Linear Algebra and Differential Equations Using MATLAB®

Martin Golubitsky
University of Houston

Michael Dellnitz
University of Paderborn

BROOKS/COLE PUBLISHING COMPANY

I(T)P® An International Thomson Publishing Company

Pacific Grove • Albany • Belmont • Bonn • Boston • Cincinnati • Detroit • Johannesburg • London
Madrid • Melbourne • Mexico City • New York • Paris • Singapore • Tokyo • Toronto • Washington

GWO A GARY W. OSTEDT BOOK

Publisher ◆ Gary W. Ostedt
Assistant Editor ◆ Carol Ann Benedict
Marketing Representative ◆ Ragu Raghavan
Marketing Team ◆ Caroline Croley, Heather Woods and
    Christina DeVeto
Production Editor ◆ Mary Vezilich
Interior Design ◆ Peter Vacek

Cover Design ◆ Laurie Albrecht
Art Editor ◆ Jennifer Mackres
Interior Illustration ◆ Roger Knox
Typesetting ◆ Eigentype Compositors
Cover Printing ◆ Phoenix Color Corporation
Printing and Binding ◆ R. R. Donnelley
    —Crawfordsville

For more information, contact:

BROOKS/COLE PUBLISHING COMPANY
511 Forest Lodge Road
Pacific Grove, CA 93950
USA

International Thomson Publishing Europe
Berkshire House 168-173
High Holborn
London WC1V 7AA
England

Thomas Nelson Australia
102 Dodds Street
South Melbourne, 3205
Victoria, Australia

Nelson Canada
1120 Birchmount Road
Scarborough, Ontario
Canada M1K 5G4

International Thomson Editores
Seneca 53
Col. Polanco
11560 México, D.F., México

International Thomson Publishing GmbH
Königswinterer Strasse 418
53227 Bonn
Germany

International Thomson Publishing Asia
60 Albert St.
#15-01 Albert Complex
Singapore 189969

International Thomson Publishing Japan
Hirakawacho Kyowa Building, 3F
2-2-1 Hirakawacho
Chiyoda-ku, Tohyo 102
Japan

Printed in the United States of America

10 9 8 7 6 5 4 3 2 1

MATLAB is a registered trademark of The MathWorks, Inc.

The cover image, "Hexnuts," is a variant of one that appeared in the book *Symmetry in Chaos*
by Michael Field and Martin Golubitsky, published by Oxford University Press, 1992.

Library of Congress Cataloging-in-Publication Data

Golubitsky, Martin, [date]
    Linear algebra and differential equations using MATLAB / Martin
Golubitsky, Michael Dellnitz.
        p.   cm.
    Includes index.
    ISBN 0-534-35425-5
    1. Algebras, Linear—Data processing.   2. Differential equations—
data processing.   3. MATLAB.   I. Dellnitz, Michael, [date].
II. Title.
QA185.D37G65   1998
512'.5'0285–dc21   98-44500

Contents

Preface

This book provides an integrated approach to linear algebra and ordinary differential equations based on computers—in this case, the software package MATLAB®.[1] We believe that computers can improve the conceptual understanding of mathematics, not just enable the completion of complicated calculations. We use computers in two ways: In linear algebra computers reduce the drudgery of calculations and enable students to focus on concepts and methods, while in differential equations computers display phase portraits graphically and enable students to focus on the qualitative information embodied in solutions rather than just on developing formulas for solutions.

We develop methods for solving both systems of linear equations and systems of (constant coefficient) linear ordinary differential equations. It is generally accepted that linear algebra methods aid in finding closed form solutions to systems of linear differential equations. The fact that the graphical solution of systems of differential equations can motivate concepts (both geometric and algebraic) in linear algebra is less often discussed. This book begins by solving linear systems of equations (through standard Gaussian elimination theory) and discussing elementary matrix theory. We then introduce ordinary differential equations (ODE)—both single equations and planar systems—to motivate the notions of eigenvectors and eigenvalues. In subsequent chapters linear, algebra and ODE theory are often mixed.

Regarding differential equations, our purpose is to introduce at the sophomore–junior level ideas from dynamical systems theory. We focus on phase portraits (and time series) rather than on techniques for finding closed form solutions. We assume that now and in the future practicing scientists and mathematicians will use ODE-solving computer programs more frequently than they will use techniques of integration. For this reason we focus on the information that is embedded in the computer graphical

---

[1] MATLAB is a registered trademark of The MathWorks Inc., Natick, MA.

approach. We discuss both typical phase portraits (Morse-Smale systems) and typical one-parameter bifurcations (both local and global). One of our goals is to provide the mathematical background that is needed when interpreting the results of computer simulation.

The Integration of Computers   Our approach assumes that students have an easier time learning with computers if the computer segments are fully integrated with the course material. Thus we interweave the instructions on how to use MATLAB with the examples and theory in the text. With ease of use in mind, we also provide a number of preloaded matrices and differential equations with the notes. Any equation in this text whose label is followed by an asterisk can be loaded into MATLAB by typing just the formula number. For the successful use of this text, it is important that students have access to computers with MATLAB and the computer files associated with these notes.

John Polking has developed an excellent graphical user interface for solving planar systems of autonomous differential equations called `pplane5`. Until Chapter 14 we use `pplane5` (and a companion code `dfield5`) instead of using the MATLAB native commands for solving ODEs. We also present an introduction to `pplane5` and the other associated software routines.

For the most part we treat the computer as a black box. We do not attempt to explain how the computer, or more precisely MATLAB, performs computations. Linear algebra structures are developed (typically) with proofs, while differential equations theorems are presented (typically) without proof and are instead motivated by computer experimentation.

There are two types of exercises included with most sections: those that should be completed using pencil and paper (called Hand Exercises) and those that should be completed with the assistance of computers (called Computer Exercises).

Ways to Use the Text   We assume as a prerequisite that students have studied one year of calculus, and we envision this course as a one-year sequence replacing the standard one-semester linear algebra and ODE courses. There is a natural one-semester Linear Systems course that can be taught using the material in this book. In this course students learn both the basics of linear algebra and the basics of linear systems of differential equations. This one-semester course covers the material in the first eight chapters. The Linear Systems course stresses eigenvalues and a baby Jordan normal form theory for 2×2 matrices and culminates in a classification of phase portraits for planar constant coefficient linear systems of differential equations. Time permitting, additional linear algebra topics from Chapters 9 and 10 may be included. Such material includes changes of coordinates for linear mappings and orthogonality, including Gram-Schmidt orthonormalization and least squares fitting of data.

We believe that by being exposed to ODE theory, a student who takes only the first semester of this sequence will gain a better appreciation of linear algebra than will a student who takes a standard one-semester introduction to linear algebra. However, a more traditional Linear Algebra course can be taught by omitting Chapter 7 and deemphasizing some of the material in Chapter 6. Then there is time in a one-semester course to cover a selection of the linear algebra topics mentioned at the end of the preceding paragraph.

A schematic diagram of the dependency relationship among chapters is given in the figure. Note that if two arrows in that figure point toward the same chapter, then material in both of the preceding chapters will be needed to complete the given chapter. Of particular importance is the observation that the material in Chapters 10, 12, 16, and 18 may be omitted without repercussion.

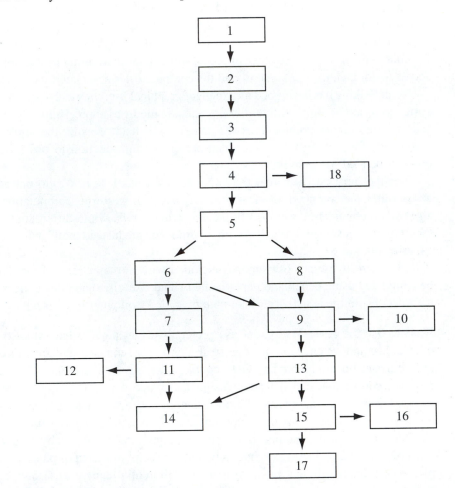

Dependency relationships among chapters

Comments on Individual Chapters

Chapters 1–3 We consider the first two chapters to be introductory material, and we attempt to cover this material as quickly as we can. Chapter 1 introduces MATLAB along with elementary remarks on vectors and matrices. In our course we ask the students to read the material in Chapter 1 and to use the computer instructions in that chapter as an entry into MATLAB. In class we cover only the material on dot product. Chapter 2 explains how to solve systems of linear equations and is required for a first course on linear algebra. The proof of the uniqueness of reduced echelon form matrices is not very illuminating for students and can be omitted in classroom discussion. Sections with material we feel can be omitted are noted by asterisks in the Table of Contents; Section 2.6 is the first example of such a section.

In Chapter 3 we introduce matrix multiplication as a notation that simplifies the presentation of systems of linear equations. We then show how matrix multiplication leads to linear mappings and how linearity leads to the principle of superposition. Multiplication of matrices is introduced as composition of linear mappings, which makes transparent the observation that multiplication of matrices is associative. The chapter ends with a discussion of inverse matrices and the role that inverses play in solving systems of linear equations. The determinant of a 2×2 matrix is introduced, and its role in determining matrix inverses is emphasized.

Chapter 4 This chapter provides a nonstandard introduction to differential equations. We begin by emphasizing that solutions to differential equations are functions (or pairs of functions for planar systems). We explain in detail the two ways that we may graph solutions to differential equations (time series and phase space) and how to go back and forth between these two graphical representations. The use of the computer is mandatory in this chapter. Chapter 4 dwells on the qualitative theory of solutions to autonomous ordinary differential equations. In one dimension we discuss the importance of knowing equilibria and their stability so that we can understand the fate of all solutions. In two dimensions we emphasize constant coefficient linear systems and the existence (numerical) of invariant directions (eigendirections). In this way we motivate the introduction of eigenvalues and eigenvectors, which are discussed in detail for 2×2 matrices. Once we know how to compute eigenvalues and eigendirections, we show how this information coupled with superposition leads to closed form solutions to initial value problems, at least when the eigenvalues are real and distinct.

We are not trying to give a thorough grounding in techniques for solving differential equations in Chapter 4; rather, we are trying to give an introduction to the ways that modern computer programs represent graphically solutions to differential equations. We include, however, a section on separation of variables for those who wish to introduce techniques for finding closed form solutions to single differential equations at this time. Our preference is to omit this section in the Linear Systems course, as well as to omit the applications in Section 4.1 of the linear growth model in one dimension to interest rates and population dynamics.

Chapter 5 In this chapter we introduce vector space theory: vector spaces, subspaces, spanning sets, linear independence, bases, dimensions, and the other basic notions in linear algebra. Since solutions to differential equations naturally reside in function spaces, we are able to illustrate that vector spaces other than $\mathbf{R}^n$ arise naturally. We have found that, depending on time, the proof of the main theorem, which appears in Section 5.6, may be omitted in a first course. The material in these chapters is mandatory in any first course on linear algebra.

Chapters 6, 7, 11, and 12   At this juncture the text divides into two tracks: one concerned with the qualitative theory of solutions to linear and nonlinear planar systems of differential equations and one mainly concerned with the development of higher dimensional linear algebra. We begin with a description of the differential equations chapters.

Chapter 6 describes closed form solutions to planar systems of constant coefficient linear differential equations in three different ways: a direct method based on eigenvalues and eigenvectors, a method based on matrix exponentials, and a related method based on similarity of matrices. Each of these methods has its virtues and vices. We note that the Jordan normal form theorem for 2×2 matrices is proved when we discuss how to solve linear planar systems using similarity of matrices.

The qualitative description of phase portraits (saddles, sinks, sources, stability, centers, etc.) for planar linear systems is presented in Chapter 7. This description depends crucially on the linear algebra of 2×2 matrices developed in Chapters 3 and 6.

In Chapter 11 we discuss the fact that nonlinear equations behave like their linear counterparts in a neighborhood of a hyperbolic equilibrium, and we verify this fact through computer experimentation. Periodic solutions to planar systems are introduced through the use of phase-amplitude equations. The chapter ends with a description of Morse-Smale nonlinear autonomous differential equations.

The discussion of planar systems of differential equations ends in Chapter 12 with a discussion of the typical bifurcations that one may expect with the variation of one parameter. These bifurcations include saddle-node bifurcations, which create a pair of equilibria, Hopf bifurcations, which create limit cycles, and global bifurcations. The discussion is framed around the breakdown of the linearization theorem at nonhyperbolic equilibria and the use of computer experimentation.

It is unusual to include the material in Chapters 11 and 12 in courses at this level. We contend, however, that the material is accessible to students who feel comfortable with computer experimentation, and it provides an alternative to learning only techniques for finding closed form solutions in a first course on differential equations. Both approaches are important, but we believe that the geometric approach based on computer simulation will be more the norm in the future than will be the determination of closed form solutions. Accordingly, we have arranged the material with the geometric theory appearing first and the techniques for finding closed form solutions appearing later.

Chapters 8, 9, 10, and 13   Material from these four chapters on linear algebra can be covered directly after Chapter 6. Chapter 8 discusses determinants, characteristic polynomials, and eigenvalues for $n \times n$ matrices. Chapter 9 presents more advanced material on linear mappings including row rank equals column rank and the matrix representation of mappings in different coordinate systems. The material in Sections 9.1 and 9.2 could be presented directly after Chapter 5, while the material in Section 9.3 explains the geometric meaning of similarity and requires the discussion of similar matrices in Chapter 6.

Orthogonal bases and orthogonal matrices, least squares and Gram-Schmidt orthonormalization, and symmetric matrices are presented in Chapter 10. This material is very important but is not required later in the text and may be omitted.

The Jordan normal form theorem for $n \times n$ matrices is presented in Chapter 13. Diagonalization of matrices with distinct real and complex eigenvalues is presented in the first two sections. The appendices, including the proof of the complete Jordan normal form theorem, are included for completeness and should be omitted in classroom presentations.

Chapter 14 In this chapter we present elements of the qualitative theory of autonomous systems of differential equations in higher dimensions. A comprehensive discussion of this topic is far beyond a course at this level. We emphasize the n-dimensional version of linearization near an equilibrium (including asymptotic stability which uses the Jordan normal form theorem of Chapter 13). We also discuss some of the complicated dynamics that appear in the solutions of systems of three and four differential equations, including quasiperiodic motion (in linear and nonlinear systems) and chaos (as represented by the Lorenz attractor). In order to explore these topics we need to introduce the MATLAB native routine ode45 for solving systems of ordinary differential equations.

Chapters 15–18 The material in the remaining chapters is more standard for a junior level course in ordinary differential equations. Chapter 15 discusses techniques for finding solutions of linear systems and higher order linear equations. This discussion requires generalized eigenvectors and hence aspects of Jordan normal form. We find solutions to higher order equations by reducing them to first-order systems. Undetermined coefficients and resonance are also presented. The popular method of Laplace transforms is discussed in Chapter 16, and the importance of this method for solving forced linear equations (including discontinuous forcing) is emphasized. The material on Laplace transforms could be presented after linearity has been discussed (after Chapter 5 or even after Chapter 3), but we do not recommend this approach.

Chapter 17 discusses linear equations with nonconstant coefficients (variation of parameters and reduction of order) and several techniques (some elementary) for solving certain nonlinear equations. These include substitution, exact differential equations, and planar Hamiltonian systems.

The text ends with a discussion of numerical techniques in Chapter 18. Euler and Runge-Kutta methods are presented. This material could, in principle, have been introduced as early as after Chapter 4; however, we include it more as an appendix than as an integrated part of the course. These topics are appropriate as assignments for independent study by interested students.

The Classroom Use of Computers

At the University of Houston we use a classroom with an IBM-compatible PC and an overhead display. Lectures are presented three hours a week using a combination of blackboard and computer display. We find it inadvisable to use the computer for more than five minutes at a time; we tend to go back and forth between standard lecture style and computer presentations. (The preloaded matrices and differential equations are important to the smooth use of the computer in class.)

We ask students to enroll in a one-hour computer lab where they can practice using the material in the text on a computer, do their homework and additional projects, and ask questions to TAs. Our computer lab has 15 Macintosh Power PCs. In addition, we ensure that MATLAB and the laode files are available on student-use computers around the campus (which is not always easy). For students who buy the student edition of MATLAB for use on their home computers, we include a CD-ROM with this book containing the required laode files. These files are also available by ftp (see page 705).

Acknowledgments   This course was first taught on a pilot basis during the 1995–96 academic year at the University of Houston. We thank the Mathematics Department and the College of Natural Sciences and Mathematics of the University of Houston for providing the resources needed to bring a course such as this to fruition. We gratefully acknowledge John Polking's help in adapting his software for our use and for allowing us access to his code so that we could write companion software for use in linear algebra.

We thank Denny Brown for his advice and his careful readings of the many drafts of this manuscript. We thank Gerhard Dangelmayr, Garret Etgen, Michael Field, Michael Friedberg, Steven Fuchs, Kimber Gross, Barbara Keyfitz, Charles Peters and David Wagner for their advice on the presentation of the material. There are many people who have helped with the production of this book and we thank them all: Laurie Albrecht, Carol Benedict, David Jansen, Roger Knox, Jennifer Mackres, Gary Ostedt, Carol Reitz, Peter Vacek, Mary Vezilich, and Helen Walden. We thank our reviewers: Bill Beckner of the University of Texas–Austin; Ted Gamelin of the University of California–Los Angeles; Cheri Shakiban, University of St. Thomas; Mo Tavakoli of Chaffney College; Robert Turner of the University of Wisconsin–Madison; and David Wagner of the University of Houston. We also thank Elizabeth Golubitsky, who has written the companion *Solutions Manual*, for her help in keeping the material accessible and in a proper order. Finally, we thank the students who stayed with this course on an experimental basis and by doing so helped to shape its form.

Martin Golubitsky
Michael Dellnitz

1

Preliminaries

The subjects of linear algebra and differential equations involve manipulating vector equations. In this chapter we introduce our notation for vectors and matrices, and we introduce MATLAB, a computer program that is designed to perform vector manipulations in a natural way.

We begin, in Section 1.1, by defining vectors and matrices and by explaining how to add and scalar multiply vectors and matrices. In Section 1.2 we explain how to enter vectors and matrices into MATLAB, and how to perform the operations of addition and scalar multiplication in MATLAB. There are many special types of matrices; these types are introduced in Section 1.3. In the concluding section we introduce the geometric interpretations of vector addition and scalar multiplication; in addition we discuss the angle between vectors through the use of the dot product of two vectors.

1.1 VECTORS AND MATRICES

In their elementary form, matrices and vectors are just lists of real numbers in different formats. An n-vector is a list of n numbers $(x_1, x_2, \ldots, x_n)$. We may write this vector as a *row* vector, as we have just done, or as a *column* vector:

$$\begin{pmatrix} x_1 \\ \vdots \\ x_n \end{pmatrix}.$$

The set of all (real-valued) n-vectors is denoted by $\mathbf{R}^n$; so points in $\mathbf{R}^n$ are called *vectors*. The sets $\mathbf{R}^n$ when n is small are very familiar. The set $\mathbf{R}^1 = \mathbf{R}$ is the real number line, the set $\mathbf{R}^2$ is the Cartesian plane, and the set $\mathbf{R}^3$ consists of points or vectors in three-dimensional space.

An $m \times n$ *matrix* is a rectangular array of numbers with m rows and n columns. A general 2×3 matrix has the form

$$A = \begin{pmatrix} a_{11} & a_{12} & a_{13} \\ a_{21} & a_{22} & a_{23} \end{pmatrix}.$$

We use the convention that matrix entries a_{ij} are indexed so that the first subscript i refers to the *row* and the second subscript j refers to the *column*. So the entry a_{21} refers to the matrix entry in the 2nd row, 1st column.

An $n \times m$ matrix A and an $n' \times m'$ matrix B are equal precisely when the sizes of the matrices are equal ($n = n'$ and $m = m'$) and when each of the corresponding entries are equal ($a_{ij} = b_{ij}$).

There is some redundancy in the use of the terms *vector* and *matrix*. For example, a row n-vector may be thought of as a $1 \times n$ matrix, and a column n-vector may be thought of as an $n \times 1$ matrix. There are situations where matrix notation is preferable to vector notation, and vice versa.

Addition and Scalar Multiplication of Vectors

There are two basic operations on vectors: addition and scalar multiplication. Let $x = (x_1, \ldots, x_n)$ and $y = (y_1, \ldots, y_n)$ be n-vectors. Then

$$x + y = (x_1 + y_1, \ldots, x_n + y_n);$$

that is, *vector addition* is defined as componentwise addition.

Similarly, *scalar multiplication* is defined as componentwise multiplication. A *scalar* is just a number. Initially we use the term *scalar* to refer to a real number, but later we sometimes use the term *scalar* to refer to a *complex* number. Suppose r is a real number; then the multiplication of a vector by the scalar r is defined as

$$rx = (rx_1, \ldots, rx_n).$$

Subtraction of vectors is defined simply as

$$x - y = (x_1 - y_1, \ldots, x_n - y_n).$$

Formally, subtraction of vectors may also be defined as

$$x - y = x + (-1)y.$$

Division of a vector x by a scalar r is defined to be

$$\frac{1}{r}x.$$

The standard difficulties concerning division by 0 still hold.

Addition and Scalar Multiplication of Matrices

Similarly, we add two $m \times n$ matrices by adding corresponding entries, and we multiply a scalar times a matrix by multiplying each entry of the matrix by that scalar. For example,

$$\begin{pmatrix} 0 & 2 \\ 4 & 6 \end{pmatrix} + \begin{pmatrix} 1 & -3 \\ 1 & 4 \end{pmatrix} = \begin{pmatrix} 1 & -1 \\ 5 & 10 \end{pmatrix}$$

and

$$4\begin{pmatrix} 2 & -4 \\ 3 & 1 \end{pmatrix} = \begin{pmatrix} 8 & -16 \\ 12 & 4 \end{pmatrix}.$$

The main restriction on adding two matrices is that the matrices must be the same size. You cannot add a 4×3 matrix to 6×2 matrix—even though they both have 12 entries.

HAND EXERCISES

In Exercises 1–3, let $x = (2, 1, 3)$ and $y = (1, 1, -1)$ and compute the given expression.

1. $x + y$ **2.** $2x - 3y$ **3.** $4x$

4. Let A be the 3×4 matrix

$$A = \begin{pmatrix} 2 & -1 & 0 & 1 \\ 3 & 4 & -7 & 10 \\ 6 & -3 & 4 & 2 \end{pmatrix}.$$

(a) For which n is a row of A a vector in $\mathbf{R}^n$?
(b) What is the 2nd column of A?
(c) Let a_{ij} be the entry of A in the ith row and the jth column. What is $a_{23} - a_{31}$?

For each pair of vectors or matrices in Exercises 5–9, decide whether addition of the members of the pair is possible; if so, perform the addition.

5. $x = (2, 1)$ and $y = (3, -1)$ **6.** $x = (1, 2, 2)$ and $y = (-2, 1, 4)$

7. $x = (1, 2, 3)$ and $y = (-2, 1)$ **8.** $A = \begin{pmatrix} 1 & 3 \\ 0 & 4 \end{pmatrix}$ and $B = \begin{pmatrix} 2 & 1 \\ 1 & -2 \end{pmatrix}$

9. $A = \begin{pmatrix} 2 & 1 & 0 \\ 4 & 1 & 0 \\ 0 & 0 & 0 \end{pmatrix}$ and $B = \begin{pmatrix} 2 & 1 \\ 1 & -2 \end{pmatrix}$

In Exercises 10 and 11, let $A = \begin{pmatrix} 2 & 1 \\ -1 & 4 \end{pmatrix}$ and $B = \begin{pmatrix} 0 & 2 \\ 3 & -1 \end{pmatrix}$ and compute the given expression.

10. $4A + B$ **11.** $2A - 3B$

1.2 MATLAB

We now use MATLAB to perform addition and scalar multiplication of vectors in two and three dimensions. This serves the purpose of introducing some basic MATLAB commands.

Entering Vectors and Vector Operations

Begin a MATLAB session. We now discuss how to enter a vector into MATLAB. The syntax is straightforward. To enter the row vector $x = (1, 2, 1)$, type[1]

    x = [1 2 1]

and MATLAB responds with

    x =
        1   2   1

It is easy to perform addition and scalar multiplication in MATLAB. Enter the row vector $y = (2, -1, 1)$ by typing

    y = [2 -1 1]

and MATLAB responds with

    y =
        2   -1   1

To add the vectors x and y, type

    x + y

and MATLAB responds with

    ans =
        3   1   2

This vector is easily checked to be the sum of the vectors x and y. Similarly, to perform a scalar multiplication, type

    2*x

which yields

    ans =
        2   4   2

MATLAB subtracts the vector y from the vector x in the natural way. Type

    x - y

to obtain

    ans =
       -1   3   0

---

[1]MATLAB has several useful line editing features. We point out two here: (a) Horizontal arrow keys ($\rightarrow$, $\leftarrow$) move the cursor one space without deleting a character, and (b) vertical arrow keys ($\uparrow$, $\downarrow$) recall previous and next command lines.

We mention two points concerning the operations that we have just performed in MATLAB:

- When we enter a vector or a number, MATLAB automatically echoes what has been entered. *This echoing can be suppressed by appending a semicolon to the line.* For example, type

  ```
  z = [-1 2 3];
  ```

  and MATLAB responds with a new line awaiting a new command. To see the contents of the vector z, just type z and MATLAB responds with

  ```
  z =
      -1    2    3
  ```

- MATLAB stores in a new vector the information obtained by algebraic manipulation. Type

  ```
  a = 2*x - 3*y + 4*z;
  ```

  Now type a to find

  ```
  a =
      -8    15    11
  ```

We see that MATLAB has created a new row vector a with the correct number of entries.

Note: In order to use the result of a calculation later in a MATLAB session, we need to name the result of that calculation. To recall the calculation 2*x - 3*y + 4*z, we needed to name that calculation, which we did by typing a = 2*x - 3*y + 4*z. Then we were able to recall the result just by typing a.

We have seen that we enter a row n-vector into MATLAB by surrounding a list of n numbers separated by spaces with square brackets. For example, to enter the 5-vector $w = (1, 3, 5, 7, 9)$, just type

```
w = [1 3 5 7 9]
```

Note that the addition of two vectors is defined only when the vectors have the same number of entries. Trying to add the 3-vector x to the 5-vector w by typing x + w in MATLAB yields the warning:

```
??? Error using ==> +
Matrix dimensions must agree.
```

In MATLAB new rows are indicated by typing ;. For example, to enter the column vector

$$z = \begin{pmatrix} -1 \\ 2 \\ 3 \end{pmatrix},$$

just type

```
z = [-1; 2; 3]
```

and MATLAB responds with

    z =
        −1
         2
         3

Note that MATLAB will not add a row vector and a column vector. Try typing x + z.

Individual entries of a vector can also be addressed. For instance, to display the first component of z, type z(1).

Entering Matrices

Matrices are entered into MATLAB row by row, with rows separated either by semicolons or by line returns. To enter the 2×3 matrix

$$A = \begin{pmatrix} 2 & 3 & 1 \\ 1 & 4 & 7 \end{pmatrix},$$

just type

    A = [2 3 1; 1 4 7]

MATLAB has very sophisticated methods for addressing the entries of a matrix. You can directly address individual entries, individual rows, and individual columns. To display the entry in the 1st row, 3rd column of A, type A(1,3). To display the 2nd column of A, type A(:,2), and to display the 1st row of A, type A(1,:). For example, to add the two rows of A and store them in the vector x, just type

    x = A(1,:) + A(2,:)

MATLAB has many operations involving matrices. We introduce them later, as needed.

COMPUTER EXERCISES

1. Enter the 3×4 matrix

$$A = \begin{pmatrix} 1 & 2 & 5 & 7 \\ -1 & 2 & 1 & -2 \\ 4 & 6 & 8 & 0 \end{pmatrix}.$$

As usual, let a_{ij} denote the entry of A in the ith row and jth column. Use MATLAB to compute the following:

(a) $a_{13} + a_{32}$

(b) Three times the 3rd column of A

(c) Twice the 2nd row of A minus the 3rd row

(d) The sum of all the columns of A

2. Verify that MATLAB adds vectors only if they are of the same type by typing

(a) x = [1 2], y = [2; 3], and x + y

(b) x = [1 2], y = [2 3 1], and x + y

In Exercises 3 and 4, let $x = (1.2, 1.4, -2.45)$ and $y = (-2.6, 1.1, 0.65)$ and use MATLAB to compute the given expression.

3. $3.27x - 7.4y$

4. $1.65x + 2.46y$

In Exercises 5 and 6, let

$$A = \begin{pmatrix} 1.2 & 2.3 & -0.5 \\ 0.7 & -1.4 & 2.3 \end{pmatrix} \quad \text{and} \quad B = \begin{pmatrix} -2.9 & 1.23 & 1.6 \\ -2.2 & 1.67 & 0 \end{pmatrix}$$

and use MATLAB to compute the given expression.

5. $-4.2A + 3.1B$

6. $2.67A - 1.1B$

1.3   SPECIAL KINDS OF MATRICES

Many matrices have special forms and hence have special names. We list them here.

- A *square* matrix is a matrix with the same number of rows and columns; that is, a square matrix is an $n \times n$ matrix.

- A *diagonal* matrix is a square matrix whose only nonzero entries are along the main diagonal; that is, $a_{ij} = 0$ if $i \neq j$. The following is a 3×3 diagonal matrix

$$\begin{pmatrix} 1 & 0 & 0 \\ 0 & 2 & 0 \\ 0 & 0 & 3 \end{pmatrix}.$$

There is a shorthand in MATLAB for entering diagonal matrices. To enter the preceding 3×3 matrix, type `diag([1 2 3])`.

- The *identity* matrix is the diagonal matrix all of whose diagonal entries are 1. The $n \times n$ identity matrix is denoted by I_n and is entered in MATLAB by typing `eye(n)`.

- A *zero* matrix is a matrix all of whose entries are 0. A zero matrix is denoted by 0. This notation is ambiguous, since there is a zero $m \times n$ matrix for every m and n. Nevertheless, this ambiguity rarely causes any difficulty. In MATLAB, to define an $m \times n$ matrix A whose entries all equal 0, just type `A = zeros(m,n)`. To define an $n \times n$ zero matrix B, type `B = zeros(n)`.

- The *transpose* of an $m \times n$ matrix A is the $n \times m$ matrix obtained from A by interchanging rows and columns. Thus the transpose of the 4×2 matrix

$$\begin{pmatrix} 2 & 1 \\ -1 & 2 \\ 3 & -4 \\ 5 & 7 \end{pmatrix}$$

is the 2×4 matrix

$$\begin{pmatrix} 2 & -1 & 3 & 5 \\ 1 & 2 & -4 & 7 \end{pmatrix}.$$

Enter this 4×2 matrix into MATLAB by typing

```
A = [2 1; -1 2; 3 -4; 5 7]
```

The transpose of a matrix A is denoted by A^t. To compute the transpose of A in MATLAB, just type A'.

- A *symmetric* matrix is a square matrix whose entries are symmetric about the main diagonal; that is, $a_{ij} = a_{ji}$. Note that a symmetric matrix is a square matrix A for which $A^t = A$.

- An *upper triangular* matrix is a square matrix all of whose entries below the main diagonal are 0; that is, $a_{ij} = 0$ if $i > j$. A *strictly upper triangular* matrix is an upper triangular matrix whose diagonal entries are also equal to 0. Similar definitions hold for *lower triangular* and *strictly lower triangular* matrices. The following 3×3 matrices are examples of upper triangular, strictly upper triangular, lower triangular, and strictly lower triangular matrices:

$$\begin{pmatrix} 1 & 2 & 3 \\ 0 & 2 & 4 \\ 0 & 0 & 6 \end{pmatrix} \quad \begin{pmatrix} 0 & 2 & 3 \\ 0 & 0 & 4 \\ 0 & 0 & 0 \end{pmatrix} \quad \begin{pmatrix} 7 & 0 & 0 \\ 5 & 2 & 0 \\ -4 & 1 & -3 \end{pmatrix} \quad \begin{pmatrix} 0 & 0 & 0 \\ 5 & 0 & 0 \\ 10 & 1 & 0 \end{pmatrix}.$$

- A square matrix A is *block diagonal* if

$$A = \begin{pmatrix} B_1 & 0 & \cdots & 0 \\ 0 & B_2 & \cdots & 0 \\ \vdots & \vdots & \ddots & \vdots \\ 0 & 0 & \cdots & B_k \end{pmatrix}$$

where each B_j is itself a square matrix. An example of a 5×5 block diagonal matrix with one 2×2 block and one 3×3 block is

$$\begin{pmatrix} 2 & 3 & 0 & 0 & 0 \\ 4 & 1 & 0 & 0 & 0 \\ 0 & 0 & 1 & 2 & 3 \\ 0 & 0 & 3 & 2 & 4 \\ 0 & 0 & 1 & 1 & 5 \end{pmatrix}.$$

HAND EXERCISES

In Exercises 1–5, decide whether or not each matrix is symmetric.

1. $\begin{pmatrix} 2 & 1 \\ 1 & 5 \end{pmatrix}$ **2.** $\begin{pmatrix} 1 & 1 \\ 0 & -5 \end{pmatrix}$ **3.** (3)   **4.** $\begin{pmatrix} 3 & 4 \\ 4 & 3 \\ 0 & 1 \end{pmatrix}$ **5.** $\begin{pmatrix} 3 & 4 & -1 \\ 4 & 3 & 1 \\ -1 & 1 & 10 \end{pmatrix}$

In Exercises 6–10, decide which of the given matrices are upper triangular and which are strictly upper triangular.

6. $\begin{pmatrix} 2 & 0 \\ -1 & -2 \end{pmatrix}$ **7.** $\begin{pmatrix} 0 & 4 \\ 0 & 0 \end{pmatrix}$ **8.** (2)   **9.** $\begin{pmatrix} 3 & 2 \\ 0 & 1 \\ 0 & 0 \end{pmatrix}$ **10.** $\begin{pmatrix} 0 & 2 & -4 \\ 0 & 7 & -2 \\ 0 & 0 & 0 \end{pmatrix}$

A general 2×2 diagonal matrix has the form $\begin{pmatrix} a & 0 \\ 0 & b \end{pmatrix}$. Thus the two unknown real numbers a and b are needed to specify each 2×2 diagonal matrix. In Exercises 11–16, how many unknown real numbers are needed to specify each of the given matrices?

11. An upper triangular 2×2 matrix   **12.** A symmetric 2×2 matrix

13. An $m \times n$ matrix   **14.** A diagonal $n \times n$ matrix

15. An upper triangular $n \times n$ matrix
Hint   Recall the summation formula:

$$1 + 2 + \cdots + k = \frac{k(k+1)}{2}.$$

16. A symmetric $n \times n$ matrix

In Exercises 17–19, determine whether each statement is true or false.

17. Every symmetric, upper triangular matrix is diagonal.

18. Every diagonal matrix is a multiple of the identity matrix.

19. Every block diagonal matrix is symmetric.

COMPUTER EXERCISES

20. Use MATLAB to compute A^t when

$$A = \begin{pmatrix} 1 & 2 & 4 & 7 \\ 2 & 1 & 5 & 6 \\ 4 & 6 & 2 & 1 \end{pmatrix}. \tag{1.3.1}$$

Use MATLAB to verify that $(A^t)^t = A$ by setting B=A' and C=B' and checking that $C = A$.

21. Use MATLAB to compute A^t when $A = (3)$ is a 1×1 matrix.

1.4 THE GEOMETRY OF VECTOR OPERATIONS

In this section we discuss the geometry of addition, scalar multiplication, and dot product of vectors. We also use MATLAB graphics to visualize these operations.

Geometry of Addition

MATLAB has an excellent graphics language that we use at various times to illustrate concepts in both two and three dimensions. In order to make the connections between ideas and graphics more transparent, we sometimes use previously developed MATLAB programs. We begin with such an example—the illustration of the parallelogram law for vector addition.

Suppose that x and y are two planar vectors. Think of these vectors as line segments from the origin to the points x and y in $\mathbf{R}^2$. We use a program written by T. A. Bryan to visualize $x + y$. In MATLAB type[2]

```
x = [1 2];
y = [-2 3];
addvec(x,y)
```

The vector x is displayed in blue, the vector y in green, and the vector $x + y$ in red. Note that $x + y$ is just the diagonal of the parallelogram spanned by x and y. A black and white version of this figure is given in Figure 1.1.

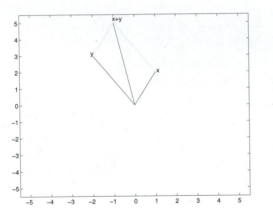

Figure 1.1
Addition of two planar vectors

The parallelogram law (the diagonal of the parallelogram spanned by x and y is $x + y$) is equally valid in three dimensions. Use MATLAB to verify this statement by typing

```
x = [1 0 2];
y = [-1 4 1];
addvec3(x,y)
```

The parallelogram spanned by x and y in $\mathbf{R}^3$ is shown in cyan; the diagonal $x + y$ is shown in blue. See Figure 1.2. To test your geometric intuition, choose several vectors x and y. Note that one vertex of the parallelogram is always the origin.

---

[2]Note that all MATLAB commands are case sensitive; upper- and lowercase must be correct.

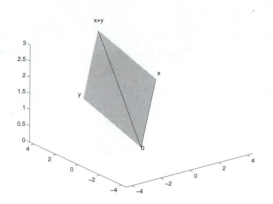

Figure 1.2
Addition of two vectors in three dimensions

Geometry of Scalar Multiplication

In all dimensions scalar multiplication just scales the length of the vector. To discuss this point we need to define the length of a vector. View an n-vector $x = (x_1, \ldots, x_n)$ as a line segment from the origin to the point x. It can be shown using the Pythagorean theorem that the *length* or *norm* of this line segment is

$$||x|| = \sqrt{x_1^2 + \cdots + x_n^2}.$$

MATLAB has the command norm for finding the length of a vector. Test this by entering the 3-vector

```
x = [1 4 2];
```

Then type

```
norm(x)
```

MATLAB responds with

```
ans =
    4.5826
```

which is indeed approximately $\sqrt{1 + 4^2 + 2^2} = \sqrt{21}$.

Now suppose $r \in \mathbf{R}$ and $x \in \mathbf{R}^n$. A calculation shows that

$$||rx|| = |r|\,||x||. \tag{1.4.1}$$

(See Exercise 17 at the end of this section.) Note also that if r is positive, then the direction of rx is the same as that of x; whereas if r is negative, then the direction of rx is opposite to the direction of x. The lengths of the vectors $3x$ and $-3x$ are each three times the length of x, but these vectors point in opposite directions. Scalar multiplication by the scalar 0 produces the zero vector, the vector whose entries are all 0.

Dot Product and Angles

The *dot product* of two n-vectors $x = (x_1, \ldots, x_n)$ and $y = (y_1, \ldots, y_n)$ is an important operation on vectors. It is defined by

$$x \cdot y = x_1 y_1 + \cdots + x_n y_n. \tag{1.4.2}$$

Note that $x \cdot x$ is just $||x||^2$, the length of x squared.

MATLAB also has a command for computing dot products of n-vectors. Type

```
x = [1 4 2];
y = [2 3 -1];
dot(x,y)
```

MATLAB responds with the dot product of x and y—namely,

```
ans =
    12
```

One of the most important facts concerning dot products is that

$$x \cdot y = 0 \quad \text{if and only if} \quad x \text{ and } y \text{ are perpendicular.} \tag{1.4.3}$$

Indeed, dot product also gives a way of numerically determining the angle between n-vectors, as follows.

Theorem 1.4.1   *Let θ be the angle between two nonzero n-vectors x and y. Then*

$$\cos \theta = \frac{x \cdot y}{||x|| \, ||y||}. \tag{1.4.4}$$

It follows that $\cos \theta = 0$ if and only if $x \cdot y = 0$. Thus (1.4.3) is valid.

Proof: Theorem 1.4.1 is just a restatement of the *law of cosines*. Recall that the law of cosines states that

$$c^2 = a^2 + b^2 - 2ab \cos \theta,$$

where a, b, and c are the lengths of the sides of a triangle and θ is the interior angle opposite the side of length c. In vector notation we can form a triangle two of whose sides are given by x and y in $\mathbf{R}^n$. The third side is just $x - y$ because $x = y + (x - y)$, as illustrated in Figure 1.3.

It follows from the law of cosines that

$$||x - y||^2 = ||x||^2 + ||y||^2 - 2||x|| \, ||y|| \cos \theta.$$

We claim that

$$||x - y||^2 = ||x||^2 + ||y||^2 - 2x \cdot y.$$

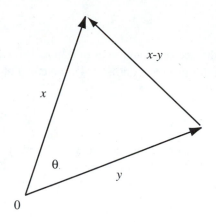

Figure 1.3
Triangle formed by vectors x and y with interior angle θ

Assuming that the claim is valid, we have

$$x \cdot y = ||x|| \, ||y|| \cos \theta,$$

which proves the theorem. Finally, compute

$$
\begin{aligned}
||x - y||^2 &= (x_1 - y_1)^2 + \cdots + (x_n - y_n)^2 \\
&= (x_1^2 - 2x_1 y_1 + y_1^2) + \cdots + (x_n^2 - 2x_n y_n + y_n^2) \\
&= (x_1^2 + \cdots + x_n^2) - 2(x_1 y_1 + \cdots + x_n y_n) + (y_1^2 + \cdots + y_n^2) \\
&= ||x||^2 - 2x \cdot y + ||y||^2
\end{aligned}
$$

to verify the claim.   ◆

Theorem 1.4.1 gives a numerically efficient method for computing the angle between vectors x and y. In MATLAB this computation proceeds by typing

```
theta = acos(dot(x,y)/(norm(x)*norm(y)))
```

where acos is the inverse cosine of a number. For example, using the 3-vectors $x = (1, 4, 2)$ and $y = (2, 3, -1)$ entered previously, MATLAB responds with

```
theta =
    0.7956
```

Remember that this answer is in radians. To convert this answer to degrees, just multiply by 360 and divide by 2π:

```
360*theta / (2*pi)
```

to obtain the answer $45.5847°$.

Area of Parallelograms

Let P be a parallelogram whose sides are the vectors v and w as in Figure 1.4. Let $|P|$ denote the area of P. As an application of dot products and (1.4.4), we calculate $|P|$. We claim that

$$|P|^2 = ||v||^2||w||^2 - (v \cdot w)^2. \tag{1.4.5}$$

We verify (1.4.5) as follows. Note that the area of P is the same as the area of the rectangle R also pictured in Figure 1.4. The side lengths of R are $||v||$ and $||w|| \sin\theta$, where θ is the angle between v and w. A computation using (1.4.4) shows that

$$\begin{aligned}
|R|^2 &= ||v||^2||w||^2 \sin^2\theta \\
&= ||v||^2||w||^2(1 - \cos^2\theta) \\
&= ||v||^2||w||^2 \left(1 - \left(\frac{v \cdot w}{||v|| \, ||w||}\right)^2\right) \\
&= ||v||^2||w||^2 - (v \cdot w)^2,
\end{aligned}$$

which establishes (1.4.5).

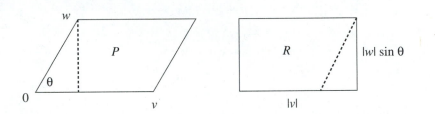

Figure 1.4
Parallelogram P beside rectangle R with same area

HAND EXERCISES

In Exercises 1–4, compute the lengths of the given vectors.

1. $x = (3, 0)$ **2.** $x = (2, -1)$

3. $x = (-1, 1, 1)$ **4.** $x = (-1, 0, 2, -1, 3)$

In Exercises 5–8, determine whether the given vectors are perpendicular.

5. $x = (1, 3)$ and $y = (3, -1)$ **6.** $x = (2, -1)$ and $y = (-2, 1)$

7. $x = (1, 1, 3, 5)$ and $y = (1, -4, 3, 0)$ **8.** $x = (2, 1, 4, 5)$ and $y = (1, -4, 3, -2)$

9. Find a real number a so that the vectors

$$x = (1, 3, 2) \quad \text{and} \quad y = (2, a, -6)$$

are perpendicular.

10. Find the lengths of the vectors $u = (2, 1, -2)$ and $v = (0, 1, -1)$ and the angle between them.

In Exercises 11–16, compute the dot product $x \cdot y$ for the given pair of vectors and the cosine of the angle between them.

11. $x = (2, 0)$ and $y = (2, 1)$ **12.** $x = (2, -1)$ and $y = (1, 2)$

13. $x = (-1, 1, 4)$ and $y = (0, 1, 3)$ **14.** $x = (-10, 1, 0)$ and $y = (0, 1, 20)$

15. $x = (2, -1, 1, 3, 0)$ and $y = (4, 0, 2, 7, 5)$

16. $x = (5, -1, 4, 1, 0, 0)$ and $y = (-3, 0, 0, 1, 10, -5)$

17. Using the definition of length, verify that formula (1.4.1) is valid.

COMPUTER EXERCISES

18. Use addvec and addvec3 to add vectors in $\mathbf{R}^2$ and $\mathbf{R}^3$. More precisely, enter pairs of 2-vectors x and y of your choosing into MATLAB, use addvec to compute $x + y$, and note the parallelogram formed by 0, x, y, and $x + y$. Similarly, enter pairs of 3-vectors and use addvec3.

19. Determine the vector of length 1 that points in the same direction as the vector

$$x = (2, 13.5, -6.7, 5.23).$$

20. Determine the vector of length 1 that points in the same direction as the vector

$$y = (2.1, -3.5, 1.5, 1.3, 5.2).$$

In Exercises 21–23, find the angle in degrees between the given pair of vectors.

21. $x = (2, 1, -3, 4)$ and $y = (1, 1, -5, 7)$

22. $x = (2.43, 10.2, -5.27, \pi)$ and $y = (-2.2, 0.33, 4, -1.7)$

23. $x = (1, -2, 2, 1, 2.1)$ and $y = (-3.44, 1.2, 1.5, -2, -3.5)$

In Exercises 24 and 25, let P be the parallelogram generated by the given vectors v and w in $\mathbf{R}^3$. Compute the area of that parallelogram.

24. $v = (1, 5, 7)$ and $w = (-2, 4, 13)$ **25.** $v = (2, -1, 1)$ and $w = (-1, 4, 3)$

Solving Linear Equations

The primary motivation for the study of vectors and matrices is based on the study of solving systems of linear equations. The algorithms that enable us to find solutions are themselves based on certain kinds of matrix manipulations. In these algorithms, matrices serve as a shorthand for calculation rather than as a basis for a theory. We will see later that these matrix manipulations lead to a rich theory of how to solve systems of linear equations, but our first step now is to see how these equations are actually solved.

We begin with a discussion in Section 2.1 of how to write systems of linear equations in terms of matrices. We also show by example how complicated it can be to write down the answers to such systems. In Section 2.2 we recall that the solution sets to systems of linear equations in two and three variables are lines and planes.

The best known and probably the most efficient method for solving systems of linear equations (especially those with a moderate to large number of unknowns) is Gaussian elimination. The idea behind this method, which is introduced in Section 2.3, is to manipulate matrices by elementary row operations to reduced echelon form. It is then possible just to look at the reduced echelon form matrix and to read off the solutions to the linear system, if any. The process of reading off the solutions is formalized in Section 2.4; see Theorem 2.4.6. Our discussion of solving linear equations is presented with equations whose coefficients are real numbers, although most of our examples have just integer coefficients. The methods work just as well with complex numbers, and this generalization is discussed in Section 2.5.

Throughout this chapter, we alternately discuss the theory and show how calculations that are tedious when done by hand can be performed easily by computer using MATLAB. The chapter ends with a proof of the uniqueness of row echelon form (a topic of theoretical importance) in Section 2.6. This section is included mainly for completeness and need not be covered on a first reading.

2.1   SYSTEMS OF LINEAR EQUATIONS AND MATRICES

It is a simple exercise to solve the system of two equations

$$
\begin{aligned}
x + y &= 7 \\
-x + 3y &= 1
\end{aligned}
\tag{2.1.1}
$$

to find that $x = 5$ and $y = 2$. One way to solve system (2.1.1) is to add the two equations, obtaining

$$4y = 8$$

and hence $y = 2$. Substituting $y = 2$ into the first equation in (2.1.1) yields $x = 5$.

This system of equations (2.1.1) can be solved in a more algorithmic fashion by solving the first equation for x as

$$x = 7 - y$$

and substituting this answer into the second equation to obtain

$$-(7 - y) + 3y = 1.$$

This equation simplifies to

$$4y = 8.$$

Now proceed as before.

Solving Larger Systems by Substitution

In contrast to solving the simple system of two equations, it is less clear how to solve a complicated system of five equations such as:

$$
\begin{aligned}
5x_1 - 4x_2 + 3x_3 - 6x_4 + 2x_5 &= 4 \\
2x_1 + x_2 - x_3 - x_4 + x_5 &= 6 \\
x_1 + 2x_2 + x_3 + x_4 + 3x_5 &= 19 \\
-2x_1 - x_2 - x_3 + x_4 - x_5 &= -12 \\
x_1 - 6x_2 + x_3 + x_4 + 4x_5 &= 4 \, .
\end{aligned}
\tag{2.1.2}
$$

The algorithmic method used to solve (2.1.1) can be expanded to produce a method, called *substitution*, for solving larger systems. We describe the substitution method as it applies to (2.1.2). Solve the first equation for x_1, obtaining

$$x_1 = \frac{4}{5} + \frac{4}{5}x_2 - \frac{3}{5}x_3 + \frac{6}{5}x_4 - \frac{2}{5}x_5. \tag{2.1.3}$$

Then substitute the right-hand side of (2.1.3) for x_1 in the remaining four equations in (2.1.2) to obtain a new system of four equations in the four variables x_2, x_3, x_4, x_5. This procedure eliminates the variable x_1. Now proceed inductively: Solve the first equation in the new system for x_2, and substitute this expression into the remaining three equations to obtain a system of three equations in three unknowns. This step eliminates the variable x_2. Continue by substitution to eliminate the variables x_3 and x_4, and arrive at a simple equation in x_5—which can then be solved. Once x_5 is known, $x_4, x_3, x_2,$ and x_1 can be found in turn.

Two Questions

- Is it realistic to expect to complete the substitution procedure without making a mistake in arithmetic?
- Will this procedure work—or will some unforeseen difficulty arise?

Almost surely, attempts to solve (2.1.2) by hand, using the substitution procedure, will lead to arithmetic errors. However, computers and software have developed to the point where solving a system such as (2.1.2) is routine. In this text we use the software package MATLAB to illustrate just how easy it has become to solve equations like (2.1.2).

The answer to the second question requires knowledge of the *theory* of linear algebra. In fact, no difficulties develop when we try to solve the particular system (2.1.2) using the substitution algorithm. We discuss why later.

Solving Equations by MATLAB

We begin by discussing the information that is needed by MATLAB to solve (2.1.2). The computer needs to know that there are five equations in five unknowns, but it does not need to keep track of the unknowns $(x_1, x_2, x_3, x_4, x_5)$ by name. Indeed, the computer just needs to know the *matrix of coefficients* in (2.1.2):

$$\begin{pmatrix} 5 & -4 & 3 & -6 & 2 \\ 2 & 1 & -1 & -1 & 1 \\ 1 & 2 & 1 & 1 & 3 \\ -2 & -1 & -1 & 1 & -1 \\ 1 & -6 & 1 & 1 & 4 \end{pmatrix} \tag{2.1.4}*$$

and the *vector* on the right-hand side of (2.1.2):

$$\begin{pmatrix} 4 \\ 6 \\ 19 \\ -12 \\ 4 \end{pmatrix}. \qquad \textbf{(2.1.5)*}$$

We now describe how to enter this information into MATLAB. To reduce the drudgery and to allow us to focus on ideas, we have entered the entries in equations that have a ∗ after their label (such as (2.1.4)*) into the laode toolbox. This information can be accessed as follows: After starting your MATLAB session, type

    e2_1_4

followed by a carriage return. This instruction tells MATLAB to load (2.1.4). The matrix of coefficients is now available in MATLAB; note that this matrix is stored in the 5×5 array A. This should appear:

```
A =
     5    -4     3    -6     2
     2     1    -1    -1     1
     1     2     1     1     3
    -2    -1    -1     1    -1
     1    -6     1     1     4
```

Indeed, comparing this result with (2.1.4), we see that they contain precisely the same information.

Since the label (2.1.5) is followed by a ∗, we can enter the vector in (2.1.5) into MATLAB by typing

    e2_1_5

Note that the right-hand side of (2.1.2) is stored in the vector b. MATLAB should respond with

```
b =
     4
     6
    19
   -12
     4
```

Now MATLAB has all the information it needs to solve the system of equations in (2.1.2). To have MATLAB solve this system, type

    x = A\b

to obtain

```
x =
    5.0000
    2.0000
    3.0000
    4.0000
    1.0000
```

This answer is interpreted as follows: The five values of the unknowns x_1, x_2, x_3, x_4, x_5 are stored in the vector x; that is,

$$x_1 = 5, \quad x_2 = 2, \quad x_3 = 3, \quad x_4 = 4, \quad x_5 = 1. \qquad (2.1.6)$$

You may verify that (2.1.6) is indeed a solution of (2.1.2) by substituting the values in (2.1.6) into the original equations.

Changing Entries in MATLAB

MATLAB also permits access to single components of x. For instance, type

```
x(5)
```

and the 5th entry of x is displayed:

```
ans =
    1.0000
```

We see that the component x(i) of x corresponds to the component x_i of the vector x, where $i = 1, 2, 3, 4, 5$. Similarly, we can access the entries of the coefficient matrix A. For instance, we type

```
A(3,4)
```

and MATLAB responds with

```
ans =
    1
```

It is also possible to change an individual entry in either a vector or a matrix. For example, if we enter

```
A(3,4) = -2
```

we obtain a new matrix A, which when displayed is

```
A =
     5    -4     3    -6     2
     2     1    -1    -1     1
     1     2     1    -2     3
    -2    -1    -1     1    -1
     1    -6     1     1     4
```

Thus the command $A(3,4) = -2$ changes the entry in the 3rd row, 4th column of A from 1 to -2. In other words, we have now entered into MATLAB the information that is needed to solve the system of equations:

$$
\begin{aligned}
5x_1 - 4x_2 + 3x_3 - 6x_4 + 2x_5 &= 4 \\
2x_1 + x_2 - x_3 - x_4 + x_5 &= 6 \\
x_1 + 2x_2 + x_3 - 2x_4 + 3x_5 &= 19 \\
-2x_1 - x_2 - x_3 + x_4 - x_5 &= -12 \\
x_1 - 6x_2 + x_3 + x_4 + 4x_5 &= 4.
\end{aligned}
$$

As expected, this change in the coefficient matrix results in a change in the solution of system (2.1.2) as well. Typing

```
x = A\b
```

now leads to the solution

```
x =
    1.9455
    3.0036
    3.0000
    1.7309
    3.8364
```

displayed to an accuracy of four decimal places.

In the next step, we can change A as follows:

```
A(2,3) = 1
```

The new system of equations is

$$
\begin{aligned}
5x_1 - 4x_2 + 3x_3 - 6x_4 + 2x_5 &= 4 \\
2x_1 + x_2 + x_3 - x_4 + x_5 &= 6 \\
x_1 + 2x_2 + x_3 - 2x_4 + 3x_5 &= 19 \\
-2x_1 - x_2 - x_3 + x_4 - x_5 &= -12 \\
x_1 - 6x_2 + x_3 + x_4 + 4x_5 &= 4.
\end{aligned}
\qquad \textbf{(2.1.7)}
$$

The command

```
x = A\b
```

now leads to the message

```
Warning: Matrix is singular to working precision.
x =
    Inf
    Inf
    Inf
    Inf
    Inf
```

Obviously, something is *wrong*; MATLAB cannot find a solution to this system of equations! Assuming that MATLAB is working correctly, we have shed light on one of our previous questions: The method of substitution described by (2.1.3) need *not* always lead to a solution, even though the method does work for system (2.1.2). Why? As we will see, this is one of the questions that is answered by the theory of linear algebra. In the case of (2.1.7), it is fairly easy to see what the difficulty is: The second and fourth equations have the forms $y = 6$ and $-y = -12$.

Warning: The MATLAB command

```
x = A\b
```

may give an error message similar to the one shown here. When this happens, one must approach the answer with caution.

HAND EXERCISES

In Exercises 1–3, find solutions to each system of linear equations.

1. $\begin{aligned} 2x - y &= 0 \\ 3x &= 6 \end{aligned}$ **2.** $\begin{aligned} 3x - 4y &= 2 \\ 2y + z &= 1 \\ 3z &= 9 \end{aligned}$ **3.** $\begin{aligned} -2x + y &= 9 \\ 3x + 3y &= -9 \end{aligned}$

4. Write the coefficient matrices for each of the systems of linear equations in Exercises 1–3.

5. Neither of the following systems of three equations in three unknowns has a unique solution—but for different reasons. Solve each system and explain why it cannot be solved uniquely.

(a) $\begin{aligned} x - y &= 4 \\ x + 3y - 2z &= -6 \\ 4x + 2y - 3z &= 1 \end{aligned}$ **(b)** $\begin{aligned} 2x - 4y + 3z &= 4 \\ 3x - 5y + 3z &= 5 \\ 2y - 3z &= -4 \end{aligned}$

6. Last year Dick was twice as old as Jane. Four years ago the sum of Dick's age and Jane's age was twice Jane's age now. How old are Dick and Jane?

Hint   Rewrite the two statements as linear equations in D—Dick's age now—and J—Jane's age now. Then solve the system of linear equations.

7. (a) Find a quadratic polynomial $p(x) = ax^2 + bx + c$ satisfying $p(0) = 1$, $p(1) = 5$, and $p(-1) = -5$.

(b) Prove that for every triple of real numbers L, M, and N, there is a quadratic polynomial satisfying $p(0) = L$, $p(1) = M$, and $p(-1) = N$.

(c) Let x_1, x_2, x_3 be three unequal real numbers and let A_1, A_2, A_3 be three real numbers. Show that finding a quadratic polynomial $q(x)$ that satisfies $q(x_i) = A_i$ is equivalent to solving a system of three linear equations.

COMPUTER EXERCISES

8. Using MATLAB, type the commands e2_1_8 and e2_1_9 to load the matrix

$$A = \begin{pmatrix} -5.6 & 0.4 & -9.8 & 8.6 & 4.0 & -3.4 \\ -9.1 & 6.6 & -2.3 & 6.9 & 8.2 & 2.7 \\ 3.6 & -9.3 & -8.7 & 0.5 & 5.2 & 5.1 \\ 3.6 & -8.9 & -1.7 & -8.2 & -4.8 & 9.8 \\ 8.7 & 0.6 & 3.7 & 3.1 & -9.1 & -2.7 \\ -2.3 & 3.4 & 1.8 & -1.7 & 4.7 & -5.1 \end{pmatrix}$$
(2.1.8)*

and the *vector*

$$b = \begin{pmatrix} 9.7 \\ 4.5 \\ 5.1 \\ 3.0 \\ -8.5 \\ 2.6 \end{pmatrix}.$$
(2.1.9)*

Solve the corresponding system of linear equations.

9. Matrices are entered in MATLAB as follows: To enter the 2×3 matrix A, type A = [-1 1 2; 4 1 2]. Enter this matrix into MATLAB; the displayed matrix should be

```
A =
    -1    1    2
     4    1    2
```

Now change the entry in the 2nd row, 1st column to -5.

10. Column vectors with n entries are viewed by MATLAB as $n \times 1$ matrices. Enter the vector b = [1; 2; -4]. Then change the 3rd entry in b to 13.

11. This problem illustrates some of the different ways that MATLAB displays numbers using the format long, the format short and the format rational commands. Use MATLAB to solve the following system of equations:

$$\begin{aligned} 2x_1 - 4.5x_2 + 3.1x_3 &= 4.2 \\ x_1 + x_2 + x_3 &= -5.1 \\ x_1 - 6.2x_2 + x_3 &= 1.3 \, . \end{aligned}$$

You may change the format of your answer in MATLAB. For example, to print your result with an accuracy of 15 digits, type format long and redisplay the answer. Similarly, to print your result as fractions, type format rational and redisplay your answer.

12. Enter the following matrix and vector into MATLAB:

```
A = [ 1 0 -1 ; 2 5 3 ; 5 -1 0];
b = [ 1; 1; -2];
```

and solve the corresponding system of linear equations by typing

```
x = A\b
```

Your answer should be

```
x =
    -0.2000
     1.0000
    -1.2000
```

Find an integer for the entry in the 2nd row, 2nd column of A so that the solution

```
x = A\b
```

is not defined.

Hint   The answer is an integer between -4 and 4.

13. The MATLAB command `rand(m,n)` defines matrices with random entries between 0 and 1. For example, the command $A = \text{rand}(5,5)$ generates a random 5×5 matrix, whereas the command $b = \text{rand}(5,1)$ generates a column vector with five random entries. Use these commands to construct several systems of linear equations and then solve them.

14. Suppose that the four substances S_1, S_2, S_3, and S_4 contain the given percentages of vitamins A, B, C, and F by weight:

Vitamin	S_1	S_2	S_3	S_4
A	25%	19%	20%	3%
B	2%	14%	2%	14%
C	8%	4%	1%	0%
F	25%	31%	25%	16%

Mix the substances S_1, S_2, S_3, and S_4 so that the resulting mixture contains precisely 3.85 grams of vitamin A, 2.30 grams of vitamin B, 0.80 gram of vitamin C, and 5.95 grams of vitamin F. How many grams of each substance have to be contained in the mixture? Discuss what happens when we require that the resulting mixture contains 2.00 grams of vitamin B instead of 2.30 grams.

2.2   THE GEOMETRY OF LOW-DIMENSIONAL SOLUTIONS

In this section we discuss how to use MATLAB graphics to solve systems of linear equations in two and three unknowns.

Linear Equations in Two Dimensions

The set of all solutions to the equation

$$2x - y = 6 \tag{2.2.1}$$

is a straight line in the xy-plane; this line has slope 2 and y-intercept equal to -6. We can use MATLAB to plot the solutions to this equation, although some understanding of the way MATLAB works is needed.

The `plot` command in MATLAB plots a sequence of points in the plane, as follows: Let X and Y be n-vectors. Then

```
plot(X,Y)
```

plots the points $(X(1), Y(1))$, $(X(2), Y(2))$, ..., $(X(n), Y(n))$ in the xy-plane.

To plot points on the line (2.2.1) we need to enter the x-coordinates of the points we wish to plot. If we want to plot 100 points, we face a tedious task, but MATLAB has a command to simplify this job. Typing

```
x = linspace(-5,5,100);
```

produces a vector x that has 100 entries, with the first entry equal to -5, the last entry equal to 5, and the remaining 98 entries equally spaced between -5 and 5. MATLAB has another command that allows us to create a vector of points x. In this command we specify the distance between points rather than the number of points; for example,

```
x = -5:0.1:5;
```

Producing x by either command is acceptable.

Typing

```
y = 2*x - 6;
```

produces a vector whose entries correspond to the y-coordinates of points on the line (2.2.1). Then typing

```
plot(x,y)
```

produces the desired plot. It is useful to label the axes on this figure, which is accomplished by typing

```
xlabel('x')
ylabel('y')
```

We can now use MATLAB to solve equation (2.1.1) graphically. Recall that (2.1.1) is

$$\begin{aligned} x + y &= 7 \\ -x + 3y &= 1. \end{aligned}$$

A solution to this system of equations is a point that lies on both lines in the system. Suppose we search for a solution to this system that has an x-coordinate between -3 and 7. We type the commands

```
x = linspace(-3,7,100);
y = 7 - x;
plot(x,y)
xlabel('x')
ylabel('y')
hold on
y = (1 + x)/3;
plot(x,y)
axis('equal')
grid
```

The MATLAB command hold on tells MATLAB to keep the present figure and to add the information that follows to that figure. The command axis('equal') instructs MATLAB to make unit distances on the x- and y-axes equal. The last MATLAB command superimposes grid lines. See Figure 2.1; you can see that the solution to this system is $(x, y) = (5, 2)$, which we already knew.

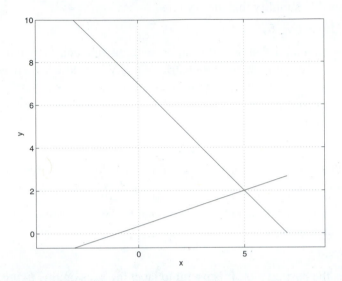

Figure 2.1
Graph of equations in (2.1.1)

There are two principles that follow from this exercise:

- Solutions to a single linear equation in two variables form a straight line.
- Solutions to two linear equations in two unknowns lie at the intersection of two straight lines in the plane.

It follows that the solution to two linear equations in two variables is a single point if the lines are not parallel. If these lines are parallel and unequal, then there are no solutions because there are no points of intersection.

Linear Equations in Three Dimensions

We begin by observing that the set of all solutions to a linear equation in three variables forms a plane. More precisely, the solutions to the equation

$$ax + by + cz = d \tag{2.2.2}$$

form a plane that is perpendicular to the vector (a, b, c)—assuming of course that the vector (a, b, c) is nonzero.

This fact is most easily proved using the *dot product*. Recall from (1.4.2) that the dot product is defined by

$$X \cdot Y = x_1 y_1 + x_2 y_2 + x_3 y_3,$$

where $X = (x_1, x_2, x_3)$ and $Y = (y_1, y_2, y_3)$. We recall from (1.4.3) the following important fact concerning dot products:

$$X \cdot Y = 0$$

if and only if the vectors X and Y are perpendicular.

Suppose that $N = (a, b, c) \neq 0$. Consider the plane that is perpendicular to the *normal vector* N and that contains the point X_0. If the point X lies in that plane, then $X - X_0$ is perpendicular to N; that is,

$$(X - X_0) \cdot N = 0. \tag{2.2.3}$$

If we use the notation

$$X = (x, y, z) \quad \text{and} \quad X_0 = (x_0, y_0, z_0),$$

then (2.2.3) becomes

$$a(x - x_0) + b(y - y_0) + c(z - z_0) = 0.$$

Setting

$$d = ax_0 + by_0 + cz_0$$

puts equation (2.2.3) into the form (2.2.2). In this way we see that the set of solutions to a single linear equation in three variables forms a plane. See Figure 2.2.

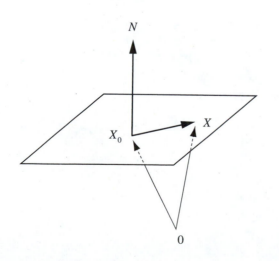

Figure 2.2
The plane containing X_0 and perpendicular to N

We now use MATLAB to visualize the planes that are solutions to linear equations. Plotting an equation in three dimensions in MATLAB follows a structure similar to the planar plots. Suppose that we wish to plot the solutions to the equation

$$-2x + 3y + z = 2. \qquad \textbf{(2.2.4)}$$

We can rewrite (2.2.4) as

$$z = 2x - 3y + 2.$$

It is this function that we actually graph by typing the commands

```
[x,y] = meshgrid(-5:0.5:5);
z = 2*x - 3*y + 2;
surf(x,y,z)
```

The first command tells MATLAB to create a square grid in the xy-plane. Grid points are equally spaced between -5 and 5 at intervals of 0.5 on both the x- and y-axes. The second command tells MATLAB to compute the z value of the solution to (2.2.4) at each grid point. The third command tells MATLAB to graph the surface containing the points (x, y, z). See Figure 2.3.

We can now see that solutions to a system of two linear equations in three unknowns consist of points that lie simultaneously on two planes. As long as the normal vectors to these planes are not parallel, the intersection of the two planes is a line in three dimensions. Indeed, consider the equations

$$-2x + 3y + z = 2$$
$$2x - 3y + z = 0.$$

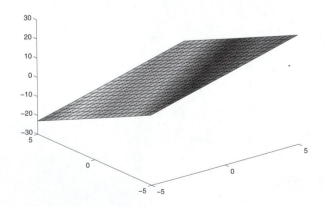

Figure 2.3
Graph of (2.2.4)

We can graph the solution using MATLAB, as follows. We continue from the previous graph by typing

```
hold on
z = −2*x + 3*y;
surf(x,y,z)
```

The result, which illustrates that the intersection of two planes in $\mathbf{R}^3$ is generally a line, is shown in Figure 2.4.

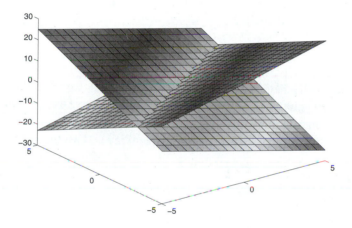

Figure 2.4
Line of intersection of two planes

We can now see geometrically that the solution to three simultaneous linear equations in three unknowns is generally a point—since generally three planes in three-space intersect in a point. To visualize this intersection, as shown in Figure 2.5, we extend the previous system of equations to

$$-2x + 3y + z = 2$$
$$2x - 3y + z = 0$$
$$-3x + 0.2y + z = 1.$$

Continuing in MATLAB type, we have

```
z = 3*x − 0.2*y + 1;
surf(x,y,z)
```

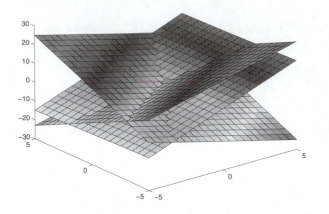

Figure 2.5
Point of intersection of three planes

Unfortunately, visualizing the point of intersection of these planes geometrically does not really help us get an accurate numerical value of the coordinates of this intersection point. However, we can use MATLAB to solve this system accurately. Denote the 3×3 matrix of coefficients by A, the vector of coefficients on the right-hand side by b, and the solution by x. Solve the system in MATLAB by typing

```
A = [ -2 3 1; 2 -3 1; -3 0.2 1];
b = [2; 0; 1];
x = A\b
```

The point of intersection of the three planes is

```
x =
    0.0233
    0.3488
    1.0000
```

Three planes in three-dimensional space need not intersect in a single point. For example, if two of the planes are parallel, they need not intersect at all. The normal vectors must point in *independent* directions to guarantee that the intersection is a point. Understanding the notion of independence (it is more complicated than just not being parallel) is part of the subject of linear algebra. MATLAB returns Inf, which we have seen previously, when these normal vectors are (approximately) dependent. For example, consider Exercise 6 at the end of this section.

Plotting Nonlinear Functions in MATLAB

Suppose that we want to plot the graph of a nonlinear function of a single variable, such as

$$y = x^2 - 2x + 3 \tag{2.2.5}$$

on the interval $[-2, 5]$ using MATLAB. There is a difficulty: How do we enter the term x^2? For example, suppose that we type

```
x = linspace(-2,5);
y = x*x - 2*x + 3;
```

Then MATLAB responds with

```
??? Error using ==> *
Inner matrix dimensions must agree.
```

The problem is that in MATLAB the variable x is a vector of 100 equally spaced points x(1), x(2), ..., x(100). What we really need is a vector consisting of entries x(1)*x(1), x(2)*x(2), ..., x(100)*x(100). MATLAB has the facility to perform this operation automatically, and the syntax for the operation is .* rather than *. So typing

```
x = linspace(-2,5);
y = x.*x - 2*x + 3;
plot(x,y)
```

produces the graph of (2.2.5) in Figure 2.6. In a similar fashion, MATLAB has the dot operations of ./, .\, and .^, as well as .*.

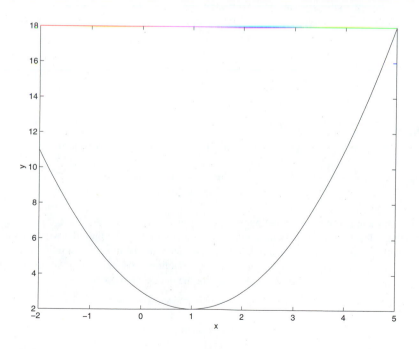

Figure 2.6
Graph of $y = x^2 - 2x + 3$

HAND EXERCISES

1. Find the equation for the plane that is perpendicular to the vector $(2, 3, 1)$ and contains the point $(-1, -2, 3)$.

2. Determine three systems of two linear equations in two unknowns so that the first system has a unique solution, the second system has an infinite number of solutions, and the third system has no solutions.

3. Write the equation of the plane through the origin containing the vectors $(1, 0, 1)$ and $(2, -1, 2)$.

4. Find a system of two linear equations in three unknowns whose solution set is the line consisting of scalar multiples of the vector $(1, 2, 1)$.

5. **(a)** Find a vector u normal to the plane $2x + 2y + z = 3$.
   (b) Find a vector v normal to the plane $x + y + 2z = 4$.
   (c) Find the cosine of the angle between the vectors u and v. Use MATLAB to find the angle in degrees.

6. Determine graphically the geometry of the set of solutions to the system of equations in the three unknowns x, y, and z:

$$
\begin{aligned}
x \phantom{{}+{}} + 3z &= 1 \\
3x \phantom{{}+{}} - z &= 1 \\
z &= 2
\end{aligned}
$$

by sketching the plane of solutions for each equation individually. Describe in words why there are no solutions to this system. (Use MATLAB graphics to verify your sketch. Note that you should enter the last equation as $z = 2 - 0*x - 0*y$ and the first two equations with $0*y$ terms. Try different views—but include `view([0 1 0])` as one view.)

COMPUTER EXERCISES

7. Use MATLAB to solve graphically the planar system of linear equations

$$
\begin{aligned}
x + 4y &= -4 \\
4x + 3y &= 4
\end{aligned}
$$

to an accuracy of two decimal points.

Hint The MATLAB command `zoom on` allows you to view the plot in a window whose axes are one-half those of the original. Each time you click with the mouse on a point, the axes' limits are halved and centered at the designated point. Coupling `zoom on` with `grid on` allows you to determine approximate numerical values for the intersection point.

8. Use MATLAB to solve graphically the planar system of linear equations

$$
\begin{aligned}
4.23x + 0.023y &= -1.1 \\
1.65x - 2.81y &= 1.63
\end{aligned}
$$

to an accuracy of two decimal points.

9. Use MATLAB to find an approximate graphical solution to the three-dimensional system of linear equations

$$
\begin{aligned}
3x - 4y + 2z &= -11 \\
2x + 2y + z &= 7 \\
-x + y - 5z &= 7.
\end{aligned}
$$

Then use MATLAB to find an exact solution.

10. Use MATLAB to determine graphically the geometry of the set of solutions to the system of equations:

$$
\begin{aligned}
x + 3y + 4z &= 5 \\
2x + y + z &= 1 \\
-4x + 3y + 5z &= 7.
\end{aligned}
$$

Attempt to use MATLAB to find an exact solution to this system and discuss the implications of your calculations.

Hint   After setting up the graphics display in MATLAB, you can use the command view([0,1,0]) to get a better view of the solution point.

11. Use MATLAB to graph the function $y = 2 - x \sin(x^2 - 1)$ on the interval $[-2, 3]$. How many relative maxima does this function have on this interval?

2.3 GAUSSIAN ELIMINATION

A general system of m *linear* equations in n unknowns has the form

$$
\begin{aligned}
a_{11}x_1 + a_{12}x_2 + \cdots + a_{1n}x_n &= b_1 \\
a_{21}x_1 + a_{22}x_2 + \cdots + a_{2n}x_n &= b_2 \\
\vdots \qquad \vdots \qquad \vdots \qquad \vdots \\
a_{m1}x_1 + a_{m2}x_2 + \cdots + a_{mn}x_n &= b_m.
\end{aligned}
\tag{2.3.1}
$$

The entries a_{ij} and b_i are constants. Our task is to find a method for solving (2.3.1) for the variables $x_1, \ldots, x_n$.

Easily Solved Equations

Some systems are easily solved. The system of three equations ($m = 3$) in three unknowns ($n = 3$),

$$
\begin{aligned}
x_1 + 2x_2 + 3x_3 &= 10 \\
x_2 - \tfrac{1}{5}x_3 &= \tfrac{7}{5} \\
x_3 &= 3
\end{aligned}
\tag{2.3.2}
$$

is one example. The third equation states that $x_3 = 3$. Substituting this value into the second equation allows us to solve the second equation as $x_2 = 2$. Finally, substituting

$x_2 = 2$ and $x_3 = 3$ into the first equation allows us to solve for $x_1 = -3$. The process that we have just described is called *back substitution*.

Next consider the system of two equations ($m = 2$) in three unknowns ($n = 3$):

$$x_1 + 2x_2 + 3x_3 = 10$$
$$x_3 = 3. \tag{2.3.3}$$

The second equation states that $x_3 = 3$. Substituting this value into the first equation leads to

$$x_1 = 1 - 2x_2.$$

We have shown that every solution to (2.3.3) has the form $(x_1, x_2, x_3) = (1 - 2x_2, x_2, 3)$ and that every vector $(1 - 2x_2, x_2, 3)$ is a solution of (2.3.3). Thus there are an infinite number of solutions to (2.3.3), and these solutions can be parameterized by one number x_2.

Equations That Have No Solutions

Note that the system of equations

$$x_1 - x_2 = 1$$
$$x_1 - x_2 = 2$$

has no solutions.

Definition 2.3.1 *A linear system of equations is* inconsistent *if the system has no solutions and* consistent *if the system does have solutions.*

As discussed in Section 2.1, (2.1.7) is an example of a linear system that MAT-LAB cannot solve. In fact, that system is inconsistent; inspect the second and fourth equations in (2.1.7).

Gaussian elimination is an algorithm for finding all solutions to a system of linear equations by reducing the given system to ones like (2.3.2) and (2.3.3), which are easily solved by back substitution. Consequently, Gaussian elimination can also be used to determine whether a system is consistent or inconsistent.

Elementary Equation Operations

There are three ways to change a system of equations without changing the set of solutions; Gaussian elimination is based on this observation. These are the three elementary operations:

1. Swap two equations.
2. Multiply a single equation by a nonzero number.

3. Add a scalar multiple of one equation to another.

We begin with an example:

$$\begin{aligned} x_1 + 2x_2 + 3x_3 &= 10 \\ x_1 + 2x_2 + x_3 &= 4 \\ 2x_1 + 9x_2 + 5x_3 &= 27. \end{aligned} \qquad (2.3.4)$$

Gaussian elimination works by eliminating variables from the equations in a fashion similar to the substitution method in the preceding section. To begin, eliminate the variable x_1 from all but the first equation, as follows. Subtract the first equation from the second, and subtract twice the first equation from the third, obtaining

$$\begin{aligned} x_1 + 2x_2 + 3x_3 &= 10 \\ -2x_3 &= -6 \\ 5x_2 - x_3 &= 7. \end{aligned} \qquad (2.3.5)$$

Next swap the second and third equations, so that the coefficient of x_2 in the new second equation is nonzero. This yields

$$\begin{aligned} x_1 + 2x_2 + 3x_3 &= 10 \\ 5x_2 - x_3 &= 7 \\ -2x_3 &= -6. \end{aligned} \qquad (2.3.6)$$

Now divide the second equation by 5 and the third equation by -2 to obtain a system of equations identical to our first example (2.3.2), which we solved by back substitution.

Augmented Matrices

The process of performing Gaussian elimination when the number of equations is greater than two or three is painful. The computer, however, can help with the manipulations. We begin by introducing the *augmented matrix*. The augmented matrix associated with (2.3.1) has m rows and $n + 1$ columns and is written as

$$\left(\begin{array}{cccc|c} a_{11} & a_{12} & \cdots & a_{1n} & b_1 \\ a_{21} & a_{22} & \cdots & a_{2n} & b_2 \\ \vdots & \vdots & & \vdots & \vdots \\ a_{m1} & a_{m2} & \cdots & a_{mn} & b_m \end{array} \right). \qquad (2.3.7)$$

The augmented matrix contains all the information that is needed to solve system (2.3.1).

Elementary Row Operations

The elementary operations used in Gaussian elimination can be interpreted as *row operations* on the augmented matrix as follows:

1. Swap two rows.
2. Multiply a single row by a nonzero number.
3. Add a scalar multiple of one row to another.

We claim that by using these elementary row operations intelligently, we can always solve a consistent linear system; indeed, we can determine when a linear system is consistent or inconsistent. The idea is to perform elementary row operations in such a way that the new augmented matrix has 0 entries below the diagonal.

We describe this process inductively. Begin with the 1st column. We assume for now that some entry in this column is nonzero. If $a_{11} = 0$, then swap two rows so that the number a_{11} is nonzero. Then divide the 1st row by a_{11} so that the leading entry in that row is 1. Now subtract a_{i1} times the 1st row from the ith row for each row i from 2 to m. The end result is that the 1st column has a 1 in the 1st row and a 0 in every row below the 1st. The result is

$$\begin{pmatrix} 1 & * & \cdots & * \\ 0 & * & \cdots & * \\ \vdots & \vdots & \vdots & \vdots \\ 0 & * & \cdots & * \end{pmatrix}.$$

Next we consider the 2nd column. We assume that some entry in that column below the 1st row is nonzero. So, if necessary, we can swap two rows below the 1st row so that the entry a_{22} is nonzero. Then we divide the 2nd row by a_{22} so that its leading nonzero entry is 1. We subtract appropriate multiples of the 2nd row from each row below the 2nd so that all the entries in the 2nd column below the 2nd row are 0. The result is

$$\begin{pmatrix} 1 & * & \cdots & * \\ 0 & 1 & \cdots & * \\ \vdots & \vdots & \vdots & \vdots \\ 0 & 0 & \cdots & * \end{pmatrix}.$$

We continue with the 3rd column. That's the idea. However, does this process always work? What happens if all the entries in a column are 0? Before answering these questions, we do experimentation with MATLAB.

Row Operations in MATLAB

In MATLAB the ith row of a matrix A is specified by A(i,:). Thus to replace the 5th row of a matrix A by twice itself, we need only type

```
A(5,:) = 2*A(5,:)
```

In general, we can replace the ith row of the matrix A by c times itself by typing

```
A(i,:) = c*A(i,:)
```

Similarly, we can divide the ith row of the matrix A by the nonzero number c by typing

```
A(i,:) = A(i,:)/c
```

The third elementary row operation is performed similarly. Suppose we want to add c times the ith row to the jth row; then we type

```
A(j,:) = A(j,:) + c*A(i,:)
```

For example, subtracting three times the 7th row from the 4th row of the matrix A is accomplished by typing

```
A(4,:) = A(4,:) - 3*A(7,:)
```

The first elementary row operation, swapping two rows, requires a different kind of MATLAB command. In MATLAB, the ith and jth rows of the matrix A are permuted by the command

```
A([i j],:) = A([j i],:)
```

So, to swap the 1st and 3rd rows of the matrix A, we type

```
A([1 3],:) = A([3 1],:)
```

Examples of Row Reduction in MATLAB

Let us see how the row operations can be used in MATLAB. As an example, we consider the augmented matrix

$$\left(\begin{array}{cccc|c} 1 & 3 & 0 & -1 & -8 \\ 2 & 6 & -4 & 4 & 4 \\ 1 & 0 & -1 & -9 & -35 \\ 0 & 1 & 0 & 3 & 10 \end{array} \right). \qquad \textbf{(2.3.8)*}$$

We enter this information into MATLAB by typing

```
e2_3_8
```

which produces the result

```
A =
     1     3     0    -1    -8
     2     6    -4     4     4
     1     0    -1    -9   -35
     0     1     0     3    10
```

We now perform Gaussian elimination on A and then solve the resulting system by back substitution. Gaussian elimination uses elementary row operations to set the entries that are in the lower left part of A to 0. These entries are indicated by numbers in the following matrix:

```
     *     *     *     *     *
     2     *     *     *     *
     1     0     *     *     *
     0     1     0     *     *
```

Gaussian elimination works inductively. Since the first entry in the matrix A is equal to 1, the first step in Gaussian elimination is to set to 0 all entries in the 1st column below the 1st row. We begin by eliminating the 2 that is the first entry in the 2nd row of A. We replace the 2nd row by the 2nd row minus twice the 1st row. To accomplish this elementary row operation, we type

```
A(2,:) = A(2,:) - 2*A(1,:)
```

and the result is

```
A =
    1    3    0   -1   -8
    0    0   -4    6   20
    1    0   -1   -9  -35
    0    1    0    3   10
```

In the next step, we eliminate the 1 from the entry in the 3rd row, 1st column of A. We do this by typing

```
A(3,:) = A(3,:) - A(1,:)
```

which yields

```
A =
    1    3    0   -1   -8
    0    0   -4    6   20
    0   -3   -1   -8  -27
    0    1    0    3   10
```

Using elementary row operations, we have now set the entries in the 1st column below the 1st row to 0.

Next we alter the 2nd column. We begin by swapping the 2nd and 4th rows so that the leading nonzero entry in the 2nd row is 1. To accomplish this swap, we type

```
A([2 4],:) = A([4 2],:)
```

and obtain

```
A =
    1    3    0   -1   -8
    0    1    0    3   10
    0   -3   -1   -8  -27
    0    0   -4    6   20
```

The next elementary row operation is the command

```
A(3,:) = A(3,:) + 3*A(2,:)
```

which leads to

```
A =
    1    3    0   -1   -8
    0    1    0    3   10
    0    0   -1    1    3
    0    0   -4    6   20
```

Now we have set all entries in the 2nd column below the 2nd row to 0.

Next we set the first nonzero entry in the 3rd row to 1 by multiplying the 3rd row by -1, obtaining

A =

1	3	0	−1	−8
0	1	0	3	10
0	0	1	−1	−3
0	0	−4	6	20

Since the leading nonzero entry in the 3rd row is 1, we next eliminate the nonzero entry in the 3rd column, 4th row. This is accomplished by the MATLAB command

```
A(4,:) = A(4,:) + 4*A(3,:)
```

Finally, we divide the 4th row by 2 to obtain

A =

1	3	0	−1	−8
0	1	0	3	10
0	0	1	−1	−3
0	0	0	1	4

By using elementary row operations, we have arrived at the system

$$
\begin{aligned}
x_1 + 3x_2 \quad\quad - x_4 &= -8 \\
x_2 \quad + 3x_4 &= 10 \\
x_3 - x_4 &= -3 \\
x_4 &= 4 ,
\end{aligned}
\tag{2.3.9}
$$

which can now be solved by back substitution. We obtain

$$
x_4 = 4, \quad x_3 = 1, \quad x_2 = -2, \quad x_1 = 2.
\tag{2.3.10}
$$

We return to the original set of equations corresponding to (2.3.8):

$$
\begin{aligned}
x_1 + 3x_2 \quad\quad - x_4 &= -8 \\
2x_1 + 6x_2 - 4x_3 + 4x_4 &= 4 \\
x_1 \quad\quad - x_3 - 9x_4 &= -35 \\
x_2 \quad\quad + 3x_4 &= 10 .
\end{aligned}
\tag{2.3.11}*
$$

We load the corresponding linear system into MATLAB by typing

```
e2_3_11
```

The information in (2.3.11) is contained in the coefficient matrix C and the right-hand side b. A direct solution is found by typing

```
x = C\b
```

which yields the same answer as in (2.3.10)—namely,

```
x =
    2.0000
   -2.0000
    1.0000
    4.0000
```

Introduction to Echelon Form

Now we discuss how Gaussian elimination works in an example in which the number of rows and the number of columns in the coefficient matrix are unequal. We consider the augmented matrix

$$\left(\begin{array}{cccccc|c} 1 & 0 & -2 & 3 & 4 & 0 & 1 \\ 0 & 1 & 2 & 4 & 0 & -2 & 0 \\ 2 & -1 & -4 & 0 & -2 & 8 & -4 \\ -3 & 0 & 6 & -8 & -12 & 2 & -2 \end{array}\right). \qquad (2.3.12)*$$

This information is entered into MATLAB by typing

```
e2_3_12
```

Again, the augmented matrix is denoted by A.

   We begin by eliminating the entry 2 in the 3rd row, 1st column. To accomplish the corresponding elementary row operation, we type

```
A(3,:) = A(3,:) - 2*A(1,:)
```

which results in

```
A =
     1     0    -2     3     4     0     1
     0     1     2     4     0    -2     0
     0    -1     0    -6   -10     8    -6
    -3     0     6    -8   -12     2    -2
```

We proceed with

```
A(4,:) = A(4,:) + 3*A(1,:)
```

to create two more 0s in the 4th row. Finally, we eliminate the -1 in the 3rd row, 2nd column by

```
A(3,:) = A(3,:) + A(2,:)
```

to arrive at

```
A =
     1     0    -2     3     4     0     1
     0     1     2     4     0    -2     0
     0     0     2    -2   -10     6    -6
     0     0     0     1     0     2     1
```

Next we set the leading nonzero entry in the 3rd row to 1 by dividing the 3rd row by 2. That is, we type

```
A(3,:) = A(3,:)/2
```

to obtain

```
A =
     1     0    -2     3     4     0     1
     0     1     2     4     0    -2     0
     0     0     1    -1    -5     3    -3
     0     0     0     1     0     2     1
```

We say that the matrix A is in (row) *echelon form* since the first nonzero entry in each row is a 1, each entry in a column below a leading 1 is 0, and the leading 1 moves to the right as you go down the matrix. In row echelon form, the entries where leading 1s occur are called *pivots*.

If we compare the structure of this matrix to the ones we have obtained previously, then we see that here we have two columns too many. Indeed, we may solve these equations by back substitution for any choice of the variables x_5 and x_6.

The idea behind back substitution is to solve the last equation for the variable corresponding to the first nonzero coefficient. In this case, we use the fourth equation to solve for x_4 in terms of x_5 and x_6, and then we substitute for x_4 in the first three equations. This process can also be accomplished by elementary row operations. Indeed, eliminating the variable x_4 from the first three equations is the same as using row operations to set the first three entries in the 4th column to 0. We can do this by typing

```
A(3,:) = A(3,:) + A(4,:);
A(2,:) = A(2,:) - 4*A(4,:);
A(1,:) = A(1,:) - 3*A(4,:)
```

Remember: By typing semicolons after the first two rows, we have told MATLAB not to print the intermediate results. Since we have not typed a semicolon after the 3rd row, MATLAB outputs

```
A =
     1     0    -2     0     4    -6    -2
     0     1     2     0     0   -10    -4
     0     0     1     0    -5     5    -2
     0     0     0     1     0     2     1
```

We proceed with back substitution by eliminating the nonzero entries in the first two rows of the 3rd column. To do this, type

```
A(2,:) = A(2,:) - 2*A(3,:);
A(1,:) = A(1,:) + 2*A(3,:)
```

which yields

$$A = \begin{pmatrix} 1 & 0 & 0 & 0 & -6 & 4 & -6 \\ 0 & 1 & 0 & 0 & 10 & -20 & 0 \\ 0 & 0 & 1 & 0 & -5 & 5 & -2 \\ 0 & 0 & 0 & 1 & 0 & 2 & 1 \end{pmatrix}$$

The augmented matrix is now in *reduced echelon form*, and the corresponding system of equations has the form

$$\begin{aligned}
x_1 \qquad\quad - 6x_5 + 4x_6 &= -6 \\
x_2 \qquad + 10x_5 - 20x_6 &= 0 \\
x_3 \qquad - 5x_5 + 5x_6 &= -2 \\
x_4 \qquad + 2x_6 &= 1.
\end{aligned} \tag{2.3.13}$$

A matrix is in reduced echelon form if it is in echelon form and if *every* entry in a column containing a pivot, other than the pivot itself, is 0.

Reduced echelon form allows us to solve directly this system of equations in terms of the variables x_5 and x_6:

$$\begin{pmatrix} x_1 \\ x_2 \\ x_3 \\ x_4 \\ x_5 \\ x_6 \end{pmatrix} = \begin{pmatrix} -6 + 6x_5 - 4x_6 \\ -10x_5 + 20x_6 \\ -2 + 5x_5 - 5x_6 \\ 1 - 2x_6 \\ x_5 \\ x_6 \end{pmatrix}. \tag{2.3.14}$$

It is important to note that every consistent system of linear equations corresponding to an augmented matrix in reduced echelon form can be solved as in (2.3.14)—and this is one reason for emphasizing reduced echelon form. We discuss the reduction to reduced echelon form in more detail in the next section.

HAND EXERCISES

In Exercises 1–3, determine whether or not the given matrix is in reduced echelon form.

1. $\begin{pmatrix} 1 & -1 & 0 & 1 \\ 0 & 1 & 0 & -6 \\ 0 & 0 & 1 & 0 \end{pmatrix}$ **2.** $\begin{pmatrix} 1 & 0 & -2 & 0 \\ 0 & 1 & 4 & 0 \\ 0 & 0 & 0 & 1 \end{pmatrix}$ **3.** $\begin{pmatrix} 0 & 1 & 0 & 3 \\ 0 & 0 & 2 & 1 \\ 0 & 0 & 0 & 0 \end{pmatrix}$

In Exercises 4–6, we list the reduced echelon form of an augmented matrix of a system of linear equations. Which columns in these augmented matrices contain pivots? Describe all solutions to these systems of equations in the form of (2.3.14).

4. $\left(\begin{array}{ccc|c} 1 & 4 & 0 & 0 \\ 0 & 0 & 1 & 5 \\ 0 & 0 & 0 & 0 \end{array} \right)$ **5.** $\left(\begin{array}{cccc|c} 1 & 2 & 0 & 0 & 0 \\ 0 & 0 & 1 & 1 & 0 \\ 0 & 0 & 0 & 0 & 1 \end{array} \right)$ **6.** $\left(\begin{array}{cccc|c} 1 & -6 & 0 & 0 & 1 \\ 0 & 0 & 1 & 0 & 9 \\ 0 & 0 & 0 & 0 & 0 \end{array} \right)$

7. (a) Consider the 2×2 matrix

$$\begin{pmatrix} a & b \\ c & 1 \end{pmatrix},$$ (2.3.15)

where $a, b, c \in \mathbf{R}$ and $a \neq 0$. Use elementary row operations to transform (2.3.15) to the matrix

$$\begin{pmatrix} 1 & \dfrac{b}{a} \\ 0 & \dfrac{a - bc}{a} \end{pmatrix}.$$

(b) Show that (2.3.15) can be transformed to the identity matrix by elementary row operations if and only if $a \neq bc$.

8. Use row reduction and back substitution to solve the system of two equations in three unknowns:

$$\begin{aligned} x_1 - x_2 + x_3 &= 1 \\ 2x_1 + x_2 - x_3 &= -1. \end{aligned}$$

Is $(1, 2, 2)$ a solution to this system? If not, is there a solution for which $x_3 = 2$?

In Exercises 9 and 10, determine the augmented matrix and all solutions for each system of linear equations.

9. $\begin{aligned} x - y + z &= 1 \\ 4x + y + z &= 5 \\ 2x + 3y - z &= 2 \end{aligned}$ **10.** $\begin{aligned} 2x - y + z + w &= 1 \\ x + 2y - z + w &= 7 \end{aligned}$

In Exercises 11–14, consider the augmented matrices representing systems of linear equations, and decide (a) if there are zero, one or infinitely many solutions, and (b) if solutions are not unique, how many variables can be assigned arbitrary values.

11. $\begin{pmatrix} 1 & 0 & 0 & | & 3 \\ 0 & 2 & 1 & | & 1 \\ 0 & 0 & 0 & | & 0 \end{pmatrix}$ **12.** $\begin{pmatrix} 1 & 2 & 0 & 0 & | & 3 \\ 0 & 1 & 1 & 0 & | & 1 \\ 0 & 0 & 0 & 0 & | & 2 \end{pmatrix}$ **13.** $\begin{pmatrix} 1 & 0 & 2 & | & 1 \\ 0 & 5 & 0 & | & 2 \\ 0 & 0 & 4 & | & 3 \end{pmatrix}$ **14.** $\begin{pmatrix} 1 & 0 & 2 & 0 & | & 3 \\ 2 & 3 & 6 & 1 & | & 16 \\ 0 & 3 & 2 & 1 & | & 10 \\ 0 & 0 & 0 & 0 & | & 0 \end{pmatrix}$

A system of m equations in n unknowns is linear if it has the form (2.3.1); any other system of equations is called *nonlinear*. In Exercises 15–19, decide whether each system of equations is linear or nonlinear.

15. $\begin{aligned} 3x_1 - 2x_2 + 14x_3 - 7x_4 &= 35 \\ 2x_1 + 5x_2 - 3x_3 + 12x_4 &= -1 \end{aligned}$ **16.** $\begin{aligned} 3x_1 + \pi x_2 &= 0 \\ 2x_1 - e x_2 &= 1 \end{aligned}$

17. $\begin{aligned} 3x_1 x_2 - x_2 &= 10 \\ 2x_1 - x_2^2 &= -5 \end{aligned}$ **18.** $\begin{aligned} 3x_1 - x_2 &= \cos(12) \\ 2x_1 - x_2 &= -5 \end{aligned}$

19. $\begin{aligned} 3x_1 - \sin(x_2) &= 12 \\ 2x_1 - x_3 &= -5 \end{aligned}$

COMPUTER EXERCISES

In Exercises 20–22, use elementary row operations and MATLAB to put each of the given matrices into row echelon form. Suppose that the matrix is the augmented matrix for a system of linear equations. Is the system consistent or inconsistent?

20. $\begin{pmatrix} 2 & 1 & 1 \\ 4 & 2 & 3 \end{pmatrix}$

21. $\begin{pmatrix} 3 & -4 & 0 & 2 \\ 0 & 2 & 3 & 1 \\ 3 & 1 & 4 & 5 \end{pmatrix}$

22. $\begin{pmatrix} -2 & 1 & 9 & 1 \\ 3 & 3 & -4 & 2 \\ 1 & 4 & 5 & 5 \end{pmatrix}$

Observation: In standard format MATLAB displays all nonzero real numbers with four decimal places while it displays zero as 0. An unfortunate consequence of this display is that when a matrix has both zero and noninteger entries, the columns will not align—which is a nuisance. You can work with rational numbers rather than decimal numbers by typing `format rational`. Then the columns will align.

23. Load the following 6×8 matrix A into MATLAB by typing e2_3_16:

$$A = \begin{pmatrix} 0 & 0 & 0 & 1 & 3 & 5 & 0 & 9 \\ 0 & 3 & 6 & -6 & -6 & -12 & 0 & 1 \\ 0 & 2 & 4 & -5 & -7 & 14 & 0 & 1 \\ 0 & 1 & 2 & 1 & 14 & 21 & 0 & -1 \\ 0 & 0 & 0 & 2 & 4 & 9 & 0 & 7 \\ 0 & 5 & 10 & -11 & -13 & 2 & 0 & 2 \end{pmatrix}. \tag{2.3.16}*$$

Use MATLAB to transform this matrix to row echelon form.

24. Use row reduction and back substitution to solve the following system of linear equations:

$$\begin{aligned} 2x_1 + 3x_2 - 4x_3 + x_4 &= 2 \\ 3x_1 - x_2 - x_3 + 2x_4 &= 4 \\ x_1 - 7x_2 + 5x_3 - x_4 &= 6. \end{aligned}$$

25. Comment: To understand the point of this exercise you must begin by typing the MATLAB command `format short e`. This command will set a format in which you can see the difficulties that sometimes arise in numerical computations.

Consider the following two 3×3-matrices:

$$A = \begin{pmatrix} 1 & 3 & 4 \\ 2 & 1 & 1 \\ -4 & 3 & 5 \end{pmatrix} \qquad B = \begin{pmatrix} 3 & 1 & 4 \\ .1 & 2 & 1 \\ 3 & -4 & 5 \end{pmatrix}. \tag{2.3.17}*$$

Note that matrix B is obtained from matrix A by interchanging the first two columns.

(a) Use MATLAB to put A into row echelon form using these transformations:

    1. Subtract two times the 1st row from the 2nd row.

    2. Add four times the 1st row to the 3rd row.

    3. Divide the 2nd row by -5.

    4. Subtract 15 times the 2nd row from the 3rd row.

(b) Put B by hand into row echelon form using these transformations:

    1. Divide the 1st row by 3.

    2. Subtract the 1st row from the 2nd row.

    3. Subtract three times the 1st row from the 3rd row.

    4. Multiply the 2nd row by $\frac{3}{5}$.

    5. Add five times the 2nd row to the 3rd row.

(c) Use MATLAB to put B into row echelon form using the same transformations as in (b).

(d) Discuss the outcome of the three transformations. Is there a difference in the results? Would you expect to see a difference? Could the difference be crucial when solving a system of linear equations?

26. Find a cubic polynomial

$$p(x) = ax^3 + bx^2 + cx + d$$

so that $p(1) = 2$, $p(2) = 3$, $p'(-1) = -1$, and $p'(3) = 1$.

2.4   REDUCTION TO ECHELON FORM

In this section we formalize our previous numerical experiments. We define more precisely the notions of echelon form and reduced echelon form matrices, and we prove that every matrix can be put into reduced echelon form using a sequence of elementary row operations. Consequently, we will have developed an algorithm for determining whether a system of linear equations is consistent or inconsistent, and for determining all solutions to a consistent system.

Definition 2.4.1   *A matrix E is in (row)* echelon form *if two conditions hold:*

(a)   *The first nonzero entry in each row of E is equal to 1. This leading entry 1 is called a* pivot.

(b)   *A pivot in the $(i + 1)$st row of E occurs in a column to the right of the column where the pivot in the ith row occurs.*

Here are three examples of matrices that are in echelon form. The pivot in each row (which is always a 1) is preceded by a $*$:

$$\begin{pmatrix} *1 & 0 & -1 & 0 & -6 & 4 & -6 \\ 0 & *1 & 4 & 0 & 0 & -2 & 0 \\ 0 & 0 & 0 & *1 & -5 & 5 & -2 \\ 0 & 0 & 0 & 0 & 0 & *1 & 0 \end{pmatrix}$$

$$\begin{pmatrix} *1 & 0 & -1 & 0 & -6 \\ 0 & *1 & 0 & 3 & 0 \\ 0 & 0 & 0 & *1 & -5 \\ 0 & 0 & 0 & 0 & 0 \end{pmatrix}$$

$$\begin{pmatrix} 0 & *1 & -1 & 14 & -6 \\ 0 & 0 & 0 & *1 & 15 \\ 0 & 0 & 0 & 0 & 0 \\ 0 & 0 & 0 & 0 & 0 \end{pmatrix}$$

Here are three examples of matrices that are *not* in echelon form.

$$\begin{pmatrix} 0 & 0 & 1 & 15 \\ 1 & -1 & 14 & -6 \\ 0 & 0 & 0 & 0 \end{pmatrix} \qquad \begin{pmatrix} 1 & -1 & 14 & -6 \\ 0 & 0 & 3 & 15 \\ 0 & 0 & 0 & 0 \end{pmatrix} \qquad \begin{pmatrix} 1 & -1 & 14 & -6 \\ 0 & 0 & 0 & 0 \\ 0 & 0 & 1 & 15 \end{pmatrix}$$

Definition 2.4.2   *Two $m \times n$ matrices are* row equivalent *if one can be transformed to the other by a sequence of elementary row operations.*

Let $A = (a_{ij})$ be a matrix with m rows and n columns. We want to show that we can perform row operations on A so that the transformed matrix is in echelon form; that is, A is row equivalent to a matrix in echelon form. If $A = 0$, then we are finished. So we assume that some entry in A is nonzero and that the first column where that nonzero entry occurs is the kth column. By swapping rows, we can assume that a_{1k} is nonzero. Next we divide the 1st row by a_{1k}, thus setting $a_{1k} = 1$. Now, using MATLAB notation, we perform the row operations

```
A(i,:) = A(i,:) - A(i,k)*A(1,:)
```

for each $i \geq 2$. This sequence of row operations leads to a matrix whose first nonzero column has a 1 in the 1st row and a 0 in each row below the 1st row.

Now we look for the next column that has a nonzero entry below the 1st row and call that column ℓ. By construction $\ell > k$. We can swap rows so that the entry in the 2nd row, ℓth column is nonzero. Then we divide the 2nd row by this nonzero element, so that the pivot in the 2nd row is 1. Again we perform elementary row operations so that all entries below the 2nd row in the ℓth column are set to 0. Now we proceed inductively until we run out of nonzero rows.

This argument proves:

Proposition 2.4.3   *Every matrix is row equivalent to a matrix in echelon form.*

More important, the previous argument provides an algorithm for transforming matrices into echelon form.

Reduction to Reduced Echelon Form

Definition 2.4.4   *A matrix E is in* reduced echelon form *if:*

(a) E is in echelon form, and

(b)   in every column of E having a pivot, every entry in that column other than the pivot is 0.

We can now prove the next theorem.

Theorem 2.4.5   *Every matrix is row equivalent to a matrix in reduced echelon form.*

Proof: Let A be a matrix. Proposition 2.4.3 states that we can transform A by elementary row operations into a matrix E in echelon form. Next we transform E into reduced echelon form by some additional elementary row operations, as follows: Choose the pivot in the last nonzero row of E. Call that row ℓ, and let k be the column where the pivot occurs. By adding multiples of the ℓth row to the rows above, we can transform each entry in the kth column above the pivot to 0. Note that none of these row operations alters the matrix before the kth column. (Also note that this process is identical to the process of back substitution.)

Again we proceed inductively by choosing the pivot in the $(\ell - 1)$st row, which is 1, and zeroing out all entries above that pivot using elementary row operations. ◆

Reduced Echelon Form in MATLAB

Preprogrammed into MATLAB is a routine to row reduce any matrix to reduced echelon form. The command is `rref`. For example, recall the 4×7 matrix A in (2.3.12) by typing e2_3_12. Put A into reduced row echelon form by typing `rref(A)` and obtaining

```
ans =
   1    0    0    0   -6    4   -6
   0    1    0    0   10  -20    0
   0    0    1    0   -5    5   -2
   0    0    0    1    0    2    1
```

Compare the result with the system of equations (2.3.13).

Solutions to Systems of Linear Equations

We introduced elementary row operations as operations that do not change solutions to the linear system. More precisely, we discussed how solutions to the original system are still solutions to the transformed system and how no new solutions are introduced by elementary row operations. This argument is most easily seen by observing that **all elementary row operations are invertible**; they can be undone.

For example, swapping two rows is undone by just swapping these rows again. Similarly, multiplying a row by a nonzero number c is undone by just dividing that same row by c. Finally, adding c times the jth row to the ith row is undone by subtracting c times the jth row from the ith row.

Thus we can make several observations about solutions to linear systems. Let E be an augmented matrix corresponding to a system of linear equations having n variables. Since an augmented matrix is formed from the matrix of coefficients by adding a column, we see that the augmented matrix has $n + 1$ columns.

Theorem 2.4.6 *Suppose that E is an $m \times (n + 1)$ augmented matrix that is in reduced echelon form. Let ℓ be the number of nonzero rows in E.*

(a) *The system of linear equations corresponding to E is inconsistent if and only if the ℓth row in E has a pivot in the $(n + 1)$st column.*

(b) *If the linear system corresponding to E is consistent, then the set of all solutions is parameterized by n − ℓ parameters.*

Proof: Suppose that the last nonzero row in E has its pivot in the $(n + 1)$st column. Then the corresponding equation is

$$0x_1 + 0x_2 + \cdots + 0x_n = 1,$$

which has no solutions. Thus the system is inconsistent.

Conversely, suppose that the last nonzero row has its pivot before the last column. Without loss of generality, we can renumber the columns—that is, we can renumber the variables x_j—so that the pivot in the ith row occurs in the ith column, where $1 \leq i \leq \ell$. Then the associated system of linear equations has the form:

$$x_1 + a_{1,\ell+1}x_{\ell+1} + \cdots + a_{1,n}x_n = b_1$$
$$x_2 + a_{2,\ell+1}x_{\ell+1} + \cdots + a_{2,n}x_n = b_2$$
$$\vdots \quad \vdots$$
$$x_\ell + a_{\ell,\ell+1}x_{\ell+1} + \cdots + a_{\ell,n}x_n = b_\ell.$$

This system can be rewritten in the form:

$$\begin{aligned}
x_1 &= b_1 - a_{1,\ell+1}x_{\ell+1} - \cdots - a_{1,n}x_n \\
x_2 &= b_2 - a_{2,\ell+1}x_{\ell+1} - \cdots - a_{2,n}x_n \\
&\vdots \quad \vdots \\
x_\ell &= b_\ell - a_{\ell,\ell+1}x_{\ell+1} - \cdots - a_{\ell,n}x_n.
\end{aligned} \qquad (2.4.1)$$

Thus each choice of the $n - \ell$ numbers $x_{\ell+1}, \ldots, x_n$ uniquely determines values of $x_1, \ldots, x_\ell$ so that $x_1, \ldots, x_n$ is a solution to this system. In particular, the system is consistent, so (a) is proved; and the set of all solutions is parameterized by $n - \ell$ numbers, so (b) is proved. ◆

Two Examples Illustrating Theorem 2.4.6

The reduced echelon form matrix

$$E = \begin{pmatrix} 1 & 5 & 0 & | & 0 \\ 0 & 0 & 1 & | & 0 \\ 0 & 0 & 0 & | & 1 \end{pmatrix}$$

is the augmented matrix of an inconsistent system of three equations in three unknowns.

The reduced echelon form matrix

$$E = \begin{pmatrix} 1 & 5 & 0 & | & 2 \\ 0 & 0 & 1 & | & 5 \\ 0 & 0 & 0 & | & 0 \end{pmatrix}$$

is the augmented matrix of a consistent system of three equations in three unknowns x_1, x_2, x_3. For this matrix $n = 3$ and $\ell = 2$. It follows from Theorem 2.4.6 that the solutions to this system are specified by one parameter. Indeed the solutions are

$$x_1 = 2 - 5x_2$$
$$x_3 = 5$$

and are specified by the one parameter x_2.

Consequences of Theorem 2.4.6

It follows from Theorem 2.4.6 that linear systems of equations with fewer equations than unknowns and with 0s on the right-hand side always have nonzero solutions. More precisely:

Corollary 2.4.7   *Let A be an $m \times n$ matrix where $m < n$. Then the system of linear equations whose augmented matrix is $(A|0)$ has a nonzero solution.*

Proof: Perform elementary row operations on the augmented matrix $(A|0)$ to arrive at the reduced echelon form matrix $(E|0)$. Since the zero vector is a solution, the associated system of equations is consistent. Now the number of nonzero rows ℓ in $(E|0)$ is less than or equal to the number of rows m in E. By assumption $m < n$ and hence $\ell < n$. It follows from Theorem 2.4.6 that solutions to the linear system are parameterized by $n - \ell \geq 1$ parameters and that there are nonzero solutions.   ◆

Recall that two $m \times n$ matrices are row equivalent if one can be transformed to the other by elementary row operations.

Corollary 2.4.8   *Let A be an $n \times n$ square matrix and let b be in $\mathbf{R}^n$. Then A is row equivalent to the identity matrix I_n if and only if the system of linear equations whose augmented matrix is $(A|b)$ has a unique solution.*

Proof: Suppose that A is row equivalent to I_n. Then, if we use the same sequence of elementary row operations, it follows that the $n \times (n + 1)$ augmented matrix $(A|b)$ is row equivalent to $(I_n|c)$ for some vector $c \in \mathbf{R}^n$. The system of linear equations that corresponds to $(I_n|c)$ is

$$
\begin{aligned}
x_1 &= c_1 \\
\vdots\ \ \vdots\ \ &\ \ \vdots \\
x_n &= c_n,
\end{aligned}
$$

which clearly has the unique solution $x = (c_1, \ldots, c_n)$. Since elementary row operations do not change the solutions of the equations, the original augmented system $(A|b)$ also has a unique solution.

Conversely, suppose that the system of linear equations associated with $(A|b)$ has a unique solution. Suppose that $(A|b)$ is row equivalent to a reduced echelon form matrix E. Suppose that the last nonzero row in E is the ℓth row. Since the system has a solution, it is consistent. Hence Theorem 2.4.6(b) implies that the solutions to the system corresponding to E are parameterized by $n - \ell$ parameters. If $\ell < n$, then the solution is not unique, so $\ell = n$.

Next observe that since the system of linear equations is consistent, it follows from Theorem 2.4.6(a) that the pivot in the nth row must occur in a column before the $(n + 1)$st. Therefore the reduced echelon matrix $E = (I_n|c)$ for some $c \in \mathbf{R}^n$. Since $(A|b)$ is row equivalent to $(I_n|c)$, it follows, by using the same sequence of elementary row operations, that A is row equivalent to I_n. ◆

Uniqueness of Reduced Echelon Form and Rank

Abstractly, our discussion of reduced echelon form has one point remaining to be proved. We know that every matrix A can be transformed by elementary row operations to reduced echelon form. Suppose, however, that we use two different sequences of elementary row operations to transform A to two reduced echelon form matrices E_1 and E_2. Can E_1 and E_2 be different? The answer is no.

Theorem 2.4.9 *For each matrix A, there is precisely one reduced echelon form matrix E that is row equivalent to A.*

The proof of Theorem 2.4.9 is given in Section 2.6. Since every matrix is row equivalent to a unique matrix in reduced echelon form, we can define the rank of a matrix as follows:

Definition 2.4.10 *Let A be an m × n matrix that is row equivalent to a reduced echelon form matrix E. Then the* rank *of A, denoted* rank *(A), is the number of nonzero rows in E.*

We make three remarks concerning the rank of a matrix:

- An echelon form matrix is always row equivalent to a reduced echelon form matrix with the same number of nonzero rows. Thus, to compute the rank of a matrix, we need only perform elementary row operations until the matrix is in echelon form.
- The rank of any matrix is easily computed in MATLAB. Enter a matrix A and type rank(A).
- The number ℓ in the statement of Theorem 2.4.6 is just the rank of E.

In particular, if the rank of the augmented matrix corresponding to a consistent system of linear equations in n unknowns has rank ℓ, then the solutions to this system are parameterized by $n - \ell$ parameters.

HAND EXERCISES

In Exercises 1 and 2, row reduce the given matrix to reduced echelon form and determine the rank of A.

1. $A = \begin{pmatrix} 1 & 2 & 1 & 6 \\ 3 & 6 & 1 & 14 \\ 1 & 2 & 2 & 8 \end{pmatrix}$
2. $B = \begin{pmatrix} 1 & -2 & 3 \\ 3 & -6 & 9 \\ 1 & -8 & 2 \end{pmatrix}$

3. The augmented matrix of a consistent system of five equations in seven unknowns has rank equal to three. How many parameters are needed to specify all solutions?

4. The augmented matrix of a consistent system of nine equations in 12 unknowns has rank equal to five. How many parameters are needed to specify all solutions?

COMPUTER EXERCISES

In Exercises 5–8, use rref on the given augmented matrices to determine whether the associated system of linear equations is consistent or inconsistent. If the equations are consistent, then determine how many parameters are needed to enumerate all solutions.

5.

$$A = \begin{pmatrix} 2 & 1 & 3 & -2 & 4 & | & 1 \\ 5 & 12 & -1 & 3 & 5 & | & 1 \\ -4 & -21 & 11 & -12 & 2 & | & 1 \\ 23 & 59 & -8 & 17 & 21 & | & 4 \end{pmatrix}$$ (2.4.2)*

6.

$$B = \begin{pmatrix} 2 & 4 & 6 & -2 & | & 1 \\ 0 & 0 & 4 & 1 & | & -1 \\ 2 & 4 & 0 & 1 & | & 2 \end{pmatrix}$$ (2.4.3)*

7.

$$C = \begin{pmatrix} 2 & 3 & -1 & | & 4 \\ 8 & 11 & -7 & | & 8 \\ 2 & 2 & -4 & | & -3 \end{pmatrix}$$ (2.4.4)*

8.

$$D = \begin{pmatrix} 2.3 & 4.66 & -1.2 & 2.11 & -2 \\ 0 & 0 & 1.33 & 0 & 1.44 \\ 4.6 & 9.32 & -7.986 & 4.22 & -10.048 \\ 1.84 & 3.728 & -5.216 & 1.688 & -6.208 \end{pmatrix}$$ **(2.4.5)***

In Exercises 9–11, compute the rank of each matrix.

9. $\begin{pmatrix} 1 & -2 \\ -3 & 6 \end{pmatrix}$ **10.** $\begin{pmatrix} 2 & 1 & 0 & 1 \\ -1 & 3 & 2 & 4 \\ 5 & -1 & 2 & -2 \end{pmatrix}$ **11.** $\begin{pmatrix} 3 & 1 & 0 \\ -1 & 2 & 4 \\ 2 & 3 & 4 \\ 4 & -1 & -4 \end{pmatrix}$

2.5   LINEAR EQUATIONS WITH SPECIAL COEFFICIENTS

In this chapter we have shown how to use elementary row operations to solve systems of linear equations. We have assumed that each linear equation in the system has the form

$$a_{j1}x_1 + \cdots + a_{jn}x_n = b_j,$$

where the a_{ji}s and the b_js are real numbers. For simplicity, in our examples we have chosen only equations with integer coefficients, such as

$$2x_1 - 3x_2 + 15x_3 = -1.$$

Systems with Nonrational Coefficients

In fact, a more general choice of coefficients for a system of two equations might have been

$$\sqrt{2}x_1 + 2\pi x_2 = 22.4$$
$$3x_1 + 36.2x_2 = e.$$ **(2.5.1)**

Suppose that we solve (2.5.1) by elementary row operations. In matrix form we have the augmented matrix

$$\begin{pmatrix} \sqrt{2} & 2\pi & 22.4 \\ 3 & 36.2 & e \end{pmatrix}.$$

Proceed with the following elementary row operations: Divide the 1st row by $\sqrt{2}$ to obtain

$$\left(\begin{array}{cc|c} 1 & \pi\sqrt{2} & 11.2\sqrt{2} \\ 3 & 36.2 & e \end{array}\right).$$

Next, subtract three times the 1st row from the 2nd row to get

$$\left(\begin{array}{cc|c} 1 & \pi\sqrt{2} & 11.2\sqrt{2} \\ 0 & 36.2 - 3\pi\sqrt{2} & e - 33.6\sqrt{2} \end{array}\right).$$

Then divide the 2nd row by $36.2 - 3\pi\sqrt{2}$, obtaining

$$\left(\begin{array}{cc|c} 1 & \pi\sqrt{2} & 11.2\sqrt{2} \\ 0 & 1 & \dfrac{e - 33.6\sqrt{2}}{36.2 - 3\pi\sqrt{2}} \end{array}\right).$$

Finally, multiply the 2nd row by $\pi\sqrt{2}$ and subtract it from the 1st row to get

$$\left(\begin{array}{cc|c} 1 & 0 & 11.2\sqrt{2} - \pi\sqrt{2}\,\dfrac{e - 33.6\sqrt{2}}{36.2 - 3\pi\sqrt{2}} \\ 0 & 1 & \dfrac{e - 33.6\sqrt{2}}{36.2 - 3\pi\sqrt{2}} \end{array}\right).$$

So

$$\begin{aligned} x_1 &= 11.2\sqrt{2} - \pi\sqrt{2}\,\frac{e - 33.6\sqrt{2}}{36.2 - 3\pi\sqrt{2}} \\ x_2 &= \frac{e - 33.6\sqrt{2}}{36.2 - 3\pi\sqrt{2}}, \end{aligned} \tag{2.5.2}$$

which is both hideous to look at and quite uninformative. It is, however, correct.

Both x_1 and x_2 are real numbers. They had to be because all of the manipulations involved addition, subtraction, multiplication, and division of real numbers—which yield real numbers.

If we want to use MATLAB to perform these calculations, we have to convert $\sqrt{2}$, π, and e to their decimal equivalents—at least up to a certain decimal place accuracy. This introduces errors, which for the moment we assume are small.

To enter A and b into MATLAB, type

```
A = [sqrt(2) 2*pi; 3 36.2];
b = [22.4; exp(1)];
```

Now type A to obtain

```
A =
    1.4142    6.2832
    3.0000   36.2000
```

As its default display, MATLAB displays real numbers to four decimal place accuracy. Similarly, type b to get

```
b =
   22.4000
    2.7183
```

Next use MATLAB to solve this system by typing

```
A\b
```

to obtain

```
ans =
   24.5417
   -1.9588
```

We can check that this answer agrees with the answer in (2.5.2) to MATLAB output accuracy by typing

```
x1 = 11.2*sqrt(2)-pi*sqrt(2)*(exp(1)-33.6*sqrt(2))/(36.2-3*pi*sqrt(2))
x2 = (exp(1)-33.6*sqrt(2))/(36.2-3*pi*sqrt(2))
```

to obtain

```
x1 =
   24.5417
```

and

```
x2 =
   -1.9588
```

More Accuracy

MATLAB can display numbers in machine precision (15 digits) rather than the standard four decimal place accuracy. To change to this display, type

```
format long
```

Now solve the system of equations (2.5.1) again by typing

```
A\b
```

and obtaining

```
ans =
   24.54169560069650
   -1.95875151860858
```

Systems with Integers and Rational Numbers

Now suppose that all of the coefficients in a system of linear equations are integers. When we add, subtract, or multiply integers, we get integers. In general, however, when we divide an integer by an integer we get a rational number rather than an integer. Indeed, since elementary row operations involve only the operations of addition, subtraction, multiplication, and division, we see that if we perform elementary row operations on a matrix with integer entries, we end up with a matrix that has rational numbers as entries.

MATLAB can display calculations using rational numbers rather than decimal numbers. To display calculations using only rational numbers, type

```
format rational
```

For example, let

$$A = \begin{pmatrix} 2 & 2 & 1 & 0 \\ 1 & 3 & -5 & 1 \\ 4 & 2 & 1 & 3 \\ 2 & 1 & -1 & 4 \end{pmatrix} \qquad (2.5.3)*$$

and let

$$b = \begin{pmatrix} 1 \\ 1 \\ -5 \\ 2 \end{pmatrix}. \qquad (2.5.4)*$$

Enter A and b into MATLAB by typing

```
e2_5_3
e2_5_4
```

Solve the system by typing

```
A\b
```

to obtain

```
ans =
   -357/41
    309/41
    137/41
    156/41
```

To display the answer in standard decimal form, type

```
format
A\b
```

obtaining

```
ans =
   -8.7073
    7.5366
    3.3415
    3.8049
```

The same logic shows that if we begin with a system of equations whose coefficients are rational numbers, then we obtain an answer consisting of rational numbers—since adding, subtracting, multiplying, and dividing rational numbers yield rational numbers. More precisely:

Theorem 2.5.1   *Let A be an $n \times n$ matrix that is row equivalent to I_n, and let b be an n vector. Suppose that all entries of A and b are rational numbers. Then there is a unique solution to the system corresponding to the augmented matrix $(A|b)$ and this solution has rational numbers as entries.*

Proof: Since A is row equivalent to I_n, Corollary 2.4.8 states that this linear system has a unique solution x. As we have just discussed, solutions are found using elementary row operations; hence the entries of x are rational numbers.   ◆

Systems with Complex Numbers

Earlier in this section we discussed why solutions to linear systems whose coefficients are rational numbers must themselves have entries that are rational numbers. We now discuss solving linear equations whose coefficients are more general than real numbers—that is, whose coefficients are complex numbers.

Recall that addition, subtraction, multiplication, and division of complex numbers yield complex numbers. Suppose that

$$a = \alpha + i\beta$$
$$b = \gamma + i\delta$$

where α, β, γ, and δ are real numbers and $i = \sqrt{-1}$. Then

$$a + b = (\alpha + \gamma) + i(\beta + \delta)$$
$$a - b = (\alpha - \gamma) + i(\beta - \delta)$$
$$ab = (\alpha\gamma - \beta\delta) + i(\alpha\delta + \beta\gamma)$$
$$\frac{a}{b} = \frac{\alpha\gamma + \beta\delta}{\gamma^2 + \delta^2} + i\frac{\beta\gamma - \alpha\delta}{\gamma^2 + \delta^2}.$$

MATLAB has been programmed to do arithmetic with complex numbers using exactly the same instructions it uses to do arithmetic with real and rational numbers. For example, we can solve the system of linear equations

$$(4 - i)x_1 + 2x_2 = 3 - i$$
$$2x_1 + (4 - 3i)x_2 = 2 + i$$

in MATLAB by typing

```
A = [4-i 2; 2 4-3i];
b = [3-i; 2+i];
A\b
```

The solution to this system of equations is

```
ans =
   0.8457 - 0.1632i
  -0.1098 + 0.2493i
```

Note: Care must be taken when entering complex numbers into arrays in MAT-LAB. For example, if you type

```
b = [3 -i; 2 +i]
```

then MATLAB responds with the 2×2 matrix

```
b =
   3.0000                   0 - 1.0000i
   2.0000                   0 + 1.0000i
```

Typing either b = [3-i; 2+i] or b = [3 - i; 2 + i] yields the desired 2×1 column vector.

All of the theorems concerning the existence and uniqueness of row echelon form—and for solving systems of linear equations—work when the coefficients of the linear system are complex numbers as opposed to real numbers. In particular:

Theorem 2.5.2 *If the coefficients of a system of n linear equations in n unknowns are complex numbers and if the coefficient matrix is row equivalent to I_n, then there is a unique solution to this system whose entries are complex numbers.*

Complex Conjugation

Let $a = \alpha + i\beta$ be a complex number. Then the *complex conjugate* of a is defined to be

$$\bar{a} = \alpha - i\beta.$$

Let $a = \alpha + i\beta$ and $c = \gamma + i\delta$ be complex numbers. Then we claim that

$$\overline{a + c} = \bar{a} + \bar{c}$$
$$\overline{ac} = \bar{a}\,\bar{c}. \tag{2.5.5}$$

To verify these statements, calculate

$$\overline{a + c} = \overline{(\alpha + \gamma) + i(\beta + \delta)} = (\alpha + \gamma) - i(\beta + \delta) = (\alpha - i\beta) + (\gamma - i\delta) = \bar{a} + \bar{c}$$

and

$$\overline{ac} = \overline{(\alpha\gamma - \beta\delta) + i(\alpha\delta + \beta\gamma)} = (\alpha\gamma - \beta\delta) - i(\alpha\delta + \beta\gamma) = (\alpha - i\beta)(\gamma - i\delta) = \bar{a}\,\bar{c}.$$

HAND EXERCISES

1. Solve the system of equations

$$\begin{aligned} x_1 - ix_2 &= 1 \\ ix_1 + 3x_2 &= -1. \end{aligned}$$

Check your answer using MATLAB.

Solve the systems of linear equations in Exercises 2 and 3, and verify that the answers are rational numbers.

2. $\begin{aligned} x_1 + x_2 - 2x_3 &= 1 \\ x_1 + x_2 + x_3 &= 2 \\ x_1 - 7x_2 + x_3 &= 3 \end{aligned}$

3. $\begin{aligned} x_1 - x_2 &= 1 \\ x_1 + 3x_2 &= -1 \end{aligned}$

COMPUTER EXERCISES

In Exercises 4–6, use MATLAB to solve each system of linear equations to four significant decimal places.

4. $\begin{aligned} 0.1x_1 + \sqrt{5}x_2 - 2x_3 &= 1 \\ -\sqrt{3}x_1 + \pi x_2 - 2.6x_3 &= 14.3 \\ x_1 - 7x_2 + \tfrac{\pi}{2}x_3 &= \sqrt{2} \end{aligned}$

5. $\begin{aligned} (4-i)x_1 + (2+3i)x_2 &= -i \\ ix_1 - 4x_2 &= 2.2 \end{aligned}$

6. $\begin{aligned} (2+i)x_1 + (\sqrt{2}-3i)x_2 - 10.66x_3 &= 4.23 \\ 14x_1 - \sqrt{5}ix_2 + (10.2-i)x_3 &= 3-1.6i \\ -4.276x_1 + 2x_2 - (4-2i)x_3 &= \sqrt{2}i \end{aligned}$

Hint When entering $\sqrt{2}i$ in MATLAB, you must type sqrt(2)*i, even though when you enter $2i$, you can type just 2i.

2.6 *UNIQUENESS OF REDUCED ECHELON FORM

In this section we prove Theorem 2.4.9, which states that every matrix is row equivalent to precisely one reduced echelon form matrix.

Proof of Theorem 2.4.9: Suppose that E and F are two $m \times n$ reduced echelon matrices that are row equivalent to A. Since elementary row operations are invertible, the two matrices E and F are row equivalent. Thus the systems of linear equations associated with the $m \times (n+1)$ matrices $(E|0)$ and $(F|0)$ must have exactly the same set of solutions. It is the fact that the solution sets of the linear equations associated with $(E|0)$ and $(F|0)$ are identical that allows us to prove that $E = F$.

Begin by renumbering the variables $x_1, \ldots, x_n$ so that the equations associated with $(E|0)$ have the form

$$
\begin{aligned}
x_1 &= -a_{1,\ell+1}x_{\ell+1} - \cdots - a_{1,n}x_n \\
x_2 &= -a_{2,\ell+1}x_{\ell+1} - \cdots - a_{2,n}x_n \\
&\vdots \qquad \vdots \\
x_\ell &= -a_{\ell,\ell+1}x_{\ell+1} - \cdots - a_{\ell,n}x_n.
\end{aligned}
\qquad (2.6.1)
$$

In this form, pivots of E occur in columns $1, \ldots, \ell$. We begin by showing that the matrix F also has pivots in columns $1, \ldots, \ell$. Moreover, there is a unique solution to these equations for *every* choice of numbers $x_{\ell+1}, \ldots, x_n$.

Suppose that the pivots of F do not occur in columns $1, \ldots, \ell$. Then there is a row in F whose first nonzero entry occurs in a column $k > \ell$. This row corresponds to an equation

$$
x_k = c_{k+1}x_{k+1} + \cdots + c_n x_n.
$$

Now consider solutions that satisfy

$$
x_{\ell+1} = \cdots = x_{k-1} = 0 \quad \text{and} \quad x_{k+1} = \cdots = x_n = 0.
$$

In the equations associated with the matrix $(E|0)$, a unique solution is associated with every number x_k, whereas in the equations associated with the matrix $(F|0)$, x_k must be 0 to be a solution. This argument contradicts the fact that the $(E|0)$ equations and the $(F|0)$ equations have the same solutions. So the pivots of F must also occur in columns $1, \ldots, \ell$, and the equations associated with F must have the form

$$
\begin{aligned}
x_1 &= -\hat{a}_{1,\ell+1}x_{\ell+1} - \cdots - \hat{a}_{1,n}x_n \\
x_2 &= -\hat{a}_{2,\ell+1}x_{\ell+1} - \cdots - \hat{a}_{2,n}x_n \\
&\vdots \qquad \vdots \\
x_\ell &= -\hat{a}_{\ell,\ell+1}x_{\ell+1} - \cdots - \hat{a}_{\ell,n}x_n,
\end{aligned}
\qquad (2.6.2)
$$

where $\hat{a}_{i,j}$ are scalars.

To complete this proof, we show that $a_{i,j} = \hat{a}_{i,j}$. These equalities are verified as follows: There is just one solution to each system (2.6.1) and (2.6.2) of the form

$$
x_{\ell+1} = 1, \quad x_{\ell+2} = \cdots = x_n = 0.
$$

These solutions are

$$
(-a_{1,\ell+1}, \ldots, -a_{\ell,\ell+1}, 1, 0, \ldots, 0)
$$

for (2.6.1) and

$$
(-\hat{a}_{1,\ell+1}, \ldots, -\hat{a}_{\ell,\ell+1}, 1, 0 \ldots, 0)
$$

for (2.6.2). It follows that $a_{j,\ell+1} = \hat{a}_{j,\ell+1}$ for $j = 1, \ldots, \ell$. Complete this proof by repeating this argument. Just inspect solutions of the form

$$x_{\ell+1} = 0, \quad x_{\ell+2} = 1, \quad x_{\ell+3} = \cdots = x_n = 0$$

through

$$x_{\ell+1} = \cdots = x_{n-1} = 0, \quad x_n = 1.$$

◆

Matrices and Linearity

In this chapter we take the first step in abstracting vectors and matrices to mathematical objects that are more than just arrays of numbers. We begin the discussion in Section 3.1 by introducing the multiplication of a matrix times a vector. Matrix multiplication simplifies the way in which we write systems of linear equations and is the way by which we view matrices as mappings. This latter point is discussed in Section 3.2.

The mappings that are produced by matrix multiplication are special and are called *linear mappings*. Some properties of linear maps are discussed in Section 3.3. One consequence of linearity is the *principle of superposition,* which enables solutions to systems of linear equations to be built out of simpler solutions. This principle is discussed in Section 3.4.

In Section 3.5 we introduce multiplication of two matrices, and we discuss properties of this multiplication in Section 3.6. Matrix multiplication is defined in terms of composition of linear mappings, which leads to an explicit formula for matrix multiplication. This dual role of multiplication of two matrices—first by formula and second as composition—enables us to solve linear equations in a conceptual way as well as in an algorithmic way. The conceptual way of solving linear equations uses matrix inverses (or inverse mappings), which is described in Section 3.7. In this section we also present important properties of matrix inversion and a method of computation of matrix inverses. There is a simple formula for computing inverses of 2×2 matrices based on determinants. The chapter ends with a discussion of determinants of 2×2 matrices in Section 3.8.

3.1   MATRIX MULTIPLICATION OF VECTORS

In Chapter 2 we discussed how matrices appear when solving systems of m linear equations in n unknowns. Given the system

$$
\begin{aligned}
a_{11}x_1 + a_{12}x_2 + \cdots + a_{1n}x_n &= b_1 \\
a_{21}x_1 + a_{22}x_2 + \cdots + a_{2n}x_n &= b_2 \\
\vdots \qquad\qquad \vdots \qquad\quad \vdots \qquad\qquad \vdots \\
a_{m1}x_1 + a_{m2}x_2 + \cdots + a_{mn}x_n &= b_m ,
\end{aligned}
\tag{3.1.1}
$$

we saw that all relevant information is contained in the $m \times n$ matrix of coefficients

$$
A = \begin{pmatrix}
a_{11} & a_{12} & \cdots & a_{1n} \\
a_{21} & a_{22} & \cdots & a_{2n} \\
\vdots & \vdots &   & \vdots \\
a_{m1} & a_{m2} & \cdots & a_{mn}
\end{pmatrix}
$$

and the n-vector

$$
b = \begin{pmatrix} b_1 \\ \vdots \\ b_n \end{pmatrix}.
$$

Matrices Times Vectors

We present multiplication of a matrix times a vector as a notational advantage that simplifies the presentation of the linear systems. It is, however, much more than that. This concept of multiplication allows us to think of matrices as mappings and these mappings tell us much about the structure of solutions to linear systems. But first we discuss the notational advantage.

Multiplying an $m \times n$ matrix A times an n-vector x produces an m-vector, as follows:

$$
Ax = \begin{pmatrix}
a_{11} & \cdots & a_{1n} \\
\vdots &   & \vdots \\
a_{m1} & \cdots & a_{mn}
\end{pmatrix}
\begin{pmatrix} x_1 \\ \vdots \\ x_n \end{pmatrix}
=
\begin{pmatrix}
a_{11}x_1 + \cdots + a_{1n}x_n \\
\vdots \\
a_{m1}x_1 + \cdots + a_{mn}x_n
\end{pmatrix}.
\tag{3.1.2}
$$

For example, when $m = 2$ and $n = 3$, the product is a 2-vector:

$$
\begin{pmatrix}
a_{11} & a_{12} & a_{13} \\
a_{21} & a_{22} & a_{23}
\end{pmatrix}
\begin{pmatrix} x_1 \\ x_2 \\ x_3 \end{pmatrix}
=
\begin{pmatrix}
a_{11}x_1 + a_{12}x_2 + a_{13}x_3 \\
a_{21}x_1 + a_{22}x_2 + a_{23}x_3
\end{pmatrix}.
\tag{3.1.3}
$$

As a specific example, compute

$$\begin{pmatrix} 2 & 3 & -1 \\ 4 & 1 & 5 \end{pmatrix} \begin{pmatrix} 2 \\ -3 \\ 4 \end{pmatrix} = \begin{pmatrix} 2 \cdot 2 + 3 \cdot (-3) + (-1) \cdot 4 \\ 4 \cdot 2 + 1 \cdot (-3) + 5 \cdot 4 \end{pmatrix} = \begin{pmatrix} -9 \\ 25 \end{pmatrix}.$$

Using (3.1.2) we have a compact notation for writing systems of linear equations. For example, as a special instance of (3.1.3),

$$\begin{pmatrix} 2 & 3 & -1 \\ 4 & 1 & 5 \end{pmatrix} \begin{pmatrix} x_1 \\ x_2 \\ x_3 \end{pmatrix} = \begin{pmatrix} 2x_1 + 3x_2 - x_3 \\ 4x_1 + x_2 + 5x_3 \end{pmatrix}.$$

In this notation we can write the system of two linear equations in three unknowns

$$\begin{aligned} 2x_1 + 3x_2 - x_3 &= 2 \\ 4x_1 + x_2 + 5x_3 &= -1 \end{aligned}$$

as the matrix equation

$$\begin{pmatrix} 2 & 3 & -1 \\ 4 & 1 & 5 \end{pmatrix} \begin{pmatrix} x_1 \\ x_2 \\ x_3 \end{pmatrix} = \begin{pmatrix} 2 \\ -1 \end{pmatrix}.$$

Indeed, the general system of linear equations (3.1.1) can be written in matrix form using matrix multiplication as

$$Ax = b,$$

where A is the $m \times n$ matrix of coefficients, x is the n-vector of unknowns, and b is the m-vector of constants on the right-hand side of (3.1.1).

Matrices Times Vectors in MATLAB

We have seen how to define matrices and vectors in MATLAB. Now we show how to multiply a matrix times a vector using MATLABmatrix vector product!in MATLAB .

Load the matrix A,

$$A = \begin{pmatrix} 5 & -4 & 3 & -6 & 2 \\ 2 & -4 & -2 & -1 & 1 \\ 1 & 2 & 1 & -5 & 3 \\ -2 & -1 & -2 & 1 & -1 \\ 1 & -6 & 1 & 1 & 4 \end{pmatrix} \qquad \text{(3.1.4)*}$$

and the vector x

$$x = \begin{pmatrix} -1 \\ 2 \\ 1 \\ -1 \\ 3 \end{pmatrix} \qquad (3.1.5)*$$

into MATLAB by typing

    e3_1_4
    e3_1_5

The multiplication Ax is performed by typing

    b = A*x

and the result should be

    b =
         2
        -8
        18
        -6
        -1

We can verify this result by solving the system of linear equations $Ax = b$. Indeed if we type

    A\b

then we get the vector x back as the answer.

HAND EXERCISES

1. Let

$$A = \begin{pmatrix} 2 & 1 \\ -1 & 4 \end{pmatrix} \quad \text{and} \quad x = \begin{pmatrix} 3 \\ -2 \end{pmatrix}.$$

Compute Ax.

2. Let

$$B = \begin{pmatrix} 3 & 4 & 1 \\ 1 & 2 & 3 \end{pmatrix} \quad \text{and} \quad y = \begin{pmatrix} 2 \\ 5 \\ -2 \end{pmatrix}.$$

Compute By.

In Exercises 3–6, decide whether or not the matrix vector product Ax can be computed; if it can, compute the product.

3. $A = \begin{pmatrix} 1 & 2 \\ 0 & -5 \end{pmatrix}$ and $x = \begin{pmatrix} 2 \\ 2 \end{pmatrix}$

4. $A = \begin{pmatrix} 1 & 2 \\ 0 & -5 \end{pmatrix}$ and $x = \begin{pmatrix} 2 \\ 2 \\ 4 \end{pmatrix}$

5. $A = \begin{pmatrix} 1 & 2 & 4 \end{pmatrix}$ and $x = \begin{pmatrix} -1 \\ 1 \\ 3 \end{pmatrix}$

6. $A = (5)$ and $x = \begin{pmatrix} 1 \\ 0 \end{pmatrix}$

7. Let

$$A = \begin{pmatrix} a_{11} & a_{12} & \cdots & a_{1n} \\ a_{21} & a_{22} & \cdots & a_{2n} \\ \vdots & \vdots & & \vdots \\ a_{m1} & a_{m2} & \cdots & a_{mn} \end{pmatrix} \quad \text{and} \quad x = \begin{pmatrix} x_1 \\ x_2 \\ \vdots \\ x_n \end{pmatrix}.$$

Denote the columns of the matrix A by

$$A_1 = \begin{pmatrix} a_{11} \\ a_{21} \\ \vdots \\ a_{m1} \end{pmatrix}, \quad A_2 = \begin{pmatrix} a_{12} \\ a_{22} \\ \vdots \\ a_{m2} \end{pmatrix}, \quad \ldots, \quad A_n = \begin{pmatrix} a_{1n} \\ a_{2n} \\ \vdots \\ a_{mn} \end{pmatrix}.$$

Show that the matrix vector product Ax can be written as

$$Ax = x_1 A_1 + x_2 A_2 + \cdots + x_n A_n,$$

where $x_j A_j$ denotes scalar multiplication (see Chapter 1).

8. Let

$$C = \begin{pmatrix} 1 & 1 \\ 2 & -1 \end{pmatrix} \quad \text{and} \quad b = \begin{pmatrix} 1 \\ 1 \end{pmatrix}.$$

Find a 2-vector z such that $Cz = b$.

9. Write the system of linear equations

$$\begin{aligned} 2x_1 + 3x_2 - 2x_3 &= 4 \\ 6x_1 \qquad\quad - 5x_3 &= 1 \end{aligned}$$

in the matrix form $Ax = b$.

10. Find all solutions to

$$\begin{pmatrix} 1 & 3 & -1 & 4 \\ 2 & 1 & 5 & 7 \\ 3 & 4 & 4 & 11 \end{pmatrix} \begin{pmatrix} x_1 \\ x_2 \\ x_3 \\ x_4 \end{pmatrix} = \begin{pmatrix} 14 \\ 17 \\ 31 \end{pmatrix}.$$

11. Let A be a 2×2 matrix. Find A so that

$$A \begin{pmatrix} 1 \\ 0 \end{pmatrix} = \begin{pmatrix} 3 \\ -5 \end{pmatrix}$$

$$A \begin{pmatrix} 0 \\ 1 \end{pmatrix} = \begin{pmatrix} 1 \\ 4 \end{pmatrix}.$$

12. Let A be a 2×2 matrix. Find A so that

$$A \begin{pmatrix} 1 \\ 1 \end{pmatrix} = \begin{pmatrix} 2 \\ -1 \end{pmatrix}$$

$$A \begin{pmatrix} 1 \\ -1 \end{pmatrix} = \begin{pmatrix} 4 \\ 3 \end{pmatrix}.$$

13. Is there an upper triangular 2×2 matrix A such that

$$A \begin{pmatrix} 1 \\ 0 \end{pmatrix} = \begin{pmatrix} 1 \\ 2 \end{pmatrix}?$$ (3.1.6)

Is there a symmetric 2×2 matrix A satisfying (3.1.6)?

COMPUTER EXERCISES

In Exercises 14 and 15 use MATLAB to compute $b = Ax$ for the given A and x.

14.

$$A = \begin{pmatrix} -0.2 & -1.8 & 3.9 & -6 & -1.6 \\ 6.3 & 8 & 3 & 2.5 & 5.1 \\ -0.8 & -9.9 & 9.7 & 4.7 & 5.9 \\ -0.9 & -4.1 & 1.1 & -2.5 & 8.4 \\ -1 & -9 & -2 & -9.8 & 6.9 \end{pmatrix} \quad \text{and} \quad x = \begin{pmatrix} -2.6 \\ 2.4 \\ 4.6 \\ -6.1 \\ 8.1 \end{pmatrix}$$ (3.1.7)*

15.

$$A = \begin{pmatrix} 14 & -22 & -26 & -2 & -77 & 100 & -90 \\ 26 & 25 & -15 & -63 & 33 & 92 & 14 \\ -53 & 40 & 19 & 40 & -27 & -88 & 40 \\ 10 & -21 & 13 & 97 & -72 & -28 & 92 \\ 86 & -17 & 43 & 61 & 13 & 10 & 50 \\ -33 & 31 & 2 & 41 & 65 & -48 & 48 \\ 31 & 68 & 55 & -3 & 35 & 19 & -14 \end{pmatrix} \quad \text{and} \quad x = \begin{pmatrix} 2.7 \\ 6.1 \\ -8.3 \\ 8.9 \\ 8.3 \\ 2 \\ -4.9 \end{pmatrix}$$ (3.1.8)*

16. Let

$$A = \begin{pmatrix} 2 & 4 & -1 \\ 1 & 3 & 2 \\ -1 & -2 & 5 \end{pmatrix} \quad \text{and} \quad b = \begin{pmatrix} 2 \\ 1 \\ 4 \end{pmatrix}.$$ (3.1.9)*

Find a 3-vector x such that $Ax = b$.

17. Let

$$A = \begin{pmatrix} 1.3 & -4.15 & -1.2 \\ 1.6 & -1.2 & 2.4 \\ -2.5 & 2.35 & 5.09 \end{pmatrix} \quad \text{and} \quad b = \begin{pmatrix} 1.12 \\ -2.1 \\ 4.36 \end{pmatrix}.$$ (3.1.10)*

Find a 3-vector x such that $Ax = b$.

18. Let A be a 3×3 matrix. Find A so that

$$A \begin{pmatrix} 2 \\ -1 \\ 1 \end{pmatrix} = \begin{pmatrix} 1 \\ 1 \\ -1 \end{pmatrix}$$

$$A \begin{pmatrix} 1 \\ -1 \\ 0 \end{pmatrix} = \begin{pmatrix} -1 \\ -2 \\ 1 \end{pmatrix}$$

$$A \begin{pmatrix} 0 \\ 2 \\ 4 \end{pmatrix} = \begin{pmatrix} 5 \\ 1 \\ 1 \end{pmatrix}.$$

Hint Rewrite these three conditions as a system of linear equations in the nine entries of A. Then solve this system using MATLAB. (Then pray that there is an easier way!)

3.2 MATRIX MAPPINGS

Having illustrated the notational advantage of using matrices and matrix multiplication, we now begin to discuss why there is also a *conceptual advantage* to matrix multiplication, a conceptual advantage that will help us understand how to solve systems of linear equations and linear differential equations.

Matrix multiplication allows us to view $m \times n$ matrices as mappings from $\mathbf{R}^n$ to $\mathbf{R}^m$. Let A be an $m \times n$ matrix and let x be an n-vector. Then

$$x \mapsto Ax$$

defines a mapping from $\mathbf{R}^n$ to $\mathbf{R}^m$.

The simplest example of a matrix mapping is given by 1×1 matrices. Matrix mappings defined from $\mathbf{R} \to \mathbf{R}$ are

$$x \mapsto ax,$$

where a is a real number. Note that the graph of this function is just a straight line through the origin (with slope a). From this example we see that matrix mappings are very special mappings indeed. In higher dimensions, matrix mappings provide a richer set of mappings; we explore here *planar* mappings—mappings of the plane into itself—using MATLAB graphics and the program map.

The simplest planar matrix mappings are *dilatations*. Let $A = cI_2$, where $c > 0$ is a scalar. When $c < 1$, vectors are contracted by a factor of c, and these mappings are examples of *contractions*. When $c > 1$, vectors are stretched or expanded by a factor of c, and these dilatations are examples of *expansions*. We now explore some more complicated planar matrix mappings.

The next planar motions that we study are those given by the matrices

$$A = \begin{pmatrix} \lambda & 0 \\ 0 & \mu \end{pmatrix}.$$

Here the matrix mapping is given by $(x, y) \mapsto (\lambda x, \mu y)$—that is, a mapping that independently stretches and/or contracts the x- and y-coordinates. Even these simple-looking mappings can move objects in the plane in a somewhat complicated fashion.

The Program map

We can use MATLAB to explore planar matrix mappings in an efficient way using the program map. Type the command

    map

and a menu appears labeled MAP Setup. The 2×2 matrix

$$\begin{pmatrix} 0 & -1 \\ 1 & 0 \end{pmatrix}$$

has been pre-entered. Click on the Proceed button. A window entitled MAP Display appears. Click on Icons and click on an icon—say, Dog. Then click in the MAP Display window and a blue dog will appear in that window. Now click on the Map button and a new version of the dog will appear in yellow, but the yellow dog is rotated about the origin counterclockwise by 90° from the blue dog. Indeed, this matrix A just rotates the plane counterclockwise by 90°. To verify this statement, just click on Map again and see that the yellow dog rotates 90° counterclockwise into the magenta dog. Of course, the magenta dog is just rotated 180° from the original blue dog. Clicking on Map again produces a fourth dog—this one in cyan. Finally one more click on the Map button rotates the cyan dog into a red dog that exactly covers the original blue dog.

Choose another icon from the Icons menu; a blue version of this icon appears in the MAP Display window. Now click on Map to see that your chosen icon is just rotated counterclockwise by 90°.

Other matrices produce different motions of the plane. You may either type the entries of a matrix in the Map Setup window and click on the Proceed button or recall one of the pre-assigned matrices listed in the menu obtained by clicking on Gallery. For example, clicking on the Contracting rotation button enters the matrix

$$\begin{pmatrix} 0.3 & -0.8 \\ 0.8 & 0.3 \end{pmatrix}.$$

This matrix rotates the plane through an angle of approximately 69.4° counterclockwise and contracts the plane by a factor of approximately 0.85. Now click on Dog in the Icons menu to bring up the blue dog again. Repeated clicking on Map rotates and contracts the dog so that dogs in a cycling set of colors slowly converge toward the origin in a spiral of dogs.

Rotations

Rotating the plane counterclockwise through an angle θ is a motion given by a matrix mapping. We show that the matrix that performs this rotation is

$$R_\theta = \begin{pmatrix} \cos\theta & -\sin\theta \\ \sin\theta & \cos\theta \end{pmatrix}. \tag{3.2.1}$$

To verify that R_θ rotates the plane counterclockwise through angle θ, let v_φ be the unit vector whose angle from the horizontal is φ; that is, $v_\varphi = (\cos\varphi, \sin\varphi)$. We can write every vector in $\mathbf{R}^2$ as rv_φ for some number $r \geq 0$. Using the trigonometric identities for the cosine and sine of the sum of two angles, we have

$$\begin{aligned} R_\theta(rv_\varphi) &= \begin{pmatrix} \cos\theta & -\sin\theta \\ \sin\theta & \cos\theta \end{pmatrix} \begin{pmatrix} r\cos\varphi \\ r\sin\varphi \end{pmatrix} \\ &= \begin{pmatrix} r\cos\theta\cos\varphi - r\sin\theta\sin\varphi \\ r\sin\theta\cos\varphi + r\cos\theta\sin\varphi \end{pmatrix} \\ &= r\begin{pmatrix} \cos(\theta + \varphi) \\ \sin(\theta + \varphi) \end{pmatrix} \\ &= rv_{\varphi+\theta}. \end{aligned}$$

This calculation shows that R_θ rotates every vector in the plane counterclockwise through angle θ.

It follows from (3.2.1) that $R_{180°} = -I_2$. So rotating a vector in the plane by $180°$ is the same as reflecting the vector through the origin. It also follows that the movement associated with the linear map $x \mapsto -cx$, where $c > 0$, may be thought of as a dilatation ($x \mapsto cx$) followed by rotation through $180°$ ($x \mapsto -x$).

We claim that combining dilatations with general rotations produces spirals. Consider the matrix

$$S = \begin{pmatrix} c\cos\theta & -c\sin\theta \\ c\sin\theta & c\cos\theta \end{pmatrix} = cR_\theta,$$

where $c < 1$. Then a calculation similar to the previous one shows that

$$S(rv_\varphi) = c(rv_{\varphi+\theta}).$$

So S rotates vectors in the plane while contracting them by the factor c. Thus multiplying a vector repeatedly by S spirals that vector into the origin. The example that we considered while using `map` is

$$\begin{pmatrix} 0.3 & -0.8 \\ 0.8 & 0.3 \end{pmatrix} \cong \begin{pmatrix} 0.85\cos(69.4°) & -0.85\sin(69.4°) \\ 0.85\sin(69.4°) & 0.85\cos(69.4°) \end{pmatrix},$$

which has the general form of S.

A Notation for Matrix Mappings

We reinforce the idea that matrices are mappings by introducing a notation for the mapping associated with an $m \times n$ matrix A. We define

$$L_A : \mathbf{R}^n \to \mathbf{R}^m$$

by

$$L_A(x) = Ax$$

for every $x \in \mathbf{R}^n$.

There are two special matrices: the $m \times n$ zero matrix O all of whose entries are 0 and the $n \times n$ identity matrix I_n whose diagonal entries are 1 and whose off-diagonal entries are 0. For instance,

$$I_3 = \begin{pmatrix} 1 & 0 & 0 \\ 0 & 1 & 0 \\ 0 & 0 & 1 \end{pmatrix}.$$

The mappings associated with these special matrices are also special. Let x be an n-vector. Then

$$Ox = 0, \tag{3.2.2}$$

where the 0 on the right-hand side of (3.2.2) is the m-vector all of whose entries are 0. The mapping L_O is the *zero mapping* —the mapping that maps every vector x to 0.

Similarly,

$$I_n x = x$$

for every vector x. It follows that

$$L_{I_n}(x) = x$$

is the *identity mapping*, because it maps every element to itself. It is for this reason that the matrix I_n is called the $n \times n$ *identity matrix*.

HAND EXERCISES

In Exercises 1–3, find a nonzero vector that is mapped to the origin by the given matrix.

1. $A = \begin{pmatrix} 0 & 1 \\ 0 & -2 \end{pmatrix}$
      2. $B = \begin{pmatrix} 1 & 2 \\ -2 & -4 \end{pmatrix}$
      3. $C = \begin{pmatrix} 3 & -1 \\ -6 & 2 \end{pmatrix}$

4. What 2×2 matrix rotates the plane about the origin counterclockwise by $30°$?

5. What 2×2 matrix rotates the plane clockwise by $45°$?

6. What 2×2 matrix rotates the plane clockwise by $90°$ while dilating it by a factor of 2?

7. Find a 2×2 matrix that reflects vectors in the (x, y)-plane across the x-axis.

8. Find a 2×2 matrix that reflects vectors in the (x, y)-plane across the y-axis.

9. Find a 2×2 matrix that reflects vectors in the (x, y)-plane across the line $x = y$.

10. The matrix

$$A = \begin{pmatrix} 1 & K \\ 0 & 1 \end{pmatrix}$$

is a *shear*. Describe the action of A on the plane for different values of K.

11. Determine a rotation matrix that maps the vectors $(3, 4)$ and $(1, -2)$ onto the vectors $(-4, 3)$ and $(2, 1)$, respectively.

12. Find a 2×3 matrix P that projects three-dimensional xyz-space onto the xy-plane.
Hint   Such a matrix will satisfy

$$P \begin{pmatrix} 0 \\ 0 \\ z \end{pmatrix} = \begin{pmatrix} 0 \\ 0 \end{pmatrix} \quad \text{and} \quad P \begin{pmatrix} x \\ y \\ 0 \end{pmatrix} = \begin{pmatrix} x \\ y \end{pmatrix}.$$

13. Show that every matrix of the form $\begin{pmatrix} a & -b \\ b & a \end{pmatrix}$ corresponds to rotating the plane through the angle θ followed by a dilatation cI_2, where

$$c = \sqrt{a^2 + b^2}$$
$$\cos \theta = \frac{a}{c}$$
$$\sin \theta = \frac{b}{c}.$$

14. Using Exercise 13, observe that the matrix $\begin{pmatrix} 3 & 4 \\ -4 & 3 \end{pmatrix}$ rotates the plane counterclockwise through an angle θ and then dilates the planes by a factor of c. Find θ and c. Use map to verify your results.

COMPUTER EXERCISES

In Exercises 15–17, use map to find vectors that are stretched and/or contracted to a multiple of themselves by the given linear mapping.
Hint   Choose a vector in the MAP Display window and apply Map several times.

15. $A = \begin{pmatrix} 2 & 0 \\ 1.5 & 0.5 \end{pmatrix}$ **16.** $B = \begin{pmatrix} 1.2 & -1.5 \\ -0.4 & 1.2 \end{pmatrix}$ **17.** $C = \begin{pmatrix} 2 & -1.25 \\ 0 & -0.5 \end{pmatrix}$

In Exercises 18–20, use Exercise 13 and map to verify that the given matrices rotate the plane through an angle θ followed by a dilatation cI_2. Find θ and c in each case.

18. $A = \begin{pmatrix} 1 & -2 \\ 2 & 1 \end{pmatrix}$ **19.** $B = \begin{pmatrix} -2.4 & -0.2 \\ 0.2 & -2.4 \end{pmatrix}$ **20.** $C = \begin{pmatrix} 2.67 & 1.3 \\ -1.3 & 2.67 \end{pmatrix}$

In Exercises 21–25, use map to help describe the planar motions of the associated linear mappings for the given 2×2 matrix.

21. $A = \begin{pmatrix} \frac{\sqrt{3}}{2} & \frac{1}{2} \\ -\frac{1}{2} & \frac{\sqrt{3}}{2} \end{pmatrix}$ **22.** $B = \begin{pmatrix} \frac{1}{2} & -\frac{1}{2} \\ \frac{1}{2} & \frac{1}{2} \end{pmatrix}$ **23.** $C = \begin{pmatrix} 0 & 1 \\ 1 & 0 \end{pmatrix}$

24. $D = \begin{pmatrix} 1 & 0 \\ 0 & 0 \end{pmatrix}$ **25.** $E = \begin{pmatrix} \frac{1}{2} & \frac{1}{2} \\ \frac{1}{2} & \frac{1}{2} \end{pmatrix}$

26. The matrix

$$A = \begin{pmatrix} 0 & -1 \\ -1 & 0 \end{pmatrix}$$

reflects the xy-plane across the diagonal line $y = -x$, while the matrix

$$B = \begin{pmatrix} -1 & 0 \\ 0 & -1 \end{pmatrix}$$

rotates the plane through an angle of $180°$. Using the program map, verify that both matrices map the vector $(1, 1)$ to its negative $(-1, -1)$. Now perform two experiments. First, using the Icons menu in map, place a dog icon at about the point $(1, 1)$ and move that dog using matrix A. Then replace the dog in its original position near $(1, 1)$ and move that dog using matrix B. Describe the difference in the results.

3.3 LINEARITY

We begin by recalling the vector operations of addition and scalar multiplication. Given two n-vectors, vector addition is defined by

$$\begin{pmatrix} x_1 \\ \vdots \\ x_n \end{pmatrix} + \begin{pmatrix} y_1 \\ \vdots \\ y_n \end{pmatrix} = \begin{pmatrix} x_1 + y_1 \\ \vdots \\ x_n + y_n \end{pmatrix}.$$

Multiplication of a scalar times a vector is defined by

$$c \begin{pmatrix} x_1 \\ \vdots \\ x_n \end{pmatrix} = \begin{pmatrix} cx_1 \\ \vdots \\ cx_n \end{pmatrix}.$$

Using (3.1.2), we can check that matrix multiplication satisfies

$$A(x + y) = Ax + Ay \qquad (3.3.1)$$
$$A(cx) = c(Ax). \qquad (3.3.2)$$

Using MATLAB, we can also verify that the identities (3.3.1) and (3.3.2) are valid for some particular choices of x, y, c, and A. For example, let

$$A = \begin{pmatrix} 2 & 3 & 4 & 1 \\ 1 & 1 & 2 & 3 \end{pmatrix}, \quad x = \begin{pmatrix} 1 \\ 5 \\ 4 \\ 3 \end{pmatrix}, \quad y = \begin{pmatrix} 1 \\ -1 \\ -1 \\ 4 \end{pmatrix}, \quad c = 5. \qquad (3.3.3)*$$

Typing e3_3_3 enters this information into MATLAB. Now type

```
z1 = A*(x+y)
z2 = A*x + A*y
```

and compare z1 and z2. The fact that they are both equal to

$$\begin{pmatrix} 35 \\ 33 \end{pmatrix}$$

verifies (3.3.1) in this case. Similarly, type

```
w1 = A*(c*x)
w2 = c*(A*x)
```

and compare w1 and w2 to verify (3.3.2).

The central idea in linear algebra is *linearity*.

Definition 3.3.1   *A mapping* $L : \mathbf{R}^n \to \mathbf{R}^m$ *is* linear *if*

(a) $L(x + y) = L(x) + L(y)$ *for all* $x, y \in \mathbf{R}^n$, *and*

(b) $L(cx) = cL(x)$ *for all* $x \in \mathbf{R}^n$ *and all scalars* $c \in \mathbf{R}$.

To better understand the meaning of Definition 3.3.1, we verify these conditions for the mapping $L : \mathbf{R}^2 \to \mathbf{R}^2$ defined by

$$L(x) = (x_1 + 3x_2, 2x_1 - x_2), \qquad (3.3.4)$$

where $x = (x_1, x_2) \in \mathbf{R}^2$. To verify Definition 3.3.1(a), let $y = (y_1, y_2) \in \mathbf{R}^2$. Then

$$\begin{aligned} L(x + y) &= L(x_1 + y_1, x_2 + y_2) \\ &= ((x_1 + y_1) + 3(x_2 + y_2), 2(x_1 + y_1) - (x_2 + y_2)) \\ &= (x_1 + y_1 + 3x_2 + 3y_2, 2x_1 + 2y_1 - x_2 - y_2). \end{aligned}$$

On the other hand,

$$L(x) + L(y) = (x_1 + 3x_2, 2x_1 - x_2) + (y_1 + 3y_2, 2y_1 - y_2)$$
$$= (x_1 + 3x_2 + y_1 + 3y_2, 2x_1 - x_2 + 2y_1 - y_2).$$

Hence

$$L(x + y) = L(x) + L(y)$$

for every pair of vectors x and y in $\mathbf{R}^2$.

Similarly, to verify Definition 3.3.1(b), let $c \in \mathbf{R}$ be a scalar and compute

$$L(cx) = L(cx_1, cx_2) = ((cx_1) + 3(cx_2), 2(cx_1) - (cx_2)).$$

Then compute

$$cL(x) = c(x_1 + 3x_2, 2x_1 - x_2) = (c(x_1 + 3x_2), c(2x_1 - x_2)),$$

from which it follows that

$$L(cx) = cL(x)$$

for every vector $x \in \mathbf{R}^2$ and every scalar $c \in \mathbf{R}$. Thus L is a linear mapping.

In fact, the mapping (3.3.4) is a matrix mapping and could have been written in the form

$$L(x) = \begin{pmatrix} 1 & 3 \\ 2 & -1 \end{pmatrix} x.$$

Hence the linearity of L could have been checked using identities (3.3.1) and (3.3.2). Indeed, matrix mappings are always linear mappings, as we now discuss.

Matrix Mappings Are Linear Mappings

Let A be an $m \times n$ matrix and recall that the matrix mapping $L_A : \mathbf{R}^n \to \mathbf{R}^m$ is defined by $L_A(x) = Ax$. We may rewrite (3.3.1) and (3.3.2) using this notation as

$$L_A(x + y) = L_A(x) + L_A(y)$$
$$L_A(cx) = cL_A(x).$$

Thus all matrix mappings are linear mappings. We will show that all linear mappings are matrix mappings (see Theorem 3.3.5), but first we discuss linearity in the simplest context of mappings from $\mathbf{R} \to \mathbf{R}$.

Linear and Nonlinear Mappings of R → R

Note that 1×1 matrices are just scalars $A = (a)$. It follows from (3.3.1) and (3.3.2) that we have shown that the matrix mappings $L_A(x) = ax$ are all linear, although this point could have been verified directly. Before showing that these are all the linear mappings of $\mathbf{R} \to \mathbf{R}$, we focus on examples of functions of $\mathbf{R} \to \mathbf{R}$ that are *not* linear.

Examples of Mappings That Are Not Linear

- $f(x) = x^2$: Calculate

$$f(x + y) = (x + y)^2 = x^2 + 2xy + y^2$$

while

$$f(x) + f(y) = x^2 + y^2.$$

The two expressions are not equal and $f(x) = x^2$ is not linear.

- $f(x) = e^x$: Calculate

$$f(x + y) = e^{x+y} = e^x e^y$$

while

$$f(x) + f(y) = e^x + e^y.$$

The two expressions are not equal and $f(x) = e^x$ is not linear.

- $f(x) = \sin x$: Recall that

$$f(x + y) = \sin(x + y) = \sin x \cos y + \cos x \sin y$$

while

$$f(x) + f(y) = \sin x + \sin y.$$

The two expressions are not equal and $f(x) = \sin x$ is not linear.

Linear Functions of One Variable

Suppose we take the opposite approach and ask what functions of $\mathbf{R} \to \mathbf{R}$ are linear. Observe that if $L : \mathbf{R} \to \mathbf{R}$ is linear, then

$$L(x) = L(x \cdot 1).$$

Since we are looking at the special case of linear mappings on $\mathbf{R}$, we note that x is a real number as well as a vector. Thus we can use Definition 3.3.1(b) to observe that

$$L(x \cdot 1) = xL(1).$$

So if we let $a = L(1)$, then we see that

$$L(x) = ax.$$

Thus linear mappings of $\mathbf{R}$ into $\mathbf{R}$ are very special mappings indeed; they are all scalar multiples of the identity mapping.

All Linear Mappings Are Matrix Mappings

We end this section by proving that every linear mapping is given by matrix multiplication, but first we state and prove two lemmas. There is a standard set of vectors that is used over and over again in linear algebra; we now define it:

Definition 3.3.2 *Let j be an integer between 1 and n. The n-vector e_j is the vector that has a 1 in the jth entry and 0s in all other entries.*

Lemma 3.3.3 *Let $L_1 \colon \mathbf{R}^n \to \mathbf{R}^m$ and $L_2 \colon \mathbf{R}^n \to \mathbf{R}^m$ be linear mappings. Suppose that $L_1(e_j) = L_2(e_j)$ for every $j = 1, \ldots, n$. Then $L_1 = L_2$.*

Proof: Let $x = (x_1, \ldots, x_n)$ be a vector in $\mathbf{R}^n$. Then

$$x = x_1 e_1 + \cdots + x_n e_n.$$

Linearity of L_1 and L_2 implies that

$$
\begin{aligned}
L_1(x) &= x_1 L_1(e_1) + \cdots + x_n L_1(e_n) \\
&= x_1 L_2(e_1) + \cdots + x_n L_2(e_n) \\
&= L_2(x).
\end{aligned}
$$

Since $L_1(x) = L_2(x)$ for all $x \in \mathbf{R}^n$, it follows that $L_1 = L_2$. ◆

Lemma 3.3.4 *Let A be an $m \times n$ matrix. Then Ae_j is the jth column of A.*

Proof: Recall the definition of matrix multiplication given in (3.1.2). In that formula, just set x_i equal to 0 for all $i \neq j$ and set $x_j = 1$. ◆

Theorem 3.3.5 *Let $L \colon \mathbf{R}^n \to \mathbf{R}^m$ be a linear mapping. Then there exists an $m \times n$ matrix A such that $L = L_A$.*

Proof: There are two steps to the proof: Determine the matrix A and verify that $L_A = L$. Let A be the matrix whose jth column is $L(e_j)$. By Lemma 3.3.4, $L(e_j) = Ae_j$; that is, $L(e_j) = L_A(e_j)$. Lemma 3.3.3 implies that $L = L_A$. ◆

Theorem 3.3.5 provides a simple way of showing that

$$L(0) = 0$$

for any linear map L. Indeed $L(0) = L_A(0) = A0 = 0$ for some matrix A. (This fact can also be proved directly from the definition of linear mapping.)

Using Theorem 3.3.5 to Find Matrices Associated with Linear Maps

The proof of Theorem 3.3.5 shows that the jth column of the matrix A associated with a linear mapping L is $L(e_j)$ viewed as a column vector. As an example, let $L : \mathbf{R}^2 \to \mathbf{R}^2$ be rotation clockwise through $90°$. Geometrically, it is easy to see that

$$L(e_1) = L\left(\begin{pmatrix} 1 \\ 0 \end{pmatrix}\right) = \begin{pmatrix} 0 \\ -1 \end{pmatrix} \quad \text{and} \quad L(e_2) = L\left(\begin{pmatrix} 0 \\ 1 \end{pmatrix}\right) = \begin{pmatrix} 1 \\ 0 \end{pmatrix}.$$

Since we know that rotations are linear maps, it follows that the matrix A associated with the linear map L is

$$A = \begin{pmatrix} 0 & 1 \\ -1 & 0 \end{pmatrix}.$$

Additional examples of linear mappings whose associated matrices can be found using Theorem 3.3.5 are given in Exercises 10–13.

HAND EXERCISES

1. Compute $ax + by$ for each case:
   (a) $a = 2, b = -3, x = (2, 4)$, and $y = (3, -1)$
   (b) $a = 10, b = -2, x = (1, 0, -1)$, and $y = (2, -4, 3)$
   (c) $a = 5, b = -1, x = (4, 2, -1, 1)$, and $y = (-1, 3, 5, 7)$

2. Let $x = (4, 7)$ and $y = (2, -1)$. Write the vector $\alpha x + \beta y$ as a vector in coordinates.

3. Let $x = (1, 2)$, $y = (1, -3)$, and $z = (-2, -1)$. Show that you can write

$$z = \alpha x + \beta y$$

for some $\alpha, \beta \in \mathbf{R}$.

Hint Set up a system of two linear equations in the unknowns α and β, and then solve this linear system.

4. Can the vector $z = (2, 3, -1)$ be written as

$$z = \alpha x + \beta y,$$

where $x = (2, 3, 0)$ and $y = (1, -1, 1)$?

5. Let $x = (3, -2)$, $y = (2, 3)$, and $z = (1, 4)$. For which real numbers α, β, and γ does

$$\alpha x + \beta y + \gamma z = (1, -2)?$$

In Exercises 6–9, determine whether or not each transformation is linear.

6. $T : \mathbf{R}^3 \to \mathbf{R}^2$ defined by $T(x_1, x_2, x_3) = (x_1 + 2x_2 - x_3, x_1 - 4x_3)$

7. $T : \mathbf{R}^2 \to \mathbf{R}^2$ defined by $T(x_1, x_2) = (x_1 + x_1 x_2, 2x_2)$

8. $T : \mathbf{R}^2 \to \mathbf{R}^2$ defined by $T(x_1, x_2) = (x_1 + x_2, x_1 - x_2 - 1)$

9. $T : \mathbf{R}^2 \to \mathbf{R}^3$ defined by $T(x_1, x_2) = (1, x_1 + x_2, 2x_2)$

10. Find the 2×3 matrix A that satisfies

$$Ae_1 = \begin{pmatrix} 2 \\ 3 \end{pmatrix}, \quad Ae_2 = \begin{pmatrix} 1 \\ -1 \end{pmatrix}, \quad \text{and} \quad Ae_3 = \begin{pmatrix} 0 \\ 1 \end{pmatrix}.$$

11. The *cross product* of two 3-vectors $x = (x_1, x_2, x_3)$ and $y = (y_1, y_2, y_3)$ is the 3-vector

$$x \times y = (x_2 y_3 - x_3 y_2, -(x_1 y_3 - x_3 y_1), x_1 y_2 - x_2 y_1).$$

Let $K = (2, 1, -1)$. Show that the mapping $L : \mathbf{R}^3 \to \mathbf{R}^3$ defined by

$$L(x) = x \times K$$

is a linear mapping. Find the 3×3 matrix A such that

$$L(x) = Ax;$$

that is, $L = L_A$.

12. Argue geometrically that rotation of the plane counterclockwise through an angle of $45°$ is a linear mapping. Find a 2×2 matrix A such that L_A rotates the plane counterclockwise by $45°$.

13. Let σ permute coordinates cyclically in $\mathbf{R}^3$; that is,

$$\sigma(x_1, x_2, x_3) = (x_2, x_3, x_1).$$

Find a 3×3 matrix A such that $\sigma = L_A$.

14. Let L be a linear map. Using the definition of linearity, prove that $L(0) = 0$.

15. Let $L_1 : \mathbf{R}^n \to \mathbf{R}^m$ and $L_2 : \mathbf{R}^n \to \mathbf{R}^m$ be linear mappings. Prove that $L : \mathbf{R}^n \to \mathbf{R}^m$ defined by

$$L(x) = L_1(x) + L_2(x)$$

is also a linear mapping. Theorem 3.3.5 states that there are matrices A, A_1, and A_2 such that

$$L = L_A \quad \text{and} \quad L_j = L_{A_j}$$

for $j = 1, 2$. What is the relationship among the matrices A, A_1, and A_2?

COMPUTER EXERCISES

16. Let

$$A = \begin{pmatrix} 0.5 & 0 \\ 0 & 2 \end{pmatrix}.$$

Use map to verify that the linear mapping L_A halves the x-component of a point while it doubles the y-component.

17. Let

$$A = \begin{pmatrix} 0 & 0.5 \\ -0.5 & 0 \end{pmatrix}.$$

Use map to determine how the mapping L_A acts on 2-vectors. Describe this action in words.

In Exercises 18 and 19, use MATLAB to verify (3.3.1) and (3.3.2).

18.

$$A = \begin{pmatrix} 1 & 2 & 3 \\ 0 & 1 & -2 \\ 4 & 0 & 1 \end{pmatrix}, \quad x = \begin{pmatrix} 3 \\ 2 \\ -1 \end{pmatrix}, \quad y = \begin{pmatrix} 0 \\ -5 \\ 10 \end{pmatrix}, \quad c = 21 \qquad \textbf{(3.3.5)*}$$

19.

$$A = \begin{pmatrix} 4 & 0 & -3 & 2 & 4 \\ 2 & 8 & -4 & -1 & 3 \\ -1 & 2 & 1 & 10 & -2 \\ 4 & 4 & -2 & 1 & 2 \\ -2 & 3 & 1 & 1 & -1 \end{pmatrix}, \quad x = \begin{pmatrix} 1 \\ 3 \\ -2 \\ 3 \\ -1 \end{pmatrix}, \quad y = \begin{pmatrix} 2 \\ 0 \\ 13 \\ -2 \\ 1 \end{pmatrix}, \quad c = -13 \qquad \textbf{(3.3.6)*}$$

3.4 THE PRINCIPLE OF SUPERPOSITION

The principle of superposition is just a restatement of the fact that matrix mappings are linear. Nevertheless, this restatement is helpful in understanding the structure of solutions to systems of linear equations.

Homogeneous Equations

A system of linear equations is *homogeneous* if it has the form

$$Ax = 0, \qquad \textbf{(3.4.1)}$$

where A is an $m \times n$ matrix and $x \in \mathbf{R}^n$. Note that homogeneous systems are consistent, since $0 \in \mathbf{R}^n$ is always a solution; that is, $A(0) = 0$.

The *principle of superposition* makes two assertions:

- Suppose that y and z in $\mathbf{R}^n$ are solutions to (3.4.1) (that is, suppose that $Ay = 0$ and $Az = 0$); then $y + z$ is a solution to (3.4.1).
- Suppose that c is a scalar; then cy is a solution to (3.4.1).

The principle of superposition is proved using the linearity of matrix multiplication. Calculate

$$A(y + z) = Ay + Az = 0 + 0 = 0$$

to verify that $y + z$ is a solution, and calculate

$$A(cy) = c(Ay) = c \cdot 0 = 0$$

to verify that cy is a solution.

We see that solutions to homogeneous systems of linear equations always satisfy the general property of superposition: Sums of solutions are solutions, and scalar multiples of solutions are solutions.

We illustrate this principle by explicitly solving the system of equations

$$\begin{pmatrix} 1 & 2 & -1 & 1 \\ 2 & 5 & -4 & -1 \end{pmatrix} \begin{pmatrix} x_1 \\ x_2 \\ x_3 \\ x_4 \end{pmatrix} = \begin{pmatrix} 0 \\ 0 \end{pmatrix}.$$

We use row reduction to show that the matrix

$$\begin{pmatrix} 1 & 2 & -1 & 1 \\ 2 & 5 & -4 & -1 \end{pmatrix}$$

is row equivalent to

$$\begin{pmatrix} 1 & 0 & 3 & 7 \\ 0 & 1 & -2 & -3 \end{pmatrix},$$

which is in reduced echelon form. Recall, using the methods of Section 2.3, that every solution to this linear system has the form

$$\begin{pmatrix} -3x_3 - 7x_4 \\ 2x_3 + 3x_4 \\ x_3 \\ x_4 \end{pmatrix} = x_3 \begin{pmatrix} -3 \\ 2 \\ 1 \\ 0 \end{pmatrix} + x_4 \begin{pmatrix} -7 \\ 3 \\ 0 \\ 1 \end{pmatrix}.$$

Superposition is verified again by observing that the form of the solutions is preserved under vector addition and scalar multiplication. For instance, suppose that

$$\alpha_1 \begin{pmatrix} -3 \\ 2 \\ 1 \\ 0 \end{pmatrix} + \alpha_2 \begin{pmatrix} -7 \\ 3 \\ 0 \\ 1 \end{pmatrix} \quad \text{and} \quad \beta_1 \begin{pmatrix} -3 \\ 2 \\ 1 \\ 0 \end{pmatrix} + \beta_2 \begin{pmatrix} -7 \\ 3 \\ 0 \\ 1 \end{pmatrix}$$

are two solutions. Then the sum has the form

$$\gamma_1 \begin{pmatrix} -3 \\ 2 \\ 1 \\ 0 \end{pmatrix} + \gamma_2 \begin{pmatrix} -7 \\ 3 \\ 0 \\ 1 \end{pmatrix},$$

where $\gamma_j = \alpha_j + \beta_j$.

We have actually proved more than superposition. We have shown in this example that every solution is a superposition of just two solutions:

$$\begin{pmatrix} -3 \\ 2 \\ 1 \\ 0 \end{pmatrix} \quad \text{and} \quad \begin{pmatrix} -7 \\ 3 \\ 0 \\ 1 \end{pmatrix}.$$

Inhomogeneous Equations

The linear system of m equations in n unknowns is written as

$$Ax = b,$$

where A is an $m \times n$ matrix, $x \in \mathbf{R}^n$, and $b \in \mathbf{R}^m$. This system is *inhomogeneous* when the vector b is nonzero. Note that if $y, z \in \mathbf{R}^n$ are solutions to the inhomogeneous equation (that is, $Ay = b$ and $Az = b$), then $y - z$ is a solution to the homogeneous equation; that is,

$$A(y - z) = Ay - Az = b - b = 0.$$

For example, let

$$A = \begin{pmatrix} 1 & 2 & 0 \\ -2 & 0 & 1 \end{pmatrix} \quad \text{and} \quad b = \begin{pmatrix} 3 \\ -1 \end{pmatrix}.$$

Then

$$y = \begin{pmatrix} 1 \\ 1 \\ 1 \end{pmatrix} \quad \text{and} \quad z = \begin{pmatrix} 3 \\ 0 \\ 5 \end{pmatrix}$$

are both solutions to the linear system $Ax = b$. It follows that

$$y - z = \begin{pmatrix} -2 \\ 1 \\ -4 \end{pmatrix}$$

is a solution to the homogeneous system $Ax = 0$, which can be checked by direct calculation.

Thus we can completely solve the inhomogeneous equation by finding one solution to the inhomogeneous equation and then adding to that solution every solution of the homogeneous equation. More precisely, suppose that we know all of the solutions w to the homogeneous equation $Ax = 0$ and one solution y to the inhomogeneous equation $Ax = b$. Then $y + w$ is another solution to the inhomogeneous equation and *every* solution to the inhomogeneous equation has this form.

An Example of an Inhomogeneous Equation

Suppose that we want to find all solutions of $Ax = b$, where

$$A = \begin{pmatrix} 3 & 2 & 1 \\ 0 & 1 & -2 \\ 3 & 3 & -1 \end{pmatrix} \quad \text{and} \quad b = \begin{pmatrix} -2 \\ 4 \\ 2 \end{pmatrix}.$$

Suppose we are told that $y = (-5, 6, 1)^t$ is a solution of the inhomogeneous equation. (This fact can be verified by a short calculation—just multiply Ay and see that the result equals b.) Next we find all solutions to the homogeneous equation $Ax = 0$ by putting A into reduced echelon form. The resulting row echelon form matrix is

$$\begin{pmatrix} 1 & 0 & \frac{5}{3} \\ 0 & 1 & -2 \\ 0 & 0 & 0 \end{pmatrix}.$$

Hence we see that the solutions of the homogeneous equation $Ax = 0$ are

$$\begin{pmatrix} -\frac{5}{3}s \\ 2s \\ s \end{pmatrix} = s \begin{pmatrix} -\frac{5}{3} \\ 2 \\ 1 \end{pmatrix}.$$

Combining these results, we conclude that all the solutions of $Ax = b$ are given by

$$\begin{pmatrix} -5 \\ 6 \\ 1 \end{pmatrix} + s \begin{pmatrix} -\frac{5}{3} \\ 2 \\ 1 \end{pmatrix}.$$

HAND EXERCISES

1. Consider the homogeneous linear equation

$$x + y + z = 0.$$

   (a) Write all solutions to this equation as a general superposition of a pair of vectors v_1 and v_2.
   (b) Write all solutions as a general superposition of a second pair of vectors w_1 and w_2.

2. Write all solutions to the homogeneous system of linear equations

$$x_1 + 2x_2 + x_4 - x_5 = 0$$
$$x_3 - 2x_4 + x_5 = 0$$

as the general superposition of three vectors.

3. (a) Find all solutions to the homogeneous equation $Ax = 0$, where

$$A = \begin{pmatrix} 2 & 3 & 1 \\ 1 & 1 & 4 \end{pmatrix}.$$

(b) Find a single solution to the inhomogeneous equation

$$Ax = \begin{pmatrix} 6 \\ 6 \end{pmatrix}.$$

(3.4.2)

(c) Use your answers in (a) and (b) to find all solutions to (3.4.2).

3.5 COMPOSITION AND MULTIPLICATION OF MATRICES

The *composition* of two matrix mappings leads to another matrix mapping from which the concept of multiplication of two matrices follows. Matrix multiplication can be introduced by formula, but then the idea is unmotivated and one is left to wonder why matrix multiplication is defined in such a seemingly awkward way.

We begin with the example of 2×2 matrices. Suppose that

$$A = \begin{pmatrix} 2 & 1 \\ 1 & -1 \end{pmatrix} \quad \text{and} \quad B = \begin{pmatrix} 0 & 3 \\ -1 & 4 \end{pmatrix}.$$

We have seen that the mappings

$$x \mapsto Ax \quad \text{and} \quad x \mapsto Bx$$

map 2-vectors to 2-vectors. So we can ask what happens when we compose these mappings. In symbols, we compute

$$L_A \circ L_B(x) = L_A(L_B(x)) = A(Bx).$$

In coordinates, we let $x = (x_1, x_2)$ and compute

$$A(Bx) = A \begin{pmatrix} 3x_2 \\ -x_1 + 4x_2 \end{pmatrix}$$
$$= \begin{pmatrix} -x_1 + 10x_2 \\ x_1 - x_2 \end{pmatrix}.$$

It follows that we can rewrite $A(Bx)$ using multiplication of a matrix times a vector as

$$A(Bx) = \begin{pmatrix} -1 & 10 \\ 1 & -1 \end{pmatrix} \begin{pmatrix} x_1 \\ x_2 \end{pmatrix}.$$

In particular, $L_A \circ L_B$ is again a linear mapping—namely, L_C, where

$$C = \begin{pmatrix} -1 & 10 \\ 1 & -1 \end{pmatrix}.$$

With this computation in mind, we define the product

$$AB = \begin{pmatrix} 2 & 1 \\ 1 & -1 \end{pmatrix} \begin{pmatrix} 0 & 3 \\ -1 & 4 \end{pmatrix} = \begin{pmatrix} -1 & 10 \\ 1 & -1 \end{pmatrix}.$$

Using the same approach, we can derive a formula for matrix multiplication of 2×2 matrices. Suppose

$$A = \begin{pmatrix} a_{11} & a_{12} \\ a_{21} & a_{22} \end{pmatrix} \quad \text{and} \quad B = \begin{pmatrix} b_{11} & b_{12} \\ b_{21} & b_{22} \end{pmatrix}.$$

Then

$$
\begin{aligned}
A(Bx) &= A \begin{pmatrix} b_{11}x_1 + b_{12}x_2 \\ b_{21}x_1 + b_{22}x_2 \end{pmatrix} \\
&= \begin{pmatrix} a_{11}(b_{11}x_1 + b_{12}x_2) + a_{12}(b_{21}x_1 + b_{22}x_2) \\ a_{21}(b_{11}x_1 + b_{12}x_2) + a_{22}(b_{21}x_1 + b_{22}x_2) \end{pmatrix} \\
&= \begin{pmatrix} (a_{11}b_{11} + a_{12}b_{21})x_1 + (a_{11}b_{12} + a_{12}b_{22})x_2 \\ (a_{21}b_{11} + a_{22}b_{21})x_1 + (a_{21}b_{12} + a_{22}b_{22})x_2 \end{pmatrix} \\
&= \begin{pmatrix} a_{11}b_{11} + a_{12}b_{21} & a_{11}b_{12} + a_{12}b_{22} \\ a_{21}b_{11} + a_{22}b_{21} & a_{21}b_{12} + a_{22}b_{22} \end{pmatrix} \begin{pmatrix} x_1 \\ x_2 \end{pmatrix}.
\end{aligned}
$$

Hence, for 2×2 matrices, we see that composition of matrix mappings defines matrix multiplication as

$$\begin{pmatrix} a_{11} & a_{12} \\ a_{21} & a_{22} \end{pmatrix} \begin{pmatrix} b_{11} & b_{12} \\ b_{21} & b_{22} \end{pmatrix} = \begin{pmatrix} a_{11}b_{11} + a_{12}b_{21} & a_{11}b_{12} + a_{12}b_{22} \\ a_{21}b_{11} + a_{22}b_{21} & a_{21}b_{12} + a_{22}b_{22} \end{pmatrix}. \tag{3.5.1}$$

Formula (3.5.1) may seem a bit formidable, but it does have structure. Suppose A and B are 2×2 matrices. Then the entry of

$$C = AB$$

in the ith row, jth column may be written as

$$a_{i1}b_{1j} + a_{i2}b_{2j} = \sum_{k=1}^{2} a_{ik}b_{kj}.$$

We shall see that an analog of this formula is available for matrix multiplications of all sizes. But to derive this formula, it is easier to develop matrix multiplication abstractly.

Lemma 3.5.1   *Let $L_1 : \mathbf{R}^n \to \mathbf{R}^m$ and $L_2 : \mathbf{R}^p \to \mathbf{R}^n$ be linear mappings. Then $L = L_1 \circ L_2 : \mathbf{R}^p \to \mathbf{R}^m$ is a linear mapping.*

Proof: Compute

$$L(x + y) = L_1 \circ L_2(x + y)$$
$$= L_1(L_2(x) + L_2(y))$$
$$= L_1(L_2(x)) + L_1(L_2(y))$$
$$= L_1 \circ L_2(x) + L_1 \circ L_2(y)$$
$$= L(x) + L(y).$$

Similarly, compute $L_1 \circ L_2(cx) = cL_1 \circ L_2(x)$. ◆

We apply Lemma 3.5.1 in the following way. Let A be an $m \times n$ matrix and let B be an $n \times p$ matrix. Then $L_A : \mathbf{R}^n \to \mathbf{R}^m$ and $L_B : \mathbf{R}^p \to \mathbf{R}^n$ are linear mappings, and the mapping $L = L_A \circ L_B : \mathbf{R}^p \to \mathbf{R}^m$ is defined and linear. Theorem 3.3.5 implies that there is an $m \times p$ matrix C such that $L = L_C$. Abstractly, we define the *matrix product AB to be C. Note that the matrix product AB is defined only when the number of columns of A is equal to the number of rows of B.*

Calculating the Product of Two Matrices

Next we discuss how to calculate the product of matrices; this discussion generalizes our discussion of the product of 2×2 matrices. Lemma 3.3.4 tells how to compute $C = AB$. The jth column of the matrix product is just

$$Ce_j = A(Be_j),$$

where $B_j \equiv Be_j$ is the jth column of the matrix B. Therefore

$$C = (AB_1 | \cdots | AB_p). \tag{3.5.2}$$

Indeed, the (i, j)th entry of C is the ith entry of AB_j; that is, the ith entry of

$$A \begin{pmatrix} b_{1j} \\ \vdots \\ b_{nj} \end{pmatrix} = \begin{pmatrix} a_{11}b_{1j} + \cdots + a_{1n}b_{nj} \\ \vdots \\ a_{m1}b_{1j} + \cdots + a_{mn}b_{nj} \end{pmatrix}.$$

It follows that the entry c_{ij} of C in the ith row, jth column is

$$c_{ij} = a_{i1}b_{1j} + a_{i2}b_{2j} + \cdots + a_{in}b_{nj} = \sum_{k=1}^{n} a_{ik}b_{kj}. \tag{3.5.3}$$

We can interpret (3.5.3) in the following way: To calculate c_{ij}, multiply the entries of the ith row of A with the corresponding entries in the jth column of B and add the results. This interpretation reinforces the idea that for the matrix product AB to be defined, the number of columns in A must equal the number of rows in B.

For example, we now perform the following multiplication:

$$\begin{pmatrix} 2 & 3 & 1 \\ 3 & -1 & 2 \end{pmatrix} \begin{pmatrix} 1 & -2 \\ 3 & 1 \\ -1 & 4 \end{pmatrix}$$

$$= \begin{pmatrix} 2 \cdot 1 + 3 \cdot 3 + 1 \cdot (-1) & 2 \cdot (-2) + 3 \cdot 1 + 1 \cdot 4 \\ 3 \cdot 1 + (-1) \cdot 3 + 2 \cdot (-1) & 3 \cdot (-2) + (-1) \cdot 1 + 2 \cdot 4 \end{pmatrix}$$

$$= \begin{pmatrix} 10 & 3 \\ -2 & 1 \end{pmatrix}.$$

Some Special Matrix Products

Let A be an $m \times n$ matrix. Then

$$OA = O$$
$$AO = O$$
$$AI_n = A$$
$$I_m A = A.$$

The first two equalities are easily checked using (3.5.3). It is not significantly more difficult to verify the last two equalities using (3.5.3), but we shall verify them using the language of linear mappings, as follows:

$$L_{AI_n}(x) = L_A \circ L_{I_n}(x) = L_A(x),$$

since $L_{I_n}(x) = x$ is the identity map. Therefore $AI_n = A$. A similar proof verifies that $I_m A = A$. Although the verification of these equalities using the notions of linear mappings may appear to be a case of overkill, the next section contains results where these notions truly simplify the discussion.

HAND EXERCISES

In Exercises 1–4, determine whether or not the matrix product AB or BA can be computed for each given pair of matrices A and B. If the product is possible, perform the computation.

1. $A = \begin{pmatrix} 1 & 0 \\ -2 & 1 \end{pmatrix}$ and $B = \begin{pmatrix} -2 & 0 \\ 3 & -1 \end{pmatrix}$

2. $A = \begin{pmatrix} 0 & -2 & 1 \\ 4 & 10 & 0 \end{pmatrix}$ and $B = \begin{pmatrix} 0 & 2 \\ 3 & -1 \end{pmatrix}$

3. $A = \begin{pmatrix} 8 & 0 & 2 & 3 \\ -3 & 0 & -10 & 3 \end{pmatrix}$ and $B = \begin{pmatrix} 0 & 2 & 5 \\ -1 & 3 & -1 \\ 0 & 1 & -5 \end{pmatrix}$

4. $A = \begin{pmatrix} 8 & -1 \\ -3 & 12 \\ 5 & -4 \end{pmatrix}$ and $B = \begin{pmatrix} 2 & 8 & 0 & -3 \\ 1 & 4 & 0 & 1 \\ -5 & 6 & 7 & -20 \end{pmatrix}$

In Exercises 5–8, compute each matrix product.

5. $\begin{pmatrix} 2 & 3 \\ 0 & 1 \end{pmatrix} \begin{pmatrix} -1 & 1 \\ -3 & 2 \end{pmatrix}$

6. $\begin{pmatrix} 1 & 2 & 3 \\ -2 & 3 & -1 \end{pmatrix} \begin{pmatrix} 2 & 3 \\ -2 & 5 \\ 1 & -1 \end{pmatrix}$

7. $\begin{pmatrix} 2 & 3 \\ -2 & 5 \\ 1 & -1 \end{pmatrix} \begin{pmatrix} 1 & 2 & 3 \\ -2 & 3 & -1 \end{pmatrix}$

8. $\begin{pmatrix} 2 & -1 & 3 \\ 1 & 0 & 5 \\ 1 & 5 & -1 \end{pmatrix} \begin{pmatrix} 1 & 7 \\ -2 & -1 \\ -5 & 3 \end{pmatrix}$

9. Determine all the 2×2 matrices B such that $AB = BA$, where A is the matrix

$$A = \begin{pmatrix} 2 & 0 \\ 0 & -1 \end{pmatrix}.$$

10. Let

$$A = \begin{pmatrix} 2 & 5 \\ 1 & 4 \end{pmatrix} \quad \text{and} \quad B = \begin{pmatrix} a & 3 \\ b & 2 \end{pmatrix}.$$

For which values of a and b does $AB = BA$?

11. Let

$$A = \begin{pmatrix} 1 & 0 & -3 \\ -2 & 1 & 1 \\ 0 & 1 & -5 \end{pmatrix}.$$

Let A^t be the transpose of the matrix A, as defined in Section 1.3. Compute AA^t.

COMPUTER EXERCISES

In Exercises 12–14, decide for the given pair of matrices A and B whether or not the product AB or BA is defined. Compute the products when possible.

12.

$$A = \begin{pmatrix} 2 & 2 & -2 \\ -4 & 4 & 0 \end{pmatrix} \quad \text{and} \quad B = \begin{pmatrix} 3 & -2 & 0 \\ 0 & -1 & 4 \\ -2 & -3 & 5 \end{pmatrix} \tag{3.5.4)*}$$

13.

$$A = \begin{pmatrix} -4 & 1 & 0 & 5 & -1 \\ 5 & -1 & -2 & -4 & -2 \\ 1 & 5 & -4 & 1 & 5 \end{pmatrix} \quad \text{and} \quad B = \begin{pmatrix} 1 & 3 & -4 & 3 & -2 & 1 \\ 0 & 3 & 2 & 3 & -1 & 4 \\ 5 & 4 & 4 & 5 & -1 & 0 \\ -4 & -3 & 2 & 4 & 1 & 4 \end{pmatrix} \tag{3.5.5)*}$$

14.

$$A = \begin{pmatrix} -2 & -2 & 4 & 5 \\ 0 & -3 & -4 & 3 \\ 1 & -3 & 1 & 1 \\ 0 & 1 & 0 & 4 \end{pmatrix} \quad \text{and} \quad B = \begin{pmatrix} 2 & 3 & -4 & 5 \\ 4 & -3 & 0 & -2 \\ -3 & -4 & -4 & -3 \\ -2 & -2 & 3 & -1 \end{pmatrix} \tag{3.5.6)*}$$

3.6   PROPERTIES OF MATRIX MULTIPLICATION

In this section we discuss the facts that matrix multiplication is associative (but not commutative) and that certain distributive properties hold. We also discuss how matrix multiplication is performed in MATLAB.

Matrix Multiplication Is Associative

Theorem 3.6.1   *Matrix multiplication is associative. That is, let A be an $m \times n$ matrix, let B be an $n \times p$ matrix, and let C be a $p \times q$ matrix. Then*

$$(AB)C = A(BC).$$

Proof: Begin by observing that composition of mappings is always associative. In symbols, let $f : \mathbf{R}^n \to \mathbf{R}^m$, $g : \mathbf{R}^p \to \mathbf{R}^n$, and $h : \mathbf{R}^q \to \mathbf{R}^p$. Then

$$\begin{aligned}
f \circ (g \circ h)(x) &= f[(g \circ h)(x)] \\
&= f[g(h(x))] \\
&= (f \circ g)(h(x)) \\
&= [(f \circ g) \circ h](x).
\end{aligned}$$

It follows that

$$f \circ (g \circ h) = (f \circ g) \circ h.$$

We can apply this result to linear mappings. Thus

$$L_A \circ (L_B \circ L_C) = (L_A \circ L_B) \circ L_C.$$

Since

$$L_{A(BC)} = L_A \circ L_{BC} = L_A \circ (L_B \circ L_C)$$

and

$$L_{(AB)C} = L_{AB} \circ L_C = (L_A \circ L_B) \circ L_C,$$

it follows that

$$L_{A(BC)} = L_{(AB)C}$$

and

$$A(BC) = (AB)C.$$

$\blacklozenge$

It is worth convincing yourself that Theorem 3.6.1 has content by verifying by hand that matrix multiplication of 2×2 matrices is associative.

Matrix Multiplication Is Not Commutative

Although matrix multiplication is associative, it is *not* commutative. This statement is trivially true when the matrix AB is defined while the matrix BA is not. Suppose, for example, that A is a 2×3 matrix and that B is a 3×4 matrix. Then AB is a 2×4 matrix, while the multiplication BA makes no sense whatsoever.

More important, suppose that A and B are both $n \times n$ square matrices. Then $AB = BA$ is generally not valid. For example, let

$$A = \begin{pmatrix} 1 & 0 \\ 0 & 0 \end{pmatrix} \quad \text{and} \quad B = \begin{pmatrix} 0 & 1 \\ 0 & 0 \end{pmatrix}.$$

Then

$$AB = \begin{pmatrix} 0 & 1 \\ 0 & 0 \end{pmatrix} \quad \text{and} \quad BA = \begin{pmatrix} 0 & 0 \\ 0 & 0 \end{pmatrix}.$$

So $AB \neq BA$. In certain cases it does happen that $AB = BA$. For example, when $B = I_n$, we have

$$AI_n = A = I_n A.$$

But these cases are rare.

Additional Properties of Matrix Multiplication

Recall that if $A = (a_{ij})$ and $B = (b_{ij})$ are both $m \times n$ matrices, then $A + B$ is the $m \times n$ matrix $(a_{ij} + b_{ij})$. We now present three properties of matrix multiplication.

- Let A and B be $m \times n$ matrices and let C be an $n \times p$ matrix. Then

$$(A + B)C = AC + BC.$$

Similarly, if D is a $q \times m$ matrix, then

$$D(A + B) = DA + DB.$$

So matrix multiplication distributes across matrix addition.

- If α and β are scalars, then

$$(\alpha + \beta)A = \alpha A + \beta A.$$

So addition distributes with scalar multiplication.

• Scalar multiplication and matrix multiplication satisfy

$$(\alpha A)C = \alpha(AC).$$

Matrix Multiplication and Transposes

Let A be an $m \times n$ matrix and let B be an $n \times p$ matrix, so that the matrix product AB is defined and AB is an $m \times p$ matrix. Note that A^t is an $n \times m$ matrix and that B^t is a $p \times n$ matrix, so that in general the product $A^t B^t$ is *not* defined. However, the product $B^t A^t$ is defined and is an $p \times m$ matrix, as is the matrix $(AB)^t$. We claim that

$$(AB)^t = B^t A^t. \tag{3.6.1}$$

We verify this claim by direct computation. The (i, k)th entry in $(AB)^t$ is the (k, i)th entry in AB. That entry is

$$\sum_{j=1}^{n} a_{kj} b_{ji}.$$

The (i, k)th entry in $B^t A^t$ is

$$\sum_{j=1}^{n} b_{ij}^t a_{jk}^t,$$

where a_{jk}^t is the (j, k)th entry in A^t and b_{ij}^t is the (i, j)th entry in B^t. It follows from the definition of transpose that the (i, k)th entry in $B^t A^t$ is

$$\sum_{j=1}^{n} b_{ji} a_{kj} = \sum_{j=1}^{n} a_{kj} b_{ji},$$

which verifies the claim.

Matrix Multiplication in MATLAB

We can now explain how matrix multiplication works in MATLAB. We load the matrices

$$A = \begin{pmatrix} -5 & 2 & 0 \\ -1 & 1 & -4 \\ -4 & 4 & 2 \\ -1 & 3 & -1 \end{pmatrix} \quad \text{and} \quad B = \begin{pmatrix} 2 & -2 & -2 & 5 & 5 \\ 4 & -5 & 1 & -1 & 2 \\ 3 & 2 & 3 & -3 & 3 \end{pmatrix} \tag{3.6.2*}$$

by typing

    e3_6_2

Now the command C = A*B asks MATLAB to compute the matrix C as the product of A and B. We obtain

```
C =
    -2     0    12   -27   -21
   -10   -11    -9     6   -15
    14    -8    18   -30    -6
     7   -15     2    -5    -2
```

We can confirm this result by another computation. As we have seen, the 4th column of C should be given by the product of A with the 4th column of B. Indeed, if we perform this computation and type

```
A*B(:,4)
```

the result is

```
ans =
   -27
     6
   -30
    -5
```

which is precisely the 4th column of C.

MATLAB also recognizes when a matrix multiplication of two matrices is not defined. For example, the product of the 3×5 matrix B with the 4×3 matrix A is not defined, and if we type B*A, then we obtain the error message:

```
??? Error using ==> *
Inner matrix dimensions must agree.
```

The size of a matrix A can be seen using the MATLAB command size. For example, the command size(A) leads to

```
ans =
     4     3
```

reflecting the fact that A is a matrix with four rows and three columns.

HAND EXERCISES

1. Let A be an $m \times n$ matrix. Show that the matrices AA^t and $A^t A$ are symmetric.

2. Let

$$A = \begin{pmatrix} 1 & 2 \\ -1 & -1 \end{pmatrix} \quad \text{and} \quad B = \begin{pmatrix} 2 & 3 \\ 1 & 4 \end{pmatrix}.$$

Compute AB and $B^t A^t$. Verify that $(AB)^t = B^t A^t$ for these matrices A and B.

3. Let

$$A = \begin{pmatrix} 0 & 1 & 0 \\ 0 & 0 & 1 \\ 0 & 0 & 0 \end{pmatrix}.$$

Compute $B = I + A + \frac{1}{2}A^2$ and $C = I + tA + \frac{1}{2}(tA)^2$.

4. Let

$$I = \begin{pmatrix} 1 & 0 \\ 0 & 1 \end{pmatrix} \quad \text{and} \quad J = \begin{pmatrix} 0 & -1 \\ 1 & 0 \end{pmatrix}.$$

(a) Show that $J^2 = -I$.

(b) Evaluate $(aI + bJ)(cI + dJ)$ in terms of I and J.

5. Recall that a square matrix C is *upper triangular* if $c_{ij} = 0$ when $i > j$. Show that the matrix product of two upper triangular $n \times n$ matrices is also upper triangular.

COMPUTER EXERCISES

In Exercises 6–8, use MATLAB to verify that $(A + B)C = AC + BC$ for the given matrices.

6. $A = \begin{pmatrix} 0 & 2 \\ 2 & 1 \end{pmatrix}$, $B = \begin{pmatrix} -2 & 1 \\ 3 & 0 \end{pmatrix}$, and $C = \begin{pmatrix} 2 & -1 \\ 1 & 5 \end{pmatrix}$

7. $A = \begin{pmatrix} 12 & -2 \\ 3 & 1 \end{pmatrix}$, $B = \begin{pmatrix} 8 & -20 \\ 3 & 10 \end{pmatrix}$, and $C = \begin{pmatrix} 10 & 2 & 4 \\ 2 & 13 & -4 \end{pmatrix}$

8. $A = \begin{pmatrix} 6 & 1 \\ 3 & 20 \\ -5 & 3 \end{pmatrix}$, $B = \begin{pmatrix} 2 & -10 \\ 5 & 0 \\ 3 & 1 \end{pmatrix}$, and $C = \begin{pmatrix} -2 & 10 \\ 12 & 10 \end{pmatrix}$

9. Use the rand(3,3) command in MATLAB to choose five pairs of 3×3 matrices A and B at random. Compute AB and BA using MATLAB to see that in general these matrix products are unequal.

10. Experimentally, find two symmetric 2×2 matrices A and B for which the matrix product AB is *not* symmetric.

3.7 SOLVING LINEAR SYSTEMS AND INVERSES

When we solve the simple equation

$$ax = b,$$

we do so by dividing by a to obtain

$$x = \frac{1}{a}b.$$

This division works as long as $a \neq 0$.

Writing systems of linear equations as

$$Ax = b$$

suggests that solutions should have the form

$$x = \frac{1}{A}b,$$

and the MATLAB command for solving linear systems

    x=A\b

suggests that there is some merit to this analogy.

Here is a better analogy. Multiplication by a has the inverse operation division by a; multiplying a number x by a and then multiplying the result by $a^{-1} = 1/a$ leaves the number x unchanged (as long as $a \neq 0$). In this sense we should write the solution to $ax = b$ as

$$x = a^{-1}b.$$

For systems of equations $Ax = b$, we wish to write solutions as

$$x = A^{-1}b.$$

In this section we consider the questions: What does A^{-1} mean? and When does A^{-1} exist? (Even in one dimension, we have seen that the inverse does not always exist, since $0^{-1} = \frac{1}{0}$ is undefined.)

Invertibility

We begin by giving a precise definition of invertibility for square matrices.

Definition 3.7.1   *The $n \times n$ matrix A is* invertible *if there is an $n \times n$ matrix B such that*

$$AB = I_n \quad and \quad BA = I_n.$$

The matrix B is called an inverse *of A. If A is not invertible, then A is* noninvertible *or* singular.

Geometrically, we can see that some matrices are invertible. For example, the matrix

$$R_{90} = \begin{pmatrix} 0 & -1 \\ 1 & 0 \end{pmatrix}$$

rotates the plane counterclockwise through 90° and is invertible. The inverse matrix of R_{90} is the matrix that rotates the plane clockwise through 90°. That matrix is

$$R_{-90} = \begin{pmatrix} 0 & 1 \\ -1 & 0 \end{pmatrix}.$$

This statement can be checked algebraically by verifying that $R_{90}R_{-90} = I_2$ and that $R_{-90}R_{90} = I_2$.

Similarly,

$$B = \begin{pmatrix} 5 & 3 \\ 2 & 1 \end{pmatrix}$$

is an inverse of

$$A = \begin{pmatrix} -1 & 3 \\ 2 & -5 \end{pmatrix},$$

because matrix multiplication shows that $AB = I_2$ and $BA = I_2$. In fact, there is an elementary formula for finding inverses of 2×2 matrices (when they exist); see (3.8.1).

On the other hand, not all matrices are invertible. For example, the zero matrix is noninvertible, since $0B = 0$ for any matrix B.

Lemma 3.7.2 *If an $n \times n$ matrix A is invertible, then its inverse is unique and is denoted by A^{-1}.*

Proof: Let B and C be $n \times n$ matrices that are inverses of A. Then

$$BA = I_n \quad \text{and} \quad AC = I_n.$$

We use the associativity of matrix multiplication to prove that $B = C$. Compute

$$B = BI_n = B(AC) = (BA)C = I_nC = C. \qquad \blacklozenge$$

We now show how to compute inverses for products of invertible matrices.

Proposition 3.7.3 *Let A and B be two invertible $n \times n$ matrices. Then AB is also invertible and*

$$(AB)^{-1} = B^{-1}A^{-1}.$$

Proof: Use associativity of matrix multiplication to compute

$$(AB)(B^{-1}A^{-1}) = A(BB^{-1})A^{-1} = AI_nA^{-1} = AA^{-1} = I_n.$$

Similarly,

$$(B^{-1}A^{-1})(AB) = B^{-1}(A^{-1}A)B = B^{-1}B = I_n.$$

Therefore AB is invertible with the desired inverse. $\qquad \blacklozenge$

Proposition 3.7.4   *Suppose that A is an invertible $n \times n$ matrix. Then A^t is invertible and*

$$(A^t)^{-1} = (A^{-1})^t.$$

Proof: We must show that $(A^{-1})^t$ is the inverse of A^t. Identity (3.6.1) implies that

$$(A^{-1})^t A^t = (AA^{-1})^t = (I_n)^t = I_n$$

and

$$A^t (A^{-1})^t = (A^{-1}A)^t = (I_n)^t = I_n.$$

Therefore $(A^{-1})^t$ is the inverse of A^t, as claimed.   ◆

Invertibility and Unique Solutions

Next we discuss the implications of invertibility for solving the inhomogeneous linear system

$$Ax = b, \tag{3.7.1}$$

where A is an $n \times n$ matrix and $b \in \mathbf{R}^n$.

Proposition 3.7.5   *Let A be an invertible $n \times n$ matrix and let b be in $\mathbf{R}^n$. Then the system of linear equations (3.7.1) has a unique solution.*

Proof: We can solve the linear system (3.7.1) by setting

$$x = A^{-1}b. \tag{3.7.2}$$

This solution is easily verified by calculating

$$Ax = A(A^{-1}b) = (AA^{-1})b = I_n b = b.$$

Next suppose that x is a solution to (3.7.1). Then

$$x = I_n x = (A^{-1}A)x = A^{-1}(Ax) = A^{-1}b.$$

So $A^{-1}b$ is the only possible solution.   ◆

Corollary 3.7.6   *An invertible matrix is row equivalent to I_n.*

Proof: Let A be an invertible $n \times n$ matrix. Proposition 3.7.5 states that the system of linear equations $Ax = b$ has a unique solution. Corollary 2.4.8 states that A is row equivalent to I_n.   ◆

The converse of Corollary 3.7.6 is also valid.

Proposition 3.7.7   *An $n \times n$ matrix A that is row equivalent to I_n is invertible.*

Proof: Form the $n \times 2n$ matrix $M = (A | I_n)$. Since A is row equivalent to I_n, there is a sequence of elementary row operations so that M is row equivalent to $(I_n | B)$. Eliminating all columns from the right half of M except the jth column yields the matrix $(A | e_j)$. The same sequence of elementary row operations results in the matrix $(A | e_j)$, which is row equivalent to $(I_n | B_j)$, where B_j is the jth column of B. It follows that B_j is the solution to the system of linear equations $Ax = e_j$ and that the matrix product

$$AB = (AB_1 | \cdots | AB_n) = (e_1 | \cdots | e_n) = I_n.$$

So $AB = I_n$.

We claim that $BA = I_n$ and hence that A is invertible. To verify this claim, form the $n \times 2n$ matrix $N = (I_n | A)$. Using the same sequence of elementary row operations again shows that N is row equivalent to $(B | I_n)$. By construction the matrix B is row equivalent to I_n. Therefore there is a unique solution to the system of linear equations $Bx = e_j$. Now eliminating all columns except the jth from the right-hand side of the matrix $(B | I_n)$ shows that the solution to the system of linear equations $Bx = e_j$ is just A_j, where A_j is the jth column of A. It follows that

$$BA = (BA_1 | \cdots | BA_n) = (e_1 | \cdots | e_n) = I_n.$$

Hence $BA = I_n$.   ◆

Theorem 3.7.8   *Let A be an $n \times n$ matrix. Then the following are equivalent:*

(a)   *A is invertible.*

(b)   *The equation $Ax = b$ has a unique solution for each $b \in \mathbf{R}^n$.*

(c)   *The only solution to $Ax = 0$ is $x = 0$.*

(d)   *A is row equivalent to I_n.*

Proof: $(a) \Rightarrow (b)$ This implication is just Proposition 3.7.5.

$(b) \Rightarrow (c)$ This implication is straightforward—just take $b = 0$ in (3.7.1).

$(c) \Rightarrow (d)$ This implication is just a restatement of Corollary 2.4.8.

$(d) \Rightarrow (a)$ This implication is just Proposition 3.7.7.   ◆

A Method for Computing Inverse Matrices

The proof of Proposition 3.7.7 gives a constructive method for finding the inverse of any invertible square matrix.

Theorem 3.7.9   *Let A be an n × n matrix that is row equivalent to I_n, and let M be the n × 2n augmented matrix*

$$M = (A|I_n).$$ (3.7.3)

Then the matrix M is row equivalent to $(I_n|A^{-1})$.

An Example

Compute the inverse of the matrix

$$A = \begin{pmatrix} 1 & 2 & 0 \\ 0 & 1 & 3 \\ 0 & 0 & 1 \end{pmatrix}.$$

Begin by forming the 3 × 6 matrix

$$M = \left(\begin{array}{ccc|ccc} 1 & 2 & 0 & 1 & 0 & 0 \\ 0 & 1 & 3 & 0 & 1 & 0 \\ 0 & 0 & 1 & 0 & 0 & 1 \end{array} \right).$$

To put M in row echelon form by row reduction, first subtract three times the 3rd row from the 2nd row, obtaining

$$\left(\begin{array}{ccc|ccc} 1 & 2 & 0 & 1 & 0 & 0 \\ 0 & 1 & 0 & 0 & 1 & -3 \\ 0 & 0 & 1 & 0 & 0 & 1 \end{array} \right).$$

Second, subtract two times the 2nd row from the 1st row to get

$$\left(\begin{array}{ccc|ccc} 1 & 0 & 0 & 1 & -2 & 6 \\ 0 & 1 & 0 & 0 & 1 & -3 \\ 0 & 0 & 1 & 0 & 0 & 1 \end{array} \right).$$

Theorem 3.7.9 implies that

$$A^{-1} = \begin{pmatrix} 1 & -2 & 6 \\ 0 & 1 & -3 \\ 0 & 0 & 1 \end{pmatrix},$$

which can be verified by matrix multiplication.

Computing the Inverse Using MATLAB

We can compute inverses using MATLAB in two ways: Either we can perform the row reduction of (3.7.3) directly, or we can use the command `inv`. We illustrate both of these methods. First type e3_7_4 to recall the matrix

$$A = \begin{pmatrix} 1 & 2 & 4 \\ 3 & 1 & 1 \\ 2 & 0 & -1 \end{pmatrix}. \tag{3.7.4}*$$

To perform the row reduction of (3.7.3) we need to form the matrix M. The MATLAB command for generating an $n \times n$ identity matrix is `eye(n)`. Therefore typing

```
M = [A eye(3)]
```

in MATLAB yields the result

```
M =
     1     2     4     1     0     0
     3     1     1     0     1     0
     2     0    -1     0     0     1
```

Now row reduce M to reduced echelon form as follows: Type

```
M(3,:) = M(3,:) - 2*M(1,:)
M(2,:) = M(2,:) - 3*M(1,:)
```

obtaining

```
M =
     1     2     4     1     0     0
     0    -5   -11    -3     1     0
     0    -4    -9    -2     0     1
```

Next type

```
M(2,:) = M(2,:)/M(2,2)
M(3,:) = M(3,:) + 4*M(2,:)
M(1,:) = M(1,:) - 2*M(2,:)
```

to get

```
M =
   1.0000        0   -0.4000   -0.2000    0.4000        0
        0   1.0000    2.2000    0.6000   -0.2000        0
        0        0   -0.2000    0.4000   -0.8000   1.0000
```

Finally, type

```
M(3,:) = M(3,:)/M(3,3)
M(2,:) = M(2,:) - M(2,3)*M(3,:)
M(1,:) = M(1,:) - M(1,3)*M(3,:)
```

to obtain

```
M =
    1.0000         0         0   -1.0000    2.0000   -2.0000
         0    1.0000         0    5.0000   -9.0000   11.0000
         0         0    1.0000   -2.0000    4.0000   -5.0000
```

Thus $C = A^{-1}$ is obtained by extracting the last three columns of M by typing

```
C = M(:,[4 5 6])
```

which yields

```
C =
   -1.0000    2.0000   -2.0000
    5.0000   -9.0000   11.0000
   -2.0000    4.0000   -5.0000
```

You may check that C is the inverse of A by typing A*C and C*A.

In fact, this entire scheme for computing the inverse of a matrix has been preprogrammed into MATLAB. Just type

```
inv(A)
```

to obtain

```
ans =
   -1.0000    2.0000   -2.0000
    5.0000   -9.0000   11.0000
   -2.0000    4.0000   -5.0000
```

We illustrate again this simple method for computing the inverse of a matrix A. For example, reload the matrix in (3.1.4) by typing e3_1_4. We get

```
A =
     5    -4     3    -6     2
     2    -4    -2    -1     1
     1     2     1    -5     3
    -2    -1    -2     1    -1
     1    -6     1     1     4
```

The command B = inv(A) stores the inverse of the matrix A in the matrix B, and we obtain the result

```
B =
   -0.0712    0.2856   -0.0862   -0.4813   -0.0915
   -0.1169    0.0585    0.0690   -0.2324   -0.0660
    0.1462   -0.3231   -0.0862    0.0405    0.0825
   -0.1289    0.0645   -0.1034   -0.2819    0.0555
   -0.1619    0.0810    0.1724   -0.1679    0.1394
```

This computation also illustrates the fact that even when the matrix A has integer entries, the inverse of A usually has noninteger entries.

Let $b = (2, -8, 18, -6, -1)$. Then we may use the inverse $B = A^{-1}$ to compute the solution of $Ax = b$. Indeed if we type

```
b = [2;-8;18;-6;-1];
x = B*b
```

then we obtain

```
x =
   -1.0000
    2.0000
    1.0000
   -1.0000
    3.0000
```

as desired (see (3.1.5)). With this computation we have confirmed the analytical results of the previous subsections.

HAND EXERCISES

1. Verify by matrix multiplication that these two matrices are inverses of each other:

$$\begin{pmatrix} 1 & 0 & 2 \\ 0 & -1 & 2 \\ 1 & 0 & 1 \end{pmatrix} \quad \text{and} \quad \begin{pmatrix} -1 & 0 & 2 \\ 2 & -1 & -2 \\ 1 & 0 & -1 \end{pmatrix}.$$

2. Let $\alpha \neq 0$ be a real number and let A be an invertible matrix. Show that the inverse of the matrix αA is given by $\frac{1}{\alpha} A^{-1}$.

3. Let $A = \begin{pmatrix} a & 0 \\ 0 & b \end{pmatrix}$ be a 2 × 2 diagonal matrix. For which values of a and b is A invertible?

4. Let A, B, and C be general $n \times n$ matrices. Simplify the expression $A^{-1}(BA^{-1})^{-1}(CB^{-1})^{-1}$.

In Exercises 5 and 6, use row reduction to find the inverse of the given matrix.

5. $\begin{pmatrix} 1 & 4 & 5 \\ 0 & 1 & -1 \\ -2 & 0 & -8 \end{pmatrix}$ $\qquad$ **6.** $\begin{pmatrix} 1 & -1 & -1 \\ 0 & 2 & 0 \\ 2 & 0 & -1 \end{pmatrix}$

7. Let A be an $n \times n$ matrix that satisfies

$$A^3 + a_2 A^2 + a_1 A + I_n = 0,$$

where $A^2 = AA$ and $A^3 = AA^2$. Show that A is invertible.

Hint Let $B = -(A^2 + a_2 A + a_1 I_n)$ and verify that $AB = BA = I_n$.

8. Let A be an $n \times n$ matrix that satisfies

$$A^m + a_{m-1} A^{m-1} + \cdots + a_1 A + I_n = 0.$$

Show that A is invertible.

9. For which values of a, b, and c is the matrix

$$A = \begin{pmatrix} 1 & a & b \\ 0 & 1 & c \\ 0 & 0 & 1 \end{pmatrix}$$

invertible? Find A^{-1} when it exists.

COMPUTER EXERCISES

In Exercises 10 and 11, use row reduction to find the inverse of the given matrix. Then confirm your results using the command inv.

10.

$$A = \begin{pmatrix} 2 & 1 & 3 \\ 1 & 2 & 3 \\ 5 & 1 & 0 \end{pmatrix} \tag{3.7.5}*$$

11.

$$B = \begin{pmatrix} 0 & 5 & 1 & 3 \\ 1 & 5 & 3 & -1 \\ 2 & 1 & 0 & -4 \\ 1 & 7 & 2 & 3 \end{pmatrix} \tag{3.7.6}*$$

12. Try to compute the inverse of the matrix

$$C = \begin{pmatrix} 1 & 0 & 3 \\ -1 & 2 & -2 \\ 0 & 2 & 1 \end{pmatrix} \tag{3.7.7}*$$

in MATLAB using the command inv. What happens? Can you explain the outcome?
    Now compute the inverse of the matrix

$$\begin{pmatrix} 1 & \epsilon & 3 \\ -1 & 2 & -2 \\ 0 & 2 & 1 \end{pmatrix}$$

for some nonzero numbers ϵ of your choice. What can be observed in the inverse if ϵ is very small? What happens when ϵ tends to 0?

3.8 DETERMINANTS OF 2 × 2 MATRICES

There is a simple way to determine whether a 2 × 2 matrix A is invertible, and there is a simple formula for finding A^{-1}. First we present the formula. Let

$$A = \begin{pmatrix} a & b \\ c & d \end{pmatrix}$$

and suppose that $ad - bc \neq 0$. Then

$$A^{-1} = \frac{1}{ad - bc} \begin{pmatrix} d & -b \\ -c & a \end{pmatrix}. \tag{3.8.1}$$

This is most easily verified by directly applying the formula for matrix multiplication. So A is invertible when $ad - bc \neq 0$. We prove below that $ad - bc$ must be nonzero when A is invertible.

    From this discussion it is clear that the number $ad - bc$ must be an important quantity for 2 × 2 matrices. We now define it:

Definition 3.8.1   *The determinant of the 2 × 2 matrix A is*

$$\det(A) = ad - bc. \tag{3.8.2}$$

Proposition 3.8.2   *As a function on 2 × 2 matrices, the determinant satisfies the following properties.*

(a)   *The determinant of an upper triangular matrix is the product of the diagonal elements.*

(b)   *The determinants of a matrix and its transpose are equal.*

(c) $\det(AB) = \det(A)\det(B)$.

Proof: Both (a) and (b) are easily verified by direct calculation. Property (c) is also verified by direct calculation—but a more extensive one. Note that

$$\begin{pmatrix} a & b \\ c & d \end{pmatrix} \begin{pmatrix} \alpha & \beta \\ \gamma & \delta \end{pmatrix} = \begin{pmatrix} a\alpha + b\gamma & a\beta + b\delta \\ c\alpha + d\gamma & c\beta + d\delta \end{pmatrix}.$$

Therefore

$$\begin{aligned}
\det(AB) &= (a\alpha + b\gamma)(c\beta + d\delta) - (a\beta + b\delta)(c\alpha + d\gamma) \\
&= (ac\alpha\beta + bc\beta\gamma + ad\alpha\delta + bd\gamma\delta) - (ac\alpha\beta + bc\alpha\delta + ad\beta\gamma + bd\gamma\delta) \\
&= bc(\beta\gamma - \alpha\delta) + ad(\alpha\delta - \beta\gamma) \\
&= (ad - bc)(\alpha\delta - \beta\gamma) \\
&= \det(A)\det(B),
\end{aligned}$$

as asserted.   ◆

Corollary 3.8.3   *A 2 × 2 matrix A is invertible if and only if* $\det(A) \neq 0$.

Proof: If A is invertible, then $AA^{-1} = I_2$. Proposition 3.8.2 implies that

$$\det(A)\det(A^{-1}) = \det(I_2) = 1.$$

Therefore $\det(A) \neq 0$. Conversely, if $\det(A) \neq 0$, then (3.8.1) implies that A is invertible.   ◆

Determinants and Area

Suppose that v and w are two vectors in $\mathbf{R}^2$ that point in different directions. Then the set of points

$$z = \alpha v + \beta w, \quad \text{where } 0 \leq \alpha, \beta \leq 1,$$

is a parallelogram that we denote by P. We denote the area of P by $|P|$. For example, the unit square S, whose corners are $(0, 0)$, $(1, 0)$, $(0, 1)$, and $(1, 1)$, is the parallelogram generated by the unit vectors e_1 and e_2.

Next let A be a 2×2 matrix and let

$$A(P) = \{Az : z \in P\}.$$

It follows from linearity (since $Az = \alpha Av + \beta Aw$) that $A(P)$ is the parallelogram generated by Av and Aw.

Proposition 3.8.4   *Let A be a 2×2 matrix and let S be the unit square. Then*

$$|A(S)| = |\det A|. \qquad (3.8.3)$$

Proof: Note that $A(S)$ is the parallelogram generated by $u_1 = Ae_1$ and $u_2 = Ae_2$, and u_1 and u_2 are the columns of A. It follows that

$$(\det A)^2 = \det(A^t)\det(A) = \det(A^t A) = \det \begin{pmatrix} u_1^t u_1 & u_1^t u_2 \\ u_2^t u_1 & u_2^t u_2 \end{pmatrix}.$$

Hence

$$(\det A)^2 = \det \begin{pmatrix} ||u_1||^2 & u_1 \cdot u_2 \\ u_1 \cdot u_2 & ||u_2||^2 \end{pmatrix} = ||u_1||^2 ||u_2||^2 - (u_1 \cdot u_2)^2.$$

Recall that (1.4.5) states that

$$|P|^2 = ||v||^2 ||w||^2 - (v \cdot w)^2,$$

where P is the parallelogram generated by v and w. Therefore $(\det A)^2 = |A(S)|^2$, and (3.8.3) is verified.   ◆

Theorem 3.8.5   *Let P be a parallelogram in $\mathbf{R}^2$ and let A be a 2×2 matrix. Then*

$$|A(P)| = |\det A| \, |P|. \qquad (3.8.4)$$

Proof: First note that (3.8.3) is a special case of (3.8.4), since $|S| = 1$. Next let P be the parallelogram generated by the (column) vectors v and w, and let $B = (v|w)$. Then $P = B(S)$. It follows from (3.8.3) that $|P| = |\det B|$. Moreover,

$$\begin{aligned} |A(P)| &= |(AB)(S)| \\ &= |\det(AB)| \\ &= |\det A| \, |\det B| \\ &= |\det A| \, |P|, \end{aligned}$$

as desired.   ◆

HAND EXERCISES

1. Find the inverse of the matrix $\begin{pmatrix} 2 & 1 \\ 3 & 2 \end{pmatrix}$.

2. Find the inverse of the shear matrix $\begin{pmatrix} 1 & K \\ 0 & 1 \end{pmatrix}$.

3. Show that the 2×2 matrix $A = \begin{pmatrix} a & b \\ c & d \end{pmatrix}$ is row equivalent to I_2 if and only if $ad - bc \neq 0$.

Hint   Prove this result separately in the two cases $a \neq 0$ and $a = 0$.

4. Let A be a 2×2 matrix having integer entries. Find a condition on the entries of A that guarantees that A^{-1} has integer entries.

5. Let A be a 2×2 matrix and assume that $\det(A) \neq 0$. Then use the explicit form for A^{-1} given in (3.8.1) to verify that

$$\det(A^{-1}) = \frac{1}{\det(A)}.$$

6. Sketch the triangle whose vertices are 0, $p = (3, 0)^t$, and $q = (0, 2)^t$. Then find the area of this triangle. Let

$$M = \begin{pmatrix} -4 & -3 \\ 5 & -2 \end{pmatrix}.$$

Sketch the triangle whose vertices are 0, Mp, and Mq. Then find the area of this triangle.

7. Cramer's rule provides a method based on determinants for finding the unique solution to the linear equation $Ax = b$ when A is an invertible matrix. More precisely, let A be an invertible 2×2 matrix and let $b \in \mathbf{R}^2$ be a column vector. Let B_j be the 2×2 matrix obtained from A by replacing the jth column of A by the vector b. Let $x = (x_1, x_2)^t$ be the unique solution to $Ax = b$. Then Cramer's rule states that

$$x_j = \frac{\det(B_j)}{\det(A)}. \tag{3.8.5}$$

Prove Cramer's rule.

Hint   Write the general system of two equations in two unknowns as

$$a_{11}x_1 + a_{12}x_2 = b_1$$
$$a_{21}x_1 + a_{22}x_2 = b_2.$$

Subtract a_{11} times the second equation from a_{21} times the first equation to eliminate x_1; then solve for x_2 and verify (3.8.5). Use a similar calculation to solve for x_1.

In Exercises 8 and 9, use Cramer's rule (3.8.5) to solve the given system of linear equations.

8. Solve $\begin{array}{l} 2x + 3y = 2 \\ 3x - 5y = 1 \end{array}$ for x.

9. Solve $\begin{array}{l} 4x - 3y = -1 \\ x + 2y = 7 \end{array}$ for y.

COMPUTER EXERCISES

10. Use MATLAB to choose five 2×2 matrices at random and compute their inverses. Do you get the impression that "typically" 2×2 matrices are invertible? Try to find a reason for this using the determinant of 2×2 matrices.

In Exercises 11–14, use the Unit square icon in the program map to test Proposition 3.8.4, as follows. Enter the given matrix A into map and map the Unit square icon. Compute $\det(A)$ by estimating the area of $A(S)$—given that S has unit area. For each matrix, use this numerical experiment to decide whether or not the matrix is invertible.

11. $A = \begin{pmatrix} 0 & -2 \\ 2 & 0 \end{pmatrix}$

12. $A = \begin{pmatrix} -0.5 & -0.5 \\ 0.7 & 0.7 \end{pmatrix}$

13. $A = \begin{pmatrix} -1 & -0.5 \\ -2 & -1 \end{pmatrix}$

14. $A = \begin{pmatrix} 0.7071 & 0.7071 \\ -0.7071 & 0.7071 \end{pmatrix}$

Solving Ordinary Differential Equations

The study of linear systems of equations in Chapter 2 provides one motivation for the study of matrices and linear algebra. Linear constant coefficient systems of ordinary differential equations are a second motivation for this study. In this chapter we show how the phase space geometry of systems of differential equations leads to the idea of *eigendirections* (or invariant directions) and *eigenvalues* (or growth rates).

We begin this chapter with a discussion of the theory and application of the simplest linear differential equations—the linear growth equation, $\dot{x} = \lambda x$. In Section 4.1 we solve the linear growth equation and discuss the fact that solutions to differential equations are functions; we emphasize this point by using MATLAB to graph solutions of x as a function of t. We also illustrate the applicability of this very simple equation with a discussion of compound interest and a simple population model.

In Section 4.2 we discuss two different ways to plot solutions of differential equations: *time series* and *phase space* plots. The first method just plots the graph of a solution as a function of time t, as discussed in Section 4.1, while the second method is based on thinking of the differential equation as describing how a point moves in space. Both methods are important: Time series are typical ways of representing results of experiments, and phase space plots are central to a geometric understanding of solutions to differential equations. In Sections 4.2 and 4.3 we introduce two MATLAB programs, dfield5 (written by John Polking) and pline, which illustrate the two methods of plotting the output of a differential equation.

In the optional Section 4.4 we present one method for solving differential equations analytically where $f(t, x)$, the right-hand side in the ODE (ordinary differential equation), is a product of a function of x and a function of t. This method is called *separation of variables* and is based on integration theory from calculus. We will see that even these simple differential equations may lead to solutions that are defined only implicitly and not in closed form.

The next two sections introduce planar constant coefficient linear differential equations. In these sections we use the program `pplane5` (also written by John Polking), which solves numerically planar systems of differential equations. In Section 4.5 we discuss uncoupled systems—two independent one-dimensional systems like those presented in Section 4.1—whose solution geometry in the plane is somewhat more complicated than might be expected. In Section 4.6 we discuss coupled linear systems and illustrate the existence and nonexistence of eigendirections.

In Section 4.7 we show how initial value problems can be solved by building the solution—through the use of superposition as discussed in Section 3.4—from simpler solutions. These simpler solutions are ones generated from real eigenvalues and eigenvectors—when they exist. In Section 4.8 we develop the theory of *eigenvalues* and *characteristic polynomials* of 2×2 matrices. (The corresponding theory for $n \times n$ matrices is developed in Chapter 8.)

The method for solving planar constant coefficient linear differential equations with real eigenvalues is summarized in Section 4.9. This method is based on the material of Sections 4.7 and 4.8. The complete discussion of the solutions of linear planar systems of differential equations is given in Chapter 6.

The chapter ends with an optional discussion of *Markov chains* in Section 4.10. Markov chains give a method for analyzing branch processes where at each time unit several outcomes are possible, each with a given probability.

4.1 A SINGLE DIFFERENTIAL EQUATION

Algebraic operations such as addition and multiplication are performed on numbers, whereas the calculus operations of differentiation and integration are performed on functions. Thus algebraic equations (such as $x^2 = 9$) are solved for numbers ($x = \pm 3$), whereas differential (and integral) equations are solved for functions.

In Chapter 2 we discussed how to solve systems of linear equations such as

$$x_1 + x_2 = 2$$
$$x_1 - x_2 = 4$$

for numbers

$$x_1 = 3 \quad \text{and} \quad x_2 = -1.$$

In this chapter we discuss how to solve some systems of differential equations for functions.

Solving a single linear equation in one unknown x is a simple task. For example, solve

$$2x = 4$$

for $x = 2$. Solving a single differential equation in one unknown function $x(t)$ is far from trivial.

Integral Calculus As a Differential Equation

Mathematically, the simplest type of differential equation is

$$\frac{dx}{dt}(t) = f(t), \tag{4.1.1}$$

where f is some continuous function. In words, this equation asks us to find all functions $x(t)$ whose derivative is $f(t)$. The fundamental theorem of calculus tells us the answer: $x(t)$ is an antiderivative of $f(t)$. Thus to find all solutions, we just integrate both sides of (4.1.1) with respect to t. Formally, using indefinite integrals, we have

$$\int \frac{dx}{dt}(t)\, dt = \int f(t)\, dt + C, \tag{4.1.2}$$

where C is an arbitrary constant. (It is tempting to put a constant of integration on both sides of (4.1.2), but two constants are not needed, because we can just combine both constants on the right-hand side of this equation.) Since the indefinite integral of dx/dt is just the function $x(t)$, we have

$$x(t) = \int f(\tau)\, d\tau + C. \tag{4.1.3}$$

In particular, finding formulas for all solutions to differential equations of the type (4.1.1) is equivalent to finding formulas for all definite integrals of the function $f(t)$. Indeed, to find solutions to differential equations like (4.1.1), we need to know all of the techniques of integration from integral calculus.

Initial Conditions and the Role of the Integration Constant C

Equation (4.1.3) tells us that there are an infinite number of solutions to the differential equation (4.1.1), each one corresponding to a different choice of the constant C. To understand how to interpret the constant C, consider the example

$$\frac{dx}{dt}(t) = \cos t.$$

Using (4.1.3), we see that the answer is

$$x(t) = \int \cos \tau\, d\tau + C = \sin t + C.$$

Note that

$$x(0) = \sin(0) + C = C.$$

Thus the constant C represents an *initial condition* for the differential equation. We return to the discussion of initial conditions several times later in this chapter.

See Exercise 15 at the end of this section for a more interesting example of this type of differential equation.

Solutions to Differential Equations Are Functions

Consider the differential equation

$$\frac{dx}{dt}(t) = tx(t). \tag{4.1.4}$$

Are the functions

$$x_1(t) = t^2 \quad \text{and} \quad x_2(t) = e^{t^2/2}$$

solutions to the differential equation (4.1.4)?

To test whether or not the function $x_1(t)$ is a solution, we compute the left- and right-hand sides of (4.1.4):

$$\text{LHS:} \quad \frac{d}{dt}x_1(t) = 2t$$
$$\text{RHS:} \quad tx_1(t) = t^3.$$

Since the left- and right-hand sides are unequal, the function $x_1(t)$ is not a solution to (4.1.4).

To test whether or not the function $x_2(t)$ is a solution, we again compute the left- and right-hand sides of (4.1.4):

$$\text{LHS:} \quad \frac{d}{dt}x_2(t) = te^{t^2/2}$$
$$\text{RHS:} \quad tx_2(t) = te^{t^2/2}.$$

Since the left- and right-hand sides are equal, the function $x_2(t)$ is a solution to (4.1.4).

Note that we have not discussed how we knew that the function $x_2(t)$ is a solution to (4.1.4). For the most part, the issue of how one finds solutions to a differential equation is discussed in later chapters, although we do determine solutions to a very important equation next.

The Linear Differential Equation of Growth and Decay

The real subject of differential equations begins when the function f on the right-hand side of (4.1.1) depends explicitly on the function x. The simplest such differential equation is

$$\frac{dx}{dt}(t) = x(t).$$

Using results from calculus, we can even solve this equation; indeed, we can even solve the slightly more complicated equation

$$\frac{dx}{dt}(t) = \lambda x(t), \tag{4.1.5}$$

where $\lambda \in \mathbf{R}$ is a constant. The differential equation (4.1.5) is *linear* since $x(t)$ appears by itself on the right-hand side. Moreover, (4.1.5) is *homogeneous* since the constant function $x(t) = 0$ is a solution.

In words, (4.1.5) asks: For which functions $x(t)$ is the derivative of $x(t)$ a scalar multiple of $x(t)$? The function

$$x(t) = e^{\lambda t}$$

is such a function since

$$\frac{dx}{dt}(t) = \frac{d}{dt}e^{\lambda t} = \lambda e^{\lambda t} = \lambda x(t).$$

More generally, the function

$$x(t) = Ke^{\lambda t} \tag{4.1.6}$$

is a solution to (4.1.5) for any real constant K. We claim that the functions (4.1.6) list all (differentiable) functions that solve (4.1.5).

To verify this claim, we let $x(t)$ be a solution to (4.1.5) and show that the ratio

$$\frac{x(t)}{e^{\lambda t}} = x(t)e^{-\lambda t}$$

is a constant (independent of t). Using the product rule and (4.1.5), we compute

$$\frac{d}{dt}\left[x(t)e^{-\lambda t}\right] = \frac{d}{dt}\left(x(t)\right)e^{-\lambda t} + x(t)\frac{d}{dt}\left(e^{-\lambda t}\right)$$
$$= (\lambda x(t))e^{-\lambda t} + x(t)(-\lambda e^{-\lambda t})$$
$$= 0.$$

Now recall that the only functions whose derivatives are identically 0 are the constant functions. Thus

$$x(t)e^{-\lambda t} = K$$

for some constant $K \in \mathbf{R}$. Hence $x(t)$ has the form (4.1.6), as claimed.

Next we discuss the role of the constant K. We write the function as $x(t)$, and we want you to think of the variable t as time. Thus $x(0)$ is the initial value of the function $x(t)$ at time $t = 0$; we say that $x(0)$ is the *initial value* of $x(t)$. From (4.1.6) we see that

$$x(0) = K$$

and that K is the initial value of the solution of (4.1.5). Henceforth, we write K as x_0 so that the notation calls attention to the special meaning of this constant.

By deriving (4.1.6), we have proved the next theorem.

Theorem 4.1.1   *There is a unique solution to the* initial value problem

$$\frac{dx}{dt}(t) = \lambda x(t)$$
$$x(0) = x_0. \tag{4.1.7}$$

That solution is

$$x(t) = x_0 e^{\lambda t}.$$

As a consequence of Theorem 4.1.1 we see that there is a qualitative difference in the behavior of solutions to (4.1.7) depending on whether $\lambda > 0$ or $\lambda < 0$. Suppose that $x_0 > 0$. Then

$$\lim_{t \to \infty} x(t) = \lim_{t \to \infty} x_0 e^{\lambda t} = \begin{cases} +\infty & \lambda > 0 \\ 0 & \lambda < 0. \end{cases} \tag{4.1.8}$$

When $\lambda > 0$, we say that the solution has *exponential growth*, and when $\lambda < 0$, we say that the solution has *exponential decay*. In either case, however, the number λ is called the *growth rate*. We can visualize this discussion by graphing the solutions in MATLAB.

Suppose we set $x_0 = 1$ and $\lambda = \pm 0.5$. Type

```
x0 = 1;
lambda = 0.5;
t = linspace(-1,4,100);
x = x0*exp(lambda*t);
plot(t,x)
hold on
xlabel('t')
ylabel('x')
lambda = -0.5;
x = x0*exp(lambda*t);
plot(t,x)
```

The result of this calculation is shown in Figure 4.1. In this way we can actually see the difference between exponential growth ($\lambda = 0.5$) and exponential decay ($\lambda = -0.5$), as discussed in (4.1.8).

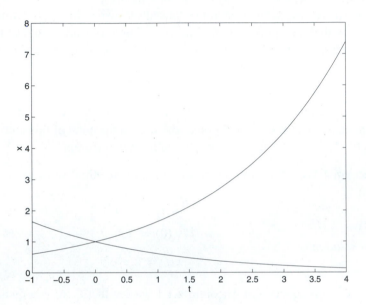

Figure 4.1
Solutions of (4.1.5) for $t \in [-1, 4]$, $x_0 = 1$, and $\lambda = \pm 0.5$

The Inhomogeneous Linear Differential Equation

It follows from (4.1.8) that solutions to the linear homogeneous differential equation (4.1.7) either are unbounded as $t \to \infty$ or approach 0. Now we consider an inhomogeneous differential equation and show that solutions can approach fixed values for increasing t that are neither 0 nor infinity.

As an example, consider the linear differential equation

$$\frac{dx}{dt} = -2x - 6. \tag{4.1.9}$$

Observe that $x(t) = 0$ is not a solution of (4.1.9), and therefore (4.1.9) is *inhomogeneous*. It is easy to verify, however, that the constant function $x(t) = -3$ is a solution.

Equation (4.1.9) can be solved by introducing a new function that transforms it into a homogeneous equation. Let

$$y(t) = x(t) + 3$$

and compute

$$\frac{dy}{dt} = \frac{dx}{dt} = -2x - 6 = -2y.$$

From Theorem 4.1.1, it follows that $y(t)$ has the form

$$y(t) = y_0 e^{-2t}.$$

Therefore

$$x(t) = y_0 e^{-2t} - 3$$

is a solution of (4.1.9) for every constant y_0. Moreover,

$$\lim_{t \to \infty} x(t) = \lim_{t \to \infty} y_0 e^{-2t} - 3 = -3,$$

which is neither 0 nor infinity.

Any equation of the form

$$\frac{dx}{dt}(t) = \lambda x(t) + \rho \qquad\qquad \textbf{(4.1.10)}$$

can be solved in a similar fashion. Just set

$$y(t) = x(t) + \frac{\rho}{\lambda}.$$

It follows from (4.1.10) that

$$\frac{dy}{dt} = \frac{dx}{dt} = \lambda x(t) + \rho = \lambda\left(y(t) - \frac{\rho}{\lambda}\right) + \rho = \lambda y(t).$$

Theorem 4.1.1 implies that $y(t) = y_0 e^{\lambda t}$ for some constant y_0 and

$$x(t) = y_0 e^{\lambda t} - \frac{\rho}{\lambda}.$$

Note that the limit of $x(t)$ as $t \to \infty$ is $-\rho/\lambda$ when $\lambda < 0$.

Some Examples of (4.1.7)

Even though the differential equation (4.1.5) is one of the simplest differential equations, it still has some use in applications. We present two here: compound interest and population dynamics.

Compound Interest

Banks pay interest on an account in the following way: At the end of each day, the bank determines the interest rate r_{day} for that day, checks the principal P in the account, and then deposits an additional $r_{day}P$. So the next day the principal in this account is $(1 + r_{day})P$. Note that if r denotes the interest rate per year, then $r_{day} = r/365$. Of course, a day is just a convenient measure for elapsed time. Before the use of computers was prevalent, banks paid interest yearly or quarterly or monthly or, in a few cases, even weekly, depending on the particular bank rules.

Observe that the more frequently interest is paid, the more money is earned. For example, if interest is paid only once at the end of a year, then the money in the account at the end of the year is $(1 + r)P$, and the amount rP is called *simple interest*. But if interest is paid twice a year, then the principal at the end of six months is $(1 + r/2)P$, and the principal at the end of the year is $(1 + r/2)^2 P$. Since

$$\left(1 + \frac{r}{2}\right)^2 = 1 + r + \frac{1}{4}r^2 > 1 + r,$$

there is more money in the account at the end of the year if the interest is compounded semiannually rather than annually. But how much is the difference and what is the maximum earning potential?

While making the calculation in the preceding paragraph, we implicitly made a number of simplifying assumptions. In particular, we assumed:

- an initial principal P_0 is deposited in the bank on January 1,
- the money is not withdrawn for one year,
- no new money is deposited in that account during the year,
- the yearly interest rate r remains constant throughout the year, and
- interest is added to the account N times during the year.

In this *model*, simple interest corresponds to $N = 1$, compound monthly interest to $N = 12$, and compound daily interest to $N = 365$.

We first answer the question: How much money is in this account after one year? After one time unit of $1/N$ year, the amount of money in the account is

$$Q_1 = \left(1 + \frac{r}{N}\right) P_0.$$

The interest rate in each time period is r/N, the yearly rate r divided by the number of time periods N. Here we have used the assumption that the interest rate remains constant throughout the year. After two time units, the principal is

$$Q_2 = \left(1 + \frac{r}{N}\right) Q_1 = \left(1 + \frac{r}{N}\right)^2 P_0,$$

and at the end of the year (that is, after N time periods),

$$Q_N = \left(1 + \frac{r}{N}\right)^N P_0. \qquad \text{(4.1.11)}$$

Here we have used the assumption that money is neither deposited nor withdrawn from our account. Note that Q_N is the amount of money in the bank after **one** year assuming that interest has been compounded N (equally spaced) times during that year, and the effective interest rate when compounding N times is

$$\left(1 + \frac{r}{N}\right)^N - 1.$$

For the curious, we can write a program in MATLAB to compute (4.1.11). Suppose we assume that the initial deposit $P_0 = \$1,000$, the simple interest rate is 6% per year, and the interest payments are made monthly. In MATLAB type

```
N  = 12;
P0 = 1000;
r  = 0.06;
QN = (1 + r/N)^N*P0
```

The answer is $QN = \$1,061.68$, and the *effective* interest rate for monthly payments is 6.16778%. For daily interest payments $N = 365$, the answer is $QN = \$1,061.83$, and the effective interest rate is 6.18313%.

To find the maximum effective interest, we ask the bank to compound interest continuously; that is, we ask the bank to compute

$$\lim_{N \to \infty} \left(1 + \frac{r}{N}\right)^N.$$

We compute this limit using differential equations. The concept of continuous interest is rephrased as follows: Let $P(t)$ be the principal at time t, where t is measured in units of years. Suppose we assume that interest is compounded N times during the year. The length of time in each compounding period is

$$\Delta t = \frac{1}{N},$$

and the change in principal during that time period is

$$\Delta P = \frac{r}{N} P = r P \Delta t.$$

It follows that

$$\frac{\Delta P}{\Delta t} = r P,$$

and, on taking the limit $\Delta t \to 0$, we have the differential equation

$$\frac{dP}{dt}(t) = r P(t).$$

Since $P(0) = P_0$, the solution of the initial value problem given in Theorem 4.1.1 shows that

$$P(t) = P_0 e^{rt}.$$

After one year ($t = 1$), we find that

$$P(1) = e^r P_0.$$

Note that

$$P(1) = \lim_{N \to \infty} Q_N,$$

and we have thus verified that

$$\lim_{N \to \infty} \left(1 + \frac{r}{N}\right)^N = e^r.$$

Thus the maximum effective interest rate is $e^r - 1$. When $r = 6\%$, the maximum effective interest rate is 6.18365%.

An Example from Population Dynamics

To provide a second interpretation of the constant λ in (4.1.5), we discuss a simplified model for population dynamics. Let $p(t)$ be the size of a population of a certain species at time t, and let r be the rate at which the population p is changing at time t. In general, r depends on the time t and is a complicated function of birth and death rates and of immigration and emigration, as well as other factors. Indeed, the rate r may well depend on the size of the population itself. (Overcrowding can be modeled by assuming that the death rate increases with the size of the population.) These population models assume that the rate of change in the size of the population dp/dt is given by

$$\frac{dp}{dt}(t) = rp(t). \tag{4.1.12}$$

They differ just on the precise form of r. In general, the rate r depends on the size of the population p as well as the time t; that is, r is a function $r(p, t)$.

The simplest population model—which we now assume—is the one in which r is assumed to be constant. Then (4.1.12) is identical to (4.1.5) after we identify p with x and r with λ. Hence we may interpret r as the growth rate for the population. The form of the solution in (4.1.6) shows that the size of a population grows exponentially if $r > 0$ and decays exponentially if $r < 0$.

The mathematical description of this simplest population model shows that the assumption of a constant growth rate leads to exponential growth (or exponential decay). Is this realistic? Surely no population will grow exponentially for all time; other factors, such as limited living space, have to be taken into account. On the other hand, exponential growth describes well the growth in human population during much of history. So this model, though surely oversimplified, gives some insight into population growth.

HAND EXERCISES

In Exercises 1–3, find solutions to the initial value problems.

1. $\frac{dx}{dt} = \sin(2t), \quad x(\pi) = 2$ **2.** $\frac{dx}{dt} = t^2, \quad x(2) = 8$ **3.** $\frac{dx}{dt} = \frac{1}{t^2}, \quad x(1) = 1$

In Exercises 4–7, determine whether or not each of the given functions $x_1(t)$ and $x_2(t)$ is a solution to each ordinary differential equation.

4. ODE: $\frac{dx}{dt} = \frac{t}{x-1}$;   functions: $x_1(t) = t + 1$ and $x_2(t) = \frac{1 + \sqrt{4t^2 + 1}}{2}$

5. ODE: $\frac{dx}{dt} = x + e^t$;   functions: $x_1(t) = te^t$ and $x_2(t) = 2e^t$

6. ODE: $\frac{dx}{dt} = x^2 + 1$;   functions: $x_1(t) = -\tan t$ and $x_2(t) = \tan t$

7. ODE: $\frac{dx}{dt} = \frac{x}{t}$;   functions: $x_1(t) = t + 1$ and $x_2(t) = 5t$

8. Solve the differential equation

$$\frac{dx}{dt} = 2x,$$

where $x(0) = 1$. At what time t_1 will $x(t_1) = 2$?

9. Solve the differential equation

$$\frac{dx}{dt} = -3x.$$

At what time t_1 will $x(t_1)$ be half of $x(0)$?

10. Bacteria grown in a culture increase at a rate proportional to the number present. If the number of bacteria doubles every 2 hours, then how many bacteria will be present after 5 hours? Express your answer in terms of x_0, the initial number of bacteria.

11. Suppose you deposit $10,000 in a bank account at an interest rate of 7.5% compounded continuously. How much money will be in your account a year and a half later? How much would you have if the interest were compounded monthly?

12. Newton's law of cooling states that the rate at which a body changes temperature is proportional to the difference between the body temperature and the temperature of the surrounding medium. That is,

$$\frac{dT}{dt} = \alpha(T - T_m), \tag{4.1.13}$$

where $T(t)$ is the temperature of the body at time t, T_m is the constant temperature of the surrounding medium, and α is the constant of proportionality. Suppose the body is in air of temperature $50°$ and the body cools from $100°$ to $75°$ in 20 minutes. What will the temperature of the body be after one hour?

Hint   Rewrite (4.1.13) in terms of $U(t) = T(t) - T_m$.

13. Let $p(t)$ be the population of group Grk at time t measured in years. Let r be the growth rate of the group Grk. Suppose that the population of Grks changes according to the differential equation (4.1.12). Find r so that the population of Grks doubles every 50 years. How large must r be so that the population doubles every 25 years?

14. You deposit $4,000 in a bank account at an interest rate of 5.5%, but after half a year the bank changes the interest rate to 4.5%. Suppose that the interest is compounded continuously. How much money will be in your account after one year?

15. As an application of (4.1.3), answer the following question (posed by R. P. Agnew): One day it started snowing at a steady rate. A snowplow started at noon and went 2 miles in the first hour and 1 mile in the second hour. Assume that the speed of the snowplow times the depth of the snow is constant. At what time did it start to snow?

To set up this problem, let $d(t)$ be the depth of the snow at time t, where t is measured in hours and $t = 0$ is noon. Since the snow is falling at a constant rate r, $d(t) = r(t - t_0)$, where t_0 is the time that it started snowing. Let $x(t)$ be the position of the snowplow along the road. The assumption that speed times the depth equals a constant k means that

$$\frac{dx}{dt}(t) = \frac{k}{d(t)} = \frac{K}{t - t_0},$$

where $K = k/r$. The information about how far the snowplow goes in the first 2 hours translates into

$$x(1) = 2 \quad \text{and} \quad x(2) = 3.$$

Now solve the problem.

COMPUTER EXERCISES

In Exercises 16–19, use MATLAB to graph the given function f on the specified interval.

16. $f(t) = t^2$ on the interval $t \in [0, 2]$

17. $f(t) = e^t - t$ on the interval $t \in [0, 3]$

18. $f(t) = \cos(2t) - t$ on the interval $t \in [2, 8]$

19. $f(t) = \sin(5t)$ on the interval $t \in [0, 6.5]$.

Hint   Use the fact that the trigonometric functions sin and cos can be evaluated in MATLAB in the same way as the exponential function—that is, by using sin and cos instead of exp.

20. Two banks each pay 7% interest per year; one compounds money daily and one compounds money continuously. What is the difference in earnings in one year in an account having $10,000?

21. There are two banks in town: Intrastate and Statewide. You plan to deposit $5,000 in one of these banks for two years. Statewide Bank's best savings account pays 8% interest per year compounded quarterly and charges $10 to open an account. Intrastate Bank's best savings account pays 7.75% interest compounded daily. Which bank will pay you the most money when you withdraw your money? Would your answer change if you had planned to keep your money in the bank for only one year?

22. In the beginning of the year 1990 the population of the United States was approximately 250,000,000 and the growth rate was estimated at 3% per year. Assuming that the growth rate does not change, during what year will the population of the United States reach 400,000,000?

4.2   GRAPHING SOLUTIONS TO DIFFERENTIAL EQUATIONS

Solutions to differential equations can be graphed in several different ways, each giving different insight into the structure of the solutions. We begin by asking what object is to be graphed. Do we first solve the differential equation and then graph the solution, or do we let the computer find the solution numerically and then graph the result? The first method assumes that we can find a formula for the solution (such as $x(t) = x_0 e^{\lambda t}$). A solution to a differential equation for which we have an explicit formula is called a *closed form* solution. Using MATLAB, we can graph closed form solutions, as we showed in Figure 4.1. The second method of graphing solutions requires having a numerical method that can *numerically integrate* the differential equation to any desired degree of accuracy.

In fact, there are rather few differential equations that can be solved in closed form (although the linear systems that we describe in this chapter are ones that can be solved in closed form). Without formulas, the first method is impossible. There are, however, several efficient algorithms for the numerical solution of (systems of) ordinary differential equations and these methods have been preprogrammed in MATLAB. In our discussions, we treat MATLAB as a *black box* numerical integration solver of ordinary differential equations.

A Single First-Order Ordinary Differential Equation

We begin our discussion of the numerical integration of differential equations with the single first-order differential equation of the form

$$\frac{dx}{dt}(t) = f(t, x(t)). \qquad (4.2.1)$$

The equation is *first order* since only the first derivative of the function $x(t)$ appears in the equation. If the second derivative appeared in the equation, then it would be a second-order equation.

Independent (t) and Dependent (x) Variables

Sometimes (4.2.1) is also written in the form

$$\frac{dx}{dt} = f(t, x), \qquad (4.2.2)$$

and in this form both t and x appear as variables. But x is a function $x(t)$ depending on t, and therefore the variable t is called the *independent* variable, while the variable x is called the *dependent* variable.

Autonomous Versus Nonautonomous

When the right-hand side f does not depend explicitly on the independent time variable t, the equation is called *autonomous*. More explicitly, in (4.2.2), the differential equation is autonomous when $f(t, x) = g(x)$. Equation (4.1.5) is an example of an autonomous differential equation since $f(t, x) = \lambda x$.

When f depends explicitly on the independent time variable t, the differential equation is called *nonautonomous*. Suppose in our example of interest rates in Section 4.1 we had assumed that the interest rate r changes over time. Then we would write $r = r(t)$, and the differential equation modeling how the principal $P(t)$ changes in time would be written as

$$\frac{dP}{dt} = r(t)P. \tag{4.2.3}$$

This equation is an example of a nonautonomous differential equation since $f(t, P) = r(t)P$. For instance, suppose that at time $t = 0$ the principal in our account is P_0 and the interest rate is 5.5%. Now suppose the bank changes the interest rate after six months to 4.5%. Then $P(t)$ is a solution to (4.2.3), where

$$r(t) = \begin{cases} 0.055 & \text{for } 0 \leq t < 0.5 \\ 0.045 & \text{for } 0.5 \leq t \end{cases}$$

and $P(0) = P_0$.

Another example of a nonautonomous differential equation is given by

$$\frac{dx}{dt} = x^2 - t. \tag{4.2.4}$$

Time Series and Phase Space Plots

There are two different methods for visualizing the result of numerical integration of differential equations of the form (4.2.1): time series plots and phase space plots. These two methods are based on interpreting the derivative dx/dt either as the slope of a tangent line or as the velocity of a particle.

A *time series* plot for a solution to (4.2.1) is found by plotting x versus t as we did for the closed form solution in Figure 4.1. However, when graphing time series of solutions, we do not need to find closed form solutions. To understand how this is done, we briefly discuss what equation (4.2.1) is actually saying about a solution $x(t)$. This equation states that the slope of the tangent line to the graph of the function $x(t)$ at time t (dx/dt) is known and equals $f(t, x(t))$. Thus we can use the right-hand side of (4.2.1) to draw the tangent lines to $x(t)$ at each point in the tx-plane. This leads to the notion of a line field. The rough idea behind the numerical integration scheme is to fit a curve $(t, x(t))$ into the tx-plane in such a way that the tangent lines to the curve match the tangent lines specified by the slope f.

A *phase space* plot is based on the *other* interpretation of a derivative as a rate of change—a velocity. We let $x(t)$ denote the position of a particle on the real line at time t. The function $f(t, x)$ denotes the velocity of that particle when the particle is at position x at time t. Thus, to view the phase space plot, we need to see the particle moving along the real line; that is, we need to see how $x(t)$ changes in t. Later we use MATLAB graphics to visualize the particle movement.

Thus time series are graphs of functions in the tx-plane, while phase space plots are graphs on the real line x. We discuss time series plots in this section and phase line plots in the next.

Line Fields

We begin our discussion of line fields (or synonymously direction fields) by focusing on the information about solutions that can be extracted directly from the equation itself. To illustrate this we consider the differential equation (4.2.4).

As mentioned, the differential equation $\dot{x} = x^2 - t$ reflects that the value of the derivative of a solution $x(t)$ at time t is given by $(x(t))^2 - t$. In other words, the slope of the tangent line to the solution is known and is given by the right-hand side of the differential equation.

We can use this information to sketch all the tangent lines at each point (t_0, x_0) of a rectangle in the tx-plane. We do this by drawing a small line segment at each point, with the slope determined by the right-hand side. In this way we obtain the *line field*. Figure 4.2 shows a line field corresponding to the differential equation (4.2.4).

By looking at the left-hand diagram in Figure 4.2, we can imagine how solution curves fit into it; indeed, we can use this line field to make freehand sketches of solutions to (4.2.4). The right-hand image in Figure 4.2 shows the solution starting at the initial condition $x(-2) = -4$, which is the point $(x, t) = (-4, -2)$.

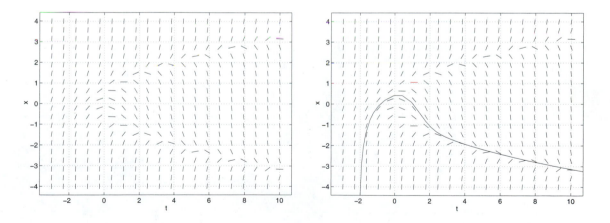

Figure 4.2
Left: Line field for (4.2.4) for $t \in [-3, 10]$ and $x \in [-4, 4]$. *Right*: A solution starting for $t = -2$ at $x(-2) = -4$.

Graphing Solutions Using dfield5

We now explain how to use MATLAB to display the graphs of solutions to the differential equation (4.2.4) for different choices of initial conditions. It is a tedious process to use MATLAB directly to both compute and graphically display these solutions. Instead, we use a program written in MATLAB by John Polking for graphing both the line field and the time series of a solution to any ordinary differential equation of the form (4.2.1). In MATLAB this program is addressed by typing

```
dfield5
```

In response, a window appears with the title DFIELD5 Setup. The differential equation under consideration is displayed in the upper big gray frame. In this case it is

$$x' = x^2 - t,$$

which is (4.2.4). Now use the left mouse button to click onto the button Proceed. Then another window having the title DFIELD5 Display appears. In this window, one should see the line field shown in Figure 4.2. We can compute the solution going through the point $(t_0, x_0) = (-2, -4)$ in the (t, x)-plane by clicking on that point with any mouse button. dfield5 should reproduce the right-hand side in Figure 4.2.

Suppose that we want to solve numerically equation (4.1.5) using dfield5. To enter this equation with $\lambda = 0.5$, we have to change the setup. Begin by clicking into the window where the right-hand side x^2−t can be found and then replace it by 0.5*x. Now use the left mouse button to click onto the button Proceed. In the window titled DFIELD5 Display, one should see the line field shown on the left in Figure 4.3.

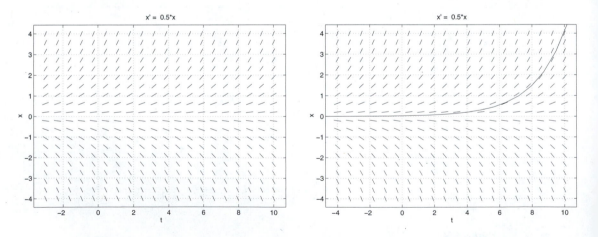

Figure 4.3
Left: Line field for $\dot{x} = 0.5x$ for $t \in [-4, 10]$ and $x \in [-4, 4]$. *Right*: A solution starting at $t = 0$ and $x > 0$ but small.

Now we may compute solutions going through a certain point (t_0, x_0) in the (t, x)-plane by clicking with any mouse button on that point. The solution is then computed first in forward time and then in backward time. For example, if we click on a point near $t = x = 0$, where $x > 0$, dfield5 produces the solution shown on the right in Figure 4.3. Note the similarity with the graph of the closed form solution in Figure 4.1 when $\lambda = 0.5$.

After we have clicked several times, it appears that all solutions diverge to either plus or minus infinity as t goes to infinity. Indeed, by (4.1.6) we know that the solutions are of the form $x(t) = x_0 e^{0.5t}$, and hence this behavior is expected for $x_0 \neq 0$. To compute a solution corresponding to the case when $x_0 = 0$, we bring up the menu DFIELD5 Options and select Keyboard input. This allows us to type in the initial values $t = -2$ and $x = 0$. The action Compute then leads to the computation of the solution $x(t) = 0$ corresponding to $x_0 = 0$.

The value of $\lambda = 0.5$ can be changed by editing the corresponding window in the DFIELD5 Setup. For instance, if we replace the 0.5 by -0.8 and push Proceed, then the current line field is replaced by the line field shown in Figure 4.4.

After we compute different solutions, it seems as though all of them converge to 0 as t goes to infinity, which agrees with (4.1.8).

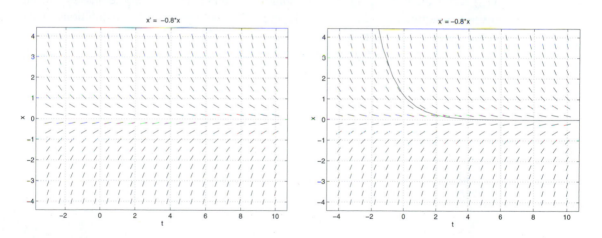

Figure 4.4
Left: Line field for $\dot{x} = -0.8x$ for $t \in [-4, 10]$ and $x \in [-4, 4]$. *Right*: A solution starting at $t = 0$ and x between 1 and 2.

Autonomous and Nonautonomous Equations in dfield5

In a sense, solutions of autonomous equations do not depend on the initial time t_0 but just on the initial position x_0. More precisely, let $x_1(t)$ be the solution to

$$\frac{dx}{dt} = f(x)$$

with initial condition $x(0) = x_0$, and let $x_2(t)$ be a solution to the same differential equation with initial condition $x(t_0) = x_0$. Then

$$x_2(t) = x_1(t - t_0). \tag{4.2.5}$$

This statement can be verified by noting that the definition of $x_2(t)$ in (4.2.5) satisfies the initial value

$$x_2(t_0) = x_1(t_0 - t_0) = x_1(0) = x_0$$

and, using the chain rule, the differential equation

$$\frac{dx_2}{dt}(t) = \frac{dx_1}{dt}(t - t_0) = f(x_1(t - t_0)) = f(x_2(t)).$$

So the solution $x_2(t)$ is the same as the solution $x_1(t)$ with just a shift in time t. In general, the same statement is *not* true for nonautonomous equations.

This difference between autonomous and nonautonomous equations can be visualized using dfield5. On the left in Figure 4.5 are the graphs of two solutions of the *autonomous* differential equation $\dot{x} = x^2 - 2x$ with initial conditions $x(0) = 1$ and $x(2) = 1$. Note that the graph of one solution is obtained from the other just by shifting by two time units. On the right in Figure 4.5 are the graphs of two solutions of the *nonautonomous* differential equation $\dot{x} = x^2 - t$ with initial conditions $x(0) = 1$ and $x(2) = 1$. Note that the graphs of the two solutions are most definitely not obtained from each other by a time shift.

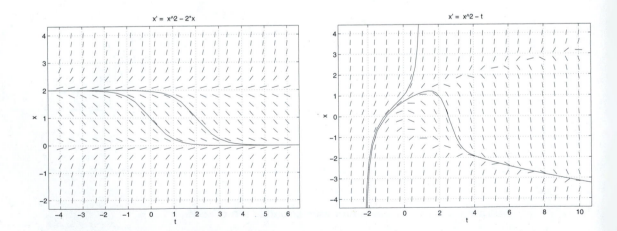

Figure 4.5
Left: Solutions of the autonomous equation $\dot{x} = x^2 - 2x$ with initial conditions $x(0) = 1$ and $x(2) = 1$. *Right*: Solutions of the nonautonomous differential equation $\dot{x} = x^2 - t$ with initial conditions $x(0) = 1$ and $x(2) = 1$.

HAND EXERCISES

In Exercises 1–4, determine whether the solution to the given ordinary differential equation with the given initial condition is increasing or decreasing at the initial point.

1. ODE: $\dot{x} = x - t$; initial condition: $x(1) = 2$

2. ODE: $\dot{x} = x - t$; initial condition: $x(2) = 1$

3. ODE: $\dot{x} = x^2 - tx$; initial condition: $x(1) = 2$

4. ODE: $\dot{x} = x^2 - tx - t$; initial condition: $x(2) = -1$

In Exercises 5–8, sketch by hand the line field of the given differential equation on the given rectangle.

5. ODE: $\dot{x} = x - t$; rectangle, $0 \le x \le 2, -1 \le t \le 1$

6. ODE: $\dot{x} = x + t$; rectangle, $-2 \le x \le 2, -1 \le t \le 2$

7. ODE: $\dot{x} = xt$; rectangle, $-1 \le x \le 1, -1 \le t \le 1$

8. ODE: $\dot{x} = x/t$; rectangle, $0 < x \le 2, 0 < t \le 2$

In Exercises 9–12, determine whether each differential equation is autonomous or nonautonomous.

9. $\dot{x} = x - t$ **10.** $\dot{x} = x^2 - x$

11. $\dot{x} = x \sin(x)$ **12.** $\dot{x} = x \cos(t)$

COMPUTER EXERCISES

In Exercises 13–15, use dfield5 to compute several solutions to the given differential equations in the specified region.

13. $\dot{x} = xt$; $t \in [0, 2], x \in [-1, 1]$ **14.** $\dot{x} = tx^2$; $t \in [0, 4], x \in [-4, 4]$

15. $\dot{x} = x - \sin(t)$; $t \in [-2, 10], x \in [-4, 4]$

16. Compute $x(2)$, where $x(t)$ is the solution to the differential equation $\dot{x} = 0.6x$ with initial condition $x(0) = 0.5$, in two different ways, as follows:

(a) Use Keyboard input in dfield5 to compute the solution with initial value $x(0) = 0.5$. Use the zoom in feature in the DFIELD5 Edit menu to compute $x(2)$ to an accuracy of two decimal places. (Drag a rectangle around the point you are interested in. Repeat this several times until you can read off the value $x(2)$ of the solution.)

(b) Use MATLAB to compute the value of the exact solution of the form (4.1.6) with initial value $x(0) = 0.5$.

(c) Is your answer obtained using dfield5 in (a) accurate to within two decimal places of the answer obtained using (b)? If not, which answer do you trust more? Why?

17. Use dfield5 to compute solutions to the differential equation $\dot{x} = x^2 - tx + 2t$. Use Keyboard input to compute the solution with initial value $x(-1) = -1$. What is the minimum value of this solution $x(t)$ on the interval $-2 \le t \le 1$? Plot solutions to this equation starting with at least six or seven different initial conditions. Then print the result.

18. Use dfield5 to compute solutions to the differential equation

$$\dot{x} = x^3 - 2t^2 x - t. \tag{4.2.6}$$

Print the line field of (4.2.6) on the intervals $t \in [-2, 2]$ and $x \in [-2, 3]$. Use the line field to draw freehand the solution to (4.2.6) starting at $(t_0, x_0) = (-2, 1)$. Then set the initial condition using

Keyboard input and print the numerically computed solution. Finally, compare your freehand drawing with the numerically computed result.

19.

(a) Draw the direction field for

$$\frac{dx}{dt} = \frac{x}{x-t}.$$ **(4.2.7)**

Assuming that $x(2) = 3$, estimate $x(137)$.

(b) Verify that

$$x(t) = t + \sqrt{t^2 - 3}$$

is the solution to (4.2.7). Then show that $x(t)$ satisfies the initial condition $x(2) = 3$.

(c) Compare your estimate of the solution to (4.2.7) obtained using dfield5 with the exact solution:

$$x(137) = 137 + \sqrt{18,766} \approx 273.989.$$

4.3   PHASE SPACE PICTURES AND EQUILIBRIA

Recall that in a *phase space* plot, the solution $x(t)$ represents the position x of a particle on a line for each time t. Phase space plots are difficult to draw, since motion must be built into the plot. However, we can view this dynamically in MATLAB using the program pline.

In this discussion, we plot only solutions of autonomous, first-order differential equations—that is, equations of the form

$$\dot{x} = g(x).$$

To begin, type

    pline

and the window with the PLINE Setup appears on the computer screen. The layout is essentially the same as in the DFIELD5 Setup described earlier. However, there are two differences:

• Since we assume that our equation is autonomous, the time variable—the independent variable—does not appear explicitly and no time interval is specified. Rather the Integration time, the period for which the solution is to be computed, has to be declared. This change is because of the convention that all numerical computations start at $t = 0$.

• We may enter equations with one free parameter. In the PLINE Setup that parameter is lambda. Moreover, we may choose the value for that parameter. The default value in the PLINE Setup is -0.8. Hence the default equation used by pline is (4.1.5) with $\lambda = -0.8$.

When we click with the left mouse button on the button Proceed, the display window with the title PLINE Display opens and the x-axis shown in Figure 4.6 becomes visible.

Similar to dfield5 we may start the numerical solution by clicking with the left mouse button on the initial value of x. It is not necessary to hit the axis precisely. For example, when we click (approximately) on 3, a colored disk becomes visible and moves to the left until it stops at a value for x that is between 0.05 and 0.06. This value can be read at the bottom of the window from the message Endpoint: 0.05.... We can also enter the initial point by choosing the option Keyboard input from the PLINE Options menu. In fact, if we enter $x = 3$ in that window and click on Compute, then the corresponding solution is computed and we obtain the message Endpoint: 0.05495. Sometimes it helps to clear all markers in the display window; this is accomplished by clicking on Clear.

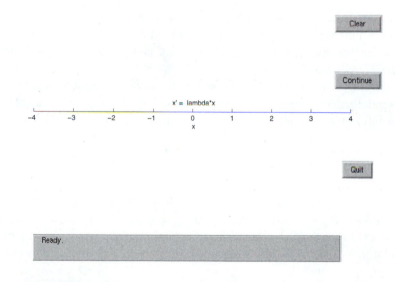

Figure 4.6
PLINE Display for $\dot{x} = \lambda x$ and $x \in [-4, 4]$

Equilibria and Dynamics

The simplest solution to a differential equation is a solution that remains constant for all time. Such solutions are called *equilibria*. Equilibria are found as follows:

Lemma 4.3.1   *Consider the autonomous differential equation*

$$\frac{dx}{dt}(t) = g(x(t)). \tag{4.3.1}$$

Then $x(t) = x_0$ is an equilibrium if and only if $g(x_0) = 0$.

Proof: Suppose that $x(t) = x_0$ is an equilibrium solution to (4.3.1). Then $g(x_0) = 0$, since $dx/dt = 0$. Conversely, suppose $g(x_0) = 0$. Then $x(t) = x_0$ is a solution to (4.3.1).   ◆

We now return to pline and the autonomous equation $g(x) = -0.8x$. If we continue the integration by pushing Continue in the display window we see again that the solution approaches 0 as t goes to infinity; that is, the solution approaches the equilibrium given by $x(t) = 0$. Correspondingly, the point that indicates the position of $x(t)$ in the display window does not move any more. Solutions that have initial conditions near 0 tend either toward 0 when $\lambda < 0$ or away from 0 when $\lambda > 0$.

An equilibrium $x(t) = x_0$ is *asymptotically stable* if all solutions $y(t)$ with initial condition $y(0)$ near x_0 have the limit x_0 as t goes to infinity. In symbols we require

$$\lim_{t \to \infty} y(t) = x_0.$$

(The definition of asymptotic stability is more complicated in higher dimensions.) The equilibrium $x(t) = x_0$ is *unstable* if trajectories starting near x_0 move away from x_0. Thus, in our example, the equilibrium $x = 0$ is *asymptotically stable* when $\lambda < 0$ and *unstable* when $\lambda > 0$.

The dynamic behavior of autonomous differential equations of the form (4.3.1) is essentially determined by the equilibria of g. We explore this statement by using pline to analyze the dynamic behavior of the ordinary differential equation

$$\dot{x} = x(\lambda - x^2), \tag{4.3.2}$$

where λ is a constant.

First, enter this equation by editing the upper box in the PLINE Setup window. Do this editing by clicking in the window and deleting the default equation, and then type x*(lambda-x^2). Next change the minimum and maximum values of x from -4 and 4 to -2 and 2. Finally, change the integration time to 1 and the value of lambda to -1.

Second, push on Proceed so that the x-axis becomes visible in the display window. On integrating the system, we find that solutions seem to behave in a similar way to the solutions of (4.1.5) for $\lambda < 0$: All the solutions approach 0 as t goes to infinity. Indeed, we can see that $x(t) = 0$ is an equilibrium solution of (4.3.2) for all values of λ. When $\lambda < 0$, numerical exploration suggests that 0 is an asymptotically stable equilibrium.

We now explore the stability properties of $x(t) = 0$ when $\lambda > 0$. We begin by changing the value of λ to $+1$. We do this in the setup window and afterward we confirm the change by pushing Proceed. If we now compute solutions of the differential equation, then we see that they come to a rest at $+1$ if we start with a positive value for $x(0)$, whereas they approach -1 if we start with a negative value for $x(0)$. Even if we begin the numerical computation very close to $x = 0$, the solutions tend to either

+1 or −1. These calculations indicate that 0 is an unstable equilibrium, but both +1 and −1 are stable equilibria of (4.3.2) when $\lambda = 1$.

We use the Keyboard input to check that $x(t) = +1$ or $x(t) = -1$ are equilibrium solutions. Start the computations with initial value $x(0) = +1$ or $x(0) = -1$ and see that the solutions remain constant in time. Alternatively, solve the equation

$$x(1 - x^2) = 0$$

to see that $x = 0, 1, -1$ are all equilibria of (4.3.2) when $\lambda = 1$.

Our numerical computations indicate that changing the value of λ from −1 to +1 changes the stability property of the equilibrium $x(t) = 0$.

We can now discuss a method for completely determining the dynamics of a single autonomous differential equation (4.3.1)—assuming that equilibria are isolated.

1. Determine all equilibria of (4.3.1) by solving the equation $g(x) = 0$.

2. Choose an initial point between each pair of consecutive equilibria and determine the direction of motion (sign of g) at that initial point. This can be done either directly or by using pline.

3. On a line plot the equilibria and connect them by arrows indicating the direction of the dynamics.

For example, the dynamics of the differential equations $\dot{x} = x(-1 - x^2)$ and $\dot{x} = x(1 - x^2)$ are shown schematically in Figure 4.7.

Hyperbolic Equilibria

Suppose that x_0 is an equilibrium for (4.3.1); that is suppose $g(x_0) = 0$. We denote the derivative of $g(x)$ with respect to x, dg/dx, by g'. The equilibrium x_0 is *hyperbolic* if $g'(x_0) \neq 0$. Assume that x_0 is a hyperbolic equilibrium, and use the tangent line

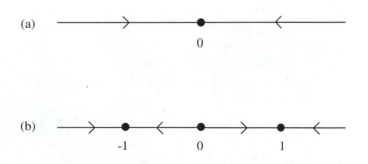

Figure 4.7
Schematic dynamics of (a) $\dot{x} = x(-1 - x^2)$ and (b) $\dot{x} = x(1 - x^2)$.

approximation to $g(x)$ near x_0 ($\Delta g = g'(x_0)\Delta x$) and the fact that $g(x_0) = 0$ to conclude that

$$g(x) \approx g'(x_0)(x - x_0).$$

It follows that if $g'(x_0) < 0$, then $g(x)$ is negative when $x > x_0$ and positive when $x < x_0$. So when $g'(x_0) < 0$, solutions of (4.3.1) starting just to the right of x_0 will move left ($g < 0$) and tend toward x_0 and solutions starting just to the left of x_0 will move right ($g > 0$) and tend toward x_0. Similarly, if $g'(x_0) > 0$, then $g(x)$ is positive when $x > x_0$ and negative when $x < x_0$. Thus solutions near x_0 tend away from x_0 when $g'(x_0) > 0$. We have shown the next theorem:

Theorem 4.3.2 (Stability of hyperbolic equilibria) *Let x_0 be a hyperbolic equilibrium for the differential equation*

$$\frac{dx}{dt} = g(x).$$

If $g'(x_0) < 0$, then the equilibrium is asymptotically stable; if $g'(x_0) > 0$, then the equilibrium is unstable.

It follows from Theorem 4.3.2 that the phase line picture near a hyperbolic equilibrium is particularly simple as the arrows beside that equilibrium either both point toward the equilibrium (as in Figure 4.7(a)) or both point away from the equilibrium (as near 0 in Figure 4.7(b)).

Comparing Phase Lines and Time Series

Phase line plots and time series graphs are different ways of presenting the same information. It is therefore important to be able to recreate one type of plot from the other. For example, let $x(t)$ be the solution to the differential equation $\dot{x} = x(1 - x^2)$ with initial condition $x(0) = 2$. Figure 4.7(b) shows the schematic phase line for all solutions to this differential equation. How can we reconstruct a (schematic) graph of the time series $x(t)$ for this solution just from the phase line?

To answer this question, note that in Figure 4.7(b) the initial condition $x(0) = 2$ lies to the right of all equilibria of this equation, and the arrow indicates that solution trajectories starting in this area move to the left; that is, they decrease to the equilibrium at $x = 1$. It follows that

$$\lim_{t \to \infty} x(t) = 1.$$

Since there are no equilibria to the right of $x = 2$, the graph of $x(t)$ must increase to infinity in *backward* time. So the graph of $x = x(t)$ is decreasing and asymptotic to $x = 1$ for large positive t. A schematic graph is given in Figure 4.8(a). Using dfield5,

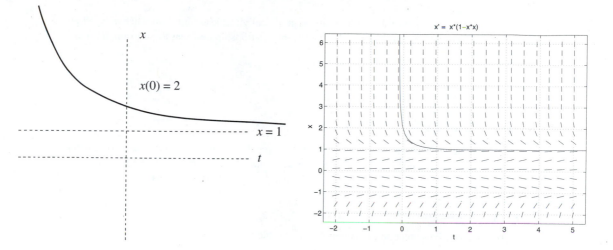

Figure 4.8
Time series for solution to $\dot{x} = x(1 - x^2)$ with $x(0) = 2$. *Left*: Sketch using asymptotic information.
Right: dfield5 computation.

we can check this description by numerically integrating the differential equation. This
graph is shown in Figure 4.8(b).

COMPUTER EXERCISES

1. Use pline to find an initial condition $x(0)$ for the ordinary differential equation (4.1.5) with
$\lambda = -0.2$ such that the corresponding solution $x(t)$ satisfies $x(20) \in [0.001, 0.002]$.

2. Use pline to find all the equilibria of the ordinary differential equation

$$\dot{x} = (x^2 - 1)(x + 2).$$

Which of the equilibria are stable and which are unstable?

3. Consider the following ordinary differential equation:

$$\dot{x} = \lambda x(1 - x).$$

Use pline to find a value for the parameter λ such that $x(t) = 1$ is a stable equilibrium.

4. The differential equation

$$\frac{dx}{dt} = x^2$$

has an equilibrium at the origin. Use pline to determine those x_0 for which solutions to the initial
value $x(0) = x_0$ tend toward the origin.

5. Consider the differential equation

$$\frac{dx}{dt} = ax^3.$$

Use pline to verify that the origin is an asymptotically stable equilibrium when $a = -1$ and is an unstable equilibrium when $a = +1$. Discuss the relationship between these examples and Theorem 4.3.2.

HAND EXERCISES

In Exercises 6–9, compute the equilibria of the given differential equation, determine whether these equilibria are asymptotically stable or unstable, and draw a schematic of the dynamics of this equation like the one in Figure 4.7. You may use pline to check your answer.

6. $\dot{x} = x^2 + 2x - 3$

7. $\dot{x} = x^3 - 2x^2 - 8x$

8. $\dot{x} = x^3 + 2x^2 - 3x$

9. $\dot{x} = x^2 + 6x + 1$

10. Let $x(t)$ be a solution to the initial value problem

$$\dot{x} = g(x)$$
$$x(0) = x_0.$$

Let $y(t) = x(-t)$. Show that $y(t)$ is a solution to the initial value problem

$$\dot{y} = -g(y)$$
$$y(0) = x_0.$$

(Thus, to integrate the solution $x(t)$ backward in time is the same as solving the differential equation $\dot{y} = -g(y)$ forward in time.)

11. Use Exercise 10 to devise a shortcut for solving Exercise 1.

12. Sketch the time series for the solution to the differential equation pictured in Figure 4.7(b) with initial condition $x(0) = \frac{1}{2}$. Use only the phase space plot given in this figure. Use dfield5 to verify your answer.

In Exercises 13–16, use each line field in Figure 4.9 to answer the following:

(a) Is the differential equation that was used to draw this figure autonomous or nonautonomous?

(b) If the differential equation is autonomous, then draw the phase line noting the values of x where equilibria occur and whether or not they are asymptotically stable or not. If the differential equation is nonautonomous, then draw the time series for solutions with initial conditions $x(0) = 0$ and $x(2) = 0$.

13. Figure 4.9(i)   **14.** Figure 4.9(ii)   **15.** Figure 4.9(iii)   **16.** Figure 4.9(iv)

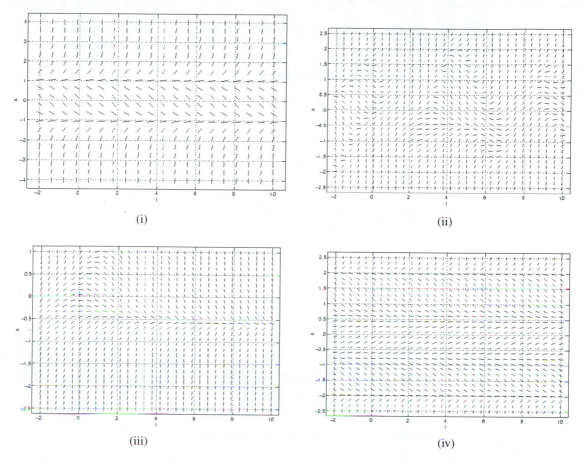

Figure 4.9
Figures for Exercises 13–16

4.4   *SEPARATION OF VARIABLES

In this section we discuss a method for finding closed form solutions to a particular class of nonautonomous first-order ordinary differential equations:

$$\frac{dx}{dt} = f(t, x).$$ **(4.4.1)**

These particular equations are *separable equations* having the form

$$\frac{dx}{dt} = g(x)h(t),$$ **(4.4.2)**

where $g, h : \mathbf{R} \to \mathbf{R}$ are continuous functions. Two special cases of (4.4.2) are $g(x) = 1$ and $h(t) = 1$.

The Special Case g(x) = 1: Integration Theory

The assumption that $g(x) = 1$ in (4.4.2) leads to the differential equation

$$\frac{dx}{dt} = h(t). \tag{4.4.3}$$

In Section 4.1 we saw that this differential equation is easily solved by direct integration. For example, if $h(t) = t^2$, then the solution to (4.4.3) with initial condition $x(2) = 5$ is found as follows: Using direct integration, we have

$$x(t) = \int t^2 dt = \frac{1}{3}t^3 + C.$$

It then follows that

$$x(2) = \frac{8}{3} + C = 5,$$

so

$$C = 5 - \frac{8}{3} = \frac{7}{3}.$$

The Special Case h(t) = 1: Autonomous Equations

The second special case in (4.4.2) leads to the autonomous differential equation

$$\frac{dx}{dt} = g(x). \tag{4.4.4}$$

Begin by noting that equilibria are special solutions to (4.4.4) that we can find by solving the equation $g(x) = 0$. More precisely, if $g(x_0) = 0$, then $x(t) = x_0$ is a constant solution to (4.4.4).

We find the nonequilibrium solutions by direct integration—but only after using change of variables in integration. To solve (4.4.4), we just divide both sides of this equation by $g(x)$, obtaining

$$\frac{1}{g(x)}\frac{dx}{dt} = 1,$$

and integrate with respect to t to get

$$\int \frac{1}{g(x)}\frac{dx}{dt}dt = \int dt + C = t + C.$$

After we substitute $y = x(t)$ and use the chain rule, the integral on the left becomes

$$\int \frac{1}{g(y)} dy.$$

Replacing y by x, we obtain

$$\int \frac{1}{g(x)} dx = t + C.$$

As a simple example, solve the growth rate equation (4.1.5):

$$\frac{dx}{dt} = \lambda x \qquad\qquad \textbf{(4.4.5)}$$

using this technique of integration. It follows that

$$\int \frac{1}{x} dx = \lambda t + C.$$

So, to solve (4.4.5), we need to know how to integrate the function $1/x$. Recalling that this integral is just $\ln |x|$, we get

$$\ln |x(t)| = \lambda t + C.$$

We solve this equation by exponentiation, obtaining

$$|x(t)| = K e^{\lambda t},$$

where $K = e^C \geq 0$. On dropping the absolute value signs, we obtain

$$x(t) = K e^{\lambda t}$$

for arbitrary K. Of course, this is precisely the solution that we found in (4.1.6).

It is worth reflecting on the information from calculus that we needed to solve (4.4.5). In Section 4.1 we found the solution to this equation by asking what function has a derivative that is a multiple of itself. We had to remember that $e^{\lambda t}$ is such a function, and then we had to prove that up to the constant K this is the only such function (recall Theorem 4.1.1). Here we needed to remember the indefinite integral of $1/x$ and then solve for x in terms of t.

We now summarize this technique for solving the autonomous differential equation (4.4.4). Let $G(x)$ be an indefinite integral of $1/g(x)$. Then, after division by $g(x)$, integrating both sides of (4.4.4) with respect to t leads to the equation

$$G(x) = t + C.$$

Then we need to solve this algebraic equation for x in terms of t. This last step is often quite difficult, as we show by example; see (4.4.8).

One additional point needs to be remembered when using this technique. The constant C is, as usual, related to an initial condition. Indeed, if we wish to solve (4.4.4) with the initial condition $x(t_0) = x_0$, then we can solve for

$$C = G(x_0) - t_0.$$

An Example of Blow-up in Finite Time

Consider the nonlinear differential equation

$$\frac{dx}{dt} = x^2 \tag{4.4.6}$$

satisfying the initial condition $x(t_0) = x_0$. Using the preceding discussion, we can solve (4.4.6) by integration. Specifically, on division we obtain

$$\frac{1}{x^2} \frac{dx}{dt} = 1,$$

and on integration with respect to t, we get

$$-\frac{1}{x} = t + C.$$

Solving for x in terms of t, we obtain

$$x(t) = -\frac{1}{t + C}.$$

Finally, use the initial condition to solve for C. On substitution,

$$x_0 = x(t_0) = -\frac{1}{t_0 + C}.$$

Hence

$$C = -\frac{1}{x_0} - t_0,$$

which is fine unless x_0 happens to equal 0. In that case, however, we have just recovered the equilibrium solution $x(t) = 0$.

On setting

$$K = \frac{1}{x_0} + t_0,$$

we obtain the solution

$$x(t) = \frac{1}{K - t}.$$

For example, if $t_0 = 0.1$ and $x_0 = 2$, then $K = 0.6$ and

$$x(t) = \frac{1}{0.6 - t}. \tag{4.4.7}$$

Example (4.4.6) shows that solutions to nonlinear differential equations possess qualitative properties that are different from those of the linear equation $\dot{x} = \lambda x$. In particular:

- Solutions can approach infinity in finite time. For example, the solution (4.4.7) with initial condition $x(0.1) = 2$ goes to infinity as t approaches 0.6.
- Solutions may not be defined for all $t \in \mathbf{R}$. Indeed, the solution (4.4.7) has a singularity at $t = 0.6$ and is defined for either $t > 0.6$ or $t < 0.6$.

We can also solve (4.4.6) using dfield5. Using Keyboard input, set the initial condition at $(x_0, t_0) = (2, 0.1)$ and obtain the trajectory in Figure 4.10 (left). Note that this solution goes to infinity in forward time while approaching $t = 0.6$. Now set the initial condition to $(x_0, t_0) = (-2.5, 1)$ and see that the solution goes to negative infinity as t approaches 0.6 from above; see Figure 4.10 (right). Finally note that both of these solutions are given by the same formula (4.4.7).

An Example That Cannot Be Solved in Closed Form

Consider the differential equation

$$\frac{dx}{dt} = \frac{x}{x - 1} \tag{4.4.8}$$

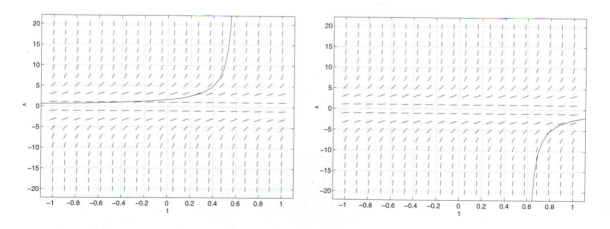

Figure 4.10
Solutions using dfield5 for $dx/dt = x^2$ with initial conditions. *Left*: $(x_0, t_0) = (2, 0.1)$; *right*: $(x_0, t_0) = (-2.5, 1)$.

with initial condition $x(1) = 2$. On division, (4.4.8) becomes

$$\left(1 - \frac{1}{x}\right)\frac{dx}{dt} = 1.$$

Integration with respect to t yields

$$x - \ln|x| = t + C.$$

Using the initial condition, we see that

$$C = 2 - \ln 2 - 1 = 1 - \ln 2.$$

Note that for t near 1, the initial condition implies that $x(t) > 0$. Hence $x(t)$ satisfies

$$x - \ln x = t + 1 - \ln 2. \tag{4.4.9}$$

Unfortunately, this equation cannot be solved explicitly for x as a function of t; that is, there is no closed form solution for $x(t)$. Nevertheless, $x(t)$ is defined *implicitly* by (4.4.9). This equation can, however, be solved numerically by `dfield5` just as easily as equations that have closed form solutions; see Exercise 17 at the end of this section.

The General Solution by Separation of Variables

The solution to the initial value problem for the separation of variables equation

$$\frac{dx}{dt} = g(x)h(t), \tag{4.4.10}$$

where $x(t_0) = x_0$, is obtained by combining the integrations of the two special cases just considered.

Note that if $g(x_0) = 0$, then $x(t) = x_0$ is an equilibrium solution to (4.4.10). So we can assume

$$g(x_0) \neq 0 \tag{4.4.11}$$

and divide (4.4.10) by $g(x)$ to obtain

$$\frac{1}{g(x)}\frac{dx}{dt} = h(t).$$

Integrating with respect to t yields

$$\int \frac{1}{g(x)}\frac{dx}{dt}dt = \int h(t)dt + C.$$

As before, on changing variables, we obtain

$$\int \frac{1}{g(x)} dx = \int h(t) dt.$$

Thus, the abstract solution to (4.4.10) can be written as

$$G(x) = H(t) + C, \tag{4.4.12}$$

where G is an indefinite integral of $1/g$, H is an indefinite integral of h, and

$$C = G(x_0) - H(t_0). \tag{4.4.13}$$

An Example Solving the Initial Value Problem

We illustrate the technique of separation of variables with an example. Find the solution of the initial value problem

$$\frac{dx}{dt} = \frac{t}{x^2},$$

where $x(1) = 2$. Here $g(x) = 1/x^2$ and $h(t) = t$, so that

$$G(x) = \int x^2 dx = \frac{1}{3} x^3 \quad \text{and} \quad H(t) = \int t dt = \frac{1}{2} t^2.$$

Since $t_0 = 1$ and $x_0 = 2$, we can use (4.4.13) to see that

$$C = G(2) - H(1) = \frac{8}{3} - \frac{1}{2} = \frac{13}{6}$$

and (4.4.12) to see that the solution $x(t)$ satisfies

$$\frac{1}{3} x(t)^3 = \frac{1}{2} t^2 + \frac{13}{6}.$$

Therefore

$$x(t) = \left(\frac{3t^2 + 13}{2} \right)^{1/3}.$$

You may check that this function is indeed a solution that satisfies the specified initial condition.

An Example Finding a General Solution

With this example, we illustrate how to use separation of variables to find all solutions of a differential equation in the form (4.4.2). Consider

$$\frac{dx}{dt} = \frac{(x+1)(t^2+1)}{t} \qquad (4.4.14)$$

for $t > 0$. Using our notation, we have

$$g(x) = x + 1 \quad \text{and} \quad h(t) = t + \frac{1}{t}.$$

Note that $g(-1) = 0$ and hence that $x(t) = -1$ is the constant equilibrium solution to (4.4.14). For all the other initial conditions, we obtain

$$G(x) = \int \frac{1}{x+1} dx = \ln|x+1| \quad \text{and} \quad H(t) = \int \left(t + \frac{1}{t}\right) dt = \frac{1}{2}t^2 + \ln|t|.$$

Thus, in this example, identity (4.4.12) implies that the solution $x(t)$ satisfies

$$\ln|x(t) + 1| = \frac{1}{2}t^2 + \ln|t| + C.$$

Hence

$$|x(t) + 1| = K|t|e^{t^2/2}$$

for some constant $K > 0$, and nonconstant solutions of (4.4.14) are in one-to-one correspondence with functions

$$x(t) = Kte^{t^2/2} - 1$$

for $K \in \mathbf{R}$. Note that setting $K = 0$ recovers the constant solution $x(t) = -1$.

HAND EXERCISES

In Exercises 1–4, decide whether or not the method of separation of variables can be applied to the given differential equation. If this method can be applied, then specify the functions g and h—but do not perform the integrations.

1. $\dfrac{dx}{dt} = x^{1/2} \cos t$

2. $\dfrac{dx}{dt} = (x + xt) \tan x$

3. $\dfrac{dx}{dt} = (x + t)(x - t)$

4. $\dfrac{dx}{dt} = 3 - 2x + 3t - 2xt$

In Exercises 5–7, solve each initial value problem by separation of variables.

5. $\dfrac{dx}{dt} = -3x; \quad x(0) = -1$

6. $\dfrac{dx}{dt} = \dfrac{t^3 - 1}{\cos x}; \quad x(\pi) = \dfrac{\pi}{2}$

7. $\dfrac{dx}{dt} = 2\sqrt{tx}; \quad x(1) = 1$

In Exercises 8–10, use separation of variables to find all the solutions of the given differential equation.

8. $\dfrac{dx}{dt} = 2x$

9. $\dfrac{dx}{dt} = 5x + 2$

10. $\dfrac{dx}{dt} = \dfrac{\sin t}{x}$

COMPUTER EXERCISES

We have studied the autonomous linear differential equation $\dot{x} = \lambda x$ when $\lambda \neq 0$, and we have shown that solutions exist for all time and tend either to 0 or to $\pm\infty$ as $t \to \pm\infty$. See (4.1.8) and Figure 4.3. In Exercises 11–16, explore the behavior of solutions $x(t)$ of the given nonlinear (and often nonautonomous) differential equation using dfield5. Where possible, describe the differences between the behaviors of solutions of the given differential equation and the linear differential equation.

Hint Here is a list of properties of solutions that are *not valid* for solutions to the linear differential equation:

(a) Solutions blow up in finite time (that is, $\lim_{t \to t_0} x(t) = \pm\infty$).

(b) Solutions do not exist for all time.

(c) Multiple solutions are bounded in both forward and backward time. (In the linear system only the zero solution $x(t) = 0$ is bounded in both forward and backward time.)

(d) Solutions stop (that is, solutions limit in either forward or backward time on a finite value of x in finite time t).

Use dfield5 to determine which of these properties of solutions are valid for the given differential equations. Also, when exploring Exercises 11–16, be prepared to use the Stop button in the DFIELD5 Display window to stop the numerical integration.

11. $\dfrac{dx}{dt} = \dfrac{1}{x}$

12. $\dfrac{dx}{dt} = \dfrac{1}{tx}$

13. $\dfrac{dx}{dt} = \sin(tx)$

14. $\dfrac{dx}{dt} = t^2 - x^3$

15. $\dfrac{dx}{dt} = x(1 - x^2)$

16. $\dfrac{dx}{dt} = \dfrac{t}{x^2}$

17. Recall that the differential equation (4.4.8), $\dot{x} = x/(x - 1)$, cannot be solved in closed form.

   (a) Use dfield5 to solve this differential equation numerically on the interval $0.8 \leq t \leq 1.2$ with initial condition $x(1) = 2$. (Set the x interval to be $[0, 4]$.)

   (b) Using this result, estimate the value of $x(1.15)$ to two decimal places.

   (c) Next change the time interval to be $0.5 \leq t \leq 1.2$. The numerically computed solution $x(t)$ behaves strangely when $t \leq 0.7$. What is the approximate value of x in this range of t? Use this observed value to explain why the numerical integration for this differential equation is badly behaved.

18. The ability to find closed form solutions to differential equations enables us to test the accuracy of numerically computed solutions. As an example where difficulties arise in numerical computations, consider the differential equation

$$\frac{dx}{dt} = \left(\frac{x}{t}\right)^2 . \tag{4.4.15}$$

   (a) Use separation of variables to find the solution $x(t)$ to the initial value problem $x(0.001) = 2/1999$ of (4.4.15).

   (b) Show that $\lim_{t \to 2^-} x(t) = \infty$ for the solution of (4.4.15) obtained in (a).

(c) For the initial condition specified in (a) compare the analytic solution of (4.4.15) to the numerical solution computed by dfield5 in the region where the maximum value of t is 2.5 and the maximum value of x is 100. How does the behavior of the numerically computed solution to (4.4.15) differ from that of the analytically computed solution?

(d) Compute the solution of (4.4.15) using the general initial condition $x(t_0) = x_0$, where $t_0, x_0 > 0$, and show that $\lim_{t \to 0^+} x(t) = 0$ for every such solution.

Comment: Part (d) hints at why there is a discrepancy between the numerically approximated solution and the analytic solution: Small errors in the approximate solution with initial condition (t_0, x_0) near $(0, 0)$ cause different solutions to be computed.

4.5 UNCOUPLED LINEAR SYSTEMS OF TWO EQUATIONS

An autonomous system of two (first-order) differential equations has the form

$$\frac{dx}{dt}(t) = f(x(t), y(t))$$

$$\frac{dy}{dt}(t) = g(x(t), y(t)),$$

(4.5.1)

where f and g are functions of the two variables x and y. Solutions to (4.5.1) are pairs of functions $(x(t), y(t))$.

As in the single equation (4.3.1), the simplest solutions are equilibria. An *equilibrium* solution to (4.5.1) is a solution where both functions $x(t)$ and $y(t)$ are constant functions; that is,

$$x(t) = x_0 \quad \text{and} \quad y(t) = y_0.$$

For equilibria, it follows that both

$$\frac{dx}{dt} = 0 \quad \text{and} \quad \frac{dy}{dt} = 0.$$

Hence, equilibria of (4.5.1) can be found by simultaneously solving the equations

$$f(x, y) = 0$$
$$g(x, y) = 0.$$

An autonomous *linear* system of ordinary differential equations has the form

$$\frac{dx}{dt}(t) = ax(t) + by(t)$$

$$\frac{dy}{dt}(t) = cx(t) + dy(t),$$

(4.5.2)

where a, b, c, and d are real constants. Note that the origin $(x(t), y(t)) = (0, 0)$ is always an equilibrium for a linear system.

We begin our discussion of linear systems of ordinary differential equations by considering uncoupled systems of the form

$$\frac{dx}{dt}(t) = ax(t)$$

$$\frac{dy}{dt}(t) = dy(t). \tag{4.5.3}$$

Since the system is *uncoupled* (that is, the equation for $\dot{x}$ does not depend on y and the equation for $\dot{y}$ does not depend on x), we can solve this system by solving each equation independently, as we did for (4.1.5):

$$\begin{aligned} x(t) &= x_0 e^{at} \\ y(t) &= y_0 e^{dt}. \end{aligned} \tag{4.5.4}$$

There are now two initial conditions that are identified by

$$x(0) = x_0 \quad \text{and} \quad y(0) = y_0.$$

Having found all the solutions to (4.5.3) in (4.5.4), we now explore the geometry of the phase plane for these uncoupled systems both analytically and by using MATLAB.

Asymptotic Stability of the Origin

As we did for the single equation (4.1.5), we ask what happens to solutions to (4.5.3) starting at (x_0, y_0) as time t increases. That is, we compute

$$\lim_{t \to \infty} (x(t), y(t)) = \lim_{t \to \infty} (x_0 e^{at}, y_0 e^{dt}).$$

This limit is $(0, 0)$ when both $a < 0$ and $d < 0$, but if either a or d is positive, then most solutions diverge to infinity, since either

$$\lim_{t \to \infty} |x(t)| = \infty \quad \text{or} \quad \lim_{t \to \infty} |y(t)| = \infty.$$

Roughly speaking, an equilibrium (x_0, y_0) is *asymptotically stable* if every trajectory $(x(t), y(t))$ beginning from an initial condition near (x_0, y_0) stays near (x_0, y_0) for all positive t, and

$$\lim_{t \to \infty} (x(t), y(t)) = (x_0, y_0).$$

The equilibrium is *unstable* if there are trajectories with initial conditions arbitrarily close to the equilibrium that move far away from that equilibrium.

At this stage, it is not clear how to determine whether the origin is asymptotically stable for a general linear system (4.5.2). However, for uncoupled linear systems we have shown that the origin is an asymptotically stable equilibrium when both $a < 0$ and $d < 0$. If either $a > 0$ or $d > 0$, then $(0, 0)$ is unstable.

Invariance of the Axes

We can make another observation for uncoupled systems. Suppose that the initial condition for an uncoupled system lies on the x-axis; that is, suppose $y_0 = 0$. Then the solution $(x(t), y(t)) = (x_0 e^{at}, 0)$ also lies on the x-axis for all time. Similarly, if the initial condition lies on the y-axis, then the solution $(0, y_0 e^{dt})$ lies on the y-axis for all time.

This invariance of the coordinate axes for uncoupled systems follows directly from (4.5.4). It turns out that many linear systems of differential equations have invariant lines; this is a topic to which we return later in this chapter.

Generating Phase Space Pictures with pplane5

How can we visualize a solution $(x(t), y(t))$ to a system of differential equations (4.5.1)? The time series approach suggests that we should graph $(x(t), y(t))$ as a function of t; that is, we should plot the curve

$$(t, x(t), y(t))$$

in three dimensions. Using MATLAB, we can plot such a graph—but such a graph by itself is difficult to interpret. Alternatively, we could graph either of the functions $x(t)$ or $y(t)$ alone as we do for solutions to single equations, but then some information is lost.

The method we prefer is the *phase space* plot obtained by thinking of $(x(t), y(t))$ as the position of a particle in the xy-plane at time t. We then graph the point $(x(t), y(t))$ in the plane as t varies. When looking at phase space plots, we naturally call solutions *trajectories*, since we can imagine that we are watching a particle moving in the plane as time changes.

We begin by considering uncoupled linear equations. As we saw, when the initial conditions are on the coordinate axes (either $(x_0, 0)$ or $(0, y_0)$), the solutions remain on the coordinate axes. For these initial conditions, the equations behave as if they were one dimensional. However, if we consider an initial condition (x_0, y_0) that is not on a coordinate axis, then even for an uncoupled system it is a little difficult to *see* what the trajectory looks like. At this point it is helpful to use the computer.

The method used to integrate planar systems of autonomous differential equations is similar to that used to integrate nonautonomous single equations in dfield5. The solution curve $(x(t), y(t))$ to (4.5.1) at a point (x_0, y_0) is tangent to the direction $(f(x_0, y_0), g(x_0, y_0))$. So the differential equation solver plots the direction field (f, g) and then finds curves that are tangent to these vectors at each point in time.

The program pplane5, written by John Polking, is the two-dimensional analog of pline. In MATLAB type

    pplane5

and the window with the PPLANE5 Setup appears. pplane5 has a number of prepro-grammed differential equations listed in a menu accessed by clicking on Gallery. To explore linear systems, choose linear system in the Gallery. (Note that the param-eters in the linear system are given by capital letters rather than lowercase a, b, c, d.)

To integrate the uncoupled linear system, set the parameters b and c equal to 0. We now have the system (4.5.3) with $a = 2$ and $d = -3$. After we push Proceed, a display window similar to dfield5 appears. The main difference is that the plane is filled by vectors (f, g) indicating directions, rather than by line segments indicating slopes.

As with dfield5 we may start the computations by clicking with a mouse button on an initial value (x_0, y_0). For example, if we click approximately onto $(x(0), y(0)) = (x_0, y_0) = (1, 1)$, then the trajectory in the upper right quadrant of Figure 4.11 is displayed.

First pplane5 draws the trajectory in forward time for $t \geq 0$, and then it draws the trajectory in backward time for $t \leq 0$. More precisely, when we click on a point (x_0, y_0) in the (x, y)-plane, pplane5 computes that part of the solution that lies inside the specified display window and that goes through this point. For linear systems

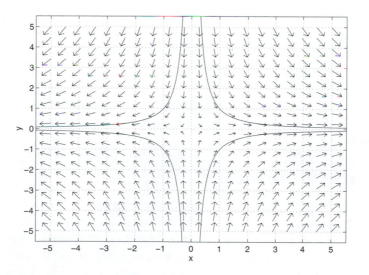

Figure 4.11
PPLANE5 Display for (4.5.3) with $a = 2$, $d = -3$, and $x, y \in [-5, 5]$. Solutions going through $(\pm 1, \pm 1)$ are shown.

there is precisely one solution that goes through a specified point in the (x, y)-plane. We prove this fact in Section 6.3.

Saddles, Sinks, and Sources for the Uncoupled System (4.5.3)

In a qualitative fashion, the trajectories of uncoupled linear systems are determined by the invariance of the coordinate axes and by the signs of the constants a and d.

Saddles: $ad < 0$

In Figure 4.11, where $a = 2 > 0$ and $d = -3 < 0$, the origin is a *saddle*. If we choose several initial values (x_0, y_0) one after another, then we find that as time increases all solutions approach the x-axis. That is, if $(x(t), y(t))$ is a solution to this system of differential equations, then $\lim_{t \to \infty} y(t) = 0$. This observation is particularly noticeable when we choose initial conditions close to the origin $(0, 0)$. On the other hand, solutions also approach the y-axis as $t \to -\infty$. These qualitative features of the phase plane are valid whenever $a > 0$ and $d < 0$.

When $a < 0$ and $d > 0$, the origin is also a saddle—but the roles of the x- and y-axes are reversed.

Sinks: $a < 1$ and $d < 0$

Now we change the parameter a to -1. After clicking on Proceed and specifying several initial conditions, we see that all solutions approach the origin as time tends to infinity. Hence—as mentioned previously, and in contrast to saddles—the equilibrium $(0, 0)$ is asymptotically stable. Observe that solutions approach the origin on trajectories that are tangent to the x-axis. Since $d < a < 0$, the trajectory decreases to 0 faster in the y direction than it does in the x direction. If we change parameters so that $a < d < 0$, then trajectories approach the origin tangent to the y-axis.

Sources: $a > 0$ and $d > 0$

We choose the constants a and d so that both are positive. In forward time, all trajectories, except the equilibrium at the origin, move toward infinity and the origin is called a *source*.

Time Series Using pplane5

We may also use pplane5 to graph the time series of the single components $x(t)$ and $y(t)$ of a solution $(x(t), y(t))$. For this we choose x vs. t from the Graph menu. After we use the mouse to select a solution curve, another window with the title PPLANE5 t-plot appears. There the time series of $x(t)$ is shown. For example, when the differential equation is a sink, we observe that this component approaches 0 as time t tends to infinity. We may also display the time series of both components $x(t)$ and $y(t)$ simultaneously by clicking on Both in the PPLANE5 t-plot window. Again we see that both $x(t)$ and $y(t)$ tend to 0 for increasing t.

We may also visualize the time series of $x(t)$ and $y(t)$ in the three-dimensional (x, y, t)-space. To see this, click on 3 D and a curve $(x(t), y(t), t)$ becomes visible. Since $x(t)$ and $y(t)$ approach 0 for $t \to \infty$, we see that this curve approaches the t-axis for increasing time t. Finally, we may look at all the different visualizations—the phase space plot, the time series for $x(t)$ and $y(t)$, and the three-dimensional representation of the solution—by clicking on the Composite button. See Figure 4.12.

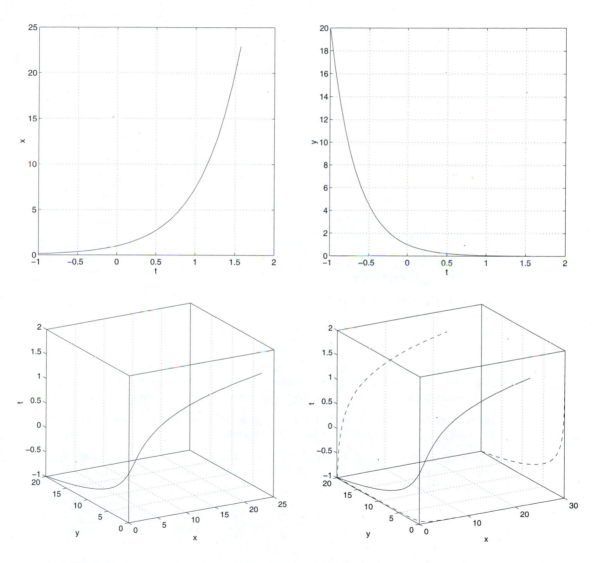

Figure 4.12
PPLANE5 Display for (4.5.3) with $a = 2, d = -3$ and $x \in [0, 25], y \in [0, 20]$. The solution going through $(1, 1)$ is shown. *Upper left*: $(t, x(t))$; *upper right*: $(t, y(t))$; *lower left*: $(x(t), y(t), t)$; *lower right*: all plots.

HAND EXERCISES

In Exercises 1 and 2, find all equilibria of each system of nonlinear autonomous differential equations.

1. $\dot{x} = x - y$
$\dot{y} = x^2 - y$

2. $\dot{x} = x^2 - xy$
$\dot{y} = x^2 + y^2 - 4$

In Exercises 3–5, consider the uncoupled system of differential equations (4.5.3). For each choice of a and d, determine whether the origin is a saddle, source, or sink.

3. $a = 1$ and $d = -1$ **4.** $a = -0.01$ and $d = -2.4$ **5.** $a = 0$ and $d = -2.3$

6. Let $(x(t), y(t))$ be the solution (4.5.4) of (4.5.3) with initial condition $(x(0), y(0)) = (x_0, y_0)$, where $x_0 \neq 0 \neq y_0$.
   (a) Show that the points $(x(t), y(t))$ lie on the curve whose equation is

$$y_0^a x^d - x_0^d y^a = 0.$$

   (b) Verify that if $a = 1$ and $d = 2$, then the solution lies on a parabola tangent to the x-axis.

7. Use the phase plane picture in Figure 4.11 to draw the time series $x(t)$ when $(x(0), y(0)) = (\frac{1}{2}, \frac{1}{2})$. Check your answer using pplane5.

COMPUTER EXERCISES

8. For the three choices of a and d in the uncoupled system of linear differential equations in Exercises 3–5, use pplane5 to compute phase portraits. Use Keyboard input to look at solutions with initial conditions on the x- and y-axes. As t increases, do solutions with these initial conditions tend toward or away from the origin?

9. Suppose a and d are both negative, so that the origin is asymptotically stable. Make several choices of $a < d < 0$ and observe that solution trajectories tend to approach the origin tangent to one of the axes. Determine which one. Try to prove that your experimental guess is always correct.

10. Suppose that $a = d < 0$. Verify experimentally using pplane5 that all trajectories approach the origin along straight lines. Try to prove this conjecture.

11. Use pplane5 to compute several solutions of the linear system by setting $a = 2$ and $d = -3$ in the region $x, y \in [-5, 5]$. Then use dfield5 to compute several solutions of the single differential equation

$$\frac{dx}{dt} = -\frac{2x}{3t}$$

for $x, t \in [-5, 5]$. Is there a relationship between solutions to the two equations? Explain what that relationship is.

4.6 COUPLED LINEAR SYSTEMS

The general linear constant coefficient system in two unknown functions x_1, and x_2 is:

$$\frac{dx_1}{dt}(t) = ax_1(t) + bx_2(t)$$

$$\frac{dx_2}{dt}(t) = cx_1(t) + dx_2(t).$$

(4.6.1)

The uncoupled systems studied in Section 4.5 are obtained by setting $b = c = 0$ in (4.6.1). We have discussed how to solve (4.6.1) by formula (4.5.4) when the system is uncoupled. We have also discussed how to visualize the phase plane for different choices of the diagonal entries a and d. At present, we cannot solve (4.6.1) by formula when the coefficient matrix is not diagonal. But we may use pplane5 to solve the initial value problems numerically for these coupled systems. We illustrate this point by solving

$$\frac{dx_1}{dt}(t) = -x_1(t) + 3x_2(t)$$

$$\frac{dx_2}{dt}(t) = 3x_1(t) - x_2(t).$$

After starting pplane5, select linear system from the Gallery and set the constants to

$$a = -1, \quad b = 3, \quad c = 3, \quad d = -1.$$

Click on Proceed. In order to have equally spaced coordinates on the x- and y-axes, do the following: In the PPLANE5 Display window click on the Edit button and then on the zoom in square command. Then, using the mouse, click on the origin.

Eigendirections

After computing several solutions, we find that for increasing time t, all the solutions seem to approach the diagonal line given by the equation $x_1 = x_2$. Similarly, in backward time t the solutions approach the antidiagonal $x_1 = -x_2$. In other words, as for the case of uncoupled systems, we find two distinguished directions in the (x, y)-plane; see Figure 4.13. Moreover, the computations indicate that these lines are invariant in the sense that solutions starting on these lines remain on them for all time. This statement can be verified numerically by using the Keyboard input in the PPLANE5 Options to choose initial conditions $(x_0, y_0) = (1, 1)$ and $(x_0, y_0) = (1, -1)$.

Definition 4.6.1 *An invariant line for a linear system of differential equations is called an* eigendirection.

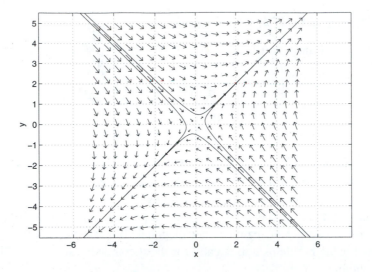

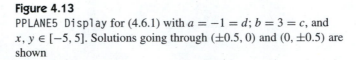

Figure 4.13
PPLANE5 Display for (4.6.1) with $a = -1 = d$; $b = 3 = c$, and
$x, y \in [-5, 5]$. Solutions going through $(\pm 0.5, 0)$ and $(0, \pm 0.5)$ are
shown

Observe that eigendirections vary if we change parameters. For example, if we set b to 1, then there are still two distinguished lines but these lines are no longer perpendicular.

For uncoupled systems, we have shown analytically that the x- and y-axes are eigendirections. The numerical computations that we just performed indicate that eigendirections exist for many coupled systems. This discussion leads naturally to two questions:

• Do eigendirections always exist?

• How can we find eigendirections?

The second question is answered in Sections 4.7 and 4.8. We can answer the first question by performing another numerical computation. In the setup window, change the parameter b to -2. Then numerically compute some solutions to see that there are no eigendirections in the phase space of this system. Observe that all solutions appear to spiral into the origin as time goes to infinity. The phase portrait is shown in Figure 4.14.

Second-Order Differential Equations

We now show analytically that certain linear systems of differential equations have no invariant lines in their phase portrait. We do this by showing that second-order differential equations can be reduced to first-order systems by a simple but important trick. Indeed, sometimes it is easier to solve a single second-order equation, and

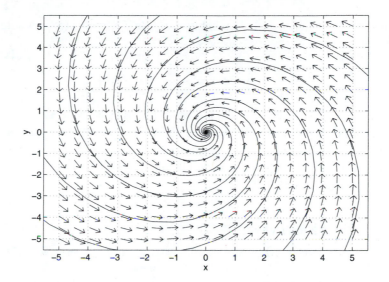

Figure 4.14
PPLANE5 Display for the linear system with $a = -1$, $b = -2$, $c = 3$, and $d = -1$

sometimes it is easier to solve the first-order system. At this stage we introduce this connection by considering the differential equation

$$\frac{d^2x}{dt^2} + x = 0. \qquad (4.6.2)$$

This differential equation states that we are looking for a function $x(t)$ whose second derivative is $-x(t)$. From calculus, we know that $x(t) = \cos t$ is such a function.

Let $y(t) = \dot{x}(t)$. Then (4.6.2) may be rewritten as a first-order coupled system in $x(t)$ and $y(t)$ as follows:

$$\begin{aligned} \dot{x} &= y \\ \dot{y} &= -x. \end{aligned} \qquad (4.6.3)$$

Observe that if $x(t)$ is a solution to (4.6.2), then

$$0 = \frac{d^2x}{dt^2} + x = \frac{dy}{dt} + x.$$

Hence

$$\dot{y} = -x.$$

Conversely, if $(x(t), y(t))$ is a solution to (4.6.3), then $x(t)$ is a solution to (4.6.2); that is,

$$\frac{d^2x}{dt^2} + x = \frac{dy}{dt} + x = -x + x = 0.$$

It follows from the discussion that $(x(t), y(t)) = (\cos t, \sin t)$ is a solution to the differential equation (4.6.3). We have shown analytically that the unit circle centered at the origin is a solution trajectory for (4.6.3). Hence (4.6.3) has no eigendirections. It may be checked using MATLAB that all solution trajectories for (4.6.3) are just circles centered at the origin.

COMPUTER EXERCISES

1. Choose the linear system in pplane5 and set $a = 0$, $b = 1$, and $c = -1$. Then find values d such that except for the origin itself all solutions appear to:
   (a) spiral into the origin;
   (b) spiral away from the origin;
   (c) form circles around the origin.

2. Choose the linear system in pplane5 and set $a = -1$, $c = 3$, and $d = -1$. Then find a value for b such that the behavior of the solutions of the system is "qualitatively" the same as for a diagonal system where a and d are negative. In particular, the origin should be an asymptotically stable equilibrium and the solutions should approach that equilibrium along a distinguished line.

3. Choose the linear system in pplane5 and set $a = d$ and $b = c$. Verify that for these systems of differential equations:
   (a) When $|a| < b$, typical trajectories approach the line $y = x$ as $t \to \infty$ and the line $y = -x$ as $t \to -\infty$.
   (b) Assume that b is positive, a is negative, and $b < -a$. With these assumptions show that the origin is a sink and that typical trajectories approach the origin tangent to the line $y = x$.

HAND EXERCISES

4. Sketch the time series $y(t)$ for the solution to the differential equation whose phase plane is pictured in Figure 4.14 with initial condition $(x(0), y(0)) = (\frac{1}{2}, \frac{1}{2})$. Check your answer using pplane5.

In Exercises 5–8, determine which of the function pairs $(x_1(t), y_1(t))$ and $(x_2(t), y_2(t))$ are solutions to the given system of ordinary differential equations.

5. The ODE is

$$\dot{x} = 2x + y$$
$$\dot{y} = 3y.$$

The pairs of functions are

$$(x_1(t), y_1(t)) = (e^{2t}, 0) \quad \text{and} \quad (x_2(t), y_2(t)) = (e^{3t}, e^{3t}).$$

6. The ODE is

$$\dot{x} = 2x - 3y$$
$$\dot{y} = x - 2y.$$

The pairs of functions are

$$(x_1(t), y_1(t)) = e^t(3, 1) \quad \text{and} \quad (x_2(t), y_2(t)) = (e^{-t}, e^{-t}).$$

7. The ODE is

$$\dot{x} = x + y$$
$$\dot{y} = -x + y.$$

The pairs of functions are

$$(x_1(t), y_1(t)) = (3e^t, -2e^t) \quad \text{and} \quad (x_2(t), y_2(t)) = e^t(\sin t, \cos t).$$

8. The ODE is

$$\dot{x} = y$$
$$\dot{y} = -\frac{1}{t^2}x + \frac{1}{t}y + 1.$$

The pairs of functions are

$$(x_1(t), y_1(t)) = (t^2, 2t) \quad \text{and} \quad (x_2(t), y_2(t)) = (2t^2, 4t).$$

4.7   THE INITIAL VALUE PROBLEM AND EIGENVECTORS

The general *constant coefficient* system of differential equations has the form

$$
\begin{aligned}
\frac{dx_1}{dt}(t) &= c_{11}x_1(t) + \cdots + c_{1n}x_n(t) \\
&\ \vdots \qquad\qquad\qquad\quad \vdots \\
\frac{dx_n}{dt}(t) &= c_{n1}x_1(t) + \cdots + c_{nn}x_n(t),
\end{aligned}
\tag{4.7.1}
$$

where the coefficients $c_{ij} \in \mathbf{R}$ are constants. Suppose that (4.7.1) satisfies the initial conditions $x_1(0) = K_1, \ldots, x_n(0) = K_n$.

Using matrix multiplication of a vector and matrix, we can rewrite these differential equations in a compact form. Consider the $n \times n$ coefficient matrix

$$
C = \begin{pmatrix}
c_{11} & c_{12} & \cdots & c_{1n} \\
c_{21} & c_{22} & \cdots & c_{2n} \\
\vdots & \vdots & & \vdots \\
c_{n1} & c_{n2} & \cdots & c_{nn}
\end{pmatrix}
$$

and the n-vectors of initial conditions and unknowns

$$X_0 = \begin{pmatrix} K_1 \\ \vdots \\ K_n \end{pmatrix} \quad \text{and} \quad X = \begin{pmatrix} x_1 \\ \vdots \\ x_n \end{pmatrix}.$$

Then (4.7.1) has the compact form

$$\frac{dX}{dt} = CX$$

$$X(0) = X_0.$$

(4.7.2)

In Section 4.6 we plotted the phase space picture of the planar system of differential equations

$$\begin{pmatrix} \dot{x} \\ \dot{y} \end{pmatrix} = C \begin{pmatrix} x(t) \\ y(t) \end{pmatrix},$$

(4.7.3)

where

$$C = \begin{pmatrix} -1 & 3 \\ 3 & -1 \end{pmatrix}.$$

In those calculations we observed that there is a solution to (4.7.3) that stayed on the main diagonal for each moment in time. Note that a vector is on the main diagonal if it is a scalar multiple of $\begin{pmatrix} 1 \\ 1 \end{pmatrix}$. Thus a solution that stays on the main diagonal for all time t must have the form

$$\begin{pmatrix} x(t) \\ y(t) \end{pmatrix} = u(t) \begin{pmatrix} 1 \\ 1 \end{pmatrix}$$

(4.7.4)

for some real-valued function $u(t)$. When a function of form (4.7.4) is a solution to (4.7.3), it satisfies

$$\dot{u}(t) \begin{pmatrix} 1 \\ 1 \end{pmatrix} = \begin{pmatrix} \dot{x}(t) \\ \dot{y}(t) \end{pmatrix} = C \begin{pmatrix} x(t) \\ y(t) \end{pmatrix} = Cu(t) \begin{pmatrix} 1 \\ 1 \end{pmatrix} = u(t)C \begin{pmatrix} 1 \\ 1 \end{pmatrix}.$$

A calculation shows that

$$C \begin{pmatrix} 1 \\ 1 \end{pmatrix} = 2 \begin{pmatrix} 1 \\ 1 \end{pmatrix}.$$

Hence

$$\dot{u}(t) \begin{pmatrix} 1 \\ 1 \end{pmatrix} = 2u(t) \begin{pmatrix} 1 \\ 1 \end{pmatrix}.$$

It follows that the function $u(t)$ must satisfy the differential equation

$$\frac{du}{dt} = 2u$$

whose solutions are

$$u(t) = \alpha e^{2t}$$

for some scalar α.

Similarly, we also saw in our MATLAB experiments that there was a solution that for all time stayed on the antidiagonal, the line $y = -x$. Such a solution must have the form

$$\begin{pmatrix} x(t) \\ y(t) \end{pmatrix} = v(t) \begin{pmatrix} 1 \\ -1 \end{pmatrix}.$$

A similar calculation shows that $v(t)$ must satisfy the differential equation

$$\frac{dv}{dt} = -4v.$$

Solutions to this equation all have the form

$$v(t) = \beta e^{-4t}$$

for some real constant β.

Thus, using matrix multiplication, we are able to prove analytically that there are solutions to (4.7.3) of exactly the type suggested by our MATLAB experiments. However, even more is true and this extension is based on the principle of superposition that was introduced for algebraic equations in Section 3.4.

Superposition in Linear Differential Equations

Consider a general linear differential equation of the form

$$\frac{dX}{dt} = CX, \tag{4.7.5}$$

where C is an $n \times n$ matrix. Suppose that $Y(t)$ and $Z(t)$ are solutions to (4.7.5) and $\alpha, \beta \in \mathbf{R}$ are scalars. Then $X(t) = \alpha Y(t) + \beta Z(t)$ is also a solution. We verify this fact using the "linearity" of d/dt. Calculate

$$\begin{aligned}
\frac{d}{dt} X(t) &= \alpha \frac{dY}{dt}(t) + \beta \frac{dZ}{dt}(t) \\
&= \alpha C Y(t) + \beta C Z(t) \\
&= C(\alpha Y(t) + \beta Z(t)) \\
&= C X(t).
\end{aligned}$$

So superposition is valid for solutions of linear differential equations.

Initial Value Problems

Suppose that we wish to find a solution to (4.7.3) satisfying the initial conditions

$$\begin{pmatrix} x(0) \\ y(0) \end{pmatrix} = \begin{pmatrix} 1 \\ 3 \end{pmatrix}.$$

Then we can use the principle of superposition to find this solution in closed form. Superposition implies that for each pair of scalars $\alpha, \beta \in \mathbf{R}$, the functions

$$\begin{pmatrix} x(t) \\ y(t) \end{pmatrix} = \alpha e^{2t} \begin{pmatrix} 1 \\ 1 \end{pmatrix} + \beta e^{-4t} \begin{pmatrix} 1 \\ -1 \end{pmatrix} \tag{4.7.6}$$

are solutions to (4.7.3). Moreover, for a solution of this form,

$$\begin{pmatrix} x(0) \\ y(0) \end{pmatrix} = \begin{pmatrix} \alpha + \beta \\ \alpha - \beta \end{pmatrix}.$$

Thus we can solve our prescribed initial value problem if we can solve the system of linear equations

$$\alpha + \beta = 1$$
$$\alpha - \beta = 3.$$

This system is solved for $\alpha = 2$ and $\beta = -1$. Thus

$$\begin{pmatrix} x(t) \\ y(t) \end{pmatrix} = 2e^{2t} \begin{pmatrix} 1 \\ 1 \end{pmatrix} - e^{-4t} \begin{pmatrix} 1 \\ -1 \end{pmatrix}$$

is the desired closed form solution.

Eigenvectors and Eigenvalues

We emphasize that just knowing that there are two lines in the plane that are invariant under the dynamics of the system of linear differential equations is sufficient information to solve these equations. So it seems appropriate to ask the question: When is there a line that is invariant under the dynamics of a system of linear differential equations? This is equivalent to asking: When is there a nonzero vector v and a nonzero real-valued function $u(t)$ such that

$$X(t) = u(t)v$$

is a solution to (4.7.5)?

Suppose that $X(t)$ is a solution to the system of differential equations $\dot{X} = CX$. Then $u(t)$ and v must satisfy

$$\dot{u}(t)v = \frac{dX}{dt} = CX(t) = u(t)Cv. \qquad (4.7.7)$$

Since u is nonzero, it follows that v and Cv must lie on the same line through the origin. Hence

$$Cv = \lambda v \qquad (4.7.8)$$

for some real number λ.

Definition 4.7.1 *A nonzero vector v satisfying (4.7.8) is called an* eigenvector *of the matrix C, and the number λ is an* eigenvalue *of the matrix C.*

Geometrically, the matrix C maps an eigenvector onto a multiple of itself—that multiple is the eigenvalue.

Note that scalar multiples of eigenvectors are also eigenvectors. More precisely:

Lemma 4.7.2 *Let v be an eigenvector of the matrix C with eigenvalue λ. Then αv is also an eigenvector of C with eigenvalue λ as long as $\alpha \neq 0$.*

Proof: By assumption, $Cv = \lambda v$ and v is nonzero. Now calculate

$$C(\alpha v) = \alpha Cv = \alpha \lambda v = \lambda(\alpha v).$$

The lemma follows from the definition of eigenvector. $\blacklozenge$

It follows from (4.7.7) and (4.7.8) that if v is an eigenvector of C with eigenvalue λ, then

$$\frac{du}{dt} = \lambda u.$$

Thus we have returned to our original linear differential equation that has solutions

$$u(t) = Ke^{\lambda t}$$

for all constants K.

We have proved the following theorem:

Theorem 4.7.3 *Let v be an eigenvector of the $n \times n$ matrix C with eigenvalue λ. Then*

$$X(t) = e^{\lambda t}v$$

is a solution to the system of differential equations $\dot{X} = CX$.

Finding eigenvalues and eigenvectors from first principles—even for 2×2 matrices—is not a simple task. We end this section with a calculation illustrating that real eigenvalues need not exist. In Section 4.8 we present a natural method for computing eigenvalues (and eigenvectors) of 2×2 matrices. We defer the discussion of how to find eigenvalues and eigenvectors of $n \times n$ matrices until Chapter 8.

An Example of a Matrix with No Real Eigenvalues

Not every matrix has *real* eigenvalues and eigenvectors. Recall the linear system of differential equations $\dot{x} = Cx$ whose phase plane is pictured in Figure 4.14. That phase plane shows no evidence of an invariant line and indeed there is none. The matrix C in that example is

$$C = \begin{pmatrix} -1 & -2 \\ 3 & -1 \end{pmatrix}.$$

We ask: Is there a value of λ and a nonzero vector (x, y) such that

$$C \begin{pmatrix} x \\ y \end{pmatrix} = \lambda \begin{pmatrix} x \\ y \end{pmatrix}? \tag{4.7.9}$$

Equation (4.7.9) implies that

$$\begin{pmatrix} -1 - \lambda & -2 \\ 3 & -1 - \lambda \end{pmatrix} \begin{pmatrix} x \\ y \end{pmatrix} = 0.$$

If this matrix is row equivalent to the identity matrix, then the only solution of the linear system is $x = y = 0$. To have a nonzero solution, the matrix

$$\begin{pmatrix} -1 - \lambda & -2 \\ 3 & -1 - \lambda \end{pmatrix}$$

must not be row equivalent to I_2. Dividing the 1st row by $-(1 + \lambda)$ leads to

$$\begin{pmatrix} 1 & \dfrac{2}{1 + \lambda} \\ 3 & -1 - \lambda \end{pmatrix}.$$

Subtracting three times the 1st row from the second produces the matrix

$$\begin{pmatrix} 1 & \dfrac{2}{1 + \lambda} \\ 0 & -(1 + \lambda) - \dfrac{6}{1 + \lambda} \end{pmatrix}.$$

This matrix is not row equivalent to I_2 when the lower right-hand entry is 0—that is, when

$$(1 + \lambda) + \frac{6}{1 + \lambda} = 0$$

or when

$$(1 + \lambda)^2 = -6,$$

which is not possible for any real number λ. This example shows that whether a given matrix has a real eigenvalue and a real eigenvector—and hence when the associated system of differential equations has a line that is invariant under the dynamics—is a subtle question.

Questions concerning eigenvectors and eigenvalues are central to much of the theory of linear algebra. We discuss this topic for 2×2 matrices in Section 4.8 and Chapter 6 and for general square matrices in Chapters 8 and 13.

HAND EXERCISES

1. Write the system of linear ordinary differential equations

$$\frac{dx_1}{dt}(t) = 4x_1(t) + 5x_2(t)$$
$$\frac{dx_2}{dt}(t) = 2x_1(t) - 3x_2(t)$$

in matrix form.

2. Show that all solutions to the system of linear differential equations

$$\frac{dx}{dt} = 3x$$
$$\frac{dy}{dt} = -2y$$

are linear combinations of the two solutions

$$U(t) = e^{3t} \begin{pmatrix} 1 \\ 0 \end{pmatrix} \quad \text{and} \quad V(t) = e^{-2t} \begin{pmatrix} 0 \\ 1 \end{pmatrix}.$$

3. Consider

$$\frac{dX}{dt}(t) = CX(t), \tag{4.7.10}$$

where

$$C = \begin{pmatrix} 2 & 3 \\ 0 & -1 \end{pmatrix}.$$

Let

$$v_1 = \begin{pmatrix} 1 \\ 0 \end{pmatrix} \quad \text{and} \quad v_2 = \begin{pmatrix} 1 \\ -1 \end{pmatrix},$$

and let

$$Y(t) = e^{2t} v_1 \quad \text{and} \quad Z(t) = e^{-t} v_2.$$

(a) Show that $Y(t)$ and $Z(t)$ are solutions to (4.7.10).
(b) Show that $X(t) = 2Y(t) - 14Z(t)$ is a solution to (4.7.10).
(c) Use the principle of superposition to verify that $X(t) = \alpha Y(t) + \beta Z(t)$ is a solution to (4.7.10).
(d) Using the general solution found in (c), find a solution $X(t)$ to (4.7.10) such that

$$X(0) = \begin{pmatrix} 3 \\ -1 \end{pmatrix}.$$

4. Find a solution to

$$\dot{X}(t) = CX(t),$$

where

$$C = \begin{pmatrix} 1 & -1 \\ -1 & 1 \end{pmatrix}$$

and

$$X(0) = \begin{pmatrix} 2 \\ 1 \end{pmatrix}.$$

Hint   Observe that

$$\begin{pmatrix} 1 \\ 1 \end{pmatrix} \quad \text{and} \quad \begin{pmatrix} 1 \\ -1 \end{pmatrix}$$

are eigenvectors of C.

5. Let

$$C = \begin{pmatrix} a & b \\ b & a \end{pmatrix}.$$

Show that

$$\begin{pmatrix} 1 \\ 1 \end{pmatrix} \quad \text{and} \quad \begin{pmatrix} 1 \\ -1 \end{pmatrix}$$

are eigenvectors of C. What are the corresponding eigenvalues?

6. Let

$$C = \begin{pmatrix} 1 & 2 \\ -3 & -1 \end{pmatrix}.$$

Show that C has no real eigenvectors.

7. Suppose that A is an $n \times n$ matrix with 0 as an eigenvalue. Show that A is not invertible.

Hint Assume that A is invertible and compute $A^{-1}Av$, where v is an eigenvector of A corresponding to the zero eigenvalue.

> **Remark:** In fact, A is invertible if all of the eigenvalues of A are nonzero. See Corollary 8.2.5

COMPUTER EXERCISES

8. Consider the matrix A and vector X_0 given by

$$A = \begin{pmatrix} 2 & 1 \\ 0 & 1 \end{pmatrix} \quad \text{and} \quad X_0 = \begin{pmatrix} 1 \\ 1 \end{pmatrix}.$$

Use map to compute $X_1 = AX_0$, $X_2 = AX_1$, $X_3 = AX_2$, and so on, by a repeated use of the Map button in the MAP Display window. What do you observe? What happens if you start the iteration process with a different choice for X_0 and, in particular, for an X_0 that is close to $\begin{pmatrix} -1 \\ 1 \end{pmatrix}$?

In Exercises 9 and 10, use map to find an (approximate) eigenvector for the given matrix.
Hint Choose a vector in map and repeatedly click on the Map button until the vector maps to a multiple of itself. You may wish to use the Rescale feature in the MAP Options. Then the length of the vector is rescaled to 1 after each use of the command Map. In this way, you can avoid overflows in the computations while still being able to see the directions where the vectors are moved by the matrix mapping. The coordinates of the new vector obtained by applying map can be viewed in the Vector input window.

9. $B = \begin{pmatrix} 2 & -2 \\ 2 & 7 \end{pmatrix}$ **10.** $C = \begin{pmatrix} 1 & 1.5 \\ 0 & -2 \end{pmatrix}$

11. Use MATLAB to verify that solutions to the system of linear differential equations

$$\frac{dx}{dt} = 2x + y$$
$$\frac{dy}{dt} = y$$

are linear combinations of the two solutions

$$U(t) = e^{2t} \begin{pmatrix} 1 \\ 0 \end{pmatrix} \quad \text{and} \quad V(t) = e^t \begin{pmatrix} -1 \\ 1 \end{pmatrix}.$$

More concretely, proceed as follows:

(a) By superposition, the general solution to the differential equation has the form $X(t) = \alpha U(t) + \beta V(t)$. Find constants α and β such that $\alpha U(0) + \beta V(0) = \begin{pmatrix} 0 \\ 1 \end{pmatrix}$.

(b) Graph the second component $y(t)$ of this solution using the MATLAB plot command.

(c) Use pplane5 to compute a solution via the Keyboard input starting at $(x(0), y(0)) = (0, 1)$, and then use the y vs t command in pplane5 to graph this solution.

(d) Compare the results of the two plots.

(e) Repeat steps (a)–(d) using the initial vector $\begin{pmatrix} 1 \\ 1 \end{pmatrix}$.

4.8   EIGENVALUES OF 2 × 2 MATRICES

We now discuss how to find eigenvalues of 2×2 matrices in a way that does not depend explicitly on finding eigenvectors. This direct method will show that eigenvalues can be complex as well as real.

We begin the discussion with a general square matrix. Let A be an $n \times n$ matrix. Recall that $\lambda \in \mathbf{R}$ is an eigenvalue of A if there is a nonzero vector $v \in \mathbf{R}^n$ for which

$$Av = \lambda v. \tag{4.8.1}$$

The vector v is called an *eigenvector*. We may rewrite (4.8.1) as

$$(A - \lambda I_n)v = 0.$$

Since v is nonzero, it follows that if λ is an eigenvalue of A, then the matrix $A - \lambda I_n$ is singular.

Conversely, suppose that $A - \lambda I_n$ is singular for some real number λ. Then Theorem 3.7.8 implies that there is a nonzero vector $v \in \mathbf{R}^n$ such that $(A - \lambda I_n)v = 0$. Hence (4.8.1) holds and λ is an eigenvalue of A. So, if we had a direct method for determining when a matrix is singular, then we would have a method for determining eigenvalues.

Characteristic Polynomials

Corollary 3.8.3 states that 2×2 matrices are singular precisely when their determinant is 0. It follows that $\lambda \in \mathbf{R}$ is an eigenvalue for the 2×2 matrix A precisely when

$$\det(A - \lambda I_2) = 0. \tag{4.8.2}$$

We can compute (4.8.2) explicitly as follows: Note that

$$A - \lambda I_2 = \begin{pmatrix} a - \lambda & b \\ c & d - \lambda \end{pmatrix}.$$

Therefore

$$\begin{aligned} \det(A - \lambda I_2) &= (a - \lambda)(d - \lambda) - bc \\ &= \lambda^2 - (a + d)\lambda + (ad - bc). \end{aligned} \tag{4.8.3}$$

Definition 4.8.1   *The* characteristic polynomial *of the matrix A is*

$$p_A(\lambda) = \det(A - \lambda I_2).$$

For an $n \times n$ matrix $A = (a_{ij})$, we define the *trace* of A to be the sum of the diagonal elements of A; that is

$$\text{tr}(A) = a_{11} + \cdots + a_{nn}. \tag{4.8.4}$$

Thus, using (4.8.3), we can rewrite the characteristic polynomial for 2×2 matrices as

$$p_A(\lambda) = \lambda^2 - \text{tr}(A)\lambda + \det(A). \tag{4.8.5}$$

As an example, consider the 2×2 matrix

$$A = \begin{pmatrix} 2 & 3 \\ 1 & 4 \end{pmatrix}. \tag{4.8.6}$$

Then

$$A - \lambda I_2 = \begin{pmatrix} 2 - \lambda & 3 \\ 1 & 4 - \lambda \end{pmatrix}$$

and

$$p_A(\lambda) = (2 - \lambda)(4 - \lambda) - 3 = \lambda^2 - 6\lambda + 5.$$

It is now easy to verify (4.8.5) for (4.8.6).

Eigenvalues

For 2×2 matrices A, $p_A(\lambda)$ is a quadratic polynomial. As we have discussed, the real roots of p_A are real eigenvalues of A. For 2×2 matrices we now generalize our first definition of eigenvalues, Definition 4.7.1, to include complex eigenvalues.

Definition 4.8.2   *An* eigenvalue *of A is a root of the characteristic polynomial* p_A.

Suppose that λ_1 and λ_2 are the roots of p_A. It follows that

$$p_A(\lambda) = (\lambda - \lambda_1)(\lambda - \lambda_2) = \lambda^2 - (\lambda_1 + \lambda_2)\lambda + \lambda_1\lambda_2. \tag{4.8.7}$$

Equating the two forms of p_A (4.8.5) and (4.8.7) shows that

$$\text{tr}(A) = \lambda_1 + \lambda_2 \tag{4.8.8}$$
$$\det(A) = \lambda_1\lambda_2. \tag{4.8.9}$$

Thus, for 2×2 matrices, the trace is the sum of the eigenvalues and the determinant is the product of the eigenvalues. In Theorems 8.2.4(b) and 8.2.9 we show that these statements are also valid for $n \times n$ matrices.

Recall that in example (4.8.6) the characteristic polynomial is

$$p_A(\lambda) = \lambda^2 - 6\lambda + 5 = (\lambda - 5)(\lambda - 1).$$

Thus the eigenvalues of A are $\lambda_1 = 1$ and $\lambda_2 = 5$, and identities (4.8.8) and (4.8.9) are easily verified for this example.

Next we consider an example with complex eigenvalues and verify that these identities are equally valid in this instance. Let

$$B = \begin{pmatrix} 2 & -3 \\ 1 & 4 \end{pmatrix}.$$

The characteristic polynomial is

$$p_B(\lambda) = \lambda^2 - 6\lambda + 11.$$

Using the quadratic formula, we see that the roots of p_B (that is, the eigenvalues of B) are

$$\lambda_1 = 3 + i\sqrt{2} \quad \text{and} \quad \lambda_2 = 3 - i\sqrt{2}.$$

Again the sum of the eigenvalues is 6, which equals the trace of B, and the product of the eigenvalues is 11, which equals the determinant of B.

Since the characteristic polynomial of 2×2 matrices is always a quadratic polynomial, it follows that 2×2 matrices have precisely two eigenvalues—including multiplicity. Properties of these eigenvalues are described as follows. The *discriminant of A* is

$$D = [\text{tr}(A)]^2 - 4\det(A). \tag{4.8.10}$$

Theorem 4.8.3 *There are three possibilities for the two eigenvalues of a 2×2 matrix A that we can describe in terms of the discriminant:*

 (a) The eigenvalues of A are real and distinct ($D > 0$).

 (b) The eigenvalues of A are a complex conjugate pair ($D < 0$).

 (c) The eigenvalues of A are real and equal ($D = 0$).

Proof: We can find the roots of the characteristic polynomial using the form of p_A given in (4.8.5) and the quadratic formula. The roots are

$$\frac{1}{2}\left(\text{tr}(A) \pm \sqrt{[\text{tr}(A)]^2 - 4\det(A)}\right) = \frac{\text{tr}(A) \pm \sqrt{D}}{2}.$$

The proof of the theorem now follows. If $D > 0$, then the eigenvalues of A are real and distinct; if $D < 0$, then the eigenvalues are complex conjugates; and if $D = 0$, then the eigenvalues are real and equal. ◆

Eigenvectors

The following lemma contains an important observation about eigenvectors:

Lemma 4.8.4   *Every eigenvalue λ of a 2 × 2 matrix A has an eigenvector v. That is, there is a nonzero vector $v \in \mathbf{C}^2$ satisfying*

$$Av = \lambda v.$$

Proof: When the eigenvalue λ is real, we know that an eigenvector $v \in \mathbf{R}^2$ exists. However, when λ is complex, we must show that there is a complex eigenvector $v \in \mathbf{C}^2$, and this we have not yet done. More precisely, we must show that if λ is a complex root of the characteristic polynomial p_A, then there is a complex vector v such that

$$(A - \lambda I_2)v = 0.$$

As we discussed in Section 2.5, finding v is equivalent to showing that the complex matrix

$$A - \lambda I_2 = \begin{pmatrix} a - \lambda & b \\ c & d - \lambda \end{pmatrix}$$

is not row equivalent to the identity matrix; see Theorem 2.5.2. Since a is real and λ is not, $a - \lambda \neq 0$. A short calculation shows that $A - \lambda I_2$ is row equivalent to the matrix

$$\begin{pmatrix} 1 & \dfrac{b}{a - \lambda} \\ 0 & \dfrac{p_A(\lambda)}{a - \lambda} \end{pmatrix}.$$

This matrix is not row equivalent to the identity matrix, since $p_A(\lambda) = 0$.   ◆

An Example of a Matrix with Real Eigenvectors

Once we know the eigenvalues of a 2 × 2 matrix, the associated eigenvectors can be found by direct calculation. For example, we showed previously that the matrix

$$A = \begin{pmatrix} 2 & 3 \\ 1 & 4 \end{pmatrix}$$

in (4.8.6) has eigenvalues $\lambda_1 = 1$ and $\lambda_2 = 5$. With this information we can find the associated eigenvectors. To find an eigenvector associated with the eigenvalue $\lambda_1 = 1$ compute

$$A - \lambda_1 I_2 = A - I_2 = \begin{pmatrix} 1 & 3 \\ 1 & 3 \end{pmatrix}.$$

It follows that $v_1 = (3, -1)^t$ is an eigenvector, since

$$(A - I_2)v_1 = 0.$$

Similarly, to find an eigenvector associated with the eigenvalue $\lambda_2 = 5$ compute

$$A - \lambda_2 I_2 = A - 5I_2 = \begin{pmatrix} -3 & 3 \\ 1 & -1 \end{pmatrix}.$$

It follows that $v_2 = (1, 1)^t$ is an eigenvector, since

$$(A - 5I_2)v_2 = 0.$$

Examples of Matrices with Complex Eigenvectors

Let

$$A = \begin{pmatrix} 0 & -1 \\ 1 & 0 \end{pmatrix}.$$

Then $p_A(\lambda) = \lambda^2 + 1$ and the eigenvalues of A are $\pm i$. To find the eigenvector $v \in \mathbf{C}^2$ whose existence is guaranteed by Lemma 4.8.4, we need to solve the complex system of linear equations $Av = iv$. We can rewrite this system as

$$\begin{pmatrix} -i & -1 \\ 1 & -i \end{pmatrix} \begin{pmatrix} v_1 \\ v_2 \end{pmatrix} = 0.$$

A calculation shows that

$$v = \begin{pmatrix} i \\ 1 \end{pmatrix} \tag{4.8.11}$$

is a solution. Since the coefficients of A are real, we can take the complex conjugate of the equation $Av = iv$ to obtain

$$A\bar{v} = -i\bar{v}.$$

Thus

$$\bar{v} = \begin{pmatrix} -i \\ 1 \end{pmatrix}$$

is the eigenvector corresponding to the eigenvalue $-i$. This comment is valid for any complex eigenvalue.

More generally, let

$$A = \begin{pmatrix} \sigma & -\tau \\ \tau & \sigma \end{pmatrix}, \tag{4.8.12}$$

where $\tau \neq 0$. Then

$$\begin{aligned} p_A(\lambda) &= \lambda^2 - 2\sigma\lambda + \sigma^2 + \tau^2 \\ &= (\lambda - (\sigma + i\tau))(\lambda - (\sigma - i\tau)), \end{aligned}$$

and the eigenvalues of A are the complex conjugates $\sigma \pm i\tau$. Thus A has no real eigenvectors. The complex eigenvectors of A are v and $\bar{v}$, where v is defined in (4.8.11).

HAND EXERCISES

1. For which values of λ is the matrix

$$\begin{pmatrix} 1 - \lambda & 4 \\ 2 & 3 - \lambda \end{pmatrix}$$

not invertible? **Note:** These values of λ are just the eigenvalues of the matrix $\begin{pmatrix} 1 & 4 \\ 2 & 3 \end{pmatrix}$.

In Exercises 2–5, compute the determinant, trace, and characteristic polynomials for the given 2×2 matrix.

2. $\begin{pmatrix} 1 & 4 \\ 0 & -1 \end{pmatrix}$ **3.** $\begin{pmatrix} 2 & 13 \\ -1 & 5 \end{pmatrix}$ **4.** $\begin{pmatrix} 1 & 4 \\ 1 & -1 \end{pmatrix}$ **5.** $\begin{pmatrix} 4 & 10 \\ 2 & 5 \end{pmatrix}$

In Exercises 6–8, compute the eigenvalues for each 2×2 matrix.

6. $\begin{pmatrix} 1 & 2 \\ 0 & -5 \end{pmatrix}$ **7.** $\begin{pmatrix} -3 & 2 \\ 1 & 0 \end{pmatrix}$ **8.** $\begin{pmatrix} 3 & -2 \\ 2 & -1 \end{pmatrix}$

9.

(a) Let A and B be 2×2 matrices. Using direct calculation, show that

$$\text{tr}(AB) = \text{tr}(BA). \tag{4.8.13}$$

(b) Now let A and B be $n \times n$ matrices. Verify by direct calculation that (4.8.13) is still valid.

COMPUTER EXERCISES

In Exercises 10–12, use the program map to guess whether the given matrix has real or complex conjugate eigenvalues. For each example, write the reasons for your guess.

10. $A = \begin{pmatrix} 0.97 & -0.22 \\ 0.22 & 0.97 \end{pmatrix}$ **11.** $B = \begin{pmatrix} 0.97 & 0.22 \\ 0.22 & 0.97 \end{pmatrix}$ **12.** $C = \begin{pmatrix} 0.4 & -1.4 \\ 1.5 & 0.5 \end{pmatrix}$

In Exercises 13–14, use the program map to guess one of the eigenvectors of the given matrix. What is the corresponding eigenvalue? Using map, can you find a second eigenvalue and eigenvector?

13. $A = \begin{pmatrix} 2 & 4 \\ 2 & 0 \end{pmatrix}$ **14.** $B = \begin{pmatrix} 2 & -1 \\ 0.25 & 1 \end{pmatrix}$

Hint Use the feature rescale in the MAP Options. Then the length of the vector is rescaled to 1 after each use of the command map. In this way you can avoid overflows in the computations while still being able to see the directions where the vectors are moved by the matrix mapping.

15. The MATLAB command eig computes the eigenvalues of matrices. Use eig to compute the eigenvalues of $A = \begin{pmatrix} 2.34 & -1.43 \\ \pi & e \end{pmatrix}$.

4.9   INITIAL VALUE PROBLEMS REVISITED

To summarize the ideas developed in this chapter, we review the method that we developed to solve the system of differential equations

$$\dot{x} = ax + by$$
$$\dot{y} = cx + dy$$

$$(4.9.1)$$

satisfying the initial conditions

$$x(0) = x_0$$
$$y(0) = y_0.$$

$$(4.9.2)$$

Begin by rewriting (4.9.1) in matrix form:

$$\dot{X} = CX,$$

$$(4.9.3)$$

where

$$C = \begin{pmatrix} a & b \\ c & d \end{pmatrix} \quad \text{and} \quad X(t) = \begin{pmatrix} x(t) \\ y(t) \end{pmatrix}.$$

Rewrite the initial conditions (4.9.2) in vector form:

$$X(0) = X_0,$$

$$(4.9.4)$$

where

$$X_0 = \begin{pmatrix} x_0 \\ y_0 \end{pmatrix}.$$

When the eigenvalues of C are *real* and *distinct*, we now know how to solve the initial value problem (4.9.3) and (4.9.4). This solution is found in four steps.

Step 1:   Find the eigenvalues λ_1 and λ_2 of C.

These eigenvalues are the roots of the characteristic polynomial as given by (4.8.5):

$$p_C(\lambda) = \lambda^2 - \text{tr}(C)\lambda + \det(C).$$

These roots may be found either by factoring p_C or by using the quadratic formula. The roots are real and distinct when the discriminant

$$D = \text{tr}(C)^2 - 4\det(C) > 0.$$

Recall (4.8.10) and Theorem 4.8.3.

Step 2:   Find the eigenvectors v_1 and v_2 of C associated with the eigenvalues λ_1 and λ_2. For $j = 1$ and $j = 2$, the eigenvector v_j is found by solving the homogeneous system of linear equations

$$(C - \lambda_j I_2)v = 0 \tag{4.9.5}$$

for one nonzero solution. Lemma 4.8.4 tells us that there is always a nonzero solution to (4.9.5), since λ_j is an eigenvalue of C.

Step 3:   Using superposition, write the *general solution* to the system of ODEs (4.9.3) as

$$X(t) = \alpha_1 e^{\lambda_1 t} v_1 + \alpha_2 e^{\lambda_2 t} v_2, \tag{4.9.6}$$

where $\alpha_1, \alpha_2 \in \mathbf{R}$. Theorem 4.7.3 tells us that for $j = 1, 2$,

$$X_j(t) = e^{\lambda_j t} v_j$$

is a solution to (4.9.3). The principle of superposition (see Section 4.7) allows us to conclude that

$$X(t) = \alpha_1 X_1(t) + \alpha_2 X_2(t)$$

is also a solution to (4.9.3) for any scalars $\alpha_1, \alpha_2 \in \mathbf{R}$. Thus, (4.9.6) is valid.
   Note that the initial condition corresponding to the general solution (4.9.6) is

$$X(0) = \alpha_1 v_1 + \alpha_2 v_2, \tag{4.9.7}$$

since $e^0 = 1$.

Step 4:   Solve the initial value problem by solving the system of linear equations

$$X_0 = \alpha_1 v_1 + \alpha_2 v_2 \tag{4.9.8}$$

for α_1 and α_2; see (4.9.7). Let A be the 2×2 matrix whose columns are v_1 and v_2; that is,

$$A = (v_1 | v_2). \tag{4.9.9}$$

Then we may rewrite (4.9.8) in the form

$$A \begin{pmatrix} \alpha_1 \\ \alpha_2 \end{pmatrix} = X_0. \tag{4.9.10}$$

We claim that the matrix $A = (v_1 | v_2)$, defined in (4.9.9), is always invertible. Recall Lemma 4.7.2, which states that if w is a nonzero multiple of v_2, then w is also an eigenvector of A associated with the eigenvalue λ_2. Since the eigenvalues λ_1 and λ_2

are distinct, it follows that the eigenvector v_1 is not a scalar multiple of the eigenvector v_2 (see Lemma 4.7.2). Therefore, the area of the parallelogram spanned by v_1 and v_2 is nonzero and the determinant of A is nonzero by Theorem 3.8.5. Corollary 3.8.3 now implies that A is invertible. Thus the unique solution to (4.9.10) is

$$\begin{pmatrix} \alpha_1 \\ \alpha_2 \end{pmatrix} = A^{-1}X_0.$$

This equation is easily solved since we have an explicit formula for A^{-1} when A is a 2×2 matrix; see (3.8.1). Indeed

$$A^{-1} = \frac{1}{\det(A)} \begin{pmatrix} d & -b \\ -c & a \end{pmatrix}.$$

An Initial Value Problem Solved by Hand

Solve the linear system of differential equations

$$\begin{aligned} \dot{x} &= 3x - y \\ \dot{y} &= 4x - 2y \end{aligned} \qquad \text{(4.9.11)}$$

with initial conditions

$$\begin{aligned} x(0) &= 2 \\ y(0) &= -3. \end{aligned} \qquad \text{(4.9.12)}$$

Rewrite the system (4.9.11) in matrix form as

$$\dot{X} = CX,$$

where

$$C = \begin{pmatrix} 3 & -1 \\ 4 & -2 \end{pmatrix}.$$

Rewrite the initial conditions (4.9.12) in vector form as

$$X(0) = X_0 = \begin{pmatrix} 2 \\ -3 \end{pmatrix}.$$

Now proceed through the four steps outlined previously.

Step 1:   Find the eigenvalues of C. The characteristic polynomial of C is

$$p_C(\lambda) = \lambda^2 - \text{tr}(C)\lambda + \det(C) = \lambda^2 - \lambda - 2 = (\lambda - 2)(\lambda + 1).$$

Therefore the eigenvalues of C are

$$\lambda_1 = 2 \quad \text{and} \quad \lambda_2 = -1.$$

Step 2:   Find the eigenvectors of C. Find an eigenvector associated with the eigenvalue $\lambda_1 = 2$ by solving the system of equations

$$(C - \lambda_1 I_2)v = \left(\begin{pmatrix} 3 & -1 \\ 4 & -2 \end{pmatrix} - \begin{pmatrix} 2 & 0 \\ 0 & 2 \end{pmatrix} \right) v = \begin{pmatrix} 1 & -1 \\ 4 & -4 \end{pmatrix} v = 0.$$

One particular solution to this system is

$$v_1 = \begin{pmatrix} 1 \\ 1 \end{pmatrix}.$$

Similarly, find an eigenvector associated with the eigenvalue $\lambda_2 = -1$ by solving the system of equations

$$(C - \lambda_2 I_2)v = \left(\begin{pmatrix} 3 & -1 \\ 4 & -2 \end{pmatrix} - \begin{pmatrix} -1 & 0 \\ 0 & -1 \end{pmatrix} \right) v = \begin{pmatrix} 4 & -1 \\ 4 & -1 \end{pmatrix} v = 0.$$

One particular solution to this system is

$$v_2 = \begin{pmatrix} 1 \\ 4 \end{pmatrix}.$$

Step 3:   Write the general solution to the system of differential equations. Using superposition, we find the general solution to the system (4.9.11) is

$$X(t) = \alpha_1 e^{2t} v_1 + \alpha_2 e^{-t} v_2 = \alpha_1 e^{2t} \begin{pmatrix} 1 \\ 1 \end{pmatrix} + \alpha_2 e^{-t} \begin{pmatrix} 1 \\ 4 \end{pmatrix},$$

where $\alpha_1, \alpha_2 \in \mathbf{R}$. Note that the initial state of this solution is

$$X(0) = \alpha_1 \begin{pmatrix} 1 \\ 1 \end{pmatrix} + \alpha_2 \begin{pmatrix} 1 \\ 4 \end{pmatrix} = \begin{pmatrix} \alpha_1 + \alpha_2 \\ \alpha_1 + 4\alpha_2 \end{pmatrix}.$$

Step 4:   Solve the initial value problem. Let

$$A = (v_1 | v_2) = \begin{pmatrix} 1 & 1 \\ 1 & 4 \end{pmatrix}.$$

The equation for the initial condition is

$$A \begin{pmatrix} \alpha_1 \\ \alpha_2 \end{pmatrix} = X_0.$$

See (4.9.9).

We can write the inverse of A by formula as

$$A^{-1} = \frac{1}{3} \begin{pmatrix} 4 & -1 \\ -1 & 1 \end{pmatrix}.$$

It follows that we solve for the coefficients α_j as

$$\begin{pmatrix} \alpha_1 \\ \alpha_2 \end{pmatrix} = A^{-1} X_0 = \frac{1}{3} \begin{pmatrix} 4 & -1 \\ -1 & 1 \end{pmatrix} \begin{pmatrix} 2 \\ -3 \end{pmatrix} = \frac{1}{3} \begin{pmatrix} 11 \\ -5 \end{pmatrix}.$$

In coordinates

$$\alpha_1 = \frac{11}{3} \quad \text{and} \quad \alpha_2 = -\frac{5}{3}.$$

The solution to the initial value problem (4.9.11) and (4.9.12) is

$$X(t) = \frac{1}{3} \left(11 e^{2t} v_1 - 5 e^{-t} v_2 \right) = \frac{1}{3} \left(11 e^{2t} \begin{pmatrix} 1 \\ 1 \end{pmatrix} - 5 e^{-t} \begin{pmatrix} 1 \\ 4 \end{pmatrix} \right).$$

Expressing the solution in coordinates, we obtain

$$x(t) = \frac{1}{3} \left(11 e^{2t} - 5 e^{-t} \right)$$

$$y(t) = \frac{1}{3} \left(11 e^{2t} - 20 e^{-t} \right).$$

An Initial Value Problem Solved Using MATLAB

Next solve the system of ODEs

$$\dot{x} = 1.7x + 3.5y$$
$$\dot{y} = 1.3x - 4.6y$$

with initial conditions

$$x(0) = 2.7$$
$$y(0) = 1.1 .$$

Rewrite this system in matrix form as

$$\dot{X} = CX,$$

where

$$C = \begin{pmatrix} 1.7 & 3.5 \\ 1.3 & -4.6 \end{pmatrix}.$$

Rewrite the initial conditions in vector form as

$$X_0 = \begin{pmatrix} 2.7 \\ 1.1 \end{pmatrix}.$$

Now proceed through the four steps outlined previously. In MATLAB begin by typing

```
C  = [1.7 3.5; 1.3 -4.6]
X0 = [2.7; 1.1]
```

Step 1:   Find the eigenvalues of C by typing

```
lambda = eig(C)
```

and obtaining

```
lambda =
    2.3543
   -5.2543
```

So the eigenvalues of C are real and distinct.

Step 2:   To find the eigenvectors of C, we need to solve two homogeneous systems of linear equations. The matrix associated with the first system is obtained by typing

```
C1 = C - lambda(1)*eye(2)
```

which yields

```
C1 =
   -0.6543     3.5000
    1.3000    -6.9543
```

We can solve the homogeneous system $(C1)x = 0$ by row reduction—but MATLAB has this process preprogrammed in the command null. So type

```
v1 = null(C1)
```

and obtain

```
v1 =
   -0.9830
   -0.1838
```

Similarly, to find an eigenvector associated with the eigenvalue λ_2, type

```
C2 = C - lambda(2)*eye(2);
v2 = null(C2)
```

and obtain

```
v2 =
   -0.4496
    0.8932
```

Step 3:   The general solution to this system of differential equations is

$$X(t) = \alpha_1 e^{2.3543t} \begin{pmatrix} -0.9830 \\ -0.1838 \end{pmatrix} + \alpha_2 e^{-5.2543t} \begin{pmatrix} -0.4496 \\ 0.8932 \end{pmatrix}.$$

Step 4:   Solve the initial value problem by finding the scalars α_1 and α_2. Form the matrix A by typing

```
A = [v1 v2]
```

Then solve for the αs by typing

```
alpha = inv(A)*X0
```

obtaining

```
alpha =
    -3.0253
     0.6091
```

Therefore the closed form solution to the initial value problem is

$$X(t) = 3.0253e^{2.3543t} \begin{pmatrix} 0.9830 \\ 0.1838 \end{pmatrix} + 0.6091e^{-5.2543t} \begin{pmatrix} -0.4496 \\ 0.8932 \end{pmatrix}.$$

HAND EXERCISES

In Exercises 1–4, find the solution to the system of differential equations $\dot{X} = CX$ satisfying $X(0) = X_0$.

1. $C = \begin{pmatrix} 1 & 1 \\ 0 & 2 \end{pmatrix}$ and $X_0 = \begin{pmatrix} 1 \\ 4 \end{pmatrix}$

2. $C = \begin{pmatrix} 2 & -3 \\ 0 & -1 \end{pmatrix}$ and $X_0 = \begin{pmatrix} 1 \\ -2 \end{pmatrix}$

3. $C = \begin{pmatrix} -3 & 2 \\ -2 & 2 \end{pmatrix}$ and $X_0 = \begin{pmatrix} -1 \\ 3 \end{pmatrix}$

4. $C = \begin{pmatrix} 2 & 1 \\ 1 & 2 \end{pmatrix}$ and $X_0 = \begin{pmatrix} 1 \\ 2 \end{pmatrix}$

5. Solve the initial value problem $\dot{X} = CX$, where $X_0 = e_1$, given that

(a) $X(t) = e^{-t} \begin{pmatrix} 1 \\ 2 \end{pmatrix}$ is a solution,   (b) $\text{tr}(C) = 3$, and

(c) C is a symmetric matrix.

COMPUTER EXERCISES

In Exercises 6 and 7, with MATLAB assistance, find the solution to the system of differential equations $\dot{X} = CX$ satisfying $X(0) = X_0$.

6. $C = \begin{pmatrix} 1.76 & 4.65 \\ 0.23 & 1.11 \end{pmatrix}$ and $X_0 = \begin{pmatrix} 0.34 \\ -0.50 \end{pmatrix}$ **7.** $C = \begin{pmatrix} 1.23 & 2\pi \\ \pi/2 & 1.45 \end{pmatrix}$ and $X_0 = \begin{pmatrix} 1.2 \\ 1.6 \end{pmatrix}$

In Exercises 8 and 9, find the solution to $\dot{X} = CX$ satisfying $X(0) = X_0$ in two different ways:

(a) Use pplane5 to find $X(0.5)$.
   Hint Use the Specify a computation interval option in the PPLANE5 Keyboard input window to compute the solution to $t = 0.5$. Then use the zoom in square feature to determine an answer to three decimal places.

(b) Next use MATLAB to find the eigenvalues and eigenvectors of C and to find a closed form solution $X(t)$. Use this formula to evaluate $X(0.5)$ to three decimal places.

(c) Do the two answers agree?

8. $C = \begin{pmatrix} 2.65 & -2.34 \\ -1.5 & -1.2 \end{pmatrix}$ and $X_0 = \begin{pmatrix} 0.5 \\ 0.1 \end{pmatrix}$ **9.** $C = \begin{pmatrix} 1.2 & 2.4 \\ 0.6 & -3.5 \end{pmatrix}$ and $X_0 = \begin{pmatrix} 0.5 \\ 0.7 \end{pmatrix}$

4.10 *MARKOV CHAINS

Markov chains provide an interesting and useful application of matrices and linear algebra. In this section we introduce Markov chains via some of the theory and two examples. The theory can be understood and applied to examples using just the background in linear algebra that we have developed in this chapter.

An Example of Cats

Consider the four-room apartment pictured in Figure 4.15. One-way passages between the rooms are indicated by arrows. For example, it is possible to go from room 1 directly to any other room, but from room 3 it is possible to go only to room 4.

Suppose that there is a cat in the apartment and that at each hour the cat is asked to move from the room that it is in to another. True to form, however, the cat chooses with equal probability to stay in the room for another hour or to move through one of the allowed passages. Suppose we let p_{ij} be the probability that the cat will move from room i to room j; in particular, p_{ii} is the probability that the cat will stay in room i. For example, when the cat is in room 1, it has four choices: It can stay in room 1 or move to any of the other rooms. Assuming that each of these choices is made with equal probability, we see that

$$p_{11} = \frac{1}{4}, \quad p_{12} = \frac{1}{4}, \quad p_{13} = \frac{1}{4}, \quad p_{14} = \frac{1}{4}.$$

It is now straightforward to verify that

$$p_{21} = \frac{1}{2}, \quad p_{22} = \frac{1}{2}, \quad p_{23} = 0, \quad p_{24} = 0,$$
$$p_{31} = 0, \quad p_{32} = 0, \quad p_{33} = \frac{1}{2}, \quad p_{34} = \frac{1}{2},$$
$$p_{41} = 0, \quad p_{42} = \frac{1}{3}, \quad p_{43} = \frac{1}{3}, \quad p_{44} = \frac{1}{3}.$$

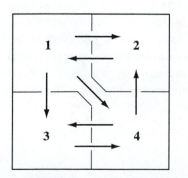

Figure 4.15
Schematic design of apartment passages

Putting these probabilities together yields the *transition matrix*

$$P = \begin{pmatrix} \frac{1}{4} & \frac{1}{4} & \frac{1}{4} & \frac{1}{4} \\ \frac{1}{2} & \frac{1}{2} & 0 & 0 \\ 0 & 0 & \frac{1}{2} & \frac{1}{2} \\ 0 & \frac{1}{3} & \frac{1}{3} & \frac{1}{3} \end{pmatrix}. \qquad \textbf{(4.10.1)*}$$

This transition matrix has the properties that all entries are nonnegative and that the entries in each row sum to 1.

Three Basic Questions

Using the transition matrix P, we discuss the answers to three questions:

A. What is the probability that a cat starting in room i will be in room j after exactly k steps? We call the movement that occurs after each hour a *step*.

B. Suppose that we put 100 cats in the apartment with some initial distribution of cats in each room. What will the distribution of cats look like after a large number of steps?

C. Suppose that a cat is initially in room i and takes a large number of steps. For how many of those steps will the cat be expected to be in room j?

A Discussion of Question A

We begin to answer Question A by determining the probability that the cat moves from room 1 to room 4 in two steps. We denote this probability by $p_{14}^{(2)}$ and compute

$$p_{14}^{(2)} = p_{11}p_{14} + p_{12}p_{24} + p_{13}p_{34} + p_{14}p_{44}; \qquad \textbf{(4.10.2)}$$

that is, the probability is the sum of the probabilities that the cat will move from room 1 to each room i and then from room i to room 4. In this case the answer is

$$p_{14}^{(2)} = \frac{1}{4} \times \frac{1}{4} + \frac{1}{4} \times 0 + \frac{1}{4} \times \frac{1}{2} + \frac{1}{4} \times \frac{1}{3} = \frac{13}{48} \approx 0.27.$$

It follows from (4.10.2) and the definition of matrix multiplication that $p_{14}^{(2)}$ is just the $(1, 4)$th entry in the matrix P^2. An induction argument shows that the probability of the cat moving from room i to room j in k steps is precisely the (i, j)th entry in the matrix P^k—which answers Question A. In particular, we can answer the question: What is the probability that the cat will move from room 4 to room 3 in four steps? With MATLAB the answer is given by typing e4_10_1 to recall the matrix P and then typing

```
P4 = P^4;
P4(4,3)
```

obtaining

```
ans =
    0.2728
```

A Discussion of Question B

We answer Question B in two parts: First we compute a formula for determining the number of cats that are expected to be in room i after k steps, and second we explore that formula numerically for large k. We begin by supposing that 100 cats are distributed in the rooms according to the initial vector $V_0 = (v_1, v_2, v_3, v_4)^t$; that is, the number of cats initially in room i is v_i. Next we denote the number of cats that are expected to be in room i after k steps by $v_i^{(k)}$. For example, we determine how many cats we expect to be in room 2 after one step. That number is

$$v_2^{(1)} = p_{12}v_1 + p_{22}v_2 + p_{32}v_3 + p_{42}v_4; \tag{4.10.3}$$

that is, $v_2^{(1)}$ is the sum of the proportion of cats in each room i that are expected to migrate to room 2 in one step. In this case the answer is

$$\frac{1}{4}v_1 + \frac{1}{2}v_2 + \frac{1}{3}v_4.$$

It now follows from (4.10.3), the definition of the transpose of a matrix, and the definition of matrix multiplication that $v_2^{(1)}$ is the 2nd entry in the vector $P^t V_0$. Indeed, it follows by induction that $v_i^{(k)}$ is the ith entry in the vector $(P^t)^k V_0$, which answers the first part of Question B.

We may rephrase the second part of Question B as follows: Let

$$V_k = (v_1^k, v_2^k, v_3^k, v_4^k)^t = (P^t)^k V_0.$$

Question B actually asks: What will the vector V_k look like for large k? To answer that question we need some results about matrices like the matrix P in (4.10.1). But first we explore the answer to this question numerically using MATLAB.

Suppose, for example, that the initial vector is

$$V_0 = \begin{pmatrix} 2 \\ 43 \\ 21 \\ 34 \end{pmatrix}. \tag{4.10.4}*$$

Typing e4_10_1 and e4_10_4 enters the matrix P and the initial vector V_0 into MATLAB. To compute V_{20}, the distribution of cats after 20 steps, type

```
Q=P'
V20 = Q^(20)*V0
```

and obtain

```
V20 =
    18.1818
    27.2727
    27.2727
    27.2727
```

Thus, after rounding to the nearest integer, we expect 27 cats to be in each of rooms 2, 3, and 4 and 18 cats to be in room 1 after 20 steps. In fact, the vector V_{20} has a remarkable feature. Compute Q*V20 in MATLAB and see that $V_{20} = P^t V_{20}$; that is, V_{20} is, to within four-digit numerical precision, an eigenvector of P^t with eigenvalue equal to 1. This computation was not a numerical accident, as we now describe. Indeed, compute V_{20} for several initial distributions V_0 of cats and see that the answer will always be the same—up to four-digit accuracy.

A Discussion of Question C

Suppose just one cat is in the apartment, and we ask how many times that cat is expected to visit room 3 in 100 steps. Suppose the cat starts in room 1; then the initial distribution of cats is one cat in room 1 and zero cats in any of the other rooms. So $V_0 = e_1$. In our discussion of Question B we saw that the 3rd entry in $(P^t)^k V_0$ gives the probability c_k that the cat will be in room 3 after k steps.

In the extreme, suppose that the probability that the cat will be in room 3 is 1 for each step k. Then the fraction of the time that the cat is in room 3 is

$$\frac{1 + 1 + \cdots + 1}{100} = 1.$$

In general, the fraction of the time f that the cat will be in room 3 during a span of 100 steps is

$$f = \frac{1}{100}(c_1 + c_2 + \cdots + c_{100}).$$

Since $c_k = (P^t)^k V_0$, we see that

$$f = \frac{1}{100}(P^t V_0 + (P^t)^2 V_0 + \cdots + (P^t)^{100} V_0). \qquad \textbf{(4.10.5)}$$

So, to answer Question C, we need a way to sum the expression for f in (4.10.5), at least approximately. This is not an easy task—although the answer itself is easy to explain. Let V be the eigenvector of P^t with eigenvalue 1 such that the sum of the entries in V is 1. The answer is: f is approximately equal to V. See Theorem 4.10.4 for a more precise statement.

In our previous calculations the vector V_{20} was seen to be (approximately) an eigenvector of P^t with eigenvalue 1. Moreover the sum of the entries in V_{20} is precisely 100. Therefore we normalize V_{20} to get V by setting

$$V = \frac{1}{100} V_{20}.$$

So the fraction of time that the cat spends in room 3 is $f \approx 0.2727$. Indeed, we expect the cat to spend approximately 27% of its time in rooms 2, 3, and 4 and about 18% of its time in room 1.

Markov Matrices

We now abstract the salient properties of our cat example. A *Markov chain* is a system with a finite number of states labeled $1, \ldots, n$ along with probabilities p_{ij} of moving from site i to site j in a single step. The Markov assumption is that these probabilities depend only on the site that you are in and not on how you got there. In our example, we assumed that the probability of the cat moving from, say, room 2 to room 4 did not depend on how the cat got to room 2 in the first place.

We make a second assumption: There is a k such that it is possible to move from any site i to any site j in exactly k steps. This assumption is *not* valid for general Markov chains, although it is valid for the cat example, since it is possible to move from any room to any other room in that example in exactly three steps. (It takes a minimum of three steps to get from room 3 to room 1 in the cat example.) To simplify our discussion we include this assumption in our definition of a Markov chain.

Definition 4.10.1 *Markov matrices are square matrices P such that*

- *(a) all entries in P are nonnegative,*
- *(b) the entries in each row of P sum to 1, and*
- *(c) there is a positive integer k such that all of the entries in P^k are positive.*

It is straightforward to verify that (a) and (b) in the definition of Markov matrices are satisfied by the transition matrix

$$P = \begin{pmatrix} p_{11} & \cdots & p_{1n} \\ \vdots & \vdots & \vdots \\ p_{n1} & \cdots & p_{nn} \end{pmatrix}$$

of a Markov chain. To verify (c) requires further discussion.

Proposition 4.10.2 *Let P be a transition matrix for a Markov chain.*

- *(a) The probability of moving from site i to site j in exactly k steps is the (i, j)th entry in the matrix P^k.*
- *(b) The expected number of individuals at site i after exactly k steps is the ith entry in the vector $V_k \equiv (P^t)^k V_0$.*
- *(c) P is a Markov matrix.*

Proof: Only minor changes in our discussion of the cat example prove (a) and (b) of the proposition.

For (c), the assumption that it is possible to move from each site i to each site j in exactly k steps means that the (i, j)th entry of P^k is positive. For that k, all of the entries of P^k are positive. In the cat example, all entries of P^3 are positive. ◆

Proposition 4.10.2 gives the answer to Question A and the first part of Question B for general Markov chains.

Let $v_i^{(0)} \geq 0$ be the number of individuals initially at site i, and let $V_0 = (v_1^{(0)}, \dots, v_n^{(0)})^t$. The total number of individuals in the initial population is

$$\#(V_0) = v_1^{(0)} + \cdots + v_n^{(0)}.$$

Theorem 4.10.3 *Let P be a Markov matrix. Then*

(a) $\#(V_k) = \#(V_0)$; *that is, the number of individuals after k steps is the same as the initial number.*

(b) $V = \lim_{k \to \infty} V_k$ *exists and $\#(V) = \#(V_0)$.*

(c) *V is an eigenvector of P^t with eigenvalue equal to 1.*

Proof: (a) By induction it is sufficient to show that $\#(V_1) = \#(V_0)$. We do this by calculating from $V_1 = P^t V_0$ that

$$\begin{aligned}
\#(V_1) &= v_1^{(1)} + \cdots + v_n^{(1)} \\
&= (p_{11} v_1^{(0)} + \cdots + p_{n1} v_n^{(0)}) + \cdots + (p_{1n} v_1^{(0)} + \cdots + p_{nn} v_n^{(0)}) \\
&= (p_{11} + \cdots + p_{1n}) v_1^{(0)} + \cdots + (p_{n1} + \cdots + p_{nn}) v_n^{(0)} \\
&= v_1^{(0)} + \cdots + v_n^{(0)},
\end{aligned}$$

since the entries in each row of P sum to 1. Thus $\#(V_1) = \#(V_0)$, as claimed.

(b) The hard part of this theorem is proving that the limiting vector V exists; we give a proof of this fact in Theorem 13.5.4. Once V exists, it follows directly from (a) that $\#(V) = \#(V_0)$.

(c) Just calculate that

$$P^t V = P^t (\lim_{k \to \infty} V_k) = P^t (\lim_{k \to \infty} (P^t)^k V_0) = \lim_{k \to \infty} (P^t)^{k+1} V_0 = \lim_{k \to \infty} (P^t)^k V_0 = V,$$

which proves (c). $\blacklozenge$

Theorem 4.10.3(b) gives the answer to the second part of Question B for general Markov chains. Next we discuss Question C.

Theorem 4.10.4 *Let P be a Markov matrix. Let V be the eigenvector of P^t with eigenvalue 1 and $\#(V) = 1$. Then after a large number of steps N, the expected number of times an individual will visit site i is $N v_i$, where v_i is the ith entry in V.*

Sketch of proof: In our discussion of Question C for the cat example, we explained why the fraction f_N of time that an individual will visit site j when starting initially at site i is the jth entry in the sum

$$f_N = \frac{1}{N}(P^t + (P^t)^2 + \cdots + (P^t)^N)e_i.$$

See (4.10.5). The proof of this theorem involves being able to calculate the limit of f_N as $N \to \infty$. There are two main ideas. First, the limit of the matrix $(P^t)^N$ exists as N approaches infinity—call that limit Q. Moreover, Q is a matrix all of whose columns equal V. Second, for large N, the sum

$$P^t + (P^t)^2 + \cdots + (P^t)^N \approx Q + Q + \cdots + Q = NQ,$$

so that the limit of the f_N is $Qe_i = V$.

The verification of these statements is beyond the scope of this text. For those interested, the idea of the proof of the second part is roughly the following. Fix k large enough so that $(P^t)^k$ is close to Q. Then when N is large, much larger than k, the sum of the first k terms in the series is nearly 0.   ◆

Theorem 4.10.4 gives the answer to Question C for a general Markov chain. It follows from Theorem 4.10.4 that for Markov chains the amount of time that an individual spends in room i is independent of the individual's initial room—at least after a large number of steps.

A complete proof of this theorem relies on a result known as the *ergodic theorem.* Roughly speaking, the ergodic theorem relates space averages with time averages. To see how this point is relevant, note that Question B deals with the issue of how a large number of individuals will be distributed in space after a large number of steps, whereas Question C deals with the issue of how the path of a single individual will be distributed in time after a large number of steps.

An Example of Umbrellas

This example focuses on the utility of answering Question C and reinforces the fact that results in Theorem 4.10.3 have the second interpretation given in Theorem 4.10.4.

Consider the problem of a man with four umbrellas. If it is raining in the morning when the man is about to leave for his office, then the man takes an umbrella from home to office, assuming that he has an umbrella at home. If it is raining in the afternoon, then the man takes an umbrella from office to home, assuming that he has an umbrella in his office. Suppose that the probability that it will rain in the morning is $p = 0.2$ and the probability that it will rain in the afternoon is $q = 0.3$, and these probabilities are independent. What percentage of days will the man get wet going from home to office; that is, what percentage of the days will the man be at home on a rainy morning with all of his umbrellas at the office?

There are five states in the system depending on the number of umbrellas that are at home. Let s_i, where $0 \le i \le 4$, be the state with i umbrellas at home and $4 - i$ umbrellas at work. For example, s_2 is the state of having two umbrellas at home and two at the office. Let P be the 5×5 transition matrix of state changes from morning to afternoon, and let Q be the 5×5 transition matrix of state changes from afternoon to morning. For example, the probability p_{23} of moving from site s_2 to site s_3 is 0, since it is not possible to have more umbrellas at home after going to work in the morning. The probability $q_{23} = q$, since the number of umbrellas at home will increase by one only if it is raining in the afternoon. The transition probabilities between all states are given in the following transition matrices:

$$P = \begin{pmatrix} 1 & 0 & 0 & 0 & 0 \\ p & 1-p & 0 & 0 & 0 \\ 0 & p & 1-p & 0 & 0 \\ 0 & 0 & p & 1-p & 0 \\ 0 & 0 & 0 & p & 1-p \end{pmatrix} \qquad Q = \begin{pmatrix} 1-q & q & 0 & 0 & 0 \\ 0 & 1-q & q & 0 & 0 \\ 0 & 0 & 1-q & q & 0 \\ 0 & 0 & 0 & 1-q & q \\ 0 & 0 & 0 & 0 & 1 \end{pmatrix}.$$

Specifically,

$$P = \begin{pmatrix} 1 & 0 & 0 & 0 & 0 \\ 0.2 & 0.8 & 0 & 0 & 0 \\ 0 & 0.2 & 0.8 & 0 & 0 \\ 0 & 0 & 0.2 & 0.8 & 0 \\ 0 & 0 & 0 & 0.2 & 0.8 \end{pmatrix} \qquad Q = \begin{pmatrix} 0.7 & 0.3 & 0 & 0 & 0 \\ 0 & 0.7 & 0.3 & 0 & 0 \\ 0 & 0 & 0.7 & 0.3 & 0 \\ 0 & 0 & 0 & 0.7 & 0.3 \\ 0 & 0 & 0 & 0 & 1 \end{pmatrix}. \qquad \textbf{(4.10.6)}^*$$

The transition matrix M from moving from state s_i on one morning to state s_j on the next morning is just $M = PQ$. We can compute this matrix using MATLAB by typing

```
e4_10_6
M = P*Q
```

obtaining

```
M =
    0.7000    0.3000         0         0         0
    0.1400    0.6200    0.2400         0         0
         0    0.1400    0.6200    0.2400         0
         0         0    0.1400    0.6200    0.2400
         0         0         0    0.1400    0.8600
```

It is easy to check using MATLAB that all entries in the matrix M^4 are nonzero. So M is a Markov matrix and we can use Theorem 4.10.4 to find the limiting distribution of states. Start with some initial condition like $V_0 = (0, 0, 1, 0, 0)^t$ corresponding to the state in which two umbrellas are at home and two at the office. Then compute the vectors $V_k = (M^t)^k V_0$ until arriving at an eigenvector of M^t with eigenvalue 1. For example, V_{70} is computed by typing V70 = M'^(70)*V0 and obtaining

V70 =
    0.0419
    0.0898
    0.1537
    0.2633
    0.4512

We interpret $V \approx V_{70}$ in the following way. Since v_1 is approximately 0.042, it follows that for approximately 4.2% of all steps the umbrellas are in state s_0. That is, approximately 4.2% of all days there are no umbrellas at home. The probability that it will rain in the morning on one of those days is 0.2. Therefore the probability of being at home in the morning when it is raining without any umbrellas is approximately 0.008.

HAND EXERCISES

1. Let P be a Markov matrix and let $w = (1, \ldots, 1)^t$. Show that the vector w is an eigenvector of P with eigenvalue 1.

In Exercises 2–4, which of the matrices are Markov matrices, and why?

2. $P = \begin{pmatrix} 0.8 & 0.2 \\ 0.2 & 0.8 \end{pmatrix}$

3. $Q = \begin{pmatrix} 0.8 & 0.2 \\ 0 & 1 \end{pmatrix}$

4. $R = \begin{pmatrix} 0.8 & 0.2 \\ -0.2 & 1.2 \end{pmatrix}$

5. The state diagram of a Markov chain is given in Figure 4.16. Assume that each arrow leaving a state has an equal probability of being chosen. Find the transition matrix for this chain.

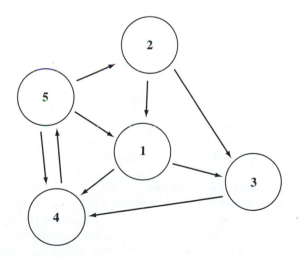

Figure 4.16
State diagram of a Markov chain

6. Suppose that P and Q are each $n \times n$ matrices whose rows sum to 1. Show that PQ is also an $n \times n$ matrix whose rows sum to 1.

COMPUTER EXERCISES

   7. Suppose that the apartment in Figure 4.15 is populated by dogs rather than cats. Suppose that dogs will actually move when told; that is, at each step a dog will move from the room that he occupies to another room.
   (a) Calculate the transition matrix PDOG for this Markov chain and verify that PDOG is a Markov matrix.
   (b) Find the probability that a dog starting in room 2 will end up in room 3 after five steps.
   (c) Find the probability that a dog starting in room 3 will end up in room 1 after four steps. Explain why your answer is correct without using MATLAB.
   (d) Suppose that the initial population consists of 100 dogs. After a large number of steps, what will be the distribution of the dogs in the four rooms?

   8. A truck rental company has locations in three cities: A, B, and C. Statistically, the company knows that the trucks rented at one location will be returned in one week to the three locations in the proportions listed in the table:

Rental Location	Returned to A	Returned to B	Returned to C
A	75%	10%	15%
B	5%	85%	10%
C	20%	20%	60%

Suppose that the company has 250 trucks. How should the company distribute the trucks so that the number of trucks available at each location remains approximately constant from one week to the next?

   9. Let

$$P = \begin{pmatrix} 0.10 & 0.20 & 0.30 & 0.15 & 0.25 \\ 0.05 & 0.35 & 0.10 & 0.40 & 0.10 \\ 0 & 0 & 0.35 & 0.55 & 0.10 \\ 0.25 & 0.25 & 0.25 & 0.25 & 0 \\ 0.33 & 0.32 & 0 & 0 & 0.35 \end{pmatrix} \qquad (4.10.7)*$$

be the transition matrix of a Markov chain.
   (a) What is the probability that an individual at site 2 will move to site 5 in three steps?
   (b) What is the probability that an individual at site 4 will move to site 1 in seven steps?
   (c) Suppose that 100 individuals are initially uniformly distributed at the five sites. How will the individuals be distributed after four steps?
   (d) Find an eigenvector of P^t with eigenvalue 1.

   10. Suppose that the probability that it will rain in the morning is $p = 0.3$ and the probability that it will rain in the afternoon is $q = 0.25$. In the man with umbrellas example, what is the probability that he will be at home with no umbrellas while it is raining?

   11. Suppose that the original man with umbrellas has only three umbrellas instead of four. What is the probability that on a given day he will get wet going to work?

CHAPTER

5

Vector Spaces

In Chapter 2 we discussed how to solve systems of m linear equations in n unknowns. We found that solutions of these equations are vectors $(x_1, \ldots, x_n) \in \mathbf{R}^n$. In Chapter 3 we discussed how the notation of matrices and matrix multiplication drastically simplifies the presentation of linear systems and how matrix multiplication leads to linear mappings. We also discussed briefly how linear mappings lead to methods for solving linear systems—superposition, eigenvectors, inverses. In Chapter 4 we discussed how to solve systems of n linear differential equations in n unknown functions. These chapters have introduced many of the ideas of linear algebra, and now we begin the task of formalizing these ideas.

Sets having the two operations of vector addition and scalar multiplication are called *vector spaces*. This concept is introduced in Section 5.1 along with the two primary examples: the set $\mathbf{R}^n$ in which solutions to systems of linear equations sit and the set C^1 of differentiable functions in which solutions to systems of ordinary differential equations sit. Solutions to systems of homogeneous linear equations form subspaces of $\mathbf{R}^n$, and solutions to systems of linear differential equations form subspaces of C^1. These issues are discussed in Sections 5.1 and 5.2.

When we *solve* a homogeneous system of equations, we write every solution as a superposition of a finite number of specific solutions. Abstracting this process is one of the main points of this chapter. Specifically, we show that every vector in many commonly occurring vector spaces (in particular, the subspaces of solutions) can be written as a *linear combination* (superposition) of a few solutions. The minimum number of solutions needed is called the *dimension* of that vector space. Sets of vectors that generate all solutions by superposition and that consist of that minimum number of vectors are called *bases*. These ideas are discussed in detail in Sections 5.3–5.5. The proof of the main theorem (Theorem 5.5.3), which gives a computable method for determining when a set is a basis, is given in Section 5.6. This proof may be omitted on a first reading, but the statement of the theorem is most important and must be understood.

5.1 VECTOR SPACES AND SUBSPACES

Vector Spaces

Vector spaces abstract the arithmetic properties of addition and scalar multiplication of vectors. In $\mathbf{R}^n$ we know how to add vectors and how to multiply vectors by scalars. Indeed, it is straightforward to verify that each of the eight properties listed in Table 5.1 is valid for vectors in $V = \mathbf{R}^n$. Remarkably, sets that satisfy these eight properties have much in common with $\mathbf{R}^n$. So we define:

Definition 5.1.1 *Let V be a set having the two operations of addition and scalar multiplication. Then V is a* vector space *if the eight properties listed in Table 5.1 hold. The elements of a vector space are called* vectors.

The vector 0 mentioned in (A3) in Table 5.1 is called the *zero vector*.

When we say that a vector space V has the two operations of addition and scalar multiplication, we mean that the sum of two vectors in V is again a vector in V and that the scalar product of a vector with a number is again a vector in V. These two properties are called *closure under addition* and *closure under scalar multiplication*.

In this discussion we focus on just two types of vector spaces: $\mathbf{R}^n$ and function spaces. The reason we make this choice is that solutions to linear equations are vectors

Table 5.1
Properties of vector spaces: suppose $u, v, w \in V$ and $r, s \in \mathbf{R}$

(A1)	Addition is commutative.	$v + w = w + v$
(A2)	Addition is associative.	$(u + v) + w = u + (v + w)$
(A3)	Additive identity 0 exists.	$v + 0 = v$
(A4)	Additive inverse $-v$ exists.	$v + (-v) = 0$
(M1)	Multiplication is associative.	$(rs)v = r(sv)$
(M2)	Multiplicative identity exists.	$1v = v$
(D1)	Distributive law for scalars	$(r + s)v = rv + sv$
(D2)	Distributive law for vectors	$r(v + w) = rv + rw$

in $\mathbf{R}^n$, whereas solutions to linear systems of differential equations are vectors of functions.

An Example of a Function Space

For example, let $\mathcal{F}$ denote the set of all functions $f : \mathbf{R} \to \mathbf{R}$. Note that functions like $f_1(t) = t^2 - 2t + 7$ and $f_2(t) = \sin t$ are in $\mathcal{F}$ since they are defined for all real numbers t, but that functions like $g_1(t) = 1/t$ and $g_2(t) = \tan t$ are not in $\mathcal{F}$ since they are not defined for all t.

We can add two functions f and g by defining the function $f + g$ to be

$$(f + g)(t) = f(t) + g(t).$$

We can also multiply a function f by a scalar $c \in \mathbf{R}$ by defining the function cf to be

$$(cf)(t) = cf(t).$$

With these operations of addition and scalar multiplication, $\mathcal{F}$ is a vector space; that is, $\mathcal{F}$ satisfies the eight vector space properties listed in Table 5.1. More precisely:

(A3) Define the zero function $\mathcal{O}$ by

$$\mathcal{O}(t) = 0 \quad \text{for all } t \in \mathbf{R}.$$

For every x in $\mathcal{F}$, the function $\mathcal{O}$ satisfies

$$(x + \mathcal{O})(t) = x(t) + \mathcal{O}(t) = x(t) + 0 = x(t).$$

Therefore $x + \mathcal{O} = x$, and $\mathcal{O}$ is the additive identity in $\mathcal{F}$.

(A4) Let x be a function in $\mathcal{F}$ and define $y(t) = -x(t)$. Then y is also a function in $\mathcal{F}$, and

$$(x + y)(t) = x(t) + y(t) = x(t) + (-x(t)) = 0 = \mathcal{O}(t).$$

Thus x has the additive inverse $-x$.

After these comments, it is straightforward to verify that the remaining six properties in Table 5.1 are satisfied by functions in $\mathcal{F}$.

Sets That Are Not Vector Spaces

It is worth considering how closure under vector addition and scalar multiplication can fail. Consider the following three examples.

(i) Let V_1 be the set that consists of just the x- and y-axes in the plane. Since $(1, 0)$ and $(0, 1)$ are in V_1 but

$$(1, 0) + (0, 1) = (1, 1)$$

is not in V_1, we see that V_1 is not closed under vector addition. On the other hand, V_1 is closed under scalar multiplication.

(ii) Let V_2 be the set of all vectors $(k, \ell) \in \mathbf{R}^2$, where k and ℓ are integers. The set V_2 is closed under addition but not under scalar multiplication since $\frac{1}{2}(1, 0) = (\frac{1}{2}, 0)$ is not in V_2.

(iii) Let $V_3 = [1, 2]$ be the closed interval in $\mathbf{R}$. The set V_3 is closed neither under addition $(1 + 1.5 = 2.5 \notin V_3)$ nor under scalar multiplication $(4 \cdot 1.5 = 6 \notin V_3)$. Hence the set V_3 is not closed under vector addition and not closed under scalar multiplication.

Subspaces

Definition 5.1.2 *Let V be a vector space. A nonempty subset $W \subset V$ is a subspace if W is a vector space using the operations of addition and scalar multiplication defined on V.*

Note that in order for a subset W of a vector space V to be a subspace it must be *closed under addition* and *closed under scalar multiplication*. That is, suppose $w_1, w_2 \in W$ and $r \in \mathbf{R}$. Then (i) $w_1 + w_2 \in W$, and (ii) $rw_1 \in W$.

The x-axis and the xz-plane are examples of subsets of $\mathbf{R}^3$ that are closed under addition and closed under scalar multiplication. Every vector on the x-axis has the form $(a, 0, 0) \in \mathbf{R}^3$. The sum of two vectors $(a, 0, 0)$ and $(b, 0, 0)$ on the x-axis is $(a + b, 0, 0)$, which is also on the x-axis. The x-axis is also closed under scalar multiplication as $r(a, 0, 0) = (ra, 0, 0)$, and the x-axis is a subspace of $\mathbf{R}^3$. Similarly, every vector in the xz-plane in $\mathbf{R}^3$ has the form $(a_1, 0, a_3)$. As in the case of the x-axis, it is easy to verify that this set of vectors is closed under addition and scalar multiplication. Thus the xz-plane is also a subspace of $\mathbf{R}^3$.

In Theorem 5.1.4 we show that every subset of a vector space that is closed under addition and scalar multiplication is a subspace. To verify this statement, we need the following lemma in which some special notation is used. Typically we use the same notation 0 to denote the real number 0 and the zero vector. In the next lemma it is convenient to distinguish the two different uses of 0, and we write the zero vector in boldface.

Lemma 5.1.3 *Let V be a vector space, and let $\mathbf{0} \in V$ be the zero vector. Then*

$$0v = \mathbf{0} \quad and \quad (-1)v = -v \cdot$$

for every vector in $v \in V$.

Proof: Let v be a vector in V and use (D1) to compute

$$0v + 0v = (0 + 0)v = 0v.$$

By (A4) the vector $0v$ has an additive inverse $-0v$. Adding $-0v$ to both sides yields

$$(0v + 0v) + (-0v) = 0v + (-0v) = \mathbf{0}.$$

Associativity of addition (A2) now implies that

$$0v + (0v + (-0v)) = \mathbf{0}.$$

A second application of (A4) implies that

$$0v + \mathbf{0} = \mathbf{0},$$

and (A3) implies that $0v = \mathbf{0}$.

Next we show that the additive inverse $-v$ of a vector v is unique; that is, if $v + a = \mathbf{0}$, then $a = -v$.

Before beginning the proof, we note that commutativity of addition (A1) together with (A3) implies that $\mathbf{0} + v = v$. Similarly (A1) and (A4) imply that $-v + v = \mathbf{0}$.

To prove uniqueness of additive inverses, add $-v$ to both sides of the equation $v + a = \mathbf{0}$, yielding

$$-v + (v + a) = -v + \mathbf{0}.$$

Properties (A2) and (A3) imply that

$$(-v + v) + a = -v,$$

but

$$(-v + v) + a = \mathbf{0} + a = a.$$

Therefore $a = -v$, as claimed.

To verify that $(-1)v = -v$, we show that $(-1)v$ is the additive inverse of v. Using (M1), (D1), and the fact that $0v = \mathbf{0}$, we calculate

$$v + (-1)v = 1v + (-1)v = (1 - 1)v = 0v = \mathbf{0}.$$

Thus $(-1)v$ is the additive inverse of v and must equal $-v$, as claimed.   ◆

Theorem 5.1.4   *Let W be a subset of the vector space V. If W is closed under addition and closed under scalar multiplication, then W is a subspace.*

Proof: We have to show that W is a vector space using the operations of addition and scalar multiplication defined on V. That is, we need to verify that the eight properties listed in Table 5.1 are satisfied. Note that properties (A1), (A2), (M1), (M2), (D1), and (D2) are valid for vectors in W since they are valid for vectors in V.

It remains to verify (A3) and (A4). Let $w \in W$ be any vector. Since W is closed under scalar multiplication, it follows that $0w$ and $(-1)w$ are in W. Lemma 5.1.3 states that $0w = 0$ and $(-1)w = -w$; it follows that 0 and $-w$ are in W. Hence, properties (A3) and (A4) are valid for vectors in W since they are valid for vectors in V.   ◆

Examples of Subspaces of $\mathbf{R}^n$

Example 5.1.5

(a) *Let V be a vector space. Then the subsets V and $\{0\}$ are always subspaces of V. A subspace $W \subset V$ is* proper *if $W \neq 0$ and $W \neq V$.*

(b) *Lines through the origin are subspaces of $\mathbf{R}^n$. Let $w \in \mathbf{R}^n$ be a nonzero vector and let $W = \{rw : r \in \mathbf{R}\}$. The set W is closed under addition and scalar multiplication and is a subspace of $\mathbf{R}^n$ by Theorem 5.1.4. The subspace W is just a line through the origin in $\mathbf{R}^n$, since the vector rw points in the same direction as w when $r > 0$ and in the exact opposite direction when $r < 0$.*

(c) *Planes containing the origin are subspaces of $\mathbf{R}^3$. To verify this point, let P be a plane through the origin and let N be a vector perpendicular to P. Then P consists of all vectors $v \in \mathbf{R}^3$ perpendicular to N; using the dot product (see (2.2.3)), we recall that such vectors satisfy the linear equation $N \cdot v = 0$. By superposition, the set of all solutions to this equation is closed under addition and scalar multiplication and is therefore a subspace by Theorem 5.1.4.*

In a sense that will be made precise, all subspaces of $\mathbf{R}^n$ can be written as the span of a finite number of vectors, generalizing Example 5.1.5(b), or as solutions to a system of linear equations, generalizing Example 5.1.5(c).

Examples of Subspaces of the Function Space $\mathcal{F}$

Let $\mathcal{P}$ be the set of all polynomials in $\mathcal{F}$. The sum of two polynomials is a polynomial, and the scalar multiple of a polynomial is a polynomial. Thus $\mathcal{P}$ is closed under addition and scalar multiplication, and $\mathcal{P}$ is a subspace of $\mathcal{F}$.

As a second example of a subspace of $\mathcal{F}$, let $\mathcal{C}^1$ be the set of all continuously differentiable functions $u : \mathbf{R} \to \mathbf{R}$. A function u is in $\mathcal{C}^1$ if u and u' exist and are continuous for all $t \in \mathbf{R}$. Here are some examples of functions in $\mathcal{C}^1$:

(i)   Every polynomial $p(t) = a_m t^m + a_{m-1} t^{m-1} + \cdots + a_1 t + a_0$ is in $\mathcal{C}^1$.

(ii)   The function $u(t) = e^{\lambda t}$ is in $\mathcal{C}^1$ for each constant $\lambda \in \mathbf{R}$.

(iii)   The trigonometric functions $u(t) = \sin(\lambda t)$ and $v(t) = \cos(\lambda t)$ are in $\mathcal{C}^1$ for each constant $\lambda \in \mathbf{R}$.

(iv) $u(t) = t^{7/3}$ is twice differentiable everywhere and is in $\mathcal{C}^1$.

Equally there are many commonly used functions that are not in $\mathcal{C}^1$. Examples include the following:

(i) $u(t) = 1/(t - 5)$ is neither defined nor continuous at $t = 5$.

(ii) $u(t) = |t|$ is not differentiable (at $t = 0$).

(iii) $u(t) = \csc(t)$ is neither defined nor continuous at $t = k\pi$ for any integer k.

The subset $\mathcal{C}^1 \subset \mathcal{F}$ is a subspace and hence a vector space. The reason is simple. If $x(t)$ and $y(t)$ are continuously differentiable, then

$$\frac{d}{dt}(x + y) = \frac{dx}{dt} + \frac{dy}{dt}.$$

Hence $x + y$ is differentiable and is in $\mathcal{C}^1$, and $\mathcal{C}^1$ is closed under addition. Similarly $\mathcal{C}^1$ is closed under scalar multiplication. Let $r \in \mathbf{R}$ and let $x \in \mathcal{C}^1$. Then

$$\frac{d}{dt}(rx)(t) = r\frac{dx}{dt}(t).$$

Hence rx is differentiable and is in $\mathcal{C}^1$.

The Vector Space $(\mathcal{C}^1)^n$

Another example of a vector space that combines the features of both $\mathbf{R}^n$ and $\mathcal{C}^1$ is $(\mathcal{C}^1)^n$. Vectors $u \in (\mathcal{C}^1)^n$ have the form

$$u(t) = (u_1(t), \ldots, u_n(t)),$$

where each coordinate function $u_j(t) \in \mathcal{C}^1$. Addition and scalar multiplication in $(\mathcal{C}^1)^n$ are defined coordinatewise—just like addition and scalar multiplication in $\mathbf{R}^n$. That is, let u, v be in $(\mathcal{C}^1)^n$ and let r be in $\mathbf{R}$; then

$$(u + v)(t) = (u_1(t) + v_1(t), \ldots, u_n(t) + v_n(t))$$
$$(ru)(t) = (ru_1(t), \ldots, ru_n(t)).$$

The set $(\mathcal{C}^1)^n$ satisfies the eight properties of vector spaces and is a vector space. Solutions to systems of n linear ordinary differential equations are vectors in $(\mathcal{C}^1)^n$.

HAND EXERCISES

1. Verify that the set V_1 consisting of all scalar multiples of $(1, -1, -2)$ is a subspace of $\mathbf{R}^3$.

2. Let V_2 be the set of all 2×3 matrices. Verify that V_2 is a vector space.

3. Let

$$A = \begin{pmatrix} 1 & 1 & 0 \\ 1 & -1 & 1 \end{pmatrix}.$$

Let V_3 be the set of vectors $x \in \mathbf{R}^3$ such that $Ax = 0$. Verify that V_3 is a subspace of $\mathbf{R}^3$. Compare V_1 with V_3.

In Exercises 4–10, you are given a vector space V and a subset W. For each pair, decide whether or not W is a subspace of V.

4. $V = \mathbf{R}^3$ and W consists of vectors in $\mathbf{R}^3$ that have a 0 in their first component.

5. $V = \mathbf{R}^3$ and W consists of vectors in $\mathbf{R}^3$ that have a 1 in their first component.

6. $V = \mathbf{R}^2$ and W consists of vectors in $\mathbf{R}^2$ for which the sum of the components is 1.

7. $V = \mathbf{R}^2$ and W consists of vectors in $\mathbf{R}^2$ for which the sum of the components is 0.

8. $V = \mathcal{C}^1$ and W consists of functions $x(t) \in \mathcal{C}^1$ satisfying $\int_{-2}^{4} x(t)dt = 0$.

9. $V = \mathcal{C}^1$ and W consists of functions $x(t) \in \mathcal{C}^1$ satisfying $x(1) = 0$.

10. $V = \mathcal{C}^1$ and W consists of functions $x(t) \in \mathcal{C}^1$ satisfying $x(1) = 1$.

In Exercises 11–15, which of the sets S are subspaces?

11. $S = \{(a, b, c) \in \mathbf{R}^3 : a \geq 0,\ b \geq 0,\ c \geq 0\}$

12. $S = \{(x_1, x_2, x_3) \in \mathbf{R}^3 : a_1 x_1 + a_2 x_2 + a_3 x_3 = 0,\ \text{where } a_1, a_2, a_3 \in \mathbf{R} \text{ are fixed}\}$

13. $S = \{(x, y) \in \mathbf{R}^2 : (x, y) \text{ is on the line through } (1, 1) \text{ with slope } 1\}$

14. $S = \{x \in \mathbf{R}^2 : Ax = 0\}$, where A is a 3×2 matrix

15. $S = \{x \in \mathbf{R}^2 : Ax = b\}$, where A is a 3×2 matrix and $b \in \mathbf{R}^3$ is a fixed nonzero vector

16. Let V be a vector space, and let W_1 and W_2 be subspaces. Show that the intersection $W_1 \cap W_2$ is also a subspace of V.

17. For which scalars a, b, c do the solutions to the equation

$$ax + by = c$$

form a subspace of $\mathbf{R}^2$?

18. For which scalars a, b, c, d do the solutions to the equation

$$ax + by + cz = d$$

form a subspace of $\mathbf{R}^3$?

19. Show that the set of all solutions to the differential equation $\dot{x} = 2x$ is a subspace of $\mathcal{C}^1$.

20. Recall from equation (4.7.6) that solutions to the system of differential equations

$$\frac{dX}{dt} = \begin{pmatrix} -1 & 3 \\ 3 & -1 \end{pmatrix} X$$

are

$$X(t) = \alpha e^{2t} \begin{pmatrix} 1 \\ 1 \end{pmatrix} + \beta e^{-4t} \begin{pmatrix} 1 \\ -1 \end{pmatrix}.$$

Use this formula for solutions to show that the set of solutions to this system of differential equations is a vector subspace of $(\mathcal{C}^1)^2$.

5.2 CONSTRUCTION OF SUBSPACES

The principle of superposition shows that the set of all solutions to a homogeneous system of linear equations is closed under addition and scalar multiplication and is a subspace. Indeed, there are two ways to describe subspaces: first as solutions to linear systems, and second as the span of a set of vectors. We shall see that solving a homogeneous linear system of equations just means writing the solution set as the span of a finite set of vectors.

Solutions to Homogeneous Systems Form Subspaces

Definition 5.2.1   *Let A be an m × n matrix. The* null space *of A is the set of solutions to the homogeneous system of linear equations*

$$Ax = 0. \tag{5.2.1}$$

Lemma 5.2.2   *Let A be an m × n matrix. Then the null space of A is a subspace of* $\mathbf{R}^n$.

Proof: Suppose that x and y are solutions to (5.2.1). Then

$$A(x + y) = Ax + Ay = 0 + 0 = 0,$$

so $x + y$ is a solution of (5.2.1). Similarly for $r \in \mathbf{R}$,

$$A(rx) = rAx = r0 = 0,$$

so rx is a solution of (5.2.1). Thus $x + y$ and rx are in the null space of A, and the null space is closed under addition and scalar multiplication. So Theorem 5.1.4 implies that the null space is a subspace of the vector space $\mathbf{R}^n$.   ◆

Solutions to Linear Systems of Differential Equations Form Subspaces

Let C be an $n \times n$ matrix, and let W be the set of solutions to the linear system of ordinary differential equations

$$\frac{dx}{dt}(t) = Cx(t). \tag{5.2.2}$$

We will see later that a solution to (5.2.2) has coordinate functions $x_j(t)$ in C^1. The principle of superposition then shows that W is a subspace of $(C^1)^n$. Suppose $x(t)$ and $y(t)$ are solutions of (5.2.2). Then

$$\frac{d}{dt}(x(t) + y(t)) = \frac{dx}{dt}(t) + \frac{dy}{dt}(t) = Cx(t) + Cy(t) = C(x(t) + y(t)),$$

so $x(t) + y(t)$ is a solution of (5.2.2) and in W. A similar calculation shows that $rx(t)$ is also in W and that $W \subset (C^1)^n$ is a subspace.

Writing Solution Subspaces As a Span

The way we solve homogeneous systems of equations gives a second method for defining subspaces. For example, consider the system

$$Ax = 0,$$

where

$$A = \begin{pmatrix} 2 & 1 & 4 & 0 \\ -1 & 0 & 2 & 1 \end{pmatrix}.$$

The matrix A is row equivalent to the reduced echelon form matrix

$$E = \begin{pmatrix} 1 & 0 & -2 & -1 \\ 0 & 1 & 8 & 2 \end{pmatrix}.$$

Therefore $x = (x_1, x_2, x_3, x_4)$ is a solution of $Ex = 0$ if and only if $x_1 = 2x_3 + x_4$ and $x_2 = -8x_3 - 2x_4$. It follows that every solution of $Ex = 0$ can be written as

$$x = x_3 \begin{pmatrix} 2 \\ -8 \\ 1 \\ 0 \end{pmatrix} + x_4 \begin{pmatrix} 1 \\ -2 \\ 0 \\ 1 \end{pmatrix}.$$

Since row operations do not change the set of solutions, it follows that every solution of $Ax = 0$ has this form. We have also shown that every solution is generated by two vectors by use of vector addition and scalar multiplication. We say that this subspace is *spanned* by the two vectors

$$\begin{pmatrix} 2 \\ -8 \\ 1 \\ 0 \end{pmatrix} \quad \text{and} \quad \begin{pmatrix} 1 \\ -2 \\ 0 \\ 1 \end{pmatrix}.$$

For example, a calculation verifies that the vector

$$\begin{pmatrix} -1 \\ -2 \\ 1 \\ -3 \end{pmatrix}$$

is also a solution of $Ax = 0$. Indeed we may write it as

$$\begin{pmatrix} -1 \\ -2 \\ 1 \\ -3 \end{pmatrix} = \begin{pmatrix} 2 \\ -8 \\ 1 \\ 0 \end{pmatrix} - 3 \begin{pmatrix} 1 \\ -2 \\ 0 \\ 1 \end{pmatrix}. \tag{5.2.3}$$

Spans

Let $v_1, \ldots, v_k$ be a set of vectors in a vector space V. A vector $v \in V$ is a *linear combination* of $v_1, \ldots, v_k$ if

$$v = r_1 v_1 + \cdots + r_k v_k$$

for some scalars $r_1, \ldots, r_k$.

Definition 5.2.3   *The set of all linear combinations of the vectors $v_1, \ldots, v_k$ in a vector space V is the* span *of $v_1, \ldots, v_k$ and is denoted by* span$\{v_1, \ldots, v_k\}$.

For example, the vector on the left-hand side in (5.2.3) is a linear combination of the two vectors on the right-hand side.

The simplest example of a span is $\mathbf{R}^n$ itself. Let $v_j = e_j$, where $e_j \in \mathbf{R}^n$ is the vector with a 1 in the jth coordinate and 0 in all other coordinates. Then every vector $x = (x_1, \ldots, x_n) \in \mathbf{R}^n$ can be written as

$$x = x_1 e_1 + \cdots + x_n e_n.$$

It follows that

$$\mathbf{R}^n = \text{span}\{e_1, \ldots, e_n\}.$$

Similarly, the set span$\{e_1, e_3\} \subset \mathbf{R}^3$ is just the $x_1 x_3$-plane, since vectors in this span are

$$x_1 e_1 + x_3 e_3 = x_1(1, 0, 0) + x_3(0, 0, 1) = (x_1, 0, x_3).$$

Proposition 5.2.4   *Let V be a vector space and let $w_1, \ldots, w_k \in V$. Then $W = \text{span}\{w_1, \ldots, w_k\} \subset V$ is a subspace.*

Proof: Suppose $x, y \in W$. Then

$$x = r_1 w_1 + \cdots + r_k w_k$$
$$y = s_1 w_1 + \cdots + s_k w_k$$

for some scalars $r_1, \ldots, r_k$ and $s_1, \ldots, s_k$. It follows that

$$x + y = (r_1 + s_1)w_1 + \cdots + (r_k + s_k)w_k$$

and

$$rx = (rr_1)w_1 + \cdots + (rr_k)w_k$$

are both in span$\{w_1, \ldots, w_k\}$. Hence $W \subset V$ is closed under addition and scalar multiplication and is a subspace by Theorem 5.1.4.   ◆

For example, let

$$v = (2, 1, 0) \quad \text{and} \quad w = (1, 1, 1) \tag{5.2.4}$$

be vectors in $\mathbf{R}^3$. Then linear combinations of the vectors v and w have the form

$$\alpha v + \beta w = (2\alpha + \beta, \alpha + \beta, \beta)$$

for real numbers α and β. Note that every one of these vectors is a solution to the linear equation

$$x_1 - 2x_2 + x_3 = 0; \tag{5.2.5}$$

that is, the 1st coordinate minus twice the 2nd coordinate plus the 3rd coordinate equals 0. Moreover, you may verify that every solution of (5.2.5) is a linear combination of the vectors v and w in (5.2.4). Thus the set of solutions to the homogeneous linear equation (5.2.5) is a subspace, and that subspace can be written as the span of all linear combinations of the vectors v and w.

In this language we see that the process of solving a homogeneous system of linear equations is just the process of finding a set of vectors that span the subspace of all solutions. Indeed we can now restate Theorem 2.4.6. Recall that a matrix A has *rank* ℓ if it is row equivalent to a matrix in echelon form with ℓ nonzero rows.

Proposition 5.2.5   *Let A be an $m \times n$ matrix with rank ℓ. Then the null space of A is the span of $n - \ell$ vectors.*

We have now seen that there are two ways to describe subspaces: as solutions of homogeneous systems of linear equations and as a span of a set of vectors, the *spanning set*. Much of linear algebra is concerned with determining how one goes from one description of a subspace to the other.

HAND EXERCISES

In Exercises 1–4, a single equation in three variables is given. For each equation write the subspace of solutions in $\mathbf{R}^3$ as the span of two vectors in $\mathbf{R}^3$.

1. $4x - 2y + z = 0$

2. $x - y + 3z = 0$

3. $x + y + z = 0$

4. $y = z$

In Exercises 5–8, each of the given matrices is in reduced echelon form. Write solutions of the corresponding homogeneous system of linear equations as a span of vectors.

5. $A = \begin{pmatrix} 1 & 2 & 0 & 1 & 0 \\ 0 & 0 & 1 & 4 & 0 \\ 0 & 0 & 0 & 0 & 1 \end{pmatrix}$
 6. $B = \begin{pmatrix} 1 & 3 & 0 & 5 \\ 0 & 0 & 1 & 2 \end{pmatrix}$

7. $A = \begin{pmatrix} 1 & 0 & 2 \\ 0 & 1 & 1 \end{pmatrix}$
 8. $B = \begin{pmatrix} 1 & -1 & 0 & 5 & 0 & 0 \\ 0 & 0 & 1 & 2 & 0 & 2 \\ 0 & 0 & 0 & 0 & 1 & 2 \end{pmatrix}$

9. Write a system of two linear equations of the form $Ax = 0$, where A is a 2×4 matrix whose subspace of solutions in $\mathbf{R}^4$ is the span of the two vectors

$$v_1 = \begin{pmatrix} 1 \\ -1 \\ 0 \\ 0 \end{pmatrix} \quad \text{and} \quad v_2 = \begin{pmatrix} 0 \\ 0 \\ 1 \\ -1 \end{pmatrix}.$$

10. Write the matrix $A = \begin{pmatrix} 2 & 2 \\ -3 & 0 \end{pmatrix}$ as a linear combination of the matrices

$$B = \begin{pmatrix} 1 & 1 \\ 0 & 0 \end{pmatrix} \quad \text{and} \quad C = \begin{pmatrix} 0 & 0 \\ 1 & 0 \end{pmatrix}.$$

11. Is $(2, 20, 0)$ in the span of $w_1 = (1, 1, 3)$ and $w_2 = (1, 4, 2)$? Answer this question by setting up a system of linear equations and solving that system by row reducing the associated augmented matrix.

In Exercises 12–15, let $W \subset C^1$ be the subspace spanned by the two polynomials $x_1(t) = 1$ and $x_2(t) = t^2$. For the given function $y(t)$ decide whether or not $y(t)$ is an element of W. Furthermore, if $y(t) \in W$, determine whether the set $\{y(t), x_2(t)\}$ is a spanning set for W.

12. $y(t) = 1 - t^2$
13. $y(t) = t^4$
14. $y(t) = \sin t$
15. $y(t) = 0.5t^2$

16. Let $W \subset \mathbf{R}^4$ be the subspace that is spanned by the vectors

$$w_1 = (-1, 2, 1, 5) \quad \text{and} \quad w_2 = (2, 1, 3, 0).$$

Find a linear system of two equations such that $W = \text{span}\{w_1, w_2\}$ is the set of solutions of this system.

17. Let V be a vector space and let $v \in V$ be a nonzero vector. Show that

$$\text{span}\{v, v\} = \text{span}\{v\}.$$

18. Let V be a vector space and let $v, w \in V$ be vectors. Show that

$$\text{span}\{v, w\} = \text{span}\{v, w, v + 3w\}.$$

19. Let $W = \text{span}\{w_1, \ldots, w_k\}$ be a subspace of the vector space V, and let $w_{k+1} \in W$ be another vector. Prove that $W = \text{span}\{w_1, \ldots, w_{k+1}\}$.

20. Let $Ax = b$ be a system of m linear equations in n unknowns, and let $r = \text{rank}(A)$ and $s = \text{rank}(A|b)$. Suppose that this system has a unique solution. What can you say about the relative magnitudes of m, n, r, and s?

5.3   SPANNING SETS AND MATLAB

In this section we discuss:

- how to find a spanning set for the subspace of solutions to a homogeneous system of linear equations using the MATLAB command `null`, and
- how to determine when a vector is in the subspace spanned by a set of vectors using the MATLAB command `rref`.

Spanning Sets for Homogeneous Linear Equations

In Chapter 2 we saw how to use Gaussian elimination, back substitution, and MATLAB to compute solutions to a system of linear equations. For systems of homogeneous equations, MATLAB provides a command to find a spanning set for the subspace of solutions. That command is `null`. For example, if we type

```
A = [2 1 4 0; −1 0 2 1]
B = null(A)
```

then we obtain

```
B =
     0.4830          0
    −0.4140     0.8729
    −0.1380    −0.2182
     0.7591     0.4364
```

The two columns of the matrix B span the set of solutions of the equation $Ax = 0$. In particular, the vector $(2, -8, 1, 0)$ is a solution to $Ax = 0$ and is therefore a linear combination of the column vectors of B. Indeed, type

```
4.1404*B(:,1)−7.2012*B(:,2)
```

and observe that this linear combination is the desired one.

Next we describe how to find the coefficients `4.1404` and `-7.2012` by showing that these coefficients themselves are solutions to another system of linear equations.

When Is a Vector in a Span?

Let $w_1, \ldots, w_k$ and v be vectors in $\mathbf{R}^n$. We now describe a method that allows us to decide whether v is in span$\{w_1, \ldots, w_k\}$. To answer this question one has to solve a system of n linear equations in k unknowns. The unknowns correspond to the coefficients in the linear combination of the vectors $w_1, \ldots, w_k$ that gives v.

Let us be more precise. The vector v is in span$\{w_1, \ldots, w_k\}$ if and only if there are constants $r_1, \ldots, r_k$ such that the equation

$$r_1 w_1 + \cdots + r_k w_k = v \tag{5.3.1}$$

is valid. Define the $n \times k$ matrix A as the one having $w_1, \dots, w_k$ as its columns; that is,

$$A = (w_1 | \cdots | w_k). \qquad (5.3.2)$$

Let r be the k-vector

$$r = \begin{pmatrix} r_1 \\ \vdots \\ r_k \end{pmatrix}.$$

Then we may rewrite equation (5.3.1) as

$$Ar = v. \qquad (5.3.3)$$

To summarize:

Lemma 5.3.1   *Let $w_1, \dots, w_k$ and v be vectors in $\mathbf{R}^n$. Then v is in $span\{w_1, \dots, w_k\}$ if and only if the system of linear equations (5.3.3) has a solution where A is the $n \times k$ defined in (5.3.2).*

To solve (5.3.3) we row reduce the augmented matrix $(A|v)$. For example, is $v = (2, 1)$ in the span of $w_1 = (1, 1)$ and $w_2 = (1, -1)$? That is, do there exist scalars r_1 and r_2 such that

$$r_1 \begin{pmatrix} 1 \\ 1 \end{pmatrix} + r_2 \begin{pmatrix} 1 \\ -1 \end{pmatrix} = \begin{pmatrix} 2 \\ 1 \end{pmatrix}?$$

As noted, we can rewrite this equation as

$$\begin{pmatrix} 1 & 1 \\ 1 & -1 \end{pmatrix} \begin{pmatrix} r_1 \\ r_2 \end{pmatrix} = \begin{pmatrix} 2 \\ 1 \end{pmatrix}.$$

We can solve this equation by row reducing the augmented matrix

$$\begin{pmatrix} 1 & 1 & | & 2 \\ 1 & -1 & | & 1 \end{pmatrix}$$

to obtain

$$\begin{pmatrix} 1 & 0 & | & \frac{3}{2} \\ 0 & 1 & | & \frac{1}{2} \end{pmatrix}.$$

So $v = \frac{3}{2}w_1 + \frac{1}{2}w_2$.

Row reduction to reduced echelon form has been preprogrammed in the MATLAB command `rref`. Consider the following example: Let

$$w_1 = (2, 0, -1, 4) \quad \text{and} \quad w_2 = (2, -1, 0, 2) \qquad \text{(5.3.4)}$$

and ask whether $v = (-2, 4, -3, 4)$ is in span$\{w_1, w_2\}$.

In MATLAB load the matrix A having w_1 and w_2 as its columns and the vector v by typing e5_3_5:

$$A = \begin{pmatrix} 2 & 2 \\ 0 & -1 \\ -1 & 0 \\ 4 & 2 \end{pmatrix} \quad \text{and} \quad v = \begin{pmatrix} -2 \\ 4 \\ -3 \\ 4 \end{pmatrix}. \qquad \text{(5.3.5)*}$$

We can solve the system of equations using MATLAB. First form the augmented matrix by typing

```
aug = [A v]
```

Then solve the system by typing `rref(aug)` to obtain

```
ans =
    1    0    3
    0    1   -4
    0    0    0
    0    0    0
```

It follows that $(r_1, r_2) = (3, -4)$ is a solution and $v = 3w_1 - 4w_2$.

Now we change the 4th entry in v slightly by typing `v(4) = 4.01`. There is no solution to the system of equations

$$Ar = \begin{pmatrix} -2 \\ 4 \\ -3 \\ 4.01 \end{pmatrix},$$

as we now show. Type

```
aug = [A v]
rref(aug)
```

which yields

```
ans =
    1    0    0
    0    1    0
    0    0    1
    0    0    0
```

This matrix corresponds to an inconsistent system; thus v is no longer in the span of w_1 and w_2.

COMPUTER EXERCISES

In Exercises 1–3, use the null command in MATLAB to find all the solutions of the linear system of equations $Ax = 0$.

1.

$$A = \begin{pmatrix} -4 & 0 & -4 & 3 \\ -4 & 1 & -1 & 1 \end{pmatrix}$$

(5.3.6)*

2.

$$A = \begin{pmatrix} 1 & 2 \\ 1 & 0 \\ 3 & -2 \end{pmatrix}$$

(5.3.7)*

3.

$$A = \begin{pmatrix} 1 & 1 & 2 \\ -1 & 2 & -1 \end{pmatrix}$$

(5.3.8)*

4. Use the null command in MATLAB to verify your answers to Exercises 5 and 6 at the end of Section 5.2.

5. Use row reduction to find the solutions to $Ax = 0$, where A is given in (5.3.6). Does your answer agree with the MATLAB answer using null? If not, explain why.

In Exercises 6–8, let $W \subset \mathbf{R}^5$ be the subspace spanned by the vectors

$$w_1 = (2, 0, -1, 3, 4), \quad w_2 = (1, 0, 0, -1, 2), \quad w_3 = (0, 1, 0, 0, -1).$$

(5.3.9)*

Use MATLAB to decide whether the given vectors are elements of W.

6. $v_1 = (2, 1, -2, 8, 3)$ **7.** $v_2 = (-1, 12, 3, -14, -1)$

8. $v_3 = (-1, 12, 3, -14, -14)$

5.4 LINEAR DEPENDENCE AND LINEAR INDEPENDENCE

An important question in linear algebra concerns finding spanning sets for subspaces having the smallest number of vectors. Let $w_1, \ldots, w_k$ be vectors in a vector space V and let $W = \text{span}\{w_1, \ldots, w_k\}$. Suppose that W is generated by a subset of these k vectors. Indeed, suppose that the kth vector is redundant in the sense that $W = \text{span}\{w_1, \ldots, w_{k-1}\}$. Since $w_k \in W$, this is possible only if w_k is a linear combination of the $k - 1$ vectors $w_1, \ldots, w_{k-1}$—that is, only if

$$w_k = r_1 w_1 + \cdots + r_{k-1} w_{k-1}.$$

(5.4.1)

Definition 5.4.1   *Let $w_1, \ldots, w_k$ be vectors in the vector space V. The set $\{w_1, \ldots, w_k\}$ is* linearly dependent *if one of the vectors w_j can be written as a linear combination of the remaining $k - 1$ vectors.*

Note that when $k = 1$, the phrase "$\{w_1\}$ is linearly dependent" means that $w_1 = 0$.
   If we set $r_k = -1$, then we may rewrite (5.4.1) as

$$r_1 w_1 + \cdots + r_{k-1} w_{k-1} + r_k w_k = 0.$$

It follows that:

Lemma 5.4.2   *The set of vectors $\{w_1, \ldots, w_k\}$ is linearly dependent if and only if there exist scalars $r_1, \ldots, r_k$ such that*

   (a)   at least one of the r_j is nonzero, and
   (b) $r_1 w_1 + \cdots + r_k w_k = 0$.

For example, the vectors $w_1 = (2, 4, 7)$, $w_2 = (5, 1, -1)$, and $w_3 = (1, -7, -15)$ are linearly dependent since $2w_1 - w_2 + w_3 = 0$.

Definition 5.4.3   *A set of k vectors $\{w_1, \ldots, w_k\}$ is* linearly independent *if none of the k vectors can be written as a linear combination of the other $k - 1$ vectors.*

Since linear independence means *not* linearly dependent, Lemma 5.4.2 can be rewritten as:

Lemma 5.4.4   *The set of vectors $\{w_1, \ldots, w_k\}$ is linearly independent if and only if whenever*

$$r_1 w_1 + \cdots + r_k w_k = 0,$$

it follows that

$$r_1 = r_2 = \cdots = r_k = 0.$$

Let e_j be the vector in $\mathbf{R}^n$ whose jth component is 1 and all of whose other components are 0. The set of vectors $e_1, \ldots, e_n$ is the simplest example of a set of linearly independent vectors in $\mathbf{R}^n$. We use Lemma 5.4.4 to verify independence by supposing that

$$r_1 e_1 + \cdots + r_n e_n = 0.$$

A calculation shows that

$$0 = r_1 e_1 + \cdots + r_n e_n = (r_1, \ldots, r_n).$$

It follows that each r_j equals 0, and the vectors $e_1, \ldots, e_n$ are linearly independent.

Deciding Linear Dependence and Linear Independence

Deciding whether a set of k vectors in $\mathbf{R}^n$ is linearly dependent or linearly independent is equivalent to solving a system of linear equations. Let $w_1, \ldots, w_k$ be vectors in $\mathbf{R}^n$, and view these vectors as column vectors. Let

$$A = (w_1 | \cdots | w_k) \tag{5.4.2}$$

be the $n \times k$ matrix whose columns are the vectors w_j. Then a vector

$$R = \begin{pmatrix} r_1 \\ \vdots \\ r_k \end{pmatrix}$$

is a solution to the system of equations $AR = 0$ precisely when

$$r_1 w_1 + \cdots + r_k w_k = 0. \tag{5.4.3}$$

If there is a nonzero solution R to $AR = 0$, then the vectors $\{w_1, \ldots, w_k\}$ are linearly dependent; if the only solution to $AR = 0$ is $R = 0$, then the vectors are linearly independent.

The preceding discussion is summarized by:

Lemma 5.4.5   *The vectors $w_1, \ldots, w_k$ in $\mathbf{R}^n$ are linearly dependent if the null space of the $n \times k$ matrix A defined in (5.4.2) is nonzero and linearly independent if the null space of A is 0.*

A Simple Example of Linear Independence with Two Vectors

The two vectors

$$w_1 = \begin{pmatrix} 2 \\ -8 \\ 1 \\ 0 \end{pmatrix} \quad \text{and} \quad w_2 = \begin{pmatrix} 1 \\ -2 \\ 0 \\ 1 \end{pmatrix}$$

are linearly independent. To see this suppose that $r_1 w_1 + r_2 w_2 = 0$. If we use the components of w_1 and w_2, this equality is equivalent to the system of four equations

$$2r_1 + r_2 = 0, \quad -8r_1 - 2r_2 = 0, \quad r_1 = 0, \quad \text{and} \quad r_2 = 0.$$

In particular, $r_1 = r_2 = 0$; hence w_1 and w_2 are linearly independent.

Using MATLAB to Decide Linear Dependence

Suppose that we want to determine whether or not the vectors

$$w_1 = \begin{pmatrix} 1 \\ 2 \\ -1 \\ 3 \\ 5 \end{pmatrix}, \quad w_2 = \begin{pmatrix} -1 \\ 1 \\ 4 \\ -2 \\ 0 \end{pmatrix}, \quad w_3 = \begin{pmatrix} 1 \\ 1 \\ -1 \\ 3 \\ 12 \end{pmatrix}, \quad w_4 = \begin{pmatrix} 0 \\ 4 \\ 3 \\ 1 \\ -2 \end{pmatrix} \qquad \textbf{(5.4.4)*}$$

are linearly dependent. After typing e5_4_4 in MATLAB, form the 5×4 matrix A by typing

```
A = [w1 w2 w3 w4]
```

Determine whether there is a nonzero solution to $AR = 0$ by typing

```
null(A)
```

The response from MATLAB is

```
ans =
    -0.7559
    -0.3780
     0.3780
     0.3780
```

showing that there is a nonzero solution to $AR = 0$ and the vectors w_j are linearly dependent. Indeed, this solution for R shows that we can solve for w_1 in terms of w_2, w_3, w_4. We can now ask whether or not w_2, w_3, w_4 are linearly dependent. To answer this question, form the matrix

```
B = [w2 w3 w4]
```

and type null(B) to obtain

```
ans =
    Empty matrix: 3-by-0
```

showing that the only solution to $BR = 0$ is the zero solution $R = 0$. Thus w_2, w_3, w_4 are linearly independent. For these particular vectors, any three of the four are linearly independent.

HAND EXERCISES

1. Let w be a vector in the vector space V. Show that the sets of vectors $\{w, 0\}$ and $\{w, -w\}$ are linearly dependent.

2. For which values of b are the vectors $(1, b)$ and $(3, -1)$ linearly independent?

3. Let

$$u_1 = (1, -1, 1), \quad u_2 = (2, 1, -2), \quad u_3 = (10, 2, -6).$$

Is the set $\{u_1, u_2, u_3\}$ linearly dependent or linearly independent?

4. For which values of b are the vectors $(1, b, 2b)$ and $(2, 1, 4)$ linearly independent?

5. Show that the polynomials $p_1(t) = 2 + t$, $p_2(t) = 1 + t^2$, and $p_3(t) = t - t^2$ are linearly independent vectors in the vector space $\mathcal{C}^1$.

6. Show that the functions $f_1(t) = \sin t$, $f_2(t) = \cos t$, and $f_3(t) = \cos\left(t + \frac{\pi}{3}\right)$ are linearly dependent vectors in $\mathcal{C}^1$.

7. Suppose that the three vectors $u_1, u_2, u_3 \in \mathbf{R}^n$ are linearly independent. Show that the set

$$\{u_1 + u_2, u_2 + u_3, u_3 + u_1\}$$

is also linearly independent.

COMPUTER EXERCISES

In Exercises 8–10, determine whether the given sets of vectors are linearly independent or linearly dependent.

8.

$$v_1 = (2, 1, 3, 4), \quad v_2 = (-4, 2, 3, 1), \quad v_3 = (2, 9, 21, 22) \qquad \text{(5.4.5)*}$$

9.

$$w_1 = (1, 2, 3), \quad w_2 = (2, 1, 5), \quad w_3 = (-1, 2, -4) \quad w_4 = (0, 2, -1) \qquad \text{(5.4.6)*}$$

10.

$$x_1 = (3, 4, 1, 2, 5), \quad x_2 = (-1, 0, 3, -2, 1), \quad x_3 = (2, 4, -3, 0, 2) \qquad \text{(5.4.7)*}$$

11. Perform the following experiments:

(a) Use MATLAB to choose randomly three column vectors in $\mathbf{R}^3$. The MATLAB commands to choose these vectors are:

```
y1 = rand(3,1)
y2 = rand(3,1)
y3 = rand(3,1)
```

Use the methods of this section to determine whether these vectors are linearly independent or linearly dependent.

(b) Now perform this exercise five times and record the number of times a linearly independent set of vectors is chosen and the number of times a linearly dependent set is chosen.

(c) Repeat the experiment in (b)—but this time randomly choose four vectors in $\mathbf{R}^3$ to be in your set.

5.5   DIMENSION AND BASES

The minimum number of vectors that span a vector space has special significance.

Definition 5.5.1   *The vector space V has* finite dimension *if V is the span of a finite number of vectors. If V has finite dimension, then the smallest number of vectors that span V is called the* dimension *of V and is denoted by* dim *V*.

For example, recall that e_j is the vector in $\mathbf{R}^n$ whose jth component is 1 and all of whose other components are 0. Let $x = (x_1, \ldots, x_n)$ be in $\mathbf{R}^n$. Then

$$x = x_1 e_1 + \cdots + x_n e_n. \tag{5.5.1}$$

Since every vector in $\mathbf{R}^n$ is a linear combination of the vectors $e_1, \ldots, e_n$, it follows that $\mathbf{R}^n = \text{span}\{e_1, \ldots, e_n\}$. Thus $\mathbf{R}^n$ is finite dimensional. Moreover, the dimension of $\mathbf{R}^n$ is at most n, since $\mathbf{R}^n$ is spanned by n vectors. It seems unlikely that $\mathbf{R}^n$ could be spanned by fewer than n vectors—but this point needs to be proved.

An Example of a Vector Space That Is Not Finite Dimensional

Next we discuss an example of a vector space that does not have finite dimension. Consider the vector space $\mathcal{C}^1$ and the subspace $\mathcal{P}$ consisting of polynomials of all degrees. We show that $\mathcal{P}$ is not the span of a finite number of vectors and hence that $\mathcal{P}$ does not have finite dimension. Let $p_1(t), p_2(t), \ldots, p_k(t)$ be a set of k polynomials and let d be the maximum degree of these k polynomials. Then every polynomial in the span of $p_1(t), \ldots, p_k(t)$ has degree less than or equal to d. In particular, $p(t) = t^{d+1}$ is a polynomial that is not in the span of $p_1(t), \ldots, p_k(t)$ and $\mathcal{P}$ is not spanned by finitely many vectors.

Bases and the Main Theorem

Definition 5.5.2   *Let $\mathcal{B} = \{w_1, \ldots, w_k\}$ be a set of vectors in a vector space W. The subset $\mathcal{B}$ is a* basis *for W if $\mathcal{B}$ is a spanning set for W with the smallest number of elements in a spanning set for W.*

It follows that if $\{w_1, \ldots, w_k\}$ is a basis for W, then $k = \dim W$. The main theorem about bases is:

Theorem 5.5.3   *A set of vectors $\mathcal{B} = \{w_1, \ldots, w_k\}$ in a vector space W is a basis for W if and only if the set $\mathcal{B}$ is linearly independent and spans W.*

Remark: The importance of Theorem 5.5.3 is that we can show that a set of vectors is a basis by verifying spanning and linear independence. We never have to check directly that the spanning set has the minimum number of vectors for a spanning set.

For example, we have shown previously that the set of vectors $\{e_1, \ldots, e_n\}$ in $\mathbf{R}^n$ is linearly independent and spans $\mathbf{R}^n$. It follows from Theorem 5.5.3 that this set is a basis and that the dimension of $\mathbf{R}^n$ is n. In particular, $\mathbf{R}^n$ cannot be spanned by fewer than n vectors.

The proof of Theorem 5.5.3 is given in Section 5.6.

Consequences of Theorem 5.5.3

We discuss two applications of Theorem 5.5.3. First, we use this theorem to derive a way of determining the dimension of the subspace spanned by a finite number of vectors. Second, we show that the dimension of the subspace of solutions to a homogeneous system of linear equation $Ax = 0$ is $n - \text{rank}(A)$, where A is an $m \times n$ matrix.

Computing the Dimension of a Span

We show that the dimension of a span of vectors can be found using elementary row operations on M.

Lemma 5.5.4 *Let $w_1, \ldots, w_k$ be k row vectors in $\mathbf{R}^n$, and let $W = span\{w_1, \ldots, w_k\} \subset \mathbf{R}^n$. Define*

$$M = \begin{pmatrix} w_1 \\ \vdots \\ w_k \end{pmatrix}$$

to be the matrix whose rows are the w_js. Then

$$\dim(W) = \text{rank}(M). \tag{5.5.2}$$

Proof: To verify (5.5.2), observe that the span of $w_1, \ldots, w_k$ is unchanged by

   (a) swapping w_i and w_j,

   (b) multiplying w_i by a nonzero scalar, and

   (c) adding a multiple of w_i to w_j.

That is, if we perform elementary row operations on M, the vector space spanned by the rows of M does not change. So we may perform elementary row operations on M until we arrive at the matrix E in reduced echelon form. Suppose that $\ell = \text{rank}(M)$; that is, suppose that ℓ is the number of nonzero rows in E. Then

$$E = \begin{pmatrix} v_1 \\ \vdots \\ v_\ell \\ 0 \\ \vdots \\ 0 \end{pmatrix},$$

where the v_j are the nonzero rows in the reduced echelon form matrix.

We claim that the vectors $v_1, \ldots, v_\ell$ are linearly independent. It then follows from Theorem 5.5.3 that $\{v_1, \ldots, v_\ell\}$ is a basis for W and that the dimension of W is ℓ. To verify the claim, suppose

$$a_1 v_1 + \cdots + a_\ell v_\ell = 0. \tag{5.5.3}$$

We show that a_i must equal 0 as follows: In the ith row, the pivot must occur in some column—say, in the jth column. It follows that the jth entry in the vector on the left-hand side of (5.5.3) is

$$0a_1 + \cdots + 0a_{i-1} + 1a_i + 0a_{i+1} + \cdots + 0a_\ell = a_i,$$

since all entries in the jth column of E other than the pivot must be 0 because E is in reduced echelon form. $\blacklozenge$

For instance, let $W = \mathrm{span}\{w_1, w_2, w_3\}$ in $\mathbf{R}^4$, where

$$w_1 = (3, -2, 1, -1), \quad w_2 = (1, 5, 10, 12), \quad w_3 = (1, -12, -19, -25). \tag{5.5.4}*$$

To compute $\dim W$ in MATLAB, type e5_5_4 to load the vectors and type

    M = [w1; w2; w3]

Row reduction of the matrix M in MATLAB leads to the reduced echelon form matrix

```
ans =
     1.0000          0     1.4706     1.1176
          0     1.0000     1.7059     2.1765
          0          0          0          0
```

indicating that the dimension of the subspace W is two, and therefore $\{w_1, w_2, w_3\}$ is not a basis of W. Alternatively, we can use the MATLAB command rank(M) to compute the rank of M and the dimension of the span W.

However, if we change one of the entries in w_3—for instance, w3(3)=-18—then indeed the command rank([w1;w2;w3]) gives the answer three, indicating that for this choice of vectors $\{w_1, w_2, w_3\}$ is a basis for $\mathrm{span}\{w_1, w_2, w_3\}$.

Solutions to Homogeneous Systems Revisited

We return to our discussions in Chapter 2 on solving linear equations. Recall that we can write all solutions to the system of homogeneous equations $Ax = 0$ in terms of a few parameters, and that the null space of A is the subspace of solutions (see Definition 5.2.1). More precisely, Proposition 5.2.5 states that the number of parameters needed is $n - \mathrm{rank}(A)$, where n is the number of variables in the homogeneous system. We claim that the dimension of the null space is exactly $n - \mathrm{rank}(A)$.

For example, consider the reduced echelon form 3×7 matrix

$$A = \begin{pmatrix} 1 & -4 & 0 & 2 & -3 & 0 & 8 \\ 0 & 0 & 1 & 3 & 2 & 0 & 4 \\ 0 & 0 & 0 & 0 & 0 & 1 & 2 \end{pmatrix} \qquad (5.5.5)$$

that has rank three. Suppose that the unknowns for this system of equations are $x_1, \ldots, x_7$. We can solve the equations associated with A by solving the first equation for x_1, the second equation for x_3, and the third equation for x_6, as follows:

$$x_1 = 4x_2 - 2x_4 + 3x_5 - 8x_7$$
$$x_3 = -3x_4 - 2x_5 - 4x_7$$
$$x_6 = -2x_7.$$

Thus all solutions to this system of equations have the form

$$\begin{pmatrix} 4x_2 - 2x_4 + 3x_5 - 8x_7 \\ x_2 \\ -3x_4 - 2x_5 - 4x_7 \\ x_4 \\ x_5 \\ -2x_7 \\ x_7 \end{pmatrix} = x_2 \begin{pmatrix} 4 \\ 1 \\ 0 \\ 0 \\ 0 \\ 0 \\ 0 \end{pmatrix} + x_4 \begin{pmatrix} -2 \\ 0 \\ -3 \\ 1 \\ 0 \\ 0 \\ 0 \end{pmatrix} + x_5 \begin{pmatrix} 3 \\ 0 \\ -2 \\ 0 \\ 1 \\ 0 \\ 0 \end{pmatrix} + x_7 \begin{pmatrix} -8 \\ 0 \\ -4 \\ 0 \\ 0 \\ -2 \\ 1 \end{pmatrix}. \qquad (5.5.6)$$

We can rewrite the right-hand side of (5.5.6) as a linear combination of four vectors w_2, w_4, w_5, w_7:

$$x_2 w_2 + x_4 w_4 + x_5 w_5 + x_7 w_7. \qquad (5.5.7)$$

This calculation shows that the null space of A, which is $W = \{x \in \mathbf{R}^7 : Ax = 0\}$, is spanned by the four vectors w_2, w_4, w_5, w_7. Moreover, this same calculation shows that the four vectors are linearly independent. From the left-hand side of (5.5.6) we see that if this linear combination sums to 0, then $x_2 = x_4 = x_5 = x_7 = 0$. It follows from Theorem 5.5.3 that dim $W = 4$.

Definition 5.5.5 *The nullity of A is the dimension of the null space of A.*

Theorem 5.5.6 *Let A be an $m \times n$ matrix. Then*

$$\text{nullity}(A) + \text{rank}(A) = n.$$

Proof: Neither the rank nor the null space of A is changed by elementary row operations, so we can assume that A is in reduced echelon form. The rank of A is the number of nonzero rows in the reduced echelon form matrix. Proposition 5.2.5 states that the null space is spanned by p vectors, where $p = n - \text{rank}(A)$. We must show that these vectors are linearly independent.

Let $j_1, \ldots, j_p$ be the columns of A that do not contain pivots. In example (5.5.5), $p = 4$ and

$$j_1 = 2, \quad j_2 = 4, \quad j_3 = 5, \quad j_4 = 7.$$

After solving for the variables corresponding to pivots, we find that the spanning set of the null space consists of p vectors in $\mathbf{R}^n$, which we label as $\{w_{j_1}, \ldots, w_{j_p}\}$; see (5.5.6). Note that the j_mth entry of w_{j_m} is 1, while the j_mth entry in all of the other $p - 1$ vectors is 0. Again, see (5.5.6) as an example that supports this statement. It follows that the set of spanning vectors is a linearly independent set. That is, suppose that

$$r_1 w_{j_1} + \cdots + r_p w_{j_p} = 0.$$

From the j_mth entry in this equation, it follows that $r_m = 0$, and the vectors are linearly independent. ◆

Theorem 5.5.6 has an interesting and useful interpretation. We have seen in the preceding subsection that the rank of a matrix A is just the number of linearly independent rows in A. In linear systems each row of the coefficient matrix corresponds to a linear equation. Thus the rank of A may be thought of as the number of independent equations in a system of linear equations. This theorem just states that the space of solutions loses a dimension for each independent equation.

HAND EXERCISES

1. Show that $\mathcal{U} = \{u_1, u_2, u_3\}$, where

$$u_1 = (1, 1, 0), \quad u_2 = (0, 1, 0), \quad u_3 = (-1, 0, 1),$$

is a basis for $\mathbf{R}^3$.

2. Let $S = \text{span}\{v_1, v_2, v_3\}$, where

$$v_1 = (1, 0, -1, 0), \quad v_2 = (0, 1, 1, 1), \quad v_3 = (5, 4, -1, 4).$$

Find the dimension of S and find a basis for S.

3. Find a basis for the null space of

$$A = \begin{pmatrix} 1 & 0 & -1 & 2 \\ 1 & -1 & 0 & 0 \\ 4 & -5 & 1 & -2 \end{pmatrix}.$$

What is the dimension of the null space of A?

4. Show that the set V of all 2×2 matrices is a vector space. Show that the dimension of V is four by finding a basis of V with four elements. Show that the space $M(m, n)$ of all $m \times n$ matrices is also a vector space. What is $\dim M(m, n)$?

5. Show that the set P_n of all polynomials of degree less than or equal to n is a subspace of $\mathcal{C}^1$. What is dim P_2? What is dim P_n?

6. Let P_3 be the vector space of polynomials of degree at most three in one variable t. Let $p(t) = t^3 + a_2 t^2 + a_1 t + a_0$, where $a_0, a_1, a_2 \in \mathbf{R}$ are fixed constants. Show that

$$\left\{ p, \frac{dp}{dt}, \frac{d^2 p}{dt^2}, \frac{d^3 p}{dt^3} \right\}$$

is a basis for P_3.

7. Let $u \in \mathbf{R}^n$ be a nonzero row vector.
   (a) Show that the $n \times n$ matrix $A = u^t u$ is symmetric and that $\mathrm{rank}(A) = 1$.
      Hint   Begin by showing that $Av^t = 0$ for every vector $v \in \mathbf{R}^n$ that is perpendicular to u and that Au^t is a nonzero multiple of u^t.
   (b) Show that the matrix $P = I_n + u^t u$ is invertible.
      Hint   Show that $\mathrm{rank}(P) = n$.

5.6   THE PROOF OF THE MAIN THEOREM

We begin the proof of Theorem 5.5.3 with two lemmas on linearly independent and spanning sets.

Lemma 5.6.1   *Let $\{w_1, \ldots, w_k\}$ be a set of vectors in a vector space V, and let W be the subspace spanned by these vectors. Then there is a linearly independent subset of $\{w_1, \ldots, w_k\}$ that also spans W.*

Proof: If $\{w_1, \ldots, w_k\}$ is linearly independent, then the lemma is proved. If not, then the set $\{w_1, \ldots, w_k\}$ is linearly dependent. If this set is linearly dependent, then at least one of the vectors is a linear combination of the others. By renumbering if necessary, we can assume that w_k is a linear combination of $w_1, \ldots, w_{k-1}$; that is,

$$w_k = a_1 w_1 + \cdots + a_{k-1} w_{k-1}.$$

Now suppose that $w \in W$. Then

$$w = b_1 w_1 + \cdots + b_k w_k.$$

It follows that

$$w = (b_1 + b_k a_1) w_1 + \cdots + (b_{k-1} + b_k a_{k-1}) w_{k-1}$$

and that $W = \mathrm{span}\{w_1, \ldots, w_{k-1}\}$. If the vectors $w_1, \ldots, w_{k-1}$ are linearly independent, then the proof of the lemma is complete. If not, continue inductively until a linearly independent subset of the w_j that also spans W is found.   ◆

The important point in proving that linear independence together with spanning implies that we have a basis is discussed in the next lemma.

Lemma 5.6.2   *Let W be an m-dimensional vector space and let $k > m$ be an integer. Then any set of k vectors in W is linearly dependent.*

Proof: Since the dimension of W is m, we know that this vector space can be written as $W = \text{span}\{v_1, \ldots, v_m\}$. Moreover, Lemma 5.6.1 implies that the vectors $v_1, \ldots, v_m$ are linearly independent. Suppose that $\{w_1, \ldots, w_k\}$ is another set of vectors, where $k > m$. We have to show that the vectors $w_1, \ldots, w_k$ are linearly dependent; that is, we must show that there exist scalars $r_1, \ldots, r_k$ not all of which are 0 that satisfy

$$r_1 w_1 + \cdots + r_k w_k = 0. \tag{5.6.1}$$

We find these scalars by solving a system of linear equations, as we now show.

The fact that W is spanned by the vectors v_j implies that

$$w_1 = a_{11} v_1 + \cdots + a_{m1} v_m$$
$$w_2 = a_{12} v_1 + \cdots + a_{m2} v_m$$
$$\vdots$$
$$w_k = a_{1k} v_1 + \cdots + a_{mk} v_m.$$

It follows that $r_1 w_1 + \cdots + r_k w_k$ equals

$$r_1 (a_{11} v_1 + \cdots + a_{m1} v_m) +$$
$$r_2 (a_{12} v_1 + \cdots + a_{m2} v_m) + \cdots +$$
$$r_k (a_{1k} v_1 + \cdots + a_{mk} v_m).$$

Rearranging terms leads to the expression

$$(a_{11} r_1 + \cdots + a_{1k} r_k) v_1 +$$
$$(a_{21} r_1 + \cdots + a_{2k} r_k) v_2 + \cdots + \tag{5.6.2}$$
$$(a_{m1} r_1 + \cdots + a_{mk} r_k) v_m.$$

Thus (5.6.1) is valid if and only if (5.6.2) sums to 0. Since the set $\{v_1, \ldots, v_m\}$ is linearly independent, (5.6.2) can equal 0 if and only if

$$a_{11} r_1 + \cdots + a_{1k} r_k = 0$$
$$a_{21} r_1 + \cdots + a_{2k} r_k = 0$$
$$\vdots$$
$$a_{m1} r_1 + \cdots + a_{mk} r_k = 0.$$

Since $m < k$, Theorem 2.4.6 implies that this system of homogeneous linear equations always has a nonzero solution $r = (r_1, \ldots, r_k)$—from which it follows that the w_i are linearly dependent.   ◆

Corollary 5.6.3   *Let V be a vector space of dimension n, and let $\{u_1, \ldots, u_k\}$ be a linearly independent set of vectors in V. Then $k \le n$.*

Proof: If $k > n$, then Lemma 5.6.2 implies that $\{u_1, \ldots, u_k\}$ is linearly dependent. Since we have assumed that this set is linearly independent, it follows that $k \le n$.   ◆

Proof of Theorem 5.5.3:   Suppose that $\mathcal{B} = \{w_1, \ldots, w_k\}$ is a basis for W. By definition, $\mathcal{B}$ spans W and $k = \dim W$. We must show that $\mathcal{B}$ is linearly independent. Suppose that $\mathcal{B}$ is linearly dependent; then Lemma 5.6.1 implies that there is a proper subset of $\mathcal{B}$ that spans W (and is linearly independent). This contradicts the fact that as a basis $\mathcal{B}$ has the smallest number of elements of any spanning set for W.

Suppose that $\mathcal{B} = \{w_1, \ldots, w_k\}$ both spans W and is linearly independent. Linear independence and Corollary 5.6.3 imply that $k \le \dim W$. Since, by definition, any spanning set of W has at least $\dim W$ vectors, it follows that $k \ge \dim W$. Thus $k = \dim W$ and $\mathcal{B}$ is a basis.   ◆

Extending Linearly Independent Sets to Bases

Lemma 5.6.1 leads to one approach to finding bases. Suppose that the subspace W is spanned by a finite set of vectors $\{w_1, \ldots, w_k\}$. Then we can throw out vectors one by one until we arrive at a linearly independent subset of the w_j. This subset is a basis for W.

We now discuss a second approach to finding a basis for a nonzero subspace W of a finite-dimensional vector space V.

Lemma 5.6.4   *Let $\{u_1, \ldots, u_k\}$ be a linearly independent set of vectors in a vector space V and assume that*

$$u_{k+1} \notin span\{u_1, \ldots, u_k\}.$$

Then $\{u_1, \ldots, u_{k+1}\}$ is also a linearly independent set.

Proof: Let $r_1, \ldots, r_{k+1}$ be scalars such that

$$r_1 u_1 + \cdots + r_{k+1} u_{k+1} = 0. \tag{5.6.3}$$

To prove independence, we need to show that all $r_j = 0$. Suppose $r_{k+1} \neq 0$. Then we can solve (5.6.3) for

$$u_{k+1} = -\frac{1}{r_{k+1}}(r_1 u_1 + \cdots + r_k u_k),$$

which implies that $u_{k+1} \in \mathrm{span}\{u_1, \ldots, u_k\}$. This contradicts the choice of u_{k+1}. So $r_{k+1} = 0$ and

$$r_1 u_1 + \cdots + r_k u_k = 0.$$

Since $\{u_1, \ldots, u_k\}$ is linearly independent, it follows that $r_1 = \cdots = r_k = 0$.   ◆

The second method for constructing a basis is:

1. Choose a nonzero vector w_1 in W.
2. If W is not spanned by w_1, then choose a vector w_2 that is not on the line spanned by w_1.
3. If $W \neq \mathrm{span}\{w_1, w_2\}$, then choose a vector $w_3 \notin \mathrm{span}\{w_1, w_2\}$.
4. If $W \neq \mathrm{span}\{w_1, w_2, w_3\}$, then choose a vector $w_4 \notin \mathrm{span}\{w_1, w_2, w_3\}$.
5. Continue until a spanning set for W is found. This set is a basis for W.

We now justify this approach to finding bases for subspaces. Suppose that W is a subspace of a finite-dimensional vector space V. For example, suppose that $W \subset \mathbf{R}^n$. Then our approach to finding a basis of W is as follows: Choose a nonzero vector $w_1 \in W$. If $W = \mathrm{span}\{w_1\}$, then we are done. If not, choose a vector $w_2 \in W - \mathrm{span}\{w_1\}$. It follows from Lemma 5.6.4 that $\{w_1, w_2\}$ is linearly independent. If $W = \mathrm{span}\{w_1, w_2\}$, then Theorem 5.5.3 implies that $\{w_1, w_2\}$ is a basis for W, $\dim W = 2$, and we are done. If not, choose $w_3 \in W - \mathrm{span}\{w_1, w_2\}$ and $\{w_1, w_2, w_3\}$ is linearly independent. The finite dimension of V implies that continuing inductively must lead to a spanning set of linear independent vectors for W—which by Theorem 5.5.3 is a basis. This discussion proves:

Corollary 5.6.5   *Every linearly independent subset of a finite-dimensional vector space V can be extended to a basis of V.*

Further Consequences of Theorem 5.5.3

We summarize here several important facts about dimensions.

Corollary 5.6.6   *Let W be a subspace of a finite-dimensional vector space V.*

(a)   *Suppose that W is a proper subspace. Then $\dim W < \dim V$.*

(b)   *Suppose that $\dim W = \dim V$. Then $W = V$.*

Proof: (a) Let dim $W = k$ and let $\{w_1, \ldots, w_k\}$ be a basis for W. Since W is a proper subspace of V, there is a vector $w \in V - W$. It follows from Lemma 5.6.4 that $\{w_1, \ldots, w_k, w\}$ is a linearly independent set. Therefore, Corollary 5.6.3 implies that $k + 1 \leq n$.

(b) Let $\{w_1, \ldots, w_k\}$ be a basis for W. Theorem 5.5.3 implies that this set is linearly independent. If $\{w_1, \ldots, w_k\}$ does not span V, then it can be extended to a basis as above. But then dim $V > $ dim W, which is a contradiction.   ◆

Corollary 5.6.7   *Let $\mathcal{B} = \{w_1, \ldots, w_n\}$ be a set of n vectors in an n-dimensional vector space V. Then the following are equivalent:*

(a) $\mathcal{B}$ is a spanning set of V,

(b) $\mathcal{B}$ is a basis for V, and

(c) $\mathcal{B}$ is a linearly independent set.

Proof: By definition, (a) implies (b) since a basis is a spanning set with the number of vectors equal to the dimension of the space. Theorem 5.5.3 states that a basis is a linearly independent set, so (b) implies (c). If $\mathcal{B}$ is a linearly independent set of n vectors, then it spans a subspace W of dimension n. It follows from Corollary 5.6.6(b) that $W = V$ and that (c) implies (a).   ◆

Subspaces of R³

We can now classify all subspaces of $\mathbf{R}^3$. They are: the origin, lines through the origin, planes through the origin, and $\mathbf{R}^3$. All of these sets were shown to be subspaces in Example 5.1.5(a–c).

To verify that these sets are the only subspaces of $\mathbf{R}^3$, note that Theorem 5.5.3 implies that proper subspaces of $\mathbf{R}^3$ have dimension equal to either one or two. (The zero-dimensional subspace is the origin and the only three-dimensional subspace is $\mathbf{R}^3$ itself.) One-dimensional subspaces of $\mathbf{R}^3$ are spanned by one nonzero vector and are just lines through the origin; see Example 5.1.5(b). We claim that all two-dimensional subspaces are planes through the origin.

Suppose that $W \subset \mathbf{R}^3$ is a subspace spanned by two noncollinear vectors w_1 and w_2. We show that W is a plane through the origin using results in Chapter 2. Observe that there is a vector $N = (N_1, N_2, N_3)$ perpendicular to $w_1 = (a_{11}, a_{12}, a_{13})$ and $w_2 = (a_{21}, a_{22}, a_{23})$. Such a vector N satisfies the two linear equations:

$$w_1 \cdot N = a_{11}N_1 + a_{12}N_2 + a_{13}N_3 = 0$$
$$w_2 \cdot N = a_{21}N_1 + a_{22}N_2 + a_{23}N_3 = 0.$$

Theorem 2.4.6 implies that a system of two linear equations in three unknowns has a nonzero solution. Let P be the plane perpendicular to N that contains the origin. We show that $W = P$ and hence that the claim is valid.

The choice of N shows that the vectors w_1 and w_2 are both in P. In fact, since P is a subspace, it contains every vector in span$\{w_1, w_2\}$. Thus $W \subset P$. If P contains just one additional vector $w_3 \in \mathbf{R}^3$ that is not in W, then the span of w_1, w_2, w_3 is three dimensional and $P = W = \mathbf{R}^3$.

HAND EXERCISES

In Exercises 1–3, you are given a pair of vectors v_1 and v_2 spanning a subspace of $\mathbf{R}^3$. Decide whether that subspace is a line or a plane through the origin. If it is a plane, then compute a vector N that is perpendicular to that plane.

1. $v_1 = (2, 1, 2)$ and $v_2 = (0, -1, 1)$ **2.** $v_1 = (2, 1, -1)$ and $v_2 = (-4, -2, 2)$

3. $v_1 = (0, 1, 0)$ and $v_2 = (4, 1, 0)$

4. The pairs of vectors

$$v_1 = (-1, 1, 0) \quad \text{and} \quad v_2 = (1, 0, 1)$$

span a plane P in $\mathbf{R}^3$. The pairs of vectors

$$w_1 = (0, 1, 0) \quad \text{and} \quad w_2 = (1, 1, 0)$$

span a plane Q in $\mathbf{R}^3$. Show that P and Q are different and compute the subspace of $\mathbf{R}^3$ that is given by the intersection $P \cap Q$.

5. Let A be a 7×5 matrix with rank$(A) = r$.
   (a) What is the largest value that r can have?
   (b) Give a condition equivalent to the system of equations $Ax = b$ having a solution.
   (c) What is the dimension of the null space of A?
   (d) If there is a solution to $Ax = b$, then how many parameters are needed to describe the set of all solutions?

6. Let

$$A = \begin{pmatrix} 1 & 3 & -1 & 4 \\ 2 & 1 & 5 & 7 \\ 3 & 4 & 4 & 11 \end{pmatrix}.$$

   (a) Find a basis for the subspace $\mathcal{C} \subset \mathbf{R}^3$ spanned by the columns of A.
   (b) Find a basis for the subspace $\mathcal{R} \subset \mathbf{R}^4$ spanned by the rows of A.
   (c) What is the relationship between dim $\mathcal{C}$ and dim $\mathcal{R}$?

7. Show that the vectors

$$v_1 = (2, 3, 1) \quad \text{and} \quad v_2 = (1, 1, 3)$$

are linearly independent. Show that the span of v_1 and v_2 forms a plane in $\mathbf{R}^3$ by showing that every linear combination is the solution to a single linear equation. Use this equation to determine the normal vector N to this plane. Verify Lemma 5.6.4 by verifying directly that v_1, v_2, and N are linearly independent vectors.

8. Let W be an infinite dimensional subspace of the vector space V. Show that V is infinite dimensional.

COMPUTER EXERCISES

9. Consider the set of vectors:

$$w_1 = (2, -2, 1), \quad w_2 = (-1, 2, 0), \quad w_3 = (3, -2, \lambda), \quad w_4 = (-5, 6, -2),$$

where λ is a real number.

(a) Find a value for λ such that the dimension of span$\{w_1, w_2, w_3, w_4\}$ is three. Then decide whether $\{w_1, w_2, w_3\}$ or $\{w_1, w_2, w_4\}$ is a basis for $\mathbf{R}^3$.

(b) Find a value for λ such that the dimension of span$\{w_1, w_2, w_3, w_4\}$ is two.

10. Find a basis for $\mathbf{R}^5$ as follows: Randomly choose vectors $x_1, x_2 \in \mathbf{R}^5$ by typing x1 = rand(5,1) and x2 = rand(5,1). Check that these vectors are linearly independent. If not, choose another pair of vectors until you find a linearly independent set. Next choose a vector x_3 at random and check that $x_1, x_2,$ and x_3 are linearly independent. If not, randomly choose another vector for x_3. Continue until you have five linearly independent vectors—which by a dimension count must be a basis and span $\mathbf{R}^5$. Verify this comment by using MATLAB to write the vector

$$\begin{pmatrix} 2 \\ 1 \\ 3 \\ -2 \\ 4 \end{pmatrix}$$

as a linear combination of $x_1, \ldots, x_5$.

11. Find a basis for the subspace of $\mathbf{R}^5$ spanned by

$$\begin{aligned} u_1 &= (1, 1, 0, 0, 1) \\ u_2 &= (0, 2, 0, 1, -1) \\ u_3 &= (0, -1, 1, 0, 2) \\ u_4 &= (1, 4, 1, 2, 1) \\ u_5 &= (0, 0, 2, 1, 3). \end{aligned} \qquad (5.6.4)*$$

CHAPTER 6

Closed Form Solutions for Planar ODEs

In this chapter we describe three different methods to find closed form solutions to planar constant coefficient systems of linear differential equations. In Section 6.1 we begin by discussing those systems of differential equations that have unique solutions to initial value problems; these systems include linear systems. Then we show how uniqueness to initial value problems implies that the space of solutions to a constant coefficient system of n linear differential equations is n dimensional. Using this observation, we present a direct method for solving planar linear systems in Section 6.2. This method extends the discussion of solutions to systems whose coefficient matrices have distinct real eigenvalues given in Section 4.9 to the cases of complex eigenvalues and equal real eigenvalues.

The matrix exponential is an elementary function that allows us to solve all initial value problems for constant coefficient linear systems, and this function is introduced in Section 6.3. In Section 6.4 we compute the matrix exponential for several special, but important, examples.

We compute matrix exponentials in two different ways. The first approach is based on changes of coordinates. The idea is to make the coefficient matrix of the differential equation as simple as possible; indeed we put the coefficient matrix in the form of one of the matrices whose exponential is computed in Section 6.4. This idea leads to the notion of *similarity* of matrices, which is discussed in Section 6.5, and leads to the second method for solving planar linear systems. Both the direct method and the method based on similarity require being able to compute the eigenvalues *and* the eigenvectors of the coefficient matrix.

Once these ideas have been introduced and discussed, we use the Cayley Hamilton theorem to derive a computable formula for all matrix exponentials of 2×2 matrices. This formula requires knowing the eigenvalues of the coefficient matrix—but not its eigenvectors; see Section 6.6.

In the last section of this chapter, Section 6.7, we consider solutions to second-order equations by reduction to first-order systems.

6.1   THE INITIAL VALUE PROBLEM

Recall that a planar autonomous system of ordinary differential equations has the form

$$\begin{aligned} \frac{dx}{dt} &= f(x, y) \\ \frac{dy}{dt} &= g(x, y). \end{aligned} \tag{6.1.1}$$

Computer experiments using pplane5 always appear to find a unique solution to initial value problems. More precisely, we are led to believe that there is just one solution to (6.1.1) satisfying the initial conditions

$$\begin{aligned} x(0) &= x_0 \\ y(0) &= y_0. \end{aligned}$$

Existence and Uniqueness of Solutions

In fact, existence and uniqueness of solutions are not always guaranteed—though they are guaranteed for a large class of differential equations, as the following theorem shows.

Theorem 6.1.1   *Suppose that the functions f and g in the system of differential equations (6.1.1) are differentiable near (x_0, y_0) and that the partial derivatives f_x, f_y, g_x, g_y are continuous near (x_0, y_0). Then there exists a unique solution to (6.1.1) with initial conditions $(x(0), y(0)) = (x_0, y_0)$.*

For example, consider the planar system of constant coefficient linear differential equations:

$$\begin{aligned} \frac{dx}{dt} &= ax + by \\ \frac{dy}{dt} &= cx + dy. \end{aligned} \tag{6.1.2}$$

In (6.1.2), $f(x, y) = ax + by$ and $g(x, y) = cx + dy$. The partial derivatives of f and g are easy to calculate; they are $f_x = a$, $f_y = b$, $g_x = c$, and $g_y = d$. As constant functions, all of these partial derivatives are continuous. Hence existence and uniqueness of solutions to the initial value problem for (6.1.2) are guaranteed by Theorem 6.1.1.

Although we have stated this theorem for just planar systems, the theorem itself is valid in all dimensions. For example, in one dimension the analog of Theorem 6.1.1 states that the differential equation

$$\dot{x} = f(x)$$

with initial condition

$$x(0) = x_0$$

has a unique solution $x(t)$ when $f'(x)$ exists and is continuous near x_0.

A discussion of the proof of this theorem is beyond the scope of this text; nevertheless, the theorem is valid and we use its consequences.

An Example of Nonuniqueness of Solutions

Indeed, uniqueness of solutions is not guaranteed even for single equations. Here is an example:

$$\frac{dx}{dt} = 3\sqrt[3]{x^2} = 3x^{2/3} \tag{6.1.3}$$

with the initial condition

$$x(0) = 0.$$

Certainly the constant function $x(t) = 0$ is a solution to (6.1.3) satisfying the initial value $x(0) = 0$. In addition, the function

$$x(t) = t^3$$

is a solution to (6.1.3). This fact is checked by direct calculation:

$$\frac{dx}{dt} = 3t^2 \quad \text{and} \quad 3x^{2/3} = 3(t^3)^{2/3} = 3t^2.$$

Example (6.1.3) shows that hypotheses like those in Theorem 6.1.1 are needed. Indeed, in (6.1.3)

$$f(x) = 3x^{2/3} \quad \text{and} \quad f'(x) = \frac{2}{x^{1/3}}.$$

Hence, f is not differentiable at $x = 0$ and the hypotheses of Theorem 6.1.1 fail.

At this point it is worth looking to see what dfield5 computes numerically. Start dfield5 and change the differential equation to

    x' = (x^2)^(1/3)

Writing the differential equation in this way guarantees that the computer does not compute square roots of negative numbers. Click on the Proceed button. In the DFIELD5 Display window attempt to click on the origin; that is, attempt to enter the initial condition $x(0) = 0$ using the mouse. You get a solution that has a shape similar to the graph of the cubic function $x = t^3$. In the DFIELD5 Options menu click on Keyboard input, and in the DFIELD5 Keyboard input window enter the values $t = 0$ and $x = 0$. After clicking on the Compute button, you see the solution $x = 0$. Now click on the Erase all solutions button in the DFIELD5 Options menu. Change the initial value of x to 0.00001 in the DFIELD5 Keyboard input window and click on Compute. You see a solution that looks like $x = t^3$; see Figure 6.1.

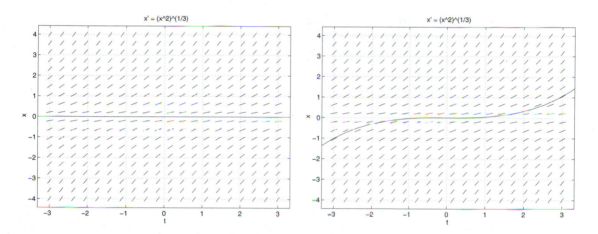

Figure 6.1
Solutions to (6.1.3) for $t \in [-3, 3]$ and $x \in [-4, 4]$. *Left:* $x(0) = 0$; *right:* $x(0) = 0.00001$.

At this stage we do not have the background to discuss the numerical method employed by dfield5.

The Initial Value Problem for Linear Systems

In this chapter we discuss how to find solutions $(x(t), y(t))$ to (6.1.2) satisfying the initial values $x(0) = x_0$ and $y(0) = y_0$. It is convenient to rewrite (6.1.2) in matrix form as

$$\frac{dX}{dt}(t) = CX(t). \tag{6.1.4}$$

The initial value problem is then stated as: Find a solution to (6.1.4) satisfying $X(0) = X_0$, where $X_0 = (x_0, y_0)^t$. Everything that we have said here works equally well for n-dimensional systems of linear differential equations. Just let C be an $n \times n$ matrix and let X_0 be an n-vector of initial conditions.

Solving the Initial Value Problem Using Superposition

In Section 4.9 we discussed how to solve (6.1.4) when the eigenvalues of C are real and distinct. Recall that when λ_1 and λ_2 are distinct real eigenvalues of C with associated eigenvectors v_1 and v_2, there are two solutions to (6.1.4) given by the explicit formulas

$$X_1(t) = e^{\lambda_1 t} v_1 \quad \text{and} \quad X_2(t) = e^{\lambda_2 t} v_2.$$

Superposition guarantees that every linear combination of these solutions

$$X(t) = \alpha_1 X_1(t) + \alpha_2 X_2(t) = \alpha_1 e^{\lambda_1 t} v_1 + \alpha_2 e^{\lambda_2 t} v_2$$

is a solution to (6.1.4). In addition, we can always choose scalars $\alpha_1, \alpha_2 \in \mathbf{R}$ to solve any given initial value problem of (6.1.4). It follows from the uniqueness of solutions to initial value problems stated in Theorem 6.1.1 that all solutions to (6.1.4) are included in this family of solutions.

We generalize this discussion so that we will be able to find closed form solutions to (6.1.4) in Section 6.2 when the eigenvalues of C are complex or are real and equal. Suppose that $X_1(t)$ and $X_2(t)$ are two solutions to (6.1.2) such that

$$v_1 = X_1(0) \quad \text{and} \quad v_2 = X_2(0)$$

are linearly independent. The existence part of Theorem 6.1.1 guarantees that such solutions exist. Then all solutions to (6.1.2) are linear combinations of these two solutions. We verify this statement as follows: Corollary 5.6.7 of Chapter 5 states that since $\{v_1, v_2\}$ is a linearly independent set in $\mathbf{R}^2$, it is also a basis of $\mathbf{R}^2$. Thus for every $X_0 \in \mathbf{R}^2$ there exist scalars r_1 and r_2 such that

$$X_0 = r_1 v_1 + r_2 v_2.$$

It follows from superposition and Theorem 6.1.1 that the solution

$$X(t) = r_1 X_1(t) + r_2 X_2(t)$$

is the unique solution whose initial condition vector is X_0.

We have proved that every solution to this linear system of differential equations is a linear combination of these two solutions; that is, we have proved that the dimension of the space of solutions to (6.1.4) is two. This proof generalizes immediately to a proof of the following theorem for $n \times n$ systems:

Theorem 6.1.2 *Let C be an n × n matrix. Suppose that $X_1(t), \ldots, X_n(t)$ are solutions to $\dot{X} = CX$ such that the vectors of initial conditions $v_j = X_j(0)$ are linearly independent in $\mathbf{R}^n$. Then the unique solution to the system (6.1.4) with initial condition $X(0) = X_0$ is*

$$X(t) = r_1 X_1(t) + \cdots + r_n X_n(t), \tag{6.1.5}$$

where $r_1, \ldots, r_n$ are scalars satisfying

$$X_0 = r_1 v_1 + \cdots + r_n v_n. \tag{6.1.6}$$

We call (6.1.5) the *general solution* to the system of differential equations $\dot{X} = CX$. When solving the initial value problem, we find a *particular solution* by specifying the scalars $r_1, \ldots, r_n$.

Corollary 6.1.3 *Let C be an n × n matrix, and let $\mathcal{X} = \{X_1(t), \ldots, X_n(t)\}$ be solutions to the differential equation $\dot{X} = CX$ such that the vectors $X_j(0)$ are linearly independent in $\mathbf{R}^n$. Then the set of all solutions to $\dot{X} = CX$ is an n-dimensional subspace of $(C^1)^n$, and $\mathcal{X}$ is a basis for the solution subspace.*

Consider a special case of Theorem 6.1.2. Suppose that the matrix C has n linearly independent eigenvectors $v_1, \ldots, v_n$ with real eigenvalues $\lambda_1, \ldots, \lambda_n$. Then the functions $X_j(t) = e^{\lambda_j t} v_j$ are solutions to $\dot{X} = CX$. Corollary 6.1.3 implies that the functions X_j form a basis for the space of solutions to this system of differential equations. Indeed, the general solution to (6.1.4) is

$$X(t) = r_1 e^{\lambda_1 t} v_1 + \cdots + r_n e^{\lambda_n t} v_n. \tag{6.1.7}$$

The particular solution that solves the initial value $X(0) = X_0$ is found by solving (6.1.6) for the scalars $r_1, \ldots, r_n$.

HAND EXERCISES

In Exercises 1–4, consider the system of differential equations

$$\frac{dx}{dt} = 65x + 42y$$
$$\frac{dy}{dt} = -99x - 64y. \tag{6.1.8}$$

1. Verify that

$$v_1 = \begin{pmatrix} 2 \\ -3 \end{pmatrix} \quad \text{and} \quad v_2 = \begin{pmatrix} -7 \\ 11 \end{pmatrix}$$

are eigenvectors of the coefficient matrix of (6.1.8) and find the associated eigenvalues.

2. Find the solution to (6.1.8) satisfying initial conditions $X(0) = (-14, 22)^t$.

3. Find the solution to (6.1.8) satisfying initial conditions $X(0) = (-3, 5)^t$.

4. Find the solution to (6.1.8) satisfying initial conditions $X(0) = (9, -14)^t$.

In Exercises 5–8, consider the system of differential equations

$$\frac{dx}{dt} = x - y$$

$$\frac{dy}{dt} = -x + y. \tag{6.1.9}$$

5. The eigenvalues of the coefficient matrix of (6.1.9) are 0 and 2. Find the associated eigenvectors.

6. Find the solution to (6.1.9) satisfying initial conditions $X(0) = (2, -2)^t$.

7. Find the solution to (6.1.9) satisfying initial conditions $X(0) = (2, 6)^t$.

8. Find the solution to (6.1.9) satisfying initial conditions $X(0) = (1, 0)^t$.

In Exercises 9–12, consider the system of differential equations

$$\frac{dx}{dt} = -y$$

$$\frac{dy}{dt} = x. \tag{6.1.10}$$

9. Show that $(x_1(t), y_1(t)) = (\cos t, \sin t)$ is a solution to (6.1.10).

10. Show that $(x_2(t), y_2(t)) = (-\sin t, \cos t)$ is a solution to (6.1.10).

11. Using Exercises 9 and 10, find a solution $(x(t), y(t))$ to (6.1.10) that satisfies $(x(0), y(0)) = (0, 1)$.

12. Using Exercises 9 and 10, find a solution $(x(t), y(t))$ to (6.1.10) that satisfies $(x(0), y(0)) = (1, 1)$.

In Exercises 13 and 14, consider the system of differential equations

$$\frac{dx}{dt} = -2x + 7y$$

$$\frac{dy}{dt} = 5y. \tag{6.1.11}$$

13. Find a solution to (6.1.11) satisfying the initial condition $(x(0), y(0)) = (1, 0)$.

14. Find a solution to (6.1.11) satisfying the initial condition $(x(0), y(0)) = (-1, 2)$.

In Exercises 15–17, consider the matrix

$$C = \begin{pmatrix} -1 & -10 & -6 \\ 0 & 4 & 3 \\ 0 & -14 & -9 \end{pmatrix}.$$

15. Verify that

$$v_1 = \begin{pmatrix} 1 \\ 0 \\ 0 \end{pmatrix}, \quad v_2 = \begin{pmatrix} 2 \\ -1 \\ 2 \end{pmatrix}, \quad v_3 = \begin{pmatrix} 6 \\ -3 \\ 7 \end{pmatrix}$$

are eigenvectors of C and find the associated eigenvalues.

16. Find a solution to the system of differential equations $\dot{X} = CX$ satisfying the initial condition $X(0) = (10, -4, 9)^t$.

17. Find a solution to the system of differential equations $\dot{X} = CX$ satisfying the initial condition $X(0) = (2, -1, 3)^t$.

18. Show that for some nonzero a the function $x(t) = at^5$ is a solution to the differential equation $\dot{x} = x^{4/5}$. Then show that there are at least two solutions to the initial value problem $x(0) = 0$ for this differential equation.

COMPUTER EXERCISES

19. Use pplane5 to investigate the system of differential equations

$$\frac{dx}{dt} = -2y$$

$$\frac{dy}{dt} = -x + y. \tag{6.1.12}$$

(a) Use pplane5 to find two independent eigendirections (and hence eigenvectors) for (6.1.12).

(b) Using (a), find the eigenvalues of the coefficient matrix of (6.1.12).

(c) Find a closed form solution to (6.1.12) satisfying the initial condition

$$X(0) = \begin{pmatrix} 4 \\ -1 \end{pmatrix}.$$

(d) Study the time series of y versus t for the solution in (c) by comparing the graph of the closed form solution obtained in (c) with the time series graph using pplane5.

6.2   CLOSED FORM SOLUTIONS BY THE DIRECT METHOD

In Section 4.9 we showed in detail how solutions are found to planar systems of constant coefficient differential equations with distinct real eigenvalues. This method was just reviewed in Section 6.1, where we saw that the crucial step in solving these systems of differential equations is where we find two linearly independent solutions. In this section we discuss how to find these two linearly independent solutions when the eigenvalues of the coefficient matrix are either complex or real and equal.

By finding these two linearly independent solutions, we find both the *general* solution of the system of differential equations $\dot{X} = CX$ and a method for solving the initial value problem

$$\frac{dX}{dt} = CX$$

$$X(0) = X_0. \tag{6.2.1}$$

We assume that C is a 2×2 matrix with eigenvalues λ_1 and λ_2. When needed, we denote the associated eigenvectors by v_1 and v_2.

Real Distinct Eigenvalues

We have discussed the case when $\lambda_1 \neq \lambda_2 \in \mathbf{R}$ on several occasions. For completeness we repeat the result. The general solution is

$$X(t) = \alpha_1 e^{\lambda_1 t} v_1 + \alpha_2 e^{\lambda_2 t} v_2. \tag{6.2.2}$$

The initial value problem is solved by finding real numbers α_1 and α_2 such that

$$X_0 = \alpha_1 v_1 + \alpha_2 v_2.$$

See Section 4.9 for a detailed discussion with examples.

Complex Conjugate Eigenvalues

Suppose that the eigenvalues of C are complex; that is, suppose that $\lambda_1 = \sigma + i\tau$ with $\tau \neq 0$ is an eigenvalue of C with eigenvector $v_1 = v + iw$, where $v, w \in \mathbf{R}^2$. We claim that $X_1(t)$ and $X_2(t)$, where

$$\begin{aligned} X_1(t) &= e^{\sigma t}(\cos(\tau t)v - \sin(\tau t)w) \\ X_2(t) &= e^{\sigma t}(\sin(\tau t)v + \cos(\tau t)w) \end{aligned} \tag{6.2.3}$$

are solutions to (6.2.1), and that the general solution to (6.2.1) is

$$X(t) = \alpha_1 X_1(t) + \alpha_2 X_2(t), \tag{6.2.4}$$

where α_1 and α_2 are real scalars.

There are several difficulties in deriving (6.2.3) and (6.2.4); these difficulties are related to using complex numbers as opposed to real numbers. In particular, in the derivation of (6.2.3) we need to define the exponential of a complex number; we begin by discussing this issue.

Euler's Formula

We find complex exponentials by using Euler's celebrated formula:

$$e^{i\theta} = \cos\theta + i\sin\theta \tag{6.2.5}$$

for any real number θ. A justification of this formula is given in Exercise 1 at the end of this section. Euler's formula allows us to differentiate complex exponentials, obtaining the expected result:

$$\frac{d}{dt}e^{i\tau t} = \frac{d}{dt}(\cos(\tau t) + i\sin(\tau t))$$

$$= \tau(-\sin(\tau t) + i\cos(\tau t))$$

$$= i\tau(\cos(\tau t) + i\sin(\tau t))$$

$$= i\tau e^{i\tau t}.$$

Euler's formula also implies that

$$e^{\lambda t} = e^{\sigma t + i\tau t} = e^{\sigma t}e^{i\tau t} = e^{\sigma t}(\cos(\tau t) + i\sin(\tau t)), \qquad \textbf{(6.2.6)}$$

where $\lambda = \sigma + i\tau$. Most important, we note that

$$\frac{d}{dt}e^{\lambda t} = \lambda e^{\lambda t}. \qquad \textbf{(6.2.7)}$$

We use (6.2.6) and the product rule for differentiation to verify (6.2.7) as follows:

$$\frac{d}{dt}e^{\lambda t} = \frac{d}{dt}\left(e^{\sigma t}e^{i\tau t}\right)$$

$$= \left(\sigma e^{\sigma t}\right)e^{i\tau t} + e^{\sigma t}\left(i\tau e^{i\tau t}\right)$$

$$= (\sigma + i\tau)e^{\sigma t + i\tau t}$$

$$= \lambda e^{\lambda t}.$$

Verification That (6.2.4) Is the General Solution

A complex vector-valued function $X(t) = X_1(t) + iX_2(t) \in \mathbf{C}^n$ consists of a *real part* $X_1(t) \in \mathbf{R}^n$ and an *imaginary part* $X_2(t) \in \mathbf{R}^n$. For such functions $X(t)$ we define

$$\dot{X} = \dot{X}_1 + i\dot{X}_2$$

and

$$CX = CX_1 + iCX_2.$$

To say that $X(t)$ is a solution to $\dot{X} = CX$ means that

$$\dot{X}_1 + i\dot{X}_2 = \dot{X} = CX = CX_1 + iCX_2. \qquad \textbf{(6.2.8)}$$

Lemma 6.2.1 *The complex vector-valued function $X(t)$ is a solution to $\dot{X} = CX$ if and only if the real and imaginary parts are real vector-valued solutions to $\dot{X} = CX$.*

Proof: Equating the real and imaginary parts of (6.2.8) implies that

$$\dot{X}_1 = CX_1 \quad \text{and} \quad \dot{X}_2 = CX_2. \qquad \blacklozenge$$

It follows from Lemma 6.2.1 that finding one complex-valued solution to a linear differential equation provides us with two real-valued solutions. Identity (6.2.7) implies that

$$X(t) = e^{\lambda_1 t} v_1$$

is a complex-valued solution to (6.2.1). Using Euler's formula, we compute the real and imaginary parts of $X(t)$ as follows:

$$
\begin{aligned}
X(t) &= e^{(\sigma + i\tau)t}(v + iw) \\
&= e^{\sigma t}(\cos(\tau t) + i\sin(\tau t))(v + iw) \\
&= e^{\sigma t}(\cos(\tau t)v - \sin(\tau t)w) + ie^{\sigma t}(\sin(\tau t)v + \cos(\tau t)w).
\end{aligned}
$$

Since the real and imaginary parts of $X(t)$ are solutions to $\dot{X} = CX$, it follows that the real-valued functions $X_1(t)$ and $X_2(t)$ defined in (6.2.3) are indeed solutions.

Returning to the case where C is a 2×2 matrix, we see that if $X_1(0) = v$ and $X_2(0) = w$ are linearly independent, then Corollary 6.1.3 implies that (6.2.4) is the general solution to $\dot{X} = CX$. The linear independence of v and w is verified using the following lemma:

Lemma 6.2.2 *Let $\lambda_1 = \sigma + i\tau$ with $\tau \neq 0$ be a complex eigenvalue of the 2×2 matrix C with eigenvector $v_1 = v + iw$, where $v, w \in \mathbf{R}^2$. Then*

$$
\begin{aligned}
Cv &= \sigma v - \tau w \\
Cw &= \tau v + \sigma w
\end{aligned}
\tag{6.2.9}
$$

and v and w are linearly independent vectors.

Proof: By assumption $Cv_1 = \lambda_1 v_1$; that is,

$$C(v + iw) = (\sigma + i\tau)(v + iw) = (\sigma v - \tau w) + i(\tau v + \sigma w). \tag{6.2.10}$$

Equating real and imaginary parts of (6.2.10) leads to the system of equations (6.2.9). Note that if $w = 0$, then $v \neq 0$ and $\tau v = 0$. Hence $\tau = 0$, contradicting the assumption that $\tau \neq 0$. So $w \neq 0$.

Note also that if v and w are linearly dependent, then $v = \alpha w$. It follows from the previous equation that

$$Cw = (\tau \alpha + \sigma)w.$$

Hence w is a real eigenvector, but the eigenvalues of C are not real and C has no real eigenvectors. $\blacklozenge$

An Example with Complex Eigenvalues

Consider an example of an initial value problem for a linear system with complex eigenvalues. Let

$$\frac{dX}{dt} = \begin{pmatrix} -1 & 2 \\ -5 & -3 \end{pmatrix} X = CX \qquad (6.2.11)$$

and

$$X_0 = \begin{pmatrix} 1 \\ 1 \end{pmatrix}.$$

The characteristic polynomial for the matrix C is

$$p_C(\lambda) = \lambda^2 + 4\lambda + 13,$$

whose roots are $\lambda_1 = -2 + 3i$ and $\lambda_2 = -2 - 3i$. So

$$\sigma = -2 \quad \text{and} \quad \tau = 3.$$

An eigenvector corresponding to the eigenvalue λ_1 is

$$v_1 = \begin{pmatrix} 2 \\ -1 + 3i \end{pmatrix} = \begin{pmatrix} 2 \\ -1 \end{pmatrix} + i \begin{pmatrix} 0 \\ 3 \end{pmatrix} = v + iw.$$

It follows from (6.2.3) that

$$X_1(t) = e^{-2t}(\cos(3t)v - \sin(3t)w)$$
$$X_2(t) = e^{-2t}(\sin(3t)v + \cos(3t)w)$$

are solutions to (6.2.11) and $X = \alpha_1 X_1 + \alpha_2 X_2$ is the general solution to (6.2.11). To solve the initial value problem we need to find α_1 and α_2 such that

$$X_0 = X(0) = \alpha_1 X_1(0) + \alpha_2 X_2(0) = \alpha_1 v + \alpha_2 w;$$

that is,

$$\begin{pmatrix} 1 \\ 1 \end{pmatrix} = \alpha_1 \begin{pmatrix} 2 \\ -1 \end{pmatrix} + \alpha_2 \begin{pmatrix} 0 \\ 3 \end{pmatrix}.$$

Therefore $\alpha_1 = \frac{1}{2}, \alpha_2 = \frac{1}{2}$, and

$$X(t) = e^{-2t} \begin{pmatrix} \cos(3t) + \sin(3t) \\ \cos(3t) - 2\sin(3t) \end{pmatrix}. \qquad (6.2.12)$$

Real and Equal Eigenvalues

There are two types of 2×2 matrices that have real and equal eigenvalues: those that are scalar multiples of the identity and those that are not. An example of a 2×2 matrix that has real and equal eigenvalues is

$$A = \begin{pmatrix} \lambda_1 & 1 \\ 0 & \lambda_1 \end{pmatrix}, \quad \lambda_1 \in \mathbf{R}. \tag{6.2.13}$$

The characteristic polynomial of A is

$$p_A(\lambda) = \lambda^2 - \text{tr}(A)\lambda + \det(A) = \lambda^2 - 2\lambda_1\lambda + \lambda_1^2 = (\lambda - \lambda_1)^2.$$

Thus the eigenvalues of A both equal λ_1.

Only One Linearly Independent Eigenvector

An important fact about the matrix A in (6.2.13) is that it has only one linearly independent eigenvector. To verify this fact, solve the system of linear equations

$$Av = \lambda_1 v.$$

In matrix form this equation is

$$0 = (A - \lambda_1 I_2)v = \begin{pmatrix} 0 & 1 \\ 0 & 0 \end{pmatrix} v.$$

A quick calculation shows that all solutions are multiples of $v_1 = e_1 = (1, 0)^t$.

In fact, this observation is valid for any 2×2 matrix that has equal eigenvalues and is not a scalar multiple of the identity, as the next lemma shows:

Lemma 6.2.3 *Let A be a 2×2 matrix. Suppose that A has two linearly independent eigenvectors both with eigenvalue λ_1. Then $A = \lambda_1 I_2$.*

Proof: Let v_1 and v_2 be two linearly independent eigenvectors of A; that is, $Av_j = \lambda_1 v_j$. It follows from linearity that $Av = \lambda_1 v$ for any linear combination $v = \alpha_1 v_1 + \alpha_2 v_2$. Since v_1 and v_2 are linearly independent and $\dim(\mathbf{R}^2) = 2$, it follows that $\{v_1, v_2\}$ is a basis of $\mathbf{R}^2$. Thus every vector $v \in \mathbf{R}^2$ is a linear combination of v_1 and v_2. Therefore A is λ_1 times the identity matrix. ◆

Generalized Eigenvectors

Suppose that C has exactly one linearly independent real eigenvector v_1 with real eigenvalue λ_1. We call w_1 a *generalized eigenvector* of C when it satisfies the system of linear equations

$$(C - \lambda_1 I_2)w_1 = v_1. \tag{6.2.14}$$

The matrix A in (6.2.13) has a generalized eigenvector. To verify this point solve the linear system

$$(A - \lambda_1 I_2) w_1 = \begin{pmatrix} 0 & 1 \\ 0 & 0 \end{pmatrix} w_1 = v_1 = \begin{pmatrix} 1 \\ 0 \end{pmatrix}$$

for $w_1 = e_2$. Note that for this matrix A, $v_1 = e_1$ and $w_1 = e_2$ are linearly independent. The next lemma shows that this observation about generalized eigenvectors is always valid.

Lemma 6.2.4   *Let C be a 2×2 matrix with both eigenvalues equal to λ_1 and with one linearly independent eigenvector v_1.*

(a) *Let w_1 be a generalized eigenvector of C. Then v_1 and w_1 are linearly independent.*

(b) *Let w be any vector such that v_1 and w are linearly independent. Then w is a nonzero scalar multiple of a generalized eigenvector of C.*

Proof: (a) If v_1 and w_1 were linearly dependent, then w_1 would be a multiple of v_1 and hence an eigenvector of C. But $C - \lambda_1 I_2$ applied to an eigenvector is 0, which is a contradiction. Therefore v_1 and w_1 are linearly independent.

(b) Let w be any vector that is linearly independent of the eigenvector v_1. It follows that $\{v_1, w\}$ is a basis for $\mathbf{R}^2$; hence

$$Cw = \alpha v_1 + \beta w \tag{6.2.15}$$

for some scalars α and β. If $\alpha = 0$, then w is an eigenvector of C, contradicting the assumption that C has only one linearly independent eigenvector. Therefore $\alpha \neq 0$.

We claim that $\beta = \lambda_1$ and we prove the claim by showing that β is an eigenvalue of C. Hence β must equal λ_1 since both eigenvalues of C equal λ_1. To see that β is an eigenvalue, define the nonzero vector

$$u = \alpha v_1 + (\beta - \lambda_1) w$$

and compute

$$Cu = \lambda_1 \alpha v_1 + (\beta - \lambda_1)(\alpha v_1 + \beta w) = \beta u.$$

So u is an eigenvector of C with eigenvalue β. It now follows from (6.2.15) that

$$(C - \lambda_1 I_2) w = \alpha v_1.$$

Therefore $w_1 = \dfrac{1}{\alpha} w$ is a generalized eigenvector of C.   ◆

Independent Solutions to Differential Equations with Equal Eigenvalues

In the equal eigenvalue, one eigenvector case, we claim that the general solution to $\dot{X} = CX$ is

$$X(t) = e^{\lambda_1 t}\left(\alpha v_1 + \beta(w_1 + tv_1)\right), \qquad (6.2.16)$$

where v_1 is an eigenvector of C and w_1 is the generalized eigenvector.

Since v_1 is an eigenvector of C with eigenvalue λ_1, the function $X_1(t) = e^{\lambda_1 t} v_1$ is a solution to $\dot{X} = CX$. Suppose we can show that $X_2(t) = e^{\lambda_1 t}(w_1 + tv_1)$ is also a solution to $\dot{X} = CX$. Then (6.2.16) is the general solution, since $X_1(0) = v_1$ and $X_2(0) = w_1$ are linearly independent by Lemma 6.2.4(a). Apply Theorem 6.1.2.

To verify that $X_2(t)$ is a solution (that is, that $\dot{X}_2 = CX_2$), calculate

$$\dot{X}_2(t) = \lambda_1 e^{\lambda_1 t}(w_1 + tv_1) + e^{\lambda_1 t} v_1 = e^{\lambda_1 t}(\lambda_1 w_1 + v_1 + t\lambda v_1)$$

and

$$CX_2(t) = e^{\lambda_1 t}(Cw_1 + tCv_1) = e^{\lambda_1 t}((v_1 + \lambda_1 w_1) + t\lambda_1 v_1)$$

using (6.2.14). Note that $X(0) = \alpha v_1 + \beta w_1$, so α and β are found by solving $X_0 = \alpha v_1 + \beta w_1$.

An Example with Equal Eigenvalues

Consider the system of differential equations

$$\frac{dX}{dt} = \begin{pmatrix} 1 & -1 \\ 9 & -5 \end{pmatrix} X \qquad (6.2.17)$$

with initial value

$$X_0 = \begin{pmatrix} 2 \\ 3 \end{pmatrix}.$$

The characteristic polynomial for the matrix $C = \begin{pmatrix} 1 & -1 \\ 9 & -5 \end{pmatrix}$ is

$$p_C(\lambda) = \lambda^2 + 4\lambda + 4 = (\lambda + 2)^2.$$

Thus $\lambda_1 = -2$ is an eigenvalue of multiplicity two. Since C is not a multiple of the identity matrix, it must have precisely one linearly independent eigenvector v_1. This eigenvector is found by solving the equation

$$0 = (C - \lambda_1 I_2)v_1 = (C + 2I_2)v_1 = \begin{pmatrix} 3 & -1 \\ 9 & -3 \end{pmatrix} v_1$$

for

$$v_1 = \begin{pmatrix} 1 \\ 3 \end{pmatrix}.$$

To find the generalized eigenvector w_1, we solve the system of linear equations

$$(C - \lambda_1 I_2)w_1 = (C + 2I_2)w_1 = \begin{pmatrix} 3 & -1 \\ 9 & -3 \end{pmatrix} w_1 = v_1 = \begin{pmatrix} 1 \\ 3 \end{pmatrix}$$

by row reducing the augmented matrix

$$\begin{pmatrix} 3 & -1 & | & 1 \\ 9 & -3 & | & 3 \end{pmatrix}$$

to obtain

$$w_1 = \begin{pmatrix} 1 \\ 2 \end{pmatrix}.$$

We may now apply (6.2.16) to find the general solution to (6.2.17):

$$X(t) = e^{-2t} \left(\alpha v_1 + \beta(w_1 + tv_1) \right).$$

We solve the initial value problem by solving

$$\begin{pmatrix} 2 \\ 3 \end{pmatrix} = X_0 = X(0) = \alpha v_1 + \beta w_1 = \begin{pmatrix} 1 & 1 \\ 3 & 2 \end{pmatrix} \begin{pmatrix} \alpha \\ \beta \end{pmatrix}$$

for $\alpha = -1$ and $\beta = 3$. So the closed form solution to this initial value problem is

$$X(t) = e^{-2t} \left(-v_1 + 3(w_1 + tv_1) \right)$$
$$= e^{-2t} \left(-\begin{pmatrix} 1 \\ 3 \end{pmatrix} + 3 \begin{pmatrix} 1+t \\ 2+3t \end{pmatrix} \right)$$
$$= e^{-2t} \begin{pmatrix} 2+3t \\ 3+9t \end{pmatrix}.$$

There is a simpler method for finding this solution—a method that does not require solving for either the eigenvector v_1 or the generalized eigenvector w_1—that we discuss in Section 6.6.

HAND EXERCISES

1. Justify Euler's formula (6.2.5) as follows: Recall the Taylor series

$$e^x = 1 + x + \frac{1}{2!}x^2 + \cdots + \frac{1}{n!}x^n + \cdots$$

$$\cos x = 1 - \frac{1}{2!}x^2 + \frac{1}{4!}x^4 + \cdots + (-1)^n \frac{1}{(2n)!}x^{2n} + \cdots$$

$$\sin x = x - \frac{1}{3!}x^3 + \frac{1}{5!}x^5 + \cdots + (-1)^n \frac{1}{(2n+1)!}x^{2n+1} + \cdots$$

Now evaluate the Taylor series $e^{i\theta}$ and separate into real and imaginary parts.

In modern language De Moivre's formula states that

$$e^{ni\theta} = \left(e^{i\theta}\right)^n.$$

In Exercises 2 and 3 use De Moivre's formula coupled with Euler's formula (6.2.5) to determine trigonometric identities for the given quantity in terms of $\cos\theta$, $\sin\theta$, $\cos\varphi$, and $\sin\varphi$.

2. $\cos(\theta + \varphi)$ 3. $\sin(3\theta)$

In Exercises 4–7, compute the general solution for the given system of differential equations.

4. $\dfrac{dX}{dt} = \begin{pmatrix} -1 & -4 \\ 2 & 3 \end{pmatrix} X$ 5. $\dfrac{dX}{dt} = \begin{pmatrix} 8 & -15 \\ 3 & -4 \end{pmatrix} X$

6. $\dfrac{dX}{dt} = \begin{pmatrix} 5 & -1 \\ 1 & 3 \end{pmatrix} X$ 7. $\dfrac{dX}{dt} = \begin{pmatrix} -4 & 4 \\ -1 & 0 \end{pmatrix} X$

6.3 SOLUTIONS USING MATRIX EXPONENTIALS

In Section 4.1 we showed that the solution of the single ordinary differential equation $\dot{x}(t) = \lambda x(t)$ with initial condition $x(0) = x_0$ is $x(t) = e^{t\lambda}x_0$; see (4.1.5). In this section we show that we may write solutions of systems of equations in a similar form. In particular, we show that the solution to the linear system of ODEs

$$\frac{dX}{dt} = CX \tag{6.3.1}$$

with initial condition

$$X(0) = X_0,$$

where C is an $n \times n$ matrix and $X_0 \in \mathbf{R}^n$, is

$$X(t) = e^{tC}X_0. \tag{6.3.2}$$

In order to make sense of the solution (6.3.2) we need to understand matrix exponentials. More precisely, since tC is an $n \times n$ matrix for each $t \in \mathbf{R}$, we need to make sense of the expression e^L, where L is an $n \times n$ matrix. For this we recall the form of the exponential function as a power series:

$$e^t = 1 + t + \frac{1}{2!}t^2 + \frac{1}{3!}t^3 + \frac{1}{4!}t^4 + \cdots .$$

In more compact notation we have

$$e^t = \sum_{k=0}^{\infty} \frac{1}{k!}t^k .$$

By analogy, define the *matrix exponential* e^L by

$$e^L = I_n + L + \frac{1}{2!}L^2 + \frac{1}{3!}L^3 + \cdots \qquad (6.3.3)$$

$$= \sum_{k=0}^{\infty} \frac{1}{k!}L^k .$$

In this formula $L^2 = LL$ is the matrix product of L with itself, and the power L^k is defined inductively by $L^k = LL^{k-1}$ for $k > 1$. Hence e^L is an $n \times n$ matrix and is the infinite sum of $n \times n$ matrices.

   Remark: The infinite series for matrix exponentials (6.3.3) does converge for all $n \times n$ matrices L, and this fact is proved in Exercises 12 and 13.

   Using (6.3.3), we can write the matrix exponential of tC for each real number t. Since $(tC)^k = t^k C^k$, we obtain

$$e^{tC} = I_n + tC + \frac{1}{2!}(tC)^2 + \frac{1}{3!}(tC)^3 + \cdots$$

$$= I_n + tC + \frac{t^2}{2!}C^2 + \frac{t^3}{3!}C^3 + \cdots . \qquad (6.3.4)$$

Next we claim that

$$\frac{d}{dt}e^{tC} = Ce^{tC} . \qquad (6.3.5)$$

We verify the claim by supposing that we can differentiate (6.3.4) term by term with respect to t. Then

$$\frac{d}{dt}e^{tC} = \frac{d}{dt}(I_n) + \frac{d}{dt}(tC) + \frac{d}{dt}\left(\frac{t^2}{2!}C^2\right) + \frac{d}{dt}\left(\frac{t^3}{3!}C^3\right) + \frac{d}{dt}\left(\frac{t^4}{4!}C^4\right) + \cdots$$

$$= 0 + C + tC^2 + \frac{t^2}{2!}C^3 + \frac{t^3}{3!}C^4 + \cdots$$

$$= C\left(I_n + tC + \frac{t^2}{2!}C^2 + \frac{t^3}{3!}C^3 + \cdots\right)$$

$$= Ce^{tC} .$$

It follows that the function $X(t) = e^{tC}X_0$ is a solution of (6.3.1) for each $X_0 \in \mathbf{R}^n$; that is,

$$\frac{d}{dt}X(t) = \frac{d}{dt}e^{tC}X_0 = Ce^{tC}X_0 = CX(t).$$

Since (6.3.3) implies that $e^{0C} = e^0 = I_n$, it follows that $X(t) = e^{tC}X_0$ is a solution of (6.3.1) with initial condition $X(0) = X_0$. This discussion shows that solving (6.3.1) in closed form is equivalent to finding a closed form expression for the matrix exponential e^{tC}.

Theorem 6.3.1 *The unique solution to the initial value problem*

$$\frac{dX}{dt} = CX$$
$$X(0) = X_0$$

is

$$X(t) = e^{tC}X_0.$$

Proof: Existence follows from the previous discussion; uniqueness follows from the n-dimensional analog of Theorem 6.1.1. ◆

Explicit Computation of Matrix Exponentials

We begin with the simplest computation of a matrix exponential.

(*a*) Let L be a multiple of the identity; that is, let $L = \alpha I_n$, where α is a real number. Then

$$e^{\alpha I_n} = e^\alpha I_n; \tag{6.3.6}$$

that is, $e^{\alpha I_n}$ is a scalar multiple of the identity. To verify (6.3.6), compute

$$e^{\alpha I_n} = I_n + \alpha I_n + \frac{\alpha^2}{2!}I_n^2 + \frac{\alpha^3}{3!}I_n^3 + \cdots = (1 + \alpha + \frac{\alpha^2}{2!} + \frac{\alpha^3}{3!} + \cdots)I_n = e^\alpha I_n.$$

(*b*) Let C be a 2×2 diagonal matrix:

$$C = \begin{pmatrix} \lambda_1 & 0 \\ 0 & \lambda_2 \end{pmatrix},$$

where λ_1 and λ_2 are real constants. Then

$$e^{tC} = \begin{pmatrix} e^{\lambda_1 t} & 0 \\ 0 & e^{\lambda_2 t} \end{pmatrix}. \tag{6.3.7}$$

To verify (6.3.7) compute

$$e^{tC} = I_2 + tC + \frac{t^2}{2!}C^2 + \frac{t^3}{3!}C^3 + \cdots$$

$$= \begin{pmatrix} 1 & 0 \\ 0 & 1 \end{pmatrix} + \begin{pmatrix} \lambda_1 t & 0 \\ 0 & \lambda_2 t \end{pmatrix} + \begin{pmatrix} \frac{t^2}{2!}\lambda_1^2 & 0 \\ 0 & \frac{t^2}{2!}\lambda_2^2 \end{pmatrix} + \cdots$$

$$= \begin{pmatrix} e^{\lambda_1 t} & 0 \\ 0 & e^{\lambda_2 t} \end{pmatrix}.$$

(c) Suppose that

$$C = \begin{pmatrix} 0 & -1 \\ 1 & 0 \end{pmatrix}.$$

Then

$$e^{tC} = \begin{pmatrix} \cos t & -\sin t \\ \sin t & \cos t \end{pmatrix}. \tag{6.3.8}$$

We begin this computation by observing that

$$C^2 = -I_2, \quad C^3 = -C, \quad \text{and} \quad C^4 = I_n.$$

Therefore, by collecting terms of odd and even power in the series expansion for the matrix exponential, we obtain

$$e^{tC} = I_2 + tC + \frac{t^2}{2!}C^2 + \frac{t^3}{3!}C^3 + \cdots$$

$$= I_2 + tC - \frac{t^2}{2!}I_2 - \frac{t^3}{3!}C + \cdots$$

$$= \left(1 - \frac{t^2}{2!} + \frac{t^4}{4!} - \frac{t^6}{6!} + \cdots\right)I_2 + \left(t - \frac{t^3}{3!} + \frac{t^5}{5!} - \frac{t^7}{7!} + \cdots\right)C$$

$$= (\cos t)I_2 + (\sin t)C$$

$$= \begin{pmatrix} \cos t & -\sin t \\ \sin t & \cos t \end{pmatrix}.$$

In this computation we have used the fact that the trigonometric functions $\cos t$ and $\sin t$ have these power series expansions:

$$\cos t = 1 - \frac{1}{2!}t^2 + \frac{1}{4!}t^4 + \cdots = \sum_{k=0}^{\infty} \frac{(-1)^k}{(2k)!}t^{2k}$$

$$\sin t = t - \frac{1}{3!}t^3 + \frac{1}{5!}t^5 + \cdots = \sum_{k=0}^{\infty} \frac{(-1)^k}{(2k+1)!}t^{2k+1}.$$

See Exercise 10 for an alternative proof of (6.3.8).

To compute the matrix exponential MATLAB matrix!exponential!in MATLAB provides the command expm. We use this command to compute the matrix exponential e^{tC} for

$$C = \begin{pmatrix} 0 & -1 \\ 1 & 0 \end{pmatrix} \quad \text{and} \quad t = \frac{\pi}{4}.$$

Type

```
C = [0, -1; 1, 0];
t = pi/4;
expm(t*C)
```

which gives the answer

```
ans =
    0.7071   -0.7071
    0.7071    0.7071
```

Indeed, this is precisely what we expect by (6.3.8), since

$$\cos\left(\frac{\pi}{4}\right) = \sin\left(\frac{\pi}{4}\right) = \frac{1}{\sqrt{2}} \approx 0.70710678.$$

(d) Let

$$C = \begin{pmatrix} 0 & 1 \\ 0 & 0 \end{pmatrix}.$$

Then

$$e^{tC} = I_2 + tC = \begin{pmatrix} 1 & t \\ 0 & 1 \end{pmatrix} \tag{6.3.9}$$

since $C^2 = 0$.

COMPUTER EXERCISES

1. Let L be the 3×3 matrix

$$L = \begin{pmatrix} 2 & 0 & -1 \\ 0 & -1 & 3 \\ 1 & 0 & 1 \end{pmatrix}.$$

Find the smallest integer m such that

$$I_3 + L + \frac{1}{2!}L^2 + \frac{1}{3!}L^3 + \cdots + \frac{1}{m!}L^m$$

is equal to e^L up to a precision of two decimal places. More exactly, use the MATLAB command expm to compute e^L and use MATLAB commands to compute the series expansion to order m. Note that the command for computing $n!$ in MATLAB is prod(1:n).

2. Use MATLAB to compute the matrix exponential e^{tC} for

$$C = \begin{pmatrix} 1 & 1 \\ 2 & -1 \end{pmatrix}$$

by choosing for t the values 1.0, 1.5, and 2.5. Does $e^C e^{1.5C} = e^{2.5C}$?

3. For the scalar exponential function e^t it is well known that for any pair of real numbers t_1 and t_2, the following equality holds:

$$e^{t_1 + t_2} = e^{t_1} e^{t_2}.$$

Use MATLAB to find two 2×2 matrices C_1 and C_2 such that

$$e^{C_1 + C_2} \neq e^{C_1} e^{C_2}.$$

HAND EXERCISES

In Exercises 4–6, compute the matrix exponential e^{tC} for the matrix.

4. $\begin{pmatrix} 0 & 1 \\ 0 & 0 \end{pmatrix}$ **5.** $\begin{pmatrix} 0 & 1 & 0 \\ 0 & 0 & 1 \\ 0 & 0 & 0 \end{pmatrix}$ **6.** $\begin{pmatrix} 0 & -2 \\ 2 & 0 \end{pmatrix}$

7. Let α and β be real numbers and let αI and βI be corresponding $n \times n$ diagonal matrices. Use properties of the scalar exponential function to show that

$$e^{(\alpha + \beta)I} = e^{\alpha I} e^{\beta I}.$$

In Exercises 8–10, we use Theorem 6.1.1, the uniqueness of solutions to initial value problems, in perhaps a surprising way.

8. Prove that

$$e^{t+s} = e^t e^s$$

for all real numbers s and t.
Hint
  (a) Fix s and verify that $y(t) = e^{t+s}$ is a solution to the initial value problem

$$\frac{dx}{dt} = x$$
$$x(0) = e^s.$$

(6.3.10)

  (b) Fix s and verify that $z(t) = e^t e^s$ is also a solution to (6.3.10).
  (c) Use Theorem 6.1.1 to conclude that $y(t) = z(t)$ for every s.

9. Let A be an $n \times n$ matrix. Prove that

$$e^{(t+s)A} = e^{tA} e^{sA}$$

for all real numbers s and t.

Hint

(a) Fix $s \in \mathbf{R}$ and $X_0 \in \mathbf{R}^n$ and verify that $Y(t) = e^{(t+s)A} X_0$ is a solution to the initial value problem

$$\frac{dX}{dt} = AX$$
$$X(0) = e^{sA} X_0.$$
(6.3.11)

(b) Fix s and verify that $Z(t) = e^{tA} \left(e^{sA} X_0 \right)$ is also a solution to (6.3.11).

(c) Use the n-dimensional version of Theorem 6.1.1 to conclude that $Y(t) = Z(t)$ for every s and every X_0.

Remark: Compare the result in this exercise with the calculation in Exercise 7.

10. Prove that

$$\exp\left(t \begin{pmatrix} 0 & -1 \\ 1 & 0 \end{pmatrix} \right) = \begin{pmatrix} \cos t & -\sin t \\ \sin t & \cos t \end{pmatrix}.$$
(6.3.12)

Hint

(a) Verify that $X_1(t) = \begin{pmatrix} \cos t \\ \sin t \end{pmatrix}$ and $X_2(t) = \begin{pmatrix} -\sin t \\ \cos t \end{pmatrix}$ are solutions to the initial value problems

$$\frac{dX}{dt} = \begin{pmatrix} 0 & -1 \\ 1 & 0 \end{pmatrix} X$$
$$X(0) = e_j$$
(6.3.13)

for $j = 1, 2$.

(b) Since $X_j(0) = e_j$, use Theorems 6.1.1 and 6.3.1 to verify that

$$X_j(t) = \exp\left(t \begin{pmatrix} 0 & -1 \\ 1 & 0 \end{pmatrix} \right) e_j.$$
(6.3.14)

(c) Show that (6.3.14) proves (6.3.12).

11. Let C be an $n \times n$ matrix. Use Theorem 6.3.1 to show that the n columns of the $n \times n$ matrix e^{tC} give a basis of solutions for the system of differential equations $\dot{X} = CX$.

Remark: The completion of Exercises 12 and 13 constitutes a proof that the infinite series definition of the matrix exponential is a convergent series for all $n \times n$ matrices.

12. Let $A = (a_{ij})$ be an $n \times n$ matrix. Define

$$\|A\|_m = \max_{1 \leq i \leq n} (|a_{i1}| + \cdots + |a_{in}|) = \max_{1 \leq i \leq n} \left(\sum_{j=1}^{n} |a_{ij}| \right).$$

That is, to compute $\|A\|_m$, first sum the absolute values of the entries in each row of A, and then take the maximum of these sums. Prove that

$$\|AB\|_m \leq \|A\|_m \|B\|_m.$$

Hint Begin by noting that

$$\|AB\|_m = \max_{1 \leq i \leq n} \left(\sum_{j=1}^{n} \left| \sum_{k=1}^{n} a_{ik} b_{kj} \right| \right) \leq \max_{1 \leq i \leq n} \left(\sum_{j=1}^{n} \sum_{k=1}^{n} |a_{ik} b_{kj}| \right) = \max_{1 \leq i \leq n} \left(\sum_{k=1}^{n} \sum_{j=1}^{n} |a_{ik} b_{kj}| \right).$$

13. Recall that an infinite series of real numbers

$$c_1 + c_2 + \cdots + c_N + \cdots$$

converges absolutely if there is a constant K such that for every N the partial sum satisfies

$$|c_1| + |c_2| + \cdots + |c_N| \leq K.$$

Let A be an $n \times n$ matrix. To prove that the matrix exponential e^A is an absolutely convergent infinite series, use Exercise 12 and the following steps: Let a_N be the (i, j)th entry in the matrix A^N, where $A^0 = I_n$.

(a) $|a_N| \leq ||A^N||_m$ (b) $||A^N||_m \leq ||A||_m^N$

(c) $|a_0| + |a_1| + \cdots + \frac{1}{N!}|a_N| \leq e^{||A||_m}$

6.4   LINEAR NORMAL FORM PLANAR SYSTEMS

There are three linear systems of ordinary differential equations that we now solve explicitly using matrix exponentials. Remarkably, in a sense to be made precise, these are the only linear planar systems. The three systems are listed in Table 6.1.

Table 6.1
Solutions to normal form ODEs with $X(0) = X_0$

Name	Equations	Closed Form Solution
(a)	$\dot{X} = \begin{pmatrix} \lambda_1 & 0 \\ 0 & \lambda_2 \end{pmatrix} X$	$X(t) = \begin{pmatrix} e^{\lambda_1 t} & 0 \\ 0 & e^{\lambda_2 t} \end{pmatrix} X_0$
(b)	$\dot{X} = \begin{pmatrix} \sigma & -\tau \\ \tau & \sigma \end{pmatrix} X$	$X(t) = e^{\sigma t} \begin{pmatrix} \cos(\tau t) & -\sin(\tau t) \\ \sin(\tau t) & \cos(\tau t) \end{pmatrix} X_0$
(c)	$\dot{X} = \begin{pmatrix} \lambda_1 & 1 \\ 0 & \lambda_1 \end{pmatrix}$	$X(t) = e^{\lambda_1 t} \begin{pmatrix} 1 & t \\ 0 & 1 \end{pmatrix} X_0$

The verification of Table 6.1(a) follows from (6.3.7), but it just reproduces earlier work in Section 4.5 where we considered uncoupled systems of two ordinary differential equations. To verify the solutions to (b) and (c), we need to prove:

Proposition 6.4.1   *Let A and B be two $n \times n$ matrices such that*

$$AB = BA. \tag{6.4.1}$$

Then

$$e^{A+B} = e^A e^B.$$

Proof: Note that (6.4.1) implies that

$$A^k B = B A^k \tag{6.4.2}$$

$$e^{tA} B = B e^{tA}. \tag{6.4.3}$$

Identity (6.4.2) is verified when $k = 2$ using associativity of matrix multiplication as follows:

$$A^2 B = AAB = ABA = BAA = BA^2.$$

The argument for general k is identical. Identity (6.4.3) follows directly from (6.4.2) and the power series definition of matrix exponentials (6.3.3).

We use Theorem 6.1.1 to complete the proof of this proposition. Recall that

$$X(t) = e^{t(A+B)} X_0$$

is the unique solution to the initial value problem

$$\frac{dX}{dt} = (A + B)X$$

$$X(0) = X_0.$$

We claim that

$$Y(t) = e^{tA} e^{tB} X_0$$

is another solution to this equation. Certainly $Y(0) = X_0$. It follows from (6.3.5) that

$$\frac{d}{dt} e^{tA} = A e^{tA} \quad \text{and} \quad \frac{d}{dt} e^{tB} = B e^{tB}.$$

Thus the product rule together with (6.4.3) implies that

$$\frac{dY}{dt} = \left(A e^{tA} \right) e^{tB} X_0 + e^{tA} \left(B e^{tB} \right) X_0$$

$$= (A + B) e^{tA} e^{tB} X_0$$

$$= (A + B) Y(t).$$

Thus

$$\frac{dY}{dt} = (A + B)Y$$

and $Y(t) = X(t)$. Since X_0 is arbitrary, it follows that

$$e^{t(A+B)} = e^{tA} e^{tB}.$$

Evaluating at $t = 1$ yields the desired result.   ◆

Verification of Table 6.1(b)

We begin by noting that the 2×2 matrix C in Table 6.1(b) is

$$C = \begin{pmatrix} \sigma & -\tau \\ \tau & \sigma \end{pmatrix} = \sigma I_2 + \tau J,$$

where

$$J = \begin{pmatrix} 0 & -1 \\ 1 & 0 \end{pmatrix}.$$

Since $I_2 J = J I_2$, it follows from Proposition 6.4.1 that

$$e^{tC} = e^{(\sigma t)I_2} e^{(\tau t)J}.$$

Thus (6.3.6) and (6.3.8) imply that

$$e^{tC} = e^{\sigma t} \begin{pmatrix} \cos(\tau t) & -\sin(\tau t) \\ \sin(\tau t) & \cos(\tau t) \end{pmatrix}, \tag{6.4.4}$$

and (b) is verified.

Verification of Table 6.1(c)

To determine the solutions to Table 6.1(c), observe that

$$C = \begin{pmatrix} \lambda_1 & 1 \\ 0 & \lambda_1 \end{pmatrix} = \lambda_1 I_2 + N,$$

where

$$N = \begin{pmatrix} 0 & 1 \\ 0 & 0 \end{pmatrix}.$$

Since $I_2 N = N I_2$, Proposition 6.4.1 implies that

$$e^{tC} = e^{(t\lambda_1)I_2} e^{tN} = e^{t\lambda} \begin{pmatrix} 1 & t \\ 0 & 1 \end{pmatrix} \tag{6.4.5}$$

by (6.3.6) and (6.3.9).

Summary

The normal form matrices in Table 6.1 are characterized by the number of linearly independent real eigenvectors. We summarize this information in Table 6.2. We show in Section 6.5 that any planar linear system of ODEs can be solved just by noting how many independent eigenvectors the corresponding matrix has; general solutions are found by transforming the equations into one of the three types of equations listed in Table 6.1.

Table 6.2

Number of linearly independent real eigenvectors

Matrix	Number of Real Eigenvectors	Reference
$\begin{pmatrix} \lambda_1 & 0 \\ 0 & \lambda_2 \end{pmatrix}$	Two linearly independent	Section 4.9
$\begin{pmatrix} \sigma & -\tau \\ \tau & \sigma \end{pmatrix}$	None	(4.8.12)
$\begin{pmatrix} \lambda_1 & 1 \\ 0 & \lambda_1 \end{pmatrix}$	One linearly independent	Lemma 6.2.3

HAND EXERCISES

1. Solve the initial value problem

$$\begin{aligned} \dot{x} &= 2x + 3y \\ \dot{y} &= -3x + 2y, \end{aligned}$$

where $x(0) = 1$ and $y(0) = -2$.

2. Solve the initial value problem

$$\begin{aligned} \dot{x} &= -2x + y \\ \dot{y} &= -2y, \end{aligned}$$

where $x(0) = 4$ and $y(0) = -1$.

3. Let A be an $n \times n$ matrix such that $A^3 = 0$. Compute e^{tC}, where $C = 2I_n + A$.

COMPUTER EXERCISES

4. Use pplane5 to plot phase plane portraits for each of the three types of linear systems (a), (b), and (c) in Table 6.1. Based on this computer exploration answer the following questions:

   (i) If a solution to that system spirals about the origin, is the system of differential equations of type (a), (b), or (c)?

   (ii) How many eigendirections are there for equations of type (c)?

   (iii) Let $(x(t), y(t))$ be a solution to one of these three types of systems, and suppose that $y(t)$ oscillates up and down infinitely often. Then $(x(t), y(t))$ is a solution for which type of system?

6.5 SIMILAR MATRICES

In Section 6.4 we discussed solutions to differential equations $\dot{X} = CX$ for three classes of matrices C; see Table 6.1. We stated that in a certain sense every 2×2 matrix can be thought of as a member of one of these families. In this section we show that every

2×2 matrix is similar to one of the matrices in that table (see Theorem 6.5.5), where similarity is defined as follows.

Definition 6.5.1   *The $n \times n$ matrices B and C are similar if there exists an invertible $n \times n$ matrix P such that*

$$C = P^{-1}BP.$$

Our present interest in similar matrices stems from the fact that if we know the solutions to the system of differential equations $\dot{Y} = CY$ in closed form, then we know the solutions to the system of differential equations $\dot{X} = BX$ in closed form. More precisely:

Lemma 6.5.2   *Suppose that B and $C = P^{-1}BP$ are similar matrices. If $Y(t)$ is a solution to the system of differential equations $\dot{Y} = CY$, then $X(t) = PY(t)$ is a solution to the system of differential equations $\dot{X} = BX$.*

Proof: Since the entries in the matrix P are constants, it follows that

$$\frac{dX}{dt} = P\frac{dY}{dt}.$$

Since $Y(t)$ is a solution to the $\dot{Y} = CY$ equation, it follows that

$$\frac{dX}{dt} = PCY.$$

Since $Y = P^{-1}X$ and $PCP^{-1} = B$,

$$\frac{dX}{dt} = PCP^{-1}X = BX.$$

Thus $X(t)$ is a solution to $\dot{X} = BX$, as claimed.   ◆

Invariants of Similarity

Lemma 6.5.3   *Let A and B be similar 2×2 matrices. Then*

$$p_A(\lambda) = p_B(\lambda)$$
$$\det(A) = \det(B)$$
$$tr(A) = tr(B)$$

and the eigenvalues of A and B are equal.

Proof: The determinant is a function on 2×2 matrices that has several important properties. Recall, in particular, from Theorem 3.8.2 that for any pair of 2×2 matrices A and B,

$$\det(AB) = \det(A)\det(B), \tag{6.5.1}$$

and for any invertible 2×2 matrix P,

$$\det(P^{-1}) = \frac{1}{\det(P)}. \tag{6.5.2}$$

Let P be an invertible 2×2 matrix so that $B = P^{-1}AP$. Using (6.5.1) and (6.5.2), we see that

$$
\begin{aligned}
p_B(\lambda) &= \det(B - \lambda I_2) \\
&= \det(P^{-1}AP - \lambda I_2) \\
&= \det(P^{-1}(A - \lambda I_2)P) \\
&= \det(A - \lambda I_2) \\
&= p_A(\lambda).
\end{aligned}
$$

Hence the eigenvalues of A and B are the same. It follows from (4.8.8) and (4.8.9) that the determinants and traces of A and B are equal. $\blacklozenge$

For example, if

$$A = \begin{pmatrix} -1 & 0 \\ 0 & 1 \end{pmatrix} \quad \text{and} \quad P = \begin{pmatrix} 1 & 2 \\ 1 & 1 \end{pmatrix},$$

then

$$P^{-1} = \begin{pmatrix} -1 & 2 \\ 1 & -1 \end{pmatrix}$$

and

$$P^{-1}AP = \begin{pmatrix} 3 & 4 \\ -2 & -3 \end{pmatrix}.$$

A calculation shows that

$$\det(P^{-1}AP) = -1 = \det(A) \quad \text{and} \quad \text{tr}(P^{-1}AP) = 0 = \text{tr}(A),$$

as stated in Lemma 6.5.3.

Similarity and Matrix Exponentials

We introduce similarity at this juncture for the following reason: If C is a matrix that is similar to B, then e^C can be computed from e^B. More precisely:

Lemma 6.5.4 *Let C and B be $n \times n$ similar matrices, and let P be an invertible $n \times n$ matrix such that*

$$C = P^{-1}BP.$$

Then

$$e^C = P^{-1}e^B P. \tag{6.5.3}$$

Proof: Note that for all powers of k we have

$$(P^{-1}BP)^k = P^{-1}B^k P.$$

Next verify (6.5.3) by computing

$$e^C = \sum_{k=0}^{\infty} \frac{1}{k!} C^k = \sum_{k=0}^{\infty} \frac{1}{k!}(P^{-1}BP)^k = \sum_{k=0}^{\infty} \frac{1}{k!} P^{-1}B^k P = P^{-1}\left(\sum_{k=0}^{\infty} \frac{1}{k!} B^k\right) P = P^{-1}e^B P. \quad \blacklozenge$$

Classification of 2 × 2 Matrices

We now classify all 2×2 matrices up to similarity.

Theorem 6.5.5 *Let C and $P = (v_1 | v_2)$ be 2×2 matrices, where the vectors v_1 and v_2 are specified below.*

(a) *Suppose that C has two linearly independent real eigenvectors v_1 and v_2 with real eigenvalues λ_1 and λ_2. Then*

$$P^{-1}CP = \begin{pmatrix} \lambda_1 & 0 \\ 0 & \lambda_2 \end{pmatrix}.$$

(b) *Suppose that C has no real eigenvectors and complex conjugate eigenvalues $\sigma \pm i\tau$, where $\tau \neq 0$. Then*

$$P^{-1}CP = \begin{pmatrix} \sigma & -\tau \\ \tau & \sigma \end{pmatrix},$$

where $v_1 + iv_2$ is an eigenvector of C associated with the eigenvalue $\lambda_1 = \sigma - i\tau$.

(c) *Suppose that C has exactly one linearly independent real eigenvector v_1 with real eigenvalue λ_1. Then*

$$P^{-1}CP = \begin{pmatrix} \lambda_1 & 1 \\ 0 & \lambda_1 \end{pmatrix},$$

where v_2 is a generalized eigenvector of C that satisfies

$$(C - \lambda_1 I_2)v_2 = v_1. \tag{6.5.4}$$

Proof: The strategy in the proof of this theorem is to determine the 1st and 2nd columns of $P^{-1}CP$ by computing (in each case) $P^{-1}CPe_j$ for $j = 1$ and $j = 2$. Note from the definition of P that

$$Pe_1 = v_1 \quad \text{and} \quad Pe_2 = v_2.$$

In addition, if P is invertible, then

$$P^{-1}v_1 = e_1 \quad \text{and} \quad P^{-1}v_2 = e_2.$$

Note that if v_1 and v_2 are linearly independent, then P is invertible.

(a)   Since v_1 and v_2 are assumed to be linearly independent, P is invertible. So we can compute

$$P^{-1}CPe_1 = P^{-1}Cv_1 = \lambda P^{-1}v_1 = \lambda e_1.$$

It follows that the 1st column of $P^{-1}CP$ is

$$\begin{pmatrix} \lambda_1 \\ 0 \end{pmatrix}.$$

Similarly, the 2nd column of $P^{-1}CP$ is

$$\begin{pmatrix} 0 \\ \lambda_2 \end{pmatrix},$$

thus verifying (a).

(b)   Lemma 6.2.2 implies that v_1 and v_2 are linearly independent and hence that P is invertible. Using (6.2.9), with τ replaced by $-\tau$, v replaced by v_1, and w replaced by w_1, we calculate

$$P^{-1}CPe_1 = P^{-1}Cv_1 = \sigma P^{-1}v_1 + \tau P^{-1}v_2 = \sigma e_1 + \tau e_2$$

and

$$P^{-1}CPe_2 = P^{-1}Cv_2 = -\tau P^{-1}v_1 + \sigma P^{-1}v_2 = -\tau e_1 + \sigma e_2.$$

Thus the columns of $P^{-1}CP$ are

$$\begin{pmatrix} \sigma \\ \tau \end{pmatrix} \quad \text{and} \quad \begin{pmatrix} -\tau \\ \sigma \end{pmatrix},$$

as desired.

   (c)   Let v_1 be an eigenvector and assume that v_2 is a generalized eigenvector satisfying (6.5.4). By Lemma 6.2.4, the vectors v_1 and v_2 exist and are linearly independent. For this choice of v_1 and v_2, compute

$$P^{-1}CPe_1 = P^{-1}Cv_1 = \lambda_1 P^{-1}v_1 = \lambda_1 e_1$$

and

$$P^{-1}CPe_2 = P^{-1}Cv_2 = P^{-1}v_1 + \lambda_1 P^{-1}v_2 = e_1 + \lambda_1 e_2.$$

Thus the two columns of $P^{-1}CP$ are

$$\begin{pmatrix} \lambda_1 \\ 0 \end{pmatrix} \quad \text{and} \quad \begin{pmatrix} 1 \\ \lambda_1 \end{pmatrix}. \qquad \blacklozenge$$

Closed Form Solutions Using Similarity

We now use Lemma 6.5.2, Theorem 6.5.5, and the explicit solutions to the normal form equations in Table 6.1 to find solutions for $\dot{X} = CX$, where C is any 2×2 matrix. The idea behind the use of similarity to solve systems of ODEs is to transform a given system into another normal form system whose solution is already known. This method is very much like the technique of change of variables used when finding indefinite integrals in calculus.

   We suppose that we are given a system of differential equations $\dot{X} = CX$ and use Theorem 6.5.5 to transform C by similarity to one of the normal form matrices listed in that theorem. We then solve the transformed equation, as we did in Section 6.4 (see Table 6.1), and use Lemma 6.5.2 to transform the solution back to the given system.

   For example, suppose that C has a complex eigenvalue $\sigma - i\tau$ with corresponding eigenvector $v + iw$. Then Theorem 6.5.5 states that

$$B = P^{-1}CP = \begin{pmatrix} \sigma & -\tau \\ \tau & \sigma \end{pmatrix},$$

where $P = (v|w)$ is an invertible matrix. From Table 6.1 the general solution to the system of equations $\dot{Y} = BY$ is

$$Y(t) = e^{\sigma t} \begin{pmatrix} \cos(\tau t) & -\sin(\tau t) \\ \sin(\tau t) & \cos(\tau t) \end{pmatrix} \begin{pmatrix} \alpha \\ \beta \end{pmatrix}.$$

Lemma 6.5.2 states that

$$X(t) = PY(t)$$

is the general solution to the $\dot{X} = CX$ system. Moreover, we can solve the initial value problem by solving

$$X_0 = PY(0) = P \begin{pmatrix} \alpha \\ \beta \end{pmatrix}$$

for α and β. In particular,

$$\begin{pmatrix} \alpha \\ \beta \end{pmatrix} = P^{-1} X_0.$$

Putting these steps together implies that

$$X(t) = e^{\sigma t} P \begin{pmatrix} \cos(\tau t) & -\sin(\tau t) \\ \sin(\tau t) & \cos(\tau t) \end{pmatrix} P^{-1} X_0 \tag{6.5.5}$$

is the solution to the initial value problem.

The Example with Complex Eigenvalues Revisited

Recall the example in (6.2.11):

$$\frac{dX}{dt} = \begin{pmatrix} -1 & 2 \\ -5 & -3 \end{pmatrix} X$$

with initial values

$$X_0 = \begin{pmatrix} 1 \\ 1 \end{pmatrix}.$$

This linear system has a complex eigenvalue $\sigma - i\tau = -2 - 3i$ with corresponding eigenvector

$$v + iw = \begin{pmatrix} 2 \\ -1 - 3i \end{pmatrix}.$$

Thus the matrix P that transforms C into normal form is

$$P = \begin{pmatrix} 2 & 0 \\ -1 & -3 \end{pmatrix} \quad \text{and} \quad P^{-1} = \frac{1}{6} \begin{pmatrix} 3 & 0 \\ -1 & -2 \end{pmatrix}.$$

It follows from (6.5.5) that the solution to the initial value problem is

$$X(t) = e^{-2t} P \begin{pmatrix} \cos(3t) & -\sin(3t) \\ \sin(3t) & \cos(3t) \end{pmatrix} P^{-1} X_0$$

$$= \frac{1}{6} e^{-2t} \begin{pmatrix} 2 & 0 \\ -1 & -3 \end{pmatrix} \begin{pmatrix} \cos(3t) & -\sin(3t) \\ \sin(3t) & \cos(3t) \end{pmatrix} \begin{pmatrix} 3 & 0 \\ -1 & -2 \end{pmatrix} \begin{pmatrix} 1 \\ 1 \end{pmatrix}.$$

A calculation gives

$$X(t) = \frac{1}{2} e^{-2t} \begin{pmatrix} 2 & 0 \\ -1 & -3 \end{pmatrix} \begin{pmatrix} \cos(3t) & -\sin(3t) \\ \sin(3t) & \cos(3t) \end{pmatrix} \begin{pmatrix} 1 \\ -1 \end{pmatrix}$$

$$= e^{-2t} \begin{pmatrix} \cos(3t) + \sin(3t) \\ \cos(3t) - 2\sin(3t) \end{pmatrix}.$$

Thus the solution to (6.2.11) that we found using similarity of matrices is identical to the solution (6.2.12) that we found by the direct method.

Solving systems with either distinct real eigenvalues or equal eigenvalues works in a similar fashion.

HAND EXERCISES

1. Suppose that the matrices A and B are similar and the matrices B and C are similar. Show that A and C are also similar matrices.

2. Use (4.8.13) to verify that the traces of similar matrices are equal.

In Exercises 3 and 4, determine whether or not the given matrices are similar, and why.

3. $A = \begin{pmatrix} 1 & 2 \\ 3 & 4 \end{pmatrix}$ and $B = \begin{pmatrix} 2 & -2 \\ -3 & 8 \end{pmatrix}$ **4.** $C = \begin{pmatrix} 2 & 2 \\ 2 & 2 \end{pmatrix}$ and $D = \begin{pmatrix} 4 & -2 \\ -2 & 4 \end{pmatrix}$

5. Let $B = P^{-1} A P$ so that A and B are similar matrices. Suppose that v is an eigenvector of B with eigenvalue λ. Show that Pv is an eigenvector of A with eigenvalue λ.

6. Which $n \times n$ matrices are similar to I_n?

7. Compute e^A, where

$$A = \begin{pmatrix} 3 & -1 \\ 1 & 1 \end{pmatrix}.$$

Check your answer using MATLAB.

6.6 *FORMULAS FOR MATRIX EXPONENTIALS

We now complete our discussion of exact solutions to planar linear systems of ODEs $\dot{X} = CX$. There are three different methods for finding closed form solutions to systems of ODEs. We have discussed two of these methods. In the first we find solutions by the direct method; that is, we find two linear independent solutions whose linear combinations form the space of solutions; see Section 6.2. In the second method we use similarity and normal form equations (whose solutions are obtained using matrix exponentials) to find closed form solutions; see Sections 6.4 and 6.5. In this section we present a third method based on computable formulas for matrix exponentials derived using the Cayley Hamilton theorem.

A Formula for the Matrix Exponential

For 2×2 matrices C with eigenvalues λ_1 and λ_2 there is a simple formula for the matrix exponential e^{tC} whose derivation depends on the Cayley Hamilton theorem. When the eigenvalues λ_1 and λ_2 of C are distinct, the formula is

$$e^{tC} = \frac{1}{\lambda_2 - \lambda_1} \left(e^{\lambda_1 t}(C - \lambda_2 I_2) - e^{\lambda_2 t}(C - \lambda_1 I_2) \right). \qquad (6.6.1)$$

When the eigenvalues are equal, the formula is

$$e^{tC} = e^{\lambda_1 t}(I_2 + tN), \qquad (6.6.2)$$

where $N = C - \lambda_1 I_2$.

Note that when we compute the matrix exponential using either (6.6.1) or (6.6.2), it is not necessary to compute the eigenvectors of C. This is a substantial simplification. But it is with the use of formula (6.6.2) that the greatest simplification occurs.

The Example with Equal Eigenvalues Revisited

Let us reconsider the system of differential equations (6.2.17):

$$\frac{dX}{dt} = \begin{pmatrix} 1 & -1 \\ 9 & -5 \end{pmatrix} X = CX$$

with initial value

$$X_0 = \begin{pmatrix} 2 \\ 3 \end{pmatrix}.$$

The eigenvalues of C are real and equal to $\lambda_1 = -2$.

We may write

$$C = \lambda_1 I_2 + N = -2I_2 + N,$$

where

$$N = \begin{pmatrix} 3 & -1 \\ 9 & -3 \end{pmatrix}.$$

It follows from (6.6.2) that

$$e^{tC} = e^{-2t} \left(I_2 + t \begin{pmatrix} 3 & -1 \\ 9 & -3 \end{pmatrix} \right) = e^{-2t} \begin{pmatrix} 1 + 3t & -t \\ 9t & 1 - 3t \end{pmatrix}. \tag{6.6.3}$$

Hence the solution to the initial value problem is

$$X(t) = e^{tC} X_0 = e^{-2t} \begin{pmatrix} 1 + 3t & -t \\ 9t & 1 - 3t \end{pmatrix} \begin{pmatrix} 2 \\ 3 \end{pmatrix} = e^{-2t} \begin{pmatrix} 2 + 3t \\ 3 + 9t \end{pmatrix}.$$

The Cayley Hamilton Theorem

The Cayley Hamilton theorem states that a matrix satisfies its own characteristic polynomial. More precisely:

Theorem 6.6.1 (Cayley Hamilton Theorem) *Let A be a 2×2 matrix and let*

$$p_A(\lambda) = \lambda^2 + a\lambda + b$$

be the characteristic polynomial of A. Then

$$p_A(A) = A^2 + aA + bI_2 = 0.$$

Proof: Suppose $B = P^{-1}AP$ and A are similar matrices. We claim that if $p_A(A) = 0$, then $p_B(B) = 0$. To verify this claim, recall from Lemma 6.5.3 that $p_A = p_B$ and calculate

$$p_B(B) = p_A(P^{-1}AP) = (P^{-1}AP)^2 + aP^{-1}AP + bI_2 = P^{-1}p_A(A)P = 0.$$

Theorem 6.5.5 classifies 2×2 matrices up to similarity. Thus we need only verify this theorem for the matrices

$$C = \begin{pmatrix} \lambda_1 & 0 \\ 0 & \lambda_2 \end{pmatrix}, \quad D = \begin{pmatrix} \sigma & -\tau \\ \tau & \sigma \end{pmatrix}, \quad E = \begin{pmatrix} \lambda_1 & 1 \\ 0 & \lambda_1 \end{pmatrix};$$

that is, we need to verify that

$$p_C(C) = 0, \quad p_D(D) = 0, \quad p_E(E) = 0.$$

Using the fact that $p_A(\lambda) = \lambda^2 - \mathrm{tr}(A)\lambda + \det(A)$, we see that

$$\begin{aligned} p_C(\lambda) &= (\lambda - \lambda_1)(\lambda - \lambda_2) \\ p_D(\lambda) &= \lambda^2 - 2\sigma\lambda + (\sigma^2 + \tau^2) \\ p_E(\lambda) &= (\lambda - \lambda_1)^2. \end{aligned}$$

It now follows that

$$p_C(C) = (C - \lambda_1 I_2)(C - \lambda_2 I_2) = \begin{pmatrix} 0 & 0 \\ 0 & \lambda_2 - \lambda_1 \end{pmatrix} \begin{pmatrix} \lambda_1 - \lambda_2 & 0 \\ 0 & 0 \end{pmatrix} = 0$$

$$p_D(D) = \begin{pmatrix} \sigma^2 - \tau^2 & -2\sigma\tau \\ 2\sigma\tau & \sigma^2 - \tau^2 \end{pmatrix} - 2\sigma \begin{pmatrix} \sigma & -\tau \\ \tau & \sigma \end{pmatrix} + (\sigma^2 + \tau^2) \begin{pmatrix} 1 & 0 \\ 0 & 1 \end{pmatrix} = 0$$

$$p_E(E) = (E - \lambda_1 I_2)^2 = \begin{pmatrix} 0 & 1 \\ 0 & 0 \end{pmatrix}^2 = 0. \qquad \blacklozenge$$

Verification of (6.6.1)

Let C be a 2×2 matrix with eigenvalues $\lambda_1 \neq \lambda_2$. Then the characteristic polynomial of C is

$$p_C(\lambda) = (\lambda - \lambda_1)(\lambda - \lambda_2).$$

We begin our verification of (6.6.1) by showing that

$$I_2 = a_1(A - \lambda_2 I_2) + a_2(A - \lambda_1 I_2), \tag{6.6.4}$$

where

$$a_1 = \frac{1}{\lambda_2 - \lambda_1} \quad \text{and} \quad a_2 = \frac{1}{\lambda_1 - \lambda_2}. \tag{6.6.5}$$

Using partial fractions, we can write

$$\frac{1}{p_C(\lambda)} = \frac{a_1}{\lambda - \lambda_1} + \frac{a_2}{\lambda - \lambda_2}, \tag{6.6.6}$$

where a_j is as in (6.6.5). Multiplying (6.6.6) by $p_C(\lambda)$ yields

$$1 = a_1(\lambda - \lambda_2) + a_2(\lambda - \lambda_1).$$

Now let v_j be the eigenvector of C corresponding to the eigenvalue λ_j and compute

$$(a_1(C - \lambda_2 I_2) + a_2(C - \lambda_1 I_2))v_1 = a_1(C - \lambda_2 I_2)v_1 = a_1(\lambda_1 - \lambda_2)v_1 = v_1.$$

Similarly

$$(a_1(C - \lambda_2 I_2) + a_2(C - \lambda_1 I_2))v_2 = v_2.$$

Since v_1 and v_2 form a basis, (6.6.4) holds by linearity.

We use the Cayley Hamilton theorem to show that

$$\begin{aligned} (C - \lambda_2 I_2)e^{tC} &= e^{\lambda_1 t}(C - \lambda_2 I_2) \\ (C - \lambda_1 I_2)e^{tC} &= e^{\lambda_2 t}(C - \lambda_1 I_2). \end{aligned} \tag{6.6.7}$$

First, assume that (6.6.7) is valid. We see that

$$e^{tC} = (a_1(C - \lambda_2 I_2) + a_2(C - \lambda_1 I_2))e^{tC}$$

by (6.6.4) and that

$$e^{tC} = a_1 e^{\lambda_1 t}(C - \lambda_2 I_2) + a_2 e^{\lambda_2 t}(C - \lambda_1 I_2)$$

by (6.6.7). Thus we have verified formula (6.6.1). To validate (6.6.7), calculate

$$e^{tC} = e^{tC - \lambda_1 t I_2 + \lambda_1 t I_2} = e^{\lambda_1 t} e^{t(C - \lambda_1 I_2)}$$

using Proposition 6.4.1 and the fact that I_2 commutes with every 2×2 matrix. Second, compute

$$
\begin{aligned}
(C - \lambda_2 I_2)e^{tC} &= e^{\lambda_1 t}(C - \lambda_2 I_2)e^{t(C - \lambda_1 I_2)} \\
&= e^{\lambda_1 t}(C - \lambda_2 I_2)(I_2 + t(C - \lambda_1 I_2) + \cdots) \\
&= e^{\lambda_1 t}(C - \lambda_2 I_2),
\end{aligned}
$$

since every other term in the infinite series contains a factor of

$$p_C(C) = (C - \lambda_2 I_2)(C - \lambda_1 I_2),$$

which vanishes by the Cayley Hamilton theorem. The second equation in (6.6.7) is proved similarly by interchanging the roles of λ_1 and λ_2.

Verification of (6.6.2)

The verification of (6.6.2) is less complicated. Since C is assumed to have a double eigenvalue λ_1, it follows that

$$N = C - \lambda_1 I_2$$

has 0 as a double eigenvalue. Hence the characteristic polynomial $p_N(\lambda) = \lambda^2$ and the Cayley Hamilton theorem implies that $N^2 = 0$. Therefore

$$e^{tC} = e^{t(C - \lambda_1 I_2) + \lambda_1 t I_2} = e^{\lambda_1 t} e^{tN} = e^{\lambda_1 t}(I_2 + tN),$$

as desired.

HAND EXERCISES

1. Solve the initial value problem

$$\frac{dX}{dt} = \begin{pmatrix} 0 & 1 \\ -2 & 3 \end{pmatrix} X,$$

where $X(0) = (2, 1)^t$.

2. Find all solutions to the linear system of ODEs

$$\frac{dX}{dt} = \begin{pmatrix} -2 & 4 \\ -1 & 1 \end{pmatrix} X.$$

3. Solve the initial value problem

$$\frac{dX}{dt} = \begin{pmatrix} 2 & 1 \\ -2 & 0 \end{pmatrix} X,$$

where $X(0) = (1, 1)^t$.

4. Let A be a 2×2 matrix. Show that

$$A^2 = \mathrm{tr}(A)A - \det(A)I_2.$$

6.7 SECOND-ORDER EQUATIONS

A second-order constant coefficient homogeneous differential equation is a differential equation of the form

$$\ddot{x} + b\dot{x} + ax = 0, \tag{6.7.1}$$

where a and b are real numbers.

Newton's Second Law

Newton's second law of motion is a second-order ordinary differential equation, and for this reason second-order equations arise naturally in mechanical systems. Newton's second law states that

$$F = ma, \tag{6.7.2}$$

where F is force, m is mass, and a is acceleration.

Newton's Second Law and Particle Motion on a Line

For a point mass moving along a line, (6.7.2) is

$$F = m\frac{d^2x}{dt^2}, \tag{6.7.3}$$

where $x(t)$ is the position of the point mass at time t. For example, suppose that a particle of mass m is falling toward the earth. If we let g be the gravitational constant and if we ignore all forces except gravitation, then the force acting on that particle is $F = -mg$. In this case Newton's second law leads to the second-order ordinary differential equation

$$\frac{d^2x}{dt^2} + g = 0. \tag{6.7.4}$$

Newton's Second Law and the Motion of a Spring

As a second example, consider the spring model pictured in Figure 6.2. Assume that the spring has zero mass and that an object of mass m is attached to the end of the spring. Let L be the natural length of the spring, and let $x(t)$ measure the distance that the spring is extended (or compressed). It follows from Newton's law that (6.7.3) is satisfied. Hooke's law states that the force F acting on a spring is

$$F = -\kappa x,$$

where κ is a positive constant. If the spring is damped by sliding friction, then

$$F = -\kappa x - \mu \frac{dx}{dt},$$

where μ is also a positive constant. Suppose, in addition, that an external force $F_{\text{ext}}(t)$ also acts on the mass and that that force is time-dependent. Then the entire force acting on the mass is

$$F = -\kappa x - \mu \frac{dx}{dt} + F_{\text{ext}}(t).$$

By Newton's second law, the motion of the mass is described by

$$m \frac{d^2x}{dt^2} + \mu \frac{dx}{dt} + \kappa x = F_{\text{ext}}(t), \tag{6.7.5}$$

which is again a second-order ordinary differential equation.

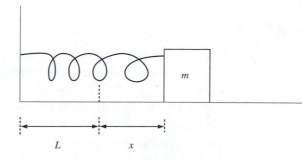

Figure 6.2
Hooke's law spring

A Reduction to a First-Order System

There is a simple trick that reduces a single linear second-order differential equation to a system of two linear first-order equations. For example, consider the linear homogeneous ordinary differential equation (6.7.1). To reduce this second-order equation to a first-order system, just set $y = \dot{x}$. Then (6.7.1) becomes

$$\dot{y} + by + ax = 0.$$

It follows that if $x(t)$ is a solution to (6.7.1) and $y(t) = \dot{x}(t)$, then $(x(t), y(t))$ is a solution to

$$
\begin{aligned}
\dot{x} &= y \\
\dot{y} &= -ax - by.
\end{aligned}
\tag{6.7.6}
$$

We can rewrite (6.7.6) as

$$\dot{X} = QX,$$

where

$$Q = \begin{pmatrix} 0 & 1 \\ -a & -b \end{pmatrix}. \tag{6.7.7}$$

Note that if $(x(t), y(t))$ is a solution to (6.7.6), then $x(t)$ is a solution to (6.7.1). Thus solving the single second-order linear equation is exactly the same as solving the corresponding first-order linear system.

The Initial Value Problem

To solve the homogeneous system (6.7.6) we need to specify two initial conditions $X(0) = (x(0), y(0))^t$. It follows that to solve the single second-order equation we need to specify two initial conditions $x(0)$ and $\dot{x}(0)$; that is, we need to specify both initial position and initial velocity.

The General Solution

There are two ways in which we can solve the second-order homogeneous equation (6.7.1). First, we know how to solve the system (6.7.6) by finding the eigenvalues and eigenvectors of the coefficient matrix Q in (6.7.7). Second, we know from the general theory of planar systems that solutions have the form $x(t) = e^{\lambda_0 t}$ for some scalar λ_0. We need only determine the values of λ_0 for which we get solutions to (6.7.1).

We now discuss the second approach. Suppose that $x(t) = e^{\lambda_0 t}$ is a solution to (6.7.1). Substituting this form of $x(t)$ in (6.7.1) yields the equation

$$\left(\lambda_0^2 + b\lambda_0 + a\right) e^{\lambda_0 t} = 0.$$

So $x(t) = e^{\lambda_0 t}$ is a solution to (6.7.1) precisely when $p_Q(\lambda_0) = 0$, where

$$p_Q(\lambda) = \lambda^2 + b\lambda + a \tag{6.7.8}$$

is the characteristic polynomial of the matrix Q in (6.7.7).

Suppose that λ_1 and λ_2 are distinct real roots of p_Q. Then the general solution to (6.7.1) is

$$x(t) = \alpha_1 e^{\lambda_1 t} + \alpha_2 e^{\lambda_2 t},$$

where $\alpha_j \in \mathbf{R}$.

An Example with Distinct Real Eigenvalues

For example, solve the initial value problem

$$\ddot{x} + 3\dot{x} + 2x = 0 \tag{6.7.9}$$

with initial conditions $x(0) = 0$ and $\dot{x}(0) = -2$. The characteristic polynomial is

$$p_Q(\lambda) = \lambda^2 + 3\lambda + 2 = (\lambda + 2)(\lambda + 1),$$

whose roots are $\lambda_1 = -1$ and $\lambda_2 = -2$. So the general solution to (6.7.9) is

$$x(t) = \alpha_1 e^{-t} + \alpha_2 e^{-2t}.$$

To find the precise solution we need to solve

$$x(0) = \alpha_1 + \alpha_2 = 0$$
$$\dot{x}(0) = -\alpha_1 - 2\alpha_2 = -2.$$

So $\alpha_1 = -2$ and $\alpha_2 = 2$, and the solution to the initial value problem (6.7.9) is

$$x(t) = -2e^{-t} + 2e^{-2t}.$$

An Example with Complex Conjugate Eigenvalues

Consider the differential equation

$$\ddot{x} - 2\dot{x} + 5x = 0. \tag{6.7.10}$$

The roots of the characteristic polynomial associated with (6.7.10) are $\lambda_1 = 1 + 2i$ and $\lambda_2 = 1 - 2i$. It follows from the discussion in the preceding section that the general solution to (6.7.10) is

$$x(t) = \mathrm{Re}\left(\alpha_1 e^{\lambda_1 t} + \alpha_2 e^{\lambda_2 t}\right),$$

where α_1 and α_2 are complex scalars. Indeed, we can rewrite this solution in real form (using Euler's formula) as

$$x(t) = e^t \left(\beta_1 \cos(2t) + \beta_2 \sin(2t) \right)$$

for real scalars β_1 and β_2.

In general, if the roots of the characteristic polynomial are $\sigma \pm i\tau$, then the general solution to the differential equation is

$$x(t) = e^{\sigma t} \left(\beta_1 \cos(\tau t) + \beta_2 \sin(\tau t) \right).$$

An Example with Multiple Eigenvalues

Note that the coefficient matrix Q of the associated first-order system in (6.7.7) is never a multiple of I_2. It follows from the preceding section that when the roots of the characteristic polynomial are real and equal, the general solution has the form

$$x(t) = \alpha_1 e^{\lambda_1 t} + \alpha_2 t e^{\lambda_2 t}.$$

Summary

It follows from this discussion that solutions to second-order homogeneous linear equations are either a linear combination of two exponentials (real unequal eigenvalues), $\alpha + \beta t$ times one exponential (real equal eigenvalues), or a time periodic function times an exponential (complex eigenvalues).

In particular, if the real part of the complex eigenvalues is 0, then the solution is time periodic. The frequency of this periodic solution is often called the *internal frequency*, a point that is made more clearly in the next example.

Solving the Spring Equation

Consider the equation for the frictionless spring without external forcing. From (6.7.5) we get

$$m\ddot{x} + \kappa x = 0, \qquad\qquad \textbf{(6.7.11)}$$

where $\kappa > 0$. The roots are $\lambda_1 = \sqrt{\kappa/m}\,i$ and $\lambda_2 = -\sqrt{\kappa/m}\,i$. So the general solution is

$$x(t) = \alpha \cos(\tau t) + \beta \sin(\tau t),$$

where $\tau = \sqrt{\kappa/m}$. Under these assumptions the motion of the spring is time periodic with period $2\pi/\tau$ or internal frequency $\tau/2\pi$. In particular, the solution satisfying initial conditions $x(0) = 1$ and $\dot{x}(0) = 0$ (the spring is extended one unit in distance and released with no initial velocity) is

$$x(t) = \cos(\tau t).$$

The graph of this function when $\tau = 1$ is given on the left in Figure 6.3.

If a small amount of friction is added, then the spring equation is

$$m\ddot{x} + \mu\dot{x} + \kappa x = 0,$$

where $\mu > 0$ is small. Since the eigenvalues of the characteristic polynomial are $\lambda = \sigma \pm i\tau$, where

$$\sigma = -\frac{\mu}{2m} < 0 \quad \text{and} \quad \tau = \sqrt{\frac{\kappa}{m} - \left(\frac{\mu}{2m}\right)^2},$$

the general solution is

$$x(t) = e^{\sigma t}(\alpha \cos(\tau t) + \beta \sin(\tau t)).$$

Since $\sigma < 0$, these solutions oscillate but damp down to 0. In particular, the solution satisfying initial conditions $x(0) = 1$ and $\dot{x}(0) = 0$ is

$$x(t) = e^{-\mu t/2m}\left(\cos(\tau t) - \frac{\mu}{2m\tau}\sin(\tau t)\right).$$

The graph of this solution when $\tau = 1$ and $\mu/2m = 0.07$ is shown on the right in Figure 6.3. Compare the solutions for the undamped and damped springs.

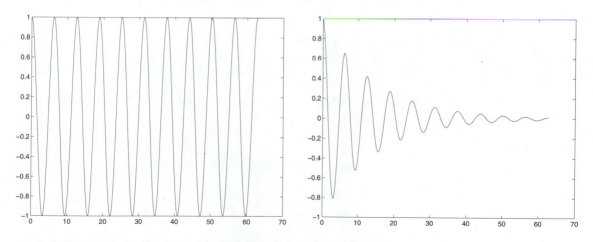

Figure 6.3
Left: Graph of solution to undamped spring equation with initial conditions $x(0) = 1$ and $\dot{x}(0) = 0$.
Right: Graph of solution to damped spring equation with the same initial conditions.

HAND EXERCISES

1. By direct integration solve the differential equation (6.7.4) for a point particle moving only under the influence of gravity. Find the solution for a particle starting at a height of 10 feet above ground with an upward velocity of 20 feet/sec. At what time will the particle hit the ground? (Recall that acceleration due to gravity is 32 feet/sec^2.)

2. By direct integration solve the differential equation (6.7.4) for a point particle moving only under the influence of gravity. Show that the solution is

$$x(t) = -\frac{1}{2}gt^2 + v_0 t + x_0,$$

where x_0 is the initial position of the particle and v_0 is the initial velocity.

In Exercises 3–5, find the general solution to the given differential equation.

3. $\ddot{x} + 2\dot{x} - 3x = 0$

4. $\ddot{x} - 6\dot{x} + 9x = 0$. In addition, find the solution to this equation that satisfies initial values $x(1) = 1$ and $\dot{x}(1) = 0$.

5. $\ddot{x} + 2\dot{x} + 2x = 0$

6. Prove that a nonzero solution to a second-order linear differential equation with constant coefficients cannot be identically equal to 0 on a nonempty interval.

7. Let $r > 0$ and $w > 0$ be constants, and let $x(t)$ be a solution to the differential equation

$$\ddot{x} + r\dot{x} + wx = 0.$$

Show that $\lim_{t \to \infty} x(t) = 0$.

In Exercises 8–10, let $x(t)$ be a solution to the second-order linear homogeneous differential equation (6.7.1). Determine whether each statement is true or false.

8. If $x(t)$ is nonconstant and time periodic, then the roots of the characteristic polynomial are purely imaginary.

9. If $x(t)$ is constant in t, then one of the roots of the characteristic polynomial is 0.

10. If $x(t)$ is not bounded, then the roots of the characteristic polynomial are equal.

11. Consider the second-order differential equation

$$\frac{d^2x}{dt^2} + a(x)\frac{dx}{dt} + b(x) = 0. \tag{6.7.12}$$

Let $y(t) = \dot{x}(t)$ and show that (6.7.12) may be rewritten as a first-order coupled system in $x(t)$ and $y(t)$ as follows:

$$\dot{x} = y$$
$$\dot{y} = -b(x) - a(x)y.$$

COMPUTER EXERCISES

12. Use pplane5 to compute solutions to the system corresponding to the spring equations with small sliding friction. Plot the time series (in x) of the solution and observe the oscillating and damping of the solution.

CHAPTER 7

Qualitative Theory of Planar ODEs

Chapter 6 discussed three methods that are used to solve planar systems of linear, constant coefficient, ordinary differential equations. The last method is based on similarity and the explicit computation of the matrix exponential for certain normal form matrices. This method depends crucially on the classification of 2×2 matrices up to similarity given in Theorem 6.5.5.

In this chapter we explore qualitative features of phase portraits for planar linear systems of differential equations using similarity. We find that the qualitative theory is completely determined by the eigenvalues and eigenvectors of the coefficient matrix—which is not surprising given that we can classify matrices up to similarity by knowing just their eigenvalues and eigenvectors. The set of planar phase portraits divides systems of linear differential equations into two camps: *hyperbolic* and *non-hyperbolic*. The hyperbolic systems consist of *saddles*, *sinks*, and *sources*, while the nonzero nonhyperbolic systems consist of *centers*, *saddle-nodes*, and *shears*.

7.1 SINKS, SADDLES, AND SOURCES

The qualitative theory of autonomous differential equations begins with the observation that many important properties of solutions to constant coefficient systems of differential equations

$$\frac{dX}{dt} = CX \qquad (7.1.1)$$

are unchanged by similarity.

Begin by noting that the origin is always an equilibrium for (7.1.1) and suppose that C is a 2×2 matrix. The origin for (7.1.1) is called a *sink* if the eigenvalues of C both have negative real part and a *source* if the eigenvalues both have positive real part. When C has one eigenvalue of each sign, the origin is called a *saddle*.

Now suppose that B is a 2×2 matrix that is similar to C. Lemma 6.5.3 states that B and C have the same eigenvalues. It follows that if the origin is a saddle for (7.1.1), then it is a saddle for $\dot{X} = BX$. Similar statements hold for sinks and sources.

Asymptotic Stability

We now discuss asymptotic stability of the origin in linear systems. Recall from our discussion in Section 4.5 that the origin is *asymptotically stable* if every trajectory $X(t)$ beginning at an initial condition near the origin stays near 0 for all positive t and

$$\lim_{t \to \infty} X(t) = 0.$$

Recall also from Lemma 6.5.2 that if $B = P^{-1}CP$, then $P^{-1}X(t)$ is a solution to $\dot{X} = BX$ whenever $X(t)$ is a solution to (7.1.1). Since P^{-1} is a matrix of constants that do not depend on t, it follows that

$$\lim_{t \to \infty} X(t) = 0 \iff \lim_{t \to \infty} P^{-1}X(t) = 0.$$

So the origin is asymptotically stable for $\dot{X} = BX$ if and only if it is asymptotically stable for (7.1.1). With this observation in hand, we prove that sinks are stable.

Theorem 7.1.1 *If the eigenvalues of C have negative real part, then the origin is an asymptotically stable equilibrium for (7.1.1). If one of the eigenvalues of C has positive real part, then the origin is unstable.*

Proof: This proof is based on the closed form of solutions given in Section 6.5. The remark preceding this theorem states that we need to prove this theorem only for differential equations up to similarity.

(a) If the eigenvalues λ_1 and λ_2 are real and there are two independent eigenvectors, then Theorem 6.5.5 states that the matrix C is similar to the diagonal matrix

$$B = \begin{pmatrix} \lambda_1 & 0 \\ 0 & \lambda_2 \end{pmatrix}.$$

The general solution to the differential equation $\dot{X} = BX$ is

$$x_1(t) = \alpha_1 e^{\lambda_1 t} \quad \text{and} \quad x_2(t) = \alpha_2 e^{\lambda_2 t}.$$

Since

$$\lim_{t\to\infty} e^{\lambda_1 t} = 0 = \lim_{t\to\infty} e^{\lambda_2 t}$$

when λ_1 and λ_2 are negative, it follows that

$$\lim_{t\to\infty} X(t) = 0$$

for all solutions $X(t)$, and the origin is asymptotically stable. Note that if one of the eigenvalues—say λ_1—is positive, then $x_1(t)$ undergoes exponential growth and the origin is unstable.

(b) If the eigenvalues of C are the complex conjugates $\sigma \pm i\tau$, where $\tau \neq 0$, then Theorem 6.5.5 states that after a similarity transformation, (7.1.1) has the form

$$\dot{X} = \begin{pmatrix} \sigma & -\tau \\ \tau & \sigma \end{pmatrix} X,$$

and solutions for this equation have the form (6.5.5); that is,

$$X(t) = e^{\sigma t} \begin{pmatrix} \cos(\tau t) & -\sin(\tau t) \\ \sin(\tau t) & \cos(\tau t) \end{pmatrix} X_0 = e^{\sigma t} R_{\tau t} X_0,$$

where $R_{\tau t}$ is a rotation matrix; recall (3.2.1). It follows that as time evolves, the vector X_0 is rotated about the origin and then expanded or contracted by the factor $e^{\sigma t}$. So when $\sigma < 0$, $\lim_{t\to\infty} X(t) = 0$ for all solutions $X(t)$. Hence the origin is asymptotically stable. Note that when $\sigma > 0$, solutions spiral away from the origin.

(c) If the eigenvalues are both equal to λ_1 and if there is only one independent eigenvector, then Theorem 6.5.5 states that after a similarity transformation, (7.1.1) has the form

$$\dot{X} = \begin{pmatrix} \lambda_1 & 1 \\ 0 & \lambda_1 \end{pmatrix} X,$$

whose solutions are

$$X(t) = e^{t\lambda} \begin{pmatrix} 1 & t \\ 0 & 1 \end{pmatrix} X_0$$

using (6.4.5). Note that the functions $e^{\lambda_1 t}$ and $t e^{\lambda_1 t}$ both have limits equal to 0 as $t \to \infty$. In the second case, use l'Hôspital's rule and the assumption that $-\lambda_1 > 0$ to compute

$$\lim_{t\to\infty} \frac{t}{e^{-\lambda_1 t}} = -\lim_{t\to\infty} \frac{1}{\lambda_1 e^{-\lambda_1 t}} = 0.$$

Hence $\lim_{t \to \infty} X(t) = 0$ for all solutions $X(t)$ and the origin is asymptotically stable. Note that initially $||X(t)||$ can grow since t is increasing. But eventually exponential decay wins out and solutions limit on the origin. Note that solutions grow exponentially when $\lambda_1 > 0$.   ◆

It is instructive to note how the time series $x_1(t)$ damps down to the origin in the three cases listed in Theorem 7.1.1. In Figure 7.1 we present the time series for the three coefficient matrices:

$$C_1 = \begin{pmatrix} -2 & 0 \\ 0 & -1 \end{pmatrix}, \quad C_2 = \begin{pmatrix} -1 & -55 \\ 55 & -1 \end{pmatrix}, \quad C_3 = \begin{pmatrix} -2 & 1 \\ 0 & -2 \end{pmatrix}.$$

In this figure we can see the exponential decay to 0 associated with the unequal real eigenvalues of C_1, the damped oscillation associated with the complex eigenvalues of C_2, and the initial growth of the time series due to the te^{-2t} term followed by exponential decay to 0 in the equal eigenvalue C_3 example.

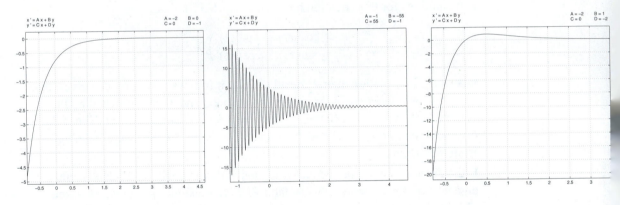

Figure 7.1
Time series for different sinks

Linear Stability

Saddles, sinks, and sources are distinguished by the stability of the origin. In Theorem 7.1.1 we showed that the origin is asymptotically stable if the eigenvalues have negative real part—that is, if the origin is a sink. There is another term that is commonly used and is synonymous with *sink*:

Definition 7.1.2   *The origin is a* linearly stable *equilibrium of (7.1.1) if the eigenvalues of C have negative real part.*

So Theorem 7.1.1 may be restated as: Linear stability implies asymptotic stability of the origin.

Sources Versus Sinks

The explicit form of solutions to planar linear systems shows that solutions with initial conditions near the origin grow exponentially in forward time when the origin of (7.1.1) is a source. We can prove this point geometrically as follows:

The phase planes of sources and sinks are almost the same; they have the same trajectories but the arrows are reversed. To verify this point, note that

$$\dot{X} = -CX \qquad\qquad\qquad (7.1.2)$$

is a sink when (7.1.1) is a source; observe that the trajectories of solutions of (7.1.1) are the same as those of (7.1.2)—just with time running backward. To verify this point, let $X(t)$ be a solution to (7.1.1); then $X(-t)$ is a solution to (7.1.2). See Figure 7.2 for plots of $\dot{X} = BX$ and $\dot{X} = -BX$, where

$$B = \begin{pmatrix} -1 & -5 \\ 5 & -1 \end{pmatrix}. \qquad\qquad\qquad (7.1.3)$$

So when we draw schematic phase portraits for sinks, we automatically know how to draw schematic phase portraits for sources. The trajectories are the same, but the arrows point in the opposite direction.

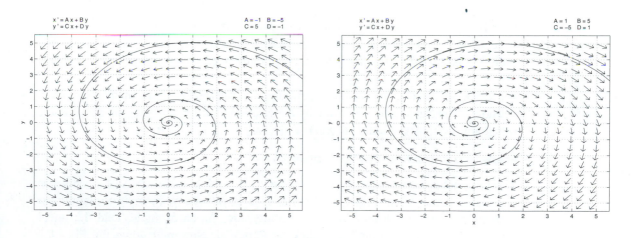

Figure 7.2
Left: Sink $\dot{X} = BX$, where B is given in (7.1.3). *Right:* Source $\dot{X} = -BX$.

Phase Portraits for Saddles

Next we discuss the phase portraits of linear saddles. Using `pplane5`, draw the phase portrait of the saddle

$$\dot{x} = 2x + y$$
$$\dot{y} = -x - 3y,$$

(7.1.4)

as in Figure 7.3. The important feature of saddles is that there are special trajectories (the eigendirections) that limit on the origin in either forward or backward time.

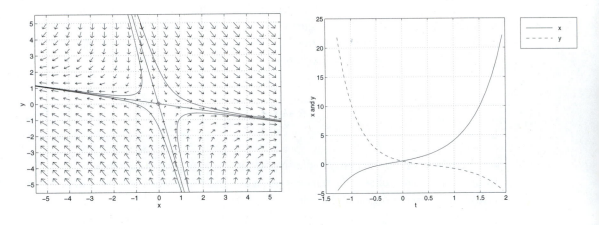

Figure 7.3
Left: Saddle phase portrait. *Right:* First quadrant solution time series.

Definition 7.1.3 *The* stable manifold *or* stable orbit *of a saddle consists of those trajectories that limit on the origin in forward time. The* unstable manifold *or* unstable orbit *of a saddle consists of those trajectories that limit on the origin in backward time.*

Let $\lambda_1 < 0$ and $\lambda_2 > 0$ be the eigenvalues of a saddle with associated eigenvectors v_1 and v_2. The stable orbits are given by the solutions $X(t) = \pm e^{\lambda_1 t} v_1$, and the unstable orbits are given by the solutions $X(t) = \pm e^{\lambda_2 t} v_2$.

Stable and Unstable Orbits Using pplane5

The program `pplane5` is set up to draw the stable and unstable orbits of a saddle on command. Although the principal use of this feature is seen when analyzing nonlinear systems, it is useful to introduce this feature here. As an example, load the linear system (7.1.4) into `pplane5` and click on Proceed. Now pull down the PPLANE5 Options menu and click on Find an equilibrium. Click the cross hairs in the PPLANE5 Display window on a point near the origin; `pplane5` responds by opening a new window—the PPLANE5 Equilibrium point data window—and by putting a small yellow circle about the origin. The circle indicates that the numerical algorithm programmed into pplane5 has detected an equilibrium near the chosen point. A new window opens and displays the message There is a saddle point at (0, 0). This window also displays the coefficient matrix (called the Jacobian for reasons discussed in Section 11.2) at the equilibrium and its eigenvalues and eigenvectors. This process numerically verifies that the origin is a saddle (a fact that could have been verified in a more straightforward way).

Now pull down the PPLANE5 Options menu again and click on Plot stable and unstable orbits. Next click on the mouse when the cross hairs are within the yellow circle, and pplane5 responds by drawing the stable and unstable orbits. The result is shown in Figure 7.3 (left). On this figure we have also plotted one trajectory from each quadrant; thus obtaining the phase portrait of a saddle. On the right of Figure 7.3 we have plotted a time series of the first quadrant solution. Note how the x time series increases exponentially to $+\infty$ in forward time and the y time series decreases in forward time while going exponentially toward $-\infty$. The two time series together give the trajectory $(x(t), y(t))$ that in forward time is asymptotic to the line given by the unstable eigendirection.

HAND EXERCISES

In Exercises 1–3, determine whether or not the equilibrium at the origin in the system of differential equations $\dot{X} = CX$ is asymptotically stable.

1. $C = \begin{pmatrix} 1 & 2 \\ 4 & 1 \end{pmatrix}$
2. $C = \begin{pmatrix} -1 & 2 \\ -4 & -1 \end{pmatrix}$
3. $C = \begin{pmatrix} 2 & 1 \\ 1 & -5 \end{pmatrix}$

In Exercises 4–9, determine whether the equilibrium at the origin in the system of differential equations $\dot{X} = CX$ is a sink, a saddle, or a source.

4. $C = \begin{pmatrix} -2 & 2 \\ 0 & -1 \end{pmatrix}$
5. $C = \begin{pmatrix} 3 & 5 \\ 0 & -2 \end{pmatrix}$
6. $C = \begin{pmatrix} 4 & 2 \\ -1 & 2 \end{pmatrix}$

7. $C = \begin{pmatrix} 8 & 0 \\ -5 & 3 \end{pmatrix}$
8. $C = \begin{pmatrix} 9 & -11 \\ -11 & 9 \end{pmatrix}$
9. $C = \begin{pmatrix} 1 & -8 \\ 2 & 1 \end{pmatrix}$

COMPUTER EXERCISES

In Exercises 10–13, use pplane5 to determine whether the origin is a saddle, sink, or source in $\dot{X} = CX$ for the given matrix C.

10. $C = \begin{pmatrix} 10 & -2.7 \\ 4.32 & 1.6 \end{pmatrix}$
11. $C = \begin{pmatrix} -10 & -2.7 \\ 4.32 & 1.6 \end{pmatrix}$

12. $C = \begin{pmatrix} -1 & 2 \\ 4.76 & 1.5 \end{pmatrix}$
13. $C = \begin{pmatrix} -2 & -2 \\ 4 & 1 \end{pmatrix}$

In Exercises 14 and 15, the given matrices B and C are similar. Observe that the phase portraits of the systems $\dot{X} = BX$ and $\dot{X} = CX$ are qualitatively the same in two steps.

(a) Use MATLAB to find the 2×2 matrix P such that $B = P^{-1}CP$. Use map to understand how the matrix P moves points in the plane.

(b) Use pplane5 to observe that P moves solutions of $\dot{X} = BX$ to the solution of $\dot{X} = CX$. Write a sentence or two describing your results.

14. $C = \begin{pmatrix} 2 & 3 \\ -1 & -3 \end{pmatrix}$ and $B = \dfrac{1}{2}\begin{pmatrix} 1 & -1 \\ -9 & -3 \end{pmatrix}$

15. $C = \begin{pmatrix} -1 & 5 \\ -5 & -1 \end{pmatrix}$ and $B = \begin{pmatrix} -1 & 0.5 \\ -50 & -1 \end{pmatrix}$

7.2 PHASE PORTRAITS OF SINKS

In this section we describe phase portraits and time series of solutions for different kinds of sinks. Sinks have coefficient matrices whose eigenvalues have negative real part. There are four types of sinks:

1. *Spiral sink*—complex eigenvalues
2. *Nodal sink*—real unequal eigenvalues
3. *Improper nodal sink*—real equal eigenvalues; one independent eigenvector
4. *Focus sink*—real equal eigenvalues; two independent eigenvectors

In the preceding section we showed that all solutions of sinks tend toward the origin in forward time. The way in which these solutions approach the origin distinguishes the different sink types; see Figure 7.1. We discuss each type of sink in turn, noting that spiral and nodal sinks are the ones that are most likely to occur.

Spiral Sinks

When the eigenvalues are complex conjugates (that is, when the coefficient matrix has no real eigenvectors), solutions spiral into the origin. This behavior can be seen from the explicit solution (6.5.5). In particular, after a similarity transformation, solutions have the form

$$X(t) = e^{\sigma t}\begin{pmatrix} \cos(\tau t) & -\sin(\tau t) \\ \sin(\tau t) & \cos(\tau t) \end{pmatrix} X_0,$$

where $\lambda = \sigma \pm \tau i$ are the eigenvalues of the matrix of the linear system. Since the real parts of the eigenvalues are assumed to be negative, $\sigma < 0$. Thus the initial vector X_0 is rotated at constant speed τ and contracted exponentially at rate σ, and the resulting trajectory forms a spiral.

Using pplane5, we compute a typical phase portrait. Load the system

$$\begin{aligned} \dot{x} &= -x + 2y \\ \dot{y} &= -5x \end{aligned} \tag{7.2.1}$$

into pplane5 and compute a trajectory. The result should be similar to Figure 7.4 (left). Note that there is no visible sign of an eigendirection in this figure. Indeed, the important feature of this phase portrait is the spiraling nature of trajectories approaching the origin. This geometric feature is typical of systems whose coefficient matrices have complex eigenvalues. In the time series, Figure 7.4 (right), note how the spiraling is realized as a damped oscillation.

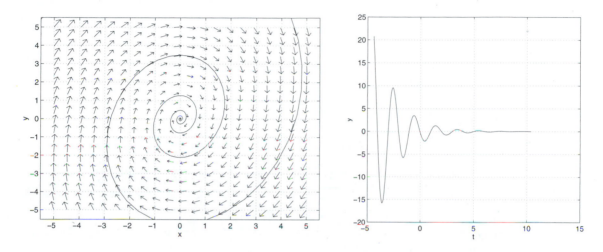

Figure 7.4

Left: Phase plane for (7.2.1) for $x, y \in [-5, 5]$. *Right:* Time series y versus t of solution.

Nodal Sinks

When the eigenvalues are real and unequal, we get a nodal sink. For example, consider the differential equation

$$\dot{x} = -x + y$$
$$\dot{y} = -2y$$

whose phase portrait is pictured in Figure 7.5 (left) along with the time series of one of the solution trajectories (right). Compare the time series of a solution to a nodal sink equation with the time series of the spiral sink solution given in Figure 7.4 (right). Note how the solution asymptotes to 0 rather than oscillating about 0.

Moreover, suppose that the eigenvalues are $\lambda_1 < \lambda_2 < 0$. Then all trajectories approach the origin in forward time tangent to the eigendirection associated with the eigenvalue λ_2. To verify this point, let v_1 and v_2 be the associated eigenvectors. Then the general solution has the form

$$X(t) = \alpha_1 e^{\lambda_1 t} v_1 + \alpha_2 e^{\lambda_2 t} v_2 = e^{\lambda_2 t}(\alpha_1 e^{(\lambda_1 - \lambda_2)t} v_1 + \alpha_2 v_2).$$

Since $\lambda_1 - \lambda_2 < 0$, in forward time $X(t)$ approaches $e^{\lambda_2 t}\alpha_2 v_2$, which is tangent to the v_2 eigendirection. The eigenvalues in our example are $\lambda_1 = -2$ and $\lambda_2 = -1$, and the eigenvector v_2 is just e_1. Indeed, note how trajectories in the phase plane Figure 7.5 (left) approach the origin tangent to the x-axis.

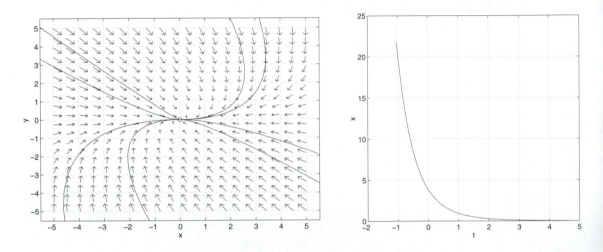

Figure 7.5
Left: Phase plane of nodal sink. *Right:* Time series of a typical trajectory.

Improper Nodes

There are two types of sinks that correspond to coefficient matrices with real equal eigenvalues: those with one independent eigenvector—an *improper node*—and those with two independent eigenvectors—a *focus*.

The phase plane of an improper node looks like the one pictured in Figure 7.6 (left) along with the time series of one of the solution trajectories (right). The equation that we use in this figure is (6.2.17). Note that trajectories approach the origin tangent to a single line—the line generated by the eigenvector. In this example, the eigenvector is approximately $(0.32, 0.95)$, which generates a line through the origin of slope approximately equal to 3.

Compare the time series of a solution to an improper nodal sink with the time series of either the spiral sink solution given in Figure 7.4 (right) or the nodal sink given in Figure 7.5 (right). Note how there is an initial excursion away from 0 followed by a simple asymptote to 0. This excursion away from 0 is typical of nodal sinks. To verify this point, let $\lambda_1 < 0$ be the eigenvalue of the coefficient matrix C. Then the general solution to this equation is

$$X(t) = e^{\lambda_1 t}(I_2 + tN)X_0,$$

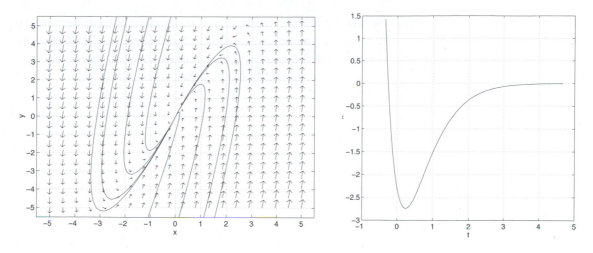

Figure 7.6

Left: Phase plane of improper nodal sink (6.2.17). *Right:* Time series of a trajectory illustrating the transient excursion away from 0.

where $N = C - \lambda_1 I_2$. The initial growth in the solution is forced by the tN term. Eventually, however, exponential decay dominates and the solution approaches 0.

In addition, solutions approach 0 in forward time tangent to the eigendirection spanned by the eigenvector v. If we choose a generalized eigenvector w so that $Nw = v$, then we can write the general solution as

$$X(t) = e^{\lambda_1 t}(I_2 + tN)(\alpha v + \beta w) = e^{\lambda_1 t}(\alpha v + \beta w + t\beta v).$$

For large $t > 0$, the solution direction $\alpha v + \beta w + t\beta v$ is dominated by $t\beta v$, and the trajectory is tangent to the v eigendirection, as claimed.

Focii

When the real equal eigenvalues correspond to a coefficient matrix C having two independent eigenvectors, we have a *focus*. Lemma 6.2.3 states that for a focus, the matrix C must be a multiple of I_2. An example of a focus is the system of differential equations

$$\begin{aligned} \dot{x} &= -0.5x \\ \dot{y} &= -0.5y. \end{aligned} \qquad (7.2.2)$$

The closed form solution to (7.2.2) is just

$$(x(t), y(t)) = e^{-0.5t}(x_0, y_0),$$

so solutions remain on straight lines for all time. The phase plane for this equation is pictured in Figure 7.7 (left), confirming that solutions stay on lines through the origin. Algebraically, the reason for this behavior is that every line through the origin is an eigendirection. There are an infinite number of eigendirections even though there are only two independent eigenvectors.

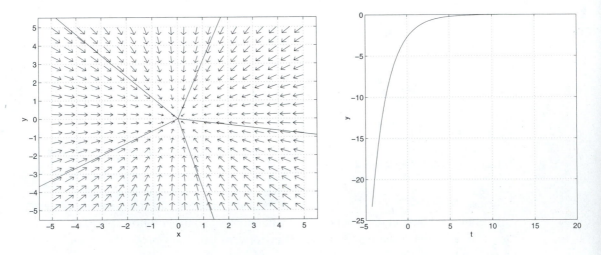

Figure 7.7
Left: Phase plane of a focus. *Right:* Time series of a focus.

Hyperbolic Systems

The simplest linear systems are the sinks, sources, and saddles. These linear systems all have eigenvalues that are neither 0 nor purely imaginary. We call the linear system $\dot{X} = CX$ *hyperbolic* when all eigenvalues of C have nonzero real parts.

Our discussion of phase portraits of hyperbolic linear systems is summarized in Table 7.1. There we reinforce the observation that the type of phase portrait is determined completely by the eigenvalues and eigenvectors of the coefficient matrix of the linear system.

Another way to determine the type of phase portrait for the *hyperbolic* linear system $\dot{X} = CX$ is through the determinant, trace, and discriminant of C. This determination can be made by answering the following four questions in order:

(Q1) **What is the determinant of C?**

$$\det(C) \begin{cases} negative \Rightarrow \text{The origin is a } saddle. \text{ Stop.} \\ zero \quad \Rightarrow \text{The system is not hyperbolic. Stop.} \\ positive \Rightarrow \text{Continue.} \end{cases}$$

(Q2) **What is the trace of C?**

$$\mathrm{tr}(C) \begin{cases} positive \Rightarrow \text{The origin is a } source. \text{ Continue.} \\ zero \quad\;\; \Rightarrow \text{The system is not hyperbolic. Stop.} \\ negative \Rightarrow \text{The origin is a } sink. \text{ Continue.} \end{cases}$$

(Q3) **What is the discriminant D of C?**

$$D \equiv \mathrm{tr}(C)^2 - 4\det(C) \begin{cases} negative \Rightarrow \text{The origin is a } spiral. \text{ Stop.} \\ positive \Rightarrow \text{The origin is a } node. \text{ Stop.} \\ zero \quad\;\; \Rightarrow \text{Continue.} \end{cases}$$

(Q4) **Is C a multiple of I_2?**

$$\begin{cases} no \;\Rightarrow \text{The origin is an } improper\ node. \text{ Stop.} \\ yes \Rightarrow \text{The origin is a } focus. \text{ Stop.} \end{cases}$$

We now verify that the answers to these four questions do indeed determine the phase portraits of hyperbolic linear systems. Recall from (4.8.9) that the determinant is the product of the eigenvalues. So if the determinant is negative, then C must have one negative eigenvalue and one positive eigenvalue, and the origin is a saddle. It also follows that C has a zero eigenvalue when $\det(C) = 0$, which contradicts hyperbolicity. When $\det(C) > 0$, C has either two real eigenvalues of the same sign or a complex conjugate pair of eigenvalues.

Next recall from (4.8.8) that the trace is the sum of the eigenvalues. Suppose the trace of C is negative. If the eigenvalues of C are real, then the two eigenvalues must

Table 7.1

Classification of planar hyperbolic equilibria

Name	Eigenvalues
Saddle	Real and of opposite sign
Spiral sink Spiral source	Complex with negative real part Complex with positive real part
Nodal sink Nodal source	Real, unequal, and negative Real, unequal, and positive
Improper nodal sink Improper nodal source	Real, equal, negative, and one eigenvector Real, equal, positive, and one eigenvector
Focus sink Focus source	Real, equal, negative, and two independent eigenvectors Real, equal, positive, and two independent eigenvectors

be negative since they have the same sign, and the origin is a sink. If the eigenvalues are a complex conjugate pair, then the trace is twice the real part of the eigenvalues. So again the origin is a sink. A similar discussion verifies that the origin is a source when the trace of C is positive. Note that $\det(C) > 0$ and $\operatorname{tr}(C) = 0$ imply that C has purely imaginary eigenvalues, which contradicts the hyperbolicity of C.

To understand the conclusions of question (Q3), recall from Theorem 4.8.3 that the eigenvalues of C are complex conjugates when the discriminant D is negative. Hence the origin is a spiral. Similarly, if $D > 0$, then the eigenvalues of C are real, and the origin is a node. Finally, if $D = 0$, then the eigenvalues of C are real and equal. The origin is an improper node when there is only one linearly independent eigenvector, and the origin is a focus when there are two linearly independent eigenvectors. More-over, when a 2×2 matrix has two equal eigenvalues and two linearly independent eigenvectors, it is a multiple of I_2.

See Figure 7.8 for a classification of phase portrait types in the determinant–trace plane.

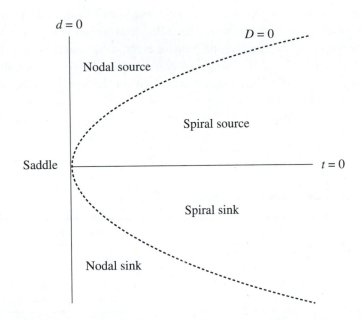

Figure 7.8
Classification of phase portraits in the t-d plane, where t is the trace, d is the determinant, and D is the discriminant

HAND EXERCISES

1. For the matrix C in Exercise 1 in Section 7.1, determine the type of phase portrait of $\dot{X} = CX$.

2. For the matrix C in Exercise 2 in Section 7.1, determine the type of phase portrait of $\dot{X} = CX$.

3. For the matrix C in Exercise 3 in Section 7.1, determine the type of phase portrait of $\dot{X} = CX$.

In Exercises 4–7, find a 2×2 matrix C so that the given statement is satisfied.

4. The differential equation $\dot{X} = CX$ has a saddle at the origin with unstable orbit in the direction $(2, 3)$.

5. The differential equation $\dot{X} = CX$ has a spiral sink at the origin where solutions decay to the origin at rate $\sigma = -0.5$.

6. The differential equation $\dot{X} = CX$ has an improper nodal source at the origin with trajectories approaching the origin tangent to the y-axis.

7. The differential equation $\dot{X} = CX$ has a nodal sink at the origin with trajectories approaching the origin tangent to the line $y = x$.

8. Each picture in Figure 7.9 is the time series of a solution to a planar system of differential equations of the form $\dot{X} = CX$. Describe the eigenvalues of C and determine the type of planar phase for each of these systems.

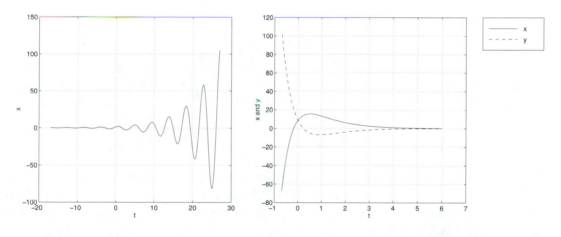

Figure 7.9
Time series for planar systems

COMPUTER EXERCISES

In Exercises 9 and 10, use pplane5 to determine the type of phase portrait for the systems of differential equations $\dot{X} = CX$, where C is the given matrix. Based on these computations answer the following questions:

(a) Is the origin asymptotically stable?
(b) How many real eigenvectors does the matrix C have?

9. $C = \begin{pmatrix} \pi & \sqrt{2} \\ -1 & 1 \end{pmatrix}$

10. $C = \begin{pmatrix} 4 & 1 \\ 6 & -1 \end{pmatrix}$

7.3   PHASE PORTRAITS OF NONHYPERBOLIC SYSTEMS

A linear constant coefficient system $\dot{X} = CX$ is not hyperbolic either when C has a zero eigenvalue ($\det(C) = 0$) or when C has purely imaginary eigenvalues ($\det(C) > 0$ and $\text{tr}(C) = 0$). These special cases are indicated by solid lines in Figure 7.8.

There are four types of nonhyperbolic planar systems:

1. *Center*—nonzero purely imaginary eigenvalues
2. *Saddle-node*—a single zero eigenvalue
3. *Shear*—a double zero eigenvalue; one independent eigenvector
4. The zero matrix itself

The dynamics in the last case are very easy to describe: All solutions are equilibria and solutions remain at the initial point for all time.

Centers

There is a new kind of solution that occurs only in centers—the time periodic solution—as we now explain. In Theorem 6.5.5 we showed that when the 2×2 matrix C has purely imaginary eigenvalues $\pm i\tau$ with $\tau \neq 0$, C is similar to the matrix

$$B = \begin{pmatrix} 0 & -\tau \\ \tau & 0 \end{pmatrix}.$$

In Table 6.1(b) we showed that solutions to $\dot{X} = BX$ are

$$X(t) = \begin{pmatrix} \cos(\tau t) & -\sin(\tau t) \\ \sin(\tau t) & \cos(\tau t) \end{pmatrix} X_0 = R_{\tau t} X_0,$$

where X_0 is the vector of initial conditions and $R_{\tau t}$ is a rotation matrix; recall (3.2.1).

Note that every solution to $\dot{X} = BX$ traverses a circle around the origin as $R_{\tau t}$ rotates the plane counterclockwise through the angle τt. As an example, we graph a solution to $\dot{X} = BX$ and its time series in Figure 7.10 when $\tau = 3$. All centers have nonzero trajectories that are either ellipses or circles, and those centers that do not have the form of B typically produce ellipses as solutions. See Exercise 16 at the end of this section.

It is instructive to see how solutions change when the real part of the eigenvalue traverses through 0. Suppose that C has eigenvalues $\sigma \pm i\tau$ and is in normal form

$$C = \begin{pmatrix} \sigma & -\tau \\ \tau & \sigma \end{pmatrix}.$$

We illustrate this change by graphing the time series (of both x and y versus t) in Figure 7.11 with $\tau = 3$ and $\sigma = -1, 0, 1$, respectively. Note that when $\sigma < 0$, the time series oscillate about 0 but also damp down to 0, whereas when $\sigma > 0$, the time series oscillate and diverge. When $\sigma = 0$, there is an exact balance leading to time periodic solutions.

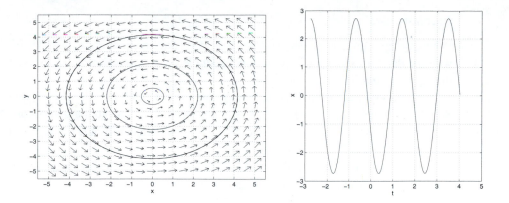

Figure 7.10
Left: Solutions to $\dot{X} = BX$ for $\tau = 3$. *Right:* Time series of a solution illustrating time periodicity.

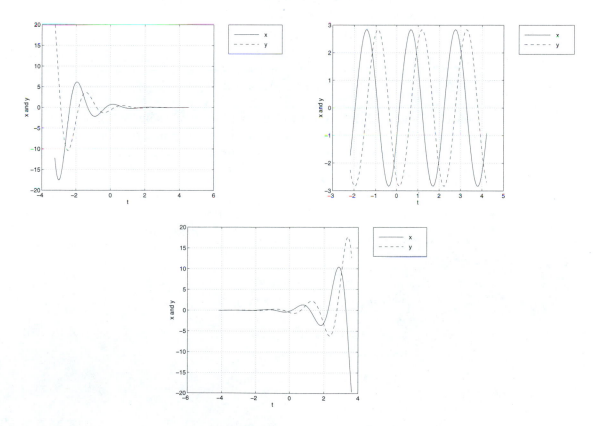

Figure 7.11
Solutions of $\dot{X} = BX$ for $\tau = 3$ and $\sigma = -1, 0, 1$

Saddle-Nodes

When the determinant of a matrix C is 0, C must have a zero eigenvalue. The simplest way for $\det(C)$ to be 0 is for C to have a single zero eigenvalue. In this case we call the origin a *saddle-node*.

So the eigenvalues of a saddle-node are $\lambda_1 = 0$ and $\lambda_2 \neq 0$. For simplicity of discussion, we assume that $\lambda_2 < 0$. Let v_1 and v_2 be the associated eigenvectors. Since $Cv_1 = 0$, it follows that saddle-nodes have a line of equilibria $X(t) = \alpha_1 v_1$ for every scalar α_1.

There is also a solution $X(t) = e^{\lambda_2 t} v_2$ that converges to the origin in forward time (since $\lambda_2 < 0$). The general solution to this system is

$$X(t) = \alpha_1 v_1 + \alpha_2 e^{\lambda_2 t} v_2$$

for scalars α_1 and α_2. Since $\lambda_2 < 0$, this solution converges on the equilibrium $\alpha_1 v_1$ in forward time. Moreover, the trajectory for any solution stays for all time on the line parallel to the vector v_2. The phase portrait for the system

$$\frac{dX}{dt} = \begin{pmatrix} -2 & 4 \\ 1 & -2 \end{pmatrix} X \tag{7.3.1}$$

is shown in Figure 7.12 (left) along with the time series of a solution (right). Note how the time series (of x versus t) asymptotes onto the x coordinate of an equilibrium.

The saddle-node is a nonhyperbolic equilibrium that sits between the hyperbolic saddle and the hyperbolic node (either the nodal sink or the nodal source), just as

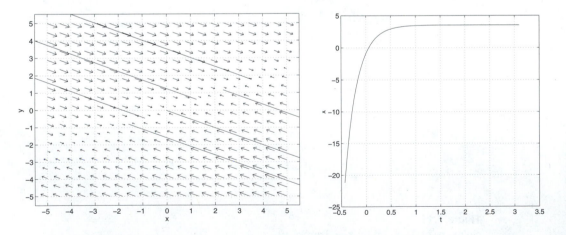

Figure 7.12
Left: Phase portrait of (7.3.1). *Right:* Time series of a solution.

the center sits between the hyperbolic spiral source and the hyperbolic spiral sink. To illustrate this point we show three time series in Figure 7.13 for the system

$$\frac{dX}{dt} = \begin{pmatrix} \mu & 0 \\ 0 & -1 \end{pmatrix} X \tag{7.3.2}$$

when $\mu < 0$, $\mu = 0$, and $\mu > 0$. See how the y time series decays exponentially to 0 in forward time in each case. But the x time series converges to 0 when $\mu < 0$ and grows exponentially when $\mu > 0$. Finally, when $\mu = 0$, the x time series is constant.

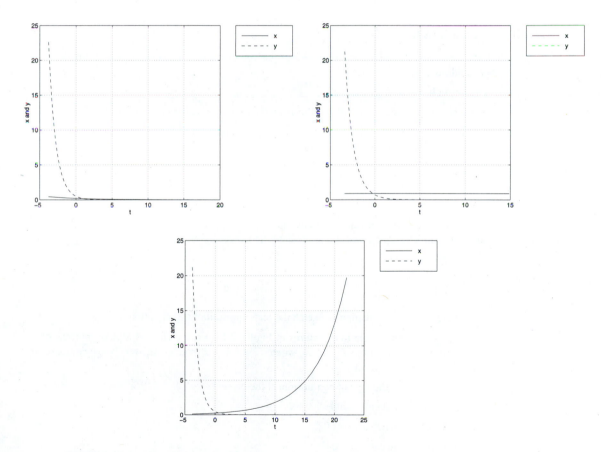

Figure 7.13
Time series for solutions of (7.3.2) when $\mu = -0.2,\ 0,\ 0.2$

Shears

The last example of a nonhyperbolic system occurs when the matrix C has two zero eigenvalues—but only one linearly independent eigenvector. In this case we call the origin a *shear*. After a similarity transformation, such a system is

$$\frac{dX}{dt} = \begin{pmatrix} 0 & 1 \\ 0 & 0 \end{pmatrix} X. \tag{7.3.3}$$

The solution is

$$(x(t), y(t)) = (x_0 + y_0 t, y_0).$$

Thus solutions move along lines parallel to the x-axis through the point (x_0, y_0) with speed y_0. These trajectories move to the right when $y_0 > 0$ and to the left when $y_0 < 0$.

HAND EXERCISES

In Exercises 1 and 2, find a 2×2 matrix C so that the differential equation $\dot{X} = CX$ satisfies the given condition.

1. The origin is a center.

2. The origin is a saddle-node with equilibria on the line generated by the vector $(-1, 2)$.

3. Recall from (6.7.11) that the undamped spring equation is $d^2x/dt^2 + \kappa x = 0$, where $\kappa > 0$.
   (a) As a first-order system this equation is

$$\dot{x} = y$$
$$\dot{y} = -\kappa x.$$

   Sketch the phase portrait of this equation.
   (b) The damped spring equation, written as a first-order system, is

$$\dot{x} = y$$
$$\dot{y} = -\kappa x - \sigma y,$$

   where $\sigma > 0$ is the damping. Sketch the phase portrait of this equation.

In Exercises 4–8, consider the four pictures in Figure 7.14. Each picture is a phase portrait of a system of differential equations $\dot{X} = CX$, where C is a 2×2 matrix. Answer the given question for each of these phase portraits.

4. What is the name of the type of equilibrium at the origin?

5. Is the origin asymptotically stable?

6. Is trace(C) positive, negative, or 0?

7. Is det(C) positive, negative, or 0?

8. Is discriminant(C) positive, negative, or 0?

In Exercises 9–15, consider the system of differential equations $\dot{X} = CX$, where C is the given matrix. For each system determine whether or not the origin is hyperbolic and the type of equilibrium at the origin (spiral sink, center, etc.).

9. $C = \begin{pmatrix} 1 & -1 \\ 2 & 1 \end{pmatrix}$

10. $C = \begin{pmatrix} 1 & 1 \\ -1 & 1 \end{pmatrix}$

11. $C = \begin{pmatrix} 3 & -1 \\ 1 & 1 \end{pmatrix}$

12. $C = \begin{pmatrix} 1 & 1 \\ 1 & 1 \end{pmatrix}$

13. $C = \begin{pmatrix} 2 & -2 \\ 4 & -2 \end{pmatrix}$

14. $C = \begin{pmatrix} 2 & 2 \\ -2 & -2 \end{pmatrix}$

15. $C = \begin{pmatrix} 1 & 1 \\ 4 & 1 \end{pmatrix}$

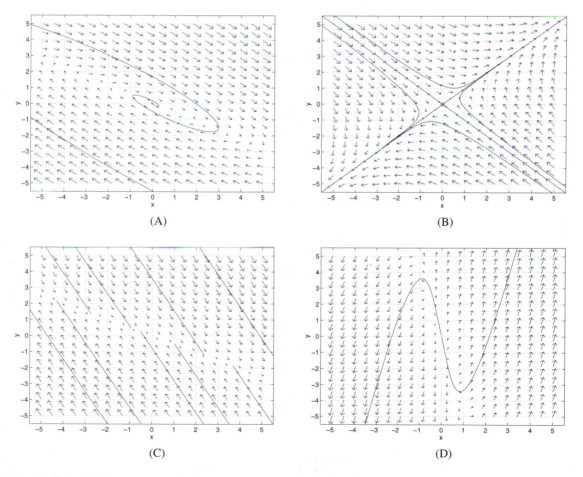

(A)

(B)

(C)

(D)

Figure 7.14
Phase portraits for planar linear systems in Exercises 4–8

COMPUTER EXERCISES

16. Consider the system of differential equations $\dot{X} = CX$, where $C = \begin{pmatrix} 2 & -10 \\ 1 & -2 \end{pmatrix}$. By hand show that this system is a center and use pplane5 to determine its phase portrait. Describe both the similarities and the differences of the phase portrait of this system and the phase portrait in Figure 7.10.

17. Consider the system of differential equations $\dot{X} = CX$, where $C = \begin{pmatrix} 2 & -4 \\ 1 & -2 \end{pmatrix}$. By hand show that this system is a shear and use pplane5 to determine its phase portrait. Describe both the similarities and the differences of the phase portrait of this system and the phase portrait of (7.3.3).

8

Determinants and Eigenvalues

In Section 3.8 we introduced determinants for 2×2 matrices A. There we showed that the determinant of A is nonzero if and only if A is invertible. In Section 4.8 we saw that the eigenvalues of A are the roots of its characteristic polynomial, and that its characteristic polynomial is just the determinant of a related matrix—namely, $p_A(\lambda) = \det(A - \lambda I_2)$.

In Section 8.1 we generalize the concept of determinants to $n \times n$ matrices, and in Section 8.2 we use determinants to show that every $n \times n$ matrix has exactly n eigenvalues—the roots of its characteristic polynomial. Properties of eigenvalues are also discussed in detail in Section 8.2. Certain details concerning determinants are deferred to Appendix 8.3.

8.1 DETERMINANTS

There are several equivalent ways to introduce determinants; none of them is easily motivated. We prefer to define determinants through the properties they satisfy rather than by formula. These properties actually enable us to compute determinants of $n \times n$ matrices where $n > 3$, which further justifies the approach. Later on, we will give an inductive formula (8.1.9) for computing the determinant.

Definition 8.1.1 *A determinant of a square $n \times n$ matrix A is a real number that satisfies the following three properties:*

(a) *If $A = (a_{ij})$ is lower triangular, then the determinant of A is the product of the diagonal entries; that is,*

$$\det(A) = a_{11} \cdots a_{nn}.$$

(b) $\det(A^t) = \det(A)$.

(c) *Let B be an $n \times n$ matrix. Then*

$$\det(AB) = \det(A)\det(B). \tag{8.1.1}$$

Theorem 8.1.2 *There exists a unique determinant function satisfying the three properties of Definition 8.1.1.*

We will show that it is possible to compute the determinant of any $n \times n$ matrix using Definition 8.1.1. Now we present a few examples:

Lemma 8.1.3 *Let A be an $n \times n$ matrix.*

(a) *Let $c \in \mathbf{R}$ be a scalar. Then $\det(cA) = c^n \det(A)$.*

(b) *If all of the entries in either a row or a column of A are 0, then $\det(A) = 0$.*

Proof:

(a) Note that Definition 8.1.1(a) implies that $\det(cI_n) = c^n$. It follows from (8.1.1) that

$$\det(cA) = \det(cI_nA) = \det(cI_n)\det(A) = c^n \det(A).$$

(b) Definition 8.1.1(b) implies that it suffices to prove this assertion when one row of A is 0. Suppose that the ith row of A is 0. Let J be an $n \times n$ diagonal matrix with a 1 in every diagonal entry except the ith diagonal entry, which is 0. A matrix calculation shows that $JA = A$. It follows from Definition 8.1.1(a) that $\det(J) = 0$ and from (8.1.1) that $\det(A) = 0$. $\blacklozenge$

Determinants of 2 × 2 Matrices

Before discussing how to compute determinants, we discuss the special case of 2×2 matrices. Recall from (3.8.2) of Section 3.8 that when

$$A = \begin{pmatrix} a & b \\ c & d \end{pmatrix},$$

we define

$$\det(A) = ad - bc. \tag{8.1.2}$$

We check that (8.1.2) satisfies the three properties in Definition 8.1.1. Observe that when A is lower triangular, then $b = 0$ and $\det(A) = ad$. So (a) is satisfied. It is straightforward to verify (b). We verified (c) in Proposition 3.8.2.

It is less obvious perhaps—but true nonetheless—that the three properties of $\det(A)$ actually force the determinant of 2×2 matrices to be given by formula (8.1.2). We begin by showing that Definition 8.1.1 implies that

$$\det \begin{pmatrix} 0 & 1 \\ 1 & 0 \end{pmatrix} = -1. \tag{8.1.3}$$

We verify this by observing that

$$\begin{pmatrix} 0 & 1 \\ 1 & 0 \end{pmatrix} = \begin{pmatrix} 1 & -1 \\ 0 & 1 \end{pmatrix} \begin{pmatrix} 1 & 0 \\ 1 & 1 \end{pmatrix} \begin{pmatrix} 1 & 0 \\ 0 & -1 \end{pmatrix} \begin{pmatrix} 1 & 1 \\ 0 & 1 \end{pmatrix}. \tag{8.1.4}$$

Hence properties (c), (a), and (b) imply that

$$\det \begin{pmatrix} 0 & 1 \\ 1 & 0 \end{pmatrix} = 1 \cdot 1 \cdot (-1) \cdot 1 = -1.$$

It is helpful to interpret the matrices in (8.1.4) as elementary row operations. Then (8.1.4) states that swapping two rows in a 2×2 matrix is the same as performing the following row operations in order:

1. Add the 2nd row to the 1st row.
2. Multiply the 2nd row by -1.
3. Add the 1st row to the 2nd row.
4. Subtract the 2nd row from the 1st row.

Suppose that $d \neq 0$. Then

$$A = \begin{pmatrix} a & b \\ c & d \end{pmatrix} = \begin{pmatrix} 1 & \dfrac{b}{d} \\ 0 & 1 \end{pmatrix} \begin{pmatrix} \dfrac{ad - bc}{d} & 0 \\ c & d \end{pmatrix}.$$

It follows from properties (c), (b), and (a) that

$$\det(A) = \frac{ad - bc}{d} \, d = ad - bc,$$

as claimed.

Now suppose that $d = 0$ and note that

$$A = \begin{pmatrix} a & b \\ c & 0 \end{pmatrix} = \begin{pmatrix} 0 & 1 \\ 1 & 0 \end{pmatrix} \begin{pmatrix} c & 0 \\ a & b \end{pmatrix}.$$

Using (8.1.3), we see that

$$\det(A) = -\det \begin{pmatrix} c & 0 \\ a & b \end{pmatrix} = -bc,$$

as desired.

We have verified that the only possible determinant function for 2×2 matrices is the determinant function defined by (8.1.2).

Row Operations Are Invertible Matrices

Proposition 8.1.4   *Let A and B be $m \times n$ matrices, where B is obtained from A by a single elementary row operation. Then there exists an invertible $m \times m$ matrix R such that $B = RA$.*

Proof: First consider multiplying the jth row of A by the nonzero constant c. Let R be the diagonal matrix whose jth entry on the diagonal is c and whose other diagonal entries are 1. Then the matrix RA is just the matrix obtained from A by multiplying the jth row of A by c. Note that R is invertible when $c \neq 0$ and that R^{-1} is the diagonal matrix whose jth entry is $1/c$ and whose other diagonal entries are 1. For example

$$\begin{pmatrix} 1 & 0 & 0 \\ 0 & 1 & 0 \\ 0 & 0 & 2 \end{pmatrix} \begin{pmatrix} a_{11} & a_{12} & a_{13} \\ a_{21} & a_{22} & a_{23} \\ a_{31} & a_{32} & a_{33} \end{pmatrix} = \begin{pmatrix} a_{11} & a_{12} & a_{13} \\ a_{21} & a_{22} & a_{23} \\ 2a_{31} & 2a_{32} & 2a_{33} \end{pmatrix}$$

multiplies the 3rd row by 2.

Next we show that the elementary row operation that swaps two rows may also be thought of as matrix multiplication. Let $R = (r_{kl})$ be the matrix that deviates from the identity matrix by changing in the four entries:

$$r_{ii} = 0$$
$$r_{jj} = 0$$
$$r_{ij} = 1$$
$$r_{ji} = 1.$$

A calculation shows that RA is the matrix obtained from A by swapping the ith and jth rows. For example,

$$\begin{pmatrix} 0 & 0 & 1 \\ 0 & 1 & 0 \\ 1 & 0 & 0 \end{pmatrix} \begin{pmatrix} a_{11} & a_{12} & a_{13} \\ a_{21} & a_{22} & a_{23} \\ a_{31} & a_{32} & a_{33} \end{pmatrix} = \begin{pmatrix} a_{31} & a_{32} & a_{33} \\ a_{21} & a_{22} & a_{23} \\ a_{11} & a_{12} & a_{13} \end{pmatrix},$$

which swaps the 1st and 3rd rows. Another calculation shows that $R^2 = I_n$ and hence that R is invertible since $R^{-1} = R$.

Finally, we claim that adding c times the ith row of A to the jth row of A can be viewed as matrix multiplication. Let $E_{k\ell}$ be the matrix all of whose entries are 0 except for the entry in the kth row and ℓth column, which is 1. Then $R = I_n + cE_{ij}$ has the property that RA is the matrix obtained by adding c times the jth row of A to the ith row. We can verify by multiplication that R is invertible and that $R^{-1} = I_n - cE_{ij}$. More precisely,

$$(I_n + cE_{ij})(I_n - cE_{ij}) = I_n + cE_{ij} - cE_{ij} - c^2 E_{ij}^2 = I_n,$$

since $E_{ij}^2 = O$ for $i \neq j$. For example,

$$(I_3 + 5E_{12})A = \begin{pmatrix} 1 & 5 & 0 \\ 0 & 1 & 0 \\ 0 & 0 & 1 \end{pmatrix} \begin{pmatrix} a_{11} & a_{12} & a_{13} \\ a_{21} & a_{22} & a_{23} \\ a_{31} & a_{32} & a_{33} \end{pmatrix}$$

$$= \begin{pmatrix} a_{11} + 5a_{21} & a_{12} + 5a_{22} & a_{13} + 5a_{23} \\ a_{21} & a_{22} & a_{23} \\ a_{31} & a_{32} & a_{33} \end{pmatrix}$$

adds five times the 2nd row to the 1st row.   ◆

Determinants of Elementary Row Matrices

Lemma 8.1.5

(a) *The determinant of a swap matrix is -1.*

(b) *The determinant of the matrix that adds a multiple of one row to another is* 1.

(c) *The determinant of the matrix that multiplies one row by c is c.*

Proof: The matrix that swaps the ith row with the jth row is the matrix whose nonzero elements are $a_{kk} = 1$, where $k \neq i, j$ and $a_{ij} = 1 = a_{ji}$. Using a similar argument as in (8.1.3), we see that the determinants of these matrices are equal to -1.

The matrix that adds a multiple of one row to another is triangular (either upper or lower) and has 1s on the diagonal. Thus property (a) in Definition 8.1.1 implies that the determinants of these matrices are equal to 1.

Finally, the matrix that multiplies the ith row by $c \neq 0$ is a diagonal matrix all of whose diagonal entries are 1 except for $a_{ii} = c$. Again property (a) implies that the determinant of this matrix is $c \neq 0$.   ◆

Computation of Determinants

We now show how to compute the determinant of any $n \times n$ matrix A using elementary row operations and Definition 8.1.1. It follows from Proposition 8.1.4 that every elementary row operation on A may be performed by premultiplying A by an elementary row matrix.

For each matrix A there is a unique reduced echelon form matrix E and a sequence of elementary row matrices $R_1, \ldots, R_s$ such that

$$E = R_s \cdots R_1 A. \tag{8.1.5}$$

It follows from Definition 8.1.1(c) that we can compute the determinant of A once we know the determinants of reduced echelon form matrices and the determinants of elementary row matrices. In particular,

$$\det(A) = \frac{\det(E)}{\det(R_1) \cdots \det(R_s)}. \tag{8.1.6}$$

It is easy to compute the determinant of any matrix in reduced echelon form using Definition 8.1.1(a) since all reduced echelon form $n \times n$ matrices are upper triangular. Lemma 8.1.5 tells us how to compute the determinants of elementary row matrices. This discussion proves the next proposition:

Proposition 8.1.6   *If a determinant function exists for $n \times n$ matrices, then it is unique.*

We still need to show that determinant functions exist when $n > 2$. More precisely, we know that the reduced echelon form matrix E is uniquely defined from A (Theorem 2.4.9), but there is more than one way to perform elementary row operations on A to get to E. Thus we can write A in the form (8.1.6) in many different ways, and these different decompositions might lead to different values for $\det A$. (They don't.)

An Example of Determinants by Row Reduction

As a practical matter we row reduce a square matrix A by premultiplying A by an elementary row matrix R_j. Thus

$$\det(A) = \frac{1}{\det(R_j)} \det(R_j A). \tag{8.1.7}$$

We use this approach to compute the determinant of the 4×4 matrix

$$A = \begin{pmatrix} 0 & 2 & 10 & -2 \\ 1 & 2 & 4 & 0 \\ 1 & 6 & 1 & -2 \\ 2 & 1 & 1 & 0 \end{pmatrix}.$$

The idea is to use (8.1.7) to keep track of the determinant while row reducing A to upper triangular form. For instance, swapping rows changes the sign of the determinant, so

$$\det(A) = - \det \begin{pmatrix} 1 & 2 & 4 & 0 \\ 0 & 2 & 10 & -2 \\ 1 & 6 & 1 & -2 \\ 2 & 1 & 1 & 0 \end{pmatrix}.$$

Adding multiples of one row to another leaves the determinant unchanged, so

$$\det(A) = - \det \begin{pmatrix} 1 & 2 & 4 & 0 \\ 0 & 2 & 10 & -2 \\ 0 & 4 & -3 & -2 \\ 0 & -3 & -7 & 0 \end{pmatrix}.$$

Multiplying a row by a scalar c corresponds to an elementary row matrix whose determinant is c. To make sure that we do not change the value of $\det(A)$, we have to divide the determinant by c as we multiply a row of A by c. So as we divide the 2nd row of the matrix by 2, we multiply the whole result by 2, obtaining

$$\det(A) = -2 \det \begin{pmatrix} 1 & 2 & 4 & 0 \\ 0 & 1 & 5 & -1 \\ 0 & 4 & -3 & -2 \\ 0 & -3 & -7 & 0 \end{pmatrix}.$$

We continue row reduction by zeroing out the last two entries in the 2nd column, obtaining

$$\det(A) = -2 \det \begin{pmatrix} 1 & 2 & 4 & 0 \\ 0 & 1 & 5 & -1 \\ 0 & 0 & -23 & 2 \\ 0 & 0 & 8 & -3 \end{pmatrix} = 46 \det \begin{pmatrix} 1 & 2 & 4 & 0 \\ 0 & 1 & 5 & -1 \\ 0 & 0 & 1 & -\frac{2}{23} \\ 0 & 0 & 8 & -3 \end{pmatrix}.$$

Thus

$$\det(A) = 46 \det \begin{pmatrix} 1 & 2 & 4 & 0 \\ 0 & 1 & 5 & -1 \\ 0 & 0 & 1 & -\frac{2}{23} \\ 0 & 0 & 0 & -\frac{53}{23} \end{pmatrix} = -106.$$

Determinants and Inverses

We end this subsection with an important observation about the determinant function. This observation generalizes to dimension n Corollary 3.8.3.

Theorem 8.1.7 *An $n \times n$ matrix A is invertible if and only if $\det(A) \neq 0$. Moreover, if A^{-1} exists, then*

$$\det A^{-1} = \frac{1}{\det A}. \tag{8.1.8}$$

Proof: If A is invertible, then

$$\det(A)\det(A^{-1}) = \det(AA^{-1}) = \det(I_n) = 1.$$

Thus $\det(A) \neq 0$ and (8.1.8) is valid. In particular, the determinants of elementary row matrices are nonzero, since they are all invertible. (This point was proved by direct calculation in Lemma 8.1.5.)

If A is singular, then A is row equivalent to a nonidentity reduced echelon form matrix E whose determinant is 0 (since E is upper triangular and its last diagonal entry is 0). So it follows from (8.1.5) that

$$0 = \det(E) = \det(R_1) \cdots \det(R_s) \det(A).$$

Since $\det(R_j) \neq 0$, it follows that $\det(A) = 0$. ◆

Corollary 8.1.8 *If the rows of an $n \times n$ matrix A are linearly dependent (for example, if one row of A is a scalar multiple of another row of A), then $\det(A) = 0$.*

An Inductive Formula for Determinants

In this subsection we present an inductive formula for the determinant; that is, we assume that the determinant is known for square $(n-1) \times (n-1)$ matrices and use this formula to define the determinant for $n \times n$ matrices. This inductive formula is called *expansion by cofactors*.

Let $A = (a_{ij})$ be an $n \times n$ matrix. Let A_{ij} be the $(n-1) \times (n-1)$ matrix formed from A by deleting the ith row and the jth column. The matrices $(-1)^{i+j} A_{ij}$ are called *cofactor* matrices of A.

Inductively we define the determinant of an $n \times n$ matrix A by

$$\det(A) = \sum_{j=1}^{n} (-1)^{1+j} a_{1j} \det(A_{1j})$$

$$= a_{11} \det(A_{11}) - a_{12} \det(A_{12}) + \cdots + (-1)^{n+1} a_{1n} \det(A_{1n}). \tag{8.1.9}$$

In Appendix 8.3 we show that the determinant function defined by (8.1.9) satisfies all properties of a determinant function. Formula (8.1.9) is also called *expansion by cofactors along the 1st row*, since the a_{1j} are taken from the 1st row of A. Since

$\det(A) = \det(A')$, it follows that if (8.1.9) is valid as an inductive definition of determinant, then expansion by cofactors along the 1st column is also valid; that is,

$$\det(A) = a_{11}\det(A_{11}) - a_{21}\det(A_{21}) + \cdots + (-1)^{n+1}a_{n1}\det(A_{n1}). \qquad \textbf{(8.1.10)}$$

We now explore some consequences of this definition, beginning with determinants of small matrices. For example, Definition 8.1.1(a) implies that the determinant of a 1×1 matrix is just

$$\det(a) = a.$$

Therefore, from (8.1.9), the determinant of a 2×2 matrix is

$$\det\begin{pmatrix} a_{11} & a_{12} \\ a_{21} & a_{22} \end{pmatrix} = a_{11}\det(a_{22}) - a_{12}\det(a_{21}) = a_{11}a_{22} - a_{12}a_{21},$$

which is just the formula for determinants of 2×2 matrices given in (8.1.2).

Similarly, we can now find a formula for the determinant of 3×3 matrices A as follows:

$$\det(A) = a_{11}\det\begin{pmatrix} a_{22} & a_{23} \\ a_{32} & a_{33} \end{pmatrix} - a_{12}\det\begin{pmatrix} a_{21} & a_{23} \\ a_{31} & a_{33} \end{pmatrix} + a_{13}\det\begin{pmatrix} a_{21} & a_{22} \\ a_{31} & a_{32} \end{pmatrix}$$

$$\textbf{(8.1.11)}$$

$$= a_{11}a_{22}a_{33} + a_{12}a_{23}a_{31} + a_{13}a_{21}a_{32} - a_{11}a_{23}a_{32} - a_{12}a_{21}a_{33} - a_{13}a_{22}a_{31}.$$

As an example, compute

$$\det\begin{pmatrix} 2 & 1 & 4 \\ 1 & -1 & 3 \\ 5 & 6 & -2 \end{pmatrix}$$

using formula (8.1.11) as

$$2(-1)(-2) + 1 \cdot 3 \cdot 5 + 4 \cdot 6 \cdot 1 - 4(-1)5 - 3 \cdot 6 \cdot 2 - (-2)1 \cdot 1$$
$$= 4 + 15 + 24 + 20 - 36 + 2 = 29.$$

There is a visual mnemonic for remembering how to compute the six terms in formula (8.1.11) for the determinant of 3×3 matrices. Write the matrix as a 3×5 array by repeating the first two columns, as shown in boldface in Figure 8.1. Then add the products of terms connected by solid lines sloping down and to the right and subtract the products of terms connected by dashed lines sloping up and to the right. **Warning:** This nice crisscross algorithm for computing determinants of 3×3 matrices does not generalize to $n \times n$ matrices.

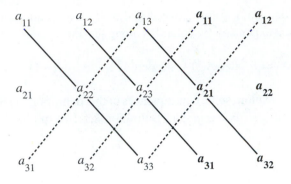

Figure 8.1
Mnemonic for computation of determinants of 3×3
matrices

When we compute determinants of $n \times n$ matrices when $n > 3$, it is usually more efficient to compute the determinant using row reduction rather than by using formula (8.1.9). In Appendix 8.3 we verify that formula (8.1.9) actually satisfies the three properties of a determinant, thus completing the proof of Theorem 8.1.2.

An interesting and useful formula for reducing the effort in computing determinants follows:

Lemma 8.1.9   *Let A be an $n \times n$ matrix of the form*

$$A = \begin{pmatrix} B & 0 \\ C & D \end{pmatrix},$$

where B is a $k \times k$ matrix and D is an $(n - k) \times (n - k)$ matrix. Then

$$\det(A) = \det(B) \det(D).$$

Proof: We prove this result using (8.1.9) coupled with induction. Assume that this lemma is valid for all $(n - 1) \times (n - 1)$ matrices of the appropriate form. Now use (8.1.9) to compute

$$\det(A) = a_{11} \det(A_{11}) - a_{12} \det(A_{12}) + \cdots \pm a_{1n} \det(A_{1n})$$
$$= b_{11} \det(A_{11}) - b_{12} \det(A_{12}) + \cdots \pm b_{1k} \det(A_{1k}).$$

Note that the cofactor matrices A_{1j} are obtained from A by deleting the 1st row and the jth column. These matrices all have the form

$$A_{1j} = \begin{pmatrix} B_{1j} & 0 \\ C_j & D \end{pmatrix},$$

where C_j is obtained from C by deleting the jth column. By induction on k,

$$\det(A_{1j}) = \det(B_{1j})\det(D).$$

It follows that

$$\det(A) = \left(b_{11}\det(B_{11}) - b_{12}\det(B_{12}) + \cdots \pm b_{1k}\det(B_{1k})\right)\det(D)$$
$$= \det(B)\det(D),$$

as desired.   ◆

Determinants in MATLAB

The determinant function has been preprogrammed in MATLAB and is quite easy to use. For example, typing e8_1_11 loads the matrix

$$A = \begin{pmatrix} 1 & 2 & 3 & 0 \\ 2 & 1 & 4 & 1 \\ -2 & -1 & 0 & 1 \\ -1 & 0 & -2 & 3 \end{pmatrix} \qquad \textbf{(8.1.12)}*$$

To compute the determinant of A, just type det(A) and obtain the answer

```
ans =
   -46
```

Alternatively, we can use row reduction techniques in MATLAB to compute the determinant of A—just to test the theory that we have developed. Note that to compute the determinant we do not need to row reduce to reduced echelon form; we need only reduce to an upper triangular matrix. This can always be done by successively adding multiples of one row to another—an operation that does not change the determinant. For example, to clear the entries in the 1st column below the 1st row, type

```
A(2,:) = A(2,:) - 2*A(1,:);
A(3,:) = A(3,:) + 2*A(1,:);
A(4,:) = A(4,:) + A(1,:)
```

obtaining

```
A =
     1     2     3     0
     0    -3    -2     1
     0     3     6     1
     0     2     1     3
```

To clear the 2nd column below the 2nd row type

```
A(3,:) = A(3,:) + A(2,:);A(4,:) = A(4,:) - A(4,2)*A(2,:)/A(2,2)
```

obtaining

```
A =
    1.0000     2.0000     3.0000          0
         0    -3.0000    -2.0000     1.0000
         0          0     4.0000     2.0000
         0          0    -0.3333     3.6667
```

Finally, to clear the entry (4, 3) type

    A(4,:) = A(4,:) −A(4,3)*A(3,:)/A(3,3)

to get

```
A =
    1.0000     2.0000     3.0000          0
         0    -3.0000    -2.0000     1.0000
         0          0     4.0000     2.0000
         0          0          0     3.8333
```

To evaluate the determinant of A, which is now an upper triangular matrix, type

    A(1,1)*A(2,2)*A(3,3)*A(4,4)

obtaining

    ans =
       −46

as expected.

HAND EXERCISES

In Exercises 1–3, compute the determinants of the each matrix.

1. $A = \begin{pmatrix} -2 & 1 & 0 \\ 4 & 5 & 0 \\ 1 & 0 & 2 \end{pmatrix}$

2. $B = \begin{pmatrix} 1 & 0 & 2 & 3 \\ -1 & -2 & 3 & 2 \\ 4 & -2 & 0 & 3 \\ 1 & 2 & 0 & -3 \end{pmatrix}$

3. $C = \begin{pmatrix} 2 & 1 & -1 & 0 & 0 \\ 1 & -2 & 3 & 0 & 0 \\ -3 & 2 & -2 & 0 & 0 \\ 1 & 1 & -1 & 2 & 4 \\ 0 & 2 & 3 & -1 & -3 \end{pmatrix}$

4. Find $\det(A^{-1})$, where $A = \begin{pmatrix} -2 & -3 & 2 \\ 4 & 1 & 3 \\ -1 & 1 & 1 \end{pmatrix}$.

5. Show that the determinants of similar $n \times n$ matrices are equal.

In Exercises 6–8, use row reduction to compute the determinant of the given matrix.

6. $A = \begin{pmatrix} -1 & -2 & 1 \\ 3 & 1 & 3 \\ -1 & 1 & 1 \end{pmatrix}$

7. $B = \begin{pmatrix} 1 & 0 & 1 & 0 \\ 0 & 1 & 0 & -1 \\ 1 & 0 & -1 & 0 \\ 0 & 1 & 0 & 1 \end{pmatrix}$

8. $C = \begin{pmatrix} 1 & 2 & 0 & 1 \\ 0 & 2 & 1 & 0 \\ -2 & -3 & 3 & -1 \\ 1 & 0 & 5 & 2 \end{pmatrix}$

9. Let

$$A = \begin{pmatrix} 2 & -1 & 0 \\ 0 & 3 & 0 \\ 1 & 5 & 3 \end{pmatrix} \quad \text{and} \quad B = \begin{pmatrix} 2 & 0 & 0 \\ 0 & -1 & 0 \\ 0 & 0 & 3 \end{pmatrix}.$$

(a) For what values of λ is $\det(\lambda A - B) = 0$?

(b) Is there a vector x for which $Ax = Bx$?

In Exercises 10 and 11, verify that the given matrix has determinant -1.

10. $A = \begin{pmatrix} 1 & 0 & 0 \\ 0 & 0 & 1 \\ 0 & 1 & 0 \end{pmatrix}$
    11. $B = \begin{pmatrix} 0 & 0 & 1 \\ 0 & 1 & 0 \\ 1 & 0 & 0 \end{pmatrix}$

12. Compute the cofactor matrices A_{13}, A_{22}, and A_{21} when $A = \begin{pmatrix} 3 & 2 & -4 \\ 0 & 1 & 5 \\ 0 & 0 & 6 \end{pmatrix}$.

13. Compute the cofactor matrices B_{11}, B_{23}, and B_{43} when $B = \begin{pmatrix} 0 & 2 & -4 & 5 \\ -1 & 7 & -2 & 10 \\ 0 & 0 & 0 & -1 \\ 3 & 4 & 2 & -10 \end{pmatrix}$.

14. Find values of λ where the determinant of the matrix

$$A_\lambda = \begin{pmatrix} \lambda - 1 & 0 & -1 \\ 0 & \lambda - 1 & 1 \\ -1 & 1 & \lambda \end{pmatrix}$$

vanishes.

15. Suppose that two $n \times p$ matrices A and B are row equivalent. Show that there is an invertible $n \times n$ matrix P such that $B = PA$.

16. Let A be an invertible $n \times n$ matrix and let $b \in \mathbf{R}^n$ be a column vector. Let B_j be the $n \times n$ matrix obtained from A by replacing the jth column of A by the vector b. Let $x = (x_1, \ldots, x_n)^t$ be the unique solution to $Ax = b$. Then Cramer's rule states that

$$x_j = \frac{\det(B_j)}{\det(A)}. \qquad (8.1.13)$$

Prove Cramer's rule.

Hint Let A_j be the jth column of A so that $A_j = Ae_j$. Show that

$$B_j = A(e_1 | \cdots | e_{j-1} | x | e_{j+1} | \cdots | e_n).$$

Using this product, compute the determinant of B_j and verify (8.1.13).

8.2 EIGENVALUES

In this section we discuss how to find eigenvalues for an $n \times n$ matrix A. This discussion parallels the discussion for 2×2 matrices given in Section 4.8. As we noted in that section, λ is a real eigenvalue of A if there exists a nonzero eigenvector v such that

$$Av = \lambda v. \qquad (8.2.1)$$

It follows that the matrix $A - \lambda I_n$ is singular, since

$$(A - \lambda I_n)v = 0.$$

Theorem 8.1.7 implies that

$$\det(A - \lambda I_n) = 0.$$

With these observations in mind, we can make the following definition:

Definition 8.2.1   *Let A be an $n \times n$ matrix. The* characteristic polynomial *of A is*

$$p_A(\lambda) = \det(A - \lambda I_n).$$

In Theorem 8.2.3 we show that $p_A(\lambda)$ is indeed a polynomial of degree n in λ. Note here that the roots of p_A are the *eigenvalues* of A. As we discussed, the real eigenvalues of A are roots of the characteristic polynomial. Conversely, if λ is a real root of p_A, then Theorem 8.1.7 states that the matrix $A - \lambda I_n$ is singular and therefore that there exists a nonzero vector v such that (8.2.1) is satisfied. Similarly, by using this extended algebraic definition of eigenvalues, we allow the possibility of complex eigenvalues. The complex analog of Theorem 8.1.7 shows that if λ is a complex eigenvalue, then there exists a nonzero complex n-vector v such that (8.2.1) is satisfied.

Example 8.2.2   *Let A be an $n \times n$ lower triangular matrix. Then the diagonal entries are the eigenvalues of A.*
    We verify this statement as follows:

$$A - \lambda I_n = \begin{pmatrix} a_{11} - \lambda & & 0 \\ & \ddots & \\ (*) & & a_{nn} - \lambda \end{pmatrix}.$$

Since the determinant of a triangular matrix is the product of the diagonal entries, it follows that

$$p_A(\lambda) = (a_{11} - \lambda) \cdots (a_{nn} - \lambda) \tag{8.2.2}$$

and hence that the diagonal entries of A are roots of the characteristic polynomial. A similar argument works if A is upper triangular.

It follows from (8.2.2) that the characteristic polynomial of a triangular matrix is a polynomial of degree n and that

$$p_A(\lambda) = (-1)^n \lambda^n + b_{n-1}\lambda^{n-1} + \cdots + b_0 \tag{8.2.3}$$

for some real constants $b_0, \ldots, b_{n-1}$. In fact, this statement is true in general.

Theorem 8.2.3 *Let A be an n × n matrix. Then p_A is a polynomial of degree n of the form (8.2.3).*

Proof: Let C be an $n \times n$ matrix whose entries have the form $c_{ij} + d_{ij}\lambda$. Then $\det(C)$ is a polynomial in λ of degree at most n. We verify this statement by induction. It is easily verified when $n = 1$, since then $C = (c + d\lambda)$ for some real numbers c and d. Then $\det(C) = c + d\lambda$, which is a polynomial of degree at most one. (It may have degree zero if $d = 0$.) So assume that this statement is true for $(n - 1) \times (n - 1)$ matrices. Recall from (8.1.9) that

$$\det(C) = (c_{11} + d_{11}\lambda) \det(C_{11}) + \cdots + (-1)^{n+1}(c_{1n} + d_{1n}\lambda) \det(C_{1n}).$$

By induction each of the determinants C_{1j} is a polynomial of degree at most $n - 1$. It follows that multiplication by $c_{1j} + d_{1j}\lambda$ yields a polynomial of degree at most n in λ. Since the sum of polynomials of degree at most n is a polynomial of degree at most n, we have verified our assertion.

Since $A - \lambda I_n$ is a matrix whose entries have the desired form, it follows that $p_A(\lambda)$ is a polynomial of degree at most n in λ. To complete the proof of this theorem we need to show that the coefficient of λ^n is $(-1)^n$. Again we verify this statement by induction. This statement is easily verified for 1×1 matrices; we assume that it is true for $(n - 1) \times (n - 1)$ matrices. Again use (8.1.9) to compute

$$\det(A - \lambda I_n) = (a_{11} - \lambda) \det(B_{11}) - a_{12} \det(B_{12}) + \cdots + (-1)^{n+1}a_{1n} \det(B_{1n}),$$

where B_{1j} are the cofactor matrices of $A - \lambda I_n$. From our previous observation, all of the terms $\det(B_{1j})$ are polynomials of degree at most $n - 1$. Thus, in this expansion, the only term that can contribute a term of degree n is

$$-\lambda \det(B_{11}).$$

Note that the cofactor matrix B_{11} is the $(n - 1) \times (n - 1)$ matrix

$$B_{11} = A_{11} - \lambda I_{n-1},$$

where A_{11} is the first cofactor matrix of the matrix A. By induction, $\det(B_{11})$ is a polynomial of degree $n - 1$ with leading term $(-1)^{n-1}\lambda^{n-1}$. Multiplying this polynomial by $-\lambda$ yields a polynomial of degree n with the correct leading term. ◆

General Properties of Eigenvalues

The *fundamental theorem of algebra* states that every polynomial of degree n has exactly n roots (counting multiplicity). For example, the quadratic formula shows that every quadratic polynomial has exactly two roots. In general, the proof of the fundamental theorem is not easy and is certainly beyond the limits of this course.

Indeed, the difficulty in proving the *fundamental theorem of algebra* is to prove that a polynomial $p(\lambda)$ of degree $n > 0$ has one (complex) root. Suppose that λ_0 is a root of $p(\lambda)$; that is, suppose that $p(\lambda_0) = 0$. Then it is easy to show that

$$p(\lambda) = (\lambda - \lambda_0)q(\lambda) \qquad (8.2.4)$$

for some polynomial q of degree $n - 1$. So once we know that p has a root, we can argue by induction to prove that p has n roots.

Recall that a polynomial need not have any real roots. For example, the polynomial $p(\lambda) = \lambda^2 + 1$ has no real roots, since $p(\lambda) > 0$ for all real λ. This polynomial does have two complex roots $\pm i = \pm\sqrt{-1}$.

However, a polynomial with real coefficients has either real roots or complex roots that come in complex conjugate pairs. To verify this statement, we need to show that if λ_0 is a complex root of $p(\lambda)$, then so is $\overline{\lambda_0}$. We claim that

$$p(\overline{\lambda}) = \overline{p(\lambda)}.$$

To verify this point, suppose that

$$p(\lambda) = c_n\lambda^n + c_{n-1}\lambda^{n-1} + \cdots + c_0,$$

where each $c_j \in \mathbf{R}$. Then

$$\overline{p(\lambda)} = \overline{c_n\lambda^n + c_{n-1}\lambda^{n-1} + \cdots + c_0} = c_n\overline{\lambda}^n + c_{n-1}\overline{\lambda}^{n-1} + \cdots + c_0 = p(\overline{\lambda}).$$

If λ_0 is a root of $p(\lambda)$, then

$$p(\overline{\lambda_0}) = \overline{p(\lambda_0)} = \overline{0} = 0.$$

Hence $\overline{\lambda_0}$ is also a root of p.

It follows that:

Theorem 8.2.4   *Every (real) $n \times n$ matrix A has exactly n eigenvalues $\lambda_1, \ldots, \lambda_n$. These eigenvalues are either real or complex conjugate pairs. Moreover,*

(a) $p_A(\lambda) = (\lambda_1 - \lambda) \cdots (\lambda_n - \lambda)$, *and*

(b) $\det(A) = \lambda_1 \cdots \lambda_n$.

Proof: Since the characteristic polynomial p_A is a polynomial of degree n with real coefficients, the first part of the theorem follows from the preceding discussion. In particular, it follows from (8.2.4) that

$$p_A(\lambda) = c(\lambda_1 - \lambda) \cdots (\lambda_n - \lambda)$$

for some constant c. Formula (8.2.3) implies that $c = 1$, which proves (a). Since $p_A(\lambda) = \det(A - \lambda I_n)$, it follows that $p_A(0) = \det(A)$. Thus (a) implies that $p_A(0) = \lambda_1 \cdots \lambda_n$, thus proving (b). ◆

The eigenvalues of a matrix do not have to be different. For example, consider the extreme case of a strictly triangular matrix A. Example 8.2.2 shows that all of the eigenvalues of A are 0.

We now discuss certain properties of eigenvalues.

Corollary 8.2.5 *Let A be an $n \times n$ matrix. Then A is invertible if and only if 0 is not an eigenvalue of A.*

Proof: The proof follows from Theorem 8.1.7 and Theorem 8.2.4(b). ◆

Lemma 8.2.6 *Let A be a singular $n \times n$ matrix. Then the null space of A is the span of all eigenvectors whose associated eigenvalue is 0.*

Proof: An eigenvector v of A has eigenvalue 0 if and only if

$$Av = 0.$$

This statement is valid if and only if v is in the null space of A. ◆

Theorem 8.2.7 *Let A be an invertible $n \times n$ matrix with eigenvalues $\lambda_1, \ldots, \lambda_n$. Then the eigenvalues of A^{-1} are $\lambda_1^{-1}, \ldots, \lambda_n^{-1}$.*

Proof: We claim that

$$p_A(\lambda) = (-1)^n \det(A) \lambda^n p_{A^{-1}} \left(\frac{1}{\lambda} \right).$$

It then follows that $1/\lambda$ is an eigenvalue for A^{-1} for each eigenvalue λ of A. This makes sense, since the eigenvalues of A are nonzero.

Compute:

$$(-1)^n \det(A) \lambda^n p_{A^{-1}} \left(\frac{1}{\lambda} \right) = (-\lambda)^n \det(A) \det \left(A^{-1} - \frac{1}{\lambda} I_n \right)$$

$$= \det(-\lambda A) \det \left(A^{-1} - \frac{1}{\lambda} I_n \right)$$

$$= \det \left(-\lambda A \left(A^{-1} - \frac{1}{\lambda} I_n \right) \right)$$

$$= \det(A - \lambda I_n)$$

$$= p_A(\lambda),$$

which verifies the claim. ◆

Theorem 8.2.8 *Let A and B be similar $n \times n$ matrices. Then*

$$p_A = p_B,$$

and hence the eigenvalues of A and B are identical.

Proof: Since B and A are similar, there exists an invertible $n \times n$ matrix S such that $B = S^{-1}AS$. It follows that

$$\det(B - \lambda I_n) = \det(S^{-1}AS - \lambda I_n) = \det(S^{-1}(A - \lambda I_n)S) = \det(A - \lambda I_n),$$

which verifies that $p_A = p_B$. ◆

Recall that the *trace* of an $n \times n$ matrix A is the sum of the diagonal entries of A; that is,

$$\text{tr}(A) = a_{11} + \cdots + a_{nn}.$$

We state without proof the following theorem:

Theorem 8.2.9 *Let A be an $n \times n$ matrix with eigenvalues $\lambda_1, \ldots, \lambda_n$. Then*

$$tr(A) = \lambda_1 + \cdots + \lambda_n.$$

It follows from Theorem 8.2.8 that the traces of similar matrices are equal.

MATLAB **Calculations**

The commands for computing characteristic polynomials and eigenvalues of square matrices are straightforward in MATLAB. In particular, for an $n \times n$ matrix A, the MATLAB command poly(A) returns the coefficients of $(-1)^n p_A(\lambda)$.

For example, reload the 4×4 matrix A of (8.1.12) by typing e8_1_11. The characteristic polynomial of A is found by typing

```
poly(A)
```

to obtain

```
ans =
    1.0000   −5.0000   15.0000   −10.0000   −46.0000
```

Thus the characteristic polynomial of A is

$$p_A(\lambda) = \lambda^4 - 5\lambda^3 + 15\lambda^2 - 10\lambda - 46.$$

The eigenvalues of A are found by typing eig(A) and obtaining

```
ans =
  -1.2224
   1.6605 + 3.1958i
   1.6605 - 3.1958i
   2.9014
```

Thus A has two real eigenvalues and one complex conjugate pair of eigenvalues. Note that MATLAB has preprogrammed not only the algorithm for finding the characteristic polynomial, but also numerical routines for finding the roots of the characteristic polynomial.

The trace of A is found by typing `trace(A)` and obtaining

```
ans =
   5
```

Using the MATLAB command `sum`, we can verify the statement of Theorem 8.2.9. Indeed `sum(v)` computes the sum of the components of the vector v, and typing

```
sum(eig(A))
```

we obtain the answer 5.0000, as expected.

HAND EXERCISES

In Exercises 1 and 2, determine the characteristic polynomial and the eigenvalues of each matrix.

1. $A = \begin{pmatrix} -9 & -2 & -10 \\ 3 & 2 & 3 \\ 8 & 2 & 9 \end{pmatrix}$

2. $B = \begin{pmatrix} 2 & 1 & -5 & 2 \\ 1 & 2 & 13 & 2 \\ 0 & 0 & 3 & -1 \\ 0 & 0 & 1 & 1 \end{pmatrix}$

3. Find a basis for the eigenspace of

$$A = \begin{pmatrix} 3 & 1 & -1 \\ -1 & 1 & 1 \\ 2 & 2 & 0 \end{pmatrix}$$

corresponding to the eigenvalue $\lambda = 2$.

4. Consider the matrix

$$A = \begin{pmatrix} -1 & 1 & 1 \\ 1 & -1 & 1 \\ 1 & 1 & -1 \end{pmatrix}.$$

(a) Verify that the characteristic polynomial of A is $p_\lambda(A) = (\lambda - 1)(\lambda + 2)^2$.

(b) Show that $(1, 1, 1)$ is an eigenvector of A corresponding to $\lambda = 1$.

(c) Show that $(1, 1, 1)$ is orthogonal to every eigenvector of A corresponding to the eigenvalue $\lambda = -2$.

5. Consider the matrix $A = \begin{pmatrix} 8 & 5 \\ -10 & -7 \end{pmatrix}$.

(a) Find the eigenvalues and eigenvectors of A.

(b) Show that the eigenvectors found in (a) form a basis for $\mathbf{R}^2$.

(c) Find the coordinates of the vector (x_1, x_2) relative to the basis in (b).

6. Find the characteristic polynomial and the eigenvalues of

$$A = \begin{pmatrix} -1 & 2 & 2 \\ 2 & 2 & 2 \\ -3 & -6 & -6 \end{pmatrix}.$$

Find eigenvectors corresponding to each of the three eigenvalues.

7. Let A be an $n \times n$ matrix. Suppose that

$$A^2 + A + I_n = 0.$$

Prove that A is invertible.

In Exercises 8 and 9, decide whether the given statements are *true* or *false*. If a statement is false, give a counterexample; if a statement is true, give a proof.

8. If the eigenvalues of a 2×2 matrix are equal to 1, then the four entries of that matrix are each less than 500.

9. The trace of the product of two $n \times n$ matrices is the product of the traces.

10. When n is odd, show that every real $n \times n$ matrix has a real eigenvalue.

COMPUTER EXERCISES

In Exercises 11 and 12, (a) use MATLAB to compute the eigenvalues, traces, and characteristic polynomials of the given matrix. (b) Use the results from (a) to confirm Theorems 8.2.7 and 8.2.9.

11.

$$A = \begin{pmatrix} -12 & -19 & -3 & 14 & 0 \\ -12 & 10 & 14 & -19 & 8 \\ 4 & -2 & 1 & 7 & -3 \\ -9 & 17 & -12 & -5 & -8 \\ -12 & -1 & 7 & 13 & -12 \end{pmatrix} \qquad (8.2.5)*$$

12.

$$B = \begin{pmatrix} -12 & -5 & 13 & -6 & -5 & 12 \\ 7 & 14 & 6 & 1 & 8 & 18 \\ -8 & 14 & 13 & 9 & 2 & 1 \\ 2 & 4 & 6 & -8 & -2 & 15 \\ -14 & 0 & -6 & 14 & 8 & -13 \\ 8 & 16 & -8 & 3 & 5 & 19 \end{pmatrix} \qquad (8.2.6)*$$

13. Use MATLAB to compute the characteristic polynomial of the matrix

$$A = \begin{pmatrix} 4 & -6 & 7 \\ 2 & 0 & 5 \\ -10 & 2 & 5 \end{pmatrix}.$$

Denote this polynomial by $p_A(\lambda) = -(\lambda^3 + p_2\lambda^2 + p_1\lambda + p_0)$. Then compute the matrix

$$B = -(A^3 + p_2A^2 + p_1A + p_0I).$$

What do you observe? In symbols, $B = p_A(A)$. Compute the matrix B for examples of other square matrices A and determine whether or not your observation was an accident.

8.3   *APPENDIX: EXISTENCE OF DETERMINANTS

The purpose of this appendix is to verify the inductive definition of determinant (8.1.9). We have already shown that if a determinant function exists, then it is unique. We also know that the determinant function exists for 1×1 matrices. So we assume by induction that the determinant function exists for $(n - 1) \times (n - 1)$ matrices, and we prove that the inductive definition gives a determinant function for $n \times n$ matrices.

Recall that A_{ij} is the cofactor matrix obtained from A by deleting the ith row and the jth column—so A_{ij} is an $(n - 1) \times (n - 1)$ matrix. The inductive definition is

$$D(A) = a_{11} \det(A_{11}) - a_{12} \det(A_{12}) + \cdots + (-1)^{n+1} a_{1n} \det(A_{1n}).$$

We use the notation $D(A)$ to remind us that we have not yet verified that this definition satisfies properties (a)–(c) of Definition 8.1.1. In this appendix we verify these properties after assuming that the inductive definition satisfies properties (a)–(c) for $(n - 1) \times (n - 1)$ matrices. For emphasis, we use the notation det to indicate the determinant of square matrices of size less than n.

Property (a) is easily verified for $D(A)$, since if A is lower triangular, then

$$D(A) = a_{11} \det(A_{11}) = a_{11} a_{22} \cdots a_{nn}$$

by induction.

Before verifying that D satisfies properties (b) and (c) of a determinant, we prove:

Lemma 8.3.1   *Let E be an elementary row matrix and let B be any $n \times n$ matrix. Then*

$$D(EB) = D(E)D(B). \tag{8.3.1}$$

Proof: We verify (8.3.1) for each of the three types of elementary row operations.

(I) Suppose that E multiplies the ith row by a nonzero scalar c. If $i > 1$, then the cofactor matrix $(EA)_{1j}$ is obtained from the cofactor matrix A_{1j} by multiplying the $(i - 1)$st row by c. By induction, $\det(EA)_{1j} = c \det(A_{1j})$ and $D(EA) = cD(A)$. On the other hand, $D(E) = \det(E_{11}) = c$. So (8.3.1) is verified in this instance. If $i = 1$, then the 1st row of EA is $(ca_{11}, \ldots, ca_{1n})$, from which it is easy to verify (8.3.1).

(II) Next suppose that E adds a multiple c of the ith row to the jth row. We note that $D(E) = 1$. When $j > 1$, we have $D(E) = \det(E_{11}) = 1$ by induction. When $j = 1$, we have $D(E) = \det(E_{11}) \pm c \det(E_{1i}) = \det(I_{n-1}) \pm c \det(E_{1i})$. But E_{1i} is strictly upper triangular and $\det(E_{1i}) = 0$. Thus $D(E) = 1$.

If $i > 1$ and $j > 1$, then the result $D(EA) = D(A) = D(E)D(A)$ follows by induction.

If $i = 1$, then

$$D(EB) = b_{11} \det((EB)_{11}) + \cdots + (-1)^{n+1} b_{1n} \det((EB)_{1n})$$
$$= D(B) + cD(C),$$

where the 1st and ith rows of C are equal.

If $j = 1$, then

$$
\begin{aligned}
D(EB) &= (b_{11} + cb_{i1}) \det(B_{11}) + \cdots + (-1)^{n+1}(b_{1n} + cb_{in}) \det(B_{1n}) \\
&= \left[b_{11} \det(B_{11}) + \cdots + (-1)^{n+1} b_{1n} \det(B_{1n}) \right] \\
&\quad + c \left[b_{i1} \det(B_{11}) + \cdots + (-1)^{n+1} b_{i1} \det(B_{1n}) \right] \\
&= D(B) + cD(C),
\end{aligned}
$$

where the 1st and ith rows of C are equal.

The hardest part of this proof is a calculation that shows that if the 1st and ith rows of C are equal, then $D(C) = 0$. By induction, we can swap the ith row with the 2nd. Hence we need to verify this fact only when $i = 2$.

(**III**) E is the matrix that swaps two rows. As we saw earlier in (8.1.4), E is the product of four matrices of types (I) and (II). It follows that $D(E) = -1$ and $D(EA) = -D(A) = D(E)D(A)$.

We now verify that if the 1st and 2nd rows of an $n \times n$ matrix C are equal, then $D(C) = 0$. This is a tedious calculation that requires some facility with indexes and summations. Rather than do this proof for general n, we present the proof for $n = 4$. This case contains all the ideas of the general proof.

We begin with the definition of $D(C)$:

$$
\begin{aligned}
D(C) = c_{11} \det \begin{pmatrix} c_{22} & c_{23} & c_{24} \\ c_{32} & c_{33} & c_{34} \\ c_{42} & c_{43} & c_{44} \end{pmatrix} - c_{12} \det \begin{pmatrix} c_{21} & c_{23} & c_{24} \\ c_{31} & c_{33} & c_{34} \\ c_{41} & c_{43} & c_{44} \end{pmatrix} \\
+ c_{13} \det \begin{pmatrix} c_{21} & c_{22} & c_{24} \\ c_{31} & c_{32} & c_{34} \\ c_{41} & c_{42} & c_{44} \end{pmatrix} - c_{14} \det \begin{pmatrix} c_{21} & c_{22} & c_{23} \\ c_{31} & c_{32} & c_{33} \\ c_{41} & c_{42} & c_{43} \end{pmatrix}.
\end{aligned}
$$

Next we expand each of the four 3×3 matrices along their 1st rows, obtaining

$$
\begin{aligned}
D(C) = c_{11} &\left(c_{22} \det \begin{pmatrix} c_{33} & c_{34} \\ c_{43} & c_{44} \end{pmatrix} - c_{23} \det \begin{pmatrix} c_{32} & c_{34} \\ c_{42} & c_{44} \end{pmatrix} + c_{24} \det \begin{pmatrix} c_{32} & c_{33} \\ c_{42} & c_{43} \end{pmatrix} \right) \\
- c_{12} &\left(c_{21} \det \begin{pmatrix} c_{33} & c_{34} \\ c_{43} & c_{44} \end{pmatrix} - c_{23} \det \begin{pmatrix} c_{31} & c_{34} \\ c_{41} & c_{44} \end{pmatrix} + c_{24} \det \begin{pmatrix} c_{31} & c_{33} \\ c_{41} & c_{43} \end{pmatrix} \right) \\
+ c_{13} &\left(c_{21} \det \begin{pmatrix} c_{32} & c_{34} \\ c_{42} & c_{44} \end{pmatrix} - c_{22} \det \begin{pmatrix} c_{31} & c_{34} \\ c_{41} & c_{44} \end{pmatrix} + c_{24} \det \begin{pmatrix} c_{31} & c_{32} \\ c_{41} & c_{42} \end{pmatrix} \right) \\
- c_{14} &\left(c_{21} \det \begin{pmatrix} c_{32} & c_{33} \\ c_{42} & c_{43} \end{pmatrix} - c_{22} \det \begin{pmatrix} c_{31} & c_{33} \\ c_{41} & c_{43} \end{pmatrix} + c_{23} \det \begin{pmatrix} c_{31} & c_{32} \\ c_{41} & c_{42} \end{pmatrix} \right).
\end{aligned}
$$

Combining the 2×2 determinants leads to

$$D(C) = (c_{11}c_{22} - c_{12}c_{21}) \det \begin{pmatrix} c_{33} & c_{34} \\ c_{43} & c_{44} \end{pmatrix} + (c_{11}c_{24} - c_{14}c_{21}) \det \begin{pmatrix} c_{32} & c_{33} \\ c_{42} & c_{43} \end{pmatrix}$$

$$+ (c_{12}c_{23} - c_{13}c_{22}) \det \begin{pmatrix} c_{31} & c_{34} \\ c_{41} & c_{44} \end{pmatrix} + (c_{13}c_{21} - c_{11}c_{23}) \det \begin{pmatrix} c_{32} & c_{34} \\ c_{42} & c_{44} \end{pmatrix}$$

$$+ (c_{13}c_{24} - c_{14}c_{23}) \det \begin{pmatrix} c_{31} & c_{32} \\ c_{41} & c_{42} \end{pmatrix} + (c_{14}c_{22} - c_{12}c_{24}) \det \begin{pmatrix} c_{31} & c_{33} \\ c_{41} & c_{43} \end{pmatrix}.$$

Supposing that

$$c_{21} = c_{11}, \quad c_{22} = c_{12}, \quad c_{23} = c_{13}, \quad c_{24} = c_{14},$$

we can easily check that $D(C) = 0$. ◆

We now return to verifying that $D(A)$ satisfies properties (b) and (c) of being a determinant. We begin by showing that $D(A) = 0$ if A has a row that is identically 0. Suppose that the zero row is the ith row and let E be the matrix that multiplies the ith row of A by c. Then $EA = A$. Using (8.3.1), we see that

$$D(A) = D(EA) = D(E)D(A) = cD(A),$$

which implies that $D(A) = 0$ since c is arbitrary.

Next we prove that $D(A) = 0$ when A is singular. Using row reduction, we can write

$$A = E_s \cdots E_1 R,$$

where the E_j are elementary row matrices and R is in reduced echelon form. Since A is singular, the last row of R is identically 0. Hence $D(R) = 0$ and (8.3.1) implies that $D(A) = 0$.

We now verify property (b). Suppose that A is singular; we show that $D(A') = D(A) = 0$. Since the row rank of A equals the column rank of A, it follows that A' is singular when A is singular. Next assume that A is nonsingular. Then A is row equivalent to I_n and we can write

$$A = E_s \cdots E_1, \tag{8.3.2}$$

where the E_j are elementary row matrices. Since

$$A' = E_1^t \cdots E_s^t$$

and $D(E) = D(E')$, property (b) follows.

We now verify property (c): $D(AB) = D(A)D(B)$. Recall that A is singular if and only if there exists a nonzero vector v such that $Av = 0$. Now if A is singular, then so is A^t. Therefore $(AB)^t = B^t A^t$ is also singular. To verify this point, let w be the nonzero vector such that $A^t w = 0$. Then $B^t A^t w = 0$. Thus AB is singular since $(AB)^t$ is singular. Thus $D(AB) = 0 = D(A)D(B)$ when A is singular. Suppose now that A is nonsingular. It follows that

$$AB = E_s \cdots E_1 B.$$

Using (8.3.1), we see that

$$D(AB) = D(E_s) \cdots D(E_1)D(B) = D(E_s \cdots E_1)D(B) = D(A)D(B),$$

as desired. We have now completed the proof that a determinant function exists.

Linear Maps and Changes of Coordinates

The first section in this chapter, Section 9.1, defines linear mappings between abstract vector spaces, shows how such mappings are determined by their values on a basis, and derives basic properties of invertible linear mappings.

The notions of *row rank* and *column rank* of a matrix are discussed in Section 9.2 along with the theorem that states that these numbers are equal to the rank of that matrix.

Section 9.3 discusses the underlying meaning of similarity—the different ways to view the same linear mapping on $\mathbf{R}^n$ in different coordinate systems or bases. This discussion makes sense only after the definitions of coordinates corresponding to bases and of changes in coordinates are given and justified. In Section 9.4 we discuss the matrix associated with a linearity transformation between two finite-dimensional vector spaces in a given set of coordinates, and we show that changes in coordinates correspond to similarity of the corresponding matrices.

9.1 LINEAR MAPPINGS AND BASES

The examples of linear mappings from $\mathbf{R}^n \to \mathbf{R}^m$ that we introduced in Section 3.3 were matrix mappings. More precisely, let A be an $m \times n$ matrix. Then

$$L_A(x) = Ax$$

defines the linear mapping $L_A : \mathbf{R}^n \to \mathbf{R}^m$. Recall that Ae_j is the jth column of A (see Lemma 3.3.4); it follows that A can be reconstructed from the vectors $Ae_1, \ldots, Ae_n$. This remark implies (Lemma 3.3.3) that linear mappings of $\mathbf{R}^n$ to $\mathbf{R}^m$ are determined by their values on the standard basis $e_1, \ldots, e_n$. Next we show that this result is valid in greater generality. We begin by defining what we mean for a mapping between vector spaces to be linear.

Definition 9.1.1   *Let V and W be vector spaces and let $L : V \to W$ be a mapping. The map L is* linear *if*

$$L(u + v) = L(u) + L(v)$$
$$L(cv) = cL(v)$$

for all $u, v \in V$ and $c \in \mathbf{R}$.

Examples of Linear Mappings

(a)   Let $v \in \mathbf{R}^n$ be a fixed vector. Use the dot product to define the mapping $L : \mathbf{R}^n \to \mathbf{R}$ by

$$L(x) = x \cdot v.$$

Then L is linear. Just check that

$$L(x + y) = (x + y) \cdot v = x \cdot v + y \cdot v = L(x) + L(y)$$

for every vector x and y in $\mathbf{R}^n$ and

$$L(cx) = (cx) \cdot v = c(x \cdot v) = cL(x)$$

for every scalar $c \in \mathbf{R}$.

(b)   The map $L : \mathcal{C}^1 \to \mathbf{R}$ defined by

$$L(f) = f'(2)$$

is linear. Indeed

$$L(f + g) = (f + g)'(2) = f'(2) + g'(2) = L(f) + L(g).$$

Similarly, $L(cf) = cL(f)$.

(c)   The map $L : \mathcal{C}^1 \to \mathcal{C}^1$ defined by

$$L(f)(t) = f(t - 1)$$

is linear. Indeed

$$L(f + g)(t) = (f + g)(t - 1) = f(t - 1) + g(t - 1) = L(f)(t) + L(g)(t).$$

Similarly, $L(cf) = cL(f)$. It may be helpful to compute $L(f)(t)$ when $f(t) = t^2 - t + 1$; that is,

$$L(f)(t) = (t-1)^2 - (t-1) + 1 = t^2 - 2t + 1 - t + 1 + 1 = t^2 - 3t + 3.$$

Constructing Linear Mappings from Bases

Theorem 9.1.2   *Let V and W be vector spaces. Let $\{v_1, \ldots, v_n\}$ be a basis for V and let $\{w_1, \ldots, w_n\}$ be n vectors in W. Then there exists a unique linear map $L : V \to W$ such that $L(v_i) = w_i$.*

Proof: Let $v \in V$ be a vector. Since $\mathrm{span}\{v_1, \ldots, v_n\} = V$, we may write v as

$$v = \alpha_1 v_1 + \cdots + \alpha_n v_n,$$

where $\alpha_1, \ldots, \alpha_n$ in $\mathbf{R}$. Moreover, $v_1, \ldots, v_n$ are linearly independent, and these scalars are uniquely defined. More precisely, if

$$\alpha_1 v_1 + \cdots + \alpha_n v_n = \beta_1 v_1 + \cdots + \beta_n v_n,$$

then

$$(\alpha_1 - \beta_1)v_1 + \cdots + (\alpha_n - \beta_n)v_n = 0.$$

Linear independence implies that $\alpha_j - \beta_j = 0$; that is, $\alpha_j = \beta_j$. We can now define

$$L(v) = \alpha_1 w_1 + \cdots + \alpha_n w_n. \qquad \textbf{(9.1.1)}$$

We claim that L is linear. Let $\hat{v} \in V$ be another vector and let

$$\hat{v} = \beta_1 v_1 + \cdots + \beta_n v_n.$$

It follows that

$$v + \hat{v} = (\alpha_1 + \beta_1)v_1 + \cdots + (\alpha_n + \beta_n)v_n$$

and hence by (9.1.1) that

$$\begin{aligned}
L(v + \hat{v}) &= (\alpha_1 + \beta_1)w_1 + \cdots + (\alpha_n + \beta_n)w_n \\
&= (\alpha_1 w_1 + \cdots + \alpha_n w_n) + (\beta_1 w_1 + \cdots + \beta_n w_n) \\
&= L(v) + L(\hat{v}).
\end{aligned}$$

Similarly,

$$\begin{aligned}
L(cv) &= L((c\alpha_1)v_1 + \cdots + (c\alpha_n)v_n) \\
&= c(\alpha_1 w_1 + \cdots + \alpha_n w_n) \\
&= cL(v).
\end{aligned}$$

Thus L is linear.

Let $M : V \to W$ be another linear mapping such that $M(v_i) = w_i$. Then

$$
\begin{aligned}
L(v) &= L(\alpha_1 v_1 + \cdots + \alpha_n v_n) \\
&= \alpha_1 w_1 + \cdots + \alpha_n w_n \\
&= \alpha_1 M(v_1) + \cdots + \alpha_n M(v_n) \\
&= M(\alpha_1 v_1 + \cdots + \alpha_n v_n) \\
&= M(v).
\end{aligned}
$$

Thus $L = M$ and the linear mapping is uniquely defined.   ◆

There are two assertions made in Theorem 9.1.2. The first is that a linear map exists mapping v_i to w_i. The second is that there is only one *linear* mapping that accomplishes this task. If we drop the constraint that the map be linear, then many mappings may satisfy these conditions. For example, find a linear map from $\mathbf{R} \to \mathbf{R}$ that maps 1 to 4. There is only one: $y = 4x$. However there are many nonlinear maps that send 1 to 4. Examples are $y = x + 3$ and $y = 4x^2$.

Finding the Matrix of a Linear Map from $\mathbf{R}^n \to \mathbf{R}^m$ Given by Theorem 9.1.2

Suppose that $V = \mathbf{R}^n$ and $W = \mathbf{R}^m$. We know that every linear map $L : \mathbf{R}^n \to \mathbf{R}^m$ can be defined as multiplication by an $m \times n$ matrix. The question that we next address is: How can we find the matrix whose existence is guaranteed by Theorem 9.1.2?

More precisely, let $v_1, \ldots, v_n$ be a basis for $\mathbf{R}^n$ and let $w_1, \ldots, w_n$ be vectors in $\mathbf{R}^m$. We suppose that all of these vectors are row vectors. Then we need to find an $m \times n$ matrix A such that $A v_i^t = w_i^t$ for all i. We find A as follows. Let $v \in \mathbf{R}^n$ be a row vector. Since the v_i form a basis, there exist scalars α_i such that

$$
v = \alpha_1 v_1 + \cdots + \alpha_n v_n.
$$

In coordinates,

$$
v^t = (v_1^t | \cdots | v_n^t) \begin{pmatrix} \alpha_1 \\ \vdots \\ \alpha_n \end{pmatrix}, \tag{9.1.2}
$$

where $(v_1^t | \cdots | v_n^t)$ is an $n \times n$ invertible matrix. By definition, (9.1.1),

$$
L(v) = \alpha_1 w_1 + \cdots + \alpha_n w_n.
$$

Thus the matrix A must satisfy

$$Av^t = (w_1^t | \cdots | w_n^t) \begin{pmatrix} \alpha_1 \\ \vdots \\ \alpha_n \end{pmatrix},$$

where $(w_1^t | \cdots | w_n^t)$ is an $m \times n$ matrix. Using (9.1.2), we see that

$$Av^t = (w_1^t | \cdots | w_n^t)(v_1^t | \cdots | v_n^t)^{-1} v^t,$$

and

$$A = (w_1^t | \cdots | w_n^t)(v_1^t | \cdots | v_n^t)^{-1} \tag{9.1.3}$$

is the desired $m \times n$ matrix.

An Example of a Linear Map from $\mathbf{R}^3$ to $\mathbf{R}^2$

As an example, we illustrate Theorem 9.1.2 and (9.1.3) by defining a linear mapping from $\mathbf{R}^3$ to $\mathbf{R}^2$ by its action on a basis. Let

$$v_1 = (1, 4, 1), \quad v_2 = (-1, 1, 1), \quad v_3 = (0, 1, 0).$$

We claim that $\{v_1, v_2, v_3\}$ is a basis of $\mathbf{R}^3$ and that there is a unique linear map for which $L(v_i) = w_i$, where

$$w_1 = (2, 0), \quad w_2 = (1, 1), \quad w_3 = (1, -1).$$

We can verify that $\{v_1, v_2, v_3\}$ is a basis of $\mathbf{R}^3$ by showing that the matrix

$$(v_1^t | v_2^t | v_3^t) = \begin{pmatrix} 1 & -1 & 0 \\ 4 & 1 & 1 \\ 1 & 1 & 0 \end{pmatrix}$$

is invertible. This can be done either in MATLAB using the `inv` command or by hand by row reducing the matrix

$$\begin{pmatrix} 1 & -1 & 0 & | & 1 & 0 & 0 \\ 4 & 1 & 1 & | & 0 & 1 & 0 \\ 1 & 1 & 0 & | & 0 & 0 & 1 \end{pmatrix}$$

to obtain

$$(v_1^t | v_2^t | v_3^t)^{-1} = \frac{1}{2} \begin{pmatrix} 1 & 0 & 1 \\ -1 & 0 & 1 \\ -3 & 2 & -5 \end{pmatrix}.$$

Now apply (9.1.3) to obtain

$$A = \frac{1}{2} \begin{pmatrix} 2 & 1 & 1 \\ 0 & 1 & -1 \end{pmatrix} \begin{pmatrix} 1 & 0 & 1 \\ -1 & 0 & 1 \\ -3 & 2 & -5 \end{pmatrix} = \begin{pmatrix} -1 & 1 & -1 \\ 1 & -1 & 3 \end{pmatrix}.$$

As a check, verify by matrix multiplication that $Av_i = w_i$, as claimed.

Properties of Linear Mappings

Lemma 9.1.3   *Let U, V, and W be vector spaces and let $L : V \to W$ and $M : U \to V$ be linear maps. Then $L \circ M : U \to W$ is linear.*

Proof: The proof of Lemma 9.1.3 is identical to that of Lemma 3.5.1.   ◆

A linear map $L : V \to W$ is *invertible* if there exists a linear map $M : W \to V$ such that $L \circ M : W \to W$ is the identity map on W and $M \circ L : V \to V$ is the identity map on V.

Theorem 9.1.4   *Let V and W be finite-dimensional vector spaces and let $v_1, \ldots, v_n$ be a basis for V. Let $L : V \to W$ be a linear map. Then L is invertible if and only if $w_1, \ldots, w_n$ is a basis for W, where $w_j = L(v_j)$.*

Proof: If $w_1, \ldots, w_n$ is a basis for W, then use Theorem 9.1.2 to define a linear map $M : W \to V$ by $M(w_j) = v_j$. Note that

$$L \circ M(w_j) = L(v_j) = w_j.$$

It follows by linearity (using the uniqueness part of Theorem 9.1.2) that $L \circ M$ is the identity of W. Similarly, $M \circ L$ is the identity map on V, and L is invertible.

Conversely, suppose that $L \circ M$ and $M \circ L$ are identity maps and that $w_j = L(v_j)$. We must show that $w_1, \ldots, w_n$ is a basis. We use Theorem 5.5.3 and verify separately that $w_1, \ldots, w_n$ are linearly independent and span W.

If there exist scalars $\alpha_1, \ldots, \alpha_n$ such that

$$\alpha_1 w_1 + \cdots + \alpha_n w_n = 0,$$

then apply M to both sides of this equation to obtain

$$0 = M(\alpha_1 w_1 + \cdots + \alpha_n w_n) = \alpha_1 v_1 + \cdots + \alpha_n v_n.$$

But the v_j are linearly independent. Therefore $\alpha_j = 0$ and the w_j are linearly independent.

To show that the w_j span W, let w be a vector in W. Since the v_j are a basis for V, there exist scalars $\beta_1, \dots, \beta_n$ such that

$$M(w) = \beta_1 v_1 + \cdots + \beta_n v_n.$$

Applying L to both sides of this equation yields

$$w = L \circ M(w) = \beta_1 w_1 + \cdots + \beta_n w_n.$$

Therefore the w_j span W. ◆

Corollary 9.1.5 *Let V and W be finite-dimensional vector spaces. Then there exists an invertible linear map $L : V \to W$ if and only if $\dim(V) = \dim(W)$.*

Proof: Suppose that $L : V \to W$ is an invertible linear map. Let $v_1, \dots, v_n$ be a basis for V, where $n = \dim(V)$. Then Theorem 9.1.4 implies that $L(v_1), \dots, L(v_n)$ is a basis for W and $\dim(W) = n = \dim(V)$.

Conversely, suppose that $\dim(V) = \dim(W) = n$. Let $v_1, \dots, v_n$ be a basis for V and let $w_1, \dots, w_n$ be a basis for W. Using Theorem 9.1.2, define the linear map $L : V \to W$ by $L(v_j) = w_j$. Theorem 9.1.4 states that L is invertible. ◆

HAND EXERCISES

1. Use the method described above to construct a linear mapping L from $\mathbf{R}^3$ to $\mathbf{R}^2$ with $L(v_i) = w_i, i = 1, 2, 3$, where

$$v_1 = (1, 0, 2), \quad v_2 = (2, -1, 1), \quad v_3 = (-2, 1, 0)$$

and

$$w_1 = (-1, 0), \quad w_2 = (0, 1), \quad w_3 = (3, 1).$$

2. Let $\mathcal{P}_n$ be the vector space of polynomials $p(t)$ of degree less than or equal to n. Show that $\{1, t, t^2, \dots, t^n\}$ is a basis for $\mathcal{P}_n$.

3. Show that

$$\frac{d}{dt} : \mathcal{P}_3 \to \mathcal{P}_2$$

is a linear mapping.

4. Show that

$$L(p) = \int_0^t p(s)ds$$

is a linear mapping of $\mathcal{P}_2 \to \mathcal{P}_3$.

5. Use Exercises 3 and 4 and Theorem 9.1.2 to show that

$$\frac{d}{dt} \circ L : \mathcal{P}_2 \to \mathcal{P}_2$$

is the identity map.

6. Let $\mathbf{C}$ denote the set of complex numbers. Verify that $\mathbf{C}$ is a two-dimensional vector space. Show that $L : \mathbf{C} \to \mathbf{C}$ is defined by

$$L(z) = \lambda z,$$

where $\lambda = \sigma + i\tau$ is a linear mapping.

7. Let $\mathcal{M}(n)$ denote the vector space of $n \times n$ matrices and let A be an $n \times n$ matrix. Let $L : \mathcal{M}(n) \to \mathcal{M}(n)$ be the mapping defined by $L(X) = AX - XA$, where $X \in \mathcal{M}(n)$. Verify that L is a linear mapping. Show that the null space of L, $\{X \in \mathcal{M} : L(X) = 0\}$, is a subspace consisting of all matrices that commute with A.

8. Let $L : \mathcal{C}^1 \to \mathbf{R}$ be defined by $L(f) = \int_0^{2\pi} f(t) \cos(t) dt$ for $f \in \mathcal{C}^1$. Verify that L is a linear mapping.

9. Let $\mathcal{P}$ be the vector space of polynomials in one variable x. Define $L : \mathcal{P} \to \mathcal{P}$ by $L(p)(x) = \int_0^x (t - 1) p(t) dt$. Verify that L is a linear mapping.

9.2 ROW RANK EQUALS COLUMN RANK

Let A be an $m \times n$ matrix. The *row space* of A is the span of the row vectors of A and is a subspace of $\mathbf{R}^n$. The *column space* of A is the span of the columns of A and is a subspace of $\mathbf{R}^m$.

Definition 9.2.1 *The* row rank *of A is the dimension of the row space of A; the* column rank *of A is the dimension of the column space of A.*

Lemma 5.5.4 states that

$$\text{row rank}(A) = \text{rank}(A).$$

We show below that row ranks and column ranks are equal. We begin by continuing the discussion of the preceding section on linear maps between vector spaces.

Null Space and Range

Each linear map between vector spaces defines two subspaces. Let V and W be vector spaces and let $L : V \to W$ be a linear map. Then

$$\text{null space}(L) = \{v \in V : L(v) = 0\} \subset V$$

and

$$\text{range}(L) = \{L(v) \in W : v \in V\} \subset W.$$

Lemma 9.2.2 *Let $L : V \to W$ be a linear map between vector spaces. Then the null space of L is a subspace of V and the range of L is a subspace of W.*

Proof: The proof that the null space of L is a subspace of V follows from linearity in precisely the same way that the null space of an $m \times n$ matrix is a subspace of $\mathbf{R}^n$. That is, if v_1 and v_2 are in the null space of L, then

$$L(v_1 + v_2) = L(v_1) + L(v_2) = 0 + 0 = 0,$$

and for $c \in \mathbf{R}$,

$$L(cv_1) = cL(v_1) = c0 = 0.$$

So the null space of L is closed under addition and scalar multiplication and is a subspace of V.

To prove that the range of L is a subspace of W, let w_1 and w_2 be in the range of L. Then, by definition, there exist v_1 and v_2 in V such that $L(v_j) = w_j$. It follows that

$$L(v_1 + v_2) = L(v_1) + L(v_2) = w_1 + w_2.$$

Therefore $w_1 + w_2$ is in the range of L. Similarly,

$$L(cv_1) = cL(v_1) = cw_1.$$

So the range of L is closed under addition and scalar multiplication and is a subspace of W.   ◆

Suppose that A is an $m \times n$ matrix and $L_A : \mathbf{R}^n \to \mathbf{R}^m$ is the associated linear map. Then the null space of L_A is precisely the null space of A, as defined in Definition 5.2.1. Moreover, the range of L_A is the column space of A. To verify this, write $A = (A_1 | \cdots | A_n)$, where A_j is the jth column of A, and let $v = (v_1, \ldots, v_n)^t$. Then $L_A(v)$ is the linear combination of columns of A:

$$L_A(v) = Av = v_1 A_1 + \cdots + v_n A_n.$$

The next theorem relates the dimensions of the null space and range to the dimension of V.

Theorem 9.2.3   *Let V and W be vector spaces with V finite dimensional, and let $L : V \to W$ be a linear map. Then*

$$\dim(V) = \dim(\textit{null space}(L)) + \dim(\textit{range}(L)).$$

Proof: Since V is finite dimensional, the null space of L is finite dimensional (since the null space is a subspace of V) and the range of L is finite dimensional (since it is spanned by the vectors $L(v_j)$, where $v_1, \ldots, v_n$ is a basis for V). Let $u_1, \ldots, u_k$ be a basis for the null space of L and let $w_1, \ldots, w_\ell$ be a basis for the range of L. Choose vectors $y_j \in V$ such that $L(y_j) = w_j$. We claim that $u_1, \ldots, u_k, y_1, \ldots, y_\ell$ is a basis for V, which proves the theorem.

To verify that $u_1, \ldots, u_k, y_1, \ldots, y_\ell$ are linear independent, suppose that

$$\alpha_1 u_1 + \cdots + \alpha_k u_k + \beta_1 y_1 + \cdots + \beta_\ell y_\ell = 0. \tag{9.2.1}$$

Apply L to both sides of (9.2.1) to obtain

$$\beta_1 w_1 + \cdots + \beta_\ell w_\ell = 0.$$

Since the w_j are linearly independent, it follows that $\beta_j = 0$ for all j. Now (9.2.1) implies that

$$\alpha_1 u_1 + \cdots + \alpha_k u_k = 0.$$

Since the u_j are linearly independent, it follows that $\alpha_j = 0$ for all j.

To verify that $u_1, \ldots, u_k, y_1, \ldots, y_\ell$ span V, let v be in V. Since $w_1, \ldots, w_\ell$ span W, it follows that there exist scalars β_j such that

$$L(v) = \beta_1 w_1 + \cdots + \beta_\ell w_\ell.$$

Note that by choice of the y_j,

$$L(\beta_1 y_1 + \cdots + \beta_\ell y_\ell) = \beta_1 w_1 + \cdots + \beta_\ell w_\ell.$$

It follows by linearity that

$$u = v - (\beta_1 y_1 + \cdots + \beta_\ell y_\ell)$$

is in the null space of L. Hence there exist scalars α_j such that

$$u = \alpha_1 u_1 + \cdots + \alpha_k u_k.$$

Thus v is in the span of $u_1, \ldots, u_k, y_1, \ldots, y_\ell$, as desired.   ◆

Row Rank and Column Rank

Recall Theorem 5.5.6, which states that the nullity plus the rank of an $m \times n$ matrix equals n. At first glance it might seem that this theorem and Theorem 9.2.3 contain the same information, but they do not. Theorem 5.5.6 is proved using a detailed analysis of solutions of linear equations based on Gaussian elimination, back substitution, and reduced echelon form, whereas Theorem 9.2.3 is proved using abstract properties of linear maps.

Let A be an $m \times n$ matrix. Theorem 5.5.6 states that

$$\text{nullity}(A) + \text{rank}(A) = n.$$

Meanwhile, Theorem 9.2.3 states that

$$\dim(\text{null space}(L_A)) + \dim(\text{range}(L_A)) = n.$$

But the dimension of the null space of L_A equals the nullity of A, and the dimension of the range of A equals the dimension of the column space of A. Therefore

$$\text{nullity}(A) + \dim(\text{column space}(A)) = n.$$

Hence the rank of A equals the column rank of A. Since rank and row rank are identical, we have proved:

Theorem 9.2.4   *Let A be an m × n matrix. Then*

$$row\ rank\ A = column\ rank\ A.$$

Since the row rank of A equals the column rank of A^t, we have:

Corollary 9.2.5   *Let A be an m × n matrix. Then*

$$rank(A) = rank(A^t).$$

HAND EXERCISES

1. The 3×3 matrix

$$A = \begin{pmatrix} 1 & 2 & 5 \\ 2 & -1 & 1 \\ 3 & 1 & 6 \end{pmatrix}$$

has rank two. Let r_1, r_2, r_3 be the rows of A and c_1, c_2, c_3 be the columns of A. Find scalars α_j and β_j such that

$$\alpha_1 r_1 + \alpha_2 r_2 + \alpha_3 r_3 = 0$$
$$\beta_1 c_1 + \beta_2 c_2 + \beta_3 c_3 = 0.$$

2. What is the largest row rank that a 5×3 matrix can have?

3. Let

$$A = \begin{pmatrix} 1 & 1 & 0 & 1 \\ 0 & -1 & 1 & 2 \\ 1 & 2 & -1 & 3 \end{pmatrix}.$$

(**a**) Find a basis for the row space of A and the row rank of A.
(**b**) Find a basis for the column space of A and the column rank of A.
(**c**) Find a basis for the null space of A and the nullity of A.
(**d**) Find a basis for the null space of A^t and the nullity of A^t.

4. Let A be a nonzero 3×3 matrix such that $A^2 = 0$. Show that $rank(A) = 1$.

5. Let B be an $m \times p$ matrix and let C be a $p \times n$ matrix. Prove that the rank of the $m \times n$ matrix $A = BC$ satisfies

$$\text{rank}(A) \leq \min\{\text{rank}(B), \ \text{rank}(C)\}.$$

COMPUTER EXERCISES

6. Let

$$A = \begin{pmatrix} 1 & 1 & 2 & 2 \\ 0 & -1 & 3 & 1 \\ 2 & -1 & 1 & 0 \\ -1 & 0 & 7 & 4 \end{pmatrix}. \tag{9.2.2}*$$

(a) Compute rank(A) and exhibit a basis for the row space of A.
(b) Find a basis for the column space of A.
(c) Find all solutions to the homogeneous equation $Ax = 0$.
(d) Does

$$Ax = \begin{pmatrix} 4 \\ 2 \\ 2 \\ 1 \end{pmatrix}$$

have a solution?

9.3   VECTORS AND MATRICES IN COORDINATES

In the last half of this chapter we discuss how similarity of matrices should be thought of as change of coordinates for linear mappings. There are three steps in this discussion:

1. Formalize the idea of coordinates for a vector in terms of basis.

2. Discuss how to write a linear map as a matrix in each coordinate system.

3. Determine how the matrices corresponding to the same linear map in two different coordinate systems are related.

The answer to the last question is simple: The matrices are related by a change of coordinates if and only if they are similar. We discuss these steps in this section for $\mathbf{R}^n$ and in Section 9.4 for general vector spaces.

Coordinates of Vectors Using Bases

Throughout we have written vectors $v \in \mathbf{R}^n$ in coordinates as $v = (v_1, \ldots, v_n)$, and we have used this notation almost without comment. From the point of view of vector space operations, we are just writing

$$v = v_1 e_1 + \cdots + v_n e_n$$

as a linear combination of the standard basis $\mathcal{E} = \{e_1, \ldots, e_n\}$ of $\mathbf{R}^n$.

More generally, each basis provides a set of coordinates for a vector space. This fact is described by the following lemma (although its proof is identical to the first part of the proof of Theorem 9.1.2:

Lemma 9.3.1   *Let $W = \{w_1, \ldots, w_n\}$ be a basis for the vector space V. Then each vector v in V can be written uniquely as a linear combination of vectors in W; that is,*

$$v = \alpha_1 w_1 + \cdots + \alpha_n w_n$$

for uniquely defined scalars $\alpha_1, \ldots, \alpha_n$.

Proof: Since W is a basis, Theorem 5.5.3 implies that the vectors $w_1, \ldots, w_n$ span V and are linearly independent. Therefore we can write v in V as a linear combination of vectors in $\mathcal{B}$. That is, there are scalars $\alpha_1, \ldots, \alpha_n$ such that

$$v = \alpha_1 w_1 + \cdots + \alpha_n w_n.$$

Next we show that these scalars are uniquely defined. Suppose that we can write v as a linear combination of the vectors in $\mathcal{B}$ in a second way; that is, suppose

$$v = \beta_1 w_1 + \cdots + \beta_n w_n$$

for scalars $\beta_1, \ldots, \beta_n$. Then

$$(\alpha_1 - \beta_1)w_1 + \cdots + (\alpha_n - \beta_n)w_n = 0.$$

Since the vectors in W are linearly independent, it follows that $\alpha_j = \beta_j$ for all j.   ◆

Definition 9.3.2   *Let $W = \{w_1, \ldots, w_n\}$ be a basis in a vector space V. Lemma 9.3.1 states that we can write $v \in V$ uniquely as*

$$v = \alpha_1 w_1 + \cdots + \alpha_n w_n. \tag{9.3.1}$$

The scalars $\alpha_1, \ldots, \alpha_n$ are the coordinates *of v relative to the basis W, and we denote the coordinates of v in the basis W by*

$$[v]_W = (\alpha_1, \ldots, \alpha_n) \in \mathbf{R}^n. \tag{9.3.2}$$

We call the coordinates of a vector $v \in \mathbf{R}^n$ relative to the standard basis the *standard coordinates* of v.

Writing Linear Maps in Coordinates As Matrices

Let V be a finite dimensional vector space of dimension n, and let $L : V \to V$ be a linear mapping. We now show how each basis of V allows us to associate an $n \times n$ matrix with L. Previously we considered this question with the standard basis on $V = \mathbf{R}^n$. We showed in Chapter 3 that we can write the linear mapping L as a matrix mapping, as follows: Let $\mathcal{E} = \{e_1, \ldots, e_n\}$ be the standard basis in $\mathbf{R}^n$. Let A be the $n \times n$ matrix whose jth column is the n vector $L(e_j)$. Then Theorem 3.3.5 shows that the linear map is given by matrix multiplication as

$$L(v) = Av.$$

Thus every linear mapping on $\mathbf{R}^n$ can be written in this matrix form.

Remark 9.3.3 *Another way to think of the jth column of the matrix A is as the coordinate vector of $L(e_j)$ relative to the standard basis—that is, as $[L(e_j)]_{\mathcal{E}}$. We denote the matrix A by $[L]_{\mathcal{E}}$; this notation emphasizes the fact that A is the matrix of L relative to the standard basis.*

We now discuss how to write a linear map L as a matrix using different coordinates.

Definition 9.3.4 *Let $\mathcal{W} = \{w_1, \ldots, w_n\}$ be a basis for the vector space V. The $n \times n$ matrix $[L]_{\mathcal{W}}$ associated with the linear map $L : V \to V$ and the basis $\mathcal{W}$ is defined as follows: The jth column of $[L]_{\mathcal{W}}$ is $[L(w_j)]_{\mathcal{W}}$—the coordinates of $L(w_j)$ relative to the basis $\mathcal{W}$.*

Note that if $V = \mathbf{R}^n$ and $\mathcal{W} = \mathcal{E}$, the standard basis of $\mathbf{R}^n$, then the definition of the matrix $[L]_{\mathcal{E}}$ is exactly the same as the matrix associated with the linear map L in Remark 9.3.3.

Lemma 9.3.5 *The coordinate vector of $L(v)$ relative to the basis $\mathcal{W}$ is*

$$[L(v)]_{\mathcal{W}} = [L]_{\mathcal{W}}[v]_{\mathcal{W}}. \tag{9.3.3}$$

Proof: The process of choosing the coordinates of vectors relative to a given basis $\mathcal{W} = \{w_1, \ldots, w_n\}$ of a vector space V is itself linear. Indeed

$$[u + v]_{\mathcal{W}} = [u]_{\mathcal{W}} + [v]_{\mathcal{W}}$$
$$[cv]_{\mathcal{W}} = c[v]_{\mathcal{W}}.$$

Thus the coordinate mapping relative to a basis W of V defined by

$$v \mapsto [v]_W \tag{9.3.4}$$

is a linear mapping of V into $\mathbf{R}^n$. We denote this linear mapping by $[\cdot]_W : V \to \mathbf{R}^n$.

It now follows that both the left-hand and right-hand sides of (9.3.3) can be thought of as linear mappings of $V \to \mathbf{R}^n$. In verifying this comment, we recall Lemma 5.1.3, which states that the composition of linear maps is linear. On the left-hand side we have the mapping

$$v \mapsto L(v) \mapsto [L(v)]_W,$$

which is the composition of the linear maps: $[\cdot]_W$ with L; see (9.3.4). The right-hand side is

$$v \mapsto [v]_W \mapsto [L]_W[v]_W,$$

which is the composition of the linear maps: multiplication by the matrix $[L]_W$ with $[\cdot]_W$.

Theorem 5.1.2 states that linear mappings are determined by their actions on a basis. Thus to verify (9.3.3), we need only verify this equality for $v = w_j$ for all j. Since $[w_j]_W = e_j$, the right-hand side of (9.3.3) is

$$[L]_W[w_j]_W = [L]_W e_j,$$

which is just the jth column of $[L]_W$. The left-hand side of (9.3.3) is the vector $[L(w_j)]_W$, which by definition is also the jth column of $[L]_W$ (see Definition 9.3.4).◆

Computations of Vectors in Coordinates in R^*n*

We divide this subsection into three parts. We consider a simple example in $\mathbf{R}^2$ algebraically in the first part and geometrically in the second. In the third part we formalize and extend the algebraic discussion to $\mathbf{R}^n$.

An Example of Coordinates in R²

How do we find the coordinates of a vector v in a basis? For example, choose a (nonstandard) basis in the plane—say,

$$w_1 = (1, 1) \quad \text{and} \quad w_2 = (1, -2).$$

Since $\{w_1, w_2\}$ is a basis, we may write the vector v as a linear combination of the vectors w_1 and w_2. Thus we can find scalars α_1 and α_2 so that

$$v = \alpha_1 w_1 + \alpha_2 w_2 = \alpha_1(1, 1) + \alpha_2(1, -2) = (\alpha_1 + \alpha_2, \alpha_1 - 2\alpha_2).$$

In standard coordinates, set $v = (v_1, v_2)$; this equation leads to the system of linear equations

$$v_1 = \alpha_1 + \alpha_2$$
$$v_2 = \alpha_1 - 2\alpha_2$$

in the two variables α_1 and α_2. As we have seen, the fact that w_1 and w_2 form a basis of $\mathbf{R}^2$ implies that these equations do have a solution. Indeed, we can write this system in matrix form as

$$\begin{pmatrix} v_1 \\ v_2 \end{pmatrix} = \begin{pmatrix} 1 & 1 \\ 1 & -2 \end{pmatrix} \begin{pmatrix} \alpha_1 \\ \alpha_2 \end{pmatrix},$$

which is solved by inverting the matrix to obtain

$$\begin{pmatrix} \alpha_1 \\ \alpha_2 \end{pmatrix} = \frac{1}{3} \begin{pmatrix} 2 & 1 \\ 1 & -1 \end{pmatrix} \begin{pmatrix} v_1 \\ v_2 \end{pmatrix}. \tag{9.3.5}$$

For example, suppose $v = (2.0, 0.5)$. Using (9.3.5), we find that $(\alpha_1, \alpha_2) = (1.5, 0.5)$; that is, we can write

$$v = 1.5w_1 + 0.5w_2,$$

and $(1.5, 0.5)$ are the *coordinates* of v in the basis $\{w_1, w_2\}$.

Using the notation in (9.3.2), we may rewrite (9.3.5) as

$$[v]_\mathcal{W} = \frac{1}{3} \begin{pmatrix} 2 & 1 \\ 1 & -1 \end{pmatrix} [v]_\mathcal{E},$$

where $\mathcal{E} = \{e_1, e_2\}$ is the standard basis.

Planar Coordinates Viewed Geometrically Using MATLAB

Next we use MATLAB to view geometrically the notion of coordinates relative to a basis $\mathcal{W} = \{w_1, w_2\}$ in the plane. Type

```
w1 = [1 1];
w2 = [1 -2];
bcoord
```

MATLAB creates a graphics window showing the two basis vectors w_1 and w_2 in red. Using the mouse, click on a point near $(2, 0.5)$ in that figure. MATLAB responds by plotting the new vector v in yellow and the parallelogram generated by $\alpha_1 w_1$ and $\alpha_2 w_2$ in cyan. The values of α_1 and α_2 are also plotted on this figure. See Figure 9.1.

Abstracting $\mathbf{R}^2$ to $\mathbf{R}^n$

Suppose that we are given a basis $\mathcal{W} = \{w_1, \ldots, w_n\}$ of $\mathbf{R}^n$ and a vector $v \in \mathbf{R}^n$. How do we find the coordinates $[v]_\mathcal{W}$ of v in the basis $\mathcal{W}$?

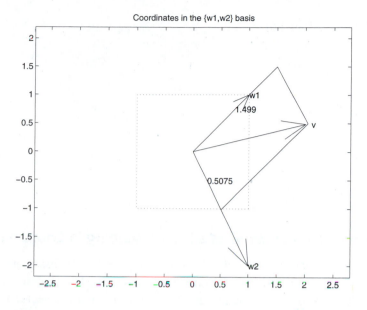

Figure 9.1
The coordinates of $v = (2.0, 0.5)$ in the basis
$w_1 = (1, 1), w_2 = (1, -2)$

For definiteness, assume that v and the w_j are row vectors. Equation (9.3.1) may be rewritten as

$$v^t = (w_1^t | \cdots | w_n^t) \begin{pmatrix} \alpha_1 \\ \vdots \\ \alpha_n \end{pmatrix}.$$

Thus

$$[v]_W = \begin{pmatrix} \alpha_1 \\ \vdots \\ \alpha_n \end{pmatrix} = P_W^{-1} v^t, \tag{9.3.6}$$

where $P_W = (w_1^t | \cdots | w_n^t)$. Since the w_j are a basis for $\mathbf{R}^n$, the columns of the matrix P_W are linearly independent, and P_W is invertible.

We may use (9.3.6) to compute $[v]_W$ with MATLAB. For example, let

$$v = (4, 1, 3)$$

and

$$w_1 = (1, 4, 7), \quad w_2 = (2, 1, 0), \quad w_3 = (-4, 2, 1).$$

Then $[v]_W$ is found by typing

```
w1 = [ 1 4 7];
w2 = [ 2 1 0];
w3 = [-4 2 1];
inv([w1' w2' w3'])*[4 1 3]'
```

The answer is

```
ans =
    0.5306
    0.3061
   -0.7143
```

Determining the Matrix of a Linear Mapping in Coordinates

Suppose that we are given the linear map $L_A : \mathbf{R}^n \to \mathbf{R}^n$ associated with the matrix A in standard coordinates and a basis $w_1, \ldots, w_n$ of $\mathbf{R}^n$. How do we find the matrix $[L_A]_W$? As above, we assume that the vectors w_j and the vector v are row vectors. Since $L_A(v) = Av^t$, we can rewrite (9.3.3) as

$$[L_A]_W[v]_W = [Av^t]_W.$$

As above, let $P_W = (w_1^t | \cdots | w_n^t)$. Using (9.3.6), we see that

$$[L_A]_W P_W^{-1} v^t = P_W^{-1} A v^t.$$

Setting

$$u = P_W^{-1} v^t,$$

we see that

$$[L_A]_W u = P_W^{-1} A P_W u.$$

Therefore

$$[L_A]_W = P_W^{-1} A P_W.$$

We have proved:

Theorem 9.3.6 *Let A be an $n \times n$ matrix and let $L_A : \mathbf{R}^n \to \mathbf{R}^n$ be the associated linear map. Let $\mathcal{W} = \{w_1, \ldots, w_n\}$ be a basis for $\mathbf{R}^n$. Then the matrix $[L_A]_W$ associated with L_A in the basis $\mathcal{W}$ is similar to A. Therefore the determinant, trace, and eigenvalues of $[L_A]_W$ are identical to those of A.*

Matrix Normal Forms in $\mathbf{R}^2$

If we are careful about how we choose the basis $\mathcal{W}$, then we can simplify the form of the matrix $[L]_{\mathcal{W}}$. Indeed, we saw examples of this process when we discussed how to find closed form solutions to linear planar systems of ODEs in Chapter 6. For example, suppose that $L : \mathbf{R}^2 \to \mathbf{R}^2$ has real eigenvalues λ_1 and λ_2 with two linearly independent eigenvectors w_1 and w_2. Then the matrix associated with L in the basis $\mathcal{W} = \{w_1, w_2\}$ is the diagonal matrix

$$[L]_{\mathcal{W}} = \begin{pmatrix} \lambda_1 & 0 \\ 0 & \lambda_2 \end{pmatrix}, \tag{9.3.7}$$

since

$$[L(w_1)]_{\mathcal{W}} = [\lambda_1 w_1]_{\mathcal{W}} = \begin{pmatrix} \lambda_1 \\ 0 \end{pmatrix} \quad \text{and} \quad [L(w_2)]_{\mathcal{W}} = [\lambda_2 w_2]_{\mathcal{W}} = \begin{pmatrix} 0 \\ \lambda_2 \end{pmatrix}.$$

In Chapter 6 we showed how to classify 2×2 matrices up to similarity (see Theorem 6.5.5) and how to use this classification to find closed form solutions to planar systems of linear ODEs (see Section 6.5). We now use the ideas of coordinates and matrices associated with bases to reinterpret the normal form result (Theorem 6.5.5) in a more geometric fashion.

Theorem 9.3.7   *Let $L : \mathbf{R}^2 \to \mathbf{R}^2$ be a linear mapping. Then in an appropriate coordinate system defined by the basis $\mathcal{W}$ below, the matrix $L_{\mathcal{W}}$ has one of the following forms:*

(a) *Suppose that L has two linearly independent real eigenvectors w_1 and w_2 with real eigenvalues λ_1 and λ_2. Then*

$$[L]_{\mathcal{W}} = \begin{pmatrix} \lambda_1 & 0 \\ 0 & \lambda_2 \end{pmatrix}.$$

(b) *Suppose that L has no real eigenvectors and complex conjugate eigenvalues $\sigma \pm i\tau$, where $\tau \neq 0$. Let $w_1 + iw_2$ be a complex eigenvector of L associated with the eigenvalue $\sigma - i\tau$. Then $\mathcal{W} = \{w_1, w_2\}$ is a basis and*

$$[L]_{\mathcal{W}} = \begin{pmatrix} \sigma & -\tau \\ \tau & \sigma \end{pmatrix}.$$

(c) *Suppose that L has exactly one linearly independent real eigenvector w_1 with real eigenvalue λ. Choose the generalized eigenvector w_2, where*

$$(L - \lambda I_2)(w_2) = w_1. \tag{9.3.8}$$

Then $\mathcal{W} = \{w_1, w_2\}$ *is a basis and*

$$[L]_{\mathcal{W}} = \begin{pmatrix} \lambda & 1 \\ 0 & \lambda \end{pmatrix}.$$

Proof: The verification of (a) was discussed in (9.3.7). The verification of (b) follows from (6.2.9) on equating w_1 with v and w_2 with w. The verification of (c) follows directly from (9.3.8) as

$$[L(w_1)]_{\mathcal{W}} = \lambda e_1 \quad \text{and} \quad [L(w_2)]_{\mathcal{W}} = e_1 + \lambda e_2. \qquad \blacklozenge$$

Visualization of Coordinate Changes in ODEs

We consider two examples. As a first example, note that the matrices

$$C = \begin{pmatrix} 1 & 0 \\ 0 & -2 \end{pmatrix} \quad \text{and} \quad B = \begin{pmatrix} 4 & -3 \\ 6 & -5 \end{pmatrix}$$

are similar matrices. Indeed, $B = P^{-1}CP$, where

$$P = \begin{pmatrix} 2 & -1 \\ 1 & -1 \end{pmatrix}. \qquad\qquad\qquad \textbf{(9.3.9)}$$

The phase portraits of the differential equations $\dot{X} = BX$ and $\dot{X} = CX$ are shown in Figure 9.2. Note that both phase portraits are pictures of the *same* saddle—just in different coordinate systems.

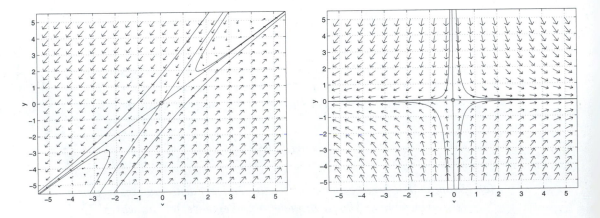

Figure 9.2
Phase planes for the saddles $\dot{X} = BX$ and $\dot{X} = CX$

As a second example, note that the matrices

$$C = \begin{pmatrix} 0 & 2 \\ -2 & 0 \end{pmatrix} \quad \text{and} \quad B = \begin{pmatrix} 6 & -4 \\ 10 & -6 \end{pmatrix}$$

are similar matrices, and both are centers. Indeed, $B = P^{-1}CP$, where P is the same matrix as in (9.3.9). The phase portraits of the differential equations $\dot{X} = BX$ and $\dot{X} = CX$ are shown in Figure 9.3. Note that both phase portraits are pictures of the *same* center—just in different coordinate systems.

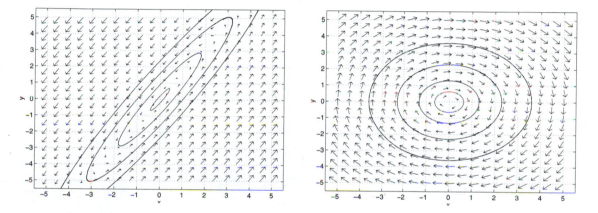

Figure 9.3
Phase planes for the centers $\dot{X} = BX$ and $\dot{X} = CX$

HAND EXERCISES

1. Let

$$w_1 = (1, 4) \quad \text{and} \quad w_2 = (-2, 1).$$

Find the coordinates of $v = (-1, 32)$ in the $\mathcal{W}$ basis.

2. Let $w_1 = (1, 2)$ and $w_2 = (0, 1)$ be a basis for $\mathbf{R}^2$. Let $L_A : \mathbf{R}^2 \to \mathbf{R}^2$ be the linear map given by the matrix

$$A = \begin{pmatrix} 2 & 1 \\ -1 & 0 \end{pmatrix}$$

in standard coordinates. Find the matrix $[L]_{\mathcal{W}}$.

3. Let E_{ij} be the 2×3 matrix whose entry in the ith row and jth column is 1 and all of whose other entries are 0.

(a) Show that

$$\mathcal{V} = \{E_{11}, E_{12}, E_{13}, E_{21}, E_{22}, E_{23}\}$$

is a basis for the vector space of 2×3 matrices.

(b) Compute $[A]_\mathcal{V}$, where

$$A = \begin{pmatrix} -1 & 0 & 2 \\ 3 & -2 & 4 \end{pmatrix}.$$

4. Verify that $\mathcal{V} = \{p_1, p_2, p_3\}$, where

$$p_1(t) = 1 + 2t, \quad p_2(t) = t + 2t^2, \quad \text{and} \quad p_3(t) = 2 - t^2$$

is a basis for the vector space of polynomials $\mathcal{P}_2$. Let $p(t) = t$ and find $[p]_\mathcal{V}$.

COMPUTER EXERCISES

5. Let

$$w_1 = (1, 0, 2), \quad w_2 = (2, 1, 4), \quad \text{and} \quad w_3 = (0, 1, -1)$$

be a basis for $\mathbf{R}^3$. Find $[v]_\mathcal{W}$, where $v = (2, 1, 5)$.

6. Let

$$\begin{aligned} w_1 &= (0.2, -1.3, 0.34, -1.1) \\ w_2 &= (0.5, -0.6, 0.7, 0.8) \\ w_3 &= (-1.0, 1.0, 2.0, 4.5) \\ w_4 &= (-5.1, 0.0, 1.6, -1.7) \end{aligned}$$

(9.3.10)*

be a basis $\mathcal{W}$ for $\mathbf{R}^4$. Find $[v]_\mathcal{W}$, where $v = (1.7, 2.3, 1.0, -5.0)$.

7. Find a basis $\mathcal{W} = \{w_1, w_2\}$ such that $[L_A]_\mathcal{W}$ is a diagonal matrix, where L_A is the linear map associated with the matrix

$$A = \begin{pmatrix} -10 & -6 \\ 18 & 11 \end{pmatrix}.$$

8. Let A be the 4×4 matrix

$$A = \begin{pmatrix} 2 & 1 & 4 & 6 \\ 1 & 2 & 1 & 1 \\ 0 & 1 & 2 & 4 \\ 2 & 1 & 1 & 5 \end{pmatrix},$$

(9.3.11)*

and let $\mathcal{W} = \{w_1, w_2, w_3, w_4\}$, where

$$\begin{aligned} w_1 &= (1, 2, 3, 4) \\ w_2 &= (0, -1, 1, 3) \\ w_3 &= (2, 0, 0, 1) \\ w_4 &= (-1, 1, 3, 0). \end{aligned}$$

(9.3.12)*

Verify that $\mathcal{W}$ is a basis of $\mathbf{R}^4$ and compute the matrix associated with A in the $\mathcal{W}$ basis.

9.4 MATRICES OF LINEAR MAPS ON A VECTOR SPACE

Returning to the general finite dimensional vector space V, suppose that

$$\mathcal{W} = \{w_1, \ldots, w_n\} \quad \text{and} \quad \mathcal{Z} = \{z_1, \ldots, z_n\}$$

are bases of V. Then we can write

$$v = \alpha_1 w_1 + \cdots + \alpha_n w_n \quad \text{and} \quad v = \beta_1 z_1 + \cdots + \beta_n z_n$$

to obtain the coordinates

$$[v]_{\mathcal{W}} = (\alpha_1, \ldots, \alpha_n) \quad \text{and} \quad [v]_{\mathcal{Z}} = (\beta_1, \ldots, \beta_n) \tag{9.4.1}$$

of v relative to the bases $\mathcal{W}$ and $\mathcal{Z}$. The question that we address is: How are $[v]_{\mathcal{W}}$ and $[v]_{\mathcal{Z}}$ related? We answer this question by finding an $n \times n$ matrix $C_{\mathcal{WZ}}$ such that

$$\begin{pmatrix} \alpha_1 \\ \vdots \\ \alpha_n \end{pmatrix} = C_{\mathcal{WZ}} \begin{pmatrix} \beta_1 \\ \vdots \\ \beta_n \end{pmatrix}. \tag{9.4.2}$$

We may rewrite (9.4.2) as

$$[v]_{\mathcal{W}} = C_{\mathcal{WZ}} [v]_{\mathcal{Z}}. \tag{9.4.3}$$

Definition 9.4.1 *Let $\mathcal{W}$ and $\mathcal{Z}$ be bases for the n-dimensional vector space V. The $n \times n$ matrix $C_{\mathcal{WZ}}$ is a transition matrix if $C_{\mathcal{WZ}}$ satisfies (9.4.3).*

Transition Mappings Defined

The next theorem presents a method for finding the transition matrix between coordinates associated with bases in an n-dimensional vector space V.

Theorem 9.4.2 *Let $\mathcal{W} = \{w_1, \ldots, w_n\}$ and $\mathcal{Z} = \{z_1, \ldots, z_n\}$ be bases for the n-dimensional vector space V. Then*

$$C_{\mathcal{WZ}} = \begin{pmatrix} c_{11} & \cdots & c_{1n} \\ \vdots & \vdots & \vdots \\ c_{n1} & \cdots & c_{nn} \end{pmatrix} \tag{9.4.4}$$

is the transition matrix, where

$$z_1 = c_{11} w_1 + \cdots + c_{n1} w_n$$
$$\vdots \tag{9.4.5}$$
$$z_n = c_{1n} w_1 + \cdots + c_{nn} w_n$$

for scalars c_{ij}.

Proof: We can restate (9.4.5) as

$$[z_j]_W = \begin{pmatrix} c_{1j} \\ \vdots \\ c_{nj} \end{pmatrix}.$$

Note that

$$[z_j]_Z = e_j$$

by definition. Since the transition matrix satisfies $[v]_W = C_{WZ}[v]_Z$ for all vectors $v \in V$, it must satisfy this relation for $v = z_j$. Therefore

$$[z_j]_W = C_{WZ}[z_j]_Z = C_{WZ}e_j.$$

It follows that $[z_j]_W$ is the jth column of C_{WZ}, which proves the theorem.   ◆

A Formula for C_{WZ} When $V = \mathbf{R}^n$

For bases in $\mathbf{R}^n$, there is a formula for finding transition matrices. Let $W = \{w_1, \ldots, w_n\}$ and $Z = \{z_1, \ldots, z_n\}$ be bases of $\mathbf{R}^n$—written as row vectors. Also, let $v \in \mathbf{R}^n$ be written as a row vector. Then (9.3.6) implies that

$$[v]_W = P_W^{-1}v^t \quad \text{and} \quad [v]_Z = P_Z^{-1}v^t,$$

where

$$P_W = (w_1^t|\cdots|w_n^t) \quad \text{and} \quad P_Z = (z_1^t|\cdots|z_n^t).$$

It follows that

$$[v]_W = P_W^{-1}P_Z[v]_Z$$

and that

$$C_{WZ} = P_W^{-1}P_Z. \tag{9.4.6}$$

As an example, consider the following bases of $\mathbf{R}^4$: Let

$$\begin{array}{ll} w_1 = [1, 4, 2, 3] & z_1 = [3, 2, 0, 1] \\ w_2 = [2, 1, 1, 4] & z_2 = [-1, 0, 2, 3] \\ w_3 = [0, 1, 5, 6] & z_3 = [3, 1, 1, 3] \\ w_4 = [2, 5, -1, 0] & z_4 = [2, 2, 3, 5]. \end{array} \tag{9.4.7*}$$

Then the matrix C_{WZ} is obtained in MATLAB by typing e9_4_7 to enter the bases and

```
inv([w1' w2' w3' w4'])*[z1' z2' z3' z4']
```

to compute C_{wz}. The answer is

```
ans =
  -8.0000    5.5000   -7.0000   -3.2500
  -0.5000    0.7500    0.0000    0.1250
   4.5000   -2.7500    4.0000    2.3750
   6.0000   -4.0000    5.0000    2.5000
```

Coordinates Relative to Two Different Bases in R²

Recall the basis $\mathcal{W}$:

$$w_1 = (1, 1) \quad \text{and} \quad w_2 = (1, -2)$$

of $\mathbf{R}^2$ that was used in a previous example. Suppose that $\mathcal{Z} = \{z_1, z_2\}$ is a second basis of $\mathbf{R}^2$. Write $v = (v_1, v_2)$ as a linear combination of the basis $\mathcal{Z}$,

$$v = \beta_1 z_1 + \beta_2 z_2,$$

obtaining the coordinates $[v]_{\mathcal{Z}} = (\beta_1, \beta_2)$.

We use MATLAB to illustrate how the coordinates of a vector v relative to two bases may be viewed geometrically. Suppose that $z_1 = (1, 3)$ and $z_2 = (1, -2)$. Then enter the two bases $\mathcal{W}$ and $\mathcal{Z}$ by typing

```
w1 = [1 1];
w2 = [1 -2];
z1 = [1 3];
z2 = [-1 2];
ccoord
```

The MATLAB program ccoord opens two graphics windows representing the $\mathcal{W}$ and $\mathcal{Z}$ planes with the basis vectors plotted in red. Clicking the left mouse button on a vector in the $\mathcal{W}$ plane simultaneously plots this vector v in both planes in yellow and the coordinates of v in the respective bases in cyan. See Figure 9.4. From this display you can visualize the coordinates of a vector relative to two different bases.

Note that the program ccoord prints the transition matrix C_{wz} in the MATLAB control window. We can verify the calculations of the program ccoord in this example by hand. Recall that (9.4.6) states that

$$
\begin{aligned}
C_{wz} &= \begin{pmatrix} 1 & 2 \\ 2 & 3 \end{pmatrix}^{-1} \begin{pmatrix} 1 & 2 \\ 4 & 1 \end{pmatrix} \\
&= \begin{pmatrix} -3 & 2 \\ 2 & -1 \end{pmatrix} \begin{pmatrix} 1 & 2 \\ 4 & 1 \end{pmatrix} \\
&= \begin{pmatrix} 5 & -4 \\ -2 & 3 \end{pmatrix}.
\end{aligned}
$$

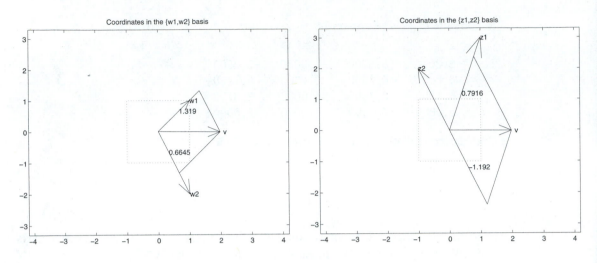

Figure 9.4
The coordinates of $v = (1.9839, -0.0097)$ in the bases $w_1 = (1, 1)$, $w_2 = (1, -2)$ and
$z_1 = (1, 3)$, $z_2 = (-1, 2)$

Matrices of Linear Maps in Different Bases

Theorem 9.4.3 *Let $L : V \to V$ be a linear mapping, and let $\mathcal{W}$ and $\mathcal{Z}$ be bases of V.
Then*

$$[L]_{\mathcal{Z}} \quad and \quad [L]_{\mathcal{W}}$$

are similar matrices. More precisely,

$$[L]_{\mathcal{W}} = C_{\mathcal{Z}\mathcal{W}}^{-1}[L]_{\mathcal{Z}}C_{\mathcal{Z}\mathcal{W}}. \tag{9.4.8}$$

Proof: For every $v \in \mathbf{R}^n$ we compute

$$
\begin{aligned}
C_{\mathcal{Z}\mathcal{W}}[L]_{\mathcal{W}}[v]_{\mathcal{W}} &= C_{\mathcal{Z}\mathcal{W}}[L(v)]_{\mathcal{W}} \\
&= [L(v)]_{\mathcal{Z}} \\
&= [L]_{\mathcal{Z}}[v]_{\mathcal{Z}} \\
&= [L]_{\mathcal{Z}}C_{\mathcal{Z}\mathcal{W}}[v]_{\mathcal{W}}.
\end{aligned}
$$

Since this computation holds for every $[v]_{\mathcal{W}}$, it follows that

$$C_{\mathcal{Z}\mathcal{W}}[L]_{\mathcal{W}} = [L]_{\mathcal{Z}}C_{\mathcal{Z}\mathcal{W}}.$$

Thus (9.4.8) is valid. $\blacklozenge$

HAND EXERCISES

1. Let

$$w_1 = (1, 2) \quad \text{and} \quad w_2 = (0, 1)$$

and

$$z_1 = (2, 3) \quad \text{and} \quad z_2 = (3, 4)$$

be two bases of $\mathbf{R}^2$. Find $C_{\mathcal{WZ}}$.

2. Let $f_1(t) = \cos t$ and $f_2(t) = \sin t$ be functions in $\mathcal{C}^1$. Let V be the two-dimensional subspace spanned by f_1, f_2; so $\mathcal{F} = \{f_1, f_2\}$ is a basis for V. Let $L : V \to V$ be the linear mapping defined by $L(f) = df/dt$. Find $[L]_{\mathcal{F}}$.

3. Let $L : V \to W$ and $M : W \to V$ be linear mappings, and assume that $\dim V > \dim W$. Show that $M \circ L : V \to V$ is not invertible.

COMPUTER EXERCISES

4. Let

$$w_1 = (0.23, 0.56) \quad \text{and} \quad w_2 = (0.17, -0.71)$$

and

$$z_1 = (-1.4, 0.3) \quad \text{and} \quad z_2 = (0.1, -0.2)$$

be two bases of $\mathbf{R}^2$ and let $v = (0.6, 0.1)$. Find $[v]_{\mathcal{W}}$, $[v]_{\mathcal{Z}}$, and $C_{\mathcal{WZ}}$.

5. Consider the matrix

$$A = \frac{1}{3}\begin{pmatrix} 1 & 1 - \sqrt{3} & 1 + \sqrt{3} \\ 1 + \sqrt{3} & 1 & 1 - \sqrt{3} \\ 1 - \sqrt{3} & 1 + \sqrt{3} & 1 \end{pmatrix} = \begin{pmatrix} 0.3333 & -0.2440 & 0.9107 \\ 0.9107 & 0.3333 & -0.2440 \\ -0.2440 & 0.9107 & 0.3333 \end{pmatrix}. \qquad \textbf{(9.4.9)*}$$

(a) Try to determine the way that the matrix A moves vectors in $\mathbf{R}^3$. For example, let

$$w_1 = (1, 1, 1)^t, \qquad w_2 = \frac{1}{\sqrt{6}}(1, -2, 1)^t, \qquad w_3 = \frac{1}{\sqrt{2}}(1, 0, -1)^t,$$

and compute Aw_j.

(b) Let $\mathcal{W} = \{w_1, w_2, w_3\}$ be the basis of $\mathbf{R}^3$ given in (a). Compute $[L_A]_{\mathcal{W}}$.

(c) Determine the way that the matrix $[L_A]_{\mathcal{W}}$ moves vectors in $\mathbf{R}^3$. For example, consider how this matrix moves the standard basis vectors e_1, e_2, and e_3. Compare this answer with that in part (a).

10

Orthogonality

In Section 10.1 we discuss orthonormal bases—bases in which each basis vector has unit length and any two basis vectors are perpendicular. We will see that the computation of coordinates in an orthonormal basis is particularly straightforward. We use orthonormality in Section 10.2 to study the geometric problem of least squares approximations (given a point v and a subspace W, find the point in W closest to v) and in Section 10.4 to study the eigenvalues and eigenvectors of symmetric matrices (the eigenvalues are real and the eigenvectors can be chosen to be orthonormal). We present two applications of least squares approximations: the Gram-Schmidt orthonormalization process for constructing orthonormal bases (Section 10.2) and regression or least squares fitting of data (Section 10.3). The chapter ends with a discussion of the QR decomposition for finding orthonormal bases in Section 10.5. This decomposition leads to an algorithm that is numerically superior to Gram-Schmidt and is the one used in MATLAB.

10.1 ORTHONORMAL BASES

In Section 9.3 we discussed how to write the coordinates of a vector in a basis. We now show that finding coordinates of vectors in certain bases is a very simple task; these bases are called orthonormal bases.

Nonzero vectors $v_1, \ldots, v_k$ in $\mathbf{R}^n$ are *orthogonal* if the dot products

$$v_i \cdot v_j = 0$$

when $i \neq j$. These vectors are *orthonormal* if they are orthogonal and of unit length; that is,

$$v_i \cdot v_i = 1.$$

The standard example of a set of orthonormal vectors in $\mathbf{R}^n$ is the standard basis $e_1, \ldots, e_n$.

Lemma 10.1.1 *Nonzero orthogonal vectors are linearly independent.*

Proof: Let $v_1, \ldots, v_k$ be a set of nonzero orthogonal vectors in $\mathbf{R}^n$ and suppose that

$$\alpha_1 v_1 + \cdots + \alpha_k v_k = 0.$$

To prove the lemma we must show that each $\alpha_j = 0$. Since $v_i \cdot v_j = 0$ for $i \neq j$,

$$\alpha_j v_j \cdot v_j = \alpha_1 v_1 \cdot v_j + \cdots + \alpha_k v_k \cdot v_j = (\alpha_1 v_1 + \cdots + \alpha_k v_k) \cdot v_j = 0 \cdot v_j = 0.$$

Since $v_j \cdot v_j = ||v_j||^2 > 0$, it follows that $\alpha_j = 0$. ◆

Corollary 10.1.2 *A set of n nonzero orthogonal vectors in $\mathbf{R}^n$ is a basis.*

Proof: Lemma 10.1.1 implies that the n vectors are linearly independent, and Corollary 5.6.7 states that n linearly independent vectors in $\mathbf{R}^n$ form a basis. ◆

Next we discuss how to find coordinates of a vector in an *orthonormal basis*—that is, a basis consisting of orthonormal vectors.

Theorem 10.1.3 *Let $V \subset \mathbf{R}^n$ be a subspace and let $\{v_1, \ldots, v_k\}$ be an orthonormal basis of V. Let $v \in V$ be a vector. Then*

$$v = \alpha_1 v_1 + \cdots + \alpha_k v_k,$$

where

$$\alpha_i = v \cdot v_i.$$

Proof: Since $\{v_1, \ldots, v_k\}$ is a basis of V, we can write

$$v = \alpha_1 v_1 + \cdots + \alpha_k v_k$$

for some scalars α_j. It follows that

$$v \cdot v_j = (\alpha_1 v_1 + \cdots + \alpha_k v_k) \cdot v_j = \alpha_j,$$

as claimed. ◆

An Example in R³

Let

$$v_1 = \frac{1}{\sqrt{3}}(1, 1, 1), \quad v_2 = \frac{1}{\sqrt{6}}(1, -2, 1), \quad v_3 = \frac{1}{\sqrt{2}}(1, 0, -1).$$

It is a straightforward calculation to verify that these vectors have unit length and are pairwise orthogonal. Let $v = (1, 2, 3)$ be a vector and determine the coordinates of v in the basis $\mathcal{V} = \{v_1, v_2, v_3\}$. Theorem 10.1.3 states that these coordinates are

$$[v]_\mathcal{V} = (v \cdot v_1, v \cdot v_2, v \cdot v_3) = (2\sqrt{3}, \frac{7}{\sqrt{6}}, -\sqrt{2}).$$

Matrices in Orthonormal Coordinates

Next we discuss how to find the matrix associated with a linear map in an orthonormal basis. Let $L : \mathbf{R}^n \to \mathbf{R}^n$ be a linear map and let $\mathcal{V} = \{v_1, \ldots, v_n\}$ be an orthonormal basis for $\mathbf{R}^n$. Then the matrix associated with L in the basis $\mathcal{V}$ is easy to calculate in terms of dot product. That matrix is

$$[L]_\mathcal{V} = (L(v_j) \cdot v_i). \tag{10.1.1}$$

To verify this claim, recall from Definition 9.3.4 that the (i, j)th entry of $[L]_\mathcal{V}$ is the ith entry in the vector $[L(v_j)]_\mathcal{V}$, which is $L(v_j) \cdot v_i$ by Theorem 10.1.3.

An Example in R²

Let $\mathcal{V} = \{v_1, v_2\} \subset \mathbf{R}^2$, where

$$v_1 = \frac{1}{\sqrt{2}} \begin{pmatrix} 1 \\ 1 \end{pmatrix} \quad \text{and} \quad v_2 = \frac{1}{\sqrt{2}} \begin{pmatrix} 1 \\ -1 \end{pmatrix}.$$

The set $\mathcal{V}$ is an orthonormal basis of $\mathbf{R}^2$. Using (10.1.1), we can find the matrix associated with the linear map

$$L_A(x) = \begin{pmatrix} 2 & 1 \\ -1 & 3 \end{pmatrix} x$$

in the basis $\mathcal{V}$ by straightforward calculation; that is, compute

$$[L]_\mathcal{V} = \begin{pmatrix} Av_1 \cdot v_1 & Av_2 \cdot v_1 \\ Av_1 \cdot v_2 & Av_2 \cdot v_2 \end{pmatrix} = \frac{1}{2} \begin{pmatrix} 5 & -3 \\ 1 & 5 \end{pmatrix}.$$

Remarks Concerning MATLAB

In the next section we prove that every vector subspace of $\mathbf{R}^n$ has an orthonormal basis (see Theorem 10.2.3), and we present a method for constructing such a basis (the Gram-Schmidt orthonormalization process). Here we note that certain commands in MATLAB produce bases for vector spaces. For those commands MATLAB always produces an orthonormal basis. For example, null(A) produces a basis for the null space of A. Take the 3×5 matrix

$$A = \begin{pmatrix} 1\ 2\ 3\ 4\ 5 \\ 0\ 1\ 2\ 3\ 4 \\ 2\ 3\ 4\ 0\ 0 \end{pmatrix}. \tag{10.1.2}*$$

Since rank(A) = 3, it follows that the null space of A is two dimensional. Typing B = null(A) in MATLAB produces

```
B =
   -0.4666          0
    0.6945     0.4313
   -0.2876    -0.3235
    0.3581    -0.6470
   -0.2984     0.5392
```

The columns of B form an orthonormal basis for the null space of A. This assertion can be checked by first typing

```
v1 = B(:,1);
v2 = B(:,2);
```

and then typing

```
norm(v1)
norm(v2)
dot(v1,v2)
A*v1
A*v2
```

which yields answers 1, 1, 0, $(0, 0, 0)^t$, $(0, 0, 0)^t$ (to within numerical accuracy). Recall that the MATLAB command norm(v) computes the norm of a vector v.

HAND EXERCISES

1. Find an orthonormal basis for the solutions to the linear equation

$$2x_1 - x_2 + x_3 = 0.$$

2.

    (a) Find the coordinates of the vector $v = (1, 4)$ in the orthonormal basis $\mathcal{V}$,

$$v_1 = \frac{1}{\sqrt{5}}(1, 2) \quad \text{and} \quad v_2 = \frac{1}{\sqrt{5}}(2, -1).$$

    (b) Let $A = \begin{pmatrix} 1 & 1 \\ 2 & -3 \end{pmatrix}$. Find $[A]_{\mathcal{V}}$.

COMPUTER EXERCISES

3. Load the matrix

$$A = \begin{pmatrix} 1 & 2 & 0 \\ 0 & 1 & 0 \\ 0 & 0 & 0 \end{pmatrix}$$

into MATLAB. Then type the command orth(A). Verify that the result is an orthonormal basis for the column space of A.

10.2   LEAST SQUARES APPROXIMATIONS

Let $W \subset \mathbf{R}^n$ be a subspace and let $x_0 \in \mathbf{R}^n$ be a vector. In this section we solve a basic geometric problem and investigate some of its consequences. The problem is: Find a vector $w_0 \in W$ that is the nearest vector in W to x_0 (see Figure 10.1).

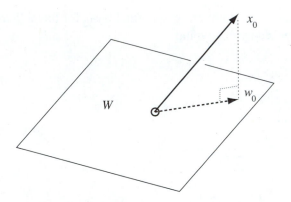

Figure 10.1
Approximation of x_0 by $w_0 \in W$ by least squares

The distance between two vectors v and w is $||v - w||$, so the geometric problem can be rephrased as follows: Find a vector $w_0 \in W$ such that

$$||x_0 - w_0|| \leq ||x_0 - w|| \quad \forall w \in W. \tag{10.2.1}$$

Condition (10.2.1) is called the *least squares approximation*. In order to see where this name comes from, we write (10.2.1) in the equivalent form

$$||x_0 - w_0||^2 \leq ||x_0 - w||^2 \quad \forall w \in W.$$

This form means that for $w = w_0$, the sum of the squares of the components of the vector $x_0 - w$ is minimal.

Before continuing, we state and prove the *Law of Pythagoras*. Let $z_1, z_2 \in \mathbf{R}^n$ be orthogonal vectors. Then

$$||z_1 + z_2||^2 = ||z_1||^2 + ||z_2||^2. \tag{10.2.2}$$

To verify (10.2.2) calculate

$$\begin{aligned}||z_1 + z_2||^2 &= (z_1 + z_2) \cdot (z_1 + z_2) = z_1 \cdot z_1 + 2z_1 \cdot z_2 + z_2 \cdot z_2 \\ &= ||z_1||^2 + 2z_1 \cdot z_2 + ||z_2||^2.\end{aligned}$$

Since z_1 and z_2 are orthogonal, $z_1 \cdot z_2 = 0$ and the Law of Pythagoras is valid.

Using (10.2.1) and (10.2.2), we can rephrase the minimum distance problem as follows.

Lemma 10.2.1 *The vector $w_0 \in W$ is the closest vector to $x_0 \in \mathbf{R}^n$ if the vector $x_0 - w_0$ is orthogonal to every vector in W (see Figure 10.1).*

Proof: Write $x_0 - w = z_1 + z_2$, where $z_1 = x_0 - w_0$ and $z_2 = w_0 - w$. By assumption, $x_0 - w_0$ is orthogonal to every vector in W, so z_1 and $z_2 \in W$ are orthogonal. It follows from (10.2.2) that

$$||x_0 - w||^2 = ||x_0 - w_0||^2 + ||w_0 - w||^2.$$

Since $||w_0 - w||^2 \geq 0$, (10.2.1) is valid, and w_0 is the vector nearest to x_0 in W. ◆

Least Squares Distance to a Line

Suppose W is as simple a subspace as possible; that is, suppose W is one dimensional with basis vector w. Since W is one dimensional, a vector $w_0 \in W$ that is closest to x_0 must be a multiple of w; that is, $w_0 = aw$. Suppose that we can find a scalar a so that $x_0 - aw$ is orthogonal to every vector in W. Then it follows from Lemma 10.2.1 that w_0 is the closest vector in W to x_0. To find a, calculate

$$0 = (x_0 - aw) \cdot w = x_0 \cdot w - aw \cdot w.$$

Then

$$a = \frac{x_0 \cdot w}{||w||^2}$$

and

$$w_0 = \frac{x_0 \cdot w}{||w||^2} w. \tag{10.2.3}$$

Observe that $||w||^2 \neq 0$ since w is a basis vector.

For example, let $x_0 = (1, 2, -1, 3) \in \mathbf{R}^4$ and $w = (0, 1, 2, 3)$. Then the vector w_0 in the space spanned by w that is nearest to x_0 is

$$w_0 = \frac{9}{14} w$$

since $x_0 \cdot w = 9$ and $||w||^2 = 14$.

Least Squares Distance to a Subspace

Similarly, using Lemma 10.2.1, we can solve the general least squares problem by solving a system of linear equations. Let $w_1, \ldots, w_k$ be a basis for W and suppose that

$$w_0 = \alpha_1 w_1 + \cdots + \alpha_k w_k$$

for some scalars α_i. We now show how to find these scalars.

Theorem 10.2.2   *Let $x_0 \in \mathbf{R}^n$ be a vector, and let $\{w_1, \ldots, w_k\}$ be a basis for the subspace $W \subset \mathbf{R}^n$. Then*

$$w_0 = \alpha_1 w_1 + \cdots + \alpha_k w_k$$

is the nearest vector in W to x_0 when

$$\begin{pmatrix} \alpha_1 \\ \vdots \\ \alpha_k \end{pmatrix} = (A^t A)^{-1} A^t x_0, \tag{10.2.4}$$

where $A = (w_1 | \cdots | w_k)$ is the $n \times k$ matrix whose columns are the basis vectors of W.

Proof: Observe that the vector $x_0 - w_0$ is orthogonal to every vector in W precisely when $x_0 - w_0$ is orthogonal to each basis vector w_j. It follows from Lemma 10.2.1 that w_0 is the closest vector to x_0 in W if

$$(x_0 - w_0) \cdot w_j = 0$$

for every j—that is, if

$$w_0 \cdot w_j = x_0 \cdot w_j$$

for every j. These equations can be rewritten as a system of equations in terms of the α_i as follows:

$$w_1 \cdot w_1 \alpha_1 + \cdots + w_1 \cdot w_k \alpha_k = w_1 \cdot x_0$$
$$\vdots \tag{10.2.5}$$
$$w_k \cdot w_1 \alpha_1 + \cdots + w_k \cdot w_k \alpha_k = w_k \cdot x_0.$$

Note that if $u, v \in \mathbf{R}^n$ are column vectors, then $u \cdot v = u^t v$. Therefore we can rewrite (10.2.5) as

$$A^t A \begin{pmatrix} \alpha_1 \\ \vdots \\ \alpha_k \end{pmatrix} = A^t x_0,$$

where A is the matrix whose columns are the w_j and x_0 is viewed as a column vector. Note that the matrix $A^t A$ is a $k \times k$ matrix.

We claim that $A^t A$ is invertible. To verify this claim, it suffices to show that the null space of $A^t A$ is 0; that is, if $A^t A z = 0$ for some $z \in \mathbf{R}^k$, then $z = 0$. First, calculate

$$||Az||^2 = Az \cdot Az = (Az)^t Az = z^t A^t A z = z^t 0 = 0.$$

It follows that $Az = 0$. Second, if we let $z = (z_1, \ldots, z_k)^t$, then the equation $Az = 0$ may be rewritten as

$$z_1 w_1 + \cdots + z_k w_k = 0.$$

Since the w_j are linearly independent, it follows that the $z_j = 0$. In particular, $z = 0$. Since $A^t A$ is invertible, (10.2.4) is valid and the theorem is proved.   ◆

Gram-Schmidt Orthonormalization Process

Suppose that $\mathcal{W} = \{w_1, \ldots, w_k\}$ is a basis for the subspace $V \subset \mathbf{R}^n$. There is a natural process by which the $\mathcal{W}$ basis can be transformed into an orthonormal basis $\mathcal{V}$ of V. This process proceeds inductively on the w_j; the orthonormal vectors $v_1, \ldots, v_k$ can be chosen so that

$$\mathrm{span}\{v_1, \ldots, v_j\} = \mathrm{span}\{w_1, \ldots, w_j\}$$

for each $j \leq k$. Moreover, the v_j are chosen using the theory of least squares that we have just discussed.

The Case $j = 2$

To gain a feeling for how the induction process works, we verify the case $j = 2$. Set

$$v_1 = \frac{1}{||w_1||} w_1, \tag{10.2.6}$$

so v_1 points in the same direction as w_1 and has unit length—that is, $v_1 \cdot v_1 = 1$. The normalization is shown in Figure 10.2.

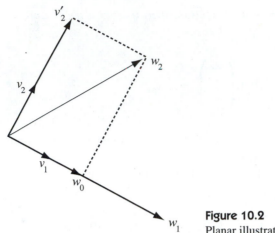

Figure 10.2
Planar illustration of Gram-Schmidt orthonormalization

Next we find a unit length vector v_2' in the plane spanned by w_1 and w_2 that is perpendicular to v_1. Let w_0 be the vector on the line generated by v_1 that is nearest to w_2. It follows from (10.2.3) that

$$w_0 = \frac{w_2 \cdot v_1}{||v_1||^2} v_1 = (w_2 \cdot v_1)v_1.$$

The vector w_0 is shown on Figure 10.2 and, as Lemma 10.2.1 states, the vector $v_2' = w_2 - w_0$ is perpendicular to v_1. That is,

$$v_2' = w_2 - (w_2 \cdot v_1)v_1 \tag{10.2.7}$$

is orthogonal to v_1.

Finally, set

$$v_2 = \frac{1}{||v_2'||} v_2' \tag{10.2.8}$$

so that v_2 has unit length. Since v_2 and v_2' point in the same direction, v_1 and v_2 are orthogonal. Note also that v_1 and v_2 are linear combinations of w_1 and w_2. Since v_1 and v_2 are orthogonal, they are linearly independent. It follows that

$$\text{span}\{v_1, v_2\} = \text{span}\{w_1, w_2\}.$$

In summary, computing v_1 and v_2 using (10.2.6), (10.2.7), and (10.2.8) yields an orthonormal basis for the plane spanned by w_1 and w_2.

The General Case

Theorem 10.2.3   *(Gram-Schmidt Orthonormalization) Let $w_1, \ldots, w_k$ be a basis for the subspace $W \subset \mathbf{R}^n$. Define v_1 as in (10.2.6) and then define inductively*

$$v'_{j+1} = w_{j+1} - (w_{j+1} \cdot v_1)v_1 - \cdots - (w_{j+1} \cdot v_j)v_j \qquad \textbf{(10.2.9)}$$

$$v_{j+1} = \frac{1}{||v'_{j+1}||} v'_{j+1}. \qquad \textbf{(10.2.10)}$$

Then $v_1, \ldots, v_k$ is an orthonormal basis of W such that for each j,

$$span\{v_1, \ldots, v_j\} = span\{w_1, \ldots, w_j\}.$$

Proof: We assume that we have constructed orthonormal vectors $v_1, \ldots, v_j$ such that

$$span\{v_1, \ldots, v_j\} = span\{w_1, \ldots, w_j\}.$$

Our purpose is to find a unit vector v_{j+1} that is orthogonal to each v_i and that satisfies

$$span\{v_1, \ldots, v_{j+1}\} = span\{w_1, \ldots, w_{j+1}\}.$$

We construct v_{j+1} in two steps. First we find a vector v'_{j+1} that is orthogonal to each of the v_i using least squares. Let w_0 be the vector in $span\{v_1, \ldots, v_j\}$ that is nearest to w_{j+1}. Theorem 10.2.2 tells us how to make this construction. Let A be the matrix whose columns are $v_1, \ldots, v_j$. Then (10.2.4) states that the coordinates of w_0 in the v_i basis are given by $(A^tA)^{-1}A^tw_{j+1}$. But since the v_is are orthonormal, the matrix A^tA is just I_k. Hence

$$w_0 = (w_{j+1} \cdot v_1)v_1 + \cdots + (w_{j+1} \cdot v_j)v_j.$$

Second, let $v'_{j+1} = w_{j+1} - w_0$ be the vector defined in (10.2.9). We claim that $v'_{j+1} = w_{j+1} - w_0$ is orthogonal to v_k for $k \leq j$ and hence to every vector in $span\{v_1, \ldots, v_j\}$. Just calculate

$$v'_{j+1} \cdot v_k = w_{j+1} \cdot v_k - w_0 \cdot v_k = w_{j+1} \cdot v_k - w_{j+1} \cdot v_k = 0.$$

Define v_{j+1} as in (10.2.10). It follows that $v_1, \ldots, v_{j+1}$ are orthonormal and that each vector is a linear combination of $w_1, \ldots, w_{j+1}$.   ◆

An Example of Orthonormalization

Let $W \subset \mathbf{R}^4$ be the subspace spanned by the vectors

$$w_1 = (1, 0, -1, 0), \quad w_2 = (2, -1, 0, 1), \quad w_3 = (0, 0, -2, 1). \qquad \textbf{(10.2.11)}$$

We find an orthonormal basis for W using Gram-Schmidt orthonormalization.

Step 1: Set

$$v_1 = \frac{1}{||w_1||} \, w_1 = \frac{1}{\sqrt{2}}(1, 0, -1, 0).$$

Step 2: Following the Gram-Schmidt process, use (10.2.9) to define

$$v_2' = w_2 - (w_2 \cdot v_1)v_1 = (2, -1, 0, 1) - \sqrt{2} \, \frac{1}{\sqrt{2}}(1, 0, -1, 0) = (1, -1, 1, 1).$$

Normalization using (10.2.10) yields

$$v_2 = \frac{1}{||v_2'||} \, v_2' = \frac{1}{2}(1, -1, 1, 1).$$

Step 3: Using (10.2.9), set

$$
\begin{aligned}
v_3' &= w_3 - (w_3 \cdot v_1)v_1 - (w_3 \cdot v_2)v_2 \\
&= (0, 0, -2, 1) - \sqrt{2} \, \frac{1}{\sqrt{2}}(1, 0, -1, 0) - \left(-\frac{1}{2}\right) \frac{1}{2}(1, -1, 1, 1) \\
&= \frac{1}{4}(-3, -1, -3, 5).
\end{aligned}
$$

Normalization using (10.2.10) yields

$$v_3 = \frac{1}{||v_3'||} \, v_3' = \frac{4}{\sqrt{44}}(-3, -1, -3, 5).$$

Hence we have constructed an orthonormal basis $\{v_1, v_2, v_3\}$ for W—namely,

$$
\begin{aligned}
v_1 &= \frac{1}{\sqrt{2}}(1, 0, -1, 0) &&\approx (0.7071, 0, -0.7071, 0) \\
v_2 &= \frac{1}{2}(1, -1, 1, 1) &&= (0.5, -0.5, 0.5, 0.5) \quad\quad\quad \textbf{(10.2.12)} \\
v_3 &= \frac{4}{\sqrt{44}}(-3, -1, -3, 5) &&\approx (-0.4523, -0.1508, -0.4523, 0.7538).
\end{aligned}
$$

HAND EXERCISES

1. Find an orthonormal basis of $\mathbf{R}^2$ by applying Gram-Schmidt orthonormalization to the vectors $w_1 = (3, 4)$ and $w_2 = (1, 5)$.

2. Find an orthonormal basis of the plane $W \subset \mathbf{R}^3$ spanned by the vectors $w_1 = (1, 2, 3)$ and $w_2 = (2, 5, -1)$ by applying Gram-Schmidt orthonormalization.

3. Let $\mathcal{W} = \{w_1, \ldots, w_k\}$ be an orthonormal basis of the subspace $W \subset \mathbf{R}^n$. Prove that $\mathcal{W}$ can be extended to an orthonormal basis $\{w_1, \ldots, w_n\}$ of $\mathbf{R}^n$.

COMPUTER EXERCISES

4. Use Gram-Schmidt orthonormalization to find an orthonormal basis for the subspace of $\mathbf{R}^5$ spanned by the vectors

$$w1 = (2, 1, 3, 5, 7), \quad w2 = (2, -1, 5, 2, 3), \quad w3 = (10, 1, -23, 2, 3). \qquad \textbf{(10.2.13)}*$$

Extend this basis to an orthonormal basis of $\mathbf{R}^5$.

10.3   LEAST SQUARES FITTING OF DATA

We begin this section by using the method of least squares to find the best straight line fit to a set of data. Later in the section we discuss best fits to other curves.

An Example of Best Linear Fit to Data

Suppose that we are given n data points (x_i, y_i) for $i = 1, \ldots, 10$. For example, consider the ten points:

$$\begin{array}{lllll} (2.0, 0.1) & (3.0, 2.7) & (1.5, -1.1) & (-1.0, -5.5) & (0.0, -3.4) \\ (3.6, 3.0) & (0.7, -2.8) & (4.1, 4.0) & (1.9, -1.9) & (5.0, 5.5). \end{array} \qquad \textbf{(10.3.1)}*$$

The ten points (x_i, y_i) are plotted in Figure 10.3 using the commands

```
e10_3_1
plot(X,Y,'o')
axis([-3,7,-8,8])
xlabel('x')
ylabel('y')
```

Next suppose that there is a linear relation between the x_i and the y_i; that is, we assume that there are constants b_1 and b_2 (that do not depend on i) for which $y_i = b_1 + b_2 x_i$ for each i. But these points are just data; errors may have been made in their measurement. So we ask: Can we find b_1^0 and b_2^0 so that the error made in fitting the data to the line $y = b_1^0 + b_2^0 x$ is minimal—that is, the error that is made in that fit is

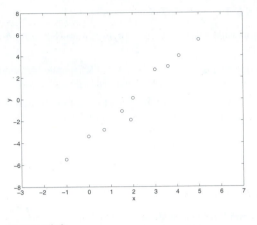

Figure 10.3
Scatter plot of data in (10.3.1)

less than or equal to the error made in fitting the data to the line $y = b_1 + b_2 x$ for any other choice of b_1 and b_2?

We begin by discussing what that error actually is. Given constants b_1 and b_2 and given a data point x_i, the difference between the data value y_i and the hypothesized value $b_1 + b_2 x_i$ is the error that is made at that data point. Next we combine the errors made at all of the data points. A standard way to combine the errors is to use the Euclidean distance

$$E(b) = \left((y_1 - (b_1 + b_2 x_1))^2 + \cdots + (y_{10} - (b_1 + b_2 x_{10}))^2 \right)^{1/2}.$$

Rewriting $E(b)$ in vector notation leads to an economy in notation and to a conceptual advantage. Let

$$X = (x_1, \ldots, x_{10})^t, \quad Y = (y_1, \ldots, y_{10})^t, \quad F_1 = (1, 1, \ldots, 1)$$

be vectors in $\mathbf{R}^{10}$. Then in coordinates,

$$Y - (b_1 F_1 + b_2 X) = \begin{pmatrix} y_1 - (b_1 + b_2 x_1) \\ \vdots \\ y_{10} - (b_1 + b_2 x_{10}) \end{pmatrix}.$$

It follows that

$$E(b) = ||Y - (b_1 F_1 + b_2 X)||.$$

The problem of making a least squares fit is to minimize E over all b_1 and b_2.

To solve the minimization problem, note that the vectors $b_1 F_1 + b_2 X$ form a two-dimensional subspace $W = \text{span}\{F_1, X\} \subset \mathbf{R}^{10}$ (at least when X is not a scalar multiple of F_1, which is almost always). Minimizing E is identical to finding a vector

$w_0 = b_1^0 F_1 + b_2^0 X \in W$ that is nearest to the vector $Y \in \mathbf{R}^{10}$. This is the least squares question that we solved in Section 10.2.

We can use MATLAB to compute the values of b_1^0 and b_2^0 that give the best linear approximation to Y. If we set the matrix $A = (F_1 | X)$, then Theorem 10.2.2 implies that the values of b_1^0 and b_2^0 are obtained using (10.2.4). In particular, type e10_3_1 to call the vectors X, Y, $F1$ into MATLAB, and then type

```
A = [F1 X];
b0 = inv(A'*A)*A'*Y
```

to obtain

```
b0(1) = -3.8597
b0(2) =  1.8845
```

Superimposing the line $y = -3.8597 + 1.8845x$ on the scatter plot in Figure 10.3 yields the plot in Figure 10.4. The total error is $E(b0) = 1.9634$ (obtained in MATLAB by typing norm(Y-(b0(1)*F1+b0(2)*X))). Compare this with the error $E(2, -4) = 2.0928$.

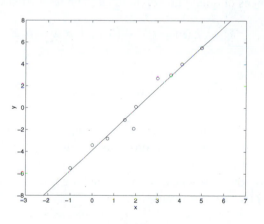

Figure 10.4
Scatter plot of data in (10.3.1) with best linear approximation

General Linear Regression

We can summarize the preceding discussion as follows: Given n data points

$$(x_1, y_1), \ldots, (x_n, y_n),$$

form the vectors

$$X = (x_1, \ldots, x_n)^t, \quad Y = (y_1, \ldots, y_n)^t, \quad F_1 = (1, \ldots, 1)^t$$

in $\mathbf{R}^n$. Find constants b_1^0 and b_2^0 so that $b_1^0 F_1 + b_2^0 X$ is a vector in $W = \mathrm{span}\{F_1, X\} \subset \mathbf{R}^n$ that is nearest to Y. Let

$$A = (F_1 | X)$$

be the $n \times 2$ matrix. This problem is solved by least squares in (10.2.4) as

$$\begin{pmatrix} b_1^0 \\ b_2^0 \end{pmatrix} = (A^t A)^{-1} A^t Y. \tag{10.3.2}$$

Least Squares Fit to a Quadratic Polynomial

Suppose that we want to fit the data (x_i, y_i) to a quadratic polynomial

$$y = b_1 + b_2 x + b_3 x^2$$

by least squares methods. We want to find constants b_1^0, b_2^0, and b_3^0 so that the error made is using the quadratic polynomial $y = b_1^0 + b_2^0 x + b_3^0 x^2$ is minimal among all possible choices of quadratic polynomials. The least squares error is

$$E(b) = ||Y - (b_1 F_1 + b_2 X + b_3 X^{(2)})||,$$

where

$$X^{(2)} = (x_1^2, \ldots, x_n^2)^t$$

and, as before, F_1 is the n vector with all components equal to 1.

We solve the minimization problem as before. In this case, the space of possible approximations to the data W is three dimensional; indeed, $W = \mathrm{span}\{F_1, X, X^{(2)}\}$. As in the case of fits to lines, we try to find a point in W that is nearest to the vector $Y \in \mathbf{R}^n$. By (10.2.4), the answer is

$$b = (A^t A)^{-1} A^t Y,$$

where $A = (F_1 | X | X^{(2)})$ is an $n \times 3$ matrix.

Suppose that we try to fit the data in (10.3.1) with a quadratic polynomial rather than a linear one. Use MATLAB as follows:

```
e10_3_1
A = [F1 X X.*X];
b = inv(A'*A)*A'*Y;
```

to obtain

```
b0(1) =    0.0443
b0(2) =    1.7054
b0(3) =   -3.8197
```

So the best parabolic fit to these data is $y = -3.8197 + 1.7054x + 0.0443x^2$. Note that the coefficient of x^2 is small, suggesting that the data are well fit by a straight line. Note also that the error is $E(b0) = 1.9098$, which is only marginally smaller than the error for the best linear fit. For comparison, in Figure 10.5 we superimpose the equation for the quadratic fit onto Figure 10.4.

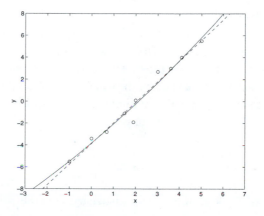

Figure 10.5
Scatter plot of data in (10.3.1) with the best linear and quadratic approximations. The best linear fit is plotted with a dashed line.

General Least Squares Fit

The approximation to a quadratic polynomial shows that least squares fits can be made to any finite-dimensional function space. More precisely, let C be a finite-dimensional space of functions, and let

$$f_1(x), \ldots, f_m(x)$$

be a basis for C. We have just considered two such spaces: $C = \text{span}\{f_1(x) = 1, f_2(x) = x\}$ for linear regression and $C = \text{span}\{f_1(x) = 1, f_2(x) = x, f_3(x) = x^2\}$ for least squares fit to a quadratic polynomial.

The general least squares fit of a data set

$$(x_1, y_1), \ldots, (x_n, y_n)$$

is the function $g_0(x) \in C$ that is nearest to the data set in the following sense: Let

$$X = (x_1, \ldots, x_n)^t \quad \text{and} \quad Y = (y_1, \ldots, y_n)^t$$

be column vectors in $\mathbf{R}^n$. For any function $g(x)$, define the column vector

$$G = (g(x_1), \ldots, g(x_n))' \in \mathbf{R}^n.$$

So G is the evaluation of $g(x)$ on the data set. Then the error

$$E(g) = ||Y - G||$$

is minimal for $g = g_0$.

More precisely, we think of the data Y as representing the (approximate) evaluation of a function on the x_i. Then we try to find a function $g_0 \in \mathcal{C}$ whose values on the x_i are as near as possible to the vector Y. This is just a least squares problem. Let $W \subset \mathbf{R}^n$ be the vector subspace spanned by the evaluations of function $g \in \mathcal{C}$ on the data points x_i—that is, the vectors G. The minimization problem is to find a vector in W that is nearest to Y. This can be solved in general using (10.2.4). That is, let A be the $n \times m$ matrix

$$A = (F_1| \cdots |F_m),$$

where $F_j \in \mathbf{R}^n$ is the column vector associated with the jth basis element of $\mathcal{C}$; that is,

$$F_j = (f_j(x_1), \ldots, f_j(x_n))' \in \mathbf{R}^n.$$

The minimizing function $g_0(x) \in \mathcal{C}$ is a linear combination of the basis functions $f_1(x), \ldots, f_n(x)$; that is,

$$g_0(x) = b_1 f_1(x) + \cdots + b_m f_m(x)$$

for scalars b_i. If we set

$$b = (b_1, \ldots, b_m) \in \mathbf{R}^m,$$

then least squares minimization states that

$$b = (A'A)^{-1}A'Y. \tag{10.3.3}$$

This equation can be solved easily in MATLAB. Enter the data as column n-vectors X and Y. Let Fj be the column vector $f_j(X)$, and then form the matrix A = [F1 F2 $\cdots$ Fm]. Finally compute

```
b = inv(A'*A)*A'*Y
```

Least Squares Fit to a Sinusoidal Function

We discuss a specific example of the general least squares formulation by considering the weather. It is reasonable to expect monthly data on the weather to vary periodically in time with a period of one year. In Table 10.1 we give average daily high and low temperatures for each month of the year for Paris and Rio de Janeiro. We attempt to fit these data with curves of the form

$$g(T) = b_1 + b_2 \cos\left(\frac{2\pi}{12}T\right) + b_3 \sin\left(\frac{2\pi}{12}T\right),$$

where T is time measured in months and b_1, b_2, b_3 are scalars. These functions are 12 periodic, which seems appropriate for weather data, and form a three-dimensional function space C. Recall the trigonometric identity

$$a \cos(\omega t) + c \sin(\omega t) = d \sin(\omega(t - \varphi)),$$

where

$$d = \sqrt{a^2 + c^2}.$$

Based on this identity we call C the space of *sinusoidal functions*. The number d is called the *amplitude* of the sinusoidal function $g(T)$.

Table 10.1
Monthly averages of daily high and low temperatures in Paris and Rio de Janeiro

Month	Paris High	Paris Low	Rio de Janeiro High	Rio de Janeiro Low	Month	Paris High	Paris Low	Rio de Janeiro High	Rio de Janeiro Low
1	55	39	84	73	7	81	64	75	63
2	55	41	85	73	8	81	64	76	64
3	59	45	83	72	9	77	61	75	65
4	64	46	80	69	10	70	54	77	66
5	68	55	77	66	11	63	46	79	68
6	75	61	76	64	12	55	41	82	71

Note that each data set consists of 12 entries—one for each month. Let $T = (1, 2, \ldots, 12)^t$ be the vector $X \in \mathbf{R}^{12}$ in the general presentation. Next let Y be the data in one of the data sets—say, the high temperatures in Paris.

Now we turn to the vectors representing basis functions in C. Let

```
F1=[1 1 1 1 1 1 1 1 1 1 1 1]'
```

be the vector associated with the basis function $f_1(T) = 1$. Let F2 and F3 be the column vectors associated with the basis functions

$$f_2(T) = \cos\left(\frac{2\pi}{12}T\right) \quad \text{and} \quad f_3(T) = \sin\left(\frac{2\pi}{12}T\right).$$

These vectors are computed by typing

```
F2 = cos(2*pi/12*T);
F3 = sin(2*pi/12*T);
```

By typing `temper`, we enter the temperatures and the vectors T, F1, F2, and F3 into MATLAB.

To find the best fit to the data by a sinusoidal function $g(T)$, we use (10.2.4). Let A be the 12×3 matrix

```
A = [F1 F2 F3];
```

The table data are entered in column vectors `ParisH` and `ParisL` for the high and low Paris temperatures and `RioH` and `RioL` for the high and low Rio de Janeiro temperatures. We can find the best least squares fit of the Paris high temperatures by a sinusoidal function $g_0(T)$ by typing

```
b = inv(A'*A)*A'*ParisH
```

obtaining

```
b(1) =  66.9167
b(2) =  -9.4745
b(3) =  -9.3688
```

The result is plotted in Figure 10.6 by typing

```
plot(T,ParisH,'o')
axis([0,13,0,100])
xlabel('time (months)')
ylabel('temperature (Fahrenheit)')
hold on
xx = linspace(0,13);
yy = b(1) + b(2)*cos(2*pi*xx/12) + b(3)*sin(2*pi*xx/12);
plot(xx,yy)
```

A similar exercise allows us to compute the best approximation to the Rio de Janeiro high temperatures, obtaining

```
b(1) =  79.0833
b(2) =   3.0877
b(3) =   3.6487
```

The value of $b(1)$ is just the mean high temperature, and not surprisingly that value is much higher in Rio than in Paris. There is yet more information contained in these approximations. For the high temperatures in Paris and Rio,

$$d_P = 13.3244 \quad \text{and} \quad d_R = 4.7798.$$

The amplitude d measures the variation of the high temperatures about the mean. It is much greater in Paris than in Rio, indicating that the difference in temperature between winter and summer is much greater in Paris than in Rio.

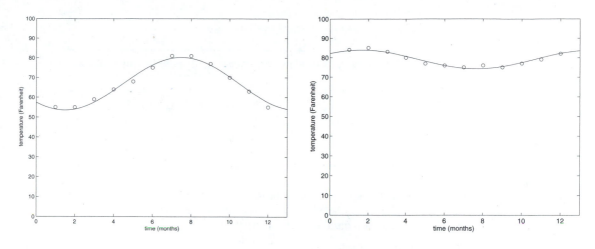

Figure 10.6
Monthly averages of daily high temperatures in Paris (*left*) and Rio de Janeiro (*right*) with best sinusoidal approximation

Least Squares Fit in MATLAB

The general formula for a least squares fit of data (10.3.3) has been preprogrammed in MATLAB. After setting up the matrix A whose columns are the vectors F_j, just type

```
b = A\Y
```

This MATLAB command can be checked on the sinusoidal fit to the Rio de Janeiro high temperature data by typing

```
b = A\RioH
```

and obtaining

```
b =
    79.0833
     3.0877
     3.6487
```

COMPUTER EXERCISES

1. World population data for each decade of this century (except 1910) are given in Table 10.2. Assume that population growth is linear $P = mT + b$, where time T is measured in decades since the year 1900 and P is measured in billions of people. These data can be recovered by typing e10_3_po.
   (a) Find m and b to give the best linear fit to these data.
   (b) Use this linear approximation to the data to make predictions of the world population in the years 1910 and 2000.
   (c) Do you expect the prediction for the year 2000 to be high, low, or on target? Explain why by graphing the data with the best linear fit superimposed and by using the differential equation population model discussed in Section 4.1.

Table 10.2
Twentieth-century world population data by decades

Year	Population (in millions)	Year	Population (in millions)
1900	1625	1950	2516
1910	n.a.	1960	3020
1920	1813	1970	3698
1930	1987	1980	4448
1940	2213	1990	5292

2. Find the best sinusoidal approximation to the monthly average low temperatures in Paris and Rio de Janeiro. How does the variation of these temperatures about the mean compare to the high temperature calculations? Was this the result you expected?

3. In Table 10.3 we present weather data from ten U.S. cities. The data are the average number of days in the year with precipitation and the percentage of sunny hours to hours when it could be sunny. Find the best linear fit to these data.

Table 10.3
Precipitation days versus sunny time for selected U.S. cities

City	Rainy Days	Sunny (%)	City	Rainy Days	Sunny (%)
Charleston	92	72	Kansas City	08	59
Chicago	121	54	Miami	114	85
Dallas	82	65	New Orleans	103	61
Denver	82	67	Phoenix	28	88
Duluth	136	52	Salt Lake City	99	59

10.4   SYMMETRIC MATRICES

Symmetric matrices have some remarkable properties that can be summarized by:

Theorem 10.4.1   *Let A be an n × n symmetric matrix. Then*

(a) every eigenvalue of A is real, and

(b) there is an orthonormal basis of $\mathbf{R}^n$ *consisting of eigenvectors of A.*

As a consequence of Theorem 10.4.1, let $\mathcal{V} = \{v_1, \ldots, v_n\}$ be an orthonormal basis for $\mathbf{R}^n$ consisting of eigenvectors of A. Indeed, suppose

$$Av_j = \lambda_j v_j,$$

where $\lambda_j \in \mathbf{R}$. Note that

$$Av_j \cdot v_i = \begin{cases} \lambda_j & i = j \\ 0 & i \neq j. \end{cases}$$

It follows from (10.1.1) that

$$[A]_\mathcal{V} = \begin{pmatrix} \lambda_1 & & 0 \\ & \ddots & \\ 0 & & \lambda_n \end{pmatrix}$$

is a diagonal matrix. So every symmetric matrix is similar to a diagonal matrix.

Hermitian Inner Products

The proof of Theorem 10.4.1 uses the *Hermitian inner product*—a generalization of dot product to complex vectors. Let $v, w \in \mathbf{C}^n$ be two complex n-vectors. Define

$$\langle v, w \rangle = v_1 \overline{w}_1 + \cdots + v_n \overline{w}_n.$$

Note that the coordinates w_i of the second vector enter this formula with a complex conjugate. However, if v and w are real vectors, then

$$\langle v, w \rangle = v \cdot w.$$

A more compact notation for the Hermitian inner product is given by matrix multiplication. Suppose that v and w are column n-vectors. Then

$$\langle v, w \rangle = v^t \overline{w}.$$

The properties of the Hermitian inner product are similar to those of dot product. We note three. Let $c \in \mathbf{C}$ be a complex scalar. Then

$$\langle v, v \rangle = ||v||^2 \geq 0$$
$$\langle cv, w \rangle = c \langle v, w \rangle$$
$$\langle v, cw \rangle = \overline{c} \langle v, w \rangle.$$

Note the complex conjugation of the complex scalar c in the preceding formula.

Let C be a complex $n \times n$ matrix. Then the main observation concerning Hermitian inner products that we use is

$$\langle Cv, w \rangle = \langle v, \overline{C}^t w \rangle.$$

This fact is verified by calculating

$$\langle Cv, w \rangle = (Cv)^t \overline{w} = (v^t C^t) \overline{w} = v^t (C^t \overline{w}) = v^t (\overline{\overline{C}^t w}) = \langle v, \overline{C}^t w \rangle.$$

So if A is a $n \times n$ real symmetric matrix, then

$$\langle Av, w \rangle = \langle v, Aw \rangle, \tag{10.4.1}$$

since $\overline{A}^t = A^t = A$.

Proof of Theorem 10.4.1(a):   Let λ be an eigenvalue of A, and let v be the associated eigenvector. Since $Av = \lambda v$ we can use (10.4.1) to compute

$$\lambda \langle v, v \rangle = \langle Av, v \rangle = \langle v, Av \rangle = \overline{\lambda} \langle v, v \rangle.$$

Since $\langle v, v \rangle = ||v||^2 > 0$, it follows that $\lambda = \overline{\lambda}$ and λ is real.   ◆

Proof of Theorem 10.4.1(b):   Let A be a real symmetric $n \times n$ matrix. We want to show that there is an orthonormal basis of $\mathbf{R}^n$ consisting of eigenvectors of A. The proof proceeds inductively on n. The theorem is trivially valid for $n = 1$, so we assume that it is valid for $n - 1$.

Theorem 8.2.4 implies that A has an eigenvalue λ_1, and Theorem 10.4.1(a) states that this eigenvalue is real. Let v_1 be a unit length eigenvector corresponding to the eigenvalue λ_1. Extend v_1 to an orthonormal basis $v_1, w_2, \ldots, w_n$ of $\mathbf{R}^n$, and let $P = (v_1 | w_2 | \cdots | w_n)$ be the matrix whose columns are the vectors in this orthonormal basis. Orthonormality and direct multiplication imply that

$$P^t P = I_n. \tag{10.4.2}$$

Therefore P is invertible; indeed $P^{-1} = P^t$.

Next let

$$B = P^{-1} A P.$$

By direct computation

$$Be_1 = P^{-1} A P e_1 = P^{-1} A v_1 = \lambda_1 P^{-1} v_1 = \lambda_1 e_1.$$

It follows that B has the form

$$B = \begin{pmatrix} \lambda_1 & * \\ 0 & C \end{pmatrix},$$

where C is an $(n - 1) \times (n - 1)$ matrix. Since $P^{-1} = P^t$, it follows that B is a symmetric matrix. To verify this point compute

$$B^t = (P^t A P)^t = P^t A^t (P^t)^t = P^t A P = B.$$

It follows that

$$B = \begin{pmatrix} \lambda_1 & 0 \\ 0 & C \end{pmatrix},$$

where C is a symmetric matrix. By induction we can choose an orthonormal basis $z_2, \ldots, z_n$ in $\{0\} \times \mathbf{R}^{n-1}$ consisting of eigenvectors of C. It follows that $e_1, z_2, \ldots, z_n$ is an orthonormal basis for $\mathbf{R}^n$ consisting of eigenvectors of B.

Finally, let $v_j = P^{-1}z_j$ for $j = 2, \ldots, n$. Since $v_1 = P^{-1}e_1$, it follows that $v_1, v_2, \ldots, v_n$ is a basis of $\mathbf{R}^n$ consisting of eigenvectors of A. We need only show that the v_j form an orthonormal basis of $\mathbf{R}^n$. This is done using (10.4.1). For notational convenience let $z_1 = e_1$ and compute

$$\langle v_i, v_j \rangle = \langle P^{-1}z_i, P^{-1}z_j \rangle = \langle P^t z_i, P^t z_j \rangle = \langle z_i, PP^t z_j \rangle = \langle z_i, z_j \rangle,$$

since $PP^t = I_n$. Thus the vectors v_j form an orthonormal basis, since the vectors z_j form an orthonormal basis. $\blacklozenge$

HAND EXERCISES

1. Let

$$A = \begin{pmatrix} a & b \\ b & d \end{pmatrix}$$

be the general real 2×2 symmetric matrix.

  (a) Prove directly using the discriminant of the characteristic polynomial that A has real eigenvalues.

  (b) Show that A has equal eigenvalues only if A is a scalar multiple of I_2.

2. Let

$$A = \begin{pmatrix} 1 & 2 \\ 2 & -2 \end{pmatrix}.$$

Find the eigenvalues and eigenvectors of A and verify that the eigenvectors are orthogonal.

COMPUTER EXERCISES

In Exercises 3–5, compute the eigenvalues and the eigenvectors of the 2×2 matrix. Then load the matrix into the program map in MATLAB and iterate. That is, choose an initial vector v_0 and use map to compute $v_1 = Av_0$, $v_2 = Av_1$, ... How does the result of iteration compare with the eigenvectors and eigenvalues that you have found?

Hint You may find it convenient to use the feature Rescale in the MAP Options. Then the norm of the vectors is rescaled to 1 after each use of the command map, and the vectors v_j will not escape from the viewing screen.

3. $A = \begin{pmatrix} 1 & 3 \\ 3 & 1 \end{pmatrix}$ **4.** $B = \begin{pmatrix} 11 & 9 \\ 9 & 11 \end{pmatrix}$ **5.** $C = \begin{pmatrix} 0.005 & -2.005 \\ -2.005 & 0.005 \end{pmatrix}$

6. Perform the same computational experiment as described in Exercises 3–5 using the matrix $A = \begin{pmatrix} 0 & 2 \\ 2 & 0 \end{pmatrix}$ and the program map. How do your results differ from the results in those exercises and why?

10.5 ORTHOGONAL MATRICES AND *QR* DECOMPOSITIONS

In this section we describe an alternative approach to Gram-Schmidt orthonormalization for constructing an orthonormal basis of a subspace $W \subset \mathbf{R}^n$. This method is called the QR decomposition and is numerically superior to Gram-Schmidt. Indeed, the QR decomposition is the method used by MATLAB to compute orthonormal bases. To discuss this decomposition we need to introduce a new type of matrices—the orthogonal matrices.

Orthogonal Matrices

Definition 10.5.1 *An $n \times n$ matrix Q is* orthogonal *if its columns form an orthonormal basis of* $\mathbf{R}^n$.

The following lemma states elementary properties of orthogonal matrices:

Lemma 10.5.2 *Let Q be an $n \times n$ matrix. Then:*

(a) *Q is orthogonal if and only if $Q^t Q = I_n$.*

(b) *Q is orthogonal if and only if $Q^{-1} = Q^t$.*

(c) *If Q_1 and Q_2 are orthogonal matrices, then the product $Q_1 Q_2$ is an orthogonal matrix.*

Proof:

(a) Let $Q = (v_1 | \cdots | v_n)$. Since Q is orthogonal, the v_j form an orthonormal basis. By direct computation note that $Q^t Q = \{(v_i \cdot v_j)\} = I_n$, since the v_j are orthonormal.

(b) This is simply a restatement of (a).

(c) Now let Q_1 and Q_2 be orthogonal. Then (a) implies

$$(Q_1 Q_2)^t (Q_1 Q_2) = Q_2^t Q_1^t Q_1 Q_2 = Q_2^t Q_2 = I_n,$$

thus proving (c). ◆

Lemma 10.5.2 together with (10.4.2) in the proof of Theorem 10.4.1(b) leads to the following result:

Proposition 10.5.3   *For each symmetric $n \times n$ matrix A, there exists an orthogonal matrix P such that $P^t A P$ is a diagonal matrix.*

Reflections Across Hyperplanes: Householder Matrices

Useful examples of orthogonal matrices are reflections across hyperplanes. An $n-1$ dimensional subspace of $\mathbf{R}^n$ is called a *hyperplane*. Let V be a hyperplane and let u be a nonzero vector normal to V. Then a *reflection* across V is a linear map $H : \mathbf{R}^n \to \mathbf{R}^n$ such that

   (a) $Hv = v$ for all $v \in V$, and

   (b) $Hu = -u$.

We claim that the matrix of a reflection across a hyperplane is orthogonal, and there is a simple formula for that matrix.

Definition 10.5.4   *A Householder matrix is an $n \times n$ matrix of the form*

$$H = I_n - \frac{2}{u^t u}\, uu^t, \qquad\qquad (10.5.1)$$

where $u \in \mathbf{R}^n$ is a nonzero vector.

This definition makes sense since $u^t u = ||u||^2$ is a number, whereas the product uu^t is an $n \times n$ matrix.

Lemma 10.5.5   *Let $u \in \mathbf{R}^n$ be a nonzero vector and let V be the hyperplane orthogonal to u. Then the Householder matrix H is a reflection across V and is orthogonal.*

Proof: By definition every vector $v \in V$ satisfies $u^t v = u \cdot v = 0$. Therefore

$$Hv = v - \frac{2}{u^t u}\, uu^t v = v$$

and

$$Hu = u - \frac{2}{u^t u}\, uu^t u = u - 2u = -u.$$

Hence H is a reflection across the hyperplane V. It also follows that $H^2 = I_n$ since $H^2 v = H(Hv) = Hv = v$ for all $v \in V$ and $H^2 u = H(-u) = u$. So H^2 acts like the identity on a basis of $\mathbf{R}^n$ and $H^2 = I_n$.

To show that H is orthogonal, we first calculate

$$H^t = I_n^t - \frac{2}{u^t u}\, (uu^t)^t = I_n - \frac{2}{u^t u}\, uu^t = H.$$

Therefore $I_n = HH = HH^t$ and $H^t = H^{-1}$. Now apply Lemma 10.5.2(b).   ◆

QR Decompositions

The Gram-Schmidt process is not used in practice to find orthonormal bases because there are other techniques available that are preferable for orthogonalization on a computer. One such procedure for the construction of an orthonormal basis is based on QR decompositions using *Householder transformations*. This method is the one implemented in MATLAB.

An $n \times k$ matrix $R = \{r_{ij}\}$ is *upper triangular* if $r_{ij} = 0$ whenever $i > j$.

Definition 10.5.6 *An $n \times k$ matrix A has a QR decomposition if*

$$A = QR, \tag{10.5.2}$$

where Q is an $n \times n$ orthogonal matrix and R is an $n \times k$ upper triangular matrix R.

QR decompositions can be used to find orthonormal bases as follows: Suppose that $\mathcal{W} = \{w_1, \ldots, w_k\}$ is a basis for the subspace $W \subset \mathbf{R}^n$. Then define the $n \times k$ matrix A that has the w_j as columns; that is,

$$A = (w_1^t | \cdots | w_k^t).$$

Suppose that $A = QR$ is a QR decomposition. Since Q is orthogonal, the columns of Q are orthonormal. So write

$$Q = (v_1^t | \cdots | v_n^t).$$

On taking transposes, we arrive at the equation $A^t = R^t Q^t$:

$$
\begin{pmatrix} w_1 \\ \vdots \\ w_k \end{pmatrix}
=
\begin{pmatrix}
r_{11} & 0 & \cdots & 0 & \cdots & 0 \\
r_{12} & r_{22} & \cdots & 0 & \cdots & 0 \\
\vdots & \vdots & \cdots & \vdots & \cdots & \vdots \\
r_{1k} & r_{2k} & \cdots & r_{kk} & \cdots & 0
\end{pmatrix}
\begin{pmatrix} v_1 \\ \vdots \\ v_n \end{pmatrix}.
$$

By equating rows in this matrix equation, we arrive at the system

$$
\begin{aligned}
w_1 &= r_{11} v_1 \\
w_2 &= r_{12} v_1 + r_{22} v_2 \\
&\;\;\vdots \\
w_k &= r_{1k} v_1 + r_{2k} v_2 + \cdots + r_{kk} v_k.
\end{aligned}
\tag{10.5.3}
$$

It now follows that $W = \text{span}\{v_1, \ldots, v_k\}$ and that $\{v_1, \ldots, v_k\}$ is an orthonormal basis for W. We have proved:

Proposition 10.5.7   *Suppose that there exist an orthogonal $n \times n$ matrix Q and an upper triangular $n \times k$ matrix RR such that the $n \times k$ matrix A has a QR decomposition*

$$A = QR.$$

Then the first k columns $v_1, \ldots, v_k$ of the matrix Q form an orthonormal basis of the subspace $W = span\{w_1, \ldots, w_k\}$, where the w_j are the columns of A. Moreover, $r_{ij} = v_i \cdot w_j$ is the coordinate of w_j in the orthonormal basis.

Conversely, we can also write down a QR decomposition for a matrix A if we have computed an orthonormal basis for the columns of A. Indeed, using the Gram-Schmidt process (Theorem 10.2.3), we have shown that QR decompositions always exist. In the remainder of this section we discuss a different way for finding QR decompositions using Householder matrices.

Construction of a *QR* Decomposition Using Householder Matrices

The QR decomposition by Householder transformations is based on the following observation:

Proposition 10.5.8   *Let $z = (z_1, \ldots, z_n) \in \mathbf{R}^n$ be nonzero and let*

$$r = \sqrt{z_j^2 + \cdots + z_n^2}.$$

Define $u = (u_1, \ldots, u_n) \in \mathbf{R}^n$ by

$$\begin{pmatrix} u_1 \\ \vdots \\ u_{j-1} \\ u_j \\ u_{j+1} \\ \vdots \\ u_n \end{pmatrix} = \begin{pmatrix} 0 \\ \vdots \\ 0 \\ z_j - r \\ z_{j+1} \\ \vdots \\ z_n \end{pmatrix}.$$

Then

$$2u^t z = u^t u,$$

and

$$Hz = \begin{pmatrix} z_1 \\ \vdots \\ z_{j-1} \\ r \\ 0 \\ \vdots \\ 0 \end{pmatrix} \qquad\qquad \textbf{(10.5.4)}$$

holds for the Householder matrix $H = I_n - \dfrac{2}{u^t u}\, uu^t$.

Proof: Begin by computing

$$u^t z = u_j z_j + z_{j+1}^2 + \cdots + z_n^2$$
$$= z_j^2 - r z_j + z_{j+1}^2 + \cdots + z_n^2$$
$$= -r z_j + r^2.$$

Next compute

$$u^t u = (z_j - r)(z_j - r) + z_{j+1}^2 + \cdots + z_n^2$$
$$= z_j^2 - 2 r z_j + r^2 + z_{j+1}^2 + \cdots + z_n^2$$
$$= 2(-r z_j + r^2).$$

Hence $2u^t z = u^t u$, as claimed.

Note that $z - u$ is the vector on the right-hand side of (10.5.4). So compute

$$Hz = \left(I_n - \frac{2}{u^t u}\, uu^t \right) z = z - \frac{2 u^t z}{u^t u}\, u = z - u$$

to see that (10.5.4) is valid. $\blacklozenge$

An inspection of the proof of Proposition 10.5.8 shows that we could have chosen

$$u_j = z_j + r$$

instead of $u_j = z_j - r$. Therefore the choice of H is not unique.

Proposition 10.5.8 allows us to determine inductively a QR decomposition of the matrix

$$A = (w_1^0 | \cdots | w_k^0),$$

where each $w_j^0 \in \mathbf{R}^n$. So A is an $n \times k$ matrix and $k \le n$.

First, set $z = w_1^0$ and use Proposition 10.5.8 to construct the Householder matrix H_1 such that

$$H_1 w_1^0 = \begin{pmatrix} r_{11} \\ 0 \\ \vdots \\ 0 \end{pmatrix} \equiv r_1.$$

Then the matrix $A_1 = H_1 A$ can be written as

$$A_1 = (r_1 | w_2^1 | \cdots | w_k^1),$$

where $w_j^1 = H_1 w_j^0$ for $j = 2, \ldots, k$.

Second, set $z = w_2^1$ in Proposition 10.5.8 and construct the Householder matrix H_2 such that

$$H_2 w_2^1 = \begin{pmatrix} r_{12} \\ r_{22} \\ 0 \\ \vdots \\ 0 \end{pmatrix} \equiv r_2.$$

Then the matrix $A_2 = H_2 A_1 = H_2 H_1 A$ can be written as

$$A_2 = (r_1 | r_2 | w_3^2 | \cdots | w_k^2),$$

where $w_j^2 = H_2 w_j^1$ for $j = 3, \ldots, k$. Observe that the 1st column r_1 is not affected by the matrix multiplication, since H_2 leaves the first component of a vector unchanged.

Proceeding inductively, in the ith step, set $z = w_i^{i-1}$ and use Proposition 10.5.8 to construct the Householder matrix H_i such that

$$H_i w_i^{i-1} = \begin{pmatrix} r_{1i} \\ \vdots \\ r_{ii} \\ 0 \\ \vdots \\ 0 \end{pmatrix} \equiv r_i,$$

and the matrix $A_i = H_i A_{i-1} = H_i \cdots H_1 A$ can be written as

$$A_i = (r_1 | \cdots | r_i | w_{i+1}^i | \cdots | w_k^i),$$

where $w_i^2 = H_i w_j^{i-1}$ for $j = i + 1, \ldots, k$.

After k steps we arrive at

$$H_k \cdots H_1 A = R,$$

where $R = (r_1 | \cdots | r_k)$ is an upper triangular $n \times k$ matrix. Since the Householder matrices $H_1, \ldots, H_k$ are orthogonal, it follows from Lemma 10.5.2(c) that the $Q^t = H_k \cdots H_1$ is orthogonal. Thus $A = QR$ is a QR decomposition of A.

Orthonormalization with MATLAB

Given a set $w_1, \ldots, w_k$ of linearly independent vectors in $\mathbf{R}^n$, the MATLAB command qr allows us to compute an orthonormal basis of the spanning set of these vectors. As mentioned earlier, the underlying technique MATLAB uses for the computation of the QR decomposition is based on Householder transformations.

The syntax of the QR decomposition in MATLAB is quite simple. For example, let $w_1 = (1, 0, -1, 0)$, $w_2 = (2, -1, 0, 1)$, and $w_3 = (0, 0, -2, 1)$ be the three vectors in (10.2.11). In Section 5.5 we computed an orthonormal basis for the subspace of $\mathbf{R}^4$ spanned by w_1, w_2, and w_3. Here we use the MATLAB command qr to find an orthonormal basis for this subspace. Let A be the matrix having the vectors w_1^t, w_2^t, and w_3^t as columns. So A is:

    A = [1 2 0; 0 −1 0; −1 0 −2; 0 1 1]

The command

    [Q R] = qr(A,0)

leads to the answer

    Q =
       −0.7071   0.5000   −0.4523
             0   −0.5000   −0.1508
        0.7071   0.5000   −0.4523
             0   0.5000   0.7538
    R =
       −1.4142   −1.4142   −1.4142
             0   2.0000   −0.5000
             0   0   1.6583

A comparison with (10.2.12) shows that the columns of the matrix Q are the elements in the orthonormal basis. The only difference is that the sign of the first vector is opposite. However, this is not surprising since we know that there is some freedom in the choice of Householder matrices, as remarked after Proposition 10.5.8.

In addition, the command qr produces the matrix R whose entries r_{ij} are the coordinates of the vectors w_j in the new orthonormal basis as in (10.5.3). For instance, the second column of R tells us that

$$w_2 = r_{12}v_1 + r_{22}v_2 + r_{32}v_3 = -1.4142v_1 + 2.0000v_2.$$

HAND EXERCISES

In Exercises 1–5, decide whether or not the given matrix is orthogonal.

1. $\begin{pmatrix} 2 & 0 \\ 0 & 1 \end{pmatrix}$

2. $\begin{pmatrix} 0 & 1 & 0 \\ 0 & 0 & 1 \\ 1 & 0 & 0 \end{pmatrix}$

3. $\begin{pmatrix} 0 & -1 & 0 \\ 0 & 0 & 1 \\ -1 & 0 & 0 \end{pmatrix}$

4. $\begin{pmatrix} \cos(1) & -\sin(1) \\ \sin(1) & \cos(1) \end{pmatrix}$

5. $\begin{pmatrix} 1 & 0 & 4 \\ 0 & 1 & 0 \end{pmatrix}$

6. Let Q be an orthogonal $n \times n$ matrix. Show that Q preserves the length of vectors; that is

$$\|Qv\| = \|v\| \quad \text{for all } v \in \mathbf{R}^n.$$

In Exercises 7–10, compute the Householder matrix H corresponding to the given vector u.

7. $u = \begin{pmatrix} 1 \\ 1 \end{pmatrix}$

8. $u = \begin{pmatrix} 0 \\ -2 \end{pmatrix}$

9. $u = \begin{pmatrix} -1 \\ 1 \\ 5 \end{pmatrix}$

10. $u = \begin{pmatrix} 1 \\ 0 \\ 4 \\ -2 \end{pmatrix}$

11. Find the matrix that reflects the plane across the line generated by the vector $(1, 2)$.

12. Prove that the rows of an $n \times n$ orthogonal matrix form an orthonormal basis for $\mathbf{R}^n$.

COMPUTER EXERCISES

In Exercises 13–16, use the MATLAB command qr to compute an orthonormal basis for each of the subspaces spanned by the given set of vectors.

13. $w_1 = (1, -1), \quad w_2 = (1, 2)$

14. $w_1 = (1, -2, 3), \quad w_2 = (0, 1, 1)$

15. $w_1 = (1, -2, 3), \quad w_2 = (0, 1, 1), \quad w_3 = (2, 2, 0)$

16. $v_1 = (1, 0, -2, 0, -1), \quad v_2 = (2, -1, 4, 2, 0), \quad v_3 = (0, 3, 5, 1, -1)$

17. Find the 4×4 Householder matrices H_1 and H_2 corresponding to the vectors

$$\begin{aligned} u_1 &= (1.04, 2, 0.76, -0.32) \\ u_2 &= (1.4, -1.3, 0.6, 1.2). \end{aligned} \qquad \text{(10.5.5)}*$$

Compute $H = H_1 H_2$ and verify that H is an orthogonal matrix.

CHAPTER 11

Autonomous Planar Nonlinear Systems

In Chapter 4 we discussed how the phase line for a single autonomous nonlinear differential equation is determined by the equilibria of the differential equation. Once the equilibria and their stability properties are known, we can find the long-time or asymptotic behavior of every solution to a single equation—even though we do not necessarily know a closed form formula for that solution.

In this chapter we discuss the corresponding and more complicated results for planar systems of nonlinear autonomous differential equations. We discuss precisely what information about planar phase portraits is needed in order to be able to tell the (asymptotic) fate of all solutions. The information that we need includes the equilibria and their type (as discussed in Chapter 7), periodic solutions and their stability type, and connecting trajectories. Once we have the needed information, we can determine the qualitative features of all solutions to a planar system—even though we cannot write a closed form formula for these solutions.

In Section 11.1 we experiment numerically with nonlinear planar systems, introducing the information that is needed to draw a qualitative phase plane portrait of a differential equation. We see that we need to know the equilibria and their type (saddle, sink, or source), time periodic solutions and their stability, and trajectories that connect these solutions (including stable and unstable orbits emanating from saddles).

In Section 11.2 we look more closely at the role of linearized systems of differential equations near a hyperbolic equilibrium. In particular, we introduce the Jacobian matrix and show numerically that solutions near an equilibrium behave like solutions to the linearized system. Then we use the results of Chapter 7 to understand specific features of the local behavior of nonlinear systems near a saddle, sink, or source in terms of the corresponding features of solutions to linear differential equations.

The analytic discussion of aspects of time periodic solutions is introduced in Section 11.3. In particular, we show how periodic solutions to certain systems of differential equations can be constructed using phase-amplitude equations in polar coordinates.

This information is then synthesized in Section 11.4 in terms of stylized phase portraits for Morse-Smale planar systems of autonomous differential equations. Morse-Smale systems are differential equations with properties that allow us to find qualitative phase plane portraits. Moreover, in a sense that we will not try to make precise, most planar autonomous systems are Morse-Smale.

11.1 INTRODUCTION

In Section 4.3 we discussed phase line pictures of a single autonomous differential equation

$$\frac{dx}{dt} = f(x).$$

We saw that knowing the equilibria and the dynamics near those equilibria was sufficient information for understanding the dynamics of all solutions. For example, if

$$f(x) = x(x - 1)(x - 2) = x^3 - 3x^2 + 2x, \tag{11.1.1}$$

then the differential equation has three equilibria located at 0, 1, and 2. From the derivative

$$f'(x) = 3x^2 - 6x + 2,$$

we see that

$$f'(0) = 2, \quad f'(1) = -1, \quad \text{and} \quad f'(2) = 2.$$

So the equilibria at $x = 0$ and $x = 2$ are unstable (since the derivative is positive at these equilibria), while the equilibrium at 1 is asymptotically stable (since the derivative is negative at $x = 1$). Recall Theorem 4.3.2. This information is summarized in the phase line picture in Figure 11.1.

Figure 11.1
Phase line plot for the one-dimensional equation (11.1.1)

The added content of Figure 11.1 is that we now know the fate of all solutions in both forward and backward time. A solution $x(t)$ with initial condition at x_0 between the equilibria 0 and 1 approaches 1 in forward time ($t \to \infty$) and 0 in backward time ($t \to -\infty$). This information is encoded in the arrows on the figure. Using dfield5, we can find the time series of (11.1.1) for an initial condition between 0 and 1; see Figure 11.2. Alternatively, this phase line tells us the equilibria and the trajectories that connect equilibria. If we assume a finite number of equilibria and that the derivative is nonzero at each equilibrium, then in one dimension there is nothing else that can happen in the dynamics.

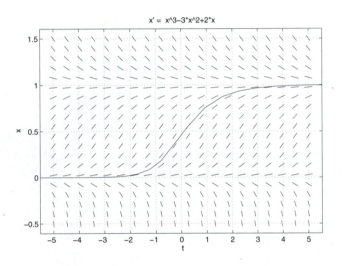

Figure 11.2
Time series for solution to (11.1.1) with initial condition between 0 and 1

Two-Dimensional Phase Planes

The purpose of this chapter is to develop a similar understanding of phase portraits for two-dimensional autonomous systems

$$
\frac{dx}{dt} = f(x, y)
$$

$$
\frac{dy}{dt} = g(x, y).
$$

(11.1.2)

As is the case in one dimension, equilibria and trajectories connecting these equilibria play a major role in determining phase portraits. In two dimensions, however, there is a new type of solution that also plays a role in determining phase plane pictures: the periodic solution. The remainder of this introduction illustrates the type of information that we want to include in two-dimensional phase portraits.

A Linear Equation

In Section 7.2 we saw that the dynamics of planar constant coefficient linear differential equations are determined by the eigenvalues and eigenvectors of the coefficient matrix. We review some of this material in Section 11.2. For example, the origin is a saddle in the system of differential equations

$$\frac{dx}{dt} = y$$

$$\frac{dy}{dt} = 2.2x + 2.1y.$$

(11.1.3)*

Note: The * after the label (11.1.3) indicates that this differential equation has been preloaded in the file e11_1_3.pps and can be addressed from pplane5 by clicking sequentially on the File button in the PPLANE5 Setup window, the Load a system from... button, and the laode toolbox button. Then click on the e11_1_3.pps file and the Done button to load the differential equation (11.1.3) into pplane5.

The fact that the origin in (11.1.3) is a saddle is seen by examining the coefficient matrix of (11.1.3):

$$C = \begin{pmatrix} 0 & 1 \\ 2.2 & 2.1 \end{pmatrix}.$$

Note that $\det(C) = -2.2 < 0$. Since the determinant is the product of the eigenvalues, the eigenvalues of C are real and of opposite sign, and the origin is a saddle. See Table 7.1.

Using MATLAB, we can determine more refined information concerning the dynamics of this equation. The two eigenvalues of C are found by entering C and typing eig(C) to obtain

```
ans =
   -0.7673
    2.8673
```

thus verifying that the eigenvalues are real and of opposite sign. Moreover, we can find the eigenvectors associated with these eigenvalues by typing

```
[V,D] = eig(C)
```

to obtain

```
V =
   -0.7934   -0.3293
    0.6087   -0.9442
D =
   -0.7673        0
        0    2.8673
```

The eigenvectors are the columns of the matrix V and this information allows us to sketch the dynamics of (11.1.3) as in Figure 11.3 (left). Indeed, we can use pplane5 to plot more accurately the phase plane picture associated with (11.1.3), as shown in Figure 11.3 (right). Note also that the eigenvalue and eigenvector information is determined in pplane5. In the PPLANE5 Display window click on the Solutions button and then on the Find an equilibrium button. Use the cross hairs to click near the origin. Then pplane5 finds the equilibrium and shows the matrix C and its eigenvalues and eigenvectors in the PPLANE5 Equilibrium point data window.

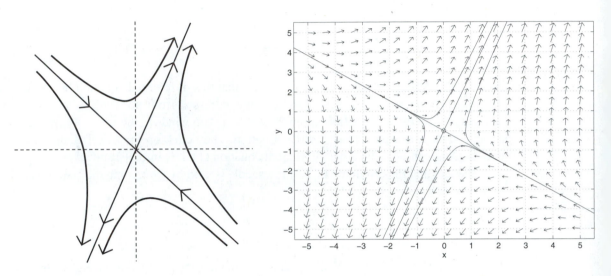

Figure 11.3
Left: Sketch of phase plane of (11.1.3) based on eigenvalues and eigenvectors of C. *Right*: Trajectories of (11.1.3) using pplane5.

The Addition of Nonlinear Terms

Next we discuss what happens to solutions when nonlinear terms are added to (11.1.3). For example, consider the system

$$\frac{dx}{dt} = y$$

$$\frac{dy}{dt} = 2.2x + 2.1y + x^2 + xy. \qquad\qquad (11.1.4)^*$$

We study the phase plane of (11.1.4) using pplane5 and find that trajectories for the nonlinear system on the square $-0.5 \le x, y \le 0.5$ look very much like those for the linear system (11.1.3)—even though we do not know how to solve the nonlinear system in closed form, as we can for the linear system. See Figure 11.4 and note the similarity with Figure 11.3 (right).

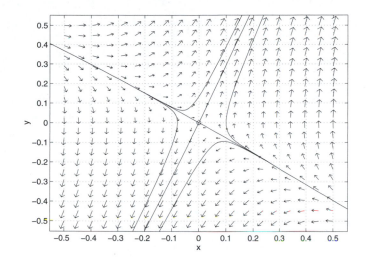

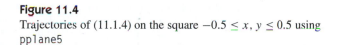

Figure 11.4
Trajectories of (11.1.4) on the square $-0.5 \le x, y \le 0.5$ using
`pplane5`

Recall that an equilibrium is *hyperbolic* if the eigenvalues of the matrix of coefficients of the linear terms have nonzero real part. There is a theorem that states that in a sufficiently small neighborhood of a hyperbolic equilibrium of a nonlinear system, the trajectories *look like* the solutions of the associated linear system. Indeed, on the small square in which these numerical calculations were performed, this theorem appears to be correct.

However, when we compute the solution trajectories to (11.1.4) on the larger square $-5 \le x, y \le 5$, it becomes clear that the phase plane picture of the nonlinear system no longer resembles the phase plane picture of the linear system. See Figure 11.5 (left). Note that there is another equilibrium, a spiral sink, surrounded by what looks like a circular trajectory. In Figure 11.5 (right) we sketch a phase portrait for this system of ODEs indicating the important information: the equilibria and type of equilibria, the periodic solutions, and the connections between these trajectories. Note that we have found these trajectories numerically even though we do not know how to solve for them in closed form.

The Importance of Phase Plane Portraits

The importance of the phase portrait is that it lets us determine the evolution of solutions starting in the square—even though we do not have a formula for these solutions. For example, we can use `Keyboard input` to start a trajectory near the origin so that its backward evolution approaches the periodic solution while its forward evolution leaves the square in the unstable direction of the saddle at the origin. In Figure 11.6 we picture the trajectory in phase space through the point $(-0.1, 0.1)$, along with the time series y versus t of this solution.

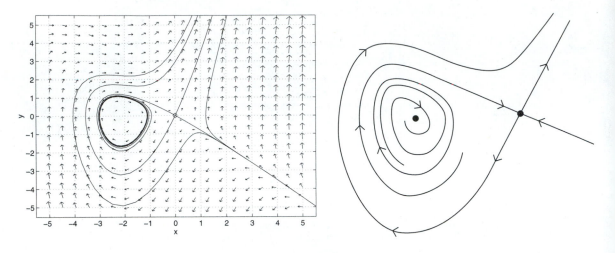

Figure 11.5
Left: Trajectories of (11.1.4) on the square $-5 \leq x, y \leq 5$ using `pplane5`. *Right*: A phase plane portrait of this equation.

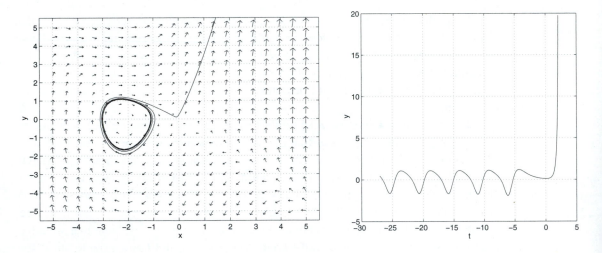

Figure 11.6
Left: Trajectory of (11.1.4) through $(-0.1, 0.1)$. *Right*: Time series y versus t of this solution.

In this chapter we discuss what kinds of dynamics we can expect from autonomous planar systems of differential equations. We show that solutions to linear systems give us much information about the structure of solutions to nonlinear systems. We also show that there are new phenomena occurring in solutions to nonlinear systems that do not occur in solutions to the linear ones. In our discussions we abandon any attempts at proofs; rather we try to survey the most important theorems with an eye to understanding how they can help with numerical explorations.

COMPUTER EXERCISES

1. Use `pline` to explore the phase line picture of the nonlinear differential equation

$$\frac{dx}{dt} = 2 + 3x - 4x^2 + x^3 - 2x^4 - x^5$$

on the interval $[-3, 1]$. Determine the number of equilibria and whether or not they are asymptotically stable.

2. Using `pplane5`, explore the phase portraits of the nonlinear systems

$$\begin{aligned} \dot{x} &= y \\ \dot{y} &= 2.2x + 2.1y - x^3 + axy, \end{aligned} \qquad \textbf{(11.1.5)*}$$

where $a = 1.6$ and $a = 1$. Observe that the phase planes of (11.1.5) resemble the phase plane of the linear system (11.1.3) in the square $-0.5 \leq x, y \leq 0.5$. In the square $-5 \leq x, y \leq 5$, discuss how these phase planes differ from each other and from the phase plane of (11.1.4) shown in Figure 11.5.

In Exercises 3–5, use `pplane5` to find a solution to (11.1.5) (with $a = 1.6$) satisfying the given properties. Plot the time series of y versus t for that solution and describe the differences in the three time series.

3. Find a solution that limits on the periodic solution in backward time and goes to infinity in the first quadrant in forward time.

4. Find a solution that limits on a spiral sink in forward time and on a periodic solution in backward time.

5. Find a solution that limits on a node in backward time and goes to infinity in the first quadrant in forward time.

11.2 EQUILIBRIA AND LINEARIZATION

Recall that equilibrium solutions of (11.1.2) are found by solving simultaneously the algebraic equations

$$\begin{aligned} f(x, y) &= 0 \\ g(x, y) &= 0. \end{aligned} \qquad \textbf{(11.2.1)}$$

In general, it is a difficult task to solve these algebraic equations explicitly, except in the simplest of cases. One example where an explicit solution may be found is (11.1.4). In that example, equilibria satisfy

$$y = 0 \quad \text{and} \quad x(2.2 + x) = 0.$$

Thus there are two equilibria, $Z_1 = (0, 0)$ and $Z_2 = (-2.2, 0)$, and these equilibria can be seen in Figure 11.5. We know from our earlier discussion that the origin Z_1 is a saddle point, but what about the other equilibrium Z_2? From the figure, it appears to be a spiral sink—dynamics that we associate with complex eigenvalues with negative real part. See Table 7.1. In this section we discuss how we can apply linear theory to nonlinear equations.

Let $F : \mathbf{R}^2 \to \mathbf{R}^2$ be the nonlinear mapping

$$F(x, y) = (f(x, y), g(x, y)).$$

Definition 11.2.1   *The* Jacobian matrix *of F at the point* (x, y) *is the* 2×2 *matrix*

$$(dF)_{(x,y)} = \begin{pmatrix} f_x(x, y) & f_y(x, y) \\ g_x(x, y) & g_y(x, y) \end{pmatrix}, \tag{11.2.2}$$

where the subscript denotes partial differentiation. For example, $f_x = \partial f / \partial x$.

Near an equilibrium, the Jacobian matrix is the best *linear* approximation to the mapping F. This statement can be formalized using Taylor series methods in two variables, but we just accept this fact here.

Definition 11.2.2   *An equilibrium is* hyperbolic *if the eigenvalues of the Jacobian matrix at the equilibrium have nonzero real part.*

Note that it is easy to compute the Jacobian matrix for linear systems. Indeed, for the linear differential equation

$$\dot{x} = ax + by$$
$$\dot{y} = cx + dy,$$

the Jacobian matrix is just

$$(dF)_{(x,y)} = \begin{pmatrix} a & b \\ c & d \end{pmatrix}$$

and is independent of x and y. So, not surprisingly, the best linear approximation to a linear system is itself.

Theorem 11.2.3   *Suppose that the system of differential equations (11.1.2) has a hyperbolic equilibrium at* $Z_0 = (x_0, y_0)$. *Then, in a sufficiently small neighborhood of* Z_0, *the phase plane for (11.1.2) is the* same *as the phase plane of the system of linear differential equations*

$$\begin{pmatrix} \dot{x} \\ \dot{y} \end{pmatrix} = (dF)_{Z_0} \begin{pmatrix} x \\ y \end{pmatrix}. \tag{11.2.3}$$

It is difficult to define precisely what we mean by the word *same*. Roughly speaking, *same* means that near Z_0 there is a nonlinear change of coordinates that transforms the nonlinear equation into the linear one.

For example, the Jacobian matrix of (11.1.4) at the equilibrium $Z_2 = (-2.2, 0)$ is

$$(dF)_{Z_2} = \begin{pmatrix} 0 & 1 \\ 2.2 + 2x + y & 2.1 + x \end{pmatrix} \Bigg|_{(x,y)=(-2.2,0)} = \begin{pmatrix} 0 & 1 \\ -2.2 & -0.1 \end{pmatrix}.$$

At the equilibrium Z_2 we may use MATLAB to show that the eigenvalues of $(dF)_{Z_2}$ are $-0.05 \pm 1.4824i$, thus verifying that the equilibrium Z_2 is a spiral sink. It is instructive to use pplane5 to show the phase plane of the linearized system. (To show the linearization, click on the Find an equilibrium point button under the Solutions button. Then in the PPLANE5 equilibrium point window click on the Display the linearization button.) The result is shown in Figure 11.7 (left); compare this result with the phase plane picture in Figure 11.5 (left) near the spiral sink.

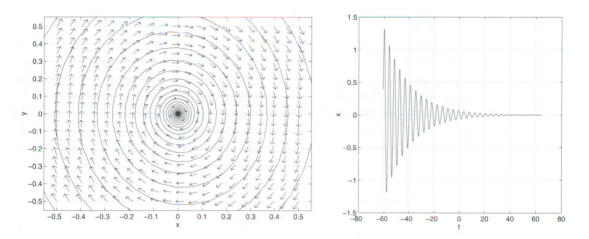

Figure 11.7
Left: Trajectory of (11.2.3) near the spiral sink Z_2. *Right*: The time series x versus t for this solution.

There is an important corollary to Theorem 11.2.3. We say that an equilibrium is *linearly stable* if all the eigenvalues of the Jacobian matrix have negative real part.

Corollary 11.2.4 *An equilibrium that is linearly stable is asymptotically stable.*

In Section 7.2 we discussed linear stability of the origin for planar linear equations. We showed (see Theorem 7.2.2) that in planar linear equations, linear stability implies that the origin is asymptotically stable. It follows from Theorem 11.2.3 that the equilibrium of the nonlinear equation is also asymptotically stable.

Remark 11.2.5 *Generalizations of Theorem 11.2.3 and Corollary 11.2.4 to n dimensions are both valid. The Jacobian matrix—the matrix of partial derivatives of the* *coordinate functions—is now an n × n matrix. In n dimensions,* hyperbolicity *of an equilibrium means that no eigenvalue of the Jacobian matrix has zero real part, and* linear stability *of an equilibrium means that the real parts of all n of the eigenvalues have negative real part. See Theorem 14.2.2.*

Important Features of Hyperbolic Equilibria

Theorem 11.2.3 states that a nonlinear autonomous system of differential equations behaves like its linearization in a small neighborhood of a hyperbolic equilibrium. To effectively utilize this theorem, we need to know the important features of phase portraits for hyperbolic linear systems. We now recall results of Section 7.2.

Saddles

Recall from Section 7.1 (see Definition 7.1.3) that planar systems of linear differential equations with saddles at the origin have invariant lines (or eigendirections) called stable and unstable orbits. One eigendirection corresponds to a negative eigenvalue and is the stable direction because solutions on that line tend to the origin in forward time. The other eigendirection corresponds to a positive eigenvalue and is the unstable direction because solutions on that line tend to the origin in backward time.

It follows from Theorem 11.2.3 that this structure is approximately recreated in nonlinear systems in small neighborhoods of saddles. In particular, there are invariant curves for nonlinear systems defined near saddle equilibria. The nonlinearities deform the invariant lines into invariant curves called *invariant manifolds*. These invariant manifolds are called *stable* and *unstable manifolds* or *stable* and *unstable orbits*; stable orbits tend to the equilibrium in forward time and unstable orbits tend to the equilibrium in backward time. See Figure 11.5 for an example.

Sinks

In (11.1.4) we have also seen that the spiraling behavior near a spiral equilibrium is unaffected by higher order terms (see Figure 11.5); this remark is also a consequence of Theorem 11.2.3. Similarly, Theorem 11.2.3 guarantees that near an improper node or a focus, higher order terms do not affect the local phase portrait of the nonlinear system.

An Example with Analytically Solvable Equilibria

As an example, consider the system of ODEs

(a)
$$\frac{dx}{dt} = 1 + x - y^2 \quad \text{and}$$

$$(11.2.4)^*$$

(b)
$$\frac{dy}{dt} = -1 + 6y + x^2 - 5y^2.$$

We find all the equilibria of this system of differential equations and their type. (The coefficients have been chosen so that the calculations are feasible. Nevertheless, it is instructive to verify the details.)

Note that $x = y^2 - 1$ is the solution to (11.2.4)(a), which we can substitute into (11.2.4)(b) to obtain

$$y^4 - 7y^2 + 6y = 0.$$

The coefficients were chosen so that this polynomial factors into

$$y(y - 1)(y - 2)(y + 3) = 0.$$

Thus there are four equilibria:

$$(-1, 0), \quad (0, 1), \quad (3, 2), \quad (8, -3).$$

We now determine the type of each equilibrium. To do this, observe that the Jacobian matrix is

$$\begin{pmatrix} 1 & -2y \\ 2x & 6 - 10y \end{pmatrix}.$$

Evaluating the Jacobian matrix at the four equilibria, we obtain

$$\begin{pmatrix} 1 & 0 \\ -2 & 6 \end{pmatrix}, \quad \begin{pmatrix} 1 & -2 \\ 0 & -4 \end{pmatrix}, \quad \begin{pmatrix} 1 & -4 \\ 6 & -14 \end{pmatrix}, \quad \begin{pmatrix} 1 & 6 \\ 16 & 36 \end{pmatrix}.$$

The determinant, trace, discriminant, and type of each of these matrices are listed in Table 11.1.

Table 11.1
Types of equilibria in example (11.2.4)

Equilibria	det(d)	trace(t)	disc(D)	Type
$(-1, 0)$	6	7	25	Nodal source
$(0, 1)$	-4	—	—	Saddle
$(3, 2)$	10	-13	129	Nodal sink
$(8, -3)$	-60	—	—	Saddle

Next we use the information about equilibria contained in Table 11.1 and pplane5 to compute numerically the phase portrait of equations (11.2.4). Using the Find the equilibrium button, put circles around the four equilibria. (Remember to make the display screen large enough—say, $-2 \le x \le 10$ and $-4 \le y \le 4$.) Also let pplane5 compute the stable and unstable orbits of the two saddles, obtaining the picture in Figure 11.8. Once this picture is drawn, we have a good idea of the phase portrait; note, in particular, the saddle sink and saddle source connections.

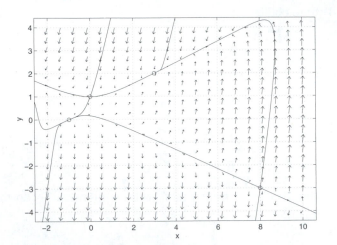

Figure 11.8
Equilibria and connections of (11.2.4)

An Example Where Equilibria Are Not Analytically Solvable

As a second example consider the system of differential equations

$$\dot{x} = x + y + x^2 - y^2 + 0.1$$
$$\dot{y} = y - 2xy + 0.5x^2 + y^2.$$

$$(11.2.5)^*$$

It is difficult to find the equilibria in (11.2.5) explicitly. We can, however, construct the phase plane of this equation with the help of pplane5. First we choose a square in which to display the direction field—say, $-5 \le x, y \le 5$; see Figure 11.9 (left). We then inspect the direction field for possible equilibria. There is one equilibrium near the origin; when we use the pplane5 search feature for equilibria by clicking on the origin, we find a spiral source. Similarly, using the search feature for equilibria by clicking along the negative x-axis leads to a saddle. Once a saddle is found, instructing pplane5 to plot the stable and unstable orbits of this saddle helps to fill in the phase plane. Some experimentation yields the four equilibria and the connecting stable and unstable orbits pictured on the right in Figure 11.9. These equilibria, which are listed by pplane5 using the List computed equilibrium points button, are:

```
(-0.1066, -0.0047)    Spiral source.
(-1.1368, -0.2110)    Saddle point.
(3.2539, 4.2672)      Saddle point.
(0.2752, -0.3372)     Saddle point.
```

It is this last picture that leads to the stylized phase plane portrait in Figure 11.10.

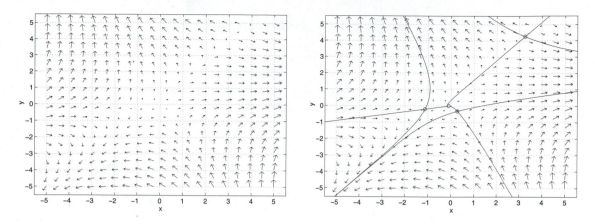

Figure 11.9
Left: Direction field of (11.2.5). *Right*: Phase plane with equilibria and stable orbits.

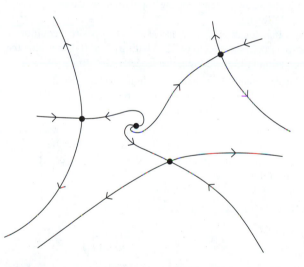

Figure 11.10
Stylized phase plane portrait of (11.2.5)

HAND EXERCISES

1. (a) Draw the phase line picture of the differential equation

$$\frac{dx}{dt} = -x^2 + 3x - 2. \tag{11.2.6}$$

(b) Let $x(t)$ be the solution to (11.2.6) satisfying $x(0) = -1$. Determine $\lim\limits_{t \to \infty} x(t)$.

(c) Let $y(t)$ be a solution to (11.2.6) satisfying $y(0) = 0.5$. Sketch the time series of this solution.

2. Recall that the simplest population growth model is

$$\frac{dp}{dt} = \lambda p,$$

where $p(t)$ is the population at time t and λ is the population growth rate. Suppose that the growth rate depends on the size of the population, as follows:

$$\lambda(p) = \lambda_0 - ap,$$

where $\lambda_0 > 0$ and $a > 0$ are constants. Suppose that $p(t)$ is a solution and $p(0) > 0$. What is the asymptotic limit of the population in forward time; that is, what is $\lim\limits_{t \to \infty} p(t)$?

3. More generally, consider the population growth model

$$\frac{dp}{dt} = \lambda(p)p, \tag{11.2.7}$$

where the growth rate depends on the population. Assume

   (i) the growth rate is positive when the population is small—that is, $\lambda(0) > 0$—and
   (ii) the growth rate decreases as the population increases—that is, $d\lambda/dp < 0$ for all p.

Consider the following:

   (a) Show that equation (11.2.7) has at most one positive equilibrium $p_e > 0$.
   (b) Let $p(t)$ be a solution to (11.2.7) with $p(0) > 0$. Determine the asymptotic limit of the population in forward time; that is, determine $\lim\limits_{t \to \infty} p(t)$.
   (c) Relate your answer in (b) to the solution to Exercise 2.

Hint   There are two possible answers to (b) depending on whether (11.2.7) has one positive equilibrium or no positive equilibria.

In Exercises 4–8, we review some of the theory of planar linear systems of differential equations that is needed in the study of planar nonlinear systems. For the given matrix C, determine whether the origin is a hyperbolic equilibrium for the differential equation $\dot{X} = CX$. When the origin is hyperbolic, what kind of system is it?

4. $C = \begin{pmatrix} 0.1 & 2 \\ -1 & -3 \end{pmatrix}$ **5.** $C = \begin{pmatrix} 4 & -2 \\ 1 & 5 \end{pmatrix}$ **6.** $C = \begin{pmatrix} 2 & -1 \\ 4 & -2 \end{pmatrix}$

7. $C = \begin{pmatrix} 1 & 3 \\ -2 & -6 \end{pmatrix}$ **8.** $C = \begin{pmatrix} 11 & -8 \\ 18 & -13 \end{pmatrix}$

9. Consider the planar nonlinear system of ODEs

$$\frac{dx}{dt} = 1 - x - y$$

$$\frac{dy}{dt} = 2xy.$$

   (a) Find all equilibria of this system.
   (b) Determine the behavior of trajectories in a small neighborhood of each equilibrium.
   (c) Sketch the phase portrait of this system.
       Hint   Check for invariant axes.
   (d) Use pplane5 to verify your sketch.

10. Consider the planar nonlinear system of ODEs

$$\frac{dx}{dt} = 1 - 2x + y + x^2 - xy$$

$$\frac{dy}{dt} = y - y^2.$$

(a) Find all equilibria of this system.
(b) Are the equilibria asymptotically stable or not?
(c) Determine the type of each equilibrium.

11. Consider the system of differential equations (11.1.5) that depends on the constant a.
(a) Find the equilibria of this system.
(b) Find the Jacobian matrices at the equilibria.
(c) For which values of a are the equilibria not hyperbolic?
(d) For those values of a where the equilibria are hyperbolic, determine whether or not these equilibria are asymptotically stable.
(e) For each nonhyperbolic equilibrium, write the linearized system and draw its phase plane.

12. Find all equilibria of the system of ODEs

$$\dot{x} = 2x - y$$
$$\dot{y} = 11x + x^2 - 3y^2.$$

Also determine the types of each of these equilibria.

13.(a) Show that the origin is the only equilibrium of

$$\begin{aligned}\dot{x} &= y - x^3 - 2xy^2 \\ \dot{y} &= -x - y^3\end{aligned} \qquad \textbf{(11.2.8)*}$$

and that the linearized system about the origin is a center. Hence the origin is not hyperbolic.

(b) Let $r = \sqrt{x^2 + y^2}$ be the distance of (x, y) from the origin. Let $(x(t), y(t))$ be a solution to (11.2.8) and show that $r(t)$ satisfies the differential equation

$$\dot{r}(t) = -r(t)^3.$$

Use this fact to show that the origin is an asymptotically stable equilibrium of (11.2.8). (Hence the phase portraits of the linearized and nonlinear systems are different on all neighborhoods of the origin and the hyperbolicity assumption in Theorem 11.2.3 is needed.)

COMPUTER EXERCISES

14. Use pplane5 to determine the phase portraits of the nonlinear system (11.2.8). Verify that pplane5 finds (approximately) periodic solutions in sufficiently small neighborhoods of the origin. This experiment, coupled with the results from Exercise 13, suggests that the numerical methods employed by pplane5 fail sufficiently close to nonhyperbolic equilibria.

15. Consider the system of differential equations

$$\dot{x} = -x - y + x^2 - y^2$$
$$\dot{y} = 0.25x - 3y - 2xy + y^2. \qquad \textbf{(11.2.9)*}$$

(a) Use pplane5 to verify that the origin is a nodal sink with eigenvalues $\lambda_1 = -2.866$ and $\lambda_2 = -1.134$.

(b) Use pplane5 on the square $-1 \leq x, y \leq 1$ to verify that trajectories starting near the origin approach the origin on curves tangent to the eigendirection corresponding to the eigenvalue λ_2.

16. Using pplane5, find all equilibria of

$$\dot{x} = 3x - 2y - 3x^2 + y^2$$
$$\dot{y} = -x + y - 3xy \qquad \textbf{(11.2.10)*}$$

on the square $-2 \leq x, y \leq 2$. Determine the type of equilibria and draw a phase portrait for this system of equations.

11.3   PERIODIC SOLUTIONS

A periodic solution to a system of differential equations

$$\frac{dX}{dt} = F(X) \qquad \textbf{(11.3.1)}$$

is a nonconstant solution such that

$$X(t) = X(t + T) \qquad \textbf{(11.3.2)}$$

for some positive real number T. The smallest positive number T satisfying (11.3.2) is called the *period* of the periodic solution. The *frequency* is the quantity $1/T$.

Uniqueness of solutions to the initial value problem has an interesting consequence concerning periodic solutions.

Lemma 11.3.1   *Let $X(t)$ be a nonconstant solution to the autonomous system (11.3.1) such that*

$$X(T) = X(0).$$

Then $X(t)$ is a periodic solution with period dividing T.

Proof: Let $Y(t) = X(t + T)$. Then $Y(t)$ is a solution to (11.3.1) since

$$\dot{Y}(t) = \dot{X}(t + T) = F(X(t + T)) = F(Y(t)).$$

This solution has initial condition $Y(0) = X(T) = X(0)$. Uniqueness of solutions with the same initial condition implies that $X(t) = Y(t)$ for all t and that $X(t) = X(t + T)$.

Since $X(t)$ is not an equilibrium, there is a smallest positive period for X and that period must divide T. ◆

In this section we study two types of planar systems that have periodic solutions, centers and phase-amplitude equations, and in both examples we prove that the system has periodic solutions. Typically, however, proving that periodic solutions exist is not a simple task; in general we rely on numerical solution to verify the existence of periodic solutions.

Nonhyperbolic Centers

Nonconstant periodic solutions occur in linear systems only when the coefficient matrix has purely imaginary eigenvalues—that is, when the origin is a center. Therefore periodic solutions occur in linear systems only when the equilibrium is nonhyperbolic, and they occur in infinite families. For example, consider the *center*

$$\frac{dX}{dt} = \begin{pmatrix} 0 & -\tau \\ \tau & 0 \end{pmatrix} X. \tag{11.3.3}$$

Solutions of (11.3.3) are

$$X(t) = \begin{pmatrix} \cos(\tau t) & -\sin(\tau t) \\ \sin(\tau t) & \cos(\tau t) \end{pmatrix} X_0.$$

All nonzero solutions of (11.3.3) are periodic with period $2\pi/\tau$. See Figure 11.11.

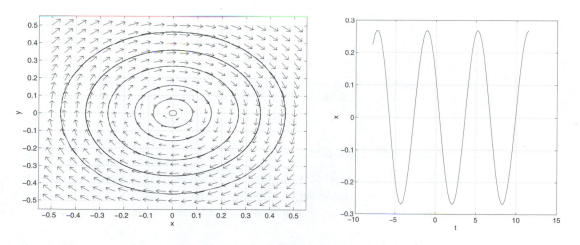

Figure 11.11
Left: Trajectories of (11.3.3) when $\tau = 3$. *Right*: A time series of one solution.

Centers are examples of nonhyperbolic equilibria, and the addition of higher order terms can change the phase portrait of a center—even near the origin. For example, consider the system

$$\dot{x} = -2y - (x^2 + y^2)x$$
$$\dot{y} = 2x - (x^2 + y^2)y. \qquad\qquad (11.3.4)^*$$

A trajectory in the phase portrait of (11.3.4) that spirals into the origin is shown in Figure 11.12 (left). Equation (11.3.4) can be modified to produce a single periodic solution as follows:

$$\dot{x} = x - 2y - (x^2 + y^2)x$$
$$\dot{y} = 2x + y - (x^2 + y^2)y. \qquad\qquad (11.3.5)^*$$

See Figure 11.12 (right). This example shows the necessity of assuming hyperbolicity as a hypothesis in Theorem 11.2.3.

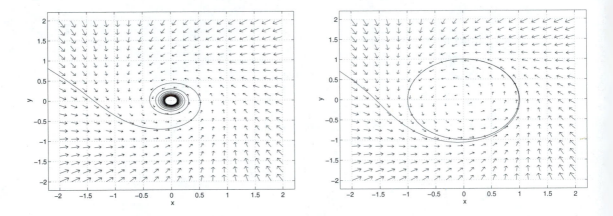

Figure 11.12
Left: A trajectory of (11.3.4) that spirals toward the origin. Note the slow convergence due to the fact that the origin is not hyperbolic. *Right*: A trajectory of (11.3.5) that spirals toward a periodic solution.

Phase-Amplitude Equations

We now show why equations (11.3.4) and (11.3.5) behave as observed in the numerical integration. We do this using polar coordinates and phase-amplitude equations. Recall that rectilinear coordinates (x, y) can be written in terms of polar coordinates (r, θ) as

$$x = r \cos \theta$$
$$y = r \sin \theta.$$

On inverting these equations, we find that

$$r^2 = x^2 + y^2$$
$$\theta = \tan^{-1}\left(\frac{y}{x}\right). \tag{11.3.6}$$

In applications, r is called the *amplitude* and θ the *phase*.

Suppose that $(x(t), y(t))$ is the solution to a planar system of ODEs of the form

$$\dot{x} = a(x^2 + y^2)x - b(x^2 + y^2)y$$
$$\dot{y} = a(x^2 + y^2)y + b(x^2 + y^2)x, \tag{11.3.7}$$

where a and b are differentiable functions of $x^2 + y^2$. Note that (11.3.4) and (11.3.5) are special cases of (11.3.7). In (11.3.4), $a(r^2) = -r^2$ and $b(r^2) = 2$, while in (11.3.5), $a(r^2) = 1 - r^2$ and $b(r^2) = 2$.

Applying the chain rule to (11.3.6), we obtain

$$\dot{r} = \frac{1}{r}(x\dot{x} + y\dot{y})$$
$$\dot{\theta} = \frac{1}{r^2}(x\dot{y} - y\dot{x}). \tag{11.3.8}$$

Substitute the values of $\dot{x}$ and $\dot{y}$ from (11.3.7) into (11.3.8) to obtain the *amplitude* equation

$$\frac{dr}{dt} = a(r^2)r \tag{11.3.9}$$

and the *phase* equation

$$\frac{d\theta}{dt} = b(r^2). \tag{11.3.10}$$

Proposition 11.3.2 *Suppose that r_0 is a positive zero of the amplitude equation, that is, $a(r_0^2) = 0$. If $b(r_0^2) \neq 0$, then the circle $r = r_0$ is the trajectory of a periodic solution to (11.3.7).*

Moreover, if $a'(r_0^2) < 0$, then the periodic solution is asymptotically stable; if $a'(r_0^2) > 0$, then the periodic solution is unstable.

Proof: The function $r(t)$ measures how far the solution $(x(t), y(t))$ to (11.3.7) is from the origin. Thus, if r_0 is a positive zero of the amplitude equation, then the distance of the solution remains constant; that is, the trajectory of $(x(t), y(t))$ in the plane lies on the circle $r = r_0$ (or $x^2 + y^2 = r^2$ in Cartesian coordinates). The solution to the phase equation corresponding to this amplitude equilibrium is $\theta(t) = b(r_0^2)t + \theta_0$. Therefore, if $b(r_0^2) \neq 0$, then the solution trajectory moves around the circle $r = r_0$ with nonzero constant speed.

To verify the statement on asymptotic stability, write the amplitude equation as

$$\dot{r} = a(r^2)r \equiv f(r).$$

The point $r_0 > 0$ is an equilibrium for the amplitude equation if $f(r_0) = 0$—that is, if $a(r_0^2) = 0$. That equilibrium is asymptotically stable if $f'(r_0) < 0$. Using the product rule and the chain rule, we see that

$$f'(r) = a'(r^2)2r^2 + a(r^2).$$

Therefore

$$f'(r_0) = 2r_0^2 a'(r_0^2).$$

It follows that $f'(r_0) < 0$ if and only if $a'(r_0^2) < 0$.

If $a'(r_0^2) < 0$, then the equilibrium of the amplitude equation is asymptotically stable and nearby trajectories to the planar periodic solution limit on that periodic solution $r = r_0$ in forward time.   ◆

The Two Examples

As noted, in both equations (11.3.4) and (11.3.5) $b(r^2) = 2$. In the first equation $a(r^2) = -r^2$, while in the second equation $a(r^2) = 1 - r^2$. In both equations the phase equation is

$$\frac{d\theta}{dt} = 2,$$

which is solved for $\theta = 2t + \theta_0$ for some initial θ_0. Thus all solutions spin around the origin counterclockwise at a constant speed. In (11.3.4) the amplitude equation is

$$\frac{dr}{dt} = -r^3,$$

and all solutions approach $r = 0$ in forward time. Thus the phase-amplitude equations together imply that all solutions to (11.3.4) spiral into the origin, even though the origin in the linearized equations is a center.

In (11.3.5) the amplitude equation is

$$\frac{dr}{dt} = r - r^3.$$

In this equation all solutions with positive initial r tend toward $r = 1$, the unit circle in the xy-plane. This is seen using the one-dimensional phase line: Note that there is only one positive equilibrium at $r = 1$ and that solutions diverge from 0 and ∞. Thus the phase-amplitude equations show that all solutions (except the origin) spiral in forward time counterclockwise into the unit circle (that is, a periodic solution).

Limit Cycles

In general, for nonlinear systems without special structure, we do not expect periodic solutions to come in continuous families. We do expect periodic solutions to be isolated—as in Figure 11.5 (left).

We call a periodic solution $x(t)$ a *limit cycle* if all nearby trajectories converge to $x(t)$ in forward time or all nearby solutions converge to $x(t)$ in backward time. Note that in phase space a limit cycle is always isolated; in particular, a limit cycle cannot be included in a continuous family of periodic solutions as happens in (11.3.3).

Theorem 11.3.3   *Suppose that the right-hand side of a planar system of differential equations is defined and smooth everywhere inside a periodic solution. Then there is an equilibrium inside the periodic solution.*

This theorem is a consequence of the celebrated Poincaré-Bendixson theorem and simplifies the search for limit cycles in the plane; limit cycles must surround at least one equilibrium. Again, the most important information for determining the dynamics of a given planar differential equation is the number and type of equilibria. Once the equilibria are known, more complicated features, such as periodic solutions and stable and unstable orbits at saddles, can be found by exploration.

HAND EXERCISES

In Exercises 1–3, use phase-amplitude equations to determine the number of limit cycles that the given system of differential equations has. Then sketch the phase portrait of this differential equation.

1. $\dot{x} = 3x - 5y - 4(x^2 + y^2)x + (x^2 + y^2)^2 x$
$\dot{y} = 5x + 3y - 4(x^2 + y^2)y + (x^2 + y^2)^2 y$

2. $\dot{x} = -5x - 3y + 4(x^2 + y^2)x + (x^2 + y^2)^2 x$
$\dot{y} = 3x - 5y + 4(x^2 + y^2)y + (x^2 + y^2)^2 y$

3. $\dot{x} = 6x - y - 5(x^2 + y^2)x - 2(x^2 + y^2)y + (x^2 + y^2)^2 x$
$\dot{y} = x + 6y - 5(x^2 + y^2)y + 2(x^2 + y^2)x + (x^2 + y^2)^2 y$

4. Use phase-amplitude equations to find a system of differential equations that has exactly three limit cycles.

5. An autonomous planar system of differential equations is a *gradient* system if there exists a real-valued differentiable function $f(x, y)$ such that

$$\dot{x} = f_x(x, y)$$
$$\dot{y} = f_y(x, y). \qquad\qquad \textbf{(11.3.11)}$$

Prove that gradient systems never have periodic solutions.
Hint   Proceed in four steps as follows:
   (a) Let $(x(t), y(t))$ be a solution to (11.3.11) and let $h(t) = f(x(t), y(t))$. So $h : \mathbf{R} \to \mathbf{R}$. Use the chain rule to show that $h'(t) \geq 0$.

(b) Show that $h'(t_0) = 0$ for some t_0 if and only if $(x(t), y(t))$ is an equilibrium of (11.3.11).

(c) Show that h is monotonic increasing along any nonconstant solution of (11.3.11).

(d) Explain why a nonconstant solution of (11.3.11) cannot be periodic.

COMPUTER EXERCISES

6. Verify that the system of differential equations

$$\begin{aligned}
\dot{x} &= x - 2y - (x^2 + y^2)x + 0.1x^2 \\
\dot{y} &= 2x + y - (x^2 + y^2)y - 0.2(y^2 - x^2)
\end{aligned} \qquad \textbf{(11.3.12)*}$$

has a single limit cycle. Compare the phase portrait of this equation with that of (11.3.5).

7. Verify that the system of differential equations

$$\begin{aligned}
\dot{x} &= x - 2y + (x^2 + y^2)^2 x - 4x^3 \\
\dot{y} &= 2x + y + (x^2 + y^2)^2 y - 7(y^2 - x^2)y
\end{aligned} \qquad \textbf{(11.3.13)*}$$

has a single limit cycle in the square $-2.5 \le x, y \le 2.5$. How many equilibria does (11.3.13) have in this square?

11.4 STYLIZED PHASE PORTRAITS

In Section 11.2 we discussed phase portraits of systems with equilibria but without periodic solutions. In Section 11.3 we discussed briefly certain planar systems having limit cycles. In this section we combine the discussions on equilibria and periodic solutions; together these discussions allow us to give a strategy for determining numerically the phase portraits of a broad class of planar autonomous differential equations—the Morse-Smale differential equations.

Planar phase portraits indicate all equilibria, periodic solutions, and connecting trajectories. The simplest phase portraits are those that describe the dynamics of differential equations satisfying the following assumptions:

- All equilibria are hyperbolic.
- All periodic solutions are limit cycles.
- There are no trajectories connecting two saddle points.

A planar system of differential equations that satisfies these three conditions is called a *Morse-Smale* system.

Definition 11.4.1 *A stylized phase portrait of a planar autonomous system of differential equations is a picture illustrating all equilibria and their type, all limit cycles, and all connecting trajectories.*

Generally, it is not an easy task to draw the stylized phase portrait of a differential equation. However, using `pplane5`, we can attempt to draw the phase portrait of a Morse-Smale system on a given rectangle in the following way:

1. Using both analysis and numerics, find the equilibria of the differential equation.
2. Mark these equilibria and their type.
3. Draw the stable and unstable orbits of saddles.
4. Determine the number and stability of limit cycles.

Putting this information together allows us to draw planar phase portraits.

The pplane5 Default Equation

As an example, we discuss the stylized phase portrait of the `pplane5` default equation:

$$\dot{x} = 2x - y + 3(x^2 - y^2) + 2xy$$
$$\dot{y} = x - 3y - 3(x^2 - y^2) + 3xy. \qquad (11.4.1)$$

Our goal is to draw the stylized phase portrait of this system of differential equations on some rectangle in the plane, and we choose the default rectangle: $-2 \le x \le 4$, $-4 \le y \le 2$.

Inspection of (11.4.1) shows that the origin is an equilibrium. By clicking on the `Find an equilibrium point` button, we see that the origin is a saddle. After we click on the `Plot stable and unstable orbits` button and then click on the mouse when the cross hairs are near the origin, `pplane5` plots the stable and unstable orbits of this saddle. See Figure 11.13.

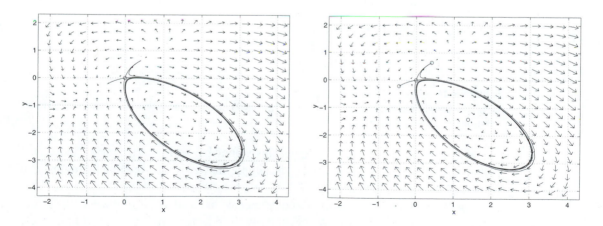

Figure 11.13
Left: Stable and unstable orbits of the saddle at the origin in (11.4.1). *Right*: Picture of (11.4.1) with equilibria added

This calculation reveals some interesting features in the phase portrait of (11.4.1). First, both stable orbits limit on the same equilibrium, and that equilibrium is near $(0.5, 0.5)$. Second, one unstable orbit limits on an equilibrium near $(-0.5, -0.2)$, while the other unstable orbit limits on what appears to be a limit cycle. Third, it follows from Theorem 11.3.3 that there must be an equilibrium inside this limit cycle—probably near $(1.3, -1.3)$. Using the information and the Find an equilibrium point button, we can locate three equilibria:

- A nodal sink at $(-0.4661, -0.2209)$
- A nodal source at $(0.4125, 0.6386)$
- A spiral source at $(1.387, -1.418)$

Entering these equilibria yields the phase portrait in Figure 11.13 (right).

We now draw the stylized phase portrait for (11.4.1), indicating the four equilibria and their type, the periodic solutions, and the connecting trajectories. See Figure 11.14 (right). A picture with additional trajectories is shown on the left of that figure.

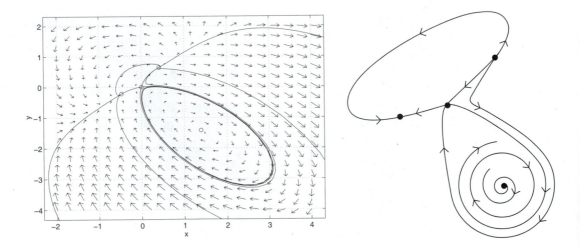

Figure 11.14
Left: Additional trajectories of (11.4.1). *Right*: Stylized phase portrait indicating equilibria, periodic solutions, and connecting trajectories.

It is reasonable to wonder whether this somewhat ad hoc process for finding the stylized phase portrait always works. The simple answer is *no*—but much progress can be made if one proceeds cautiously.

For example, one numerical difficulty appears in the analysis of the default system (11.4.1). Clicking inside the limit cycle shows a trajectory that spirals to a limit cycle in forward time, as expected. In backward time, however, the trajectory spirals toward the spiral source—but never makes it—and pplane5 indicates the possible existence of a second periodic solution. In fact, no such periodic solution exists. This point can be clarified using the zoom feature near the spiral source. These numerical calculations verify that the default system (11.4.1) is Morse-Smale.

COMPUTER EXERCISES

1. Use `pplane5` to determine the stylized phase portrait for the system

$$\dot{x} = 2x - y + 3\cos(x + y)$$
$$\dot{y} = x - 3y - 2xy \qquad\qquad (11.4.2)*$$

on the square $-10 \le x, y \le 10$. Indicate all equilibria (and their types) and all periodic solutions.

2. Use `pplane5` to determine the number of equilibria and their types for the system

$$\dot{x} = 2x - y + 3\cos(x + y)$$
$$\dot{y} = x - y - 2\sin(x - y) \qquad\qquad (11.4.3)*$$

in the rectangle $-2 \le x \le 4, -3 \le y \le 6$.

3. Use `pplane5` to determine the number of equilibria and their types, the number of limit cycles, and the connecting trajectories for the system

$$\dot{x} = 2 - y^2 - ((x^2 - 2)^2 + (y^2 - 2)^2 - 1)(x^2 - 2) - 0.2x$$
$$\dot{y} = x^2 - 2 - ((x^2 - 2)^2 + (y^2 - 2)^2 - 1)(y^2 - 2) \qquad\qquad (11.4.4)*$$

in the square $-3 \le x, y \le 3$.
   (a) Draw a qualitative phase portrait for this system.
   (b) Characterize all solutions to this differential equation that stay within the square $-3 \le x, y \le 3$ for all forward and backward time t.
   (c) Draw the time series in x versus t for the trajectory whose initial condition is $(0, 0)$.

4. Draw the stylized phase portrait of (11.3.13).
In Exercises 5 and 6, answer the following four questions for the given system of differential equations:
   (a) How many equilibria does the system of differential equations have? Any equilibrium that you find using `pplane5` may be considered an actual equilibrium, but use analysis to prove that you have found them all.
   (b) How many limit cycles does the system of differential equations have? Base your answer on numerical exploration using `pplane5`.
   (c) By hand (and not necessarily to scale) sketch the phase portrait.
   (d) Indicate which initial conditions have solutions that stay bounded in forward time.

5.

$$\dot{x} = 0.04x - y - 3y^2 + 2.5xy$$
$$\dot{y} = x - 3x^2 + 2x^2y \qquad\qquad (11.4.5)*$$

6.

$$\dot{x} = 0.05x - y - 3y^2 + 2xy$$
$$\dot{y} = x - 3x^2 + x^2y \qquad\qquad (11.4.6)*$$

CHAPTER

12

Bifurcation Theory

Many applications are modeled by autonomous systems of differential equations that contain parameters. As these parameters change, the stylized phase portraits of the differential equations may also change; parameter values where these changes occur are called *bifurcation* values. In this chapter we discuss how bifurcations occur. To frame the discussion we introduce two systems of differential equations: the Volterra-Lotka equations modeling the population evolution of two species (Section 12.1) and the CSTR equations modeling a single exothermic (or heat-producing) chemical reaction (Section 12.3). In these models we use scaling to identify the essential parameters, and we illustrate changes that can take place in phase portraits as a parameter is varied. The information concerning changes in phase portraits is summarized in *bifurcation diagrams*, which are introduced by simple examples in Section 12.2. Bifurcation diagrams are used in Section 12.3 to summarize the results of numerical explorations on the CSTR.

In Chapter 11 we showed that stylized phase portraits of planar Morse-Smale systems can often be drawn by a combination of analysis and computer. In this chapter we observe that bifurcations occur at parameter values where phase portraits of systems of differential equations are not Morse-Smale. There are three ways that an autonomous planar system of differential equations can fail to be Morse-Smale:

- There is a nonhyperbolic equilibrium.
- There is a periodic solution that is not a limit cycle.
- There is a trajectory that connects a saddle to itself or that connects two different saddles.

Typically, in a system of differential equations that depends on a single parameter ρ, there are isolated values in ρ where the differential equation fails to be Morse-Smale, and at these bifurcation values the phase portraits do actually change.

Section 12.2 discusses the typical bifurcations associated with nonhyperbolic equilibria: saddle-node bifurcations (where two equilibria are created) and Hopf bifurcations (where limit cycles are created). These bifurcations are called *local bifurcations* because the changes in the phase portraits occur in a small neighborhood of the nonhyperbolic equilibrium.

We also describe typical ways in which the remaining two failures in Morse-Smale occur. Typically, when periodic solutions are not limit cycles, two periodic solutions collide and disappear (Section 12.4). Other global bifurcations occur when there are saddle-saddle connections. Homoclinic bifurcations occur when there is a connection from a saddle point to itself; typically such bifurcations occur when a limit cycle disappears (Section 12.2). Heteroclinic bifurcations occur when a trajectory connects two different saddles (Section 12.4).

Additional details concerning saddle-node and Hopf bifurcations are given in Sections 12.5 and 12.6.

12.1   TWO-SPECIES POPULATION MODELS

Suppose that two species coexist in an isolated environment and that we want to understand how the populations of both species evolve in time. Denote the population of the first species at time t by $x(t) \geq 0$ and the population of the second species by $y(t) \geq 0$. We make one general assumption about models of population growth: If at any time t the population of a species is 0 (that is, the population is *extinct*), then the population of that species is 0 for all subsequent time. In short, extinct populations remain extinct.

If the evolution of the population of two species is modeled by a first-order system of autonomous equations, then the model has the form

$$\frac{dx}{dt} = xf_1(x, y) \tag{12.1.1}$$

$$\frac{dy}{dt} = yf_2(x, y), \tag{12.1.2}$$

where $x, y \geq 0$. The presence of the factor x in (12.1.1) and the factor y in (12.1.2) ensures that extinct populations remain extinct. The factors f_j are the *growth rates* of the species.

If $f_1(x, y) = \mu_1$ and $f_2(x, y) = \mu_2$ are constants, then the system (12.1.1), (12.1.2) is an uncoupled system of linear equations, and the two species evolve independently. In this model, the population of species j either grows exponentially ($\mu_j > 0$) or decays exponentially ($\mu_j < 0$) or remains constant ($\mu_j = 0$).

Predator–Prey Equations

Four assumptions go into the development of predator–prey models:

(a) In the absence of predators, the prey population grows.

(b) In the absence of prey, the predator population shrinks.

(c) In the presence of predators, the growth rate of the prey population decreases.

(d) In the presence of prey, the growth rate of the predator population increases.

Let x denote the size of the prey population and y denote the size of the predator population. The simplest model that incorporates these four assumptions is:

$$\dot{x} = x(\mu_1 + \sigma_1 y)$$
$$\dot{y} = y(\mu_2 + \sigma_2 x),$$

(12.1.3)*

where $\mu_1, \mu_2, \sigma_1, \sigma_2$ are constants. Assumptions (a)–(d) can be restated simply as

(a) $\mu_1 > 0$

(b) $\mu_2 < 0$

(c) $\sigma_1 < 0$

(d) $\sigma_2 > 0$

Under these assumptions, what can we expect the fate of predator–prey populations to be? Remarkably, most solutions to the differential equations (12.1.3) are periodic solutions. We test this statement using pplane5. Loading system e12_1_3.pps into pplane5, we find the predator–prey equations with default values

$$\mu_1 = 0.4, \quad \mu_2 = -0.3, \quad \sigma_1 = -0.01, \quad \sigma_2 = 0.005.$$

Press the Proceed button and plot several trajectories. You should find a picture like that in Figure 12.1. In this figure there is an equilibrium at $(x, y) = (60, 40)$ surrounded by periodic trajectories.

Note that if a solution to the predator–prey equation is time periodic, then the predator population and the prey population both vary periodically with the same period. The assumptions (a)–(d) that went into forming the predator–prey equations (12.1.3) are reflected in the time periodic solutions drawn in Figure 12.1. To see this correspondence, note that when the predator and the prey populations are near their minimum values, the prey population increases while the predator population remains nearly constant (assumption (a)). After a while, however, first the predator growth rate and then the predator population increase (assumption (d)), and the prey population begins to fall (assumption (c)). Then the prey population falls further and, as it does, the predator growth rate decreases (assumption (b)). Finally, the predator population falls until both populations are near their minimum values, and the whole process repeats itself exactly—at least in this model.

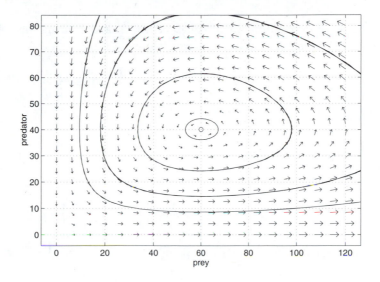

Figure 12.1
Phase portrait for predator–prey equations (12.1.3)

Scaling and Reduction in the Number of Parameters

From the discussion in Chapter 11 on Morse-Smale differential equations, it should seem surprising that a system of nonlinear differential equations produces an equilibrium surrounded by a continuous family of periodic solutions—just as the linear center does. You might think that the values of the constants $\mu_1, \mu_2, \sigma_1, \sigma_2$ were chosen just so this unlikely situation would occur. In fact, this is not the case, as a little numerical experimentation with different values of $\mu_1, \mu_2, \sigma_1, \sigma_2$ shows.

It is also surprising that a differential equation depending on four parameters has similar phase portraits for every one of these parameters (assuming the appropriate sign restrictions on the parameters). We continue our discussion of (12.1.3) by using scaling arguments to show that three of these parameters are not essential.

The idea behind scaling is simple. We think of $x(t)$ as measuring the number of some species at time t. But what is the measure? Are we counting the number of individuals or the number of thousands of individuals or even the number of millions of individuals? For example, we say that the population of the United States is 275 million people. Similarly, when speaking of population growth, we choose a time unit. Is population growth measured per year or per month or, in the case of bacteria, per minute? In general, we can change the units of population and time without changing the type of solutions. We make this change in scale by setting

$$X(t) = \alpha x(\gamma t) \quad \text{and} \quad Y(t) = \beta y(\gamma t),$$

where α, β, and γ are positive scaling constants.

Next we derive differential equations for X and Y using the chain rule and the differential equations (12.1.3). Observe that

$$\frac{dX}{dt}(t) = \alpha\gamma\frac{dx}{dt}(\gamma t) \quad \text{and} \quad \frac{dY}{dt}(t) = \beta\gamma\frac{dy}{dt}(\gamma t),$$

and from (12.1.3) that

$$\frac{dX}{dt}(t) = \alpha\gamma\frac{dx}{dt}(\gamma t)$$

$$= \alpha\gamma x(\gamma t)(\mu_1 + \sigma_1 y(\gamma t))$$

$$= \gamma X(t)\left(\mu_1 + \frac{\sigma_1}{\beta}Y(t)\right).$$

A similar equation holds for $\dot{Y}$. Dropping the explicit dependence on t, we have

$$\frac{dX}{dt} = X\left(\gamma\mu_1 + \frac{\gamma\sigma_1}{\beta}Y\right)$$

$$\frac{dY}{dt} = Y\left(\gamma\mu_2 + \frac{\gamma\sigma_2}{\alpha}X\right).$$

We now make a judicious choice of the three constants α, β, and γ. We choose

$$\alpha = \gamma\sigma_2, \quad \beta = -\gamma\sigma_1, \quad \gamma = -\frac{1}{\mu_2}. \tag{12.1.4}$$

These choices lead to the system of differential equations

$$\begin{aligned} \dot{X} &= X(\mu - Y) \\ \dot{Y} &= Y(-1 + X), \end{aligned} \tag{12.1.5*}$$

where $\mu = |\mu_1/\mu_2| > 0$. It is easy to check that the scaled predator–prey equations (12.1.5) have an equilibrium at $(X, Y) = (1, \mu)$ that is a center. It is also easy to check using pplane5 that for each μ the phase portrait of (12.1.5) appears to have periodic solutions surrounding this center.

The Volterra-Lotka Equations

Next we show that if the population model (12.1.3) is extended to a more complicated model, then the family of periodic solutions in the predator–prey equations disappears. To understand better the extension that we have in mind for the predator–prey equations, we discuss first the single species logistic equation.

The *logistic equation* is

$$\frac{dx}{dt} = x(\mu_1 + \rho_1 x).$$

In this model the growth rate of the population is assumed to depend on the size of the population. Typically, it is assumed that the growth rate decreases as the population increases; that is, it is assumed that $\rho_1 < 0$. It follows that the logistic model for two independent species is the uncoupled system

$$\dot{x} = x(\mu_1 + \rho_1 x)$$
$$\dot{y} = y(\mu_2 + \rho_2 y),$$

where $\mu_j > 0$, $\rho_j < 0$.

The *Volterra-Lotka* equations are the simplest extension of the uncoupled logistic system to interacting species and have the form

$$\dot{x} = x(\mu_1 + \rho_1 x + \sigma_1 y)$$
$$\dot{y} = y(\mu_2 + \sigma_2 x + \rho_2 y), \tag{12.1.6}$$

where σ_1 and σ_2 are nonzero constants. Note that when $\sigma_1 = \sigma_2 = 0$, system (12.1.6) reduces to the uncoupled logistic system.

These equations model three different types of situations. If $\sigma_1 > 0$, then the population growth rate of the first species increases as the population of the second species increases. If $\sigma_2 < 0$, then the population growth rate of the second species slows as the population of the first species increases. Suppose that the signs of σ_1 and σ_2 are different; that is, suppose $\sigma_1 < 0$ and $\sigma_2 > 0$. Then the second species acts like predators (the more of the second species there is, the lower the growth rate of the first species), and the first species acts like prey (the more of the first species there is, the higher the growth rate of the second species). In this case the Volterra-Lotka equations are a generalized predator–prey two-species model. Indeed, if $\rho_1 = \rho_2 = 0$ in (12.1.6), then we recover the predator–prey equations (12.1.3). Thus we may think of Volterra-Lotka equations as a small perturbation of the predator–prey equations when ρ_1 and ρ_2 are small. The two other cases in (12.1.6) correspond to *competing* species ($\sigma_1, \sigma_2 < 0$) and *cooperating* species ($\sigma_1, \sigma_2 > 0$).

Predator–Prey Volterra-Lotka Equations

What kind of population dynamics are predicted by equations (12.1.6)? The answer depends on the exact values of the six parameters μ_j, ρ_j, σ_j. As in our previous discussion of the predator–prey equations (12.1.3), we assume that $\mu_1 > 0$ and $\mu_2 < 0$. To simplify the notation, we perform the same scalings as we did when transforming (12.1.3) to (12.1.5), and obtain

$$\dot{x} = x(\mu + \rho x - y)$$
$$\dot{y} = y(-1 + x + \eta y),$$

where $\mu > 0$, ρ, η are constants.

We wish to compare the dynamics of these equations with those of the predator–prey model (12.1.3). To facilitate this comparison, we assume that $\eta = 0$ and study the system of equations

$$\dot{x} = x(\mu + \rho x - y)$$
$$\dot{y} = y(-1 + x).$$

(12.1.7)*

There is one equilibrium where neither species is extinct (that is, where $x > 0$ and $y > 0$). For population dynamics the equilibrium $(x, y) = (1, \mu + \rho)$ is a relevant solution to (12.1.7) only when $\mu + \rho > 0$, since we care about solutions only when $x \geq 0$, $y \geq 0$. To understand the dynamics near this equilibrium, we compute the Jacobian matrix

$$\begin{pmatrix} \mu + 2\rho x - y & -x \\ y & -1 + x \end{pmatrix} = \begin{pmatrix} \rho & -1 \\ \mu + \rho & 0 \end{pmatrix}.$$

The determinant of this matrix is $\mu + \rho > 0$ and the trace is ρ. Fix $\mu > 0$. It follows that for ρ near 0 and negative, this equilibrium is a spiral sink, and for ρ near 0 and positive, this equilibrium is a spiral source. The phase portraits for these two cases are shown in Figure 12.2. So we see that as long as $\rho \neq 0$, the continuous family of periodic solutions found in the simpler predator–prey equations (12.1.3) disappears.

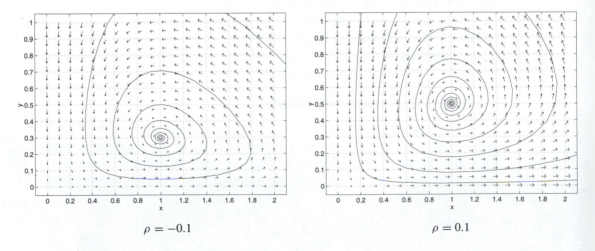

$\rho = -0.1$ $\rho = 0.1$

Figure 12.2
Phase portraits for Volterra-Lotka predator–prey equations (12.1.7) with $\mu = 0.4$

HAND EXERCISES

1. Consider the two predator–prey equations

$$\begin{aligned}\dot{x} &= x(2 - 10y)\\ \dot{y} &= y(-4 + y)\end{aligned}$$ (12.1.8)

and

$$\begin{aligned}\dot{x} &= x(2 - 0.1y)\\ \dot{y} &= y(-4 + y).\end{aligned}$$ (12.1.9)

In one of these equations the predators eat the prey 100 times more frequently than in the other. Which one is which?

2. Show that the predator–prey equations (12.1.8) and (12.1.9) both scale to the same equation (12.1.5). For each equation what are the values of the scaling coefficients α, β, and γ defined in (12.1.4)? What is the common value of μ?

3. Consider the following special case of the Volterra-Lotka equations:

$$\begin{aligned}\dot{x} &= x(\mu + 2x - y)\\ \dot{y} &= y(1 + x - 0.25y).\end{aligned}$$

For what value of μ is there an equilibrium whose linearization is a center?

4. Consider the Volterra-Lotka equations when $\mu_1, \mu_2 > 0$, $\rho_1, \rho_2 < 0$, and $\sigma_1 < 0, \sigma_2 > 0$. Scaling these equations leads to the system

$$\begin{aligned}\dot{x} &= x(\mu + \rho x - y)\\ \dot{y} &= y(1 + x + \eta y),\end{aligned}$$ (12.1.10)*

where $\mu > 0$ and $\rho, \eta < 0$.

(a) There are four possible equilibria depending on whether or not x and y vanish. Find these equilibria and their type (saddle, sink, source, or nonhyperbolic).

Hint Use the equations for the equilibrium where both coordinates are nonzero,

$$\mu + \rho x - y = 0 \quad \text{and} \quad 1 + x + \eta y = 0,$$

to show that the Jacobian matrix at that equilibrium is

$$(df) = \begin{pmatrix} \rho x & -x \\ y & \eta y \end{pmatrix}.$$

(b) Set $\rho = -1.5$ and $\eta = -1$ and discuss the phase portraits of these equations for different values of μ. Verify your answer using pplane5.

COMPUTER EXERCISES

5. Choose values for the constants in the predator–prey equations (12.1.3), and verify numerically that there is a unique equilibrium surrounded by periodic solutions in the first quadrant of the phase portrait for these equations.

6. Choose values for the constants in the Volterra-Lotka predator–prey equations (12.1.7), and verify numerically that there are no periodic solutions in the first quadrant for these equations.

7. Use `pplane5` to describe the (numerical) differences between solutions to the differential equations (12.1.8) and (12.1.9). In particular, compare the maximum number of prey in one equation with the maximum number of prey in the other equation when you start with the same number of predators and prey in each equation.

8. Foxes (the predators) and rabbits (the prey) coexist in an isolated area. In equilibrium, there are 200 foxes and 10,000 rabbits. When isolated from the foxes, the rabbit population doubles every year. When isolated from the rabbits, the fox population decreases at the rate of 10% per year. Supposing that the fox–rabbit populations are modeled by the predator–prey equations (12.1.3), answer the following questions:

(**a**) What are the values of $\mu_1, \mu_2, \sigma_1, \sigma_2$ in (12.1.3)?

(**b**) After a storm 1,200 rabbits and 110 foxes remain in the area. According to this model, what will the fox population be after three years?

   Hint   Use `Specify a computation interval` on the `PPLANE5 Keyboard input` menu.

(**c**) After the storm, what is the maximum number of rabbits that this model predicts will inhabit the area?

(**d**) How many years will it take the rabbit population to reach this maximum number for the first time?

12.2   EXAMPLES OF BIFURCATIONS

In Section 12.1 we illustrated how modeling leads to systems of differential equations that depend on external parameters. In this section we discuss some of the expected ways in which phase portraits of systems of differential equations change as a single parameter is varied. These ways include changes in the number of equilibria and in the number of limit cycles; they are called *bifurcations*. We use simple differential equations to illustrate three different kinds of bifurcation: *saddle-node* bifurcations where two equilibria collide and disappear, *Hopf* bifurcations where limit cycles are created, and *homoclinic* bifurcations where limit cycles disappear. We also show how to use *bifurcation diagrams* to summarize these changes.

Saddle-Node Bifurcations on the Line

Consider the differential equation

$$\dot{x} = \rho - x^2 \equiv f(x, \rho). \tag{12.2.1}$$

We discuss how the phase line to (12.2.1) changes as ρ increases through 0. A simple numerical experiment using `pline` shows that the phase lines for (12.2.1) when $\rho = -1, 0, 1$ are those given in Figure 12.3. From this figure we see that a pair of equilibria is created as ρ increases through 0.

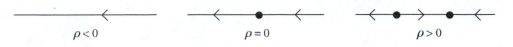

Figure 12.3
Phase lines for the differential equation (12.2.1)

These phase lines can be verified analytically by solving the algebraic equation $f(x, \rho) = 0$ and obtaining

$$x^2 = \rho. \tag{12.2.2}$$

When $\rho < 0$, there are no (real) solutions to (12.2.2); when $\rho = 0$ the only solution to (12.2.2) is $x = 0$; and when $\rho > 0$ there are two solutions to (12.2.2) given by $x = \pm\sqrt{\rho}$. It follows that (12.2.1) has no equilibria when $\rho < 0$, a single equilibrium when $\rho = 0$, and two equilibria when $\rho > 0$.

Additionally, we can determine the stability of the equilibria using Theorem 4.3.2. Observe that

$$f_x(x, \rho) = -2x.$$

Therefore $f_x < 0$ at $x = \sqrt{\rho}$ and $f_x > 0$ at $x = -\sqrt{\rho}$. Theorem 4.3.2 implies that the equilibrium at $x = -\sqrt{\rho}$ is unstable and that the equilibrium at $x = \sqrt{\rho}$ is stable.

We summarize the information about equilibria of (12.2.1) in the bifurcation diagram in Figure 12.4. This bifurcation diagram is formed as follows: Graph the points where $f(x, \rho) = 0$ in the ρx-plane; that is, graph the parabola $\rho = x^2$. On that graph we use a solid line when the equilibrium of (12.2.1) is stable and a dashed line when the equilibrium is unstable.

We have shown that at $\rho = 0$ the differential equation (12.2.1) undergoes a saddle-node bifurcation where a pair of equilibria is created. Moreover, one of this pair is stable and one is unstable. Next we consider a more complicated example:

$$\dot{x} = \rho - x^3 + 3x \equiv f(x, \rho). \tag{12.2.3}$$

We can solve for the equilibria of this differential equation by solving the algebraic equation $f(x, \rho) = 0$ for

$$\rho = x^3 - 3x. \tag{12.2.4}$$

Graphing (12.2.4) in the ρx-plane yields the bifurcation diagram in Figure 12.5. Note that two equilibria appear at $\rho = -2$ as ρ increases and that two equilibria disappear at $\rho = 2$. Thus there are two saddle-node bifurcations for (12.2.3): one at $\rho = -2$ and one at $\rho = 2$.

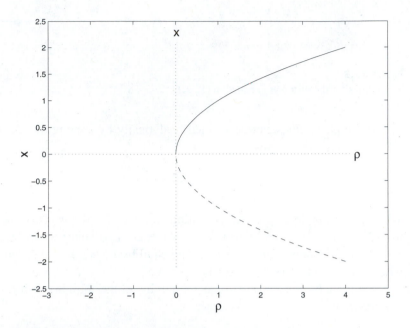

Figure 12.4
Bifurcation diagram for the differential equation (12.2.1)

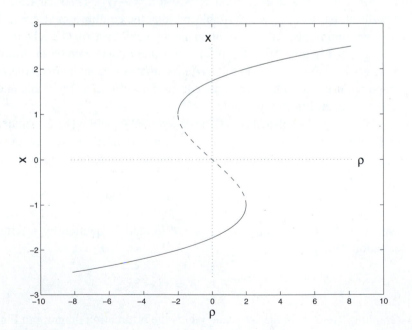

Figure 12.5
Bifurcation diagram for the differential equation (12.2.3)

In Figure 12.5 we have again plotted stable equilibria using solid lines and unstable equilibria using dashed lines. The stability can be checked analytically by computing

$$f_x(x, \rho) = -3x^2 + 3 = 3(1 - x^2).$$

It follows that $f_x < 0$ at equilibria where $|x| > 1$ and $f_x > 0$ at equilibria where $|x| < 1$. So the equilibria at x values between -1 and 1 are unstable and the equilibria at $x > 1$ and $x < -1$ are stable. The phase lines are those given in Figure 12.6. These calculations can be verified using pline.

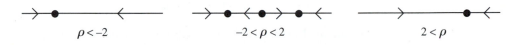

$\rho < -2$ $-2 < \rho < 2$ $2 < \rho$

Figure 12.6
Phase lines for the differential equation (12.2.3)

The Detection of One-Dimensional Saddle-Node Bifurcations

In differential equations of the form

$$\dot{x} = f(x, \rho)$$

we have seen that at points (x_0, ρ_0) where saddle-node bifurcations occur, two equilibria collide as the parameter ρ is varied, and one of these equilibria is stable while the other one is unstable. Therefore the following restrictions on f must hold:

$$f(x_0, \rho_0) = 0$$
$$f_x(x_0, \rho_0) = 0. \qquad (12.2.5)$$

The first condition just states that the point (x_0, ρ_0) is an equilibrium of the differential equation. The second condition follows from continuity; there are two equilibria nearby, one of which is asymptotically stable $f_x < 0$ and the other is unstable $f_x > 0$. In between, which means at the point (x_0, ρ_0), we must have $f_x = 0$.

For example, we can find points where saddle-node bifurcations might occur in the differential equation

$$\dot{x} = 2x^2 + \rho x + 2$$

by solving the two equations given in (12.2.5). Set

$$f(x, \rho) = 2x^2 + \rho x + 2$$

and differentiate to obtain

$$f_x(x, \rho) = 4x + \rho.$$

On setting $f_x = 0$, we find

$$\rho = -4x.$$

Substituting this result into the equation $f = 0$ yields

$$-2x^2 + 2 = 0,$$

which can be solved for $x = \pm 1$. There are two possible points where saddle-node bifurcations can occur:

$$(x_0, \rho_0) = (1, -4) \quad \text{and} \quad (x_0, \rho_0) = (-1, 4).$$

It can be checked using pline that saddle-node bifurcations do actually occur at both points.

Conditions (12.2.5) are necessary conditions for the existence of a saddle-node bifurcation; by themselves they are not sufficient. For example, consider

$$f(x, \rho) = x^2 - \rho^2.$$

Since $f(0, 0) = 0$ and $f_x(0, 0) = 0$, the origin satisfies (12.2.5). But the bifurcation is not a saddle-node bifurcation because there are two equilibria for every nonzero value of ρ.

Saddle-Node Bifurcations in the Plane

The transition from no equilibria to two equilibria can occur as a parameter is varied in differential equations with any number of variables. Here we consider one example of a planar system of differential equations where a saddle-node bifurcation occurs. Consider the system of differential equations

$$\begin{align}
\dot{x} &= x^2 - \rho + y \\
\dot{y} &= -y(x^2 + 1).
\end{align} \qquad \textbf{(12.2.6)*}$$

From the $\dot{y}$ equation we see that equilibria occur only when $y = 0$. Substituting $y = 0$ into the $\dot{x}$ equation yields the equation $x^2 = \rho$. Thus there are no equilibria for $\rho < 0$, a single equilibrium at the origin for $\rho = 0$, and two equilibria at

$$X_\pm = (\pm\sqrt{\rho}, 0)$$

for $\rho > 0$. The bifurcation diagram for the system of differential equations (12.2.6) is identical to the one for the single equation (12.2.1) given in Figure 12.3.

In Figure 12.7, the phase portraits of (12.2.6) are given for two values of ρ—namely, $\rho = -1$ and $\rho = 1$. Note that when $\rho = 0$, the origin is a nonhyperbolic equilibrium, and when $\rho = 1$, the equilibrium at $(-1, 0)$ is a nodal sink while the equilibrium at $(1, 0)$ is a saddle. Thus this saddle-node bifurcation creates a saddle and

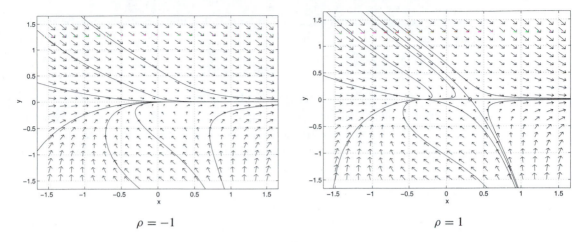

$$\rho = -1 \qquad\qquad\qquad \rho = 1$$

Figure 12.7
Phase planes for the differential equation (12.2.6)

a nodal sink. In all planar saddle-node bifurcations one of the equilibria is a saddle and the other is either a nodal source or a nodal sink (see Exercise 6 at the end of this section).

It is instructive to determine the type of equilibria in (12.2.6) by computing the Jacobian matrix. Let $F(x, y)$ denote the right-hand side of (12.2.6) and calculate

$$dF = \begin{pmatrix} 2x & 1 \\ -2xy & -x^2 - 1 \end{pmatrix} \quad \text{and} \quad (dF)_{X_\pm} = \begin{pmatrix} \pm 2\sqrt{\rho} & 1 \\ 0 & -\rho - 1 \end{pmatrix}. \qquad \textbf{(12.2.7)}$$

Observe that when $\rho = 0$, the linearized differential equation, at the origin, is the nonhyperbolic saddle-node

$$\dot{X} = \begin{pmatrix} 0 & 1 \\ 0 & -1 \end{pmatrix} X$$

whose eigenvalues are 0 and -1. See Section 7.3. We can also check using (12.2.7) that when $\rho > 0$, the linearization at X_- has two negative real eigenvalues ($-2\sqrt{\rho}$ and $-\rho - 1$) and is a node, while the linearization at X_+ has one positive ($2\sqrt{\rho}$) and one negative ($-\rho - 1$) eigenvalue and is a saddle.

Detection of Two-Dimensional Saddle-Node Bifurcations

Saddle-node bifurcations occur in planar systems at equilibria where a saddle and a node coalesce. Consider the planar system of differential equations

$$\dot{X} = F(X, \rho).$$

At a saddle-node bifurcation point (X_0, ρ_0), the following conditions must be satisfied:

$$F(X_0, \rho_0) = 0$$
$$\det(J) = 0, \tag{12.2.8}$$

where $J = (dF)_{(X_0, \rho_0)}$ satisfies $\operatorname{tr}(J) \neq 0$. The first equation just states that (X_0, ρ_0) is an equilibrium, and the second equation implies that the Jacobian has a zero eigenvalue. Since $\operatorname{tr}(J) \neq 0$, J has a nonzero eigenvalue and is a saddle-node matrix.

We can use (12.2.8) to find possible saddle-node bifurcation points in planar systems. For example, consider the system

$$\dot{x} = x^2 + y^2 - \rho$$
$$\dot{y} = x + y - 2.$$

The two parts of (12.2.8) lead to three equations:

$$x^2 + y^2 - \rho = 0$$
$$x + y - 2 = 0$$
$$\det \begin{pmatrix} 2x & 2y \\ 1 & 1 \end{pmatrix} = 0.$$

These equations can be solved for $(x_0, y_0, \rho_0) = (1, 1, 2)$. Indeed, for $\rho \approx 2$, pplane5 can be used to show that there are two equilibria near $(1, 1)$ (a saddle and a node) when $\rho > 2$ and no equilibria near $(1, 1)$ when $\rho < 2$.

Hopf Bifurcation in a Planar System

Another type of bifurcation that can occur in planar systems (though not in single autonomous equations) is the creation of a limit cycle from an equilibrium. This type of bifurcation is commonly called *Hopf bifurcation*. We begin our discussion with a numerical example.

Load the planar system of differential equations

$$\dot{x} = y$$
$$\dot{y} = -x + \rho y - y^3 \tag{12.2.9}*$$

into pplane5. It is straightforward to see that the origin is an equilibrium of (12.2.9) for all values of ρ; indeed, the origin is the only equilibrium of (12.2.9). Now plot the phase portraits for $\rho = -1$ and $\rho = 1$. These plots are shown in Figure 12.8.

The Jacobian matrix of (12.2.9) at the origin is

$$\begin{pmatrix} 0 & 1 \\ -1 & \rho \end{pmatrix}.$$

It follows that the origin goes from being a spiral sink to being a spiral source as ρ increases through 0. It also follows that the origin is a center at $\rho = 0$. The main

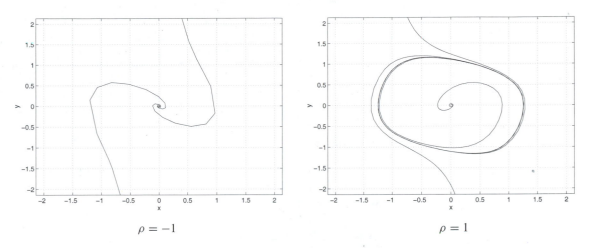

$$\rho = -1 \qquad\qquad\qquad \rho = 1$$

Figure 12.8
Phase planes for the differential equation (12.2.9)

new feature of the phase portraits is the existence of an asymptotically stable periodic
solution when $\rho > 0$. Hopf bifurcations can also produce unstable periodic solutions
(see (12.2.11) for an example).

Bifurcation Diagrams for Hopf Bifurcation

There are two salient points concerning Hopf bifurcation. The first point is that a spiral
sink equilibrium loses stability and becomes a spiral source as a parameter is varied (or
vice versa). The second point is that a limit cycle (either stable or unstable) is produced
by this change in stability. We draw schematic bifurcation diagrams in Figures 12.9
and 12.10 to indicate this change. On this diagram stable equilibria are indicated by a
solid line and unstable equilibria are indicated by a dashed line, as before. Stable limit
cycles are indicated by a heavy dotted line and unstable limit cycles are indicated by
a dot-dashed line. Note that the bifurcation diagram in Figure 12.9 describes the Hopf
bifurcation in the system of differential equations (12.2.9).

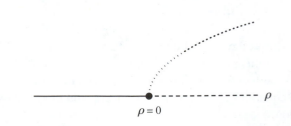

$$\rho = 0$$

Figure 12.9
Schematic bifurcation diagram for the
differential equation (12.2.9) with a branch
of stable limit cycles emanating from a point
of Hopf bifurcation indicated by a dotted line

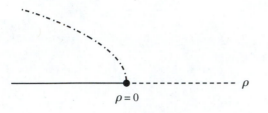

Figure 12.10

Schematic bifurcation diagram with a branch of unstable limit cycles emanating from a point of Hopf bifurcation indicated by a dot-dashed line

Detection of Hopf Bifurcations

A necessary condition for producing a limit cycle from an equilibrium by varying an external parameter is the existence of an equilibrium whose linearization is a center. Thus (X_0, ρ_0) is a possible point of Hopf bifurcation for the planar system

$$\dot{X} = F(X, \rho)$$

if the following conditions are satisfied:

$$\begin{aligned} F(X_0, \rho_0) &= 0 \\ \text{tr}(J) &= 0 \\ \det(J) &> 0, \end{aligned} \qquad \textbf{(12.2.10)}$$

where $J = (dF)_{(X_0, \rho_0)}$.

We can use (12.2.10) to find possible points of Hopf bifurcation. For example, consider the system

$$\begin{aligned} \dot{x} &= 2y - \rho + x^2 - 2y^2 \\ \dot{y} &= -x + y^2. \end{aligned} \qquad \textbf{(12.2.11)*}$$

A point of Hopf bifurcation in (12.2.11) must satisfy the three equations

$$\begin{aligned} 2y - \rho + x^2 - 2y^2 &= 0 \\ -x + y^2 &= 0 \\ 2(x + y) &= 0, \end{aligned}$$

where, in addition, $\det(J) = 4xy + 2 - 4y > 0$. It follows from the third equation, $\text{tr}(J) = 0$, that $y = -x$. On substitution for y in the first two equations, we find

$$\begin{aligned} -2x - \rho - x^2 &= 0 \\ -x + x^2 &= 0, \end{aligned}$$

where $\det(J) = -4x^2 + 4x + 2$. From the second equation it follows that $x = 0$ or $x = 1$. So there are two solutions to this system of equations: $(x_0, y_0, \rho_0) = (0, 0, 0)$ and $(x_1, y_1, \rho_1) = (1, -1, -3)$. At both of these points the determinant of the Jacobian matrix is 2. Hence the linearized equations at both point are centers, and there are two possible points of Hopf bifurcation.

Numerical exploration of (12.2.11) using `pplane5` shows that there are unstable limit cycles emanating from both of these points; hence there are exactly two points of Hopf bifurcation in this system of equations.

Conventions for Bifurcation Diagrams

In all bifurcation diagrams we indicate:

- stable equilibria by a solid line,
- unstable equilibria by a dashed line,
- stable limit cycles by a dotted line,
- unstable limit cycles by a dot-dashed line, and
- bifurcation points by black circles.

For example, the information contained in the Hopf bifurcation of (12.2.9) is summarized by the bifurcation diagram in Figure 12.9. In that figure, a curve of equilibria loses stability at $\rho = 0$ as ρ increases. A branch of limit cycles emanates from the branch of equilibria at the bifurcation point, which is indicated by a black dot. The stable limit cycles are indicated by the dotted line.

Homoclinic Bifurcations

There is a second way that planar systems of differential equations can have periodic solutions appear or disappear as parameters are varied. This change is called a *homoclinic bifurcation*. For example, consider the system of differential equations

$$\begin{aligned} \dot{x} &= \rho x - y \\ \dot{y} &= x + x^2 + xy. \end{aligned} \qquad \text{(12.2.12)*}$$

Begin by noting that the origin is an equilibrium of (12.2.12) for every value of ρ and the linearized equation about this equilibrium is

$$\dot{X} = \begin{pmatrix} \rho & -1 \\ 1 & 0 \end{pmatrix} X.$$

It follows that (12.2.12) has a Hopf bifurcation at the origin, since the origin changes from a spiral sink to a spiral source as ρ increases through 0. Calculations show that (12.2.12) has a second equilibrium at

$$X_0 = -\frac{1}{1 + \rho}(1, \rho)$$

and that this equilibrium is a saddle when $\rho > -1$.

Use `pplane5` to plot the phase plane portraits of (12.2.12) for $\rho = -0.07$, $\rho = 0.07$, and $\rho = 0.14$. (These phase portraits are most easily drawn by having `pplane5` find the saddle point and then plot the stable and unstable orbits.) The phase portraits are displayed in Figure 12.11.

From the numerical calculations we see that a standard Hopf bifurcation to a stable limit cycle occurs as ρ increases through 0—but that limit cycle disappears by the time $\rho = 0.14$. We discuss homoclinic bifurcations in more detail in Section 12.2. For the moment, just note that such transitions are possible. In bifurcation diagrams, a homoclinic bifurcation is indicated by having a branch of periodic solutions terminate with a black dot. In Figure 12.12 we show the bifurcation diagram for the system of differential equations (12.2.12) for ρ between -0.1 and 0.15. Note that there are branches for two equilibria and a branch of limit cycles beginning at a Hopf bifurcation and ending at a homoclinic bifurcation. In this particular bifurcation diagram we have

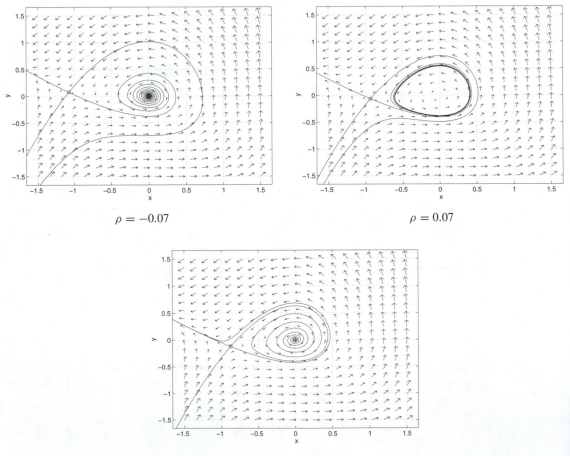

$$\rho = -0.07 \qquad\qquad \rho = 0.07$$

$$\rho = 0.14$$

Figure 12.11
Phase planes for the differential equation (12.2.12)

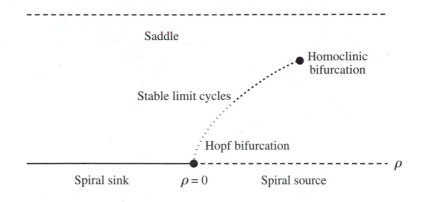

Figure 12.12
Schematic bifurcation diagram for the differential equation (12.2.12)
indicating a branch of stable limit cycles ending in a homoclinic bifurcation

labeled all pieces of information including the equilibria and their types, the bifurcations and their types, and the limit cycles.

Homoclinic bifurcations are global bifurcations; the change in the phase portrait does not occur in a small neighborhood of an equilibrium. For this reason we cannot detect homoclinic bifurcations analytically as we did for saddle-node bifurcations in (12.2.5) and (12.2.8) and Hopf bifurcations in (12.2.10).

Reading Bifurcation Diagrams

Planar systems of differential equations that model applications often lead to complicated bifurcation diagrams. We will see an example of this complexity in Section 12.3 when we study a model for a continuous flow stirred tank chemical reactor. Now we discuss the information contained in the hypothetical bifurcation pictured in Figure 12.13. In words we describe the six bifurcations contained in this hypothetical bifurcation diagram.

For $\rho < \rho_1$, there are two stable equilibria. There is a saddle-node bifurcation at ρ_1 where two additional equilibria—one stable and one unstable—are created. At ρ_2 there is a second saddle-node bifurcation where two of the four equilibria disappear.

As ρ increases past ρ_3, a homoclinic bifurcation produces an unstable limit cycle that disappears at a Hopf bifurcation at ρ_4. Note that the equilibrium becomes unstable at ρ_4. Hopf bifurcations also occur at ρ_5 and ρ_6. A stable limit cycle is created at ρ_5 and disappears at ρ_6.

As you can see, a bifurcation diagram is a shorthand method for recording an exceptional amount of information. In subsequent sections we develop some of the theory needed to understand the changes that are to be expected at each bifurcation.

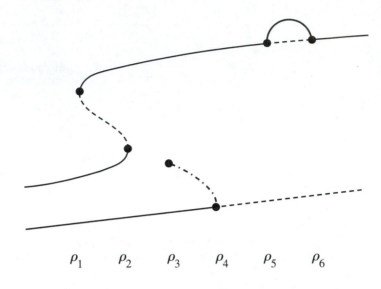

$$\rho_1 \qquad \rho_2 \qquad \rho_3 \qquad \rho_4 \qquad \rho_5 \qquad \rho_6$$

Figure 12.13
A hypothetical schematic bifurcation diagram for a planar system of
differential equations with six bifurcation points

HAND EXERCISES

1. Show that the differential equation

$$\dot{x} = \rho + x^2$$

undergoes a saddle-node bifurcation at $\rho = 0$, where two equilibria disappear as ρ increases
through 0. Draw the bifurcation diagram for this differential equation.

2. Show that the differential equation

$$\dot{x} = \rho - x^2 + 2x$$

undergoes a saddle-node bifurcation at $\rho = -1$, where two equilibria appear as ρ increases
through -1. Draw the bifurcation diagram for this differential equation.

3. Sketch the bifurcation diagram of the one-parameter family of differential equations

$$\dot{x} = 3x^4 - 8x^3 - 6x^2 + 24x - \rho. \tag{12.2.13}$$

What is the maximum number of equilibria for (12.2.13)? For which values of ρ does this
maximum number occur?
Hint Graph the function $\rho = 3x^4 - 8x^3 - 6x^2 + 24x$ by finding its maxima and minima.

4. The unforced Van der Pol equation is

$$\begin{aligned} \dot{x} &= \rho x - y - x^3 \\ \dot{y} &= x. \end{aligned} \tag{12.2.14}*$$

Verify analytically that the origin is the only equilibrium of this equation and that the only value of
ρ where that equilibrium could be a point of Hopf bifurcation is $\rho = 0$. Use pplane5 to verify that
there is a unique limit cycle when $\rho > 0$.

5. At what point (x_0, y_0) is it possible for the differential equation

$$\dot{x} = -4 + 3x - 4y + x^2 - 2y^2$$
$$\dot{y} = -21 - (1 + \rho)y + 7x^2 - 3xy \qquad \textbf{(12.2.15)*}$$

to have a Hopf bifurcation when $\rho = 0$?

COMPUTER EXERCISES

6. Change the sign of the $\dot{y}$ equation in (12.2.6), obtaining the system of differential equations

$$\dot{x} = x^2 - \rho + y$$
$$\dot{y} = y(x^2 + 1).$$

Show that a saddle-node bifurcation occurs at $\rho = 0$ in this system by plotting phase portraits for $\rho = -1$ and $\rho = 1$. How do the phase portraits of this system differ from those of (12.2.6) given in Figure 12.7?

7. Change the sign of the y^3 term in the $\dot{y}$ equation in (12.2.9) and obtain the system of differential equations

$$\dot{x} = y$$
$$\dot{y} = -x + \rho y + y^3.$$

Use pplane5 to plot the phase portraits of this system when $\rho = -1$ and $\rho = 1$. Describe the differences between these phase portraits and those in Figure 12.8. Draw the bifurcation diagram.

8. Consider the system of differential equations

$$\dot{x} = -1 + \rho + 5x + 2.5y + x^2 - x^3$$
$$\dot{y} = -5 - x + 4y - y^2. \qquad \textbf{(12.2.16)*}$$

(a) Verify that (12.2.16) has an equilibrium at $X_0 = (-1, 2)$ when $\rho = -1$.
(b) Show that the linearization of (12.2.16) at X_0 is a center when $\rho = -1$.
(c) Let ρ be a number close to -1. Use pplane5 to determine whether (12.2.16) has a spiral sink when $\rho < -1$ or when $\rho > -1$.
(d) Determine whether limit cycles exist for $\rho < -1$ or for $\rho > -1$.
(e) Find an approximate value of ρ at which (12.2.16) has a second Hopf bifurcation.

9. Consider the differential equation

$$\dot{x} = y$$
$$\dot{y} = -\rho + y + x^2 + xy. \qquad \textbf{(12.2.17)*}$$

(a) Show analytically that (12.2.17) has two equilibria at $(\pm\sqrt{\rho}, 0)$ when $\rho > 0$.
(b) Show analytically that the equilibrium at $(\sqrt{\rho}, 0)$ is always a saddle, while the equilibrium at $(-\sqrt{\rho}, 0)$ is a spiral sink when $1 < \rho < 97$.
(c) Use pplane5 to show that the phase portrait of (12.2.17) has a homoclinic trajectory between $\rho = 1.7$ and $\rho = 1.8$.

In Exercises 10–12, consider the system of differential equations:

$$\dot{x} = y - \mu x - x^2$$
$$\dot{y} = -x - \mu y + x^2. \qquad \textbf{(12.2.18)*}$$

10. Show that (12.2.18) undergoes a Hopf bifurcation near the origin when $\mu = 0$. Sketch the qualitative phase portraits on either side of the Hopf bifurcation.

11. Show that another bifurcation takes place in (12.2.18) near $\mu \approx -0.1$. Sketch the new phase portrait that appears near this bifurcation.

12. For each of the three different phase portraits examined in Exercises 10 and 11 identify the region of initial conditions in the square $-4 \leq x, y \leq 4$ where solutions stay within the square in forward time and the region where solutions stay within the square in both forward and backward time.

13. Use pplane5 to determine whether there are small amplitude periodic solutions to (12.2.15) at $\rho > 0$ or at $\rho < 0$.

12.3 THE CONTINUOUS FLOW STIRRED TANK REACTOR

Perhaps the simplest chemical engineering model of a chemical reaction is the CSTR, the continuous flow stirred tank chemical reactor. In the CSTR we imagine that a reactant flows into a vessel and undergoes a single heat-producing reaction to form inert products. The concentration of the reactant inside the vessel is denoted by c, and the temperature of the fluid inside the vessel is denoted by T. The assumption that the tank is well stirred is interpreted mathematically to mean that c and T are constant everywhere in the vessel. The CSTR model describes how c and T evolve in time.

The CSTR is pictured schematically in Figure 12.14. We assume that the reactant flows into a unit volume vessel at a constant rate r and that the product flows out of the vessel at the same rate. The fluid that flows into the vessel is called the *feed*. We assume that the temperature of the feed is held constant at T_f and that the reactant concentration of the feed is c_f. We also assume that the vessel is surrounded by a coolant whose temperature T_c is held constant.

The Reactionless CSTR

Ignoring the effects of the chemical reaction, the model equations describing the time evolution of the temperature and concentration of the reactant in the vessel are

$$\frac{dc}{dt} = r(c_f - c)$$

$$\frac{dT}{dt} = r(T_f - T) + k(T_c - T), \qquad \textbf{(12.3.1)}$$

where k is a lumped parameter that depends on a variety of physical quantities including heat transfer area and specific heat.

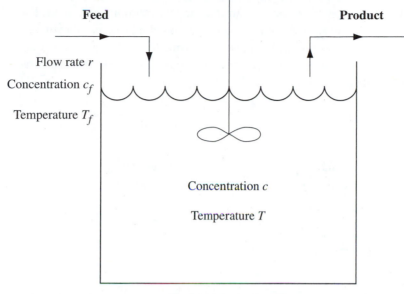

Feed   **Product**

Flow rate r

Concentration c_f

Temperature T_f

Concentration c

Temperature T

Coolant temperature T_c

Figure 12.14
Schematic diagram for CSTR

So far, the model in (12.3.1) is an inhomogeneous uncoupled system of linear differential equations. We have assumed that the concentration of the reactant grows exponentially to the feed concentration at rate r. The fate of the temperature inside the vessel is less clear, since we have assumed that the temperature grows exponentially to the feed temperature at rate r and simultaneously grows exponentially to the coolant temperature at rate k (that is just Newton's law of cooling). At this point, we have not modeled the effects of the chemical reaction on the concentration and temperature inside the vessel—though we have modeled how the external variables such as feed temperature and concentration and the coolant temperature affect the concentration and temperature inside the vessel. For this reason, this model is called the *reactionless* CSTR. Before continuing with the modeling, we nondimensionalize the variables, as engineers prefer to do.

In this model, we have assumed implicitly that concentration and temperature are positive quantities. For concentration this is clearly a reasonable assumption. For temperature this means that we are measuring temperature from absolute zero. Suppose that we let

$$x(t) = \frac{c(t/k) - c_f}{c_f} \quad \text{and} \quad y(t) = \frac{T(t/k) - T_f}{T_f}$$

be scaled concentration and temperature. We have normalized x and y to measure deviation from the feed concentration and temperature, and we scale c and T by c_f and T_f so that the state variables x and y are nondimensional; they have no physical units attached to them. In addition, we scale time by the lumped rate constant k. (Chemical engineers do not usually scale time in the equations, since k is a parameter that can be controlled in experiments.) Next, let

$$\eta = \frac{T_c - T_f}{T_f} \quad \text{and} \quad \rho = \frac{r}{k}$$

be the nondimensionalized coolant temperature and flow rate, respectively. Note that the fact that all physical quantities were assumed to be positive leads to the restrictions

$$x, y, \eta > -1 \quad \text{and} \quad \rho > 0.$$

Observe that

$$\frac{dx}{dt}(t) = \frac{1}{kc_f} \frac{dc}{dt}\left(\frac{t}{k}\right) \quad \text{and} \quad \frac{dy}{dt}(t) = \frac{1}{kT_f} \frac{dT}{dt}\left(\frac{t}{k}\right).$$

Coupling this observation with (12.3.1) yields the equations for the reactionless CSTR in nondimensionalized form:

$$\dot{x} = -\rho x \tag{12.3.2}$$
$$\dot{y} = -(\rho + 1)y + \eta. \tag{12.3.3}$$

The solution to (12.3.2) is

$$x(t) = e^{-\rho t} x_0$$

and, in the absence of a reaction, the scaled concentration decays exponentially to $x = 0$—that is, to the feed concentration.

The solution to (12.3.3) is slightly more complicated to obtain in closed form, as this is an example of an inhomogeneous linear equation. Using superposition, we can find the general solution to the inhomogeneous equation by finding one solution to the inhomogeneous equation and adding in all solutions to the homogeneous equation. In this case, it is easy to find a constant solution to the inhomogeneous equation. Just set

$$y = \frac{\eta}{\rho + 1}, \tag{12.3.4}$$

and check that this constant is a solution to (12.3.3). As we know,

$$y(t) = e^{-(\rho + 1)t} K$$

is the general solution to the homogeneous equation, and

$$y(t) = e^{-(\rho+1)t} K + \frac{\eta}{\rho + 1}$$

is the general solution to (12.3.3). Thus temperature decays exponentially to (12.3.4), the coolant temperature in nondimensionalized form scaled by the nondimensionalized flow rate.

The CSTR

Next we consider the effects of the reaction. For thermodynamic reasons, which we do not discuss here, we assume that the reaction depletes the reactant at a temperature-dependent rate proportional to concentration times the *Arrhenius* term $A(y)$, where

$$A(y) = \exp\left(\frac{\gamma y}{y + 1}\right)$$

and γ is the *activation energy*. Since the reaction is heat producing, we also assume that the temperature of the reactant increases at a rate proportional to $cA(y)$. Since the concentration c is proportional to $x + 1$, these assumptions lead to the nonlinear system of differential equations

$$\begin{aligned} \dot{x} &= -\rho x - Z(x + 1)A(y) \\ \dot{y} &= -(\rho + 1)y + \eta + hZ(x + 1)A(y), \end{aligned} \qquad \textbf{(12.3.5)}^*$$

where Z and h are proportionality constants. In this form the equations have two state variables x, y and five constants ρ, η, γ, Z, h. (The CSTR equation in e12_3_5.pps has $\gamma = 4$ so that there are just four parameters.)

Numerical Solution of the CSTR

It is a difficult problem to determine the phase portraits for the CSTR equations for all values of the five constants. Here we use pplane5 to indicate some of the different phase portraits that can occur in these equations. In order to simplify the task, we fix values for all parameters except the nondimensionalized flow rate ρ. We set

$$\gamma = 4, \quad \eta = -0.75, \quad Z = 0.5, \quad h = 3. \qquad \textbf{(12.3.6)}$$

Numerical experimentation on the rectangle $-0.75 \le x \le 0.5$, $-0.75 \le y \le 0.75$ shows the following: There are three possible equilibria: one at high temperature ($y \sim 0.1$), one at medium temperature ($y \sim -0.1$), and one at low temperature ($y \sim -0.5$). For these parameter values, the equilibrium at medium temperature is always a saddle and the equilibrium at low temperature is always a sink. We present the phase portraits for five values of the nondimensionalized flow rate ρ; the results are as follows:

- $\rho = 0.495$: The only equilibrium is the low-temperature equilibrium and it is a nodal sink.

- $\rho = 0.520$: There are three equilibria. The high-temperature spiral source and the medium-temperature saddle appear as ρ is increased. Note that the stable orbit of the saddle connects directly to the spiral source.

- $\rho = 0.545$: The high-temperature equilibrium is a spiral sink, and the stable orbit of the saddle connects to a limit cycle. At this parameter value, there are two stable equilibria.

- $\rho = 0.570$: There are still three equilibria, but the time-periodic solution has disappeared. The unstable orbit of the saddle now connects to the high-temperature spiral sink. The low-temperature sink is a spiral sink, and there are two stable spiral sinks.

- $\rho = 0.720$: The only remaining equilibrium is the high-temperature spiral sink. The saddle and the low-temperature sink have coalesced and disappeared. (Presumably the low-temperature spiral sink became a nodal sink before it coalesced with the saddle.)

These calculations show that the phase portrait of the CSTR changes between successive values of ρ; that is, we have shown that there are at least four bifurcation values in the CSTR as ρ is varied. The five different phase planes are shown in Figure 12.15.

Bifurcations in the CSTR

The observed bifurcations in the CSTR divide into local and global bifurcations, as we now discuss.

As we saw in Section 12.2, local bifurcations are of two types: steady state (saddle-node) and Hopf. For example, in Figure 12.15 we see that as the parameter ρ is decreased from 0.52 to 0.495, the middle- and high-temperature equilibria collide and disappear at a saddle-node bifurcation. (see Exercise 1 for further verification of this fact.) Similarly, between the values of $\rho = 0.570$ and $\rho = 0.720$, the low- and middle-temperature equilibria collide and disappear. We presume that a saddle-node bifurcation has also occurred in this parameter regime.

A different bifurcation occurs as ρ increases from 0.520 and 0.545. In this range, the high-temperature spiral changes from a source to a sink. For this change to occur, there must be a parameter value where the high-temperature equilibrium is a center. Since a time-periodic solution is found in the phase portrait of the CSTR at $\rho = 0.545$, we conclude that a Hopf bifurcation has occurred in this parameter region.

Both of these local bifurcations occur at parameter values where there is a nonhyperbolic equilibrium. As we noted, Hopf bifurcation occurs at a parameter value where a center is present, while steady-state bifurcation occurs at a parameter value where an equilibrium has a Jacobian matrix with a zero eigenvalue—typically a saddle-node.

A global bifurcation, one in which the bifurcation in phase portraits occurs away from equilibria, is present in the CSTR between $\rho = 0.545$ and $\rho = 0.570$. In this

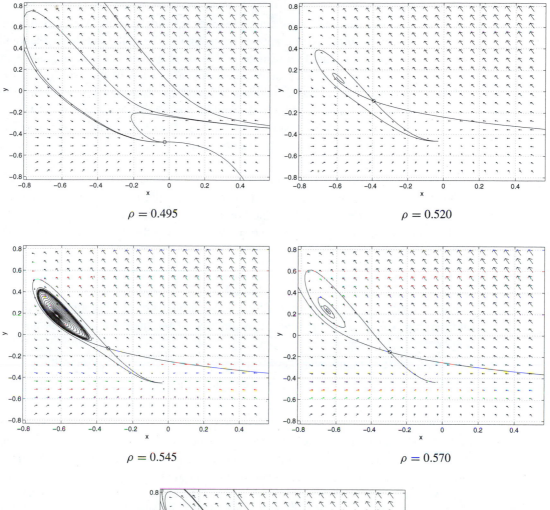

$\rho = 0.495$ $\rho = 0.520$

$\rho = 0.545$ $\rho = 0.570$

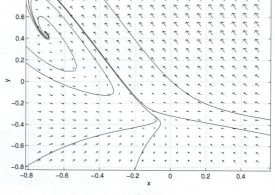

$\rho = 0.720$

Figure 12.15

Phase portraits of CSTR (12.3.5) with $\gamma = 4$, $\eta = -0.75$, $Z = 0.5$, $h = 3$

region, the periodic solution grows until it touches the saddle point. At that point, one branch of the stable orbit and one branch of the unstable orbit of the saddle are identical. Thus there is a trajectory that limits on the saddle point in both forward and backward time. As we saw in Section 12.2, such trajectories are called *homoclinic* trajectories and such bifurcations are called *homoclinic bifurcations*. Note the similarity between the phase portraits in Figure 12.11 and the middle phase portraits in Figure 12.15.

There are other global bifurcations that we have not seen in these equations. Two periodic solutions can collide and both disappear (just like the saddle-node steady-state bifurcation where two equilibria collide and disappear). This bifurcation is discussed in Section 12.4. Another global bifurcation occurs when the stable orbit of one saddle point coincides with the unstable orbit of another saddle point. This bifurcation is called a *saddle-saddle* connection. The homoclinic bifurcation is a special case of a saddle-saddle connection where the two saddle points are the same; see Section 12.2.

Bifurcation Diagrams

The discussion of transitions in the CSTR can be summarized by the use of a *bifurcation diagram*, as introduced in Section 12.2. In a bifurcation diagram we graph the equilibria *and* periodic solutions of equations as a function of the parameter ρ. In bifurcation diagrams:

- A saddle-node bifurcation appears as a turning point in a branch of equilibria.
- A Hopf bifurcation appears as a change in stability along a branch of equilibria.
- A homoclinic bifurcation is noted by a sudden ending of a branch of periodic solutions.

In the CSTR equations we have numerical evidence for two saddle-node bifurcations, a Hopf bifurcation, and a homoclinic bifurcation. This information is illustrated in the bifurcation diagram in Figure 12.16. Bifurcation diagrams contain an enormous amount of information, and it takes practice to learn to read them. For example, suppose ρ is fixed to be 0.545 in Figure 12.16. Then we may surmise that there are three equilibria of the CSTR equations at that value of ρ—the low- and high-temperature equilibria are asymptotically stable while the middle temperature one is not—and an unstable limit cycle (that surrounds the high-temperature equilibrium).

COMPUTER EXERCISES

1. Consider the CSTR equations (12.3.5) with parameters (12.3.6). Set $\rho = 0.5$ and use pplane5 to verify that there are two equilibria: a low-temperature nodal sink and a middle-temperature saddle-node equilibrium. Then change ρ to 0.505 and verify that the saddle-node has split into two equilibria: a saddle and a spiral source.

2. Use pplane5 to explore the CSTR equations (12.3.5) with parameters:

$$h = 2, \quad Z = 0.5, \quad \eta = -0.5, \quad \gamma = 5$$

for various values of the parameter ρ between 0.46 and 0.51. Describe all equilibria and all periodic solutions that you find, and how they change as ρ is varied.

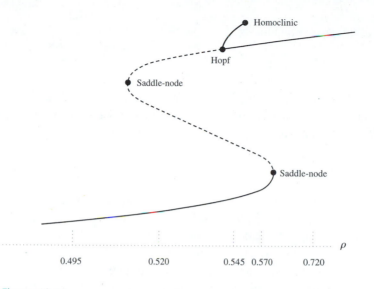

Figure 12.16
Schematic bifurcation diagram for the CSTR

12.4   THE REMAINING GLOBAL BIFURCATIONS

We have seen one kind of global bifurcation: the homoclinic bifurcation. There are two other global bifurcations that are likely to occur in planar systems of differential equations depending on one parameter: a saddle-node bifurcation of periodic solutions and a heteroclinic trajectory connecting two different saddle points.

Saddle-Node Bifurcations of Periodic Solutions

Typically, planar systems have a periodic solution that is not a limit cycle when two limit cycles collide and disappear—very much like the collision and destruction of two equilibria in a saddle-node bifurcation. We illustrate this kind of collision using phase-amplitude equations and pplane5. Indeed, the collision of periodic solutions corresponds to a saddle-node bifurcation of equilibria in the amplitude equation.

Let $X = (x, y)$ and consider the system of differential equations

$$
\begin{aligned}
\dot{x} &= a(x^2 + y^2, \rho)x - 3y \\
\dot{y} &= a(x^2 + y^2, \rho)y + 3x,
\end{aligned}
\qquad \textbf{(12.4.1)*}
$$

where $r^2 = x^2 + y^2$, and

$$
a(r^2, \rho) = \rho - (r^2 - 1)^2. \qquad \textbf{(12.4.2)}
$$

Recall from (11.3.9) and (11.3.10) that in polar coordinates (r, θ) the amplitude equation is

$$
\dot{r} = a(r^2, \rho)r
$$

and the phase equation is

$$\dot{\theta} = 3.$$

Recall that zeros of the amplitude equation $a(r^2, \rho) = 0$ correspond to periodic solutions of (12.4.1). When $0 < \rho < 1$, there are two zeros of the amplitude equation (12.4.2) that occur at

$$r^2 = 1 \pm \sqrt{\rho}.$$

When $\rho < 0$, there are no zeros of the amplitude equation. It follows that the amplitude equations undergo a saddle-node bifurcation at $(r, \rho) = (1, 0)$ and that two periodic solutions of (12.4.1) collide and disappear. This assertion can be checked using pplane5. See Figure 12.17. Note that when $\rho = 0$, the circle $r = 1$ is the trajectory of a periodic solution but that periodic solution is *not* a limit cycle. You can see this point by noting that trajectories starting at points with $r > 1$ asymptote in forward time to the periodic solution and that solutions starting at points with $r < 1$ move away from the unit circle in forward time.

In a sense that we will not make precise, the annihilation of two limit cycles in the plane through collision typically looks like the scenario shown in Figure 12.17.

Saddle-Saddle Connections: Heteroclinic Trajectories

In Section 12.2 we discussed the consequences of having a single trajectory connect a saddle point to itself. As we have seen, these homoclinic trajectories lead to the disappearance of a limit cycle.

When a single planar trajectory connects two different saddle points, the stable orbit of one saddle must become the unstable orbit of the other saddle. This connecting trajectory is called a *heteroclinic* trajectory.

We begin our discussion of heteroclinic connections with a simple example:

$$\begin{aligned} \dot{x} &= x^2 - x \\ \dot{y} &= y(0.5 - x) + \rho x. \end{aligned} \qquad \textbf{(12.4.3)}^*$$

When $\rho = 0$, the only equilibria of these equations are $(0, 0)$ and $(1, 0)$. Note that when $y = 0$, it follows that $\dot{y} = 0$. Hence the x-axis is an invariant line for the dynamics of this equation. Moreover, a short calculation shows that both of these equilibria are saddle points and that the x-axis is the stable orbit for $(0, 0)$ and the unstable orbit for $(1, 0)$. It follows that the line segment $(0, 1)$ on the x-axis is a single trajectory that connects the two equilibria. This is an example of a heteroclinic trajectory; see Figure 12.18. The time series for the heteroclinic trajectory is given in Figure 12.19. Note the similarity to the time series in Figure 11.2 of the one-dimensional equation (11.1.1) discussed in the introduction to Chapter 11. (In that example the trajectory

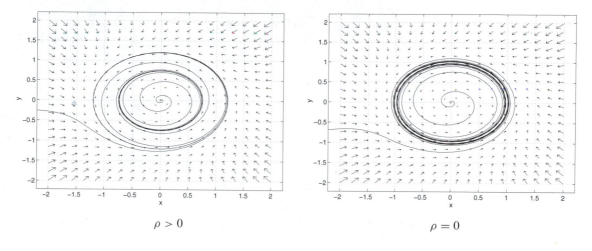

$\rho > 0$ $\rho = 0$

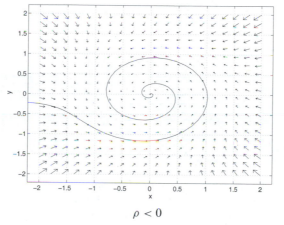

$\rho < 0$

Figure 12.17

Phase portraits for (12.4.1). Note that there are two limit cycles when $\rho > 0$, one periodic solution when $\rho = 0$, and no periodic solutions when $\rho < 0$.

connects 0 in negative time to 1 in positive time, whereas in the present example the heteroclinic trajectory connects 1 in negative time to 0 in positive time.)

When $\rho \neq 0$, the x-axis is no longer invariant for the differential equation. The two saddles remain, but there is no longer a trajectory connecting them. In Figure 12.18 we use pplane5 to plot the stable and unstable orbits when $\rho = \pm 0.1$. From this figure we can see how the stable orbit of the origin swings through the saddle at $(1, 0)$ as the parameter ρ is varied. This is typical of the way that a heteroclinic trajectory appears and disappears as a parameter is varied. The fleeting existence of a heteroclinic trajectory shows how the global fate of stable and unstable orbits emanating from a saddle point can change in a drastic fashion.

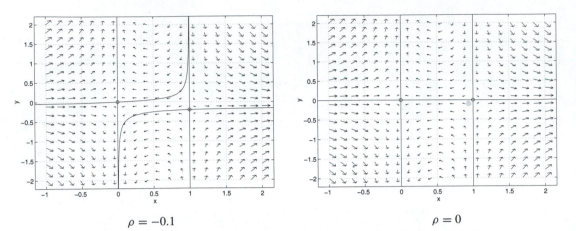

$\rho = -0.1$ $\rho = 0$

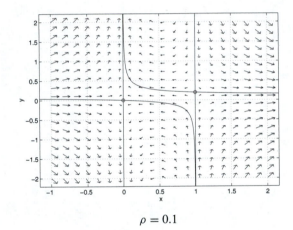

$\rho = 0.1$

Figure 12.18
Phase portraits for (12.4.3). Note the heteroclinic trajectory when $\rho = 0$.

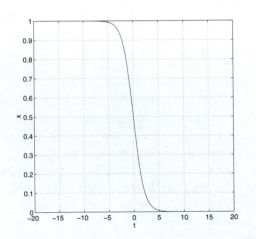

Figure 12.19
Time series for heteroclinic trajectory in
(12.4.3) when $\rho = 0$

A Classification of One-Parameter Bifurcations

The following is a complete list of expected transitions between planar Morse-Smale systems of different types in systems with one parameter:

- Two equilibria coalesce and disappear in a saddle-node bifurcation.
- A single equilibrium changes from a spiral source to a spiral sink (or vice versa) and spawns a limit cycle in a Hopf bifurcation.
- Two limit cycles coalesce in a saddle-node bifurcation of periodic solutions.
- Stable and unstable orbits from distinct saddle points coalesce and form a heteroclinic trajectory.
- Limit cycles grow until they touch a saddle and form a homoclinic trajectory.

COMPUTER EXERCISES

1. Use pplane5 to analyze the phase portraits for the system

$$\dot{x} = (-\rho + 3(0.5x^2 + y^2) - (x^2 + y^2)^2)x - 2y$$
$$\dot{y} = (-\rho + 3(x^2 + y^2) - (1.3x^2 + y^2)^2)y + x \qquad (12.4.4)^*$$

for $\rho = 0.7$ and $\rho = 1.1$ on the square $-2 \le x, y \le 2$. Is there a bifurcation between these two values of ρ? If so, what kind of bifurcation is it, and what is the bifurcation value of ρ to within ± 0.02?

In Exercises 2–4, consider the system of differential equations:

$$\dot{x} = -y + \rho x + 0.1x^2 - 0.5y^5$$
$$\dot{y} = x + \rho y + 0.1y^3 - x^4 \qquad (12.4.5)^*$$

in the square $-1.5 \le x, y \le 1.5$.

2. Verify numerically that (12.4.5) appears to have a limit cycle for $\rho = -0.02$.

3. Verify numerically that (12.4.5) does not appear to have a limit cycle when $\rho = -0.02$.

4. Determine numerically, to two significant figures, the value of ρ where (12.4.5) has a homoclinic bifurcation.

In Exercises 5 and 6, consider the system of differential equations:

$$\dot{x} = y$$
$$\dot{y} = -x + \rho y + \tfrac{1}{2}\cos(x)y. \qquad (12.4.6)^*$$

5. How many periodic solutions does (12.4.6) appear to have when $\rho = 0$? Be careful when answering this question. First use pplane5 to determine the number of limit cycles on the square $-\ell \le x, y \le \ell$ when $\ell = 2$. Then compute with $\ell = 10$ and $\ell = 40$.

6. Find the number of limit cycles to (12.4.6) on the square with $\ell = 7$ when $\rho = 0.06$ and when $\rho = 0.07$. Has a bifurcation occurred between these two values of ρ? If so, what kind of bifurcation?

In Exercises 7–9, consider the system of differential equations:

$$\dot{x} = y - x + (x - \rho)^2$$
$$\dot{y} = y - x^3 + 3x \qquad (12.4.7)^*$$

in the square $-5 \le x, y \le 5$.

7. How many equilibria does (12.4.7) have in this square when $\rho = 1.7$ and of what type are they? This question can be answered either numerically using pplane5 (easy) or analytically (more difficult, but possible).

8. What type of bifurcation occurs in (12.4.7) as ρ is varied from $\rho = 1.7$ to $\rho = 1.85$?

9. What type of bifurcation occurs in (12.4.7) as ρ is varied from $\rho = 1.85$ to $\rho = 2.1$?

10. Embed system 11.4.5 in the family of systems of differential equations

$$\begin{aligned} \dot{x} &= \rho x - y - 3y^2 + 2.5xy \\ \dot{y} &= x - 3x^2 + 2x^2 y. \end{aligned} \tag{12.4.8}$$

Find seven bifurcations that occur in (12.4.8) as ρ varies from -0.1 to 0.3.

In Exercises 11–13, consider the system of differential equations:

$$\begin{aligned} \dot{x} &= y \\ \dot{y} &= -x - 0.1y + x^3 + \rho x^2. \end{aligned} \tag{12.4.9}*$$

11. How many equilibria does (12.4.9) have for each value of ρ and of what type are they?

12. How many bifurcations occur in (12.4.9) as ρ is varied from $\rho = 0.1$ to $\rho = 0.2$?

13. What types of bifurcation occur in (12.4.9) as ρ is varied from $\rho = 0.1$ to $\rho = 0.2$?

12.5   *SADDLE-NODE BIFURCATIONS REVISITED

The simplest steady-state bifurcations—and the ones that are most likely to occur—are the saddle-node bifurcations. In Section 12.2 we found necessary conditions for the existence of saddle-node bifurcations in one dimension (12.2.5) and in two dimensions (12.2.8). In this section we describe necessary and sufficient conditions for the existence of saddle-node bifurcations.

Saddle-Node Bifurcations in One Dimension

Consider the differential equation

$$\dot{x} = f(x, \rho) \tag{12.5.1}$$

that depends on the single parameter ρ. The equilibria are found by solving the algebraic equation

$$f(x, \rho) = 0. \tag{12.5.2}$$

We can study how the equilibria in phase portraits for (12.5.1) change when ρ is varied by plotting solutions to (12.5.2) in the ρx-plane. The solutions to (12.5.2) form the bifurcation diagram of (12.5.1).

Recall from (12.2.5) that

$$f(x_0, \rho_0) = 0$$
$$f_x(x_0, \rho_0) = 0$$

(12.5.3)

are necessary conditions for a saddle-node bifurcation to occur at (x_0, ρ_0). The simplest example of a saddle-node bifurcation is

$$f(x, \rho) = ax^2 + \rho,$$

(12.5.4)

where a is a nonzero constant. The exact bifurcation diagram for (12.5.4) depends on the sign of a; see Figure 12.20. Note that the bifurcation diagram is just a parabola pointing either to the left or to the right. The general saddle-node bifurcation is one whose bifurcation diagram looks "parabolic-like." Theorem 12.5.1 makes this idea precise.

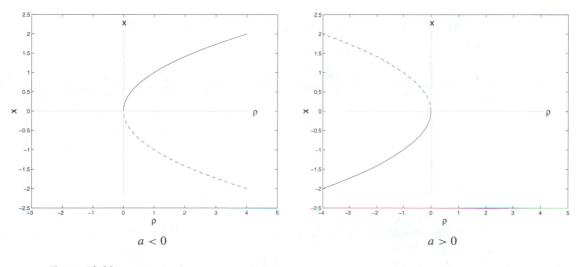

Figure 12.20
Bifurcation diagrams for (12.5.4)

Theorem 12.5.1   *Suppose that the differential equation (12.5.1) has an equilibrium at* (x_0, ρ_0) *satisfying the bifurcation conditions (12.5.3) and the nondegeneracy conditions*

$$f_\rho(x_0, \rho_0) \neq 0$$
$$f_{xx}(x_0, \rho_0) \neq 0.$$

(12.5.5)

Then the bifurcation diagram $f = 0$ *looks like the diagram in Figure 12.20, where*

$$a = \frac{f_{xx}(x_0, \rho_0)}{f_\rho(x_0, \rho_0)}.$$

Proof: The proof of this theorem uses the implicit function theorem and implicit differentiation. (Technically, we need to assume that the second derivatives of $f(x, \rho)$ are continuous before proceeding.) Since $f_\rho(x_0, \rho_0)$ is assumed to be nonzero, the implicit function theorem guarantees the existence of a function $R(x)$ such that $R(x_0) = \rho_0$ and

$$f(x, R(x)) \equiv 0 \tag{12.5.6}$$

for all x near x_0. It follows that the bifurcation diagram $f(x, \rho) = 0$ is the graph $\rho = R(x)$. We claim that $R'(x_0) = 0$ and $R''(x_0) \neq 0$. Thus the Taylor series of the function $R(x)$ is

$$R(x) = \rho_0 + \tfrac{1}{2} R''(x_0)(x - x_0)^2 + \cdots$$

for all x near x_0. It follows that the bifurcation diagram $f = 0$ is parabolic as in Figure 12.20—opening either to the left or to the right depending on the sign of $R''(x_0)$.

We use implicit differentiation to verify these claims. Differentiating (12.5.6) twice yields

$$f_x(x, R(x)) + f_\rho(x, R(x)) R'(x) \equiv 0 \tag{12.5.7}$$

and

$$f_{xx}(x, R(x)) + 2 f_{x\rho}(x, R(x)) R'(x) + f_{\rho\rho}(x, R(x)) R'(x)^2 + f_\rho(x, R(x)) R''(x) \equiv 0. \tag{12.5.8}$$

Evaluating (12.5.7) at $x = x_0$ yields

$$f_x(x_0, \rho_0) + f_\rho(x_0, \rho_0) R'(x_0) = 0.$$

Since, by assumption, $f_x(x_0, \rho_0) = 0$ and $f_\rho(x_0, \rho_0) \neq 0$, it follows that

$$R'(x_0) = 0.$$

Evaluating (12.5.8) at $x = x_0$ now yields

$$f_{xx}(x_0, \rho_0) + f_\rho(x_0, \rho_0) R''(x_0) = 0.$$

The nondegeneracy assumptions (12.5.5) imply that

$$R''(x_0) = -\frac{f_{xx}(x_0, \rho_0)}{f_\rho(x_0, \rho_0)} \neq 0.$$

Thus $a = -R''(x_0)$. So when $R''(x_0) > 0$, the number a is negative and the bifurcating branch is supercritical, as in Figure 12.20. Similarly, when $R''(x_0) < 0$, the number a is positive and the bifurcating branch is subcritical.   ◆

An Example of a One-Dimensional Saddle-Node Bifurcation

Consider the following example of a saddle-node bifurcation. Let

$$f(x, \rho) = -x^3 + 3x^2 - \rho x - 4. \tag{12.5.9}$$

A quick check shows that $x = 2$ is an equilibrium for (12.5.9) when $\rho = 0$. Moreover,

$$f(2, 0) = 0, \quad f_x(2, 0) = 0, \quad f_\rho(2, 0) = -2 \neq 0, \quad f_{xx}(2, 0) = -6 \neq 0.$$

It follows from Theorem 12.5.1 that f has a saddle-node bifurcation when $\rho = 0$ at $x = 2$. Indeed, since $a = \frac{-6}{-2} = 3 > 0$, it follows that there are two equilibria near $x = 2$ when $\rho < 0$ and none when $\rho > 0$. See Figure 12.20. This fact can be verified using pline. When ρ is small and positive, there is one stable equilibrium near $x = -1$. When ρ is small and negative, there are three equilibria: the stable one near $x = -1$ and two new equilibria near $x = 2$ formed from the saddle-node bifurcation.

Saddle-Node Bifurcations in Two Dimensions

In higher dimensions, steady-state bifurcations occur at parameter values where the Jacobian matrix has a zero eigenvalue. In one dimension, the 1×1 Jacobian matrix is the number f_x. Such a Jacobian matrix has a zero eigenvalue precisely when f_x vanishes—which is just the condition in (12.5.3).

In two dimensions, the Jacobian matrix J is a saddle-node matrix when it has a single zero eigenvalue—that is, when $\det(J) = 0$ and $\text{tr}(J) \neq 0$. Now consider the planar system of differential equations depending on a parameter ρ. We write this system as

$$\dot{X} = F(X, \rho), \tag{12.5.10}$$

where $X = (x, y) \in \mathbf{R}^2$ and $F = (g, h)$ in coordinates. Sufficient conditions for the existence of a saddle-node bifurcation in (12.5.10) are as follows:

1. Equation (12.5.10) has an equilibrium at (X_0, ρ_0), so

$$F(X_0, \rho_0) = 0.$$

2. The Jacobian matrix

$$J = (dF)_{(X_0, \rho_0)} = \begin{pmatrix} g_x(X_0, \rho_0) & g_y(X_0, \rho_0) \\ h_x(X_0, \rho_0) & h_y(X_0, \rho_0) \end{pmatrix}$$

has a zero eigenvalue and a nonzero eigenvalue. Let w be a nonzero vector in the null space of J^t. Assume

$$w \cdot F_\rho(X_0, \rho_0) \neq 0, \tag{12.5.11}$$

where $F_\rho = (g_\rho, h_\rho)$. (In one dimension, (12.5.11) just states that $f_\rho(X_0, \rho_0)$ is nonzero; that is, the first nondegeneracy condition in (12.5.5) is valid.)

3. Let $v = (v_1, v_2)$ be a nonzero vector in the null space of J. The following is an analog in two dimensions for the second nondegeneracy condition in (12.5.5), $f_{xx}(x_0, \rho_0) \neq 0$. Define F_0 to be the vector

$$F_0 = v_1^2 \begin{pmatrix} g_{xx} \\ h_{xx} \end{pmatrix} + 2v_1 v_2 \begin{pmatrix} g_{xy} \\ h_{xy} \end{pmatrix} + v_2^2 \begin{pmatrix} g_{yy} \\ h_{yy} \end{pmatrix}$$

evaluated at (X_0, ρ_0). The nondegeneracy condition in two dimensions is

$$w \cdot F_0 \neq 0. \tag{12.5.12}$$

A steady-state bifurcation is a *saddle-node bifurcation* if (12.5.11) and (12.5.12) are valid. At a saddle-node bifurcation two equilibria coalesce as ρ is varied, and the bifurcation diagram resembles that in Figure 12.20.

An Example of a Two-Dimensional Saddle-Node Bifurcation

As an example, consider the system of differential equations

$$\begin{aligned} \dot{x} &= \rho + x + 3y - xy &\equiv g(x, y) \\ \dot{y} &= -\rho + 2x + 6y + 3x^2 &\equiv h(x, y). \end{aligned} \tag{12.5.13}*$$

At $\rho = 0$, (12.5.13) has an equilibrium at the origin with Jacobian matrix

$$J = \begin{pmatrix} 1 & 3 \\ 2 & 6 \end{pmatrix}.$$

Since $\det(J) = 0$, the origin is a bifurcation point. Define the vectors v and w by

$$Jv = 0 \quad \text{and} \quad J^t w = 0,$$

so

$$v = \begin{pmatrix} 3 \\ -1 \end{pmatrix} \quad \text{and} \quad w = \begin{pmatrix} 2 \\ -1 \end{pmatrix}.$$

Next compute

$$F_\rho = \begin{pmatrix} 1 \\ -1 \end{pmatrix}.$$

Therefore

$$w \cdot F_\rho = \begin{pmatrix} 2 \\ -1 \end{pmatrix} \cdot \begin{pmatrix} 1 \\ -1 \end{pmatrix} = 3 \neq 0,$$

and (12.5.11) is satisfied.
    Next compute

$$g_{xx} = 0, \quad g_{xy} = -1, \quad g_{yy} = 0, \quad h_{xx} = 6, \quad h_{xy} = 0, \quad h_{yy} = 0.$$

Hence

$$F_0 = v_1^2 \begin{pmatrix} 0 \\ 6 \end{pmatrix} + 2v_1 v_2 \begin{pmatrix} -1 \\ 0 \end{pmatrix} = \begin{pmatrix} 6 \\ 54 \end{pmatrix}.$$

and

$$w \cdot F_0 = \begin{pmatrix} 2 \\ -1 \end{pmatrix} \cdot \begin{pmatrix} 6 \\ 54 \end{pmatrix} = -42 \neq 0.$$

So (12.5.12) is satisfied. These conditions show that there is a saddle-node bifurcation at the origin and that two equilibria coalesce as ρ varies through 0.

    Some experimentation will convince you that it is complicated to find a closed form analytic solution for the equilibria of (12.5.13). We can, however, verify the existence of a saddle-node bifurcation using pplane5. The phase plane portraits are shown in Figure 12.21.

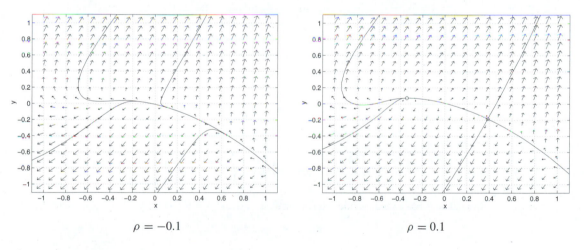

$$\rho = -0.1 \qquad\qquad\qquad \rho = 0.1$$

Figure 12.21
Phase plane portrait for (12.5.13) on the square $-1 \leq x, y \leq 1$. Note that there are no equilibria when $\rho = -0.1$ and two equilibria when $\rho = 0.1$.

HAND EXERCISES

1. Let

$$f_1(x, \rho) = \rho - x^3$$
$$f_2(x, \rho) = \rho x - x^2.$$

(a) Verify that both f_1 and f_2 satisfy the bifurcation conditions (12.5.3) at $x = 0$ when $\rho = 0$.
(b) Show that neither f_1 nor f_2 has a saddle-node bifurcation at $x = 0$ when $\rho = 0$.
(c) Draw the bifurcation diagrams $f_1(x, \rho) = 0$ and $f_2(x, \rho) = 0$ and explain why the conclusion of Theorem 12.5.1 is not satisfied for either f_1 or f_2.

2. Find all saddle-node bifurcations in the family of differential equations

$$\dot{x} = x^3 - 5x^2 + 3x + \rho \qquad \qquad \textbf{(12.5.14)}$$

analytically. Draw the bifurcation diagram.

3. Let

$$f(x, \rho) = x^4 - 6x^3 + 8x^2 + 6(x - \rho) - 3.$$

Show that f has a saddle-node bifurcation at $x = 3$ when $\rho = 1$. Are there two equilibria near $x = 3$ when $\rho < 1$ or when $\rho > 1$?

4. Let

$$g(x, y, \rho) = \rho - 2x + y + 2x^2 - y^3$$
$$h(x, y, \rho) = 2\rho + 4x - 2y - xy + y^2.$$

Show that $F = (g, h)$ has a saddle-node bifurcation at $(x, y) = 0$ and $\rho = 0$.

5. Let $X = (x, y)$ and

$$F(X, \rho) = \rho q + \begin{pmatrix} 1 & 0 \\ 0 & 0 \end{pmatrix} X + Q(X),$$

where

$$q = \begin{pmatrix} q_1 \\ q_2 \end{pmatrix} \quad \text{and} \quad Q(X) = \begin{pmatrix} \alpha x^2 + \beta xy + \gamma y^2 \\ \delta x^2 + \epsilon xy + \varphi y^2 \end{pmatrix}.$$

Show that F has a saddle-node bifurcation at $X = 0$ and $\rho = 0$ precisely when $q_2 \neq 0$ and $\varphi \neq 0$.

12.6   *HOPF BIFURCATIONS REVISITED

As discussed in Section 12.2, Hopf bifurcation occurs at an equilibrium when the Jacobian J has a pair of complex conjugate, purely imaginary eigenvalues—that is, when $\text{tr}(J) = 0$ and $\det(J) > 0$. In two dimensions, J is a center and the phase portrait of the linearized equation consists of a continuous family of periodic solutions. The Hopf bifurcation theorem states that under reasonable assumptions this family of periodic solutions persists in the nonlinear equations.

Since J is invertible, the (two-dimensional) implicit function theorem guarantees that there is a curve of equilibria with one equilibrium for each value ρ near the bifurcation value. Because of this, it makes sense to ask how the real parts of the eigenvalues of the Jacobian are changing as ρ varies. The basic assumption that we make about Hopf bifurcation is that the real parts of the eigenvalues of the Jacobian go through 0 with nonzero speed. We can rewrite this condition algebraically as follows: Let $X(\rho)$ denote the curve of equilibria and $J_{X(\rho)}$ the corresponding Jacobian matrix. Then the real part of the complex conjugate eigenvalues of the Jacobian matrix is

$$\frac{1}{2}\text{tr}(J_{X(\rho)}).$$

The condition that the real parts of the eigenvalues cross through 0 with nonzero speed is

$$\frac{d}{d\rho}\text{tr}(J_{X(\rho)}) \neq 0. \tag{12.6.1}$$

Condition (12.6.1) is called the *eigenvalue crossing condition*.

Two Examples

As a first example, consider the linear system

$$\begin{aligned}\dot{x} &= \rho x - y \\ \dot{y} &= x + \rho y.\end{aligned} \tag{12.6.2}*$$

It is easy to check that $\text{tr}(J) = 2\rho$ and that (12.6.1) is satisfied. Note that the origin is a spiral sink when $\rho < 0$ and a spiral source when $\rho > 0$. When $\rho = 0$, the origin is a center, and there is a continuous family of periodic solutions surrounding this center. See Figure 12.22.

As a second example, consider the predator–prey Volterra-Lotka equations (12.1.7):

$$\begin{aligned}\dot{x} &= x(\mu + \rho x - y) \\ \dot{y} &= y(-1 + x).\end{aligned} \tag{12.6.3}$$

We fix $\mu > 0$ and vary the parameter ρ. Our previous calculations showed that (12.6.3) has a center when $\rho = 0$. Indeed, these equations have a nontrivial equilibrium at $(x_0, y_0) = (1, \mu + \rho)$, so $X(\rho) = (1, \mu + \rho)$.

The Jacobian matrix of (12.6.3) is

$$J_{(x,y)} = \begin{pmatrix} \mu + 2\rho x - y & -x \\ y & -1 + x \end{pmatrix}.$$

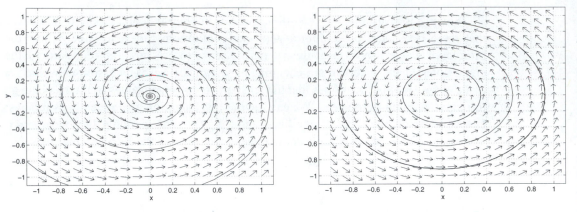

$$\rho = -0.1 \qquad\qquad \rho = 0$$

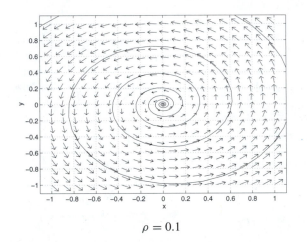

$$\rho = 0.1$$

Figure 12.22
Phase planes for (12.6.2)

Evaluating J along the equilibria $X(\rho)$ yields

$$J_{(1,\mu+\rho)} = \begin{pmatrix} \rho & -1 \\ \mu + \rho & 0 \end{pmatrix}.$$

It follows that

$$\mathrm{tr}(J_{X(\rho)}) = \rho,$$

and hence

$$\frac{d}{d\rho}\mathrm{tr}(J_{X(\rho)}) = 1 \neq 0.$$

So the eigenvalue crossing condition is satisfied for equation (12.6.3).

Theorem 12.6.1 (Simple Hopf bifurcation)   *Suppose that the system of differential equations (12.5.10) has a point of Hopf bifurcation at (X_0, ρ_0). Suppose, in addition, that the eigenvalue crossing condition (12.6.1) is satisfied. Then there exists a unique branch of periodic solutions to (12.5.10) emanating from (X_0, ρ_0).*

Moreover, every periodic solution to (12.5.10) that occurs for a parameter value ρ near ρ_0 and in phase space near X_0 is included in this branch of periodic solutions.

Recall that the predator–prey equations are obtained from (12.6.3) by setting $\rho = 0$. We have seen in our numerical analysis of the predator–prey equations that for all values of $\mu > 0$ there is a center surrounded by a continuous family of periodic solutions. (We have not proved this fact, but it is true.) Since the eigenvalue crossing condition is satisfied for the predator–prey Volterra-Lotka equations, it follows from the simple Hopf bifurcation theorem that there are no other periodic solutions near the bifurcation point. Indeed, the numerical computations shown in Figure 12.2 support this conclusion.

A Significant Effect of Nonlinearity

The most important point about the periodic solutions guaranteed by Theorem 12.6.1 is that they need not occur at the same parameter value. For example, in the CSTR there is an isolated limit cycle for each value of ρ on one side of the bifurcation value. See the numerical calculations of the CSTR in Figure 12.15 for $\rho = 0.545$, where an isolated limit cycle is observed, and for $\rho = 0.520$, where no periodic solution is seen.

Indeed, the CSTR equations behave more typically than the Volterra-Lotka equations. What we expect to happen at a Hopf bifurcation point is summarized by the bifurcation diagrams of Figures 12.23 and 12.24. In these figures we see a single branch of periodic solutions emanating from the Hopf bifurcation point either *supercritically* (for values of ρ greater than the one where Hopf bifurcation occurs) or *subcritically* (for values of ρ less than the critical Hopf bifurcation value).

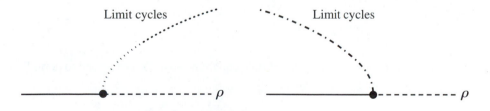

Figure 12.23
Typical bifurcation diagrams for Hopf bifurcation

Exchange of Stability

Assume that the equilibrium solution is stable at values below the bifurcation value and that the eigenvalue crossing condition (12.6.1) holds. Then typically the periodic

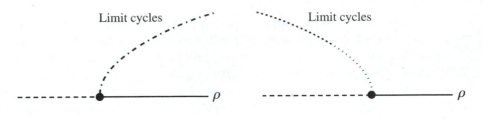

Figure 12.24
Typical bifurcation diagrams for Hopf bifurcation

solutions appear above the bifurcation value and are asymptotically stable, or they appear below the bifurcation value and are unstable. Which one of these scenarios actually occurs depends on the sign of a certain coefficient computed from the second- and third-order terms of the differential equation. This coefficient plays the same role in Hopf bifurcation that condition (12.5.12) plays in saddle-node bifurcations. The coefficient is much more complicated than (12.5.12), and we do not attempt to give its explicit form here.

If, on the other hand, the equilibrium solution is unstable at values below the bifurcation value, then the periodic solutions are unstable when supercritical and stable when subcritical. See Figure 12.24. In particular, the periodic solutions are stable only when they appear on the opposite side of criticality from the stable equilibrium. This phenomenon is called *exchange of stability*.

Hopf Bifurcation in Phase-Amplitude Equations

We can understand how typical Hopf bifurcations behave by looking at phase-amplitude equations of ODEs. The system is

$$\frac{d}{dt}\begin{pmatrix} x \\ y \end{pmatrix} = (\rho + a(x^2 + y^2))\begin{pmatrix} x \\ y \end{pmatrix} + \begin{pmatrix} -y \\ x \end{pmatrix}. \qquad \textbf{(12.6.4)*}$$

In amplitude r and phase θ, (12.6.4) becomes

$$\dot{r} = (\rho + ar^2)r$$
$$\dot{\theta} = 1.$$

The nontrivial zeros of the amplitude equation—which correspond to periodic solutions in the original system—are

$$\rho = -ar^2.$$

It follows that there is one periodic trajectory of (12.6.4) for $\rho > 0$ when $a < 0$ and for $\rho < 0$ when $a > 0$. Phase portraits for (12.6.4) when $a = -1$ are presented in Figure 12.25. Note the existence of a stable limit cycle when $\rho > 0$ and the slowness of the convergence of trajectories to the origin when $\rho = 0$.

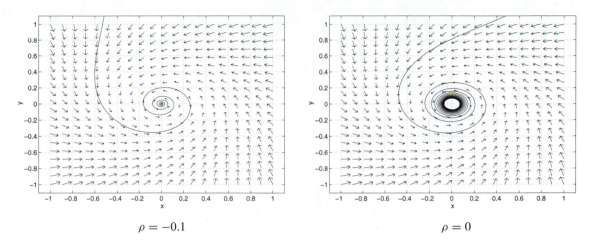

$$\rho = -0.1 \qquad\qquad \rho = 0$$

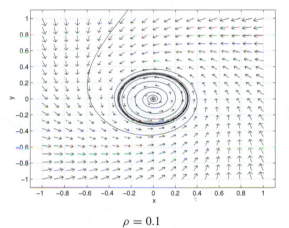

$$\rho = 0.1$$

Figure 12.25
Phase planes for (12.6.4) with $a = -1$

HAND EXERCISES

In Exercises 1 and 2, find points of Hopf bifurcation and check for these points whether the eigenvalue crossing condition (12.6.1) is satisfied.

1.
$$\dot{x} = \rho x - y - x^3$$
$$\dot{y} = x$$

2.
$$\dot{x} = y - \rho x - x^2$$
$$\dot{y} = -x - \rho y + x^2$$

Matrix Normal Forms

In this chapter we generalize to $n \times n$ matrices the theory of matrix normal forms presented in Chapter 6 for 2×2 matrices. In this theory we ask: What is the simplest form that a matrix can have up to *similarity*? After we present several preliminary results, the theory culminates in the Jordan normal form theorem, Theorem 13.4.2.

The first of the matrix normal form results—every matrix with n distinct real eigenvalues can be diagonalized—is presented in Section 13.1. The basic idea is that when a matrix has n distinct real eigenvalues, it has n linearly independent eigenvectors. In Section 13.2 we discuss matrix normal forms when the matrix has n distinct eigenvalues some of which are complex. When an $n \times n$ matrix has fewer than n linearly independent eigenvectors, it must have multiple eigenvalues and generalized eigenvectors. This topic is discussed in Section 13.3. The Jordan normal form theorem is introduced in Section 13.4 and describes similarity of matrices when the matrix has fewer than n independent eigenvectors. The proof is given in Appendix 13.6.

We introduced Markov matrices in Section 4.10. One of the theorems discussed there has a proof that relies on the Jordan normal form theorem, and we prove this theorem in Appendix 13.5.

13.1 REAL DIAGONALIZABLE MATRICES

An $n \times n$ matrix is *real diagonalizable* if it is similar to a diagonal matrix. More precisely, an $n \times n$ matrix A is real diagonalizable if there exists an invertible $n \times n$ matrix S such that

$$D = S^{-1}AS$$

is a diagonal matrix. In this section we investigate when a matrix is diagonalizable. In this discussion we assume that all matrices have real entries.

We begin with the observation that not all matrices are real diagonalizable. We saw in Example 8.2.2 that the diagonal entries of the diagonal matrix D are the eigenvalues of D. Theorem 8.2.8 states that similar matrices have the same eigenvalues. Thus, if a matrix is real diagonalizable, then it must have real eigenvalues. It follows, for example, that the 2×2 matrix

$$\begin{pmatrix} 0 & -1 \\ 1 & 0 \end{pmatrix}$$

is not real diagonalizable, since its eigenvalues are $\pm i$.

However, even if a matrix A has real eigenvalues, it need not be diagonalizable. For example, the only matrix similar to the identity matrix I_n is the identity matrix itself. To verify this point, calculate

$$S^{-1} I_n S = S^{-1} S = I_n.$$

Suppose that A is a matrix all of whose eigenvalues are equal to 1. If A is similar to a diagonal matrix D, then D must have all of its eigenvalues equal to 1. Since the identity matrix is the only diagonal matrix with all eigenvalues equal to 1, $D = I_n$. So, if A is similar to a diagonal matrix, it must itself be the identity matrix. Consider, however, the 2×2 matrix

$$A = \begin{pmatrix} 1 & 1 \\ 0 & 1 \end{pmatrix}.$$

Since A is triangular, it follows that both eigenvalues of A are equal to 1. Since A is not the identity matrix, it cannot be diagonalizable. More generally, if N is a nonzero strictly upper triangular $n \times n$ matrix, then the matrix $I_n + N$ is not diagonalizable.

These examples show that complex eigenvalues are always obstructions to real diagonalization and multiple real eigenvalues are sometimes obstructions to diagonalization. Indeed,

Theorem 13.1.1   *Let A be an $n \times n$ matrix with n distinct real eigenvalues. Then A is real diagonalizable.*

There are two ideas in the proof of Theorem 13.1.1, and they are summarized in the following lemmas.

Lemma 13.1.2   *Let $\lambda_1, \ldots, \lambda_k$ be distinct real eigenvalues for an $n \times n$ matrix A. Let v_j be eigenvectors associated with the eigenvalue λ_j. Then $\{v_1, \ldots, v_k\}$ is a linearly independent set.*

Proof: We prove the lemma by using induction on k. When $k = 1$, the proof is simple, since $v_1 \neq 0$. So we can assume that $\{v_1, \ldots, v_{k-1}\}$ is a linearly independent set.

Let $\alpha_1, \ldots, \alpha_k$ be scalars such that

$$\alpha_1 v_1 + \cdots + \alpha_k v_k = 0. \tag{13.1.1}$$

We must show that all $\alpha_j = 0$.

Begin by multiplying both sides of (13.1.1) by A to obtain

$$\begin{aligned}
0 &= A(\alpha_1 v_1 + \cdots + \alpha_k v_k) \\
&= \alpha_1 A v_1 + \cdots + \alpha_k A v_k \\
&= \alpha_1 \lambda_1 v_1 + \cdots + \alpha_k \lambda_k v_k.
\end{aligned} \tag{13.1.2}$$

Now subtract λ_k times (13.1.1) from (13.1.2) to obtain

$$\alpha_1(\lambda_1 - \lambda_k)v_1 + \cdots + \alpha_{k-1}(\lambda_{k-1} - \lambda_k)v_{k-1} = 0.$$

Since $\{v_1, \ldots, v_{k-1}\}$ is a linearly independent set, it follows that

$$\alpha_j(\lambda_j - \lambda_k) = 0$$

for $j = 1, \ldots, k - 1$. Since all of the eigenvalues are distinct, $\lambda_j - \lambda_k \neq 0$ and $\alpha_j = 0$ for $j = 1, \ldots, k - 1$. Substituting this information into (13.1.1) yields $\alpha_k v_k = 0$. Since $v_k \neq 0$, α_k is also equal to 0. $\blacklozenge$

Lemma 13.1.3 *Let A be an $n \times n$ matrix. Then A is real diagonalizable if and only if A has n real linearly independent eigenvectors.*

Proof: Suppose that A has n linearly independent eigenvectors $v_1, \ldots, v_n$. Let $\lambda_1, \ldots, \lambda_n$ be the corresponding eigenvalues of A; that is, $A v_j = \lambda_j v_j$. Let $S = (v_1 | \cdots | v_n)$ be the $n \times n$ matrix whose columns are the eigenvectors v_j. We claim that $D = S^{-1} A S$ is a diagonal matrix. Compute

$$D = S^{-1} A S = S^{-1} A(v_1 | \cdots | v_n) = S^{-1}(A v_1 | \cdots | A v_n) = S^{-1}(\lambda_1 v_1 | \cdots | \lambda_n v_n).$$

It follows that

$$D = (\lambda_1 S^{-1} v_1 | \cdots | \lambda_n S^{-1} v_n).$$

Note that

$$S^{-1} v_j = e_j,$$

since

$$S e_j = v_j.$$

Therefore

$$D = (\lambda_1 e_1 | \cdots | \lambda_n e_n)$$

is a diagonal matrix.

Conversely, suppose that A is a real diagonalizable matrix. Then there exists an invertible matrix S such that $D = S^{-1}AS$ is diagonal. Let $v_j = Se_j$. We claim that $\{v_1, \ldots, v_n\}$ is a linearly independent set of eigenvectors of A.

Since D is diagonal, $De_j = \lambda_j e_j$ for some real number λ_j. It follows that

$$Av_j = SDS^{-1}v_j = SDS^{-1}Se_j = SDe_j = \lambda_j Se_j = \lambda_j v_j.$$

So v_j is an eigenvector of A. Since the matrix S is invertible, its columns are linearly independent. Since the columns of S are v_j, the set $\{v_1, \ldots, v_n\}$ is a linearly independent set of eigenvectors of A, as claimed.   ◆

Proof of Theorem 13.1.1:   Let $\lambda_1, \ldots, \lambda_n$ be the distinct eigenvalues of A and let $v_1, \ldots, v_n$ be the corresponding eigenvectors. Lemma 13.1.2 implies that $\{v_1, \ldots, v_n\}$ is a linearly independent set in $\mathbf{R}^n$ and therefore a basis. Lemma 13.1.3 implies that A is diagonalizable.   ◆

Diagonalization Using MATLAB

Let

$$A = \begin{pmatrix} -6 & 12 & 4 \\ 8 & -21 & -8 \\ -29 & 72 & 27 \end{pmatrix}. \tag{13.1.3}*$$

We use MATLAB to answer the questions: Is A real diagonalizable and, if it is, can we find the matrix S such that $S^{-1}AS$ is diagonal? We can find the eigenvalues of A by typing eig(A). MATLAB's response is

```
ans =
   -2.0000
   -1.0000
    3.0000
```

Since the eigenvalues of A are real and distinct, Theorem 13.1.1 states that A can be diagonalized; that is, there is a matrix S such that

$$S^{-1}AS = \begin{pmatrix} -1 & 0 & 0 \\ 0 & -2 & 0 \\ 0 & 0 & 3 \end{pmatrix}.$$

The proof of Lemma 13.1.3 tells us how to find the matrix S. We need to find the eigenvectors v_1, v_2, v_3 associated with the eigenvalues $-1, -2, 3$, respectively. Then the matrix $(v_1|v_2|v_3)$ whose columns are the eigenvectors is the matrix S. To verify this construction we first find the eigenvectors of A by typing

```
v1 = null(A+eye(3));
v2 = null(A+2*eye(3));
v3 = null(A−3*eye(3));
```

Now type S = [v1 v2 v3] to obtain

```
S =
      0.8729      0.7071           0
      0.4364      0.0000      0.3162
     -0.2182      0.7071     -0.9487
```

Finally, check that $S^{-1}AS$ is the desired diagonal matrix by typing inv(S)*A*S to obtain

```
ans =
     -1.0000      0.0000           0
      0.0000     -2.0000     -0.0000
      0.0000           0      3.0000
```

It is cumbersome to use the null command to find eigenvectors, and MATLAB has been preprogrammed to do these computations automatically. We can use the eig command to find the eigenvectors and eigenvalues of a matrix A, as follows: Type

```
[S,D] = eig(A)
```

and MATLAB responds with

```
S =
     -0.7071      0.8729     -0.0000
     -0.0000      0.4364     -0.3162
     -0.7071     -0.2182      0.9487
D =
     -2.0000           0           0
           0     -1.0000           0
           0           0      3.0000
```

The matrix S is the transition matrix whose columns are the eigenvectors of A, and the matrix D is a diagonal matrix whose jth diagonal entry is the eigenvalue of A corresponding to the eigenvector in the jth column of S.

HAND EXERCISES

1. Let $A = \begin{pmatrix} 0 & 3 \\ 3 & 0 \end{pmatrix}$.

   (a) Find the eigenvalues and eigenvectors of A.

   (b) Find an invertible matrix S such that $S^{-1}AS$ is a diagonal matrix D. What is D?

2. The eigenvalues of

$$A = \begin{pmatrix} -1 & 2 & -1 \\ 3 & 0 & 1 \\ -3 & -2 & -3 \end{pmatrix}$$

are $2, -2, -4$. Find the eigenvectors of A for each of these eigenvalues, and find a 3×3 invertible matrix S so that $S^{-1}AS$ is diagonal.

3. Let

$$A = \begin{pmatrix} -1 & 4 & -2 \\ 0 & 3 & -2 \\ 0 & 4 & -3 \end{pmatrix}.$$

Find the eigenvalues and eigenvectors of A, and find an invertible matrix S so that $S^{-1}AS$ is diagonal.

4. Let A and B be similar $n \times n$ matrices.
 (a) Show that if A is invertible, then B is invertible.
 (b) Show that $A + A^{-1}$ is similar to $B + B^{-1}$.

5. Let A and B be $n \times n$ matrices. Suppose that A is real diagonalizable and that B is similar to A. Show that B is real diagonalizable.

6. Let A be an $n \times n$ real diagonalizable matrix. Show that $A + \alpha I_n$ is also real diagonalizable.

7. Let A be an $n \times n$ matrix with a real eigenvalue λ and associated eigenvector v. Assume that all other eigenvalues of A are different from λ. Let B be an $n \times n$ matrix that commutes with A; that is, $AB = BA$. Show that v is also an eigenvector for B.

8. Let A be an $n \times n$ matrix with distinct real eigenvalues, and let B be an $n \times n$ matrix that commutes with A. Using the result of Exercise 7, show that there is a matrix S that simultaneously diagonalizes A and B; that is, $S^{-1}AS$ and $S^{-1}BS$ are both diagonal matrices.

9. Let A be an $n \times n$ matrix all of whose eigenvalues equal ± 1. Show that if A is diagonalizable, then $A^2 = I_n$.

10. Let A be an $n \times n$ matrix all of whose eigenvalues equal 0 and 1. Show that if A is diagonalizable, then $A^2 = A$.

COMPUTER EXERCISES

11. Consider the 4×4 matrix

$$C = \begin{pmatrix} 12 & 48 & 68 & 88 \\ -19 & -54 & -57 & -68 \\ 22 & 52 & 66 & 96 \\ -11 & -26 & -41 & -64 \end{pmatrix}. \qquad \textbf{(13.1.4)*}$$

Use MATLAB to show that the eigenvalues of C are real and distinct. Find a matrix S so that $S^{-1}CS$ is diagonal.

In Exercises 12 and 13, use MATLAB to decide whether or not the given matrix is real diagonalizable.

12.

$$A = \begin{pmatrix} -2.2 & 4.1 & -1.5 & -0.2 \\ -3.4 & 4.8 & -1.0 & 0.2 \\ -1.0 & 0.4 & 1.9 & 0.2 \\ -14.5 & 17.8 & -6.7 & 0.6 \end{pmatrix} \qquad \textbf{(13.1.5)*}$$

13.

$$B = \begin{pmatrix} 1.9 & 2.2 & 1.5 & -1.6 & -2.8 \\ 0.8 & 2.6 & 1.5 & -1.8 & -2.0 \\ 2.6 & 2.8 & 1.6 & -2.1 & -3.8 \\ 4.8 & 3.6 & 1.5 & -3.1 & -5.2 \\ -2.1 & 1.2 & 1.7 & -0.2 & 0.0 \end{pmatrix} \qquad \textbf{(13.1.6)*}$$

13.2 SIMPLE COMPLEX EIGENVALUES

Theorem 13.1.1 states that a matrix A with real unequal eigenvalues may be diagonalized. It follows that in an appropriately chosen basis (the basis of eigenvectors), matrix multiplication by A acts as multiplication by these real eigenvalues. Moreover, geometrically, multiplication by A stretches or contracts vectors in eigendirections (depending on whether the eigenvalue is greater than or less than 1 in absolute value).

The purpose of this section is to show that a similar kind of diagonalization is possible when the matrix has distinct complex eigenvalues; see Theorems 13.2.1 and 13.2.2. We show that multiplication by a matrix with complex eigenvalues corresponds to multiplication by complex numbers. We also show that multiplication by complex eigenvalues corresponds geometrically to rotations as well as expansions and contractions.

The Geometry of Complex Eigenvalues: Rotations and Dilatations

Real 2×2 matrices are the smallest real matrices where complex eigenvalues can possibly occur. In Theorem 6.5.5(b) we discussed the classification of such matrices up to similarity. Recall that if the eigenvalues of a 2×2 matrix A are $\sigma \pm i\tau$, then A is similar to the matrix

$$\begin{pmatrix} \sigma & -\tau \\ \tau & \sigma \end{pmatrix}. \qquad \textbf{(13.2.1)}$$

Moreover, the basis in which A has the form (13.2.1) is found as follows. Let $v = w_1 + iw_2$ be the eigenvector of A corresponding to the eigenvalue $\sigma - i\tau$. Then $\{w_1, w_2\}$ is the desired basis.

Geometrically, multiplication of vectors in $\mathbf{R}^2$ by (13.2.1) is the same as a rotation followed by a dilatation. More specifically, let $r = \sqrt{\sigma^2 + \tau^2}$. So the point (σ, τ)

lies on the circle of radius r about the origin, and there is an angle θ such that $(\sigma, \tau) = (r \cos \theta, r \sin \theta)$. Now we can rewrite (13.2.1) as

$$\begin{pmatrix} \sigma & -\tau \\ \tau & \sigma \end{pmatrix} = r \begin{pmatrix} \cos \theta & -\sin \theta \\ \sin \theta & \cos \theta \end{pmatrix} = r R_\theta,$$

where R_θ is rotation counterclockwise through angle θ. From this discussion we see that geometrically complex eigenvalues are associated with rotations followed by either stretching $(r > 1)$ or contracting $(r < 1)$.

As an example, consider the matrix

$$A = \begin{pmatrix} 2 & 1 \\ -2 & 0 \end{pmatrix}. \tag{13.2.2}$$

The characteristic polynomial of A is $p_A(\lambda) = \lambda^2 - 2\lambda + 2$. Thus the eigenvalues of A are $1 \pm i$, and $\sigma = 1$ and $\tau = 1$ for this example. An eigenvector associated with the eigenvalue $1 - i$ is $v = (1, -1 - i)^t = (1, -1)^t + i(0, -1)^t$. Therefore

$$B = S^{-1} A S = \begin{pmatrix} 1 & -1 \\ 1 & 1 \end{pmatrix} \quad \text{where} \quad S = \begin{pmatrix} 1 & 0 \\ -1 & -1 \end{pmatrix},$$

as can be checked by direct calculation. Moreover, we can rewrite

$$B = \sqrt{2} \begin{pmatrix} \dfrac{\sqrt{2}}{2} & -\dfrac{\sqrt{2}}{2} \\ \dfrac{\sqrt{2}}{2} & \dfrac{\sqrt{2}}{2} \end{pmatrix} = \sqrt{2} R_{\pi/4}.$$

So, in an appropriately chosen coordinate system, multiplication by A rotates vectors counterclockwise by $45°$ and then expands the result by a factor of $\sqrt{2}$. (See Exercise 3 at the end of this section.)

The Algebra of Complex Eigenvalues: Complex Multiplication

We have shown that the normal form (13.2.1) can be interpreted geometrically as a rotation followed by a dilatation. There is a second algebraic interpretation of (13.2.1), and this interpretation is based on multiplication by complex numbers.

Let $\lambda = \sigma + i\tau$ be a complex number and consider the matrix associated with complex multiplication—that is, the linear mapping

$$z \mapsto \lambda z \tag{13.2.3}$$

on the complex plane. By identifying real and imaginary parts, we can rewrite (13.2.3) as a real 2×2 matrix in the following way: Let $z = x + iy$. Then

$$\lambda z = (\sigma + i\tau)(x + iy) = (\sigma x - \tau y) + i(\tau x + \sigma y).$$

Now identify z with the vector (x, y)—that is, the vector whose first component is the real part of z and whose second component is the imaginary part. With this identification, the complex number λz is identified with the vector $(\sigma x - \tau y, \tau x + \sigma y)$. So, in real coordinates and in matrix form, (13.2.3) becomes

$$\begin{pmatrix} x \\ y \end{pmatrix} \mapsto \begin{pmatrix} \sigma x - \tau y \\ \tau x + \sigma y \end{pmatrix} = \begin{pmatrix} \sigma & -\tau \\ \tau & \sigma \end{pmatrix} \begin{pmatrix} x \\ y \end{pmatrix}.$$

That is, the matrix corresponding to multiplication of $z = x + iy$ by the complex number $\lambda = \sigma + i\tau$ is the one that multiplies the vector $(x, y)^t$ by the normal form matrix (13.2.1).

Direct Agreement Between the Two Interpretations of (13.2.1)

We have shown that matrix multiplication by (13.2.1) may be thought of either algebraically as multiplication by a complex number (an eigenvalue) or geometrically as a rotation followed by a dilatation. We now show how to go directly from the algebraic interpretation to the geometric interpretation.

Euler's formula, (6.2.5), states that

$$e^{i\theta} = \cos\theta + i\sin\theta$$

for any real number θ. It follows that we can write a complex number $\lambda = \sigma + i\tau$ in polar form as

$$\lambda = re^{i\theta},$$

where $r^2 = \lambda\bar{\lambda} = \sigma^2 + \tau^2$, $\sigma = r\cos\theta$, and $\tau = r\sin\theta$.

Now consider multiplication by λ in polar form. Write $z = se^{i\varphi}$ in polar form and compute

$$\lambda z = re^{i\theta}se^{i\varphi} = rse^{i(\varphi+\theta)}.$$

It follows from polar form that multiplication of z by $\lambda = re^{i\theta}$ rotates z through an angle θ and dilates the result by the factor r. Thus Euler's formula directly relates the geometry of rotations and dilatations to the algebra of multiplication by a complex number.

Normal Form Matrices with Distinct Complex Eigenvalues

In the first parts of this section we have discussed a geometric and an algebraic approach to matrix multiplication by 2×2 matrices with complex eigenvalues. We now turn our attention to classifying $n \times n$ matrices that have distinct eigenvalues, whether these eigenvalues are real or complex. We will see that there are two ways to frame this classification: one algebraic (using complex numbers) and one geometric (using rotations and dilatations).

Algebraic Normal Forms: The Complex Case

Let A be an $n \times n$ matrix with real entries and n distinct eigenvalues $\lambda_1, \ldots, \lambda_n$. Let v_j be an eigenvector associated with the eigenvalue λ_j. By methods that are entirely analogous to those in Section 13.1, we can diagonalize the matrix A over the complex numbers. The resulting theorem is analogous to Theorem 13.1.1.

More precisely, the $n \times n$ matrix A is *complex diagonalizable* if there is a complex $n \times n$ matrix T such that

$$T^{-1}AT = \begin{pmatrix} \lambda_1 & 0 & \cdots & 0 \\ 0 & \lambda_2 & \cdots & 0 \\ \vdots & \vdots & \ddots & \vdots \\ 0 & 0 & \cdots & \lambda_n \end{pmatrix}.$$

Theorem 13.2.1 *Let A be an $n \times n$ matrix with n distinct eigenvalues. Then A is complex diagonalizable.*

The proof of Theorem 13.2.1 follows from a theoretical development virtually word for word the same as that used to prove Theorem 13.1.1. Beginning from the theory that we have developed so far, the difficulty in proving this theorem lies in the need to base the theory of linear algebra on complex scalars rather than real scalars. We will not pursue that development here.

As in Theorem 13.1.1, the proof of Theorem 13.2.1 shows that the complex matrix T is the matrix whose columns are the eigenvectors v_j of A; that is,

$$T = (v_1 | \cdots | v_n).$$

Finally, we mention that the computation of inverse matrices with complex entries is the same as that of matrices with real entries. That is, row reduction of the $n \times 2n$ matrix $(T|I_n)$ leads, when T is invertible, to the matrix $(I_n|T^{-1})$.

Two Examples

As a first example, consider the normal form 2×2 matrix (13.2.1) that has eigenvalues λ and $\bar{\lambda}$, where $\lambda = \sigma + i\tau$. Let

$$B = \begin{pmatrix} \sigma & -\tau \\ \tau & \sigma \end{pmatrix} \quad \text{and} \quad C = \begin{pmatrix} \lambda & 0 \\ 0 & \bar{\lambda} \end{pmatrix}.$$

Since the eigenvalues of B and C are identical, Theorem 13.2.1 implies that there is a 2×2 complex matrix T such that

$$C = T^{-1}BT. \tag{13.2.4}$$

Moreover, the columns of T are the complex eigenvectors v_1 and v_2 associated with the eigenvalues λ and $\bar{\lambda}$.

It can be checked that the eigenvectors of B are $v_1 = (1, -i)^t$ and $v_2 = (1, i)^t$. If we set

$$T = \begin{pmatrix} 1 & 1 \\ -i & i \end{pmatrix},$$

then it is a straightforward calculation to verify that $C = T^{-1}BT$.

As a second example, consider the matrix

$$A = \begin{pmatrix} 4 & 2 & 1 \\ 2 & -3 & 1 \\ 1 & -1 & -3 \end{pmatrix}. \tag{13.2.5}*$$

Using MATLAB, we find the eigenvalues of A by typing eig(A):

```
ans =
   4.6432
  -3.3216 + 0.9014i
  -3.3216 - 0.9014i
```

We can diagonalize (over the complex numbers) using MATLAB; indeed MATLAB is programmed to do these calculations over the complex numbers. Type [T,D] = eig(A) and obtain

```
T =
   0.9604     -0.1299 + 0.1587i   -0.1299 - 0.1587i
   0.2632      0.0147 - 0.5809i    0.0147 + 0.5809i
   0.0912      0.7788 - 0.1173i    0.7788 + 0.1173i
D =
   4.6432           0                    0
        0     -3.3216 + 0.9014i          0
        0           0            -3.3216 - 0.9014i
```

This calculation can be checked by typing inv(T)*A*T to see that the diagonal matrix D appears. One can also check that the columns of T are eigenvectors of A.

Note that the development here does not depend on the matrix A having real entries. Indeed, this diagonalization can be completed using $n \times n$ matrices with complex entries—and MATLAB can handle such calculations.

Geometric Normal Forms: Block Diagonalization

There is a second normal form theorem based on the geometry of rotations and dilatations for real $n \times n$ matrices A. In this normal form we determine all matrices A that have distinct eigenvalues—up to similarity by real $n \times n$ matrices S. The normal form results in matrices that are block diagonal with either 1×1 blocks or 2×2 blocks of the form (13.2.1) on the diagonal.

A real $n \times n$ matrix is in *real block diagonal form* if it is a block diagonal matrix

$$\begin{pmatrix} B_1 & 0 & \cdots & 0 \\ 0 & B_2 & \cdots & 0 \\ \vdots & \vdots & \ddots & \vdots \\ 0 & 0 & \cdots & B_m \end{pmatrix}, \tag{13.2.6}$$

where each B_j is either a 1×1 block

$$B_j = \lambda_j$$

for some real number λ_j or a 2×2 block

$$B_j = \begin{pmatrix} \sigma_j & -\tau_j \\ \tau_j & \sigma_j \end{pmatrix}, \tag{13.2.7}$$

where σ_j and $\tau_j \neq 0$ are real numbers. A matrix is *real block diagonalizable* if it is similar to a real block diagonal form matrix.

Note that the real eigenvalues of a real block diagonal form matrix are just the real numbers λ_j that occur in the 1×1 blocks. The complex eigenvalues are the eigenvalues of the 2×2 blocks B_j and are $\sigma_j \pm i\tau_j$.

Theorem 13.2.2   *Every $n \times n$ matrix A with n distinct eigenvalues is real block diagonalizable.*

We need two preliminary results.

Lemma 13.2.3   *Let $\lambda_1, \ldots, \lambda_q$ be distinct (possibly complex) eigenvalues of an $n \times n$ matrix A. Let v_j be a (possibly complex) eigenvector associated with the eigenvalue λ_j. Then $v_1, \ldots, v_q$ are linearly independent in the sense that if*

$$\alpha_1 v_1 + \cdots + \alpha_q v_q = 0 \tag{13.2.8}$$

for (possibly complex) scalars α_j, then $\alpha_j = 0$ for all j.

Proof: The proof is identical in spirit with the proof of Lemma 13.1.2. Proceed by induction on q. When $q = 1$, the lemma is trivially valid, as $\alpha v = 0$ for $v \neq 0$ implies that $\alpha = 0$, even when $\alpha \in \mathbf{C}$ and $v \in \mathbf{C}^n$.

By induction assume the lemma is valid for $q - 1$. Now apply A to (13.2.8), obtaining

$$\alpha_1 \lambda_1 v_1 + \cdots + \alpha_q \lambda_q v_q = 0.$$

Subtract this identity from λ_q times (13.2.8) and obtain

$$\alpha_1 (\lambda_1 - \lambda_q) v_1 + \cdots + \alpha_{q-1} (\lambda_{q-1} - \lambda_q) v_{q-1} = 0.$$

By induction

$$\alpha_j(\lambda_j - \lambda_q) = 0$$

for $j = 1, \ldots, q - 1$. Since the λ_j are distinct, it follows that $\alpha_j = 0$ for $j = 1, \ldots,$ $q - 1$. Hence (13.2.8) implies that $\alpha_q v_q = 0$; since $v_q \neq 0$, $\alpha_q = 0$.   ◆

Lemma 13.2.4   *Let* $\mu_1, \ldots, \mu_k$ *be distinct real eigenvalues of an* $n \times n$ *matrix A, and let* $v_1, \overline{v}_1 \ldots, v_\ell, \overline{v}_\ell$ *be distinct complex conjugate eigenvalues of A. Let* $v_j \in \mathbf{R}^n$ *be eigenvectors associated with* μ_j, *and let* $w_j = w_j^r + iw_j^i$ *be eigenvectors associated with the eigenvalues* v_j. *Then the* $k + 2\ell$ *vectors*

$$v_1, \ldots, v_k, w_1^r, w_1^i, \ldots, w_\ell^r, w_\ell^i$$

in $\mathbf{R}^n$ *are linearly independent.*

Proof: Let $w = w^r + iw^i$ be a vector in $\mathbf{C}^n$ and let β^r and β^i be real scalars. Then

$$\beta^r w^r + \beta^i w^i = \beta w + \overline{\beta}\overline{w}, \tag{13.2.9}$$

where $\beta = \frac{1}{2}(\beta^r - i\beta^i)$. Identity (13.2.9) is verified by direct calculation.

Suppose now that

$$\alpha_1 v_1 + \cdots + \alpha_k v_k + \beta_1^r w_1^r + \beta_1^i w_1^i + \cdots + \beta_\ell^r w_\ell^r + \beta_\ell^i w_\ell^i = 0 \tag{13.2.10}$$

for real scalars α_j, β_j^r, and β_j^i. Using (13.2.9), we can rewrite (13.2.10) as

$$\alpha_1 v_1 + \cdots + \alpha_k v_k + \beta_1 w_1 + \overline{\beta}_1 \overline{w}_1 + \cdots + \beta_\ell w_\ell + \overline{\beta}_\ell \overline{w}_\ell = 0,$$

where $\beta_j = \frac{1}{2}(\beta_j^r - i\beta_j^i)$. Since the eigenvalues

$$\mu_1, \ldots, \mu_k, v_1, \overline{v}_1 \ldots, v_\ell, \overline{v}_\ell$$

are all distinct, we may apply Lemma 13.2.3 to conclude that $\alpha_j = 0$ and $\beta_j = 0$. It follows that $\beta_j^r = 0$ and $\beta_j^i = 0$ as well, thus proving linear independence.   ◆

Proof of Theorem 13.2.2:   Let μ_j for $j = 1, \ldots, k$ be the real eigenvalues of A, and let $v_j, \overline{v}_j$ for $j = 1, \ldots, \ell$ be the complex eigenvalues of A. Since the eigenvalues are all distinct, it follows that $k + 2\ell = n$.

Let v_j and $w_j = w_j^r + iw_j^i$ be eigenvectors associated with the eigenvalues μ_j and $\overline{v}_j$. It follows from Lemma 13.2.4 that the n real vectors

$$v_1, \ldots, v_k, w_1^r, w_1^i, \ldots, w_\ell^r, w_\ell^i \tag{13.2.11}$$

are linearly independent and hence form a basis for $\mathbf{R}^n$.

We now show that A is real block diagonalizable. Let S be the $n \times n$ matrix whose columns are the vectors in (13.2.11). Since these vectors are linearly independent, S is invertible. We claim that $S^{-1}AS$ is real block diagonal. This statement is verified by direct calculation.

First, note that $Se_j = v_j$ for $j = 1, \ldots, k$ and compute

$$(S^{-1}AS)e_j = S^{-1}Av_j = \mu_j S^{-1}v_j = \mu_j e_j.$$

It follows that the first k columns of $S^{-1}AS$ are 0 except for the diagonal entries, and those diagonal entries equal $\mu_1, \ldots, \mu_k$.

Second, note that $Se_{k+1} = w_1^r$ and $Se_{k+2} = w_1^i$. Write the complex eigenvalues as

$$v_j = \sigma_j + i\tau_j.$$

Since $Aw_1 = \bar{v}_1 w_1$, it follows that

$$\begin{aligned} Aw_1^r + iAw_1^i &= (\sigma_1 - i\tau_1)(w_1^r + iw_1^i) \\ &= (\sigma_1 w_1^r + \tau_1 w_1^i) + i(-\tau_1 w_1^r + \sigma_1 w_1^i). \end{aligned}$$

Equating real and imaginary parts leads to

$$\begin{aligned} Aw_1^r &= \sigma_1 w_1^r + \tau_1 w_1^i \\ Aw_1^i &= -\tau_1 w_1^r + \sigma_1 w_1^i. \end{aligned} \tag{13.2.12}$$

Using (13.2.12), compute

$$(S^{-1}AS)e_{k+1} = S^{-1}Aw_1^r = S^{-1}(\sigma_1 w_1^r + \tau_1 w_1^i) = \sigma_1 e_{k+1} + \tau_1 e_{k+2}.$$

Similarly,

$$(S^{-1}AS)e_{k+2} = S^{-1}Aw_1^i = S^{-1}(-\tau_1 w_1^r + \sigma_1 w_1^i) = -\tau_1 e_{k+1} + \sigma_1 e_{k+2}.$$

Thus the kth and $(k+1)$st columns of $S^{-1}AS$ have the desired diagonal block in the kth and $(k+1)$st rows, and they have all other entries equal to 0.

The same calculation is valid for the complex eigenvalues $v_2, \ldots, v_\ell$. Thus $S^{-1}AS$ is real block diagonal, as claimed. ◆

MATLAB Calculations of Real Block Diagonal Form

Let C be the 4×4 matrix

$$C = \begin{pmatrix} 1 & 0 & 2 & 3 \\ 2 & 1 & 4 & 6 \\ -1 & -5 & 1 & 3 \\ 1 & 4 & 7 & 10 \end{pmatrix}. \tag{13.2.13}*$$

Using MATLAB, enter C by typing e13_2_14 and find the eigenvalues of C by typing eig(C) to obtain

```
ans =
    0.5855 + 0.8861i
    0.5855 - 0.8861i
   -0.6399
   12.4690
```

We see that C has two real and two complex conjugate eigenvalues. To find the complex eigenvectors associated with these eigenvalues, type

```
    [T,D] = eig(C)
```

MATLAB responds with

```
T =
   -0.0787 + 0.0899i   -0.0787 - 0.0899i    0.0464            0.2209
    0.0772 + 0.2476i    0.0772 - 0.2476i    0.0362            0.4803
   -0.5558 - 0.5945i   -0.5558 + 0.5945i   -0.8421           -0.0066
    0.3549 + 0.3607i    0.3549 - 0.3607i    0.5361            0.8488
D =
    0.5855 + 0.8861i         0                   0                 0
         0             0.5855 - 0.8861i          0                 0
         0                   0             -0.6399                 0
         0                   0                   0           12.4690
```

The 4×4 matrix T has the eigenvectors of C as columns. The jth column is the eigenvector associated with the jth diagonal entry in the diagonal matrix D.

To find the matrix S that puts C in real block diagonal form, we need to take the real and imaginary parts of the eigenvectors corresponding to the complex eigenvalues and the real eigenvectors corresponding to the real eigenvalues. In this case, type

```
    S = [real(T(:,1)) imag(T(:,1)) T(:,3) T(:,4)]
```

to obtain

```
S =
   -0.0787    0.0899    0.0464    0.2209
    0.0772    0.2476    0.0362    0.4803
   -0.5558   -0.5945   -0.8421   -0.0066
    0.3549    0.3607    0.5361    0.8488
```

Note that the 1st and 2nd columns are the real and imaginary parts of the complex eigenvector. Check that inv(S)*C*S is the matrix in complex diagonal form

```
ans =
    0.5855    0.8861    0.0000    0.0000
   -0.8861    0.5855    0.0000   -0.0000
    0.0000    0.0000   -0.6399    0.0000
   -0.0000   -0.0000   -0.0000   12.4690
```

as proved in Theorem 13.2.2.

HAND EXERCISES

1. Consider the 2×2 matrix

$$A = \begin{pmatrix} 3 & 1 \\ -2 & 1 \end{pmatrix},$$

whose eigenvalues are $2 \pm i$ and whose associated eigenvectors are

$$\begin{pmatrix} 1 - i \\ 2i \end{pmatrix} \quad \text{and} \quad \begin{pmatrix} 1 + i \\ -2i \end{pmatrix}.$$

Find a complex 2×2 matrix T such that $C = T^{-1}AT$ is complex diagonal and a real 2×2 matrix S so that $B = S^{-1}AS$ is in real block diagonal form.

2. Let

$$A = \begin{pmatrix} 2 & 5 \\ -2 & 0 \end{pmatrix}.$$

Find a complex 2×2 matrix T such that $T^{-1}AT$ is complex diagonal and a real 2×2 matrix S so that $S^{-1}AS$ is in real block diagonal form.

COMPUTER EXERCISES

3. Use map to verify that the normal form matrices (13.2.1) are just rotations followed by dilatations. In particular, use map to study the normal form matrix

$$A = \begin{pmatrix} 1 & -1 \\ 1 & 1 \end{pmatrix}.$$

Then compare your results with the similar matrix

$$B = \begin{pmatrix} 2 & 1 \\ -2 & 0 \end{pmatrix}.$$

4. Consider the 2×2 matrix

$$A = \begin{pmatrix} -0.8318 & -1.9755 \\ 0.9878 & 1.1437 \end{pmatrix}.$$

(a) Use MATLAB to find the complex conjugate eigenvalues and eigenvectors of A.

(b) Find the real block diagonal normal form of A and describe geometrically the motion of this normal form on the plane.

(c) Using map, describe geometrically how A maps vectors in the plane to vectors in the plane.

In Exercises 5–8, find a square real matrix S so that $S^{-1}AS$ is in real block diagonal form and a complex square matrix T so that $T^{-1}AT$ is in complex diagonal form.

5.

$$A = \begin{pmatrix} 1 & 2 & 4 \\ 2 & -4 & -5 \\ 1 & 10 & -15 \end{pmatrix} \qquad \textbf{(13.2.14)*}$$

6.

$$A = \begin{pmatrix} -15.1220 & 12.2195 & 13.6098 & 14.9268 \\ -28.7805 & 21.8049 & 25.9024 & 28.7317 \\ 60.1951 & -44.9512 & -53.9756 & -60.6829 \\ -44.5122 & 37.1220 & 43.5610 & 47.2927 \end{pmatrix}$$

(13.2.15)*

7.

$$A = \begin{pmatrix} 2.2125 & 5.1750 & 8.4250 & 15.0000 & 19.2500 & 0.5125 \\ -1.9500 & -3.9000 & -6.5000 & -7.4000 & -12.0000 & -2.9500 \\ 2.2250 & 3.9500 & 6.0500 & 0.9000 & 1.5000 & 1.0250 \\ -0.2000 & -0.4000 & 0 & 0.1000 & 0 & -0.2000 \\ -0.7875 & -0.8250 & -1.5750 & 1.0000 & 2.2500 & 0.5125 \\ 1.7875 & 2.8250 & 4.5750 & 0 & 4.7500 & 5.4875 \end{pmatrix}$$

(13.2.16)*

8.

$$A = \begin{pmatrix} -12 & 15 & 0 \\ 1 & 5 & 2 \\ -5 & 1 & 5 \end{pmatrix}$$

(13.2.17)*

13.3 MULTIPLICITY AND GENERALIZED EIGENVECTORS

The difficulty in generalizing the results in the preceding two sections to matrices with multiple eigenvalues stems from the fact that these matrices may not have enough (linearly independent) eigenvectors. In this section we present the basic examples of matrices with a deficiency of eigenvectors, as well as the definitions of algebraic and geometric multiplicity. These matrices are the building blocks of the Jordan normal form theorem—the theorem that classifies all matrices up to similarity.

Deficiency in Eigenvectors for Real Eigenvalues

An example of deficiency in eigenvectors is given by the following $n \times n$ matrix:

$$J_n(\lambda_0) = \begin{pmatrix} \lambda_0 & 1 & 0 & \cdots & 0 & 0 \\ 0 & \lambda_0 & 1 & \cdots & 0 & 0 \\ 0 & 0 & \lambda_0 & \cdots & 0 & 0 \\ \vdots & \vdots & \vdots & \ddots & \vdots & \vdots \\ 0 & 0 & 0 & \cdots & \lambda_0 & 1 \\ 0 & 0 & 0 & \cdots & 0 & \lambda_0 \end{pmatrix},$$

(13.3.1)

where $\lambda_0 \in \mathbf{R}$. Note that $J_n(\lambda_0)$ has all diagonal entries equal to λ_0, all superdiagonal entries equal to 1, and all other entries equal to 0. Since $J_n(\lambda_0)$ is upper triangular, all n eigenvalues of $J_n(\lambda_0)$ are equal to λ_0. However, $J_n(\lambda_0)$ has only one linearly independent eigenvector. To verify this assertion let

$$N = J_n(\lambda_0) - \lambda_0 I_n.$$

Then v is an eigenvector of $J_n(\lambda_0)$ if and only if $Nv = 0$. Therefore $J_n(\lambda_0)$ has a unique linearly independent eigenvector if:

Lemma 13.3.1   *Nullity* $(N) = 1$.

Proof: In coordinates the equation $Nv = 0$ is

$$
\begin{pmatrix}
0 & 1 & 0 & \cdots & 0 & 0 \\
0 & 0 & 1 & \cdots & 0 & 0 \\
0 & 0 & 0 & \cdots & 0 & 0 \\
\vdots & \vdots & \vdots & \ddots & \vdots & \vdots \\
0 & 0 & 0 & \cdots & 0 & 1 \\
0 & 0 & 0 & \cdots & 0 & 0
\end{pmatrix}
\begin{pmatrix}
v_1 \\ v_2 \\ v_3 \\ \vdots \\ v_{n-1} \\ v_n
\end{pmatrix}
=
\begin{pmatrix}
v_2 \\ v_3 \\ v_4 \\ \vdots \\ v_n \\ 0
\end{pmatrix}
= 0.
$$

Thus $v_2 = v_3 = \cdots = v_n = 0$, and the solutions are all multiples of e_1. Therefore the nullity of N is 1.   ◆

Note that we can express matrix multiplication by N as

$$
\begin{aligned}
Ne_1 &= 0 \\
Ne_j &= e_{j-1}, \quad j = 2, \ldots, n.
\end{aligned}
\tag{13.3.2}
$$

Note that (13.3.2) implies that $N^n = 0$.

The $n \times n$ matrix N motivates the following definition:

Definition 13.3.2   *Let λ_0 be an eigenvalue of A. The* algebraic multiplicity *of λ_0 is the number of times that λ_0 appears as a root of the characteristic polynomial $p_A(\lambda)$. The* geometric multiplicity *of λ_0 is the number of linearly independent eigenvectors of A having eigenvalue equal to λ_0.*

Abstractly, the geometric multiplicity is

$$
\text{nullity}(A - \lambda_0 I_n).
$$

Our previous calculations show that the matrix $J_n(\lambda_0)$ has an eigenvalue λ_0 with algebraic multiplicity equal to n and geometric multiplicity equal to 1.

Lemma 13.3.3   *The algebraic multiplicity of an eigenvalue is greater than or equal to its geometric multiplicity.*

Proof: For ease of notation we prove this lemma for only real eigenvalues; the proof for complex eigenvalues is similar. Let A be an $n \times n$ matrix and let λ_0 be a real eigenvalue of A. Let k be the geometric multiplicity of λ_0, and let $v_1, \ldots, v_k$ be k

linearly independent eigenvectors of A with eigenvalue λ_0. We can extend $\{v_1, \ldots, v_k\}$ to be a basis $\mathcal{V} = \{v_1, \ldots, v_n\}$ of $\mathbf{R}^n$. In this basis, the matrix of A is

$$[A]_\mathcal{V} = \begin{pmatrix} \lambda_0 I_k & (*) \\ 0 & B \end{pmatrix}.$$

The matrices A and $[A]_\mathcal{V}$ are similar matrices. Therefore they have the same characteristic polynomials and the same eigenvalues with the same algebraic multiplicities. It follows from Lemma 8.1.9 that the characteristic polynomial of A is

$$p_A(\lambda) = p_{[A]_\mathcal{V}}(\lambda) = (\lambda - \lambda_0)^k p_B(\lambda).$$

Hence λ_0 appears as a root of $p_A(\lambda)$ at least k times, and the algebraic multiplicity of λ_0 is greater than or equal to k. The same proof works when λ_0 is a complex eigenvalue—but all vectors chosen must be complex rather than real. ◆

Deficiency in Eigenvectors with Complex Eigenvalues

An example of a real matrix with complex conjugate eigenvalues having geometric multiplicity less than algebraic multiplicity is the $2n \times 2n$ block matrix

$$\widehat{J}_n(\lambda_0) = \begin{pmatrix} B & I_2 & 0 & \cdots & 0 & 0 \\ 0 & B & I_2 & \cdots & 0 & 0 \\ 0 & 0 & B & \cdots & 0 & 0 \\ \vdots & \vdots & \vdots & \ddots & \vdots & \vdots \\ 0 & 0 & 0 & \cdots & B & I_2 \\ 0 & 0 & 0 & \cdots & 0 & B \end{pmatrix}, \tag{13.3.3}$$

where $\lambda_0 = \sigma + i\tau$ and B is the 2×2 matrix

$$B = \begin{pmatrix} \sigma & -\tau \\ \tau & \sigma \end{pmatrix}.$$

Lemma 13.3.4   *Let λ_0 be a complex number. Then the algebraic multiplicity of the eigenvalue λ_0 in the $2n \times 2n$ matrix $\widehat{J}_n(\lambda_0)$ is n and the geometric multiplicity is 1.*

Proof: We begin by showing that the eigenvalues of $J = \widehat{J}_n(\lambda_0)$ are λ_0 and $\overline{\lambda_0}$, each with algebraic multiplicity n. The characteristic polynomial of J is $p_J(\lambda) = \det(J - \lambda I_{2n})$. From Lemma 8.1.9 and induction, we see that $p_J(\lambda) = p_B(\lambda)^n$. Since the eigenvalues of B are λ_0 and $\overline{\lambda_0}$, we have proved that the algebraic multiplicity of each of these eigenvalues in J is n.

Next we compute the eigenvectors of J. Let $Jv = \lambda_0 v$ and let $v = (v_1, \ldots, v_n)$, where each $v_j \in \mathbf{C}^2$. Observe that $(J - \lambda_0 I_{2n})v = 0$ if and only if

$$Qv_1 + v_2 = 0$$

$$\vdots$$

$$Qv_{n-1} + v_n = 0$$
$$Qv_n = 0,$$

where $Q = B - \lambda_0 I_2$. Using the fact that $\lambda_0 = \sigma + i\tau$, we have

$$Q = B - \lambda_0 I_2 = -\tau \begin{pmatrix} i & 1 \\ -1 & i \end{pmatrix}.$$

Hence

$$Q^2 = 2\tau^2 i \begin{pmatrix} i & 1 \\ -1 & i \end{pmatrix} = -2\tau i Q.$$

Thus

$$0 = Q^2 v_{n-1} + Qv_n = -2\tau i Qv_{n-1},$$

from which it follows that $Qv_{n-1} + v_n = v_n = 0$. Similarly, $v_2 = \cdots = v_{n-1} = 0$. Since there is only one nonzero complex vector v_1 (up to a complex scalar multiple) satisfying

$$Qv_1 = 0,$$

it follows that the geometric multiplicity of λ_0 in the matrix $\widehat{J}_n(\lambda_0)$ equals 1.   ◆

Definition 13.3.5   *The real matrices $J_n(\lambda_0)$ when $\lambda_0 \in \mathbf{R}$ and $\widehat{J}_n(\lambda_0)$ when $\lambda_0 \in \mathbf{C}$ are real Jordan blocks. The matrices $J_n(\lambda_0)$ when $\lambda_0 \in \mathbf{C}$ are (complex) Jordan blocks.*

Generalized Eigenvectors and Generalized Eigenspaces

What happens when $n \times n$ matrices have fewer that n linearly independent eigenvectors? Answer: The matrices gain generalized eigenvectors.

Definition 13.3.6   *A vector $v \in \mathbf{C}^n$ is a* generalized eigenvector *for the $n \times n$ matrix A with eigenvalue λ if*

$$(A - \lambda I_n)^k v = 0 \tag{13.3.4}$$

for some positive integer k. The smallest integer k for which (13.3.4) is satisfied is called the index *of the generalized eigenvector v.*

   Note: Eigenvectors are generalized eigenvectors with index equal to 1.
   Let λ_0 be a real number and let $N = J_n(\lambda_0) - \lambda_0 I_n$. Recall that (13.3.2) implies that $N^n = 0$. Hence every vector in $\mathbf{R}^n$ is a generalized eigenvector for the matrix $J_n(\lambda_0)$.

So $J_n(\lambda_0)$ provides a good example of a matrix whose lack of eigenvectors (there is only one independent eigenvector) is made up for by generalized eigenvectors (there are n independent generalized eigenvectors).

Let λ_0 be an eigenvalue of the $n \times n$ matrix A and let $A_0 = A - \lambda_0 I_n$. For simplicity, assume that λ_0 is real. Note that

$$\text{null space}(A_0) \subset \text{null space}(A_0^2) \subset \cdots \subset \text{null space}(A_0^k) \subset \cdots \subset \mathbf{R}^n.$$

Therefore the dimensions of the null spaces are bounded above by n, and there must be a smallest k such that

$$\dim \text{null space}(A_0^k) = \dim \text{null space}(A_0^{k+1}).$$

It follows that

$$\text{null space}(A_0^k) = \text{null space}(A_0^{k+1}). \tag{13.3.5}$$

Lemma 13.3.7 *Let λ_0 be a real eigenvalue of the $n \times n$ matrix A and let $A_0 = A - \lambda_0 I_n$. Let k be the smallest integer for which (13.3.5) is valid. Then*

$$\text{null space}(A_0^k) = \text{null space}(A_0^{k+j})$$

for every integer $j > 0$.

Proof: We can prove the lemma by induction on j if we can show that

$$\text{null space}(A_0^{k+1}) = \text{null space}(A_0^{k+2}).$$

Since null space$(A_0^{k+1}) \subset$ null space(A_0^{k+2}), we need to show that

$$\text{null space}(A_0^{k+2}) \subset \text{null space}(A_0^{k+1}).$$

Let $w \in$ null space(A_0^{k+2}). It follows that

$$A^{k+1} A w = A^{k+2} w = 0,$$

so $Aw \in$ null space$(A_0^{k+1}) =$ null space(A_0^k), by (13.3.5). Therefore

$$A^{k+1} w = A^k(Aw) = 0,$$

which verifies that $w \in$ null space(A_0^{k+1}).   ◆

Let V_{λ_0} be the set of all generalized eigenvectors of A with eigenvalue λ_0. Let k be the smallest integer satisfying (13.3.5). Then Lemma 13.3.7 implies that

$$V_{\lambda_0} = \text{null space}(A_0^k) \subset \mathbf{R}^n$$

is a subspace called the *generalized eigenspace* of A associated with the eigenvalue λ_0. It will follow from the Jordan normal form theorem (see Theorem 13.4.2) that the dimension of V_{λ_0} is the algebraic multiplicity of λ_0.

An Example of Generalized Eigenvectors

Find the generalized eigenvectors of the 4×4 matrix

$$A = \begin{pmatrix} -24 & -58 & -2 & -8 \\ 15 & 35 & 1 & 4 \\ 3 & 5 & 7 & 4 \\ 3 & 6 & 0 & 6 \end{pmatrix} \tag{13.3.6)*}$$

and their indices. When we find generalized eigenvectors of a matrix A, the first two steps are:

1. Find the eigenvalues of A.

2. Find the eigenvectors of A.

After entering A into MATLAB by typing e13_3_6, we type eig(A) and find that all the eigenvalues of A equal 6. Without additional information, there could be one, two, three, or four linearly independent eigenvectors of A corresponding to the eigenvalue 6. In MATLAB we determine the number of linearly independent eigenvectors by typing null(A-6*eye(4)) and obtaining

```
ans =
     0.8892          0
    -0.4446     0.0000
    -0.0262     0.9701
    -0.1046    -0.2425
```

We now know that (numerically) there are two linearly independent eigenvectors. The next step is to find the number of independent generalized eigenvectors of index 2. To complete this calculation, we find a basis for the null space of $(A - 6I_4)^2$ by typing null((A-6*eye(4))^2), obtaining

```
ans =
     1     0     0     0
     0     1     0     0
     0     0     1     0
     0     0     0     1
```

Thus, for this example, all generalized eigenvectors that are not eigenvectors have index 2.

HAND EXERCISES

In Exercises 1–4, determine the eigenvalues and their geometric and algebraic multiplicities for the given matrix.

1. $A = \begin{pmatrix} 2 & 0 & 0 & 0 \\ 0 & 3 & 1 & 0 \\ 0 & 0 & 3 & 0 \\ 0 & 0 & 0 & 4 \end{pmatrix}$

2. $B = \begin{pmatrix} 2 & 0 & 0 & 0 \\ 0 & 2 & 0 & 0 \\ 0 & 0 & 3 & 1 \\ 0 & 0 & 0 & 3 \end{pmatrix}$

3. $C = \begin{pmatrix} -1 & 1 & 0 & 0 \\ 0 & -1 & 0 & 0 \\ 0 & 0 & -1 & 0 \\ 0 & 0 & 0 & 1 \end{pmatrix}$

4. $D = \begin{pmatrix} 2 & -1 & 0 & 0 \\ 1 & 2 & 0 & 0 \\ 0 & 0 & 2 & 1 \\ 0 & 0 & 0 & 2 \end{pmatrix}$

In Exercises 5–8, find a basis consisting of the eigenvectors for the given matrix supplemented by generalized eigenvectors. Choose the generalized eigenvectors with lowest index possible.

5. $A = \begin{pmatrix} 1 & -1 \\ 1 & 3 \end{pmatrix}$

6. $B = \begin{pmatrix} -2 & 0 & -2 \\ -1 & 1 & -2 \\ 0 & 1 & -1 \end{pmatrix}$

7. $C = \begin{pmatrix} -6 & 31 & -14 \\ -1 & 6 & -2 \\ 0 & 2 & 1 \end{pmatrix}$

8. $D = \begin{pmatrix} 5 & 1 & 0 \\ -3 & 1 & 1 \\ -12 & -4 & 0 \end{pmatrix}$

COMPUTER EXERCISES

In Exercises 9 and 10, use MATLAB to find the eigenvalues and their algebraic and geometric multiplicities for the given matrix.

9.

$$A = \begin{pmatrix} 2 & 3 & -21 & -3 \\ 2 & 7 & -41 & -5 \\ 0 & 1 & -5 & -1 \\ 0 & 0 & 4 & 4 \end{pmatrix} \tag{13.3.7}*$$

10.

$$B = \begin{pmatrix} 179 & -230 & 0 & 10 & -30 \\ 144 & -185 & 0 & 8 & -24 \\ 30 & -39 & -1 & 3 & -9 \\ 192 & -245 & 0 & 9 & -30 \\ 40 & -51 & 0 & 2 & -7 \end{pmatrix} \tag{13.3.8}*$$

13.4 THE JORDAN NORMAL FORM THEOREM

The question that we discussed in Sections 13.1 and 13.2 is: Up to similarity, what is the simplest form that a matrix can have? We have seen that if A has real distinct eigenvalues, then A is real diagonalizable. That is, A is similar to a diagonal matrix whose diagonal entries are the real eigenvalues of A. Similarly, if A has distinct real and complex eigenvalues, then A is complex diagonalizable. That is, A is similar either to a diagonal matrix whose diagonal entries are the real and complex eigenvalues of A or to a real block diagonal matrix.

In this section we address the question of simplest form when a matrix has multiple eigenvalues. In much of this discussion we assume that A is an $n \times n$ matrix with only real eigenvalues. Lemma 13.1.3 shows that if the eigenvectors of A form a basis, then A is diagonalizable. Indeed, for A to be diagonalizable, there must be a basis of eigenvectors of A. It follows that if A is not diagonalizable, then A must have fewer than n linearly independent eigenvectors.

The prototypical examples of matrices having fewer eigenvectors than eigenvalues are the matrices $J_n(\lambda)$ for λ real (see (13.3.1)) and $\widehat{J}_n(\lambda)$ for λ complex (see (13.3.3)).

Definition 13.4.1   *A matrix is in* Jordan normal form *if it is block diagonal and the matrix in each block on the diagonal is a Jordan block—that is, $J_\ell(\lambda)$ for some integer ℓ and some real or complex number λ.*

A matrix is in real Jordan normal form *if it is block diagonal and the matrix in each block on the diagonal is a real Jordan block—that is, either $J_\ell(\lambda)$ for some integer ℓ and some real number λ or $\widehat{J}_\ell(\lambda)$ for some integer ℓ and some complex number λ.*

The main theorem about Jordan normal form is:

Theorem 13.4.2 (Jordan normal form)   *Let A be an $n \times n$ matrix. Then A is similar to a Jordan normal form matrix and to a real Jordan normal form matrix.*

This theorem is proved by constructing a basis $\mathcal{V}$ for $\mathbf{R}^n$ so that the matrix $S^{-1}AS$ is in Jordan normal form, where S is the matrix whose columns consist of vectors in $\mathcal{V}$. The algorithm for finding the basis $\mathcal{V}$ is complicated and is found in Appendix 13.6. In this section we construct $\mathcal{V}$ only in the special and simpler case where each eigenvalue of A is real and is associated with exactly one Jordan block.

More precisely, let $\lambda_1, \ldots, \lambda_s$ be the distinct eigenvalues of A and let

$$A_j = A - \lambda_j I_n.$$

The eigenvectors corresponding to λ_j are the vectors in the null space of A_j, and the generalized eigenvectors are the vectors in the null space of A_j^k for some k. The dimension of the null space of A_j is precisely the number of Jordan blocks of A associated with the eigenvalue λ_j. So the assumption that we make here is

$$\mathrm{nullity}(A_j) = 1$$

for $j = 1, \ldots, s$.

Let k_j be the integer whose existence is specified by Lemma 13.3.7. Since, by assumption, there is only one Jordan block associated with the eigenvalue λ_j, it follows that k_j is the algebraic multiplicity of the eigenvalue λ_j.

To find a basis in which the matrix A is in Jordan normal form, we proceed as follows: First, let w_{jk_j} be a vector in

$$\text{null space}(A_j^{k_j}) - \text{null space}(A_j^{k_j-1}).$$

Define the vectors w_{ji} by

$$w_{j,k_j-1} = A_j w_{j,k_j}$$

$$\vdots$$

$$w_{j,1} = A_j w_{j,2}.$$

Second, when λ_j is real, let the k_j vectors $v_{ji} = w_{ji}$, and when λ_j is complex, let the $2k_j$ vectors v_{ji} be defined by

$$v_{j,2i-1} = \text{Re}(w_{ji})$$
$$v_{j,2i} = \text{Im}(w_{ji}).$$

Let $\mathcal{V}$ be the set of vectors $v_{ji} \in \mathbf{R}^n$. We show in Appendix 13.6 that the set $\mathcal{V}$ consists of n vectors and is a basis of $\mathbf{R}^n$. Let S be the matrix whose columns are the vectors in $\mathcal{V}$. Then $S^{-1}AS$ is in Jordan normal form.

The Cayley Hamilton Theorem

As a corollary of the Jordan normal form theorem, we prove the Cayley Hamilton theorem, which states that a *square matrix satisfies its characteristic polynomial.* More precisely:

Theorem 13.4.3 (Cayley Hamilton)   *Let A be a square matrix and let $p_A(\lambda)$ be its characteristic polynomial. Then*

$$p_A(A) = 0.$$

Proof: Let A be an $n \times n$ matrix. The characteristic polynomial of A is

$$p_A(\lambda) = \det(A - \lambda I_n).$$

Suppose that $B = P^{-1}AP$ is a matrix similar to A. Theorem 8.2.8 states that $p_B = p_A$. Therefore

$$p_B(B) = p_A(P^{-1}AP) = P^{-1}p_A(A)P.$$

So if the Cayley Hamilton theorem holds for a matrix similar to A, then it is valid for the matrix A. Moreover, using the Jordan normal form theorem, we may assume that A is in Jordan normal form.

Suppose that A is block diagonal; that is,

$$A = \begin{pmatrix} A_1 & 0 \\ 0 & A_{2,} \end{pmatrix},$$

where A_1 and A_2 are square matrices. Then

$$p_A(\lambda) = p_{A_1}(\lambda) p_{A_2}(\lambda).$$

This observation follows directly from Lemma 8.1.9. Since

$$A^k = \begin{pmatrix} A_1^k & 0 \\ 0 & A_2^k \end{pmatrix},$$

it follows that

$$p_A(A) = \begin{pmatrix} p_A(A_1) & 0 \\ 0 & p_A(A_2) \end{pmatrix} = \begin{pmatrix} p_{A_1}(A_1) p_{A_2}(A_1) & 0 \\ 0 & p_{A_1}(A_2) p_{A_2}(A_2) \end{pmatrix}.$$

It now follows from this calculation that if the Cayley Hamilton theorem is valid for Jordan blocks, then $p_{A_1}(A_1) = 0 = p_{A_2}(A_2)$. So $p_A(A) = 0$, and the Cayley Hamilton theorem is valid for all matrices.

A direct calculation shows that Jordan blocks satisfy the Cayley Hamilton theorem. To begin, suppose that the eigenvalue of the Jordan block is real. Note that the characteristic polynomial of the Jordan block $J_n(\lambda_0)$ in (13.3.1) is $(\lambda - \lambda_0)^n$. Indeed, $J_n(\lambda_0) - \lambda_0 I_n$ is strictly upper triangular and $(J_n(\lambda_0) - \lambda_0 I_n)^n = 0$. If λ_0 is complex, then either repeat this calculation using the complex Jordan form or show by direct calculation that $(A - \lambda_0 I_n)(A - \overline{\lambda_0} I_n)$ is strictly upper triangular when $A = \widehat{J}_n(\lambda_0)$ is the real Jordan form of the Jordan block in (13.3.3).   ◆

An Example

Consider the 4×4 matrix

$$A = \begin{pmatrix} -147 & -106 & -66 & -488 \\ 604 & 432 & 271 & 1992 \\ 621 & 448 & 279 & 2063 \\ -169 & -122 & -76 & -562 \end{pmatrix}. \tag{13.4.1)*}$$

Using MATLAB, we can compute the characteristic polynomial of A by typing

```
poly(A)
```

The output is

```
ans =
    1.0000   -2.0000  -15.0000   -0.0000   -0.0000
```

Note that since A is a matrix of integers, we know that the coefficients of the characteristic polynomial of A must be integers. Thus the characteristic polynomial is exactly

$$p_A(\lambda) = \lambda^4 - 2\lambda^3 - 15\lambda^2 = \lambda^2(\lambda - 5)(\lambda + 3).$$

So $\lambda_1 = 0$ is an eigenvalue of A with algebraic multiplicity two, and $\lambda_2 = 5$ and $\lambda_3 = -3$ are simple eigenvalues of multiplicity one.

We can find eigenvectors of A corresponding to the simple eigenvalues by typing

```
v2 = null(A-5*eye(4));
v3 = null(A+3*eye(4));
```

At this stage we do not know how many linearly independent eigenvectors have eigenvalue 0. There are either one or two linearly independent eigenvectors, and we determine which by typing null(A) and obtaining

```
ans =
   -0.1818
    0.6365
    0.7273
   -0.1818
```

So MATLAB tells us that there is just one linearly independent eigenvector having 0 as an eigenvalue. There must be a generalized eigenvector in V_0. Indeed, the null space of A^2 is two dimensional, and this fact can be checked by typing

```
null2 = null(A^2)
```

obtaining

```
null2 =
    0.2193   -0.2236
   -0.5149   -0.8216
   -0.8139    0.4935
    0.1561    0.1774
```

Choose one of these vectors—say, the first vector—to be v_{12} by typing

```
v12 = null2(:,1);
```

Since the algebraic multiplicity of the eigenvalue 0 is two, we choose the fourth basis vector to be $v_{11} = Av_{12}$. In MATLAB we type

```
v11 = A*v12
```

obtaining

```
v11 =
   -0.1263
    0.4420
    0.5051
   -0.1263
```

Since v11 is nonzero, we have found a basis for V_0. We can now put the matrix A in Jordan normal form by setting

```
S = [v11 v12 v2 v3];
J = inv(S)*A*S
```

to obtain

```
J =
   -0.0000    1.0000    0.0000   -0.0000
    0.0000    0.0000    0.0000   -0.0000
   -0.0000   -0.0000    5.0000    0.0000
    0.0000   -0.0000   -0.0000   -3.0000
```

We have discussed a Jordan normal form example only when the eigenvalues are real and multiple. The case when the eigenvalues are complex and multiple first occurs when $n = 4$. A sample complex Jordan block when the matrix has algebraic multiplicity two eigenvalues $\sigma \pm i\tau$ of geometric multiplicity one is

$$\begin{pmatrix} \sigma & -\tau & 1 & 0 \\ \tau & \sigma & 0 & 1 \\ 0 & 0 & \sigma & -\tau \\ 0 & 0 & \tau & \sigma \end{pmatrix}.$$

Numerical Difficulties

When a matrix has multiple eigenvalues, numerical difficulties can arise when using the MATLAB command eig(A), as we now explain.

Let $p(\lambda) = \lambda^2$. Solving $p(\lambda) = 0$ is very easy—in theory—since $\lambda = 0$ is a double root of p. Suppose, however, that we want to solve $p(\lambda) = 0$ numerically. Then numerical errors lead to solving the equation

$$\lambda^2 = \epsilon,$$

where ϵ is a small number. Note that if $\epsilon > 0$, the solutions are $\pm\sqrt{\epsilon}$; whereas if $\epsilon < 0$, the solutions are $\pm i\sqrt{|\epsilon|}$. Since numerical errors are machine dependent, ϵ can be of either sign. The numerical process of finding double roots of a characteristic polynomial (that is, double eigenvalues of a matrix) is similar to numerically solving the equation $\lambda^2 = 0$, as we shall see.

For example, on a *Sun SPARCstation 10* using MATLAB version 4.2c, the eigen-values of the 4×4 matrix A in (13.4.1) (in format long) obtained using eig(A) are

```
ans =
  5.00000000001021
 -0.00000000000007 + 0.00000023858927i
 -0.00000000000007 - 0.00000023858927i
 -3.00000000000993
```

That is, MATLAB computes two complex conjugate eigenvalues

$$\pm 0.00000023858927i,$$

which corresponds to an ϵ of -5.692483975913288e-14. On a *IBM* compatible 486 computer using MATLAB version 4.2, the same computation yields eigenvalues

```
ans=
  4.99999999999164
  0.00000057761008
 -0.00000057760735
 -2.99999999999434
```

That is, on this computer MATLAB computes two real, near 0, eigenvalues

$$\pm 0.00000057761,$$

which corresponds to an ϵ of 3.336333121e-13. These errors are within round-off error in double precision computation.

A consequence of these kinds of error, however, is that when a matrix has multiple eigenvalues, we cannot use the command [V,D] = eig(A) with confidence. On the *Sun SPARCstation*, this command yields a matrix

```
V =
  -0.1652        0.0000 - 0.1818i   0.0000 + 0.1818i   -0.1642
   0.6726       -0.0001 + 0.6364i  -0.0001 - 0.6364i    0.6704
   0.6962       -0.0001 + 0.7273i  -0.0001 - 0.7273i    0.6978
  -0.1888        0.0000 - 0.1818i   0.0000 + 0.1818i   -0.1915
```

that suggests that A has two complex eigenvectors corresponding to the "complex" pair of near 0 eigenvalues. The *IBM* compatible yields the matrix

```
V =
  -0.1652    0.1818   -0.1818   -0.1642
   0.6726   -0.6364    0.6364    0.6704
   0.6962   -0.7273    0.7273    0.6978
  -0.1888    0.1818   -0.1818   -0.1915
```

indicating that MATLAB has found two real eigenvectors corresponding to the near 0 real eigenvalues. Note that the two eigenvectors corresponding to the eigenvalues 5 and −3 are correct on both computers.

HAND EXERCISES

1. Write two different 4×4 Jordan normal form matrices all of whose eigenvalues equal 2 for which the geometric multiplicity is two.

2. How many different 6×6 Jordan form matrices have all eigenvalues equal to 3? (We say that two Jordan form matrices are the same if they have the same number and type of Jordan block, though not necessarily in the same order along the diagonal.)

3. A 5×5 matrix A has three eigenvalues equal to 4 and two eigenvalues equal to -3. List the possible Jordan normal forms for A (up to similarity). Suppose that you can ask your computer to compute the nullity of precisely two matrices. Can you devise a strategy for determining the Jordan normal form of A? Explain your answer.

4. An 8×8 real matrix A has three eigenvalues equal to 2, two eigenvalues equal to $1 + i$, and one 0 eigenvalue. List the possible Jordan normal forms for A (up to similarity). Suppose that you can ask your computer to compute the nullity of precisely two matrices. Can you devise a strategy for determining the Jordan normal form of A? Explain your answer.

In Exercises 5–10, find the Jordan normal forms for each matrix.

5. $A = \begin{pmatrix} 2 & 4 \\ 1 & 1 \end{pmatrix}$

6. $B = \begin{pmatrix} 9 & 25 \\ -4 & -11 \end{pmatrix}$

7. $C = \begin{pmatrix} -5 & -8 & -9 \\ 5 & 9 & 9 \\ -1 & -2 & -1 \end{pmatrix}$

8. $D = \begin{pmatrix} 0 & 1 & 0 \\ 0 & 0 & 1 \\ 1 & 1 & -1 \end{pmatrix}$

9. $E = \begin{pmatrix} 2 & 0 & -1 \\ 2 & 1 & -1 \\ 1 & 0 & 0 \end{pmatrix}$

10. $F = \begin{pmatrix} 3 & -1 & 2 \\ -1 & 2 & -1 \\ -1 & 1 & 0 \end{pmatrix}$

11. Compute e^{tJ}, where $J = \begin{pmatrix} 2 & 0 & 0 \\ 0 & -1 & 1 \\ 0 & 0 & -1 \end{pmatrix}$.

12. Compute e^{tJ}, where $J = \begin{pmatrix} 2 & 1 & 0 & 0 & 0 \\ 0 & 2 & 0 & 0 & 0 \\ 0 & 0 & 3 & 1 & 0 \\ 0 & 0 & 0 & 3 & 1 \\ 0 & 0 & 0 & 0 & 3 \end{pmatrix}$.

13. An $n \times n$ matrix N is *nilpotent* if $N^k = 0$ for some positive integer k.
   (a) Show that the matrix N defined in (13.3.2) is nilpotent.
   (b) Show that all eigenvalues of a nilpotent matrix equal 0.
   (c) Show that any matrix similar to a nilpotent matrix is also nilpotent.
   (d) Let N be a matrix all of whose eigenvalues are 0. Use the Jordan normal form theorem to show that N is nilpotent.

14. Let A be a 3×3 matrix. Use the Cayley Hamilton theorem to show that A^{-1} is a linear combination of I_3, A, and A^2. That is, there exist real scalars a, b, and c such that

$$A^{-1} = aI_3 + bA + cA^2.$$

COMPUTER EXERCISES

In Exercises 15–20, (a) determine the real Jordan normal form for the given matrix A, and (b) find the matrix S so that $S^{-1}AS$ is in real Jordan normal form.

15.

$$A = \begin{pmatrix} -3 & -4 & -2 & 0 \\ -9 & -39 & -16 & -7 \\ 18 & 64 & 27 & 10 \\ 15 & 86 & 34 & 18 \end{pmatrix}$$ (13.4.2)*

16.

$$A = \begin{pmatrix} 9 & 45 & 18 & 8 \\ 0 & -4 & -1 & -1 \\ -16 & -69 & -29 & -12 \\ 25 & 123 & 49 & 23 \end{pmatrix}$$ (13.4.3)*

17.

$$A = \begin{pmatrix} -5 & -13 & 17 & 42 \\ -10 & -57 & 66 & 187 \\ -4 & -23 & 26 & 77 \\ -1 & -9 & 9 & 32 \end{pmatrix}$$ (13.4.4)*

18.

$$A = \begin{pmatrix} 1 & 0 & -9 & 18 \\ 12 & -7 & -26 & 77 \\ 5 & -2 & -13 & 32 \\ 2 & -1 & -4 & 11 \end{pmatrix}$$ (13.4.5)*

19.

$$A = \begin{pmatrix} -1 & -1 & 1 & 0 \\ -3 & 1 & 1 & 0 \\ -3 & 2 & -1 & 1 \\ -3 & 2 & 0 & 0 \end{pmatrix}$$ (13.4.6)*

20.

$$A = \begin{pmatrix} 0 & 0 & -1 & 2 & 2 \\ 1 & -2 & 0 & 2 & 2 \\ 1 & -1 & -1 & 2 & 2 \\ 0 & 0 & 0 & 1 & 2 \\ 0 & 0 & 0 & -1 & 3 \end{pmatrix}$$ (13.4.7)*

13.5   *APPENDIX: MARKOV MATRIX THEORY

In this appendix we use the Jordan normal form theorem to study the asymptotic dynamics of transition matrices such as those of Markov chains introduced in Section 4.10. The basic result is the following theorem:

Theorem 13.5.1   *Let A be an $n \times n$ matrix and assume that all eigenvalues λ of A satisfy $|\lambda| < 1$. Then for every vector $v_0 \in \mathbf{R}^n$,*

$$\lim_{k \to \infty} A^k v_0 = 0. \tag{13.5.1}$$

Proof: Suppose that A and B are similar matrices; that is, $B = SAS^{-1}$ for some invertible matrix S. Then $B^k = SA^k S^{-1}$, and for any vector $v_0 \in \mathbf{R}^n$ (13.5.1) is valid if and only if

$$\lim_{k \to \infty} B^k v_0 = 0.$$

Thus, when proving this theorem, we may assume that A is in Jordan normal form.

Suppose that A is in block diagonal form; that is, suppose

$$A = \begin{pmatrix} C & 0 \\ 0 & D \end{pmatrix},$$

where C is an $\ell \times \ell$ matrix and D is a $(n - \ell) \times (n - \ell)$ matrix. Then

$$A^k = \begin{pmatrix} C^k & 0 \\ 0 & D^k \end{pmatrix}.$$

So for every vector $v_0 = (w_0, u_0) \in \mathbf{R}^\ell \times \mathbf{R}^{n-\ell}$, (13.5.1) is valid if and only if

$$\lim_{k \to \infty} C^k v_0 = 0 \quad \text{and} \quad \lim_{k \to \infty} D^k v_0 = 0.$$

So, when proving this theorem, we may assume that A is a Jordan block.

Consider the case of a simple Jordan block. Suppose that $n = 1$ and that $A = (\lambda)$, where λ is either real or complex. Then

$$A^k v_0 = \lambda^k v_0.$$

It follows that (13.5.1) is valid precisely when $|\lambda| < 1$. Next, suppose that A is a nontrivial Jordan block. For example, let

$$A = \begin{pmatrix} \lambda & 1 \\ 0 & \lambda \end{pmatrix} = \lambda I_2 + N,$$

where $N^2 = 0$. It follows by induction that

$$A^k v_0 = \lambda^k v_0 + k\lambda^{k-1} N v_0 = \lambda^k v_0 + k\lambda^k \frac{1}{\lambda} N v_0.$$

Thus (13.5.1) is valid precisely when $|\lambda| < 1$. The reason for this convergence is as follows: The first term converges to 0 as before, but the second term is the product of three terms k, λ^k, and $\frac{1}{\lambda} N v_0$. The first increases to infinity, the second decreases to 0, and the third is constant independent of k. In fact, geometric decay (λ^k when $|\lambda| < 1$) always beats polynomial growth. Indeed,

$$\lim_{m \to \infty} m^j \lambda^m = 0 \qquad (13.5.2)$$

for any integer j. This fact can be proved using l'Hôspital's rule and induction.

So we see that when A has a nontrivial Jordan block, convergence is subtler than when A has only simple Jordan blocks, since initially the vectors $A v_0$ grow in magnitude. For example, suppose that $\lambda = 0.75$ and $v_0 = (1, 0)^t$. Then $A^8 v_0 = (0.901, 0.075)^t$ is the first vector in the sequence $A^k v_0$ whose norm is less than 1; that is, $A^8 v_0$ is the first vector in the sequence closer to the origin than v_0.

It is also true that (13.5.1) is valid for any Jordan block A and for all v_0 precisely when $|\lambda| < 1$. To verify this fact we use the binomial theorem. We can write a nontrivial Jordan block as $\lambda I_n + N$, where $N^{k+1} = 0$ for some integer k. We just discussed the case $k = 1$. In this case

$$(\lambda I_n + N)^m = \lambda^m I_n + m\lambda^{m-1} N + \binom{m}{2} \lambda^{m-2} N^2 + \cdots + \binom{m}{k} \lambda^{m-k} N^k,$$

where

$$\binom{m}{j} = \frac{m!}{j!(m-j)!} = \frac{m(m-1)\cdots(m-j+1)}{j!}.$$

To verify that

$$\lim_{m \to \infty} (\lambda I_n + N)^m = 0$$

we need only verify that each term

$$\lim_{m \to \infty} \binom{m}{j} \lambda^{m-j} N^j = 0.$$

Such terms are the product of three terms:

$$m(m-1)\cdots(m-j+1) \quad \text{and} \quad \lambda^m \quad \text{and} \quad \frac{1}{j!\lambda^j} N^j.$$

The first term has polynomial growth to infinity dominated by m^j, the second term decreases to 0 geometrically, and the third term is constant independent of m. The desired convergence to 0 follows from (13.5.2). ◆

Definition 13.5.2 *The $n \times n$ matrix A has a* dominant *eigenvalue $\lambda_0 > 0$ if λ_0 is a simple eigenvalue and all other eigenvalues λ of A satisfy $|\lambda| < \lambda_0$.*

Theorem 13.5.3 *Let P be a Markov matrix. Then 1 is a dominant eigenvalue of P.*

Proof: Recall from Definition 4.10.1 that a Markov matrix is a square matrix P whose entries are nonnegative, whose rows sum to 1, and for which a power P^k has all positive entries. To prove this theorem we must show that all eigenvalues λ of P satisfy $|\lambda| \leq 1$ and that 1 is a simple eigenvalue of P.

Let λ be an eigenvalue of P and let $v = (v_1, \ldots, v_n)^t$ be an eigenvector corresponding to the eigenvalue λ. We prove that $|\lambda| \leq 1$. Choose j so that $|v_j| \geq |v_i|$ for all i. Since $Pv = \lambda v$, we can equate the jth coordinates of both sides of this equality, obtaining

$$p_{j1}v_1 + \cdots + p_{jn}v_n = \lambda v_j.$$

Therefore

$$|\lambda||v_j| = |p_{j1}v_1 + \cdots + p_{jn}v_n| \leq p_{j1}|v_1| + \cdots + p_{jn}|v_n|,$$

since the p_{ij} are nonnegative. It follows that

$$|\lambda||v_j| \leq (p_{j1} + \cdots + p_{jn})|v_j| = |v_j|,$$

since $|v_i| \leq |v_j|$ and rows of P sum to 1. Since $|v_j| > 0$, it follows that $\lambda \leq 1$.

Next we show that 1 is a simple eigenvalue of P. Recall, or just calculate directly, that the vector $(1, \ldots, 1)^t$ is an eigenvector of P with eigenvalue 1. Now let $v = (v_1, \ldots, v_n)^t$ be an eigenvector of P with eigenvalue 1. Let $Q = P^k$ so that all entries of Q are positive. Observe that v is an eigenvector of Q with eigenvalue 1, and hence that all rows of Q also sum to 1.

To show that 1 is a simple eigenvalue of Q, and therefore of P, we must show that all coordinates of v are equal. Using the previous estimates (with $\lambda = 1$), we obtain

$$|v_j| = |q_{j1}v_1 + \cdots + q_{jn}v_n| \leq q_{j1}|v_1| + \cdots + q_{jn}|v_n| \leq |v_j|. \tag{13.5.3}$$

Hence

$$|q_{j1}v_1 + \cdots + q_{jn}v_n| = q_{j1}|v_1| + \cdots + q_{jn}|v_n|.$$

This equality is valid only if all of the v_i are nonnegative or if all are nonpositive. Without loss of generality, we assume that all $v_i \geq 0$. It follows from (13.5.3) that

$$v_j = q_{j1}v_1 + \cdots + q_{jn}v_n.$$

Since $q_{ji} > 0$, this inequality can hold only if all of the v_i are equal.   ◆

Theorem 13.5.4

(a)   Let Q be an $n \times n$ matrix with dominant eigenvalue $\lambda > 0$ and associated eigenvector v. Let v_0 be any vector in $\mathbf{R}^n$. Then

$$\lim_{k \to \infty} \frac{1}{\lambda^k} Q^k v_0 = cv$$

for some scalar c.

(b)   Let P be a Markov matrix and let v_0 be a nonzero vector in $\mathbf{R}^n$ with all entries nonnegative. Then

$$\lim_{k \to \infty} (P^t)^k v_0 = V,$$

where V is the eigenvector of P^t with eigenvalue 1 such that the sum of the entries in V is equal to the sum of the entries in v_0.

Proof:

(a)   After a similarity transformation, if needed, we can assume that Q is in Jordan normal form. More precisely, we can assume that

$$\frac{1}{\lambda} Q = \begin{pmatrix} 1 & 0 \\ 0 & A \end{pmatrix},$$

where A is an $(n-1) \times (n-1)$ matrix with all eigenvalues μ satisfying $|\mu| < 1$. Suppose $v_0 = (c_0, w_0) \in \mathbf{R} \times \mathbf{R}^{n-1}$. It follows from Theorem 13.5.1 that

$$\lim_{k \to \infty} \frac{1}{\lambda^k} Q^k v_0 = \lim_{k \to \infty} \left(\frac{1}{\lambda} Q \right)^k v_0 = \lim_{k \to \infty} \begin{pmatrix} c_0 & 0 \\ 0 & A^k w_0 \end{pmatrix} = c_0 e_1.$$

Since e_1 is the eigenvector of Q with eigenvalue λ, part (a) is proved.

(b)   Theorem 13.5.3 states that a Markov matrix has a dominant eigenvalue equal to 1. The Jordan normal form theorem implies that the eigenvalues of P^t are equal to the eigenvalues of P with the same algebraic and geometric multiplicities. It follows that 1 is also a dominant eigenvalue of P^t. It follows from part (a) that

$$\lim_{k \to \infty} (P^t)^k v_0 = cV$$

for some scalar c. But Theorem 4.10.3 implies that the sum of the entries in v_0 equals the sum of the entries in cV, which by assumption equals the sum of the entries in V. Thus $c = 1$.   ◆

HAND EXERCISES

1. Let A be an $n \times n$ matrix. Suppose that

$$\lim_{k \to \infty} A^k v_0 = 0$$

for every vector $v_0 \in \mathbf{R}^n$. Then the eigenvalues λ of A all satisfy $|\lambda| < 1$.

13.6 *APPENDIX: PROOF OF JORDAN NORMAL FORM

We prove the Jordan normal form theorem under the assumption that the eigenvalues of A are all real. The proof for matrices having both real and complex eigenvalues proceeds along similar lines.

Let A be an $n \times n$ matrix, let $\lambda_1, \ldots, \lambda_s$ be the distinct eigenvalues of A, and let $A_j = A - \lambda_j I_n$.

Lemma 13.6.1   *The linear mappings A_i and A_j commute.*

Proof: Just compute

$$A_i A_j = (A - \lambda_i I_n)(A - \lambda_j I_n) = A^2 - \lambda_i A - \lambda_j A + \lambda_i \lambda_j I_n$$

and

$$A_j A_i = (A - \lambda_j I_n)(A - \lambda_i I_n) = A^2 - \lambda_j A - \lambda_i A + \lambda_j \lambda_i I_n.$$

So $A_i A_j = A_j A_i$, as claimed.   ◆

Let V_j be the generalized eigenspace corresponding to eigenvalue λ_j.

Lemma 13.6.2 $A_i : V_j \to V_j$ is invertible when $i \neq j$.

Proof: Recall from Lemma 13.3.7 that $V_j = \text{null space}(A_j^k)$ for some $k \geq 1$. Suppose that $v \in V_j$. We first verify that $A_i v$ is also in V_j. Using Lemma 13.6.1, just compute

$$A_j^k A_i v = A_i A_j^k v = A_i 0 = 0.$$

Therefore $A_i v \in \text{null space}(A_j^k) = V_j$.

Let B be the linear mapping $A_i | V_j$. It follows from Theorem 9.2.3 that

$$\text{nullity}(B) + \dim \text{range}(B) = \dim(V_j).$$

Now $w \in \text{null space}(B)$ if $w \in V_j$ and $A_i w = 0$. Since $A_i w = (A - \lambda_i I_n)w = 0$, it follows that $Aw = \lambda_i w$. Hence

$$A_j w = (A - \lambda_j I_n)w = (\lambda_i - \lambda_j)w$$

and

$$A_j^k w = (\lambda_i - \lambda_j)^k w.$$

Since $\lambda_i \neq \lambda_j$, it follows that $A_j^k w = 0$ only when $w = 0$. Hence the nullity of B is 0. We conclude that

$$\dim \text{range}(B) = \dim(V_j).$$

Thus B is invertible, since the domain and range of B are the same space.   ◆

Lemma 13.6.3 *Nonzero vectors taken from different generalized eigenspaces V_j are linearly independent. More precisely, if $w_j \in V_j$ and*

$$w = w_1 + \cdots + w_s = 0,$$

then $w_j = 0$.

Proof: Let $V_j = \text{null space}(A_j^{k_j})$ for some integer k_j. Let $C = A_2^{k_2} \circ \cdots \circ A_s^{k_s}$. Then

$$0 = Cw = Cw_1,$$

since $A_j^{k_j} w_j = 0$ for $j = 2, \ldots, s$. But Lemma 13.6.2 implies that $C|V_1$ is invertible. Therefore $w_1 = 0$. Similarly, all of the remaining w_j have to vanish.   ◆

Lemma 13.6.4 *Every vector in $\mathbf{R}^n$ is a linear combination of vectors in the generalized eigenspaces V_j.*

Proof: Let W be the subspace of $\mathbf{R}^n$ consisting of all vectors of the form $z_1 + \cdots + z_s$, where $z_j \in V_j$. We need to verify that $W = \mathbf{R}^n$. Suppose that W is a proper subspace. Then choose a basis $w_1, \ldots, w_t$ of W and extend this set to a basis $\mathcal{W}$ of $\mathbf{R}^n$. In this basis the matrix $[A]_{\mathcal{W}}$ has block form; that is,

$$[A]_{\mathcal{W}} = \begin{pmatrix} A_{11} & A_{12} \\ 0 & A_{22} \end{pmatrix},$$

where A_{22} is an $(n - t) \times (n - t)$ matrix. The eigenvalues of A_{22} are eigenvalues of A. Since all of the distinct eigenvalues and eigenvectors of A are accounted for in W (that is, in A_{11}), we have a contradiction. So $W = \mathbf{R}^n$, as claimed.   ◆

Lemma 13.6.5   *Let V_j be a basis for the generalized eigenspaces V_j, and let V be the union of the sets V_j. Then V is a basis for $\mathbf{R}^n$.*

Proof: We first show that the vectors in V span $\mathbf{R}^n$. It follows from Lemma 13.6.4 that every vector in $\mathbf{R}^n$ is a linear combination of vectors in V_j. But each vector in V_j is a linear combination of vectors in V_j. Hence the vectors in V span $\mathbf{R}^n$.

Second, we show that the vectors in V are linearly independent. Suppose that a linear combination of vectors in V sums to 0. We can write this sum as

$$w_1 + \cdots + w_s = 0,$$

where w_j is the linear combination of vectors in V_j. Lemma 13.6.3 implies that each $w_j = 0$. Since V_j is a basis for V_j, it follows that the coefficients in the linear combinations w_j must all be 0. Hence the vectors in V are linearly independent.

Finally, it follows from Theorem 5.5.3 that V is a basis.   ◆

Lemma 13.6.6   *In the basis V of $\mathbf{R}^n$ guaranteed by Lemma 13.6.5, the matrix $[A]_V$ is block diagonal; that is,*

$$[A]_V = \begin{pmatrix} A_{11} & 0 & 0 \\ 0 & \ddots & 0 \\ 0 & 0 & A_{ss} \end{pmatrix},$$

where all of the eigenvalues of A_{jj} equal λ_j.

Proof: It follows from Lemma 13.6.1 that $A : V_j \to V_j$. Suppose that $v_j \in V_j$. Then Av_j is in V_j and Av_j is a linear combination of vectors in V_j. The block diagonalization of $[A]_V$ follows. Since $V_j = \text{null space}(A_j^{k_j})$, it follows that all eigenvalues of A_{jj} equal λ_j.   ◆

Lemma 13.6.6 implies that to prove the Jordan normal form theorem, we must find a basis in which the matrix A_{jj} is in Jordan normal form. So, without loss of generality, we may assume that all eigenvalues of A equal λ_0 and then find a basis in which A is in Jordan normal form. Moreover, we can replace A by the matrix $A - \lambda_0 I_n$, a matrix all of whose eigenvalues are 0. So, without loss of generality, we assume that A is an $n \times n$ matrix all of whose eigenvalues are 0. We now sketch the remainder of the proof of Theorem 13.4.2.

Let k be the smallest integer such that $\mathbf{R}^n = \text{null space}(A^k)$, and let

$$s = \dim \text{null space}(A^k) - \dim \text{null space}(A^{k-1}) > 0.$$

Let $z_1, \ldots, z_{n-s}$ be a basis for null space(A^{k-1}), and extend this set to a basis for null space(A^k) by adjoining the linearly independent vectors $w_1, \ldots, w_s$. Let

$$W_k = \text{span}\{w_1, \ldots, w_s\}.$$

It follows that $W_k \cap$ null space(A^{k-1}) = $\{0\}$.

We claim that the ks vectors $\mathcal{W} = \{w_{j\ell} = A^\ell(w_j)\}$, where $0 \leq \ell \leq k - 1$ and $1 \leq j \leq s$, are linearly independent. We can write any linear combination of the vectors in $\mathcal{W}$ as $y_k + \cdots + y_1$, where $y_j \in A^{k-j}(W_k)$. Suppose that

$$y_k + \cdots + y_1 = 0.$$

Then $A^{k-1}(y_k + \cdots + y_1) = A^{k-1}y_k = 0$. Therefore y_k is in W_k and in null space(A^{k-1}). Hence $y_k = 0$. Similarly, $A^{k-2}(y_{k-1} + \cdots + y_1) = A^{k-2}y_{k-1} = 0$. But $y_{k-1} = A\hat{y}_k$, where $\hat{y}_k \in W_k$ and $\hat{y}_k \in$ null space(A^{k-1}). Hence $\hat{y}_k = 0$ and $y_{k-1} = 0$. Similarly, all of the $y_j = 0$. It follows from $y_j = 0$ that a linear combination of the vectors $A^{k-j}(w_1), \ldots, A^{k-j}(w_s)$ is 0; that is,

$$0 = \beta_1 A^{k-j}(w_1) + \cdots + \beta_s A^{k-j}(w_s) = A^{k-j}(\beta_1 w_1 + \cdots + \beta_s w_s).$$

Applying A^{j-1} to this expression, we see that

$$\beta_1 w_1 + \cdots + \beta_s w_s$$

is in W_k and in the null space(A^{k-1}). Hence

$$\beta_1 w_1 + \cdots + \beta_s w_s = 0.$$

Since the w_j are linearly independent, each $\beta_j = 0$, thus verifying the claim.

Next we find the largest integer m so that

$$t = \dim \text{null space}(A^m) - \dim \text{null space}(A^{m-1}) > 0.$$

Proceed as above. Choose a basis for null space(A^{m-1}) and extend to a basis for null space(A^m) by adjoining the vectors $x_1, \ldots, x_t$. Adjoin the mt vectors $A^\ell x_j$ to the set $\mathcal{V}$ and verify that these vectors are all linearly independent. Repeat the process. Eventually, we arrive at a basis for $\mathbf{R}^n =$ null space(A^k).

In this basis the matrix $[A]_\mathcal{V}$ is block diagonal; indeed, each of the blocks is a Jordan block, since

$$A(w_{j\ell}) = \begin{cases} w_{j(\ell-1)} & 0 < \ell \leq k - 1 \\ 0 & \ell = 1 \end{cases}.$$

Note the resemblance to (13.3.2).

Higher Dimensional Systems

In Chapter 4 we saw that equilibria and their stabilities completely determine the phase line dynamics for single autonomous first-order differential equations. The stability of the equilibria determines the direction in which one equilibrium is connected to the next.

In Chapter 11 we discussed the extent to which equilibria and their stability determine the phase planes of systems of two autonomous systems of ordinary differential equations. When an equilibrium is hyperbolic, nonlinear planar systems behave much like their linearizations—at least in a small neighborhood of the equilibrium. However, away from equilibria, planar systems can have dynamically more interesting states: limit cycles. We have seen that qualitatively we can understand the global dynamics of most planar systems (the Morse-Smale ones) if we can find their equilibria, their periodic solutions, and their connecting orbits. Analytically, this is an impossible problem to solve in closed form—but numerically this kind of calculation is often tractable.

In this chapter we discuss briefly the dynamics of systems of three or more autonomous first-order differential equations. The situation is very complicated—even on the qualitative level. On the positive side, nonlinear systems do behave like their linear counterparts in a neighborhood of hyperbolic equilibria, and the dynamics of the linearized systems can be understood as a consequence of the Jordan normal form theorem. These issues are discussed in Sections 14.1 and 14.2. On the negative side, in Sections 14.5 and 14.6 we see that the dynamics of nonlinear systems away from equilibria are just much more complicated than their planar counterparts. In particular, we see that quasiperiodic motion may be expected (first in linear nonhyperbolic four-dimensional systems and then in nonlinear three-dimensional systems). Finally, we show that even complicated "chaotic" motion may be expected in three dimensions (the Lorenz equations).

The discussion in Sections 14.5 and 14.6 is predicated on being able to solve numerically systems of differential equations with more than two equations. To do this, we must use the MATLAB differential equations solver ode45 directly. We introduce this solver in Section 14.3 by solving certain one-dimensional differential equations. In this section we also discuss how to store functions in MATLAB m-files. In Section 14.4 we use ode45 to solve several sample differential equations in three and four dimensions, and we display the results using MATLAB graphics.

The discussion in this chapter continues an important theme: What information can we learn about the dynamics and solutions of nonlinear systems of ordinary differential equations from numerical simulation? Indeed, what can mathematics say that helps in interpreting numerically obtained solutions? We will see that even on the qualitative level, the situation is very complicated—complicated enough to guarantee that closed form solutions are, in general, not an option. It should be noted, however, that closed form solutions do exist for many particular types of equations (not least of which are the linear constant coefficient systems). In later chapters we discuss some of the techniques of integration that allow us to solve certain special kinds of differential equations in closed form.

14.1 LINEAR SYSTEMS IN JORDAN NORMAL FORM

In this section we discuss one method for finding closed form solutions to linear constant coefficient systems of ODEs based on Jordan normal form. Let

$$\frac{dX}{dt} = AX, \tag{14.1.1}$$

where $A = (a_{ij})$ is a constant $n \times n$ matrix and X is an n vector. In Theorem 6.3.1 we showed that the general solution to (14.1.1) is

$$X(t) = e^{tA}X_0,$$

where $X_0 = X(0)$ is the initial condition. In Chapter 6 we used this theorem to solve planar systems of ODEs in closed form. Now we apply matrix exponentials to solve (14.1.1) for an arbitrary matrix A—at least in principle. There are three ideas behind finding solutions to planar systems:

(a) Exponentials of similar matrices are similar—more precisely,

$$e^{P^{-1}AP} = P^{-1}e^{A}P,$$

where P is any invertible matrix. See Lemma 6.5.4.

(b) Every matrix A is similar to a matrix in Jordan normal form; in two dimensions the Jordan normal forms are

$$B_1 = \begin{pmatrix} \lambda & 0 \\ 0 & \mu \end{pmatrix}, \quad B_2 = \begin{pmatrix} \sigma & -\tau \\ \tau & \sigma \end{pmatrix} \quad \text{and} \quad B_3 = \begin{pmatrix} \lambda & 1 \\ 0 & \lambda \end{pmatrix}.$$

The first matrix corresponds to two real eigenvalues λ and μ with linearly independent eigenvectors; the second matrix corresponds to complex eigenvalues $\sigma \pm i\tau$; and the third matrix is the only nontrivial Jordan block in two dimensions.

(c) It is possible to compute the exponentials of matrices in Jordan normal form. Indeed,

$$e^{tB_1} = \begin{pmatrix} e^{\lambda t} & 0 \\ 0 & e^{\mu t} \end{pmatrix}$$

$$e^{tB_2} = e^{\sigma t} \begin{pmatrix} \cos(\tau t) & -\sin(\tau t) \\ \sin(\tau t) & \cos(\tau t) \end{pmatrix}$$

$$e^{tB_3} = e^{\lambda t} \begin{pmatrix} 1 & t \\ 0 & 1 \end{pmatrix}.$$

We use the same approach for solving linear systems in higher dimensions. But in order to compute exponentials of matrices in Jordan normal form, we need to verify one additional remark.

Lemma 14.1.1 *Let the matrix A be block diagonal, that is, let*

$$A = \begin{pmatrix} C_1 & 0 \\ 0 & C_2 \end{pmatrix},$$

where C_1 is a $\ell \times \ell$ matrix and C_2 is an $(n - \ell) \times (n - \ell)$ matrix. Then

$$e^A = \begin{pmatrix} e^{C_1} & 0 \\ 0 & e^{C_2} \end{pmatrix}.$$

Proof: This lemma is proved by noting that

$$A^k = \begin{pmatrix} C_1^k & 0 \\ 0 & C_2^k \end{pmatrix}$$

and applying the power series definition of matrix exponential; see (6.3.3). ◆

Matrix Exponentials of Jordan Normal Forms

We compute matrix exponentials in two steps: First we consider Jordan blocks corresponding to real eigenvalues, and second we consider Jordan blocks with complex eigenvalues. In this section we discuss mainly solutions to linear systems already in

Jordan normal form. That is, we assume that a similarity transformation has already been performed that put the coefficient matrix into Jordan normal form. In the next chapter, Section 15.1, we consider solving linear systems in their original coordinates.

Real Eigenvalues

A Jordan block with a real simple eigenvalue is just a 1×1 matrix (λ). This matrix corresponds to the single equation $\dot{x} = \lambda x$ whose solution is $x(t) = e^{\lambda t} x_0$.

The $k \times k$ Jordan block with real eigenvalue λ is

$$\lambda I_k + N,$$

where the only nonzero entries in N are the 1s on the superdiagonal. Recall that (13.3.2) implies that $N^k = 0$. Since I_k and N commute, Proposition 6.4.1 implies that $e^{t(I_n+N)} = e^{tI_n}e^{tN}$. We can compute the solution to the differential equation as

$$
\begin{aligned}
X(t) &= e^{\lambda t} e^{tN} X_0 \\
&= e^{\lambda t} \left(I_k + tN + \cdots + \frac{1}{k!} t^{k-1} N^{k-1} \right) X_0.
\end{aligned}
\tag{14.1.2}
$$

Note that N^2 is the matrix with 1s on the second diagonal above the main diagonal and 0s elsewhere. Similarly, N^3 is the matrix with 1s on the third diagonal above the main diagonal and 0s elsewhere; and so on.

An Example of a Real Eigenvalue Jordan Block Equation

Consider the case of a $k = 3$ Jordan block with eigenvalue $\lambda = -0.5$. Then

$$
X(t) = e^{-0.5t}
\begin{pmatrix}
1 & t & \dfrac{t^2}{2} \\
0 & 1 & t \\
0 & 0 & 1
\end{pmatrix}
X_0.
$$

We can now solve the initial value problem in closed form. For example, suppose $X_0 = (0.5, 1.5, 2)$ and let $X(t) = (x_1(t), x_2(t), x_3(t))$. Then the solution of this initial value problem is

$$
\begin{aligned}
x_1(t) &= e^{-0.5t}(0.5 + 1.5t + t^2) \\
x_2(t) &= e^{-0.5t}(1.5 + 2t) \\
x_3(t) &= 2e^{-0.5t}.
\end{aligned}
\tag{14.1.3}
$$

In Figure 14.1 we graph the time series for x_1 as a function of t. Note the initial growth in x_1 (due to the quadratic polynomial factor in x_1) before exponential decay dominates the quadratic growth. See Exercise 8 at the end of this section.

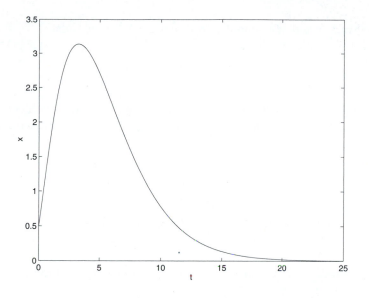

Figure 14.1
First component $x_1(t)$ of solution to Jordan block equation for $k = 3$
when $X(0) = (0.5, 1.5, 2)$

Complex Eigenvalues

If A is the Jordan block corresponding to a simple complex eigenvalue $\lambda = \sigma + i\tau$,
then (14.1.1) is the two dimensional system

$$\frac{dX}{dt} = \begin{pmatrix} \sigma & -\tau \\ \tau & \sigma \end{pmatrix} X$$

whose solutions have the form

$$X(t) = e^{\sigma t} \begin{pmatrix} \cos(\tau t) & -\sin(\tau t) \\ \sin(\tau t) & \cos(\tau t) \end{pmatrix} X_0.$$

The general Jordan block corresponding to a multiplicity k complex eigenvalue
$\sigma + i\tau$ has the form

$$J + M, \qquad\qquad \textbf{(14.1.4)}$$

where J is a block diagonal $2k \times 2k$ matrix with k identical 2×2 matrices

$$\begin{pmatrix} \sigma & -\tau \\ \tau & \sigma \end{pmatrix}$$

on the diagonal and M is a $2k \times 2k$ block form matrix whose nonzero blocks have I_2
on the superdiagonal.

Since J and M commute, Proposition 6.4.1 implies that

$$e^{t(J+M)} = e^{tJ}e^{tM}.$$

As in the case of real eigenvalues, $M^k = 0$. Now since tJ is block diagonal, we can use Lemma 14.1.1 to determine that e^{tJ} is also block diagonal with blocks

$$e^{\sigma t}\begin{pmatrix} \cos(\tau t) & -\sin(\tau t) \\ \sin(\tau t) & \cos(\tau t) \end{pmatrix}.$$

Putting these two observations together allows us to solve explicitly ODE systems whose coefficient matrix is a single Jordan block with complex eigenvalues.

An Example of a Complex Eigenvalue Jordan Block Equation

Find the solution to the differential equation

$$\frac{dX}{dt} = \left(\begin{array}{cc|cc|cc} 2 & 1 & 1 & 0 & 0 & 0 \\ -1 & 2 & 0 & 1 & 0 & 0 \\ \hline 0 & 0 & 2 & 1 & 1 & 0 \\ 0 & 0 & -1 & 2 & 0 & 1 \\ \hline 0 & 0 & 0 & 0 & 2 & 1 \\ 0 & 0 & 0 & 0 & -1 & 2 \end{array}\right) X$$

with initial condition

$$X_0 = (1, 3, -2, 0, 0, 4).$$

Using the notation of (14.1.4), we have

$$J = \left(\begin{array}{cc|cc|cc} 2 & 1 & 0 & 0 & 0 & 0 \\ -1 & 2 & 0 & 0 & 0 & 0 \\ \hline 0 & 0 & 2 & 1 & 0 & 0 \\ 0 & 0 & -1 & 2 & 0 & 0 \\ \hline 0 & 0 & 0 & 0 & 2 & 1 \\ 0 & 0 & 0 & 0 & -1 & 2 \end{array}\right) \quad \text{and} \quad M = \left(\begin{array}{cc|cc|cc} 0 & 0 & 1 & 0 & 0 & 0 \\ 0 & 0 & 0 & 1 & 0 & 0 \\ \hline 0 & 0 & 0 & 0 & 1 & 0 \\ 0 & 0 & 0 & 0 & 0 & 1 \\ \hline 0 & 0 & 0 & 0 & 0 & 0 \\ 0 & 0 & 0 & 0 & 0 & 0 \end{array}\right).$$

It follows that

$$X(t) = e^{tJ} e^{tM} X_0$$

$$
= e^{2t}
\begin{pmatrix}
\cos t & -\sin t & 0 & 0 & 0 & 0 \\
\sin t & \cos t & 0 & 0 & 0 & 0 \\
0 & 0 & \cos t & -\sin t & 0 & 0 \\
0 & 0 & \sin t & \cos t & 0 & 0 \\
0 & 0 & 0 & 0 & \cos t & -\sin t \\
0 & 0 & 0 & 0 & \sin t & \cos t
\end{pmatrix}
\begin{pmatrix}
1 & 0 & t & 0 & \frac{1}{2}t^2 & 0 \\
0 & 1 & 0 & t & 0 & \frac{1}{2}t^2 \\
0 & 0 & 1 & 0 & t & 0 \\
0 & 0 & 0 & 1 & 0 & t \\
0 & 0 & 0 & 0 & 1 & 0 \\
0 & 0 & 0 & 0 & 0 & 1
\end{pmatrix}
X_0
$$

$$
= e^{2t}
\begin{pmatrix}
\cos t & -\sin t & t\cos t & -t\sin t & \frac{1}{2}t^2\cos t & -\frac{1}{2}t^2\sin t \\
\sin t & \cos t & t\sin t & t\cos t & \frac{1}{2}t^2\sin t & \frac{1}{2}t^2\cos t \\
0 & 0 & \cos t & -\sin t & t\cos t & -t\sin t \\
0 & 0 & \sin t & \cos t & t\sin t & t\cos t \\
0 & 0 & 0 & 0 & \cos t & -\sin t \\
0 & 0 & 0 & 0 & \sin t & \cos t
\end{pmatrix}
\begin{pmatrix}
1 \\ 3 \\ -2 \\ 0 \\ 0 \\ 4
\end{pmatrix}.
$$

Finally,

$$
X(t) = e^{2t}
\begin{pmatrix}
\cos t - 3\sin t - 2t\cos t - 2t^2\sin t \\
\sin t + 3\cos t - 2t\sin t + 2t^2\cos t \\
-2\cos t - 4t\sin t \\
-2\sin t + 4t\cos t \\
-4\sin t \\
4\cos t
\end{pmatrix}.
$$

Moreover, this example is typical of all such Jordan block equations.

Solutions When Matrices Are Not in Jordan Normal Form

Let A and $B = S^{-1}AS$ be similar $n \times n$ matrices. Recall Lemma 6.5.2, which states that $X(t) = SY(t)$ is a solution to $\dot{X} = AX$ if and only if $Y(t)$ is a solution to

$$\frac{dY}{dt} = BY. \tag{14.1.5}$$

Suppose that B is in Jordan normal form. Then we know how to solve (14.1.5) using matrix exponentials and we can solve the original equation $\dot{X} = AX$ in principle by multiplying $Y(t)$ by S. This approach is clean in theory, but messy in practice.

As an example, consider the differential equation $\dot{X} = AX$, where

$$A = \begin{pmatrix} 12 & 4 & 3 & 6 \\ 5 & 22 & 6 & -6 \\ -34 & -69 & -22 & 7 \\ -5 & 16 & 3 & -10 \end{pmatrix} \qquad \textbf{(14.1.6)*}$$

with initial condition $X_0 = (1, -1, 0, 1)^t$. Use eig(A) in MATLAB to verify that the eigenvalues of A are $-1, -1, 2, 2$. Since the nullities of $A + I_n$ and $A - 2I_n$ both equal 1, it follows that both double eigenvalues have nontrivial Jordan blocks, and

$$B = \begin{pmatrix} \begin{pmatrix} -1 & 1 \\ 0 & -1 \end{pmatrix} & 0 \\ 0 & \begin{pmatrix} 2 & 1 \\ 0 & 2 \end{pmatrix} \end{pmatrix}$$

is the Jordan normal form of A. As noted previously, we can solve (14.1.5) using matrix exponentials, obtaining

$$Y(t) = \begin{pmatrix} e^{-t}\begin{pmatrix} 1 & t \\ 0 & 1 \end{pmatrix} & 0 \\ 0 & e^{2t}\begin{pmatrix} 1 & t \\ 0 & 1 \end{pmatrix} \end{pmatrix} Y_0.$$

To find a formula for $X(t)$ we need to find the matrix S such that $B = S^{-1}AS$. Then $X(t) = SY(t)$, where $Y_0 = S^{-1}X_0$. Using MATLAB, we can find vectors

$$v_2 \in \text{null space}((A + I_n)^2) - \text{null space}(A + I_n)$$
$$v_4 \in \text{null space}((A - 2I_n)^2) - \text{null space}(A - 2I_n).$$

Then set $v_1 = (A + I_n)v_2$, $v_3 = (A - 2I_n)v_4$, and $S = (v_1|v_2|v_3|v_4)$. Indeed,

```
S =
   -0.3580   -0.1181    0.2039   -0.0775
    0.3580   -0.2399    0.1019   -0.2936
   -0.3580    0.9559   -0.6116    0.9459
    0.7160   -0.1218   -0.1019   -0.1142
```

and

```
Y0 =
    5.5285
  -16.7608
   16.0879
   29.4307
```

So

$$X(t) = S \left(\begin{array}{cc} e^{-t} \begin{pmatrix} 1 & t \\ 0 & 1 \end{pmatrix} & 0 \\ 0 & e^{2t} \begin{pmatrix} 1 & t \\ 0 & 1 \end{pmatrix} \end{array} \right) \begin{pmatrix} 5.5285 \\ -16.7608 \\ 16.0879 \\ 29.4307 \end{pmatrix}. \qquad \textbf{(14.1.7)}$$

The General Functional Form of Solutions

The previous calculations suggest how to prove the following lemma.

Lemma 14.1.2   *Components of solutions to the system of differential equations $\dot{X} = AX$, where A is an $n \times n$ matrix, are linear combinations of the functions $t^j e^{\lambda t}$, where λ is a real eigenvalue of A, and $t^j e^{\sigma t} \cos(\tau t)$ and $t^j e^{\sigma t} \sin(\tau t)$, where $\sigma \pm i\tau$ is a complex eigenvalue of A. The exponent $j \leq k$, where k is the size of the largest Jordan block associated with the given eigenvalue.*

Proof: Formula (14.1.2) verifies this statement when A is in Jordan normal form with real eigenvalues. A similar formula holds with J and M when A has complex eigenvalues. Solutions to the general matrix equation are just linear combinations of solutions to the corresponding Jordan normal form equations. Indeed, as noted in the discussion leading to (14.1.5), solutions to the general system equation are obtained from the Jordan normal form equations by multiplying these special solutions by the constant matrix S. This multiplication does not alter the statement that the coordinate functions are linear combinations of the given functions. $\qquad \blacklozenge$

In Section 15.1 we discuss alternative methods for finding solutions to systems of constant coefficient linear differential equations. All of these methods require some effort to carry out by hand.

HAND EXERCISES

1. Find the solution to the system

$$\dot{x}_1 = 2x_1 + x_2$$
$$\dot{x}_2 = 2x_2$$
$$\dot{x}_3 = x_3 - x_4$$
$$\dot{x}_4 = x_3 + x_4$$

satisfying $X_0 = (-1, 1, 2, -3)^t$.

2. Find the solution to the system

$$\dot{x}_1 = -2x_1 + x_2 + x_3$$
$$\dot{x}_2 = -x_1 - 2x_2 + x_4$$
$$\dot{x}_3 = -2x_3 + x_4$$
$$\dot{x}_4 = -x_3 - 2x_4$$

satisfying $X_0 = (1, 2, -1, -3)^t$.

3. Let $X(t)$ be the solution to

$$\frac{dX}{dt} = \begin{pmatrix} -\frac{1}{2} & 1 & 0 \\ 0 & -\frac{1}{2} & 1 \\ 0 & 0 & -\frac{1}{2} \end{pmatrix} X$$

satisfying $X_0 = (0, 1, 1)^t$. Show that $X(2)$ is farther from the origin than $X(0)$.

4. Consider the differential equation $\dot{X} = JX$, where J is the single Jordan block 2×2 matrix with double eigenvalue 0. Show that most solutions are unbounded in both forward and backward time.

5. Consider the differential equation $\dot{X} = JX$, where J is the single Jordan block 4×4 matrix with double eigenvalues i and $-i$. Show that most solutions spiral away from the origin in both forward and backward time.

6. Find e^J, where

$$J = \begin{pmatrix} 1 & -2 & 1 & 0 \\ 2 & 1 & 0 & 1 \\ 0 & 0 & 1 & -2 \\ 0 & 0 & 2 & 1 \end{pmatrix}.$$

7. Find e^{tJ}, where

$$J = \begin{pmatrix} -1 & 1 & 0 & 0 & 0 \\ 0 & -1 & 1 & 0 & 0 \\ 0 & 0 & -1 & 0 & 0 \\ 0 & 0 & 0 & 2 & 1 \\ 0 & 0 & 0 & 0 & 2 \end{pmatrix}.$$

COMPUTER EXERCISES

8. Use calculus to compute the maximum value of the function $x_1(t)$ defined in (14.1.3) and graphed in Figure 14.1.

In Exercises 9 and 10, solve the initial value problem $\dot{X} = AX$ for the given matrix A and initial vector X_0 in the form of (14.1.7) by converting A to Jordan normal form.

9.

$$A = \begin{pmatrix} -3 & -7 & -2 \\ 11 & 19 & 5 \\ -32 & -51 & -13 \end{pmatrix} \quad \text{and} \quad X_0 = (1, 0, -1)^t \qquad \text{(14.1.8)*}$$

10.

$$A = \begin{pmatrix} 103 & 138 & 124 & -215 \\ -67 & -85 & -68 & 142 \\ 22 & 27 & 20 & -47 \\ 19 & 27 & 27 & -39 \end{pmatrix} \quad \text{and} \quad X_0 = (1, 2, 0, 3)^t \qquad \text{(14.1.9)*}$$

In Exercises 11 and 12, components of solutions to differential equation $\dot{X} = AX$ are linear combinations of functions of the form $t^j e^{\lambda t}$ when λ is a real eigenvalue of A and $t^j e^{\sigma t} \cos(\tau t)$ and

$t^j e^{\sigma t} \sin(\tau t)$ when $\lambda = \sigma \pm i\tau$ is a complex eigenvalue of A. Determine the precise values of j and λ for each of the given matrices. Do not attempt to write down the closed form solutions to these systems of differential equations.

11.

$$A = \begin{pmatrix} -3 & 77 & -124 & -225 \\ 10 & -1 & 69 & 105 \\ -26 & -47 & -112 & -148 \\ 18 & 24 & 89 & 124 \end{pmatrix} \qquad (14.1.10)*$$

12.

$$A = \begin{pmatrix} 29 & 27 & -4 & -7 & -3 & -20 \\ -26 & -23 & 5 & 9 & 3 & 23 \\ 12 & 26 & -4 & 32 & 7 & -24 \\ 12 & 12 & -3 & -1 & -1 & -13 \\ -8 & -10 & 4 & -3 & 1 & 16 \\ -4 & -8 & 2 & -9 & -2 & 10 \end{pmatrix} \qquad (14.1.11)*$$

14.2   QUALITATIVE THEORY NEAR EQUILIBRIA

In this section we discuss asymptotic stability and hyperbolicity for linear systems of differential equations. We show, in general, that each linear constant coefficient system divides coordinates into three groups: stable, center, and unstable. Then, in analogy with the planar case, we discuss how solutions to nonlinear autonomous systems look very much like solutions to linear differential equations, at least in a neighborhood of a hyperbolic equilibrium.

Asymptotic Stability

Consider the linear system of ODEs $\dot{X} = AX$. In the case $n = 2$ we showed that the origin is an asymptotically stable equilibrium precisely when the eigenvalues of A have negative real part. In this subsection we show that the same result holds for general n.

Definition 14.2.1   *The origin is a* linearly stable *equilibrium for the system of differential equations* $\dot{X} = AX$ *if the n eigenvalues of A all have negative real part.*

Using this definition, we have:

Theorem 14.2.2   *The origin is an asymptotically stable equilibrium for the system of differential equations* $\dot{X} = AX$ *if and only if the origin is linearly stable.*

Proof: The proof of this theorem follows directly from Lemma 14.1.2. Each coordinate function of each solution of $\dot{X} = AX$ is a linear combination of functions of the form $t^j e^{\lambda t}$, where λ is a real eigenvalue of A, and $t^j e^{\sigma t} \cos(\tau t)$ and $t^j e^{\sigma t} \sin(\tau t)$, where $\lambda = \sigma \pm i\tau$ is a complex eigenvalue of A. Observe that if $\lambda < 0$, then

$$\lim_{t \to \infty} t^k e^{\lambda t} = 0,$$

since exponentials decay to 0 faster than polynomials grow to infinity. A similar argument holds when the multiple eigenvalue is complex since

$$\lim_{t \to \infty} t^k e^{\sigma t} \cos(\tau t) = 0 = \lim_{t \to \infty} t^k e^{\sigma t} \sin(\tau t)$$

when $\sigma < 0$.   ◆

Stable, Unstable, and Center Coordinates

Recall from Chapter 7 (Theorem 9.4.3) that two matrices are similar precisely when there is a linear change of coordinates that transforms one matrix into the other. Recall from Theorem 9.4.3 that when we assume that a matrix A is in Jordan normal form in the system of differential equations $\dot{X} = AX$, we are just assuming that we have made a preliminary change of coordinates to put that matrix in normal form.

We can always group the eigenvalues of a matrix into three classes: those that have negative real part, those that have positive real part, and those that have zero real part; and we can always arrange the Jordan normal form so that:

(i)   blocks corresponding to eigenvalues with negative real part come first,

(ii)   blocks corresponding to eigenvalues with positive real part come next, and

(iii)   blocks corresponding to eigenvalues with zero real part come last.

As a consequence of this discussion we have proved:

Proposition 14.2.3   *Every $n \times n$ matrix B is similar to a block diagonal $n \times n$ matrix*

$$A = \begin{pmatrix} S & 0 & 0 \\ 0 & U & 0 \\ 0 & 0 & C \end{pmatrix}, \qquad \text{(14.2.1)}$$

where

(a)   *S is a Jordan normal form $k_S \times k_S$ matrix all of whose eigenvalues have negative real part;*

(b)   *U is a Jordan normal form $k_U \times k_U$ matrix all of whose eigenvalues have positive real part; and*

(c)   *C is a Jordan normal form $k_C \times k_C$ matrix all of whose eigenvalues have zero real part;*

and $k_S + k_U + k_C = n$.

Using Proposition 14.2.3, we can give a qualitative description of every constant coefficient linear system of differential equations.

Theorem 14.2.4   *Let A be a Jordan normal form matrix in the block diagonal form (14.2.1). Let $X = (x, y, z) \in \mathbf{R}^n$, where $x \in \mathbf{R}^s$, $y \in \mathbf{R}^u$, and $z \in \mathbf{R}^c$. Then every solution to the differential equation $\dot{X} = AX$ has the form*

$$X(t) = (x(t), y(t), z(t)),$$

where

   (a) *$x(t)$ decays to 0 exponentially fast in forward time and grows to infinity exponentially fast in backward time;*

   (b) *$y(t)$ grows to infinity exponentially fast in forward time and decays to 0 exponentially fast in backward time; and*

   (c) *$z(t)$ can be either bounded or grow or decay.*

Moreover, if the Jordan blocks of C are all trivial (that is, 1×1 zero blocks or 2×2 blocks with complex conjugate purely imaginary eigenvalues), then $z(t)$ remains bounded in both forward and backward time.

Proof: The block diagonal form of A decouples the differential equation into three subsystems:

$$\dot{x} = Sx$$
$$\dot{y} = Uy$$
$$\dot{z} = Cz.$$

Since the eigenvalues of S all have negative real part, Theorem 14.2.2 proves (a). Since the eigenvalues of U all have positive real part, we can run time backward and obtain the differential equation $\dot{y} = -Uy$, where all of the eigenvalues of $-U$ have negative real part. Again, we can use Theorem 14.2.2 to prove (b). There is nothing to prove in part (c). However, to prove the last statement, observe that if the Jordan blocks in C are trivial, then the blocks are either 1×1 zero blocks or 2×2 blocks $\begin{pmatrix} 0 & -\tau \\ \tau & 0 \end{pmatrix}$. Solutions to the subblock equations are either equilibria (in the case of a 1×1 block) or circles (in the case of the 2×2 block). The result follows, since the superposition of bounded solutions is bounded.   ◆

Definition 14.2.5   *The x-coordinates in Theorem 14.2.4 are the* stable *coordinates, the y-coordinates are the* unstable *coordinates, and the z-coordinates are the* center *coordinates.*

Theorem 14.2.4 implies that for each linear constant coefficient system of n equations, we can group coordinates of $\mathbf{R}^n$ into stable, unstable, and center coordinates for that differential equation. Moreover, we need to know just the eigenvalues of the coefficient matrix to make this grouping.

Linearizations near Hyperbolic Equilibria

Consider the autonomous system of differential equations

$$\frac{dX}{dt} = f(X), \tag{14.2.2}$$

where $X \in \mathbf{R}^n$ and $f : \mathbf{R}^n \to \mathbf{R}^n$. We restate here for higher dimensional systems of differential equations some of the ideas and terminology that we introduced previously for planar systems. The point $X_0 \in \mathbf{R}^n$ is an equilibrium or steady-state solution if $f(X_0) = 0$. As we have seen, $X_0 = 0$ is an equilibrium of any linear system, where $f(X) = AX$ for some $n \times n$ matrix A.

Suppose that X_0 is an equilibrium for (14.2.2). Let

$$f(X) = (f_1(X), \ldots, f_n(X))$$

define the coordinate functions of f as a function of $X = (x_1, \ldots, x_n)$. Recall that the *Jacobian matrix* at X_0 is the $n \times n$ matrix of partial derivatives of f and is denoted by $(df)_{X_0}$. More precisely,

$$(df)_{X_0} = \begin{pmatrix} \frac{\partial f_1}{\partial x_1}(X_0) & \cdots & \frac{\partial f_1}{\partial x_n}(X_0) \\ \vdots & & \vdots \\ \frac{\partial f_n}{\partial x_1}(X_0) & \cdots & \frac{\partial f_n}{\partial x_n}(X_0) \end{pmatrix}.$$

An equilibrium X_0 of (14.2.2) is *hyperbolic* if the n eigenvalues of $(df)_{X_0}$ all have nonzero real part. We now state the analog of Theorem 11.2.3.

Theorem 14.2.6 *Suppose that the system of differential equations (14.2.2) has a hyperbolic equilibrium at X_0. Then, in a sufficiently small neighborhood of X_0, the phase space for (14.2.2) is the same as the phase space of the system of linear differential equations*

$$\frac{dX}{dt} = (df)_{X_0} X. \tag{14.2.3}$$

As before, it is difficult to define precisely what we mean by the word *same*. Roughly speaking, *same* means that near X_0 there is a nonlinear change of coordinates that transforms the nonlinear equation into the linear one.

Asymptotically Stable Equilibria in Nonlinear Systems

We illustrate this theorem and consider the linear system $\dot{X} = AX$, where A is given by

$$A = \begin{pmatrix} -0.1 & 0.1 & -0.2 & 0.1 \\ -0.8 & -0.5 & 1.9 & 0 \\ 2.4 & -3.9 & -0.7 & 0.1 \\ 0.3 & 0.1 & -0.2 & -0.3 \end{pmatrix}. \tag{14.2.4)*}$$

We can enter A into MATLAB and compute the eigenvalues of A by typing

```
e12_2_4
eig(A)
```

and we obtain

```
ans =
  -0.5725+ 2.8194i
  -0.5725- 2.8194i
  -0.0550
  -0.4000
```

In particular, all eigenvalues of A have negative real parts, and Theorem 14.2.2 implies that the origin is an asymptotically stable equilibrium.

Theorem 14.2.6 implies that the origin remains asymptotically stable for any system of differential equations

$$\dot{X} = AX + N(X), \tag{14.2.5}$$

where $N(X)$ consists of only higher order terms. For example, suppose

$$N(X) = \begin{pmatrix} 2x_1^2 - x_1 x_2 \\ -x_4^3 \\ -x_3^2 \\ x_4 x_1 \end{pmatrix}. \tag{14.2.6}$$

Then the Jacobian matrix of the right-hand side in (14.2.5) at $X_0 = 0$ is given by A, and Theorem 14.2.6 guarantees that solutions to the nonlinear equation (14.2.5) that have an initial condition near enough to the origin tend toward the origin in forward time. It would be nice to test the consequences of this theorem numerically, but until now we have not discussed how to integrate solutions to differential equations in more than two dimensions. We do this in Section 14.4 using a three-dimensional equation.

Saddles, Sources, and Sinks in Higher Dimensions

Suppose that one eigenvalue λ of the $n \times n$ matrix A has positive real part. Then *almost all* solutions to the linear system $\dot{X} = AX$ tend to infinity in forward time. This statement can be proved by appealing to Theorem 14.2.4.

It follows from Theorem 14.2.6 that if X_0 is an equilibrium and one eigenvalue of $(df)_{X_0}$ has positive real part, then almost all solutions near the equilibrium X_0 will leave small neighborhoods of that equilibrium in forward time. Thus, in this case the equilibrium is not asymptotically stable.

Note that if $(df)_{X_0}$ has an eigenvalue with positive real part and an eigenvalue with negative real part, then most solutions with initial conditions near X_0 will leave small neighborhoods of X_0 in both forward and backward times. Such equilibria are called *saddles*.

Equilibria are called *sources* when all the eigenvalues of the corresponding Jacobian have positive real parts and *sinks* when all the eigenvalues of the corresponding Jacobian have negative real parts.

HAND EXERCISES

In Exercises 1 and 2, compute the Jacobian matrices of the given system of differential equations at the given point.

1. $\begin{cases} \dot{x} = -10 + 2x - z - 3x^2 + y^2 \\ \dot{y} = x + y + 2xz \\ \dot{z} = -27 + z + 10xy - x^2 \end{cases}$ and $(x_0, y_0, z_0) = (2, -1, 4)$

2. $\begin{cases} \dot{x} = 2 + 2z + x^2 - z^3 \\ \dot{y} = x + xyz \\ \dot{z} = -10 + y + 10xy + x^3 \end{cases}$ and $(x_0, y_0, z_0) = (1, 2, -2)$

3. Verify that $(1, 3, -2)$ is an equilibrium for the differential equation in Exercise 1. Is this equilibrium asymptotically stable or not?

4. Verify that $(1, 1, -1)$ is an equilibrium for the differential equation in Exercise 2. Is this equilibrium asymptotically stable or not?

5. Consider the system of ODEs $\dot{X} = F(X)$, where

$$F(x, y, z) = (x^2 - z, y^2 - z, 2x + y - 3).$$

Find all equilibria of this system and determine the number of stable and unstable directions for each of these equilibria.

6. Consider the system of ODEs $\dot{X} = F(X)$, where

$$F(x, y, z) = (x - 2y - x^2, y + z^2, -z - x^2).$$

Find all equilibria of this system and determine the number of stable and unstable directions for each of these equilibria.

COMPUTER EXERCISES

In Exercises 7–9, determine whether the origin is an asymptotically stable equilibrium for the system of differential equations $\dot{x} = Cx$.

7.

$$C = \begin{pmatrix} 1 & -3 \\ 2 & -5 \end{pmatrix}$$

8.

$$C = \begin{pmatrix} 94 & 174 & 132 \\ -37 & -68 & -53 \\ -16 & -29 & -24 \end{pmatrix} \qquad \textbf{(14.2.7)}*$$

9.

$$C = \begin{pmatrix} -0.28 & -0.04 & -0.22 & -0.42 \\ 0.16 & -0.62 & 0.14 & 0.04 \\ 0.06 & -0.12 & -0.06 & 0.14 \\ 0.04 & -0.28 & 0.06 & -0.34 \end{pmatrix} \qquad \textbf{(14.2.8)}*$$

In Exercises 10–12, determine the number of stable, unstable, and center directions for the system of differential equations $\dot{x} = Cx$.

10.

$$C = \begin{pmatrix} 1 & 2 & 3 & 4 \\ 5 & 4 & 3 & 2 \\ 9 & 8 & 7 & 6 \\ 1 & 2 & 6 & 7 \end{pmatrix} \qquad \textbf{(14.2.9)}*$$

11.

$$C = \begin{pmatrix} 1 & 3 & 6 & 9 & 11 \\ -2 & 4 & 6 & -8 & 0 \\ 1 & 3 & 5 & 7 & 9 \\ 0 & 2 & 5 & 8 & 0 \\ 1 & 4 & -2 & 4 & 10 \end{pmatrix} \qquad \textbf{(14.2.10)}*$$

12.

$$C = \begin{pmatrix} 1 & 3 & 5 & -11 & 20 & 5 \\ -5 & -13 & -11 & 23 & -49 & -12 \\ 13 & 47 & 41 & -82 & 167 & 41 \\ 4 & 35 & 27 & -48 & 94 & 23 \\ 8 & 10 & 6 & -16 & 45 & 11 \\ -30 & -8 & -2 & 28 & -110 & -27 \end{pmatrix} \qquad \textbf{(14.2.11)}*$$

14.3   MATLAB ode45 IN ONE DIMENSION

Previously, we have used `dfield5`, `pline`, and `pplane5` to compute solutions to one- and two-dimensional ordinary differential equations numerically. We now wish to compute solutions to higher dimensional systems of ordinary differential equations, and to do this we use the MATLAB command `ode45`. MATLAB provides the command `ode45`, among others, for solving initial value problems, and we illustrate how this command works on several examples. In this section we illustrate the command on a simple one-dimensional example, and in the next section we compute solutions to three- and four-dimensional systems.

A Simple One-Dimensional Example

Suppose that we want to compute numerically the solution of the initial value problem

$$\frac{dx}{dt}(t) = x(t) + t$$
$$x(t_0) = x_0$$

(14.3.1)

on the time interval $t \in [t_0, t_e]$. Then the *data* of this problem are

- the numbers x_0, t_0, and t_e, and
- the function $f(t, x) = x + t$.

We know how to assign specific values for t_0, x_0, and t_e in MATLAB; we will call these variables t0, x0, and te. Before proceeding, we need to understand how to make a function $f(t, x)$ available for computations in MATLAB.

The Construction of m-files in MATLAB

Functions that are used by MATLAB are defined via *m-files*. We now show how to construct m-files with two examples.

Suppose that we want the function $g : \mathbf{R}^2 \to \mathbf{R}$, defined by

$$g(y, z) = yz^2 + \sin y,$$

(14.3.2)*

to be available in MATLAB. Then we specify g by writing a MATLAB m-file that contains the information defining this function. The m-file gexam.m that defines the function $g(y, z)$ in (14.3.2) has two lines:

```
function g = gexam(y,z)
g = y*z^2 + sin(y);
```

The first line states that

- this m-file contains a function with the name gexam,
- the input arguments are y and z, and
- the output variable is g.

The value of the function g at the point (y,z) is given in the second line

```
g = ... ;
```

Note that this line must end with a semicolon.

Remark: When MATLAB is used under Windows or on a PowerMac, m-files can be created using the New entry in the File menu. On Unix-based systems we must create the m-file using a separate editor. In any case make sure that the new m-file is stored in the directory where MATLAB has been started.

As an exercise, create the m-file gexam.m. The function g is now available in MATLAB; when we type gexam(1,2), we obtain the answer 4.8415, which is indeed $g(1, 2) = 4 + \sin(1)$.

We store function m-files with the other laode files using the prefix f (for function) followed by the equation number. So the m-file gexam.m is stored as f14_3_2.m. Indeed, compare the result of typing f14_3_2(1,2) with gexam(1,2).

Both the input arguments and the output variable in an m-file can consist of vectors. For instance, suppose that we want to define the function $h : \mathbf{R}^3 \to \mathbf{R}^2$, where $u = (u_1, u_2, u_3)' \in \mathbf{R}^3$ and

$$h(u) = \begin{pmatrix} u_1 u_2 \\ u_2 - u_1 u_3 \end{pmatrix}. \qquad \textbf{(14.3.3)*}$$

As a second exercise, write the m-file hexam.m as follows:

```
function h = hexam(u)
h = [u(1)*u(2), u(2)-u(1)*u(3)]';
```

The command w = hexam([1,2,3]) gives the answer

```
w =
    2
   -1
```

which is indeed $h(1, 2, 3)$. This m-file has been stored as f14_3_3.m.

The MATLAB Command ode45

We now return to the initial value problem (14.3.1). The function

$$f(t, x) = x + t \qquad \textbf{(14.3.4)*}$$

on the right-hand side is available to MATLAB in the m-file f14_3_4.m, which is stored with the other laode files:

```
function f = f14_3_4(t,x)
f = x + t;
```

The MATLAB command

```
ode45('f14_3_4',[t0 te],x0)
```

computes a numerical approximation to the solution to the initial value problem (14.3.1) on the interval [t0,te]. For instance, we can use ode45 to solve the initial value problem

$$\frac{dx}{dt} = x + t$$
$$x(1) = 2$$

(14.3.5)

for $t \in [1, 3]$. In this case, $t_0 = 1$, $t_e = 3$, and $x_0 = 2$. After we type the command

```
[t,x] = ode45('f14_3_4',[1 3],2);
```

the approximate solution is stored in the vectors t and x. The components of the vector t are the values of t for which the solution has been computed, and the components of the vector x are the values of the solution x at the values t. Indeed, typing t and x, we see that both vectors have length 45. The length of the vectors t and x is chosen by ode45 so as to guarantee a certain accuracy in the solution. We discuss this issue in greater detail below.

We can visualize the solution by plotting its time series:

```
plot(t,x)
xlabel('t')
ylabel('x')
```

The result is shown in Figure 14.2.

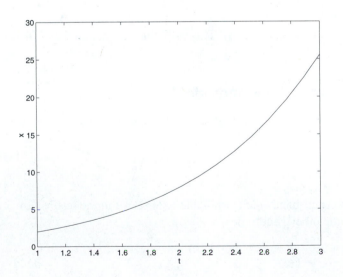

Figure 14.2
Approximation of the solution of (14.3.5) obtained by ode45 in
MATLAB

In fact, the initial value problem (14.3.5) has a simple closed form solution

$$x(t) = 4e^{t-1} - t - 1,$$

which can be verified directly by differentiation. The existence of this closed form solution allows us to check the accuracy of the numerical calculations made by ode45. Indeed, graphically we cannot distinguish the result obtained using ode45 from the exact solution. To verify this point, type

```
x_exact = 4*exp(t-1)-t-1;
hold on
plot(t,x_exact,'r')
```

and observe that the graph of the exact solution, which is in red, exactly covers the plot of the numerically computed solution.

Accuracy of ode45

We now discuss the error made by ode45 in more detail. Using ode45, we obtained an approximation of the solution of the initial value problem (14.3.5) at the times $t_k = $ t(k) for $k = 1, 2, \ldots, 45$. The exact solution at these points is

$$x(t_k) = 4e^{t_k-1} - t_k - 1,$$

and the ode45 approximation to the solution is $x_k = $ x(k). The routine ode45 automatically computes the solution subject to satisfying two constraints: *absolute error* and *relative error*.

The *absolute error* is just the absolute value of the difference between the numerically computed solution and the exact solution; that is,

$$\epsilon_{\text{abs}}(k) = |x_k - x(t_k)|.$$

We can visualize the absolute error by plotting ϵ_{abs} versus t, as follows:

```
err_abs = abs(x-x_exact);
plot(t,err_abs)
xlabel('t')
ylabel('absolute error')
```

The MATLAB command abs(v) generates a vector containing the absolute values of the components of the vector v. The result is presented in Figure 14.3. Note that even though the absolute error oscillates, it is always less than $6 \cdot 10^{-6}$, which is why we could not distinguish the graph of the exact solution from the graph of the numerically computed solution given in Figure 14.2.

Having an absolute error of 10^{-6} in a numerically computed solution might seem quite good—unless we happened to be computing a solution whose size is 10^{-7}. Then

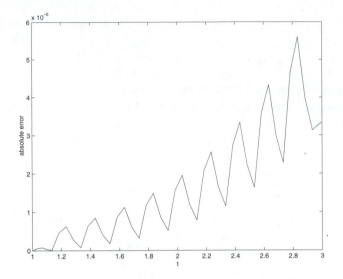

Figure 14.3

Absolute error in the approximation of the solution of (14.3.5) obtained by ode45 in MATLAB with default error bound 1e-6

the numerical error would be ten times the size of the solution itself, which is a huge error. For this reason, numerical analysts like to use another measure for success.

The *relative error* between the approximation and the exact solution is the absolute error normalized by the size of the exact solution; that is,

$$\epsilon_{rel}(k) = \frac{\epsilon_{abs}(k)}{|x(t_k)|} = \frac{|x_k - x(t_k)|}{|x(t_k)|}.$$

The numbers $\epsilon_{rel}(k)$ must be uniformly small in order to guarantee a good numerical approximation to the actual solution. Using the fact that we have a formula for the exact solution, we can compute the numbers ϵ_{rel} and plot them in MATLAB by typing

```
err_rel = abs(x-x_exact)./abs(x_exact);
plot(t,err_rel)
xlabel('t')
ylabel('relative error')
```

The result is shown in Figure 14.4. We see that the relative error oscillates but is always smaller than $3 \cdot 10^{-7}$.

Default error bounds are preset in the command ode45. Unless otherwise instructed, ode45 *attempts* to find a numerical approximation whose absolute error is everywhere less than 10^{-6} and whose relative error is everywhere less than 10^{-3}. (There is an interesting mathematical question concerning how these bounds are actually satisfied, since ode45 does not, in fact, know the exact solution.) We can use ode45 to

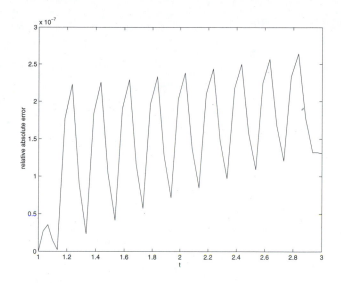

Figure 14.4
Relative error in the approximation of the solution of (14.3.5)
obtained by ode45 in MATLAB with default error bound 1e−3

compute an approximation of the solution with an even smaller error. To set this smaller error, we add an argument to the call of ode45 by typing

```
options = odeset('RelTol',1e-8);
[t,x]=ode45('f14_3_4',[1 3],2,options);
```

These instructions compute an approximation for which the relative error is smaller than 10^{-8}. (Type odeset in MATLAB in order to see a complete list of options. Type options to see a list of the currently specified options.) Hence, when we perform this calculation, we expect to obtain an even better result than before. Indeed, proceeding as above, we obtain the relative error shown in Figure 14.5. In an attempt to guarantee the reduced error, ode45 generates 61 time steps during the computation.

COMPUTER EXERCISES

For each function specified in Exercises 1–6, write an m-file that makes that function available in MATLAB.

1. $f1 : \mathbf{R} \to \mathbf{R}$, where $f1(t) = \sin(t) - t^3$

2. $f2 : \mathbf{R}^2 \to \mathbf{R}$, where $f2(x, y) = x(1 - y)$

3. $f3 : \mathbf{R}^2 \to \mathbf{R}^2$, where $f3(x, y) = \begin{pmatrix} xy - 1 \\ x + 2y \end{pmatrix}$

4. $f4 : \mathbf{R}^3 \to \mathbf{R}$, where $f4(u) = u_1 - u_2^2 + u_3$ and $u = (u_1, u_2, u_3)^t$

5. $f5 : \mathbf{R}^3 \to \mathbf{R}^2$, where $f5(u) = \begin{pmatrix} u_1 - 2u_3 \\ u_1 u_2 u_3 \end{pmatrix}$ and $u = (u_1, u_2, u_3)^t$

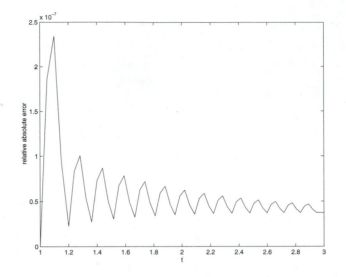

Figure 14.5
Relative error in the approximation of the solution of (14.3.5)
obtained by ode45 in MATLAB with 1e−8 error bound

6. $f6 : \mathbf{R}^3 \to \mathbf{R}^3$, where $f6(u) = \begin{pmatrix} u_3 \\ u_1 u_3 \\ \sin u_2 \end{pmatrix}$ and $u = (u_1, u_2, u_3)^t$

7. Verify that $x(t) = e^{1/2(t^2-4)}$ is a solution to the initial value problem

$$\dot{x} = tx$$
$$x(2) = 1.$$

Use ode45 to compute the solution to this initial value problem on the interval [2, 3] to within an accuracy of 10^{-6} and graphically compare this answer with the graph of the exact solution. Find the values $x(2.5)$ and $x(2.75)$.
Hint Set the exact times t where the ODE solver evaluates time by the command tspan = 2:0.01:3; and insert tspan instead of the interval [2,3] in ode45. Use help ode45 in MATLAB for additional information.

8. Verify that $x(t) = e^{(1-\cos t)}$ is a solution to the initial value problem

$$\dot{x} = x \sin t$$
$$x(0) = 1.$$

Use ode45 to compute the solution to this initial value problem on the interval [0, 15] to within an accuracy of 10^{-4} and graphically compare this answer with the graph of the exact solution. Find the values $x(2)$ and $x(3)$.

14.4   HIGHER DIMENSIONAL SYSTEMS USING ode45

In this section we discuss how to use ode45 to find solutions to linear and nonlinear systems of differential equations in three dimensions, and how to plot the results of these calculations. The same ideas work in principle in any numbers of dimensions. Specifically, we compute the solutions of a nonlinear system and its linearization at a hyperbolic equilibrium. In this example we test numerically the conclusions of Theorem 14.2.6 on linearized stability for nonlinear systems.

A Three-Dimensional Linear Example

We now use ode45 to numerically compute solutions of the system $\dot{X} = AX$, where

$$A = \begin{pmatrix} -0.25 & 3.00 & 0 \\ -3.00 & -0.25 & 0 \\ 0 & 0 & -0.2 \end{pmatrix}. \tag{14.4.1}$$

The function

$$f(X) = AX \tag{14.4.2}*$$

that is on the right-hand side of this linear system of differential equations is stored in the m-file f14_4_2.m. The following lines are included in that m-file:

```
function f = f14_4_2(t,x)
A = [ −0.25 3.0 0; −3 −0.25 0; 0 0 −0.2];
f = A*x;
```

Observe that the first argument of the function f14_4_2 has to be t even though this variable does not explicitly occur on the right-hand side.

The eigenvalues of A are $-0.25 \pm 3i$ and -0.2. It follows that in the $x_1 x_2$-plane solutions spiral into the origin, and along the x_3-axis solutions decay exponentially to the origin. By superposition most solutions spiral around the x_3-axis while decaying into the origin. We test this prediction using ode45.

We approximate the solution starting in $X_0 = (2, -1, -1)$ using ode45 on the time interval [0, 100] by typing

```
[t,x] = ode45('f14_4_2',[0 100],[2,−1,−1]');
```

Note that when we use ode45, the initial condition must be entered as a *column* vector. (The reason is that this vector is an input argument of the function f14_4_2, and in this function this argument is multiplied by the matrix A.) When the computation is complete, the approximation of $(t, x_1(t), x_2(t), x_3(t))$ is stored in the column vector t and in the three columns of the matrix x. Indeed, if we type size(t) or size(x), then we see that t is a vector of length 897 and x is a matrix with 897 rows and 3 columns.

Next we consider how to view the solutions graphically. As in one and two dimensions, there are two possibilities. First, we can visualize the three time series $(t, x_1(t))$, $(t, x_2(t))$, and $(t, x_3(t))$. This can be done by typing

```
subplot(3,1,1)
plot(t,x(:,1))
ylabel('x1')
subplot(3,1,2)
plot(t,x(:,2))
ylabel('x2')
subplot(3,1,3)
plot(t,x(:,3))
ylabel('x3')
xlabel('t')
```

We then obtain the result shown in Figure 14.6. This figure shows the expected oscillation in the x_1- and x_2-coordinates and the exponential decay in the x_3-coordinate.

In drawing Figure 14.6 we have introduced another MATLAB graphics command subplot(m,n,p). The subplot command activates one subfigure in an $m \times n$ matrix of subfigures. In this case $m = 3$ and $n = 1$, so that we produce three subfigures arranged vertically. The number p indicates which subfigure is the active subfigure—the subfigure to which the plot command refers.

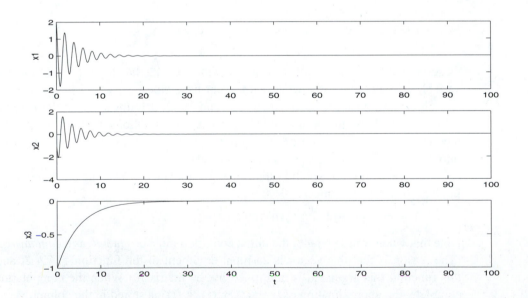

Figure 14.6
Time series showing convergence to the origin for the solution of the linear system $\dot{X} = AX$, where A is as in (14.4.1), with initial condition $X_0 = (2, -1, -1)$

The second possibility for the graphical representation of the solution is the phase space plot. Here we visualize the curve $(x_1(t), x_2(t), x_3(t))$ in three-dimensional space by typing

```
clf
plot3(x(:,1),x(:,2),x(:,3))
xlabel('x1')
ylabel('x2')
zlabel('x3')
```

The result is shown in Figure 14.7. Note that we begin by using the MATLAB graphics command clf to clear all previous graphics.

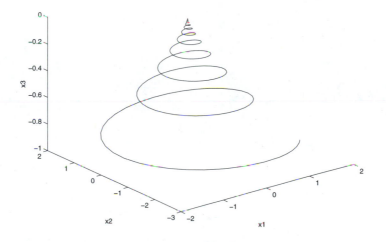

Figure 14.7

Phase space plot showing convergence to the origin for the solution of the linear system $\dot{X} = AX$, where A is as in (14.4.1), with initial condition $X_0 = (2, -1, -1)$

A Three-Dimensional Nonlinear System

We now solve the nonlinear differential equation

$$\dot{X} = AX + (2x_1^2 - x_1 x_2, -x_3^3, -x_2^2)^t \tag{14.4.3}*$$

using ode45, where A is the matrix given in (14.4.1). The m-file for this differential equation is f14_4_3.m. For completeness, this m-file is:

```
function f = f14_4_3(t,x)
A = [ -0.25 3.0 0; -3 -0.25 0; 0 0 -0.2];
f = A*x + [2*x(1)^2-x(1)*x(2); -x(3)^3; -x(2)^2];
```

The theory in Section 14.2 guarantees that the origin is asymptotically stable, and we now verify this statement numerically. Typing

```
[t,x] = ode45('f14_4_3',[0 100],[0.2,-0.1,-0.1]');
```

numerically solves this system of ODEs. The phase space picture, given in Figure 14.8, shows convergence to the origin of the nonlinear system, as expected. Note the similarity of this figure with the phase space picture of the linear system given in Figure 14.7. These numerical computations are surely in agreement with the conclusion of Theorem 14.2.6.

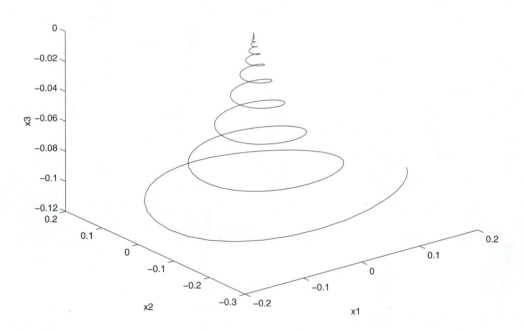

Figure 14.8
Time series showing convergence to the origin for the solution of the nonlinear system (14.4.3) with initial condition $X_0 = (2, -1, -1)$

Periodic Solutions in Three Dimensions

Since planar autonomous nonlinear systems produce limit cycles as solutions, it should come as no surprise that nonlinear three-dimensional systems can also have periodic solutions.

An example of a system of differential equations having a limit cycle as a solution is

$$
\begin{aligned}
\dot{x}_1 &= -x_1 - x_2 + x_1 x_3 \\
\dot{x}_2 &= x_1 - x_2 + x_2 x_3 \\
\dot{x}_3 &= 1 + x_3 - x_1^2 - x_2^2 - x_3^3.
\end{aligned}
\qquad (14.4.4)*
$$

Using the m-file

```
function f = f14_4_4(t,x)
f = [-x(1) - x(2) + x(1)*x(3);
     x(1) - x(2) + x(2)*x(3);
     1  + x(3) - x(1)^2 - x(2)^2 - x(3)^3];
```

numerically integrate (14.4.4) with initial condition $X_0 = (0.5, 0.4, 0.3)$ by typing

```
[t,x] = ode45('f14_4_4', [0 50], [0.5,0.4,0.3]);
```

Using subplot, we can plot the three time series, obtaining the result in Figure 14.9. After an initial transient, each component of the solution settles into a periodic motion with the same period.

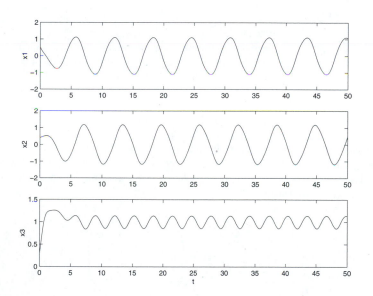

Figure 14.9
Time series showing convergence to a limit cycle for the solution of the nonlinear system (14.4.4) with initial condition $X_0 = (0.5, 0.4, 0.3)$

In the three-dimensional phase space x_1, x_2, x_3 this solution converges to a simple closed curve or a deformed "circle." See Figure 14.10, which is reproduced using the MATLAB commands

```
plot3(x(:,1),x(:,2),x(:,3))
xlabel('x1')
ylabel('x2')
zlabel('x3')
```

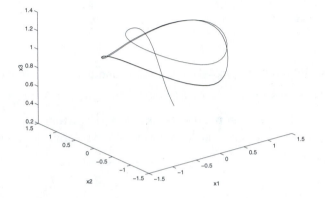

Figure 14.10
Phase space picture showing convergence to a deformed
"circle" for the solution of the nonlinear system (14.4.4)
with initial condition $X_0 = (0.5, 0.4, 0.3)$

COMPUTER EXERCISES

In Exercises 1 and 2, use linear algebra to decide whether or not the origin is an asymptotically stable equilibrium for each system of ODEs $\dot{X} = AX$. If the origin is unstable, find an initial condition such that the corresponding solution approaches the origin as t tends to infinity. Verify these calculations using ode45.

1.

$$A = \begin{pmatrix} -5 & 1 & -3 & 0 \\ -2 & 3 & -3 & 0 \\ 4 & 11 & -5 & 0 \\ 2 & -5 & 3 & -2 \end{pmatrix} \qquad (14.4.5)*$$

2.

$$A = \begin{pmatrix} 0 & 1 & -3 & 0 \\ -2 & 3 & -3 & 0 \\ 4 & 11 & -5 & 0 \\ 2 & -5 & 3 & -2 \end{pmatrix} \qquad (14.4.6)*$$

In Exercises 3–6, investigate numerically the behavior of solutions of the ODEs in a neighborhood of the given equilibrium X_0. Based on your observations decide whether or not the equilibrium is asymptotically stable.

3. Explore the system (14.4.7) near the equilibrium $X_0 = (0, 0, 0)$:

$$\begin{aligned} \dot{x}_1 &= -x_1 - x_2 + x_2 x_3 \\ \dot{x}_2 &= x_1 - x_2 + x_1^2 \\ \dot{x}_3 &= -x_3 + x_1 x_2. \end{aligned} \qquad \text{(14.4.7)*}$$

4. Explore the system (14.4.8) near the equilibrium $X_0 = (0, 0, 0)$:

$$\begin{aligned} \dot{x}_1 &= 10(x_2 - x_1) \\ \dot{x}_2 &= 28x_1 - x_2 - x_1 x_3 \\ \dot{x}_3 &= -\tfrac{8}{3} x_3 + x_1 x_2. \end{aligned} \qquad \text{(14.4.8)*}$$

5. Explore the system (14.4.9) near the equilibrium $X_0 = (1, 0, 2)$:

$$\begin{aligned} \dot{x}_1 &= 3x_1 + x_3 - 3 - x_1 x_3 \\ \dot{x}_2 &= x_1 - 1 + x_2 x_3 \\ \dot{x}_3 &= x_2 + x_3 - 2 + x_1 x_2 - x_2 x_3. \end{aligned} \qquad \text{(14.4.9)*}$$

6. Explore the system (14.4.10) near the equilibrium $X_0 = (0, 0, 2)$:

$$\begin{aligned} \dot{x}_1 &= -x_1 - x_3 + 3 - \cos(x_1) \\ \dot{x}_2 &= 1 - \cos(x_1) - x_3 \sin(x_2) \\ \dot{x}_3 &= 2 - x_3 - \sin(x_2) \cos(x_1). \end{aligned} \qquad \text{(14.4.10)*}$$

7. Use MATLAB to verify the conclusion of Theorem 14.2.6 for the nonlinear differential equation in (14.2.5). Choose the matrix A as in (14.2.4) and the higher order terms $N(X)$ as in (14.2.6).

14.5   QUASIPERIODIC MOTIONS AND TORI

In Chapter 4 we saw that in single autonomous differential equations the only asymptotically stable solutions are steady-state solutions. In Chapter 11 we saw that in two-dimensional autonomous systems the asymptotically stable solutions include limit cycles as well as equilibria. In this section and the next we show that in higher dimensions asymptotically stable solutions can be even more complicated. Formulas for these complicated solutions cannot be found analytically; therefore we use the MATLAB command ode45 to investigate these solutions.

In this section we introduce quasiperiodic two-frequency solutions in three stages. First, we discuss quasiperiodic motion in linear four-dimensional systems; second, we discuss asymptotically stable quasiperiodic motion in four dimensions; and third, we discuss asymptotically stable quasiperiodic motion in three dimensions. We also show that two-frequency quasiperiodic motions fill out a torus (the surface of a doughnut) as opposed to a point (an equilibrium) or a circle (a periodic solution).

A Linear Torus in Four Dimensions

We know that the origin is a center in a linear planar system when the eigenvalues are purely imaginary complex conjugates $\pm \tau i$. All trajectories in a center (except for the origin) lie on ellipses (or circles) surrounding the origin. We now ask what the geometry of solution trajectories is in four-dimensional linear systems with two pairs of complex conjugate purely imaginary eigenvalues $\pm\tau_1 i$ and $\pm\tau_2 i$.

Suppose we start the discussion with a Jordan normal form matrix

$$
B = \begin{pmatrix} 0 & -0.1 & 0 & 0 \\ 0.1 & 0 & 0 & 0 \\ 0 & 0 & 0 & \sqrt{23} \\ 0 & 0 & -\sqrt{23} & 0 \end{pmatrix}.
$$

The associated linear system decouples into two planar systems

$$
\begin{aligned} \dot{x}_1 &= -0.1x_2 \\ \dot{x}_2 &= 0.1x_1 \end{aligned} \quad \text{and} \quad \begin{aligned} \dot{x}_3 &= \sqrt{23}x_4 \\ \dot{x}_4 &= -\sqrt{23}x_3. \end{aligned}
$$

Since each of these systems is a center (in normal form), the phase plane of each system consists of concentric circles.

Suppose that $(x_1(t), x_2(t), x_3(t), x_4(t))$ is a solution to the four-dimensional system. Then $(x_1(t), x_2(t))$ is a solution to the first planar system and $(x_3(t), x_4(t))$ is a solution to the second planar system. Since we know what phase portraits and time series for centers look like, we conclude that each of the time series $x_j(t)$ is a periodic function (in fact just a sum of a cosine and a sine function). The only difference between these functions is the frequency that is 0.1 for the first two time series and $\sqrt{23} = 4.7958$ for the third and fourth time series.

We ask: Is this description accurate for the general four-dimensional linear system with two pairs of purely imaginary complex conjugate eigenvalues? The answer is *no*, and the answer is interesting. The general solution to such an equation lives on a torus (the surface of a doughnut), and the general time series is quasiperiodic with two frequencies. Linear algebra tells us that by simply changing coordinates we can put the matrix into Jordan normal form. So if the information about solutions that we have just described is accurate, the complication must come from viewing the solutions in a different coordinate system.

We do this as follows: Let

$$
P = \begin{pmatrix} -2 & 1 & 3 & 4 \\ -1 & 2 & 2 & 1 \\ 1 & 4 & 1 & 0 \\ 0 & 0 & 2 & 1 \end{pmatrix}
$$

and note that $\det(P) = 27$ so that P is invertible. Now let $A = PBP^{-1}$ so that A is just a matrix whose Jordan normal form is B. A calculation (using MATLAB) shows that

$$A = \begin{pmatrix} 10.6722 & -16.0417 & 5.4028 & -12.2597 \\ 5.3472 & -8.1375 & 2.7569 & -3.6597 \\ 2.1574 & -3.5194 & 1.1954 & -0.3144 \\ 5.3287 & -7.9931 & 2.6644 & -3.7301 \end{pmatrix}. \qquad \textbf{(14.5.1)*}$$

By typing

```
e14_5_1
eig(A)
```

we can check that the eigenvalues of A are $\pm 0.1i$ and $\pm\sqrt{23}i$.

We now compute the solution to the corresponding linear system of ODEs $\dot{X} = AX$, where the function A*x is stored in the m-file f14_5_1 on the time interval $[0, 100]$ with initial conditions $X_0 = (0.2, 0.6, -0.5, 0.1)$. Type

```
[t,x] = ode45('f14_5_1',[0 100],[0.2,0.6,-0.5,0.1]');
```

Using the MATLAB instruction subplot described in Section 14.4, we can graph the four time series. The results are given in Figure 14.11. Note how the time series oscillate

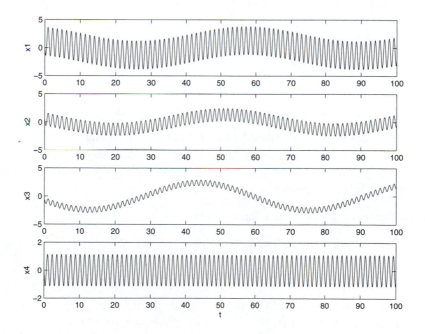

Figure 14.11

Time series showing quasiperiodic two-frequency motion for the solution of the linear system (14.5.1) with initial condition $X_0 = (0.2, 0.6, -0.5, 0.1)$

both on a short scale and on a long scale—one scale corresponding to each frequency. These trajectories are called *quasiperiodic* or *two-frequency* motions. Trajectories with many frequencies can be found in yet higher dimensional linear systems.

The phase space portrait can be viewed in three dimensions (by just ignoring one coordinate). The projections of this trajectory onto the $x_1 x_3 x_4$-hyperplane and the $x_1 x_2 x_4$-hyperplane are given in Figure 14.12. There we can see the torus. For linear systems with two pairs of complex conjugate purely imaginary eigenvalues, almost all solutions lie on tori. To verify this statement, experiment with different initial conditions and see Exercise 1 at the end of this section.

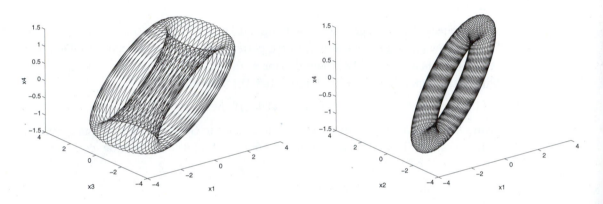

Figure 14.12
Phase space projections showing quasiperiodic motion on a torus of the solution of the linear system (14.5.1) with initial condition $X_0 = (0.2, 0.6, -0.5, 0.1)$

Asymptotically Stable Quasiperiodic Motion

In Section 12.6 we discussed Hopf bifurcation of planar autonomous systems that leads, by varying a parameter through a center, to an asymptotically stable periodic trajectory. Here we discuss how we can also find asymptotically stable quasiperiodic motions in a similar way.

Motion on a Torus in Four Dimensions

Consider the system of four differential equations

$$\dot{X} = (A + \epsilon I_4)X - ||X||^2 X, \qquad \qquad (14.5.2)^*$$

where A has two pairs of complex conjugate purely imaginary eigenvalues and $\epsilon > 0$. We see that $X = 0$ is an equilibrium for (14.5.2) and the Jacobian matrix is $A + \epsilon I_4$ whose eigenvalues are $\lambda + \epsilon$, where λ is an eigenvalue of A. Since all of the eigenvalues of the Jacobian have positive real part, the origin is a source in four dimensions. On the other hand, the term $-||X||^2 X$ in (14.5.2) always drives solutions toward the origin. The two forces balance and result in one asymptotically stable invariant torus. The reasoning here is similar to that used to find limit cycles in planar phase-amplitude equations.

Let A be the matrix (14.5.1) and solve numerically the corresponding differential equation (14.5.2) using the preloaded m-file f14_5_2.m. That m-file is:

```
function f = f14_5_2(t,x)
A = [ 10.6722 -16.0417   5.4028 -12.2597;
       5.3472  -8.1375   2.7569  -3.6597;
       2.1574  -3.5194   1.1954  -0.3144;
       5.3287  -7.9931   2.6644  -3.7301];
f = (A+0.1*eye(4))*x - norm(x)^2*x;
```

To solve this differential equation with initial conditions $X_0 = (0.2, 0.6, -0.5, 0.1)$, type

```
[t,x] = ode45('f14_5_2',[0 100],[0.2,0.6,-0.5,0.1]');
```

The time series for this nonlinear system are given in Figure 14.13. Note the initial transient that is present before the solution settles down to a quasiperiodic motion on a torus. The phase space picture in the $x_1 x_3 x_4$-hyperplane is given in Figure 14.14. Here the transient is more visible.

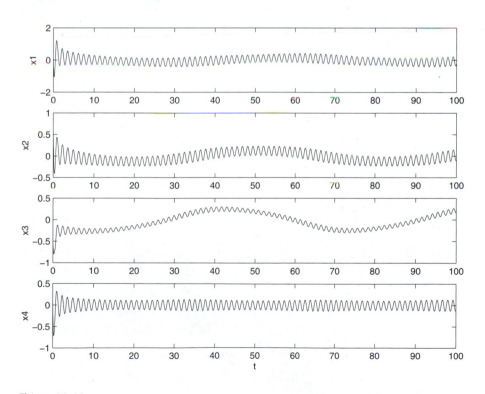

Figure 14.13
Time series showing quasiperiodic two-frequency motion for the solution of the nonlinear system (14.5.2) with initial condition $X_0 = (0.2, 0.6, -0.5, 0.1)$

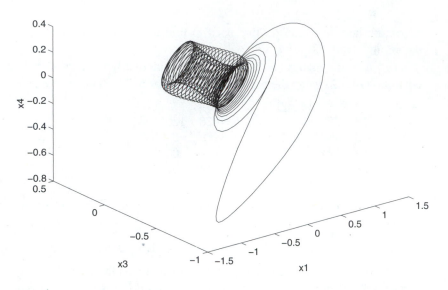

Figure 14.14

Phase space projections showing motion on a torus for the solution of the nonlinear system (14.5.2) with initial condition $X_0 = (0.2, 0.6, -0.5, 0.1)$

A Torus in Three Dimensions

Until now, the way that we have constructed two-frequency quasiperiodic solutions to systems of ODEs is based on having two independent frequencies present in a linear system. It is not obvious—yet it is true—that solutions to nonlinear systems of three differential equations can also have two-frequency quasiperiodic solutions. The theory that leads to this example is beyond the scope of this book; nevertheless, we now have the numerical techniques to see (visually) that such solutions exist.

Consider the autonomous nonlinear system of ODEs:[1]

$$\begin{aligned}
\dot{x}_1 &= (x_3 - 0.7)x_1 - 3.5x_2 \\
\dot{x}_2 &= 3.5x_1 + (x_3 - 0.7)x_2 \\
\dot{x}_3 &= 0.6 + x_3 - 0.33x_3^3 - (x_1^2 + x_2^2)(1 + .025x_3).
\end{aligned} \qquad \text{(14.5.3)*}$$

The m-file for this system of equations is f14_5_3.m and contains

```
function f = f14_5_3(t,x)
f = [(x(3)−0.7)*x(1) − 3.5*x(2);
     3.5*x(1) + (x(3)−0.7)*x(2);
     0.6 + x(3) − x(3)^3/3 − (x(1)^2+x(2)^2)*(1+0.25*x(3))];
```

The differential equation (14.5.3) is solved by typing

```
[t,x] = ode45('f14_5_3',[0 100],[0.1,0.03,0.001]');
```

---

[1]This system of equations is taken from W. F. Langford, *Numerical Studies of Torus Bifurcations*, ISNM **70**, Birkhäuser, 1984.

The time series for the system (14.5.3) are given in Figure 14.15, and the three dimensional phase space picture is given in Figure 14.16.

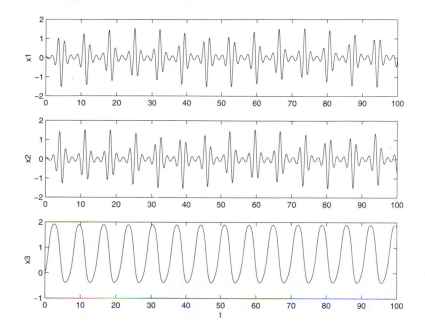

Figure 14.15
Time series showing quasiperiodic two-frequency motion for the solution of the nonlinear system (14.5.3) with initial condition $X_0 = (0.1, 0.03, 0.001)$

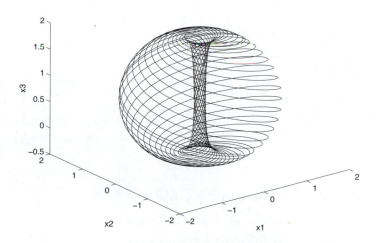

Figure 14.16
Phase space showing motion on a torus for the solution trajectory of the nonlinear system (14.5.3) with initial condition $X_0 = (0.1, 0.03, 0.001)$

COMPUTER EXERCISES

1. Let A be the 4×4 torus example given in (14.5.1). Choose three different initial conditions to the system of ODEs $\dot{X} = AX$ and use ode45 to compute solutions with these initial conditions. (To decrease the length of time needed by the computer to do these calculations, shorten the time period to, say, $[0, 30]$.)

    (a) Based on these calculations, verify that most initial conditions lead to quasiperiodic toroidal motions. Display both time series and three-dimensional phase portraits of your solutions.

    (b) Take as initial condition the real part of one of the complex eigenvectors of A. (You will need to use MATLAB to find the eigenvalues and eigenvectors of A.) What kind of phase space motion do you see now? How does this solution differ from the toroidal motions obtained in (a)? Use Jordan normal forms to explain why your numerical answer is correct.

In Exercises 2–5, use MATLAB to find out whether the system of differential equations $\dot{X} = AX$ for the given matrix A has quasiperiodic solutions.

2.
$$A = \begin{pmatrix} 4.9666 & 2.2833 & 0.8000 & 5.3666 \\ -0.9889 & -0.0944 & -1.6000 & -1.7889 \\ 2.9889 & 4.0944 & -0.4000 & 1.7889 \\ -8.9443 & -4.4721 & 0 & -4.4721 \end{pmatrix} \qquad \textbf{(14.5.4)}^*$$

3.
$$A = \begin{pmatrix} 2.7666 & 0.2833 & 1.4000 & 4.5666 \\ 2.4111 & 1.9056 & -1.8000 & -0.1889 \\ -0.4111 & 2.0944 & -0.2000 & 0.1889 \\ -7.9443 & -2.4721 & -1.0000 & -4.4721 \end{pmatrix} \qquad \textbf{(14.5.5)}^*$$

4.
$$A = \begin{pmatrix} 0.7130 & 24.8184 & 32.0740 & 2.8959 & 15.3610 \\ 0.0552 & 17.0732 & 21.1395 & 6.6120 & 11.0843 \\ -0.6168 & -12.0764 & -16.0165 & -1.6151 & -7.3997 \\ 1.0410 & -3.3312 & -2.0820 & -3.3312 & -3.1230 \\ -0.5205 & 1.6656 & 1.0410 & 1.6656 & 1.5615 \end{pmatrix} \qquad \textbf{(14.5.6)}^*$$

5.
$$A = \begin{pmatrix} -10.2870 & 40.8184 & 30.0740 & 19.8959 & 38.3610 \\ -7.4448 & 25.0732 & 16.1395 & 15.8620 & 26.0843 \\ 6.8833 & -20.0764 & -11.0165 & -10.3652 & -21.3998 \\ 1.0410 & -3.3312 & -2.0820 & -3.3312 & -3.1230 \\ -0.5205 & 1.6656 & 1.0410 & 1.1656 & 0.5615 \end{pmatrix} \qquad \textbf{(14.5.7)}^*$$

14.6   CHAOS AND THE LORENZ EQUATION

Classifying all the kinds of solutions that can occur asymptotically in autonomous systems of first-order differential equations is a difficult task and is very much a topic of current research. Until now we have discussed three types of solutions: equilibria, limit cycles, and quasiperiodic motions. The purpose of the next example, the Lorenz equations, is to illustrate that there are still other types of asymptotic behavior that can occur in solutions to autonomous ordinary differential equations in three dimensions. This type of solution is called *chaotic*, and what distinguishes chaotic solutions from the previously discussed solutions is *sensitive dependence on initial conditions*.

The prototypical example of chaos is the *Lorenz system*. The Lorenz system consists of three first-order (almost linear) ordinary differential equations (there are just two quadratic terms):

$$\begin{aligned}
\dot{x}_1 &= \sigma(x_2 - x_1) \\
\dot{x}_2 &= \rho x_1 - x_2 - x_1 x_3 \\
\dot{x}_3 &= -\beta x_3 + x_1 x_2,
\end{aligned} \qquad \textbf{(14.6.1)*}$$

where σ, ρ, and β are real constants. We consider here solutions to (14.6.1) when

$$\sigma = 10, \quad \beta = \frac{8}{3}, \quad \rho = 28.$$

The right-hand side of (14.6.1) is stored in the m-file f14_6_1.m:

```
function f = f14_6_1(t,x)
sigma = 10;   beta  = 8/3;   rho   = 28;
f      = [sigma*(x(2)-x(1));
          rho*x(1)-x(2)-x(1)*x(3);
          -beta*x(3)+x(1)*x(2)];
```

Compute a solution of the Lorenz system starting at $X_0 = (5, 5, 30)$ by typing

```
[t,x]=ode45('f14_6_1',[0 40],[5,5,30]');
```

The time series of x_1 is shown in Figure 14.17(a). This time series looks bizarre and no apparent regularity can be seen. Moreover, the motion is not just irregular, it is also sensitive to the choice of the initial conditions.

Sensitive Dependence on Initial Conditions

In Figure 14.17(b) we illustrate sensitive dependence by showing a solution of the Lorenz system with initial conditions very close to the first one—namely, at $X_0 = (5.01, 5, 30)$. In the beginning the solutions behave in a similar way, but by $t \approx 10$ the behavior is completely different. Even for smaller differences in the initial conditions, the phenomenon of sensitivity to initial conditions is still present, although the significant difference in the trajectories occurs at a later time (see Figure 14.18).

The consequence of sensitive dependence of solutions on initial conditions in the Lorenz system is long-term unpredictable behavior. Typically, in experiments, we know initial conditions only to within some (hopefully) small error. If these errors get magnified, as they do in the Lorenz system, then it is impossible to make accurate long-term predictions. This lack of predictability is the defining feature of *chaotic behavior*. So we must ask: Is chaotic behavior typical in solutions to autonomous

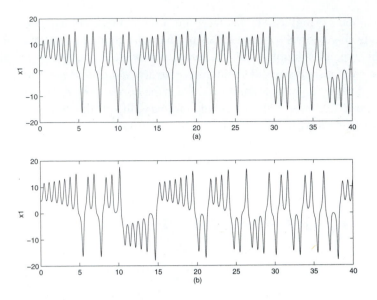

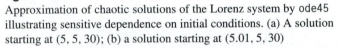

Figure 14.17
Approximation of chaotic solutions of the Lorenz system by ode45 illustrating sensitive dependence on initial conditions. (a) A solution starting at $(5, 5, 30)$; (b) a solution starting at $(5.01, 5, 30)$

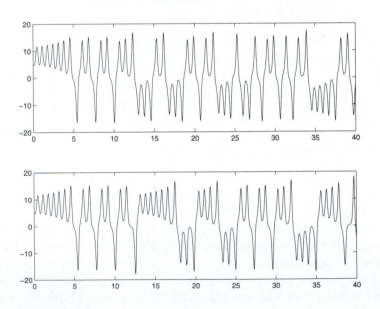

Figure 14.18
Approximation of chaotic solutions of the Lorenz system by ode45 illustrating sensitive dependence on initial conditions. (a) A solution starting at $(5.001, 5, 30)$; (b) a solution starting at $(5.0001, 5, 30)$

systems of differential equations? The answer is yes. Not every three-dimensional system of differential equations exhibits chaos—but many do.

In Figure 14.19 we show a phase space plot of the solution starting at $(5, 5, 30)$. To reproduce this figure with the correct scale and viewpoint, type

```
[t,x]=ode45('f14_6_1',[0 40],[5,5,30]);
plot3(x(:,1),x(:,2),x(:,3))
axis([-20,20,-20,20,0,60])
view(125,20)
```

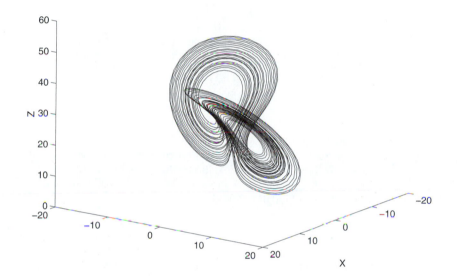

Figure 14.19

Phase space plot of a solution of the Lorenz system starting at $(5, 5, 30)$

We emphasize that the existence of sensitive dependence on initial conditions does *not* depend on the numerical algorithm used in the numerical integration nor does it depend on the computer that is used. However, different numerical algorithms and even different computers generally give different numerical results. Finally, we note that the phase space picture of the Lorenz attractor, as shown in Figure 14.19, will seem the same to the eye regardless of the choice of numerical algorithm and computer, even though the time series have readily observable differences of the type shown in Figure 14.18.

Remark: A dynamic simulation of a solution to the Lorenz equations in phase space can also be seen in MATLAB: Just type lorenz.

It should be noted that even though quasiperiodic motion is geometrically compli-cated (leading to trajectories lying on a torus), it does not exhibit sensitive dependence on initial conditions. More precisely, the time series of asymptotically stable quasiperi-odic solutions remain almost unchanged after small changes in initial conditions. To

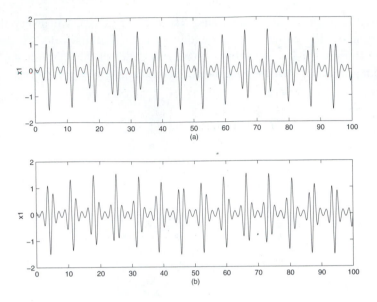

Figure 14.20

Time series for a quasiperiodic two-frequency solution of
the nonlinear system (14.5.3) illustrating a lack of sensitive
dependence on initial conditions: (a) $X_0 = (0.1, 0.03, 0.001)$ and
(b) $X_0 = (0.11, 0.031, 0.0015)$

verify this statement, we return to the numerical solution of (14.5.3). In Figure 14.20
we plot the time series for the x_1 component of solutions with nearly identical initial
conditions and note that the two time series are nearly identical.

A Summary of Observed Three-Dimensional Dynamics

We now summarize the types of attracting solutions that we have seen in autonomous
three-dimensional systems of differential equations. The states we have studied are stable equilibria (or sinks), attracting limit cycles, attracting quasiperiodic two-frequency
motions, and chaotic dynamics such as seen in the Lorenz equations. Each of these
solution types has well-defined characteristics that can be observed either in time series plots or in three-dimensional phase space plots. This information is summarized
in Table 14.1.

In Figure 14.6 we plotted the time series of a solution to a linear equation $\dot{X} = AX$
where the 3×3 matrix A had eigenvalues with negative real part. These time series showed convergence to the equilibrium at the origin by becoming horizontal as t
increased. There was transient oscillation caused by a complex conjugate pair of eigenvalues in A. The three-dimensional phase portrait Figure 14.7 indicates convergence
of this solution trajectory to a point. These types of time series and phase portraits are

Table 14.1

Summary of observed asymptotic dynamics in three dimensions

Asymptotic Solution Type	Time Series	Phase Space
Sink	Horizontal line	Point
Limit cycle	Periodic oscillation	"Circle"
Two-frequency quasiperiodic	Modulated periodic oscillation	Torus
Chaotic	Bounded irregular oscillation **with** sensitive dependence on initial conditions	Complicated surface **not** sensitive to initial conditions

equally valid for nonlinear systems near a sink, as was illustrated in Figure 14.8 for the nonlinear system (14.4.3).

In example (14.4.4) we saw a solution to an autonomous three-dimensional system of differential equations approach a limit cycle. For such examples, the time series converge to a periodic function, as in Figure 14.9. In phase space, such solutions converge to a closed curve, like a circle; see Figure 14.10. In general, the closed curve corresponding to a periodic solution can be quite complicated—but the time series still consists of periodic functions.

The time series for two-frequency quasiperiodic solutions was illustrated for a four-dimensional linear system in Figure 14.11. Note the "almost periodic" behavior on short time scales coupled with long time modulations. The short period oscillation is caused by the larger frequency and the long time modulation by the smaller frequency. Such solutions are found in linear systems when there are two complex conjugate pairs of purely imaginary eigenvalues with incommensurate frequencies. Therefore two-frequency motions can appear only in linear systems with four or more variables. However, two-frequency quasiperiodic motion can occur in three-dimensional nonlinear systems, as illustrated in the time series plots of (14.5.3) given in Figure 14.15. In phase space, the images are even more interesting because, after an initial transient, the solution fills out the surface of a torus; see Figure 14.16.

The Lorenz system (14.6.1) illustrates the possibility of yet more complicated motions occurring in three dimensions. The time series in Figures 14.17 and 14.18 illustrate the phenomenon of sensitive dependence on initial conditions, where the numerical values of solutions starting at two nearby initial conditions seem to be unrelated after numerically integrating the equations for a relatively short length of time. Nevertheless, the characteristic three-dimensional phase space picture of the Lorenz equations shown in Figure 14.19 is reproduced by almost all nearby initial conditions; see Exercise 4.

Other types of solutions are possible in three dimensions. Such solutions are not regular in the sense that they limit on a point, a circle, or a torus, and they do not exhibit sensitive dependence on initial conditions. Although the existence of these other solutions is quite an interesting topic, we do not pursue it here.

COMPUTER EXERCISES

1. Compute the three equilibria of the Lorenz system (14.6.1) and use ode45 to decide whether or not these are stable.

2. Fix the parameters $\sigma = 10$ and $\beta = \frac{8}{3}$. Then use ode45 to investigate the behavior of the Lorenz system (14.6.1) for $\rho = 0.8$, $\rho = 6$, $\rho = 20$, and $\rho = 26$.

3. Use ode45 to analyze the dynamical behavior of a variant of the *Chua circuit*,

$$\dot{x} = \alpha \left(y - m_0 x - \frac{1}{3} m_1 x^3 \right)$$
$$\dot{y} = x - y + z$$
$$\dot{z} = -\beta y.$$

In the computations, fix $\alpha = 18$, $\beta = 33$, $m_0 = -0.2$, and $m_1 = 0.01$. You will have to write your own m-file to complete this exercise.

4. Numerically integrate the Lorenz equations (14.6.1) for the standard parameter values $\sigma = 10$, $\beta = \frac{8}{3}$, and $\rho = 28$ using the initial condition $X_0 = (5.01, 5, 30)$. Compare the phase space plot of this solution with that of Figure 14.19. (You will need to use the same viewpoint and axis range as was used in that figure.) Experiment with several different choices of initial conditions.

In Exercises 5–12, determine whether the given solution is asymptotic to an equilibrium, a limit cycle, a two-frequency torus, exhibits sensitive dependence on initial conditions, or has a limiting behavior that has not been described previously.

5. Solve the system (14.6.2) with initial conditions $X_0 = (0.10, 0.20, 0.25)^t$:

$$\dot{x}_1 = 1 - x_1 - x_2^2$$
$$\dot{x}_2 = -x_2 - x_3^2 \qquad (14.6.2)^*$$
$$\dot{x}_3 = x_3 + x_1 x_2 - x_3^3.$$

6. Solve the system (14.6.3) with initial conditions $X_0 = (0.10, 0.24, 0.14)^t$:

$$\dot{x}_1 = (x_3 - 1.3)x_1 - 3.5x_2$$
$$\dot{x}_2 = 3.5x_1 + (x_3 - 1.3)x_2 \qquad (14.6.3)^*$$
$$\dot{x}_3 = 0.6 + x_3 - \frac{1}{3}x_3^3 - (x_1^2 + x_2^2)(1 + \frac{1}{4}x_3).$$

7. Solve the system (14.6.4) with initial conditions $X_0 = (0.10, 0.23, 0.15)^t$:

$$\dot{x}_1 = x_1 - (x_1^2 + 1.5x_2^2 + 0.6x_3^2)x_1$$
$$\dot{x}_2 = x_2 - (0.6x_1^2 + x_2^2 + 1.5x_3^2)x_2 \qquad (14.6.4)^*$$
$$\dot{x}_3 = x_3 - (1.5x_1^2 + 0.6x_2^2 + x_3^2)x_3.$$

8. Solve the system (14.6.5) with initial conditions $X_0 = (0.10, 0.23, 0.15)^t$:

$$\dot{x}_1 = x_1 - (x_1^2 + 0.5x_2^2 + 0.7x_3^2)x_1$$
$$\dot{x}_2 = x_2 - (0.7x_1^2 + x_2^2 + 0.5x_3^2)x_2 \qquad \textbf{(14.6.5)*}$$
$$\dot{x}_3 = x_3 - (0.5x_1^2 + 0.7x_2^2 + x_3^2)x_3.$$

9. Solve the system (14.6.6) with initial conditions $X_0 = (0.10, 0.11, 0.15)^t$:

$$\dot{x}_1 = -x_2 - x_3$$
$$\dot{x}_2 = x_1 + 0.2x_2 \qquad \textbf{(14.6.6)*}$$
$$\dot{x}_3 = 0.2 + x_3(x_1 - 5.7).$$

10. Solve the system (14.6.7) with initial conditions $X_0 = (0.1, 0.2, -0.2)^t$:

$$\dot{x}_1 = 0.65 + x_1 - x_1^{10} - (x_3^2 + x_2^2)(1 + 0.25x_1)$$
$$\dot{x}_2 = 3.5x_3 + (x_1 - 0.7)x_2 \qquad \textbf{(14.6.7)*}$$
$$\dot{x}_3 = (x_1 - 0.7)x_3 - 3.5x_2.$$

11. Solve the system (14.6.8) with initial conditions $X_0 = (2.0, 0.5, 1.0)^t$:

$$\dot{x}_1 = -x_2 - x_3$$
$$\dot{x}_2 = x_1 + 0.2x_2 \qquad \textbf{(14.6.8)*}$$
$$\dot{x}_3 = 0.2 + x_3(x_1 - 1).$$

12. Solve the system (14.6.9) with initial conditions $X_0 = (3.0084, 3.0983, 2.7673)^t$:

$$\dot{x}_1 = -1.5x_1 + (x_3 - 0.2)x_2$$
$$\dot{x}_2 = -1.5x_2 + x_1x_3 \qquad \textbf{(14.6.9)*}$$
$$\dot{x}_3 = 1 - x_1x_2.$$

Linear Differential Equations

Primarily our discussions of differential equations have focused on two issues: generalizations of solutions of the differential equation $\dot{x} = \lambda x$ to systems of equations and the qualitative interpretation of numerically obtained phase portraits for autonomous nonlinear systems. Except for finding closed form solutions of systems of equations $\dot{X} = AX$ (in the plane, Chapter 6, or in Jordan normal form, Section 14.1), we have solved in closed form virtually no other differential equation. In this chapter and the next two we focus on finding closed form solutions to a variety of differential equations. This chapter and the next discuss constant coefficient linear equations, both homogeneous and inhomogeneous, while Chapter 17 discusses nonconstant coefficient and nonlinear equations.

The system of differential equations $\dot{X} = AX$ is a constant coefficient, first-order, homogeneous, linear system of ordinary differential equations. We begin this chapter (Section 15.1) by discussing how to solve the systems $\dot{X} = AX$ in the given coordinates (rather than by first transforming A to Jordan normal form, as we did in Section 14.1). For example, we present a formula for computing e^{tA} for any matrix A. Later we solve some linear equations that are neither homogeneous nor first order.

In Section 15.2 we discuss how to solve higher order linear differential equations by reducing them to first-order systems. This section generalizes the reduction of second-order equations to planar systems described in Section 6.7. We will see that generally it is easier to solve higher order equations in closed form than to solve first-order systems. Section 15.4 discusses the solution of an inhomogeneous higher order equation by undetermined coefficients. The method of undetermined coefficients is most easily understood in the language of linear differential operators; this language is introduced in Section 15.3.

The chapter ends with a discussion of resonance in Section 15.5. Here we use explicit closed form solutions found by the method of undetermined coefficients to understand a physically motivated phenomenon that occurs in solutions to forced second-order equations.

15.1 SOLVING SYSTEMS IN ORIGINAL COORDINATES

In Section 14.1 we discussed one method for solving systems of first-order constant coefficient linear differential equations. We saw that such systems can be solved by putting the coefficient matrix A in Jordan normal form and then computing the exponential of the Jordan normal form matrix.

In this section we describe a second and a third approach to solving linear systems. The second method is based on finding the generalized eigenvectors that put A into Jordan normal form and then computing solutions directly using this information, while the third method is based on deriving a formula for the exponential e^{tA} in original coordinates. The advantage of the second method is that it is not necessary to perform the similarity that transforms the matrix A to Jordan normal form. The advantage of the third method is that it is not necessary even to compute the eigenvectors of A. Be forewarned, however, that all of these methods require substantial calculations.

Let A be an $n \times n$ matrix. All methods for solving the system

$$\dot{X} = AX \tag{15.1.1}$$

begin by finding the eigenvalues of A. This can be done either analytically (sometimes) or numerically (using MATLAB). Then the methods diverge. In the first and second methods, we need to find the eigenvectors and, if need be, the generalized eigenvectors of A; in the third method, we need to perform tedious calculations involving partial fractions and matrix multiplications. With either method, the calculations simplify enormously when the eigenvalues are simple. Indeed, this simplification also occurs in the Jordan normal form method of Section 14.1.

A Method Based on Eigenvectors

In this method we find a basis for solutions of (15.1.1). First we review the simpler case when there is a basis of eigenvectors, and then we consider (part of) the case when there is a deficiency of eigenvectors.

A Complete Set of Eigenvectors

The simplest case in solving (15.1.1) occurs when A has a basis of eigenvectors $v_1, \ldots, v_n$ corresponding to the (not necessarily distinct) eigenvalues $\lambda_1, \ldots, \lambda_n$. We showed how to find a basis for the solutions of (15.1.1) in Section 6.2, but we review the results here. Each eigenvector v_j generates the solution $X_j(t) = e^{\lambda_j t} v_j$ to (15.1.1), and the general solution is

$$X(t) = \alpha_1 X_1(t) + \cdots + \alpha_n X_n(t), \qquad \textbf{(15.1.2)}$$

where the scalars α_j are real when λ_j is real and complex when λ_j is not real.

The initial value problem is then solved by finding scalars α_j so that

$$X_0 = X(0) = \alpha_1 v_1 + \cdots + \alpha_n v_n. \qquad \textbf{(15.1.3)}$$

The solution of (15.1.3) is a well-understood linear algebra problem.

Complex Eigenvalues

The only complication occurs when some of the eigenvalues are complex. If λ is a complex eigenvalue of the real matrix A, then so is $\bar{\lambda}$. If v is a complex eigenvector corresponding to λ, then $\bar{v}$ is the complex eigenvector corresponding to $\bar{\lambda}$. With this choice of eigenvector, $\bar{\alpha}$ is the scalar corresponding to the eigenvector $\bar{v}$, where α is the scalar corresponding to the eigenvector v.

More precisely, let $\lambda = \sigma + i\tau$ be an eigenvalue of A and let $v = u + iw$ be a corresponding eigenvector. We claim that

$$
\begin{aligned}
X_1(t) &= e^{\sigma t}(\cos(\tau t)u - \sin(\tau t)w) \quad \text{and} \\
X_2(t) &= e^{\sigma t}(\sin(\tau t)u + \cos(\tau t)w)
\end{aligned}
\qquad \textbf{(15.1.4)}
$$

are solutions of the homogeneous equation (15.1.1). Verifying that X_1 and X_2 are solutions proceeds as follows: The solutions corresponding to the eigenvalue λ are

$$\alpha e^{\lambda t} v + \bar{\alpha} e^{\bar{\lambda} t} \bar{v} = 2\,\mathrm{Re}\left(\alpha e^{\lambda t} v\right)$$

for all complex scalars α. If we set $\alpha = \frac{1}{2}$, then using Euler's formula, we obtain the solution

$$\mathrm{Re}\left(e^{\lambda t} v\right) = e^{\sigma t}\mathrm{Re}\left((\cos(\tau t) + i\sin(\tau t))(u + iw)\right) = X_1(t).$$

Similarly, setting $\alpha = -\frac{1}{2}i$ leads to the solution $X_2(t)$.

Two Examples with a Complete Set of Eigenvectors

Next we consider two examples. The first has distinct eigenvalues, some of which are complex, while the second has real eigenvalues, one of which is multiple.

(***a***) Find all the solutions of the linear system of ODEs

$$\begin{pmatrix} \dfrac{dx_1}{dt} \\[2mm] \dfrac{dx_2}{dt} \\[2mm] \dfrac{dx_3}{dt} \end{pmatrix} = \begin{pmatrix} 0 & 3 & 1 \\ 4 & 1 & -1 \\ 2 & 7 & -5 \end{pmatrix} \begin{pmatrix} x_1 \\ x_2 \\ x_3 \end{pmatrix}. \qquad (15.1.5)*$$

We load the coefficient matrix A of (15.1.5) by typing e15_1_5 and compute its eigenvalues and eigenvectors using the command

```
[V D] = eig(A)
```

This leads to

```
V =
   -0.5774          0.3757 + 0.0411i     0.3757 - 0.0411i
   -0.5774         -0.3757 - 0.0411i    -0.3757 + 0.0411i
   -0.5774         -0.4579 + 0.7103i    -0.4579 - 0.7103i
D =
    4.0000               0                    0
         0         -4.0000 + 2.0000i          0
         0                0             -4.0000 - 2.0000i
```

The matrix A has the three distinct eigenvalues: $\rho = 4$, $\lambda = \sigma + i\tau = -4 + 2i$, and $\bar{\lambda} = \sigma - i\tau = -4 - 2i$. We can use these data to find all the solutions of (15.1.5). Every solution of (15.1.5) is a linear combination of the functions

$$X_1(t) = e^{4t} \begin{pmatrix} 1 \\ 1 \\ 1 \end{pmatrix}$$

$$\widehat{X}_2(t) = e^{-4t} \left[\cos(2t) \begin{pmatrix} 0.3757 \\ -0.3757 \\ -0.4579 \end{pmatrix} - \sin(2t) \begin{pmatrix} 0.0411 \\ -0.0411 \\ 0.7103 \end{pmatrix} \right]$$

$$\widehat{X}_3(t) = e^{-4t} \left[\sin(2t) \begin{pmatrix} 0.3757 \\ -0.3757 \\ -0.4579 \end{pmatrix} + \cos(2t) \begin{pmatrix} 0.0411 \\ -0.0411 \\ 0.7103 \end{pmatrix} \right].$$

In fact, it is possible to obtain a simpler form for these solutions by using different eigenvectors associated with the eigenvalues $-4 \pm 2i$. Rescaling V(:,2) by V(:,2)/V(1,2) yields the answer

```
ans =
    1.0000
   -1.0000 - 0.0000i
   -1.0000 + 2.0000i
```

and therefore $(1, -1, -1 + 2i)^t$ is also an eigenvector belonging to the eigenvalue $-4 + 2i$. When we use this eigenvector, the two solutions in (15.1.4) take the form

$$X_2(t) = e^{-4t} \begin{pmatrix} \cos(2t) \\ -\cos(2t) \\ -\cos(2t) - 2\sin(2t) \end{pmatrix}, \quad X_3(t) = e^{-4t} \begin{pmatrix} \sin(2t) \\ -\sin(2t) \\ -\sin(2t) + 2\cos(2t) \end{pmatrix}.$$

Suppose that we wish to find the solution satisfying the initial condition $X_0 = (1, 2, 3)^t$. Then we need to find scalars α_1, α_2, and α_3 that solve the system of linear equations

$$X_0 = \alpha_1 X_1(0) + \alpha_2 X_2(0) + \alpha_3 X_3(0).$$

In coordinates, this linear system is

$$\begin{pmatrix} 1 \\ 2 \\ 3 \end{pmatrix} = \begin{pmatrix} 1 & 1 & 0 \\ 1 & -1 & 0 \\ 1 & -1 & 2 \end{pmatrix} \begin{pmatrix} \alpha_1 \\ \alpha_2 \\ \alpha_3 \end{pmatrix}.$$

This linear system is easily solved by hand to obtain $\alpha_1 = \frac{3}{2}$, $\alpha_2 = -\frac{1}{2}$, and $\alpha_3 = \frac{1}{2}$. The closed form solution to this initial value problem is then

$$X(t) = \frac{1}{2} \begin{pmatrix} 3e^{4t} + e^{-4t}(\sin(2t) - \cos(2t)) \\ 3e^{4t} + e^{-4t}(-\sin(2t) + \cos(2t)) \\ 3e^{4t} + e^{-4t}(\sin(2t) + 3\cos(2t)) \end{pmatrix}. \tag{15.1.6}$$

(b) As a second example, consider the system $\dot{X} = AX$, where

$$A = \begin{pmatrix} -2 & -2 & -4 \\ 0 & 0 & 4 \\ 0 & 2 & 2 \end{pmatrix}. \tag{15.1.7}*$$

By inspection, we see that the three eigenvalues of A are -2 and the two eigenvalues of the matrix $\begin{pmatrix} 0 & 4 \\ 2 & 2 \end{pmatrix}$. A quick calculation shows that the eigenvalues of this 2×2 matrix are -2 and 4. So the eigenvalues of A are

$$\lambda_1 = \lambda_2 = -2 \quad \text{and} \quad \lambda_3 = 4.$$

There are three linearly independent eigenvectors corresponding to these eigenvalues:

$$v_1 = \begin{pmatrix} 1 \\ 0 \\ 0 \end{pmatrix}, \quad v_2 = \begin{pmatrix} 0 \\ -2 \\ 1 \end{pmatrix}, \quad v_3 = \begin{pmatrix} 1 \\ -1 \\ -1 \end{pmatrix}.$$

Therefore every solution of the equation $\dot{X} = AX$ has the form

$$X(t) = \alpha_1 e^{-2t} \begin{pmatrix} 1 \\ 0 \\ 0 \end{pmatrix} + \alpha_2 e^{-2t} \begin{pmatrix} 0 \\ -2 \\ 1 \end{pmatrix} + \alpha_3 e^{4t} \begin{pmatrix} 1 \\ -1 \\ -1 \end{pmatrix}$$

with appropriate constants $\alpha_1, \alpha_2, \alpha_3$.

A Deficiency in Eigenvectors

When working with Jordan normal forms, we have to find bases consisting of general-ized eigenvectors. As we know, this is a difficult problem. We describe a method here for finding closed form solutions to linear systems when the eigenvalues are real and there is exactly one Jordan block associated with each eigenvalue. This assumption is equivalent to assuming that the null space of $A - \lambda I_n$ is one dimensional, and under this assumption the computation of generalized eigenvalues is a tractable linear algebra problem (as we saw in Section 13.4).

Suppose that λ is a real eigenvalue of A with algebraic multiplicity k and with one linearly independent eigenvector w. Suppose that there exist linearly independent vectors w_j $(j = 1, \ldots, k)$ such that

$$\begin{aligned} (A - \lambda I_n)w_1 &= 0 \\ (A - \lambda I_n)w_2 &= w_1 \\ &\vdots \\ (A - \lambda I_n)w_k &= w_{k-1}. \end{aligned} \tag{15.1.8}$$

We use this information to find k linearly independent solutions to $\dot{X} = AX$. See Theorem 15.1.1.

The theory of Section 14.1 tells us that solutions of (15.1.1) have the form

$$X(t) = e^{\lambda t}(\alpha_1(t)w_1 + \alpha_2(t)w_2 + \cdots + \alpha_k(t)w_k),$$

where each $\alpha_j(t)$ is a polynomial of degree at most $k - 1$. See Lemma 14.1.2. Using the product rule, compute

$$\frac{dX}{dt} = e^{\lambda t}\left(\lambda(\alpha_1 w_1 + \cdots + \alpha_k w_k) + \frac{d\alpha_1}{dt}w_1 + \cdots + \frac{d\alpha_k}{dt}w_k\right).$$

On the other hand, from (15.1.8),

$$\begin{aligned} AX(t) &= e^{\lambda t}(\alpha_1 \lambda w_1 + \alpha_2(w_1 + \lambda w_2) + \cdots + \alpha_k(w_{k-1} + \lambda w_k)) \\ &= e^{\lambda t}(\lambda(\alpha_1 w_1 + \cdots + \alpha_k w_k) + \alpha_2 w_1 + \cdots + \alpha_k w_{k-1}). \end{aligned}$$

It follows that $X(t)$ is a solution if and only if

$$\frac{d\alpha_1}{dt} w_1 + \cdots + \frac{d\alpha_k}{dt} w_k = \alpha_2 w_1 + \cdots + \alpha_k w_{k-1}. \qquad (15.1.9)$$

Since $w_1, \ldots, w_k$ are linearly independent vectors, (15.1.9) is equivalent to

$$\frac{d\alpha_1}{dt} = \alpha_2$$

$$\vdots$$

$$\frac{d\alpha_{k-1}}{dt} = \alpha_k$$

$$\frac{d\alpha_k}{dt} = 0.$$

Next, choose j such that $1 \leq j \leq k$ and set

$$\alpha_{j+1} = \cdots = \alpha_k = 0$$

and

$$\alpha_j(t) = 1, \quad \alpha_{j-1}(t) = t, \quad \alpha_{j-2}(t) = \frac{t^2}{2}, \quad \ldots, \quad \alpha_1(t) = \frac{t^{j-1}}{(j-1)!}.$$

Then the function

$$X_j(t) = e^{\lambda t} \left(\frac{t^{j-1}}{(j-1)!} w_1 + \frac{t^{j-2}}{(j-2)!} w_2 + \cdots + t w_{j-1} + w_j \right) \qquad (15.1.10)$$

is a solution of (15.1.1). Moreover, since $X_j(0) = w_j$, the solutions $X_1(t), \ldots, X_k(t)$ are linearly independent. We have proved:

Theorem 15.1.1   *Let $w_1, \ldots, w_k$ be a set of linearly independent generalized eigenvectors satisfying (15.1.8). Then the functions $X_1(t), \ldots, X_k(t)$ in (15.1.10) are a linearly independent set of solutions to $\dot{X} = AX$.*

For example, when $k = 3$, we have found three linearly independent solutions:

$$X_1(t) = e^{\lambda t} w_1$$
$$X_2(t) = e^{\lambda t} (t w_1 + w_2)$$
$$X_3(t) = e^{\lambda t} \left(\frac{t^2}{2} w_1 + t w_2 + w_3 \right).$$

An Example

Consider the system $\dot{X} = AX$, where

$$A = \begin{pmatrix} 6 & 6 & 6 \\ 5 & 11 & -1 \\ 1 & -5 & 7 \end{pmatrix}.$$ **(15.1.11)***

The eigenvalues of A (using eig(A)) are

$$\lambda_1 = 0 \quad \text{and} \quad \lambda_2 = \lambda_3 = 12.$$

In this case (using [V,D] = eig(A)) we find that there are just two linearly independent eigenvectors,

$$v_1 = \begin{pmatrix} 2 \\ -1 \\ -1 \end{pmatrix}, \quad v_2 = \begin{pmatrix} 0 \\ 1 \\ -1 \end{pmatrix}.$$

It follows that the algebraic multiplicity of $\lambda_2 = 12$ is two but the geometric multiplicity is just one. Typing

    null((A-12*eye(3))^2)

shows us that $w_2 = (2, 1, 1)^t$ is a generalized eigenvector of A. Set $w_1 = (A - 12I_3)w_2 = (0, 8, -8)^t$. Then all solutions of (15.1.11) are linear combinations of

$$X_1(t) = \begin{pmatrix} 2 \\ -1 \\ -1 \end{pmatrix}, \quad X_2(t) = e^{12t} \begin{pmatrix} 0 \\ 1 \\ -1 \end{pmatrix}, \quad X_3(t) = e^{12t} \left[t \begin{pmatrix} 0 \\ 8 \\ -8 \end{pmatrix} + \begin{pmatrix} 2 \\ 1 \\ 1 \end{pmatrix} \right].$$

The Direct Computation of the Matrix Exponential e^{tA}

Explicitly solving a linear system of n ODEs when the coefficient matrix A has fewer than n linearly independent eigenvectors is a difficult problem. Several methods have been developed to find closed form solutions when eigenvector deficiencies exist; all of these methods require many calculations and are difficult to carry out using only hand calculation. We now describe one of these methods—the direct computation of the exponential e^{tA} by a method based on the Cayley Hamilton theorem (Theorem 13.4.3), which was proved as a corollary to the Jordan normal form theorem.

A Method Based on Cayley Hamilton and Partial Fractions

Let $\lambda_1, \ldots, \lambda_k$ be the distinct eigenvalues of the $n \times n$ matrix A. Let m_j be the algebraic multiplicity of the eigenvalue λ_j. Therefore the characteristic polynomial of A is

$$p_A(\lambda) = (\lambda_1 - \lambda)^{m_1} \cdots (\lambda_k - \lambda)^{m_k}.$$

We begin with a preliminary discussion of p_A based on partial fractions. In particular, we claim that

$$\frac{1}{p_A(\lambda)} = \frac{a_1(\lambda)}{(\lambda_1 - \lambda)^{m_1}} + \cdots + \frac{a_k(\lambda)}{(\lambda_k - \lambda)^{m_k}}, \qquad \textbf{(15.1.12)}$$

where $a_j(\lambda)$ is a polynomial with $\deg(a_j) \leq m_j - 1$. Partial fractions state that $1/p_A(\lambda)$ can be written as sums of expressions of the form

$$\frac{c_{j1}}{(\lambda_j - \lambda)} + \cdots + \frac{c_{jm_j}}{(\lambda_j - \lambda)^{m_j}}$$

for scalars c_{ji}. Putting these terms over a common denominator proves (15.1.12).

Define the polynomials

$$P_j(\lambda) = \frac{p_A(\lambda)}{(\lambda_j - \lambda)^{m_j}}. \qquad \textbf{(15.1.13)}$$

For instance, if $p_A(\lambda) = (\lambda_1 - \lambda)(\lambda_2 - \lambda)^2(\lambda_3 - \lambda)$ with $\lambda_1 = -2$, $\lambda_2 = 5$, and $\lambda_3 = 4$, then

$$P_1(\lambda) = \frac{p_A(\lambda)}{-2 - \lambda} = (5 - \lambda)^2(4 - \lambda)$$

$$P_2(\lambda) = \frac{p_A(\lambda)}{(5 - \lambda)^2} = (-2 - \lambda)(4 - \lambda)$$

$$P_3(\lambda) = \frac{p_A(\lambda)}{4 - \lambda} = (-2 - \lambda)(5 - \lambda)^2.$$

We can now state one method for computing e^{tA}.

Theorem 15.1.2 *Let A be an $n \times n$ matrix with distinct eigenvalues $\lambda_1, \ldots, \lambda_k$ and algebraic multiplicities $m_1, \ldots, m_k$. Then*

$$e^{tA} = \sum_{\ell=1}^{k} \left[e^{\lambda_\ell t} a_\ell(A) P_\ell(A) \sum_{j=0}^{m_\ell - 1} \frac{1}{j!} t^j (A - \lambda_\ell I_n)^j \right].$$

Note that this matrix exponential consists of functions that are linear combinations of $t^j e^{\lambda_\ell t}$, where $j \leq m_\ell - 1$. These are just the type of terms that appeared in our discussion of solutions of equations in Jordan normal form in Section 14.1.

Proof: Multiplying (15.1.12) by $p_A(\lambda)$ yields the identity

$$1 = a_1(\lambda) P_1(\lambda) + \cdots + a_k(\lambda) P_k(\lambda). \qquad \textbf{(15.1.14)}$$

Identity (15.1.14) is valid for every number λ. This is possible only if, for each $j > 0$, the sum of all coefficients of terms with λ^j is 0. Therefore we can substitute A into (15.1.14) and obtain

$$I_n = a_1(A)P_1(A) + \cdots + a_k(A)P_k(A). \tag{15.1.15}$$

We now compute

$$e^{tA} = e^{\lambda_\ell t} e^{t(A - \lambda_\ell I_n)} = e^{\lambda_\ell t} \sum_{j=0}^{\infty} \frac{1}{j!} t^j (A - \lambda_\ell I_n)^j.$$

Multiplying this identity by $a_\ell(A)P_\ell(A)$ yields

$$a_\ell(A)P_\ell(A)e^{tA} = e^{\lambda_\ell t} a_\ell(A) \sum_{j=0}^{\infty} \frac{1}{j!} t^j P_\ell(A)(A - \lambda_\ell I_n)^j.$$

Now we use the Cayley Hamilton theorem to observe that

$$P_\ell(A)(A - \lambda_\ell I_n)^{m_\ell} = (-1)^{m_\ell} p_A(A) = 0.$$

Hence $P_\ell(A)(A - \lambda_\ell I_n)^j = 0$ for all $j \geq m_\ell$, and therefore

$$a_\ell(A)P_\ell(A)e^{tA} = e^{\lambda_\ell t} a_\ell(A) \sum_{j=0}^{m_\ell - 1} \frac{1}{j!} t^j P_\ell(A)(A - \lambda_\ell I_n)^j.$$

Finally, we sum over the index ℓ and use (15.1.15) to conclude that

$$e^{tA} = \sum_{\ell=1}^{k} a_\ell(A)P_\ell(A)e^{tA} = \sum_{\ell=1}^{k} \left[e^{\lambda_\ell t} a_\ell(A)P_\ell(A) \sum_{j=0}^{m_\ell - 1} \frac{1}{j!} t^j (A - \lambda_\ell I_n)^j \right],$$

which proves the theorem.   ◆

As a special case, consider a matrix A that has n distinct eigenvalues. Then $m_\ell - 1 = 0$ for each ℓ, implying that the polynomial $a_\ell(\lambda)$ has degree zero and is a constant independent of λ. Since $0! = 1$, $t^0 = 1$, and $(A - \lambda_\ell I_n)^0 = I_n$, we obtain:

Corollary 15.1.3   *Let A be an $n \times n$ matrix with distinct eigenvalues $\lambda_1, \ldots, \lambda_n$. Then*

$$e^{tA} = \sum_{\ell=1}^{n} a_\ell e^{\lambda_\ell t} P_\ell(A).$$

An Example with Distinct Eigenvalues

We revisit the example in (15.1.5)—that is,

$$A = \begin{pmatrix} 0 & 3 & 1 \\ 4 & 1 & -1 \\ 2 & 7 & -5 \end{pmatrix}.$$

As we discussed previously, the eigenvalues of this matrix are $\lambda_1 = 4$, $\lambda_2 = -4 + 2i$, and $\lambda_3 = -4 - 2i$. Therefore the characteristic polynomial of A is

$$p_A(\lambda) = (4 - \lambda)(-4 + 2i - \lambda)(-4 - 2i - \lambda),$$

and

$$P_1(\lambda) = (-4 + 2i - \lambda)(-4 - 2i - \lambda)$$
$$P_2(\lambda) = (4 - \lambda)(-4 - 2i - \lambda)$$
$$P_3(\lambda) = (4 - \lambda)(-4 + 2i - \lambda).$$

Moreover, using partial fractions, we can write

$$\frac{1}{p_A(\lambda)} = \frac{\frac{1}{68}}{4 - \lambda} + \frac{-\frac{1}{8+32i}}{-4 + 2i - \lambda} + \frac{-\frac{1}{8-32i}}{-4 - 2i - \lambda}.$$

So

$$a_1(\lambda) = \frac{1}{68}, \quad a_2(\lambda) = -\frac{1}{8 + 32i}, \quad \text{and} \quad a_3(\lambda) = -\frac{1}{8 - 32i}.$$

Next, using MATLAB, compute

$$P_1(A) = ((-4 + 2i)I_3 - A)((-4 - 2i)I_3 - A) = \begin{pmatrix} 34 & 34 & 0 \\ 34 & 34 & 0 \\ 34 & 34 & 0 \end{pmatrix}$$

$$P_2(A) = \quad (4I_3 - A)((-4 - 2i)I_3 - A) \quad = \begin{pmatrix} -2 - 8i & 10 + 6i & -8 + 2i \\ 2 + 8i & -10 - 6i & 8 - 2i \\ 18 + 4i & -22 + 14i & 4 - 18i \end{pmatrix}$$

$$P_3(A) = \quad (4I_3 - A)((-4 + 2i)I_3 - A) \quad = \begin{pmatrix} -2 + 8i & 10 - 6i & -8 - 2i \\ 2 - 8i & -10 + 6i & 8 + 2i \\ 18 - 4i & -22 - 14i & 4 + 18i \end{pmatrix}.$$

Corollary 15.1.3 states that

$$e^{tA} = e^{4t}a_1(A)P_1(A) + e^{-4t+2it}a_2(A)P_2(A) + e^{-4t-2it}a_3(A)P_3(A).$$

Using MATLAB, we obtain

$$
e^{tA} = \frac{1}{2}\, e^{4t} \begin{pmatrix} 1 & 1 & 0 \\ 1 & 1 & 0 \\ 1 & 1 & 0 \end{pmatrix} + \frac{1}{4}\, e^{-4t+2it} \begin{pmatrix} 1 & -1+i & -i \\ -1 & 1-i & i \\ -1+2i & -1-3i & 2+i \end{pmatrix}
$$

$$
+ \frac{1}{4}\, e^{-4t-2it} \begin{pmatrix} 1 & -1-i & i \\ -1 & 1+i & -i \\ -1-2i & -1+3i & 2-i \end{pmatrix}.
$$

Therefore

$$
e^{tA} = \frac{1}{2}\, e^{4t} \begin{pmatrix} 1 & 1 & 0 \\ 1 & 1 & 0 \\ 1 & 1 & 0 \end{pmatrix} + \frac{1}{2}\, e^{-4t} \left(\cos(2t) \begin{pmatrix} 1 & -1 & 0 \\ -1 & 1 & 0 \\ -1 & -1 & 2 \end{pmatrix} + \sin(2t) \begin{pmatrix} 0 & 1 & -1 \\ 0 & -1 & 1 \\ 2 & -3 & 1 \end{pmatrix} \right).
$$

A distinct advantage of this direct method for computing e^{tA} comes when solving an initial value problem. Then we can write the solution directly as $X(t) = e^{tA} X_0$. For example, the solution to the initial value problem when $X_0 = (1, 2, 3)^t$ is

$$
X(t) = e^{tA} X_0 = \frac{1}{2}\, e^{4t} \begin{pmatrix} 3 \\ 3 \\ 3 \end{pmatrix} + \frac{1}{2}\, e^{-4t} \left(\cos(2t) \begin{pmatrix} -1 \\ 1 \\ 3 \end{pmatrix} + \sin(2t) \begin{pmatrix} -1 \\ 1 \\ -1 \end{pmatrix} \right).
$$

Compare this result with (15.1.6).

An Example with Multiple Eigenvalues

As an example, recall the matrix (15.1.11):

$$
A = \begin{pmatrix} 6 & 6 & 6 \\ 5 & 11 & -1 \\ 1 & -5 & 7 \end{pmatrix}.
$$

The eigenvalues of A (using eig(A)) are

$$
\lambda_1 = 0 \quad \text{and} \quad \lambda_2 = \lambda_3 = 12.
$$

Therefore the characteristic polynomial is

$$
p_A(\lambda) = (0 - \lambda)(12 - \lambda)^2 = -\lambda(12 - \lambda)^2.
$$

Using partial fractions, we write

$$
\frac{1}{p_A(\lambda)} = \frac{1}{-\lambda(12 - \lambda)^2} = \frac{\frac{1}{144}}{-\lambda} + \frac{-\frac{1}{6} + \frac{1}{144}\lambda}{(12 - \lambda)^2}.
$$

It follows that

$$a_1(\lambda) = \frac{1}{144} \quad \text{and} \quad a_2(\lambda) = -\frac{1}{6} + \frac{1}{144}\lambda$$

and

$$P_1(\lambda) = (12 - \lambda)^2 \quad \text{and} \quad P_2(\lambda) = -\lambda.$$

Thus, using MATLAB, we calculate

$$a_1(A) = \frac{1}{144}I_3 \quad \text{and} \quad a_2(A) = -\frac{1}{6}I_3 + \frac{1}{144}A = \frac{1}{144}\begin{pmatrix} -18 & 6 & 6 \\ 5 & -13 & -1 \\ 1 & -5 & -17 \end{pmatrix}$$

and

$$P_1(A) = (12I_3 - A)^2 = \begin{pmatrix} 72 & -72 & -72 \\ -36 & 36 & 36 \\ -36 & 36 & 36 \end{pmatrix} \quad \text{and} \quad P_2(A) = -A.$$

From Theorem 15.1.2 (again with the help of MATLAB), it follows that

$$e^{tA} = e^{0t}a_1(A)P_1(A) + e^{12t}a_2(A)P_2(A) + te^{12t}a_2(A)P_2(A)(A - 12I_3)$$

$$= \frac{1}{4}\begin{pmatrix} 2 & -2 & -2 \\ -1 & 1 & 1 \\ -1 & 1 & 1 \end{pmatrix} + \frac{1}{4}e^{12t}\begin{pmatrix} 2 & 2 & 2 \\ 1 & 3 & -1 \\ 1 & -1 & 3 \end{pmatrix} + te^{12t}\begin{pmatrix} 0 & 0 & 0 \\ 2 & 2 & 2 \\ -2 & -2 & -2 \end{pmatrix}.$$

HAND EXERCISES

In Exercises 1 and 4, consider the system of differential equations $\dot{X} = AX$ corresponding to the given matrix A and construct a complete set of linearly independent solutions.

1. $A = \begin{pmatrix} 2 & 1 \\ 0 & 2 \end{pmatrix}$

2. $A = \begin{pmatrix} 1 & 0 & 0 \\ 0 & -1 & 2 \\ 0 & 0 & -1 \end{pmatrix}$

3. $A = \begin{pmatrix} 1 & -1 \\ 1 & 3 \end{pmatrix}$

4. $A = \begin{pmatrix} -4 & -9 & 0 \\ 0 & 5 & 0 \\ -9 & -8 & 5 \end{pmatrix}$

In Exercises 5–8, use Corollary 15.1.3 to compute e^{tA} for each matrix.

5. $A = \begin{pmatrix} 0 & 1 \\ 1 & 0 \end{pmatrix}$

6. $A = \begin{pmatrix} 2 & 1 \\ 1 & 2 \end{pmatrix}$

7. $A = \begin{pmatrix} 3 & 2 \\ 0 & 1 \end{pmatrix}$

8. $A = \begin{pmatrix} 1 & -3 & 1 \\ 0 & -2 & -5 \\ 0 & 0 & 3 \end{pmatrix}$

COMPUTER EXERCISES

In Exercises 9 and 10, solve the initial value problem for the given system of ODE $\dot{X} = AX$ with initial condition $X(0) = X_0$.

9. $A = \begin{pmatrix} 1 & 2 & -1 \\ -1 & 1 & 1 \\ 2 & 2 & -2 \end{pmatrix}$ and $X_0 = \begin{pmatrix} 1 \\ 2 \\ -1 \end{pmatrix}$

10. $A = \begin{pmatrix} -3 & 0 & 2 \\ 2 & -1 & -1 \\ -4 & 0 & 3 \end{pmatrix}$ and $X_0 = \begin{pmatrix} 0 \\ 1 \\ 1 \end{pmatrix}$

In Exercises 11–14, use Theorem 15.1.2 and MATLAB to compute e^{tA} for the given matrix.

11. $A = \begin{pmatrix} 0 & 1 \\ 0 & 0 \end{pmatrix}$

12. $A = \begin{pmatrix} -1 & 2 \\ 0 & -1 \end{pmatrix}$

13. $A = \begin{pmatrix} 1 & 1 & 2 \\ 1 & 1 & 0 \\ 0 & 0 & 0 \end{pmatrix}$

14. $A = \begin{pmatrix} -1 & 2 & -6 \\ 0 & 2 & 0 \\ 0 & -1 & 2 \end{pmatrix}$

15.2   HIGHER ORDER EQUATIONS

The *order* of an ordinary differential equation is the highest order of differentiation that appears in the equation. For instance, the differential equation

$$\frac{d^3x}{dt^3} + 2\frac{dx}{dt} + 5tx = \sin t$$

has order three.

In this section we begin the study of solutions to differential equations of the form

$$a_n\frac{d^n x}{dt^n} + \cdots + a_1\frac{dx}{dt} + a_0 x = g(t), \tag{15.2.1}$$

where $a_j \in \mathbf{R}$ are constants with $a_n \neq 0$ and $g(t)$ is a continuous function. These equations are nth order and linear. The constants a_j are called the *coefficients* of the differential equation (15.2.1). After dividing (15.2.1) by a_n, we may assume that $a_n = 1$.

In analogy to systems of linear equations, we call (15.2.1) *homogeneous* if $g(t) = 0$ for all t. Otherwise the equation is *inhomogeneous*. In particular, the principle of superposition is valid for the homogeneous equations.

Reduction of Equations to First-Order Systems

In principle, we can find solutions to (15.2.1) by finding solutions to a corresponding linear system of first-order ODEs with constant coefficients. We have previously discussed this reduction in the special case of second-order equations to planar systems in Section 6.7. Let

$$y_1(t) = x(t), \quad y_2(t) = \frac{dx}{dt}(t), \quad \ldots, \quad y_n(t) = \frac{d^{n-1}x}{dt^{n-1}}(t). \tag{15.2.2}$$

With these functions, (15.2.1) (with $a_n = 1$) can be rewritten as

$$\frac{dy_n}{dt} + a_{n-1}y_n + \cdots + a_1 y_2 + a_0 y_1 = g(t).$$

Hence (15.2.1) is equivalent to

$$\frac{dy_j}{dt} = y_{j+1} \qquad\qquad (j = 1, \ldots, n-1)$$

$$\frac{dy_n}{dt} = g(t) - a_0 y_1 - \cdots - a_{n-1}y_n,$$

which is a linear constant coefficient system of ODEs. More explicitly, define the $n \times n$ coefficient matrix Q by

$$Q = \begin{pmatrix} 0 & 1 & 0 & \cdots & 0 \\ 0 & 0 & 1 & \cdots & 0 \\ \vdots & \vdots & \vdots & & \vdots \\ 0 & 0 & 0 & \cdots & 1 \\ -a_0 & -a_1 & -a_2 & \cdots & -a_{n-1} \end{pmatrix}, \tag{15.2.3}$$

and introduce the vectors

$$Y(t) = (y_1(t), \ldots, y_n(t))^t \quad \text{and} \quad G(t) = (0, \ldots, 0, g(t))^t.$$

Then we arrive at the system of ODEs

$$\frac{dY}{dt} = QY + G(t). \tag{15.2.4}$$

Let us summarize.

Proposition 15.2.1   *The function $x(t)$ is a solution to the order n ODE (15.2.1) if and only if the vector of functions $Y(t) = (y_1(t), \ldots, y_n(t))^t$ is a solution of the first-order system (15.2.4), where*

$$y_1(t) = x(t) \quad and \quad y_{j+1}(t) = \frac{d^j x}{dt^j}(t)$$

for $j = 1, 2, \ldots, n-1$.

The Homogeneous Equation

Now consider specifically the case when (15.2.1) is homogeneous. As before, we assume that $a_n = 1$ and solve

$$\frac{d^n x}{dt^n} + a_{n-1}\frac{d^{n-1}x}{dt^{n-1}} + \cdots + a_1\frac{dx}{dt} + a_0 x = 0. \tag{15.2.5}$$

Proposition 15.2.1 implies that each solution of (15.2.5) is the first component in a solution to the $n \times n$ system

$$\frac{dY}{dt} = QY, \tag{15.2.6}$$

where Q is the matrix defined in (15.2.3). From the discussion in Section 14.1, it follows that we can find all solutions to (15.2.5) if we know the eigenvalues and the Jordan block structure of Q. In particular, the first component of solutions to (15.2.6) is always just a linear combination of the functions $t^j e^{\lambda t}$, where λ is an eigenvalue of Q. The only question that remains is: Which powers t^j can actually occur? Certainly, j must be less than the multiplicity of the eigenvalue λ. It also follows from Proposition 15.2.1 that there are precisely n linearly independent functions that are solutions to (15.2.5).

We can use this abstract information to find a shortcut for solving (15.2.5). Suppose that $x(t) = e^{\lambda t}$. Then we can compute the left-hand side of (15.2.5) with this $x(t)$, obtaining

$$e^{\lambda t}\left(\lambda^n + a_{n-1}\lambda^{n-1} + \cdots + a_1\lambda + a_0\right). \tag{15.2.7}$$

Thus $x(t)$ is a solution to (15.2.5) precisely when λ is a root of

$$p(\lambda) = \lambda^n + a_{n-1}\lambda^{n-1} + \cdots + a_1\lambda + a_0. \tag{15.2.8}$$

We call $p(\lambda)$ the *characteristic polynomial* of the nth-order equation (15.2.5). The roots of the characteristic polynomial are called *eigenvalues*. This terminology is explained in Section 15.3.

It follows immediately that there is one solution associated with each eigenvalue of the characteristic polynomial $p(\lambda)$ in (15.2.8). If the eigenvalues $\lambda = \sigma \pm i\tau$ are complex, then there are two solutions to (15.2.5)—namely, $e^{\sigma t}\cos(\tau t)$ and $e^{\sigma t}\sin(\tau t)$.

Theorem 15.2.2 *Let $\lambda_1, \ldots, \lambda_k$ be the distinct eigenvalues of the characteristic polynomial $p(\lambda)$ of (15.2.8) with multiplicities $m_1, \ldots, m_k$. Then*

$$\{t^j e^{\lambda_\ell} : 1 \le \ell \le k, 0 \le j < m_\ell\}$$

is a basis of solutions to (15.2.5).

Proof: The previous calculations show that if λ is an eigenvalue of $p(\lambda)$, then $e^{\lambda t}$ is a solution to (15.2.5). The difficulty in proving this theorem is in showing that $t^j e^{\lambda t}$ is also a solution to (15.2.5) for any $j < m$, where m is the multiplicity of the eigenvalue λ. We prove this theorem in two steps. First, we show in Lemma 15.2.3 that the eigenvalues of p are just the eigenvalues of the characteristic polynomial of the coefficient matrix Q in (15.2.3). Second, we show in Lemma 15.2.4 that every eigenvalue of Q is associated with precisely one Jordan block and that every eigenvector of Q has a nonvanishing first component. Then the result follows from Theorem 15.1.1.   ◆

Lemma 15.2.3   *The characteristic polynomial of the matrix Q in (15.2.3) is*

$$p_Q(\lambda) = (-1)^n p(\lambda).$$

Proof: We prove the lemma by induction. For $n = 1$, the matrix Q is just $(-a_0)$. Hence

$$p_Q(\lambda) = \det(Q - \lambda I_1) = -a_0 - \lambda = -(\lambda + a_0),$$

as desired.

Now suppose that the result is valid for all matrices of order $n - 1$. Then we may expand the determinant by cofactors along the 1st column (see (8.1.10)) and obtain

$$p_Q(\lambda) = -\lambda \det \begin{pmatrix} -\lambda & 1 & \cdots & 0 \\ 0 & -\lambda & \cdots & 0 \\ \vdots & \vdots & & \vdots \\ 0 & 0 & \cdots & 1 \\ -a_1 & -a_2 & \cdots & -\lambda - a_{n-1} \end{pmatrix} + (-1)^n a_0 \det \begin{pmatrix} 1 & 0 & \cdots & 0 & 0 \\ -\lambda & 1 & \cdots & 0 & 0 \\ \vdots & \vdots & & \vdots & \vdots \\ 0 & 0 & \cdots & 1 & 0 \\ 0 & 0 & \cdots & -\lambda & 1 \end{pmatrix}.$$

By induction, the first determinant is

$$(-1)^{n-1}(\lambda^{n-1} + a_{n-1}\lambda^{n-2} + \cdots + a_2\lambda + a_1),$$

and by direct calculation the second determinant is 1. Therefore

$$\begin{aligned} p_Q(\lambda) &= -\lambda(-1)^{n-1}(\lambda^{n-1} + a_{n-1}\lambda^{n-2} + \cdots + a_2\lambda + a_1) + (-1)^n a_0 \\ &= (-1)^n(\lambda^n + a_{n-1}\lambda^{n-1} + \cdots + a_1\lambda + a_0). \end{aligned}$$

This proves the lemma.   ◆

Lemma 15.2.4   *Let λ be an eigenvalue of Q. Then, up to scaling, there is a unique eigenvector v corresponding to the eigenvalue λ given by*

$$
v = \begin{pmatrix} 1 \\ \lambda \\ \lambda^2 \\ \vdots \\ \lambda^{n-1} \end{pmatrix}.
$$

In particular, the geometric multiplicity of each eigenvalue of Q is one, and there is exactly one Jordan block of Q associated with the eigenvalue λ.

Proof: Let λ be an eigenvalue of Q and let $v = (v_1, \ldots, v_n)^t$ be a corresponding eigenvector. Writing $Qv = \lambda v$ in coordinates, we obtain

$$
\begin{pmatrix} v_2 \\ v_3 \\ \vdots \\ v_n \\ -\sum_{j=0}^{n-1} a_j v_{j+1} \end{pmatrix} = \begin{pmatrix} \lambda v_1 \\ \lambda v_2 \\ \vdots \\ \lambda v_{n-1} \\ \lambda v_n \end{pmatrix}.
$$

Note that if $v_1 = 0$, then $v_2 = v_3 = \cdots = v_n = 0$, which contradicts the fact that v is an eigenvector. Hence $v_1 \neq 0$ and we may rescale v so that $v_1 = 1$. It follows that

$$
\begin{aligned}
v_2 &= \lambda v_1 & &= \lambda \\
v_3 &= \lambda v_2 & &= \lambda^2 \\
&\ \ \vdots \\
v_n &= \lambda v_{n-1} & &= \lambda^{n-1},
\end{aligned}
$$

as desired.   ◆

We restate the results in Theorem 15.2.2 in real form.

Theorem 15.2.5   *Let λ be an eigenvalue of (15.2.8) with multiplicity m.*

(a)   If λ is real, then $t^j e^{\lambda t}$ is a solution of (15.2.5) for $0 \le j < m$.

(b)   If $\lambda = \sigma + i\tau$ is complex, then $t^j e^{\sigma t} \cos(\tau t)$ and $t^j e^{\sigma t} \sin(\tau t)$ are solutions of (15.2.5) for $0 \le j < m$.

(c)   Every solution of (15.2.5) is a linear combination of the functions constructed in (a) and (b).

Examples of Fourth-Order Equations

Consider the linear homogeneous ordinary differential equation

$$\frac{d^4x}{dt^4} + 2\frac{d^2x}{dt^2} + x = 0. \tag{15.2.9}$$

This type of equation arises if one considers small deformations of an elastic beam under pressure where the pressure is acting at both ends of the beam in the direction of the beam. In this case $x(t)$ represents the resulting deformation of the beam at the spatial point t.

First, we find all eigenvalues of the characteristic polynomial by solving

$$\lambda^4 + 2\lambda^2 + 1 = (\lambda^2 + 1)^2 = 0.$$

This fourth-order polynomial has two roots—namely $\pm i$—each of algebraic multiplicity two. By Theorem 15.2.5, the general solution of (15.2.9) has the form

$$x(t) = c_1 \cos(t) + c_2 \sin(t) + c_3 t \cos(t) + c_4 t \sin(t).$$

Second, if we consider a beam that is hinged at both of its ends—say, at $t = 0$ and $t = \pi$—then the corresponding solution has to satisfy

$$x(0) = x(\pi) = 0 \quad \text{and} \quad \frac{d^2x}{dt^2}(0) = \frac{d^2x}{dt^2}(\pi) = 0.$$

These conditions on the solution $x(t)$ lead to a homogeneous system of four linear equations in the four unknowns c_1, c_2, c_3, c_4. This system can easily be solved by hand, and it follows that c_1, c_3, and c_4 have to vanish, whereas c_2 is arbitrary. Thus, under these conditions a small deformation of the beam is given by $x(t) = c_2 \sin(t)$.

In general, however, we cannot expect to find the roots of a fourth-order characteristic polynomial by inspection. For example, suppose we add a first-order term to (15.2.9), obtaining the differential equation

$$\frac{d^4x}{dt^4} + 2\frac{d^2x}{dt^2} - \frac{dx}{dt} + x = 0. \tag{15.2.10}$$

The characteristic polynomial of (15.2.10) is

$$p(\lambda) = \lambda^4 + 2\lambda^2 - \lambda + 1.$$

We can use MATLAB to find the roots of $p(\lambda)$, as follows: Polynomials are entered into MATLAB by entering the coefficients in an array from highest order to lowest. So we enter $p(\lambda)$ into MATLAB by typing

```
p = [1 0 2 -1 1];
```

To find the roots of $p(\lambda)$, just type the command `roots(p)`. MATLAB responds with

```
ans =
  -0.3438 + 1.3584i
  -0.3438 - 1.3584i
   0.3438 + 0.6254i
   0.3438 - 0.6254i
```

By Theorem 15.2.5, the general solution of (15.2.10) has the form

$$x(t) = (c_1 \cos(1.3584t) + c_2 \sin(1.3584t))e^{-0.3438t}$$
$$+(c_3 \cos(0.6254t) + c_4 \sin(0.6254t))e^{0.3438t}.$$

Note that the typical solution to the fourth-order differential equation (15.2.10) contains sine and cosine terms with two different frequencies. This is not surprising since the characteristic polynomial has two distinct complex conjugate pairs of roots. See Section 14.5.

The Initial Value Problem for Higher Order Equations

For motivation, recall the introductory mechanical examples. The motions of mass points are described by second-order ODEs. To determine a *specific* motion we need to specify the position of the point together with its velocity at a certain time t_0. In other words, to obtain existence and uniqueness of solutions to second-order equations, both

$$x(t_0) \quad \text{and} \quad \frac{dx}{dt}(t_0)$$

have to be specified.

Theorem 15.2.6   *The initial value problem*

$$\frac{d^n x}{dt^n} + a_{n-1}\frac{d^{n-1}x}{dt^{n-1}} + \cdots + a_1\frac{dx}{dt} + a_0 x = 0$$

$$\frac{d^{n-1}x}{dt^{n-1}}(t_0) = x_{n-1}, \ldots, \quad \frac{dx}{dt}(t_0) = x_1, \quad x(t_0) = x_0$$

(15.2.11)

has a unique solution $x(t)$).

Proof: The result is an immediate consequence of Proposition 15.2.1, since the corresponding solution of the system of linear differential equations exists and is unique (see Theorem 6.3.1).   ◆

Inhomogeneous Higher Order Equations

Linearity implies that to find the general solution to the inhomogeneous equation (15.2.1) we need to find one solution to that equation and all solutions to the corresponding homogeneous equation (15.2.5). Sections 15.4 and 15.5 are devoted to finding specific solutions to inhomogeneous equations of various types using the method of undetermined coefficients. Our study of inhomogeneous equations continues in Chapter 16, where we use Laplace transforms to study discontinuous forcing, and in Section 17.4, where we discuss the method of reduction of order.

HAND EXERCISES

In Exercises 1–4, rewrite the given higher order equation as a first-order system.

1. $\ddot{x} + 5\dot{x} + x = 0$

2. $\ddot{x} - 10x = 0$

3. $\dfrac{d^3x}{dt^3} - 4\dfrac{d^2x}{dt^2} - 10x = 0$

4. $\dfrac{d^4x}{dt^4} + 2\dfrac{d^2x}{dt^2} - 3\dfrac{dx}{dt} - 2x = 0$

In Exercises 5–9, find the general solution to the given homogeneous linear differential equation.

5. $\ddot{x} - 3\dot{x} - 4x = 0$

6. $\ddot{x} + 2\dot{x} + 2x = 0$

7. $\ddot{x} - 6\dot{x} + 9x = 0$

8. $\dfrac{d^3x}{dt^3} - 3\dfrac{d^2x}{dt^2} + 3\dfrac{dx}{dt} - x = 0$

9. $\dfrac{d^3x}{dt^3} + 4\dfrac{dx}{dt} = 0$

10. Find a solution to the differential equation

$$\ddot{x} - 3\dot{x} + 2x = 0$$

satisfying $x(0) = 0$ and $x(\ln 2) = 4$.

COMPUTER EXERCISES

In Exercises 11–16, use MATLAB to find the roots of the characteristic polynomial for the given homogeneous linear differential equation, and then find the general solution to that equation.

11. $\dfrac{d^3x}{dt^3} - 6\dfrac{d^2x}{dt^2} + 11\dfrac{dx}{dt} - 6x = 0$

12. $\dfrac{d^3x}{dt^3} + \dfrac{d^2x}{dt^2} + 10x = 0$

13. $\dfrac{d^3x}{dt^3} + \dfrac{d^2x}{dt^2} - 4\dfrac{dx}{dt} - 4x = 0$

14. $\dfrac{d^4x}{dt^4} - 2\dfrac{d^3x}{dt^3} + \dfrac{d^2x}{dt^2} + 8\dfrac{dx}{dt} - 20x = 0$

15. $\dfrac{d^4x}{dt^4} - 3\dfrac{d^2x}{dt^2} - 4x = 0$

16. $\dfrac{d^4x}{dt^4} - 4\dfrac{d^3x}{dt^3} + 14\dfrac{d^2x}{dt^2} - 4\dfrac{dx}{dt} + 13x = 0$

15.3   LINEAR DIFFERENTIAL OPERATORS

In Section 15.4 we describe a powerful method for solving certain second-order linear differential equations. To describe this method, it is convenient to introduce the notion of *linear differential operators*.

We denote by D the simplest differential operator; that is,

$$D = \frac{d}{dt}.$$

From differential calculus we know that D acts linearly on (differentiable) functions; that is,

$$D(x(t) + y(t)) = Dx(t) + Dy(t)$$
$$D(cx(t)) = cDx(t),$$

where $c \in \mathbf{R}$. Thus we say that D is a linear differential operator.

Higher order derivatives can be written in terms of D; that is,

$$\frac{d^2x}{dt^2} = \frac{d}{dt}\left(\frac{dx}{dt}\right) = D(Dx) = D^2x,$$

where D^2 is just the composition of D with itself. Similarly,

$$\frac{d^nx}{dt^n} = D^nx.$$

It follows that $D^2, \ldots, D^n$ are all compositions of linear operators and therefore each is linear. We can even form a polynomial in D by taking linear combinations of the D^k. For example,

$$D^4 - 3D^3 + D^2 - 5D + 10 \tag{15.3.1}$$

is a differential operator. We use the following polynomial notation to denote these operators. Let $q(\lambda)$ be the polynomial

$$q(\lambda) = \lambda^4 - 3\lambda^3 + \lambda^2 - 5\lambda + 10.$$

Then we denote the linear operator in (15.3.1) by $q(D)$. For example,

$$q(D)(\sin(2t)) = 16\sin(2t) + 24\cos(2t) - 4\sin(2t) - 10\cos(2t) + 10\sin(2t)$$
$$= 22\sin(2t) + 14\cos(2t).$$

With this notation in mind, we can reformulate much of the discussion of higher order equations in terms of linear differential operators. Begin by rewriting the homogeneous equation (15.2.5)

$$\frac{d^nx}{dt^n} + a_{n-1}\frac{d^{n-1}x}{dt^{n-1}} + \cdots + a_1\frac{dx}{dt} + a_0x = (D^n + a_{n-1}D^{n-1} + \cdots + a_1D + a_0)x = 0$$

as

$$p(D)x = 0, \tag{15.3.2}$$

where $p(\lambda)$ is the characteristic polynomial of (15.2.5).

We think of the *differential operator* $p(D)$ as operating on functions (that are sufficiently differentiable).

Lemma 15.3.1   *The differential operator $p(D)$ is linear; that is,*

$$p(D)(x + y) = p(D)x + p(D)y$$
$$p(D)(cx) = cp(D)x$$

for all sufficiently differentiable functions x and y and all scalars c.

The proof is left as an exercise; see Exercise 9 at the end of this section.

Using the linearity of these differential operators allows us to reformulate certain aspects of Section 15.2 in this new language.

(a)   Solutions to the homogeneous equation (15.3.2) are just functions in the null space of $p(D)$.

(b)   Using operator notation, we can simplify (15.2.7) as

$$p(D)\left(e^{\lambda t}\right) = p(\lambda)e^{\lambda t}. \tag{15.3.3}$$

It follows from (15.3.3) that the functions $e^{\lambda t}$ are eigenvectors of the operator $p(D)$ with eigenvalue $p(\lambda)$. Usually the functions $e^{\lambda t}$ are called *eigenfunctions*. Perversely, we follow convention and reserve the term *eigenvalue* for just those λ that are roots of the characteristic polynomial—that is, those values of λ for which $p(\lambda) = 0$.

(c)   We can rewrite the inhomogeneous equation as

$$p(D)x = g.$$

Showing that the inhomogeneous equation is solvable is equivalent to showing that the function $g(t)$ is in the range of the operator $p(D)$.

Superposition and the Inhomogeneous Equation

Lemma 15.3.2   *Let $p(\lambda)$ be a polynomial, let $g_1(t)$, $g_2(t)$ be continuous functions, and let α_1, α_2 be scalars. We can find a particular solution $x_p(t)$ to the inhomogeneous differential equation*

$$p(D)x = \alpha_1 g_1(t) + \alpha_2 g_2(t)$$

by first-finding solutions $x_j(t)$ to $p(D)x_j = g_j$ and then setting

$$x_p(t) = \alpha_1 x_1(t) + \alpha_2 x_2(t).$$

Proof: The proof follows directly from the linearity of $p(D)$. Just compute

$$p(D)(\alpha_1 x_1 + \alpha_2 x_2) = \alpha_1 p(D)x_1 + \alpha_2 p(D)x_2 = \alpha_1 g_1 + \alpha_2 g_2.$$

Thus the particular solution is a superposition of the solutions x_1 and x_2.   ◆

The Method of Elimination

In Section 15.2 we showed how to solve an nth-order constant coefficient linear differential equation by converting that equation to a constant coefficient first-order system of differential equations. We now show that the process is reversible—we can solve a first-order system of n equations by finding solutions to an associated nth-order equation. This procedure is called the *method of elimination*. We first discuss this method abstractly using the language of differential operators and then discuss the pragmatic implementation of the method.

Theorem 15.3.3   *Let A be an $n \times n$ matrix and let $p_A(\lambda)$ be the characteristic polynomial of A. Let $X(t) = (x_1(t), \ldots, x_n(t))^t$ be a solution to the system of ODEs*

$$\frac{dX}{dt} = AX.$$

Then each coordinate function $x_j(t)$ satisfies the nth-order differential equation

$$p_A(D)x_j = 0. \tag{15.3.4}$$

Proof: The proof of this theorem follows from the Cayley Hamilton theorem, as follows. Rewrite the differential equation using operator notation as

$$DX = AX,$$

where DX indicates differentiation of the vector $X(t)$ by d/dt and AX indicates multiplication of the vector $X(t)$ by the matrix A. Since the coefficients of the matrix A are constants (independent of t), it follows that $DAX = ADX$. Hence

$$D^2 X = D(AX) = A(DX) = A^2 X.$$

Hence $D^k X = A^k X$ by induction, and $p(D)X = p(A)X$ for any polynomial $p(\lambda)$ by linearity. The Cayley Hamilton theorem (Theorem 13.4.3) states that $p_A(A) = 0$. Hence $p_A(D)X = 0$. So, in coordinates, $p_A(D)x_j = 0$.   ◆

Implementation of the Method of Elimination

Consider the first-order system of differential equations

$$\begin{aligned} \dot{x} &= 2x - 3y \\ \dot{y} &= -5x + 4y. \end{aligned} \qquad \textbf{(15.3.5)}$$

We can eliminate y from the second equation in (15.3.5) by solving for y in the first equation, differentiating, and substituting, as follows:

$$\begin{aligned} y &= \frac{1}{3}(2x - \dot{x}) \\[2mm] \dot{y} &= \frac{1}{3}(2\dot{x} - \ddot{x}). \end{aligned} \qquad \textbf{(15.3.6)}$$

On substituting (15.3.6) into the second equation in (15.3.5), we find

$$\frac{1}{3}(2\dot{x} - \ddot{x}) = -5x + \frac{4}{3}(2x - \dot{x}).$$

Simplification leads to the differential equation

$$\ddot{x} - 6\dot{x} - 7x = 0.$$

Since the characteristic polynomial of the coefficient matrix of (15.3.5) is $\lambda^2 - 6\lambda - 7 = (\lambda - 7)(\lambda + 1)$, this equation is the one predicted by Theorem 15.3.3.

Since the roots of the characteristic polynomial are 7 and -1, it follows that $x(t)$ has the form

$$x(t) = \alpha e^{7t} + \beta e^{-t}.$$

We can solve for $y(t)$ using (15.3.6) and obtain

$$y(t) = \frac{1}{3}(2x - \dot{x}) = -\frac{5}{3}\alpha e^{7t} + \beta e^{-t}.$$

Thus the general solution to the first-order system is

$$\begin{pmatrix} x(t) \\ y(t) \end{pmatrix} = \frac{1}{3}\alpha e^{7t} \begin{pmatrix} 3 \\ -5 \end{pmatrix} + \beta e^{-t} \begin{pmatrix} 1 \\ 1 \end{pmatrix}.$$

Note that the vectors $(3, 5)^t$ and $(1, 1)^t$ are eigenvectors of the coefficient matrix of the system (15.3.5).

Note that the second half of the method of elimination—in the previous example where we back substituted for the function $y(t)$—does not always work. For example, if the system of differential equations decouples, then the second half of the method will fail. See Exercise 8 for an example.

Thus the method of elimination provides another alternative to computing solutions to first-order systems of differential equations; however, this procedure is not carried out easily for systems of more than two or three equations.

HAND EXERCISES

In Exercises 1–4, apply the given linear differential operator $p(D)$ to the functions (a) $\sin(3t)$ and (b) $t^2 e^{-t}$.

1. $p(D) = D + 1$ **2.** $p(D) = D^2 - D$

3. $p(D) = D^4 - 2D^2 + 5$ **4.** $p(D) = 2D^3 - D^2 + 10D - 1$

In Exercises 5–7, write the homogeneous differential equation $p(D)x = 0$ using d/dt notation, factor the characteristic polynomial, and then find the general solution.

5. $p(D) = D + 1$ **6.** $p(D) = -D^2 - 5$

7. $p(D) = -4D^5 + 4D^4 - 8D^2$

8. Attempt to use the method of elimination to solve the following system:

$$\begin{aligned} \dot{x} &= 2x + z \\ \dot{y} &= 3y + z \\ \dot{z} &= z. \end{aligned}$$

Use the characteristic equation (15.3.4) to solve for $x(t)$ and then attempt to use the system of differential equations to solve for $y(t)$ and $z(t)$. What goes wrong?

9. Prove Lemma 15.3.1.

In Exercises 10–13, apply the method of elimination to find the general solution of each system of differential equations.

10. $\begin{aligned} \dot{x} &= y \\ \dot{y} &= 2x + y \end{aligned}$ **11.** $\begin{aligned} \dot{x} &= x - y \\ \dot{y} &= x + y \end{aligned}$

12. $\begin{aligned} \dot{x} &= x + y \\ \dot{y} &= 4x - 2y \end{aligned}$ **13.** $\begin{aligned} \dot{x} &= 2x - y \\ \dot{y} &= y \end{aligned}$

COMPUTER EXERCISES

In Exercises 14–16, write the homogeneous differential equation $p(D)x = 0$ using d/dt notation, use MATLAB to find the roots of the characteristic polynomial, and then find the general solution.

14. $p(D) = 8D^2 - 4D + 8$ **15.** $p(D) = D^4 + 4D^2 - 6D$

16. $p(D) = D^3 - 6$

15.4   UNDETERMINED COEFFICIENTS

In this section we find solutions to inhomogeneous linear differential equations, such as the second-order equation

$$\ddot{x} + b\dot{x} + ax = g(t), \tag{15.4.1}$$

where $g(t)$ is thought of as a forcing term. To find all solutions to the inhomogeneous equation (15.4.1), we need to find just one solution to (15.4.1) and then add to that solution all solutions to the homogeneous equation—which we know how to solve. If the forcing term $g(t)$ is sufficiently nice, then there is an elegant way to solve (15.4.1) called the method of *undetermined coefficients*.

An Illustrative Example

Consider the differential equation

$$\ddot{x} + 3\dot{x} + 2x = t. \tag{15.4.2}$$

To solve (15.4.2) we must find one solution to the inhomogeneous equation and add to that particular solution the general solution of the homogeneous equation.

The general solution to the homogeneous equation is easily found using the techniques of Section 15.2; that is, the characteristic polynomial of the homogeneous equation is

$$p(\lambda) = \lambda^2 + 3\lambda + 2 = (\lambda + 2)(\lambda + 1).$$

So the roots of $p(\lambda)$ are -2 and -1. It follows that the general solution to the homogeneous equation is

$$x_h(t) = \alpha_1 e^{-2t} + \alpha_2 e^{-t}.$$

Therefore, to solve (15.4.2) in general we must find just one solution to (15.4.2). Can we guess the answer? The answer is yes in this case. Since differentiation just lowers the degree of a polynomial, we can guess that there is a particular solution $x(t)$ to (15.4.2) that is a polynomial of degree one, that is, $x(t)$ has the form

$$y(t) = d_1 t + d_2$$

for constants d_1 and d_2. If we substitute $y(t)$ into the left-hand side of (15.4.2), we obtain

$$\left(\frac{d^2}{dt^2} + 3\frac{d}{dt} + 2 \right) y(t) = 0 + 3d_1 + 2(d_1 t + d_2) = 2d_1 t + (3d_1 + 2d_2).$$

Since we want the result of this differentiation to be t, we must choose d_1 and d_2 to solve the linear equations

$$2d_1 = 1 \quad \text{and} \quad 3d_1 + 2d_2 = 0.$$

The solution to this linear system is $d_1 = \frac{1}{2}$ and $d_2 = -\frac{3}{4}$. Therefore we get a particular solution to (15.4.2)—namely, $x_p(t) = \frac{1}{2}t - \frac{3}{4}$. It follows that the general solution to (15.4.2) is

$$x(t) = x_h(t) + x_p(t) = \alpha_1 e^{-2t} + \alpha_2 e^{-t} + \frac{1}{2}t - \frac{3}{4}.$$

Why Did the Guess Work?

What lies at the heart of undetermined coefficients is having a method for choosing a subspace of functions in which a particular solution resides. We call this subspace the *trial space*. In example (15.4.2) the trial space is the two-dimensional subspace

$$d_1 t + d_2.$$

The idea behind finding a trial subspace is the elimination of the inhomogeneity in (15.4.2) using the fact that $g(t) = t$ is itself a solution to some homogeneous differential equation. In example (15.4.2), $g(t) = t$ satisfies the differential equation

$$\frac{d^2 y}{dt^2} = 0.$$

It follows that any solution x of (15.4.2) has to satisfy

$$0 = \frac{d^2}{dt^2} t = \frac{d^2}{dt^2} (\ddot{x} + 3\dot{x} + 2x) = \frac{d^4 x}{dt^4} + 3\frac{d^3 x}{dt^3} + 2\frac{d^2 x}{dt^2}. \tag{15.4.3}$$

The characteristic polynomial of the homogeneous equation (15.4.3) is

$$\lambda^4 + 3\lambda^3 + 2\lambda^2 = \lambda^2(\lambda + 1)(\lambda + 2),$$

and its zeros are

$$\lambda_1 = \lambda_2 = 0, \quad \lambda_3 = -1, \quad \lambda_4 = -2.$$

Hence, the general solution of (15.4.3) is

$$x(t) = c_1 + c_2 t + c_3 e^{-t} + c_4 e^{-2t}.$$

Since we want to find a particular solution of the inhomogeneous equation, we need not consider terms that are solutions of the homogeneous equation. That is, we can set $c_3 = c_4 = 0$ and try to find a solution of the form

$$x(t) = c_1 + c_2 t,$$

which explains more precisely why our guess of a trial space worked.

The Method of Undetermined Coefficients

The method used in the previous example works for many differential equations. We use the notation for linear differential operators developed in Section 15.3 to discuss how the previous example generalizes to a large family of equations. In fact, we can find a particular solution of the nth-order inhomogeneous differential equation

$$p(D)x = g, \qquad\qquad\qquad\qquad \textbf{(15.4.4)}$$

where $g(t)$ is sufficiently differentiable and

$$p(D) = \frac{d^n}{dt^n} + a_{n-1}\frac{d^{n-1}}{dt^{n-1}} + \cdots + a_1\frac{d}{dt} + a_0,$$

as follows: We divide the process into three steps.

Step 1. Find an annihilator of $g(t)$.   Find a linear differential operator

$$q(D) = D^k + b_{k-1}D^{k-1} + \cdots + b_1 D + b_0$$

such that

$$q(D)g = 0. \qquad\qquad\qquad\qquad \textbf{(15.4.5)}$$

This differential operator is called the *annihilator* of g.

Remark: When $g(t)$ is a linear combination of functions, it is often simpler to solve a separate equation for each function in the linear combination, as discussed in Lemma 15.3.2.

It follows that if $x(t)$ is a solution to the inhomogeneous equation (15.4.4), then $x(t)$ is also a solution to the homogeneous equation

$$q(D)p(D)x = q(D)g = 0. \qquad\qquad\qquad\qquad \textbf{(15.4.6)}$$

Note that the roots of the characteristic polynomial pq for (15.4.6) are just the union of the roots of p and the roots of q.

We could take the trial space to be the space of solutions to (15.4.6), but in general that space is too large, as it contains all solutions to the homogeneous equation $p(D)x = 0$.

Step 2: Find the trial space. Compute the general solution to (15.4.6) and set to 0 coefficients of solutions of the original homogeneous equation $p(D)x = 0$, obtaining a subspace of trial functions

$$\hat{y}(t) = c_1 y_1(t) + \cdots + c_k y_k(t).$$

Note that if p and q have no roots in common, then the trial space is precisely the general solution of equation (15.4.5). If p and q have common roots, then the situation is more complicated. See (15.4.14) for an example.

Step 3: Find the particular solution. Substitute the trial function y into (15.4.4) and find constants $c_1, \ldots, c_k$ so that y is a particular solution to (15.4.4).

When Undetermined Coefficients Works

In fact, it is not always possible to satisfy Step 1. In Step 1 we may not be able to find a constant coefficient homogeneous linear differential equation that has $g(t)$ as a solution. However, Theorem 15.2.5 shows that all functions $g(t)$ that are linear combinations of the functions

$$t^j e^{\lambda t}, \quad t^j e^{\sigma t} \sin(\tau t), \quad t^j e^{\sigma t} \cos(\tau t),$$

for $j = 0, 1, \ldots$, are solutions to some homogeneous linear differential equation (perhaps of high order). So we can use undetermined coefficients to find particular solutions for a large class of possible forcing terms g. And when this method can be used, it is relatively straightforward to implement.

A Second Example

Consider the differential equation

$$\ddot{x} + 2\dot{x} + 2x = \cos(3t). \tag{15.4.7}$$

The characteristic polynomial of (15.4.7) is $p(\lambda) = \lambda^2 + 2\lambda + 2$; the associated eigenvalues are $\lambda = -1 \pm i$.

Step 1: The function $g(t) = \cos(3t)$ is a solution to the differential equation

$$\ddot{y} + 9y = 0. \tag{15.4.8}$$

So the differential operator $q(D) = D^2 + 9$ is an annihilator of $g(t) = \cos(3t)$.

Step 2: The roots of $q(\lambda)$ are $\pm 3i$ and they are distinct from those of $p(\lambda)$. Hence the general solution of (15.4.8),

$$y(t) = c_1 \cos(3t) + c_2 \sin(3t), \tag{15.4.9}$$

is the trial space in which to look for a particular solution of (15.4.7).

Step 3:   Substituting (15.4.9) into (15.4.7) yields

$$-9(c_1 \cos(3t) + c_2 \sin(3t)) + 6(c_2 \cos(3t) - c_1 \sin(3t))$$
$$+2(c_1 \cos(3t) + c_2 \sin(3t)) = \cos(3t).$$

That is,
$$(-7c_1 + 6c_2) \cos(3t) + (-6c_1 - 7c_2) \sin(3t) = \cos(3t).$$

Hence we have found a particular solution when the coefficients c_1 and c_2 satisfy

$$-7c_1 + 6c_2 = 1$$
$$-6c_1 - 7c_2 = 0.$$

The solution of this system is
$$c_1 = -\frac{7}{85}, \quad c_2 = \frac{6}{85}.$$

Thus

$$x_p(t) = \frac{1}{85}(6 \sin(3t) - 7 \cos(3t))$$

is a particular solution of (15.4.7).

A Third Example Using Superposition

Find a particular solution of the differential equation

$$\frac{d^3x}{dt^3} - \frac{dx}{dt} + 2x = e^{-2t} \sin t + 1. \tag{15.4.10}$$

The characteristic polynomial of the homogeneous equation associated with (15.4.10) is $p(\lambda) = \lambda^3 - \lambda + 2$, and the roots are (approximately) $-1.52, 0.76 \pm 0.86i$. This (numerical) information is found using MATLAB by typing roots([1 0 -1 2]), obtaining

```
ans =
  -1.5214
   0.7607 + 0.8579i
   0.7607 - 0.8579i
```

Using Lemma 15.3.2, we can find a particular solution to (15.4.10) by adding together solutions to

$$\frac{d^3x_1}{dt^3} - \frac{dx_1}{dt} + 2x_1 = 1 \tag{15.4.11}$$

$$\frac{d^3x_2}{dt^3} - \frac{dx_2}{dt} + 2x_2 = e^{-2t} \sin t. \tag{15.4.12}$$

The solution x_1 to (15.4.11) is found by inspection—the annihilator of $g_1(t) = 1$ is just D, and the trial space consists of the constants. By inspection, the answer is

$$x_1(t) = \frac{1}{2}.$$

To solve (15.4.12) for x_2 we proceed with the three steps associated with undetermined coefficients.

Step 1: The right-hand side of (15.4.12), $g_2(t) = e^{-2t} \sin t$, is a solution of the linear differential equation whose characteristic polynomial has roots $-2 \pm i$. Thus an annihilator for g_2 is

$$\frac{d^2 y}{dt^2} + 4\frac{dy}{dt} + 5y = 0. \tag{15.4.13}$$

To verify this point, observe that the characteristic polynomial of (15.4.13) is

$$q(\lambda) = \lambda^2 + 4\lambda + 5,$$

which has roots $-2 \pm i$.

Step 2: Since the roots of p and q are disjoint sets, the trial space is the general solution of (15.4.13)—namely,

$$y(t) = c_1 e^{-2t} \cos t + c_2 e^{-2t} \sin t.$$

We look for a particular solution to (15.4.12) in this function subspace.

Step 3: We compute

$$Dy = e^{-2t}\left((-2c_1 + c_2)\cos t - (c_1 + 2c_2)\sin t\right)$$
$$D^2 y = e^{-2t}\left((3c_1 - 4c_2)\cos t + (4c_1 + 3c_2)\sin t\right)$$
$$D^3 y = e^{-2t}\left((-2c_1 + 11c_2)\cos t - (11c_1 + 2c_2)\sin t\right).$$

Substituting into (15.4.12) leads to

$$e^{-2t}\left((2c_1 + 10c_2)\cos t + (-10c_1 + 2c_2)\sin t\right) = e^{-2t}\sin t.$$

Hence c_1 and c_2 satisfy

$$2c_1 + 10c_2 = 0$$
$$-10c_1 + 2c_2 = 1.$$

The solution is $c_1 = -\frac{5}{52}$ and $c_2 = \frac{1}{52}$, and the particular solution to (15.4.10) is

$$x_p(t) = \frac{1}{2} - \frac{5}{52} e^{-2t} \cos t + \frac{1}{52} e^{-2t} \sin t.$$

An Example Where p and q Have Common Roots

Consider the first-order differential equation

$$\frac{dx}{dt} - x = e^t. \qquad\qquad \textbf{(15.4.14)}$$

The characteristic polynomial of the homogeneous equation is $p(\lambda) = \lambda - 1$ whose root is $\lambda = 1$. An annihilator of $g(t) = e^t$ is $q(D) = D - 1$. Hence $q(\lambda)$ also has the root $\lambda = 1$. Thus to solve this differential equation by undetermined coefficients, we apply $q(D)$ to both sides of the equation, obtaining

$$q(D)\left(\frac{dx}{dt} - x\right) = \ddot{x} - 2\dot{x} + x = 0.$$

Since $\lambda = 1$ is a double root for the characteristic polynomial of this equation, the general solution is

$$x(t) = c_1 e^t + c_2 t e^t.$$

Setting to 0 the solution to the original homogeneous equation (that is, setting $c_1 = 0$), we find that the trial space for the inhomogeneous equation (15.4.14) is

$$y(t) = c_2 t e^t.$$

Substituting $y(t)$ into (15.4.14) yields

$$c_2(e^t + t e^t) - c_2 t e^t = e^t.$$

It follows that $c_2 = 1$ and that $x_p(t) = t e^t$ is a particular solution. The general solution to (15.4.14) is

$$x(t) = \alpha e^t + t e^t.$$

HAND EXERCISES

In Exercises 1–4, find annihilators for each function.

1. $g(t) = e^{4t} \cos(5t)$ **2.** $g(t) = e^t - 1$

3. $g(t) = t^2 e^{2t}$ **4.** $g(t) = t \cos(5t) + 5 \cos(t)$

In Exercises 5–11, use the method of undetermined coefficients to find particular solutions to the given differential equations.

5. $(D^2 - 3D + 2)x = \sin(t)$ **6.** $\ddot{x} + 2\dot{x} + x = t + e^t$

7. $(D^3 + 6D^2 + 9D + 4)x = e^{-t}$

8. $(D^2 + D - 2)x = 3te^t$

9. $\ddot{x} + x = t \sin t$

10. $(D^3 + D)x = 6t^2 + \sin t$

11. $(D^2 + 2D + 2)x = 8e^{-t} \sin t$

In Exercises 12–15, solve the given initial value problems.

12. $\ddot{x} - x = 1$, where $x(0) = (Dx)(0) = 0$

13. $(D^3 - 8)x = e^{2t}$, where $x(0) = (Dx)(0) = (D^2x)(0) = 0$

14. $(D^2 + D - 6)x = 0$, where $x(0) = (Dx)(0) = 0$

15. $(D^2 - 1)x = 1$, where $x(1) = (Dx)(1) = 0$

15.5   PERIODIC FORCING AND RESONANCE

Periodic Forcing

A linear second-order differential equation is *periodically forced* if it has the form

$$\ddot{x} + b\dot{x} + ax = g(t),$$

where $g(t)$ is periodic in time; that is, $g(t + T) = g(t)$ for some period T. The simplest kind of forcing is *sinusoidal* forcing; that is, $g(t) = \sin(\omega t + t_0)$, where $\omega/2\pi$ is the *forcing frequency* and t_0 is a *phase*. We simplify the discussion by choosing the phase t_0 so that $g(t) = \cos(\omega t)$ and $\omega > 0$.

Suppose, in addition, that the homogeneous equation itself has periodic solutions with *internal frequency* $\tau/2\pi$. Such a differential equation is $\ddot{x} + \tau^2 x = 0$; by rescaling time we can assume that $\tau = 1$. Thus the homogeneous equation is

$$\ddot{x} + x = 0,$$

and the forced equation is

$$\ddot{x} + x = \cos(\omega t), \tag{15.5.1}$$

where $\omega > 0$.

We focus on three features of solutions to this equation:

(a) Generally, solutions to (15.5.1) are quasiperiodic, having two frequencies—just like the torus solutions mentioned in Section 14.5.

(b) When the forcing frequency is near the internal frequency, the solution has beats.

(c) *Resonance* occurs when the forcing frequency equals the internal frequency.

The Periodically Forced Undamped Spring

One of the simplest examples of a model differential equation with both a forcing frequency and an internal frequency is the periodically forced undamped spring.

The motion of a forced undamped spring is described by the spring equation given in Chapter 6, (6.7.5), with $\mu = 0$. To simplify the computations, we set the mass to be $m = 1$ and suppose that the spring constant is $\kappa = 1$. Then, with periodic forcing $g(t) = \cos(\omega t)$, we obtain the differential equation (15.5.1).

Closed Form Solutions to (15.5.1) by Undetermined Coefficients

We can apply the method of undetermined coefficients to solve (15.5.1), but there are two cases: $\omega = 1$ and $\omega \neq 1$.

The Case $\omega \neq 1$

We use undetermined coefficients to find a particular solution to the inhomogeneous equation (15.5.1). The forcing term $\cos(\omega t)$ is a solution to the differential equation

$$\ddot{y} + \omega^2 y = 0. \tag{15.5.2}$$

Thus the annihilator of $\cos(\omega t)$ is $q(D) = D^2 + \omega^2$. The eigenvalues of the homogeneous equation associated with (15.5.1) are $\pm i$, and the eigenvalues of $q(\lambda)$ are $\pm \omega i$. Thus, when $\omega \neq 1$, we can use undetermined coefficients by choosing the general solution to (15.5.2) as the trial space for (15.5.1). So we set

$$y(t) = c_1 \cos(\omega t) + c_2 \sin(\omega t).$$

Next, substitute y into (15.5.1), obtaining

$$(1 - \omega^2)(c_1 \cos(\omega t) + c_2 \sin(\omega t)) = \cos(\omega t).$$

This equality holds when

$$c_1 = \frac{1}{1 - \omega^2} \quad \text{and} \quad c_2 = 0.$$

Since the general solution of the homogeneous equation associated with (15.5.1) is

$$\alpha \cos(t) + \beta \sin(t),$$

the general solution to (15.5.1) is

$$x(t) = \frac{1}{1 - \omega^2} \cos(\omega t) + \alpha \cos(t) + \beta \sin(t)$$

when $\omega \neq 1$.

The Case $\omega = 1$

When $\omega = 1$, the roots of $p(\lambda)$ and $q(\lambda)$ both equal $\pm i$. Thus, to use undetermined coefficients, we must first find the general solution to

$$p(D)q(D)x = 0.$$

This solution is

$$x(t) = \alpha_1 \cos t + \alpha_2 \sin t + c_1 t \cos t + c_2 t \sin t.$$

On setting the solutions to the homogeneous equation associated with (15.5.1) to 0 (that is, on setting $\alpha_1 = \alpha_2 = 0$), we find that the trial space for (15.5.1) is

$$y(t) = c_1 t \cos t + c_2 t \sin t.$$

Substituting $y(t)$ into (15.5.1) yields

$$\ddot{y} + y \equiv -2c_1 \sin t + 2c_2 \cos t = \cos t.$$

Therefore $c_1 = 0$ and $c_2 = \frac{1}{2}$, and a particular solution of (15.5.1) is

$$x_p(t) = \frac{1}{2} t \sin t.$$

The general solution can be written as

$$x(t) = \frac{1}{2} t \sin t + \alpha \cos t + \beta \sin t.$$

To summarize: The general closed form solution to (15.5.1) is

$$x(t) = \alpha \cos t + \beta \sin t + \begin{cases} \dfrac{1}{1-\omega^2} \cos(\omega t) & 0 < \omega \neq 1 \\[2mm] \dfrac{1}{2} t \sin t & \omega = 1 \end{cases} \qquad \textbf{(15.5.3)}$$

Types of Solutions to (15.5.1)

We use the closed form solution (15.5.3) to (15.5.1) to identify three different types of solutions: quasiperiodic two-frequency motion, beats, and resonance.

Quasiperiodic Two-Frequency Motion

In our discussion we now specify initial conditions. In particular, the solution to (15.5.1) with initial conditions $x(0) = 1$ and $\dot{x}(0) = 0$ is

$$x(t) = \cos t + \begin{cases} \dfrac{1}{1 - \omega^2}(\cos(\omega t) - \cos t) & \omega \neq 1 \\ \dfrac{1}{2}t \sin t & \omega = 1 \end{cases} \qquad \textbf{(15.5.4)}$$

In particular, when $\omega \neq 1$, the solution (15.5.4) is just a linear combination of two periodic functions with different frequencies:

$$x(t) = -\frac{\omega^2}{1 - \omega^2} \cos t + \frac{1}{1 - \omega^2} \cos(\omega t). \qquad \textbf{(15.5.5)}$$

In this way it is straightforward to see that the motion is quasiperiodic with two frequencies.

When ω is far from 1, the solution is very close to the solution to the undamped unforced spring equation. Indeed, note that when the frequency ω is large, it follows from (15.5.5) that $x(t)$ is approximately equal to $\cos t$. See Figure 15.1.

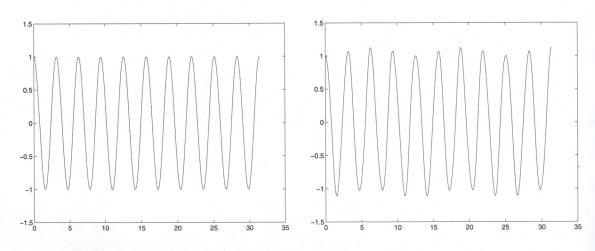

Figure 15.1
Solutions to (*left*) the unforced undamped spring equation and (*right*) the forced spring equation when $\omega = 4.5$

Beats

Using the trigonometric identity

$$\cos A - \cos B = -2 \sin \left(\frac{A + B}{2} \right) \sin \left(\frac{A - B}{2} \right),$$

we find that the solution $x(t)$ is approximately

$$x(t) \approx \frac{1}{1-\omega^2}(\cos(\omega t) - \cos(t)) = \frac{2}{\omega^2 - 1}\sin\left(\frac{\omega + 1}{2}t\right)\sin\left(\frac{\omega - 1}{2}t\right) \quad \textbf{(15.5.6)}$$

when ω is close to 1. Note that the first sine term on the right-hand side of (15.5.6) has period about 2π, while the second sine term has a large period of $4\pi/(\omega - 1)$. For example, when $\omega = 1.05$, this fact leads to periodic behavior of period about 2π and a modulation of period about 80π. See Figure 15.2.

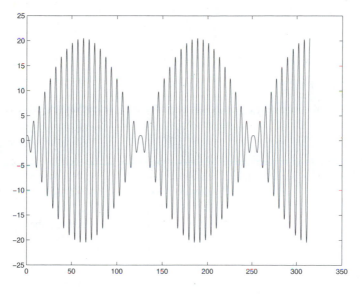

Figure 15.2
Beats in the solution $x(t)$ to (15.5.4) when $\omega = 1.05$ and $0 \le t \le 100\pi$

Resonance

When $\omega = 1$, it follows from (15.5.3) that every solution to (15.5.1) is unbounded as t goes to infinity. This phenomenon is called *resonance* and is due to the fact that the internal frequency is the same as the forcing frequency. The solution to (15.5.1) for $\omega = 1$ is shown in Figure 15.3 (right). Resonance shows that when the forcing frequency equals the internal frequency, the forcing amplifies the internal dynamics.

Note that when ω is close to 1, the solution follows the solution for $\omega = 1$ for some length of time. See Figure 15.3, where the solutions for $\omega = 1.05$ and $\omega = 1$ are given. But it is only when $\omega = 1$ exactly that unbounded growth or resonance actually occurs, while the nearby solution has long-term modulation or beats.

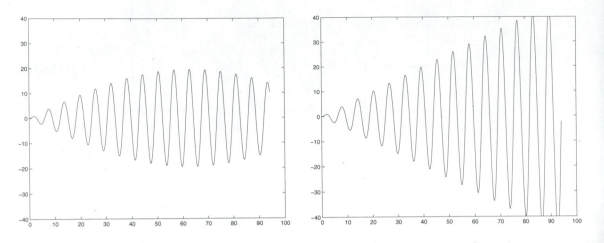

Figure 15.3
Two solutions $x(t)$ from (15.5.4) when $\omega = 1.05$ and $\omega = 1$ for $0 \le t \le 30\pi$

HAND EXERCISES

1. Consider the following equation for the periodically forced undamped spring:

$$\ddot{x} + \kappa x = A \cos(\omega t).$$

(Here the constant of the spring κ and the amplitude A of the forcing are assumed to be positive.) Determine the general solution of this equation depending on the constants κ, A, and ω.
Hint Proceed as in the text and distinguish between the two different cases where $\omega \ne \sqrt{\kappa}$ and $\omega = \sqrt{\kappa}$.

2. The following equation describes the behavior of a damped spring that is forced periodically:

$$\ddot{x} + 2\dot{x} + 2x = \sin(2t).$$

Find real constants γ and δ so that

$$x_p(t) = \gamma \cos(2t) + \delta \sin(2t)$$

is a particular solution of this second-order equation. Write down the general solution and discuss the behavior of solutions if time t is going to infinity.

3. Let $x_\omega(t)$ be the solution to (15.5.1) given in (15.5.4). Show that

$$\lim_{\omega \to 1} x_\omega(t) = x_1(t)$$

for every real number t.

In Exercises 4–7, decide whether or not beats or even resonance occurs in the given differential equations.

4. $\ddot{x} + 4x = \cos(2t)$ **5.** $\ddot{x} + 16x = \sin(3.9t)$

6. $\ddot{x} + 9x = \cos(t)$ **7.** $\ddot{x} + 9x = \cos(-3t)$

8. Recall the sinusoidally forced second-order equation (15.5.1):

$$\ddot{x} + x = \cos(\omega t).$$

The second-order differential operator that annihilates $\cos(\omega t)$ is $q(D) = D^2 + \omega^2$. It follows that any solution to (15.5.1) must satisfy the fourth-order equation

$$q(D)(\ddot{x} + x) = 0. \tag{15.5.7}$$

Compute the characteristic polynomial of (15.5.7) and its roots. Since the roots typically form two complex conjugate pairs, observe that there is a connection between the quasiperiodic solutions of this section and those of Section 14.5. What is the special property of these roots that leads to resonance and why?

16

Laplace Transforms

Laplace transforms are used to solve forced linear differential equations. In Section 15.4 we used the method of undetermined coefficients to solve forced equations when the forcing term $g(t)$ is of a special form—namely, when $g(t)$ is a linear combination of the functions $t^j e^{\lambda t}$, $t^j e^{\sigma t} \sin(\tau t)$, and $t^j e^{\sigma t} \cos(\tau t)$. Although Laplace transforms can be used to solve such systems as well, it is usually more efficient to use the method of undetermined coefficients when that method is applicable. The strength of the method of Laplace transforms is that it can be used to solve forced linear differential equations when the forcing term is more general. Specifically, Laplace transforms can be used to solve forced equations when the forcing term is either discontinuous (a step function) or an impulse function (a Dirac delta function).

The idea behind the *Laplace transform method* is that it is possible to transform any function so that there is a simple relationship between the transform of that function and the transform of its derivative. It is this observation, coupled with linearity, that leads to a useful and elegant method for solving linear, higher order, inhomogeneous differential equations. In Section 16.1 we discuss this property and the Laplace transform method for solving differential equations. In Section 16.2 we formally introduce the Laplace transform (as an improper integral), show that this definition satisfies the basic properties introduced in Section 16.1, and compute the Laplace transform for a variety of functions including step functions and Dirac delta functions.

When we use the method of Laplace transforms to solve linear differential equations, it is immediately clear that this method demands the computation of partial fraction expansions. Some of the details of partial fraction expansions are discussed in Section 16.3.

In Section 16.4 we solve several linear differential equations with discontinuous forcing using the methods developed in the first three sections. RLC circuits provide an excellent example of a physical problem modeled by second-order linear forced differential equations, and this model is discussed in Section 16.5. Discontinuous periodic forcing (AC current) also occurs in these models, and their analysis is discussed in that section.

16.1   THE METHOD OF LAPLACE TRANSFORMS

In this section we describe the basic properties of Laplace transforms and show how these properties lead to a method for solving forced equations. We also discuss the kind of information that we need to know about Laplace transforms in order to solve a general second-order forced equation.

The Basic Properties

Suppose that there is an operation or transform that transforms functions x defined on a variable t to functions $\mathcal{L}[x]$ defined on a variable s in such a way that:

$$(a)\ \mathcal{L} \text{ is linear,}$$
$$(b)\ \mathcal{L} \text{ is invertible, and} \qquad \textbf{(16.1.1)}$$
$$(c)\ \mathcal{L}[\dot{x}](s) = s\mathcal{L}[x](s) - x(0).$$

Note that (16.1.1)(c) relates the transform of the derivative of a function to the transform of the function itself. We call a transform that satisfies (16.1.1) the *Laplace transform*, because there is only one transform that satisfies these three properties.

The Laplace Transform of e^{at}

We can use properties (16.1.1) to compute the Laplace transform of the function $x(t) = e^{at}$. To perform this calculation we need only recall that e^{at} is the unique solution to the differential equation $\dot{x} = ax$ with initial value $x(0) = 1$. In essence, we can compute the Laplace transform of e^{at} without actually having defined the Laplace transform.

It follows from the differential equation $\dot{x} = ax$ and (16.1.1)(a) that

$$\mathcal{L}[\dot{x}](s) = \mathcal{L}[ae^{at}](s) = a\mathcal{L}[e^{at}](s)$$

and from (16.1.1)(c) that

$$\mathcal{L}[\dot{x}](s) = s\mathcal{L}[e^{at}](s) - 1.$$

Equating these two expressions for $\mathcal{L}[\dot{x}]$ leads to

$$s\mathcal{L}[e^{at}](s) - 1 = a\mathcal{L}[e^{at}](s)$$

and hence to

$$\mathcal{L}[e^{at}](s) = \frac{1}{s-a}. \tag{16.1.2}$$

Solving an Inhomogeneous Equation by Laplace Transforms

Properties (16.1.1) and formula (16.1.2) allow us to solve the initial value problem

$$
\begin{aligned}
(a) &\quad \dot{x} - 3x = e^{2t} \\
(b) &\quad x(0) = 1.
\end{aligned}
\tag{16.1.3}
$$

Before proceeding, note that (16.1.3) can be solved directly using the method of undetermined coefficients. Indeed, undetermined coefficients show that $x_p(t) = 2e^{2t}$ is a particular solution to (16.1.3)(a). Since the general solution to the homogeneous equation is αe^{3t}, the general solution to the inhomogeneous equation (16.1.3)(a) is

$$x(t) = 2e^{2t} + \alpha e^{3t}.$$

Setting $\alpha = -1$ solves the initial value problem (16.1.3)(b).

We illustrate the method of Laplace transforms by solving (16.1.3) using Laplace transforms. Apply the transform $\mathcal{L}$ to both sides of (16.1.3), obtaining

$$\mathcal{L}[\dot{x} - 3x](s) = \mathcal{L}[e^{2t}](s).$$

From (16.1.2) with $a = 2$, we see that

$$\mathcal{L}[e^{2t}](s) = \frac{1}{s-2}.$$

Using (16.1.1)(c, a), we see that

$$\mathcal{L}[\dot{x} - 3x](s) = (s-3)\mathcal{L}[x](s) - 1.$$

Hence

$$(s-3)\mathcal{L}[x](s) = 1 + \frac{1}{s-2}.$$

Next we solve for $\mathcal{L}[x](s)$, obtaining

$$\mathcal{L}[x](s) = \frac{1}{s-3} + \frac{1}{(s-2)(s-3)}.$$

Using partial fractions, we find

$$\mathcal{L}[x](s) = \frac{2}{s-3} - \frac{1}{s-2}.$$

We describe the method of partial fractions in more detail in Section 16.3. The particular uses of partial fractions in this section all follow from Exercise 3 at the end of this section.

Finally, from (16.1.2) we see that

$$\mathcal{L}[2e^{3t} - e^{2t}](s) = \frac{2}{s-3} - \frac{1}{s-2},$$

and using (16.1.1)(b), we conclude that

$$x(t) = 2e^{3t} - e^{2t}$$

is the unique solution to the initial value problem (16.1.3).

Three Steps in Solving Equations by Laplace Transforms

We can summarize the method for solving ordinary differential equations by Laplace transforms in three steps. In this summary it will be useful to have defined the inverse Laplace transform.

Definition 16.1.1   *The* inverse Laplace transform *of a function $Y(s)$ is the function $y(t)$ satisfying $\mathcal{L}[y(t)](s) = Y(s)$ and is denoted by $\mathcal{L}^{-1}[Y(s)]$.*

The summary of the Laplace transform method is:

1. Compute the Laplace transforms of both sides of the differential equation in (16.1.6) using the linearity of Laplace transforms, the formulas for Laplace transforms of derivatives (16.1.1)(c), and specific Laplace transforms such as in (16.1.2).
2. Explicitly solve the transformed equation for $\mathcal{L}[x(t)](s) = Y(s)$.
3. Find a function $x(t)$ whose Laplace transform is $Y(s)$; that is, find $\mathcal{L}^{-1}[Y(s)]$.

An Example of a First-Order Forced Equation

As an example, we use this three-step method to solve the initial value problem:

$$\dot{x} - x = 2$$
$$x(0) = 4. \tag{16.1.4}$$

The first step is to apply the Laplace transform to both sides of the differential equation. Using formula (16.1.1)(c) and the linearity of $\mathcal{L}$, we obtain

$$\frac{2}{s} = \mathcal{L}[2] = \mathcal{L}[\dot{x}] - \mathcal{L}[x] = (s-1)\mathcal{L}[x] - 4.$$

Note that $e^{0t} = 1$ so that $\mathcal{L}[1] = 1/(s-0) = 1/s$.

The second step requires solving this equation explicitly for $\mathcal{L}[x]$, obtaining

$$\mathcal{L}[x] = \frac{\frac{2}{s} + 4}{s - 1} = \frac{2 + 4s}{s(s - 1)} = \frac{6}{s - 1} - \frac{2}{s}. \tag{16.1.5}$$

Note that we have simplified the right-hand side by use of partial fractions.

The third step requires using (16.1.2) to see that

$$x(t) = \mathcal{L}^{-1}\left[\frac{6}{s - 1} - \frac{2}{s}\right](t) = 6\mathcal{L}^{-1}\left[\frac{1}{s - 1}\right](t) - 2\mathcal{L}^{-1}\left[\frac{1}{s}\right](t) = 6e^t - 2.$$

Laplace Transforms of Second-Order Equations

We end this section with a discussion of the type of information that we need to solve initial value problems of second-order inhomogeneous linear equations by the method of Laplace transforms. Consider the differential equation

$$\begin{aligned}
\ddot{x} + a\dot{x} + bx &= g(t) \\
x(0) &= x_0 \\
\dot{x}(0) &= \dot{x}_0,
\end{aligned} \tag{16.1.6}$$

where $a, b \in \mathbf{R}$ are constants. The function $g(t)$ is called the *forcing* term.

A Second-Order Homogeneous Example

Consider a simple example of an unforced equation:

$$\begin{aligned}
\ddot{x} + 3\dot{x} + 2x &= 0 \\
x(0) &= 1 \\
\dot{x}(0) &= 2.
\end{aligned} \tag{16.1.7}$$

To solve (16.1.7) we begin by applying (16.1.1)(c) twice to obtain

$$\begin{aligned}
\mathcal{L}[\ddot{x}](s) &= s\mathcal{L}[\dot{x}](s) - \dot{x}(0) \\
&= s(s\mathcal{L}[x](s) - x(0)) - \dot{x}(0) \\
&= s^2\mathcal{L}[x](s) - sx(0) - \dot{x}(0).
\end{aligned}$$

Where there is no ambiguity we drop the argument (s) in $\mathcal{L}[x](s)$; this change should make subsequent formulas easier to read. For instance, the preceding equation is

$$\mathcal{L}[\ddot{x}] = s^2\mathcal{L}[x] - sx(0) - \dot{x}(0). \tag{16.1.8}$$

Now apply (16.1.1) and (16.1.8) to the differential equation (16.1.7) to obtain

$$\begin{aligned}
0 &= \mathcal{L}[\ddot{x} + 3\dot{x} + 2x] \\
&= (s^2\mathcal{L}[x] - sx(0) - \dot{x}(0)) + 3(s\mathcal{L}[x] - x(0)) + 2\mathcal{L}[x] \\
&= (s^2 + 3s + 2)\mathcal{L}[x] - (sx(0) + \dot{x}(0) + 3x(0)) \\
&= (s^2 + 3s + 2)\mathcal{L}[x] - (s + 5).
\end{aligned}$$

Next, use this equation and partial fractions to solve for $\mathcal{L}[x]$ as

$$\mathcal{L}[x] = \frac{s+5}{(s+1)(s+2)} = 4\frac{1}{s+1} - 3\frac{1}{s+2}.$$

Using linearity and (16.1.2), we see that

$$\mathcal{L}[4e^{-t} - 3e^{-2t}] = 4\frac{1}{s+1} - 3\frac{1}{s+2}.$$

Hence

$$x(t) = 4e^{-t} - 3e^{-2t}$$

is the solution to (16.1.7).

Information Needed to Solve Second-Order Equations

What information about Laplace transforms do we need to solve the initial value problem (16.1.6) in general? We can answer this question just by taking the Laplace transform of (16.1.6) and solving for $\mathcal{L}[x]$. Using (16.1.8) and (16.1.1)(c), compute

$$\begin{aligned}
\mathcal{L}[g(t)] &= \mathcal{L}[\ddot{x} + a\dot{x} + bx] \\
&= \mathcal{L}[\ddot{x}] + a\mathcal{L}[\dot{x}] + b\mathcal{L}[x] \\
&= (s^2\mathcal{L}[x] - sx_0 - \dot{x}_0) + a(s\mathcal{L}[x] - x_0) + b\mathcal{L}[x] \\
&= (s^2 + as + b)\mathcal{L}[x] - (x_0 s + ax_0 + \dot{x}_0).
\end{aligned}$$

Then solve for $\mathcal{L}[x]$ as

$$\mathcal{L}[x] = \frac{x_0 s + ax_0 + \dot{x}_0}{s^2 + as + b} + \frac{G(s)}{s^2 + as + b}, \tag{16.1.9}$$

where $G = \mathcal{L}[g(t)]$. The third step in solving (16.1.6) by Laplace transforms is to compute the inverse Laplace transform of the right-hand side of (16.1.9).

To find the inverse Laplace transform of the first term, we use partial fractions. For example, suppose that the polynomial $s^2 + as + b$ has real distinct roots r_1 and r_2. Then we can rewrite the first term as

$$\frac{c_1}{s - r_1} + \frac{c_2}{s - r_2}$$

for some real constants c_1 and c_2. We can now use (16.1.2) to find the inverse Laplace transform of this first term.

If this polynomial has a double real root or a complex conjugate pair of roots, then we need to find inverse Laplace transforms of functions like

$$\frac{1}{(s - r_1)^2} \quad \text{and} \quad \frac{1}{(s - B)^2 + C^2} \quad \text{and} \quad \frac{s}{(s - B)^2 + C^2}. \tag{16.1.10}$$

Finding the inverse Laplace transform for the second term is in general more difficult, since it depends on the hitherto unspecified function $g(t)$. As it happens, this inverse Laplace transform can be computed for a number of important functions, as we discuss in the next section.

To summarize: In order to use the method of Laplace transforms successfully to solve forced second-order linear equations, we must be able to

- compute partial fraction expansions,
- compute the inverse Laplace transforms for functions in (16.1.10), and
- compute the inverse Laplace transforms of the second term in (16.1.9), which involves the forcing function $g(t)$.

HAND EXERCISES

1. Find the Laplace transform of $x(t) = e^{3t} - 2e^{-4t}$.

2. Find the Laplace transform of $x(t) = 14e^{-6t} + 3e^{7t}$.

3. Given distinct roots r_1 and r_2, the method of partial fractions states that

$$\frac{1}{(s - r_1)(s - r_2)} = \frac{a_1}{s - r_1} + \frac{a_2}{s - r_2} \qquad \textbf{(16.1.11)}$$

for some scalars a_1 and a_2. By putting the right side of (16.1.11) over a common denominator, verify that a_1 and a_2 are found by solving the system of linear equations

$$a_1 + a_2 = 0$$
$$r_2 a_1 + r_1 a_2 = -1.$$

In Exercises 4 and 5, use partial fractions to find a function $x(t)$ whose Laplace transform is the given function $Y(s)$.

Hint   In your calculations, you may use the result of Exercise 3.

4. $Y(s) = \dfrac{1}{s^2 - 4s + 3}$

5. $Y(s) = \dfrac{s - 1}{s^2 + 5s + 6}$

In Exercises 6 and 7, use Laplace transforms to compute the solution to the given initial value problem.

6. $\ddot{x} + 3\dot{x} + 2x = 0$; $x(0) = 1$, $\dot{x}(0) = -1$

7. $\ddot{x} - 3\dot{x} - 4x = 0$; $x(0) = 2$, $\dot{x}(0) = -1$

8. Let α, β, a, b, c be real numbers. Show that the solution of the initial value problem

$$\ddot{x} + a\dot{x} + bx = c; \quad x(0) = \alpha, \quad \dot{x}(0) = \beta,$$

has the Laplace transform

$$\mathcal{L}[x] = \frac{\alpha s^2 + (\beta + a\alpha)s + c}{s(s^2 + as + b)}.$$

9. Identity (16.1.8) may be generalized to

$$\mathcal{L}\left[\frac{d^k y}{dt^k}\right] = s^k \mathcal{L}[y] - s^{k-1} y(0) - \cdots - \frac{d^{k-1} y}{dt^{k-1}}(0). \qquad \textbf{(16.1.12)}$$

Use induction to verify (16.1.12).

16.2   LAPLACE TRANSFORMS AND THEIR COMPUTATION

This section divides into three parts. In the first part, we give an explicit formula for the Laplace transform and verify that this formula satisfies properties (16.1.1). In the second part, we compute a table of Laplace transforms for a number of special functions including step functions and impulse functions. In the third part, we discuss two properties of Laplace transforms that are useful when computing inverse Laplace transforms of functions of the type that appear when solving forced second-order equations.

Definition 16.2.1   *Let $x(t)$ be a function that is defined for $0 \le t < \infty$. Then the function $\mathcal{L}[x](s)$ is defined by*

$$\mathcal{L}[x](s) = \int_0^\infty e^{-st} x(t) \, dt \qquad \textbf{(16.2.1)}$$

and is called the Laplace transform *of $x(t)$.*

Note that the Laplace transform is an improper integral—which implies that some care must be taken when discussing its properties. Indeed, for some functions the Laplace transform does not exist for all values of s. What usually happens is that the Laplace transform exists for all values of s larger than some number a. As an example, we compute (16.1.2) directly using Definition 16.2.1. Let $x(t) = e^{at}$ and compute

$$\mathcal{L}[x](s) = \int_0^\infty e^{(a-s)t} \, dt = \lim_{h \to \infty} \frac{1}{s-a}\left(1 - e^{(a-s)h}\right) = \begin{cases} \dfrac{1}{s-a} & \text{for } s > a \\ \infty & \text{for } s \le a. \end{cases} \qquad \textbf{(16.2.2)}$$

The Laplace Transform Is Linear

The first step in verifying properties (16.1.1) is to show that the Laplace transform $\mathcal{L}$ is linear—that is, to show that (16.1.1)(a) holds.

Proposition 16.2.2   *Suppose $x(t)$ and $y(t)$ are functions for which the Laplace transforms $\mathcal{L}[x](s)$ and $\mathcal{L}[y](s)$ exist for $s > a$, and let c be real. Then for all $s > a$,*

$$\mathcal{L}[x + y] = \mathcal{L}[x] + \mathcal{L}[y]$$
$$\mathcal{L}[cx] = c\mathcal{L}[x]. \qquad \textbf{(16.2.3)}$$

Proof: Using properties of integrals, we obtain for each $s > a$,

$$\mathcal{L}[x + y](s) = \int_0^\infty e^{-st}(x(t) + y(t))\, dt$$

$$= \int_0^\infty e^{-st}x(t)\, dt + \int_0^\infty e^{-st}y(t)\, dt$$

$$= \mathcal{L}[x](s) + \mathcal{L}[y](s).$$

A similar computation verifies that $\mathcal{L}$ commutes with scalar multiplication.   ◆

The Derivative Property of Laplace Transforms Is Valid

Next we verify the important derivative property (16.1.1)(c) of the Laplace transform.

Proposition 16.2.3   *Suppose that x is a differentiable function such that $|x(t)| \le Me^{at}$ for constants $a > 0$ and $M > 0$. Then*

(a)   *the Laplace transform $\mathcal{L}[x](s)$ exists for $s > a$, and*

(b) $\mathcal{L}[\dot{x}](s) = s\mathcal{L}[x](s) - x(0)$ *for $s > a$.*

Proof:

(a)   Observe that by the assumption on the growth of $x(t)$ we have for $s > a$,

$$\lim_{h \to \infty} \left| \int_0^h e^{-st}x(t)\, dt \right| \le \lim_{h \to \infty} \int_0^h Me^{(a-s)t}\, dt = \lim_{h \to \infty} \left. \frac{Me^{(a-s)t}}{a - s} \right|_0^h = \frac{M}{s - a} < \infty.$$

Hence the Laplace transform $\mathcal{L}[x](s)$ exists for $s > a$.

(b)   Note that

$$\lim_{h \to \infty} |e^{-sh}x(h)| \le \lim_{h \to \infty} |Me^{(a-s)h}| = 0.$$

Now use the definition of improper integrals and integration by parts to compute

$$\mathcal{L}[\dot{x}](s) = \int_0^\infty e^{-st}\dot{x}(t)\, dt$$

$$= \lim_{h \to \infty} \int_0^h e^{-st}\dot{x}(t)\, dt$$

$$= \lim_{h \to \infty} \left(\left. e^{-st}x(t) \right|_0^h - \int_0^h (-s)e^{-st}x(t)\, dt \right)$$

$$= \lim_{h \to \infty} \left(e^{-sh}x(h) - x(0) \right) + s \lim_{h \to \infty} \int_0^h e^{-st}x(t)\, dt$$

$$= -x(0) + s\mathcal{L}[x](s)$$

for $s > a$.   ◆

Finally, we remark that (16.1.1)(b) is valid but that its proof is beyond the scope of this course. One method of proof is to develop an explicit integral formula for the inverse Laplace transform.

A Table of Laplace Transforms

In order to solve differential equations using the method of Laplace transforms, we need to have a table of Laplace transforms readily available. We now compute the Laplace transform for the functions in Table 16.1. These functions include the discontinuous step function

$$H_c(t) = \begin{cases} 0 & \text{for } 0 \le t < c \\ 1 & \text{for } t \ge c, \end{cases}$$

where c is a positive constant, and the hitherto undefined Dirac delta function $\delta_c(t)$. See Figure 16.1 for a graph of H_1.

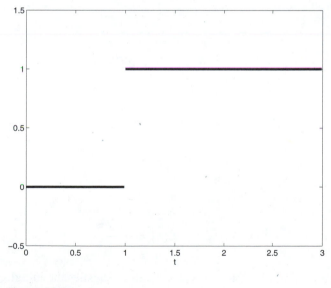

Figure 16.1
Graph of $H_1(t)$

We derive some of these formulas by using the Laplace transform formulas for derivatives (16.1.1)(c) and (16.1.8). (We could compute these transforms directly by integration from Definition 16.2.1, but the method we choose is simpler and illustrates the power of the derivative formulas.) Note that we have already computed the Laplace transforms of the first two functions in Table 16.1 when deriving (16.1.2).

Table 16.1
Laplace transforms

Function $y(t)$	Laplace Transform $\mathcal{L}[y](s)$
1	$\dfrac{1}{s}$
e^{at}	$\dfrac{1}{s-a}$
$\cos(\tau t)$	$\dfrac{s}{s^2+\tau^2}$
$\sin(\tau t)$	$\dfrac{\tau}{s^2+\tau^2}$
t^n	$\dfrac{n!}{s^{n+1}}$
$H_c(t)$	$\dfrac{1}{s}e^{-cs}$
$\delta_c(t)$	e^{-cs}

The Laplace Transforms: $\mathcal{L}[\cos(\tau t)]$ and $\mathcal{L}[\sin(\tau t)]$

To compute $\mathcal{L}[x]$, where $x(t) = \cos(\tau t)$, observe that $x(t)$ is a solution to the initial value problem

$$\ddot{x} + \tau^2 x = 0$$
$$x(0) = 1$$
$$\dot{x}(0) = 0.$$

It follows from (16.1.8) that

$$s^2\mathcal{L}[x] - s + \tau^2\mathcal{L}[x] = 0.$$

Hence

$$\mathcal{L}[x] = \frac{s}{s^2+\tau^2},$$

as desired. Similarly, note that $x(t) = \sin(\tau t)$ satisfies the initial value problem

$$\ddot{x} + \tau^2 x = 0$$
$$x(0) = 0$$
$$\dot{x}(0) = \tau.$$

The computation of the Laplace transform for $\sin(\tau t)$ proceeds identically as the computation of $\mathcal{L}[\cos(\tau t)]$.

The Laplace Transform: $\mathcal{L}[t^n]$

Next we compute $\mathcal{L}[t^n]$ by applying (16.1.1)(c) n times; that is,

$$\mathcal{L}[t^n] = \frac{n}{s}\mathcal{L}[t^{n-1}] = \cdots = \frac{n!}{s^n}\mathcal{L}[1] = \frac{n!}{s^{n+1}}.$$

The Laplace Transform: $\mathcal{L}[H_c(t)]$

We compute the Laplace transform of the step function $H_c(t)$ by direct integration:

$$\mathcal{L}[H_c](s) = \int_0^\infty e^{-st} H_c(t)\, dt = \int_c^\infty e^{-st}\, dt = \begin{cases} \frac{1}{s}e^{-cs} & \text{for } s > 0 \\ \infty & \text{for } s \le 0. \end{cases}$$

The Laplace Transform of Dirac Delta Functions

Suppose that a strong external force is applied at time $t = c$, but only for a very short time h. Then the forcing function is approximated by

$$g_h(t) = \begin{cases} \dfrac{1}{h} & c \le t \le c + h \\ 0 & \text{otherwise} \end{cases} \tag{16.2.4}$$
$$= \frac{1}{h}\left(H_c(t) - H_{c+h}(t)\right).$$

The graph of g_h is given in Figure 16.2. It is not clear how to take the limit of $g_h(t)$ as $h \to 0$ in (16.2.4), since this limit leads to a "function" that is infinity at c and 0 elsewhere. Despite this difficulty, the limit is called the *Dirac delta function* and is denoted by $\delta_c(t)$. We compute the Laplace transform of the $\delta_c(t)$ by first computing the Laplace transform of the approximating function $g_h(t)$ and then taking the limit of the Laplace transforms of the approximation as $h \to 0$.

The Laplace transform of the approximating function $g_h(t)$ is found using the linearity of $\mathcal{L}$ and Table 16.1, that is,

$$\mathcal{L}[g_h] = \frac{1}{h}\left(\mathcal{L}[H_c] - \mathcal{L}[H_{c+h}]\right) = \frac{1}{h}\left(\frac{1}{s}e^{-cs} - \frac{1}{s}e^{-(c+h)s}\right) = \frac{1}{s}e^{-cs}\frac{1 - e^{-sh}}{h}.$$

We claim that

$$\lim_{h \to 0}\left(\frac{1 - e^{-sh}}{h}\right) = s,$$

from which it follows that

$$\lim_{h \to 0}\mathcal{L}[g_h](s) = e^{-cs}.$$

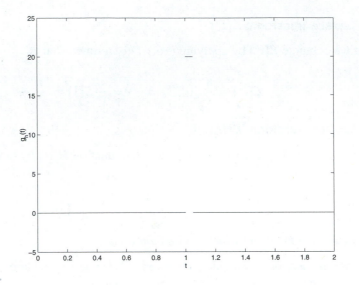

Figure 16.2
Graph of the approximating function $g_h(t)$ of (16.2.4) for $c = 1$ and $h = 0.05$

This result justifies writing

$$\mathcal{L}[\delta_c] = e^{-cs}. \tag{16.2.5}$$

To verify the claim, let $f(y) = e^{-sy}$ and recall that

$$f'(0) = \lim_{h \to 0} \frac{f(h) - f(0)}{h} = \lim_{h \to 0} \frac{e^{-sh} - 1}{h}.$$

Since $f'(0) = -s$, it follows that

$$\lim_{h \to 0} \left(\frac{1 - e^{-sh}}{h} \right) = -f'(0) = s,$$

as claimed.

Additional Techniques for Laplace Transforms

At the end of Section 16.1 we showed that when solving second-order differential equations by the method of Laplace transforms, we must compute inverse Laplace transforms of functions like

$$\frac{1}{(s - B)^2 + C^2} \quad \text{and} \quad \frac{s - B}{(s - B)^2 + C^2} \tag{16.2.6}$$

and

$$\frac{1}{(s-B)^2 + C^2} G(s) \quad \text{and} \quad \frac{s-B}{(s-B)^2 + C^2} G(s), \qquad \textbf{(16.2.7)}$$

where $G(s)$ is the Laplace transform of the forcing function $g(t)$. See (16.1.10).

Inverse Laplace Transforms of Shifted Functions

The next proposition enables us to calculate inverse Laplace transforms for functions of type (16.2.6).

Proposition 16.2.4   *Suppose that $Y(s) = \mathcal{L}[y(t)](s)$ exists. Then*

$$\begin{aligned}
(a) \ \mathcal{L}[e^{at} y(t)](s) \ &= Y(s-a) \\
(b) \ \mathcal{L}^{-1}[Y(s-a)](t) \ &= e^{at} y(t).
\end{aligned}$$

Proof: To verify (a) compute

$$\mathcal{L}[e^{at} y(t)](s) = \int_0^\infty e^{-st} e^{at} y(t) \, dt = \int_0^\infty e^{-(s-a)t} y(t) \, dt = \mathcal{L}[y(t)](s-a) = Y(s-a).$$

Part (b) follows by applying the inverse Laplace transform to both sides of (a).   ◆

For example, suppose that we want to compute the inverse Laplace transform of

$$\frac{s-1}{s^2 + 2s + 5}. \qquad \textbf{(16.2.8)}$$

The first step in this computation is to write the denominator as the sum of two squares and to shift the numerator; that is,

$$\frac{s-1}{s^2 + 2s + 5} = \frac{(s+1) - 2}{(s+1)^2 + 4} = \frac{s+1}{(s+1)^2 + 4} - 2\frac{1}{(s+1)^2 + 4}. \qquad \textbf{(16.2.9)}$$

We can now find the inverse Laplace transform of the right-hand side of (16.2.9). Except for the shift of s to $s+1$, the rearranged expression (16.2.9) is one whose inverse Laplace transform can be read from Table 16.1. More precisely, let

$$Y(s) = \frac{s}{s^2 + 4} - 2\frac{1}{s^2 + 4},$$

so that the right-hand side of (16.2.9) is $Y(s+1)$. From Table 16.1 the inverse Laplace transform of $Y(s)$ is

$$\mathcal{L}^{-1}[Y(s)](t) = \mathcal{L}^{-1}\left[\frac{s}{s^2 + 4} - 2\frac{1}{s^2 + 4}\right](t) = \cos(2t) - \sin(2t) = y(t). \quad \textbf{(16.2.10)}$$

Using Proposition 16.2.4(b), we can compute the inverse Laplace transform of $Y(s + 1)$ as

$$\mathcal{L}^{-1}[Y(s + 1)](t) = e^{-t} y(t);$$

that is,

$$\mathcal{L}^{-1}\left[\frac{s - 1}{s^2 + 2s + 5}\right](t) = e^{-t}(\cos(2t) - \sin(2t)).$$

Another Useful Property of the Laplace Transform

Next we examine the Laplace transform of a function that is multiplied by the jump function $H_c(t)$ and by doing so we can compute the inverse Laplace transform of functions like (16.2.7) when $G(s)$ is the Laplace transform of a discontinuous forcing function.

Proposition 16.2.5 *Suppose that* $Y(s) = \mathcal{L}[y(t)](s)$ *exists. Then*

(a) $\mathcal{L}[H_c(t)y(t - c)](s) = e^{-cs} Y(s)$

(b) $\mathcal{L}^{-1}[e^{-cs} Y(s)](t) = H_c(t)y(t - c).$

Proof: To verify (a) compute

$$\mathcal{L}[H_c(t)y(t - c)](s) = \int_0^\infty e^{-st} H_c(t)y(t - c)\, dt = \int_c^\infty e^{-st} y(t - c)\, dt.$$

On substituting $t + c$ for t, we obtain

$$\mathcal{L}[H_c(t)y(t - c)](s) = \int_0^\infty e^{-s(t+c)} y(t)\, dt = e^{-cs} \int_0^\infty e^{-st} y(t)\, dt = e^{-cs} Y(s).$$

Part (b) follows by applying the inverse Laplace transform to both sides of (a). ◆

As an application of Proposition 16.2.5(b), we find that

$$\mathcal{L}^{-1}\left[e^{-s}\frac{s}{s^2 + 4}\right](t) = H_1(t)\mathcal{L}^{-1}\left[\frac{s}{s^2 + 4}\right](t - 1) = H_1(t)\cos(2(t - 1)).$$

HAND EXERCISES

In Exercises 1–4, compute the Laplace transform for each function $x(t)$.

1. $x(t) = 4\cos(t - 1)$ **2.** $x(t) = \sin(3(t - 2))$ **3.** $x(t) = (t - 3)^2$ **4.** $x(t) = te^{-2t}$

In Exercises 5–9, use Laplace transforms to compute the solution to the given initial value problem.

5. $\ddot{x} + 2\dot{x} + 5x = 0$; $x(0) = 2$, $\dot{x}(0) = -6$

6. $\ddot{x} + 4\dot{x} + 20x = 0$; $x(0) = 1$, $\dot{x}(0) = -6$

7. $\ddot{x} - 6\dot{x} + 13x = 1$; $x(0) = 0$, $\dot{x}(0) = 1$

8. $\ddot{x} + 2\dot{x} - 3x = 1$; $x(0) = 2$, $\dot{x}(0) = 0$

9. $\ddot{x} + 2\dot{x} - 8x = e^t$; $x(0) = 1$, $\dot{x}(0) = 2$

10. Suppose that $z(t) = t y(t)$. Using (16.2.1), show that

$$\mathcal{L}[z](s) = -\frac{d}{ds}\mathcal{L}[y](s).$$

16.3   PARTIAL FRACTIONS

In Section 16.1 we saw that expansion into partial fractions is a necessary tool when applying the method of Laplace transforms. In the simplest case partial fractions work as follows. Assume that $p(s)$ and $q(s)$ are two polynomials such that:

   (a)   the degree of $p(s)$ is less than or equal to the degree d of $q(s)$, and

   (b) $q(s) = (s - r_1) \cdots (s - r_d)$ has no multiple roots.

The roots $r_1, \ldots, r_d$ of $q(s)$ may be either real or complex. Then the expansion into partial fractions of $p(s)/q(s)$ has the form

$$\frac{p(s)}{q(s)} = \frac{c_1}{s - r_1} + \frac{c_2}{s - r_2} + \cdots + \frac{c_d}{s - r_d}, \tag{16.3.1}$$

where $c_1, \ldots, c_d$ are scalars determined from p and q.

There is a simple way to compute the constant c_j. Define the degree $d - 1$ polynomial:

$$q_j(s) = \frac{q(s)}{s - r_j} = (s - r_1) \cdots (s - r_{j-1})(s - r_{j+1}) \cdots (s - r_d).$$

Multiply both sides of (16.3.1) by $s - r_j$ and evaluate at $s = r_j$ to obtain

$$c_j = \frac{p(r_j)}{q_j(r_j)}.$$

For example, compute the partial fraction expansion for

$$\frac{s^2 + 3s - 5}{s(s - 1)(s + 2)} = \frac{c_1}{s} + \frac{c_2}{s - 1} + \frac{c_3}{s + 2}.$$

In this example, $r_1 = 0$, $r_2 = 1$, and $r_3 = -2$. The relevant polynomials are

$$p(s) = s^2 + 3s - 5, \quad q_1(s) = (s - 1)(s + 2), \quad q_2(s) = s(s + 2), \quad q_3(s) = s(s - 1).$$

It follows that

$$c_1 = \frac{p(0)}{q_1(0)} = \frac{5}{2}, \quad c_2 = \frac{p(1)}{q_2(1)} = -\frac{1}{3}, \quad c_3 = \frac{p(-2)}{q_3(-2)} = -\frac{7}{6}.$$

Partial Fractions with Complex Roots

Suppose that the denominator $q(s)$ has complex conjugate roots r and $\bar{r}$. When r is a complex simple root of $q(s)$, the partial fraction expansion of $p(s)/q(s)$ contains the two terms

$$\frac{c}{s-r} + \frac{d}{s-\bar{r}} \,,$$

where $c, d \in \mathbf{C}$. Together these two terms must be real-valued, and it follows that $d = \bar{c}$. Therefore the expansion is

$$\frac{c}{s-r} + \frac{\bar{c}}{s-\bar{r}}$$

for some complex scalar c. These terms combine as

$$\frac{c(s-\bar{r}) + \bar{c}(s-r)}{s^2 - 2r_1 s + r\bar{r}} \,,$$

where $r = r_1 + ir_2$. With inverse Laplace transforms in mind, we prefer to write this expression as

$$\frac{2c_1(s-r_1) - 2c_2 r_2}{(s-r_1)^2 + r_2^2} \tag{16.3.2}$$

where $c = c_1 + ic_2$.

In the third part of Section 16.2 we saw how to compute inverse Laplace transforms of functions like those in (16.3.2).

Partial Fractions Using MATLAB

The MATLAB command residue can be used to determine partial fraction expansions. We begin by discussing how polynomials are defined in MATLAB. The polynomial $q(s) = a_d s^d + \cdots + a_1 s + a_0$ is stored in MATLAB by the vector q = [ad $\cdots$ a1 a0] consisting of the coefficients of $q(s)$ in *descending* order. For instance, in MATLAB, the polynomial $q(s) = 2s^3 - 3s + 5$ is identified with the vector $(2, 0, -3, 5)$.

Suppose that the two vectors p and q represent the two polynomials $p(s)$ (of degree less than D) and $q(s)$ (of degree d). Both the vector of roots $r = (r_1, \ldots, r_d)$ and the vectors of scalars $c = (c_1, \ldots, c_d)$ are determined using the command

```
[c,r] = residue(p,q)
```

To illustrate this command, find the partial fraction expansion of

$$\frac{4s+2}{s^2-s}$$

(which we computed in (16.1.5)) by typing

```
    p = [4 2];
    q = [1 -1 0];
 [c,r] = residue(p,q)
```

MATLAB responds with

```
 c =
       6
      -2
 r =
       1
       0
```

Note that this result agrees with our previous calculation in (16.1.5).

Observe that the situation where the polynomial $q(s)$ has complex roots is not excluded. Indeed, let $p(s) = s + 1$ and $q(s) = s^3 + s^2 - 4s + 6$, and type

```
    p = [1 1];
    q = [1 1 -4 6];
 [c,r] = residue(p,q)
```

to obtain the answer

```
 c =
   -0.1176
    0.0588 - 0.2647i
    0.0588 + 0.2647i
 r =
   -3.0000
    1.0000 + 1.0000i
    1.0000 - 1.0000i
```

In particular, $q(s)$ has the three roots -3, $1 + i$, and $1 - i$, and we have the expansion

$$\frac{s + 1}{s^3 + s^2 - 4s + 6} = -\frac{0.1176}{s + 3} + \frac{0.0588 - 0.2647i}{s - (1 + i)} + \frac{0.0588 + 0.2647i}{s - (1 - i)}. \qquad \textbf{(16.3.3)}$$

The Return to Real Form in Partial Fractions

We can return to a representation in real numbers by combining the terms corresponding to complex conjugate roots. When r is a complex simple root of $q(s)$, the partial fraction expansion of $p(s)/q(s)$ contains the two terms

$$\frac{c}{s - r} + \frac{\bar{c}}{s - \bar{r}}$$

for some complex scalar c. These terms combine to give

$$\frac{2c_1(s - r_1) - 2c_2 r_2}{(s - r_1)^2 + r_2^2},$$

where $c = c_1 + ic_2$, as in (16.3.2).

We can write a MATLAB m-file to perform the computations in (16.3.2) as follows:

```
function [cr,rr] = realform(c,r)
cr = [2*real(c), -2*imag(c)*imag(r)];
rr = [real(r), imag(r)^2];
```

This m-file is accessed using the command

```
[num,denom] = realform(c(i),r(i))
```

where i is the index corresponding to the complex conjugate root of $q(s)$. For example, if we consider the expansion in (16.3.3), then we type

```
[num,denom] = realform(c(2),r(2))
```

yielding the answer

```
num =
      0.1176      0.5294
denom =
      1.0000      1.0000
```

This output corresponds to the expression

$$\frac{\text{num}(1)(s - \text{denom}(1)) + \text{num}(2)}{(s - \text{denom}(1))^2 + \text{denom}(2)}.$$

Combining the second and third term on the right-hand side leads to

$$\frac{s+1}{s^3 + s^2 - 4s + 6} = -\frac{0.1176}{s+3} + \frac{0.1176(s-1) + 0.5294}{(s-1)^2 + 1}.$$

Repeating a Calculation Using MATLAB

The partial fraction expansion of (16.2.8), that is,

$$\frac{s-1}{s^2 + 2s + 5}$$

is found by typing

```
      p = [2 -2];
      q = [1 2 5];
[c,r] = residue(p,q)
```

We obtain

```
c =
      0.5000 + 0.5000i
      0.5000 - 0.5000i
r =
      -1.0000 + 2.0000i
      -1.0000 - 2.0000i
```

Hence we have the expansion

$$\frac{s-1}{s^2+2s+5} = \frac{0.5+0.5i}{s-(-1+2i)} + \frac{0.5-0.5i}{s-(-1-2i)}.$$

Now we use `realform` to return to a representation avoiding complex numbers. Type

    [num,denom] = realform(c(1),r(1))

to obtain

    num =
         1   -2
    denom =
        -1   4

which corresponds to (16.2.9).

HAND EXERCISES

In Exercises 1 and 2, use partial fractions to find a function $x(t)$ whose Laplace transform is the given function $Y(s)$.

1. $Y(s) = \dfrac{10}{s^3+s^2-6s}$

2. $Y(s) = \dfrac{s+2}{s^3-s}$

COMPUTER EXERCISES

In Exercises 3–5, use the MATLAB command `residue` to compute the expansion into partial fractions of $p(s)/q(s)$ for the given polynomials p and q.

3. $p(s) = 2(s-1)$ and $q(s) = s^2 - 3s + 2$

4. $p(s) = s^3 - 6s^2 - 45s + 50$ and $q(s) = s^4 - 8s^3 - 21s^2 + 8s + 20$

5. $p(s) = 3(s-1)$ and $q(s) = s^3 - s^2 + 4s - 4$

6. The `residue` command in MATLAB also works when the degree of the numerator $p(s)$ is greater than the degree of the denominator $q(s)$. In this case the answer has the form "partial fractions + polynomial." The `residue` command stores the polynomial part in a vector k. To explore this feature, enter the vectors p = [2,0,−2,0] and q = [1,0,−4], and type the command

    [c,r,k] = residue(p,q)

Explain the result of this calculation by comparing it to the expansion

$$\frac{2(s^3-s)}{s^2-4} = \frac{3}{s-2} + \frac{3}{s+2} + 2s.$$

7. Use the command `residue` as in Exercise 6 to compute an expansion into partial fractions for

$$\frac{s^3-4s^2+s+6}{s^2-3s+2}.$$

16.4 DISCONTINUOUS FORCING

In Section 15.5 we saw how to find solutions to a second-order ordinary differential equation modeling the dynamics of a periodically forced undamped spring. In particular, we studied equation (15.5.1), a variant of which is reproduced here:

$$\ddot{x} + 4x = g(t)$$
$$x(0) = 1 \qquad\qquad \textbf{(16.4.1)}$$
$$\dot{x}(0) = 0,$$

where $g(t) = \cos(at)$.

We now assume that for a certain period of time there is no external force acting on the spring and that, at some point, the situation changes. After that time the spring is subjected to an external forcing term. Specifically, we study the solution of (16.4.1), where $g(t)$ is assumed to have the simplest type of forcing with these properties; that is,

$$g(t) = H_1(t). \qquad\qquad \textbf{(16.4.2)}$$

So until time $t = 1$ there is no force on the spring, and after time $t = 1$ there is a unit force acting on the spring.

The first step in solving the initial value problem (16.4.1) is to apply the Laplace transform to both sides of the equation to obtain

$$s^2 \mathcal{L}[x] - s + 4\mathcal{L}[x] = \mathcal{L}[H_1] = \frac{1}{s}e^{-s}.$$

Here we have used the initial conditions $x(0) = 1$ and $\dot{x}(0) = 0$.

In the second step we solve this equation for $\mathcal{L}[x]$ and obtain

$$\mathcal{L}[x] = \frac{1}{s^2+4}\left(\frac{1}{s}e^{-s} + s\right) = \frac{1}{s(s^2+4)}e^{-s} + \frac{s}{s^2+4}.$$

This result is simplified by using partial fractions to obtain

$$\frac{1}{s(s^2+4)} = \frac{1}{4}\left(\frac{1}{s} - \frac{s}{s^2+4}\right).$$

Therefore

$$\mathcal{L}[x] = \frac{1}{4}\left(\frac{1}{s}e^{-s}\right) - \frac{1}{4}\frac{s}{s^2+4}e^{-s} + \frac{s}{s^2+4}.$$

In the third step we use the inverse Laplace transform to solve for $x(t)$. In particular,

$$x = \frac{1}{4}\mathcal{L}^{-1}\left[\frac{1}{s}e^{-s}\right] - \frac{1}{4}\mathcal{L}^{-1}\left[\frac{s}{s^2+4}e^{-s}\right] + \mathcal{L}^{-1}\left[\frac{s}{s^2+4}\right].$$

Using Table 16.1, we obtain

$$x(t) = \frac{1}{4}H_1(t) - \frac{1}{4}\mathcal{L}^{-1}\left[\frac{s}{s^2+4}e^{-s}\right](t) + \cos(2t).$$

It remains to find the inverse Laplace transform of $s/(s^2+4)e^{-s}$. This calculation can be done using Proposition 16.2.5, and we obtain

$$x(t) = \frac{1}{4}H_1(t) - \frac{1}{4}H_1(t)\cos(2(t-1)) + \cos(2t)$$

$$= \frac{1}{4}H_1(t)(1 - \cos(2(t-1))) + \cos(2t).$$

Note that the function

$$H_1(t)(1 - \cos(2(t-1)))$$

is differentiable, even though the step function $H_1(t)$ is discontinuous. See Exercise 1 at the end of this section.

An Example with Impulse Forcing

We are now in a position to solve initial value problems having forcing functions that are impulse functions. Indeed, as an example we solve

$$\dot{x} + x = \delta_2(t)$$
$$x(0) = 1 \qquad\qquad \textbf{(16.4.3)}$$

using Laplace transforms. Indeed, an application of $\mathcal{L}$ to (16.4.3) leads to

$$s\mathcal{L}[x] - 1 + \mathcal{L}[x] = e^{-2s},$$

and therefore

$$\mathcal{L}[x] = \frac{e^{-2s} + 1}{s+1} = e^{-2s}\frac{1}{s+1} + \frac{1}{s+1}.$$

Now we can use Proposition 16.2.5 combined with Table 16.1 to see that

$$x(t) = \mathcal{L}^{-1}\left[e^{-2s}\frac{1}{s+1}\right](t) + \mathcal{L}^{-1}\left[\frac{1}{s+1}\right](t) = H_2(t)e^{-(t-2)} + e^{-t}.$$

Not surprisingly, this solution has a jump discontinuity at $t = 2$ (see Figure 16.3); that is, the jump discontinuity occurs at the time when the impulse force is applied.

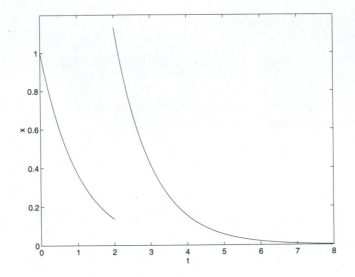

Figure 16.3
The solution of the initial value problem (16.4.3)

Grand Finale

We end this section by using Laplace transforms to solve the initial value problem

$$\ddot{x} + 4\dot{x} + 5x = \delta_1(t) + 2H_3(t)$$
$$x(0) = 0 \qquad \text{(16.4.4)}$$
$$\dot{x}(0) = 1.$$

When solving (16.4.4), we use most of the techniques that we have introduced so far.

Applying the Laplace transform to both sides of the differential equation in (16.4.4) and using the initial conditions lead to

$$s^2\mathcal{L}[x] - 1 + 4s\mathcal{L}[x] + 5\mathcal{L}[x] = e^{-s} + 2\frac{e^{-3s}}{s}.$$

Solving this equation for the Laplace transform of x yields

$$\mathcal{L}[x] = \frac{e^{-s} + 2\frac{e^{-3s}}{s} + 1}{s^2 + 4s + 5} = \frac{e^{-s} + 1}{s^2 + 4s + 5} + \frac{2e^{-3s}}{s(s^2 + 4s + 5)}.$$

To find the inverse Laplace transform of this equation, we need to compute the partial fraction expansion of

$$\frac{p(s)}{q(s)} = \frac{1}{s(s^2 + 4s + 5)}.$$

Defining p=[1] and q=[1 4 5 0], we use the command [c,r] = residue(p,q) to obtain

```
c =
  -0.1000 + 0.2000i
  -0.1000 - 0.2000i
   0.2000
r =
  -2.0000 + 1.0000i
  -2.0000 - 1.0000i
        0
```

Hence the roots of $s^2 + 4s + 5$ are $-2 \pm i$, and we have

$$\frac{1}{s(s^2 + 4s + 5)} = \frac{-0.1 + 0.2i}{s - (-2 + i)} + \frac{-0.1 - 0.2i}{s - (-2 - i)} + \frac{0.2}{s}.$$

We combine the complex conjugate terms by typing

```
[num,denom] = realform(c(1),r(1))
```

and obtain

```
num =
  -0.2000    -0.4000
denom =
    -2      1
```

The partial fraction result is

$$\frac{1}{s(s^2 + 4s + 5)} = -\frac{0.2(s + 2) + 0.4}{(s + 2)^2 + 1} + \frac{0.2}{s}.$$

Now we can rewrite $\mathcal{L}[x]$ as

$$\mathcal{L}[x] = \frac{e^{-s}}{(s + 2)^2 + 1} + \frac{1}{(s + 2)^2 + 1}$$

$$+ 2e^{-3s}\left(-0.2\frac{s + 2}{(s + 2)^2 + 1} - 0.4\frac{1}{(s + 2)^2 + 1} + 0.2\frac{1}{s}\right).$$

We use Proposition 16.2.4(a) to see that

$$\mathcal{L}^{-1}\left[\frac{1}{(s + 2)^2 + 1}\right] = e^{-2t}\sin(t)$$

$$\mathcal{L}^{-1}\left[\frac{s + 2}{(s + 2)^2 + 1}\right] = e^{-2t}\cos(t),$$

and Proposition 16.2.5(b) to see that

$$\mathcal{L}^{-1}\left[e^{-s}\frac{1}{(s+2)^2+1}\right] = H_1(t)e^{-2(t-1)}\sin(t-1)$$

$$\mathcal{L}^{-1}\left[e^{-3s}\frac{s+2}{(s+2)^2+1}\right] = H_3(t)e^{-2(t-3)}\cos(t-3)$$

$$\mathcal{L}^{-1}\left[e^{-3s}\frac{1}{(s+2)^2+1}\right] = H_3(t)e^{-2(t-3)}\sin(t-3).$$

Now we can write down the solution $x(t)$ as

$$\begin{aligned}
x(t) &= H_1(t)e^{-2(t-1)}\sin(t-1) + e^{-2t}\sin(t) \\
&\quad - 0.4H_3(t)e^{-2(t-3)}\cos(t-3) - 0.8H_3(t)e^{-2(t-3)}\sin(t-3) + 0.4H_3(t) \\
&= H_1(t)e^{-2(t-1)}\sin(t-1) + e^{-2t}\sin(t) \\
&\quad - 0.4H_3(t)e^{-2(t-3)}(\cos(t-3) + 2\sin(t-3)) + 0.4H_3(t).
\end{aligned}$$

The solution is shown in Figure 16.4. The two changes in the behavior of the solution occurring at $t = 1$ and at $t = 3$ are readily observed.

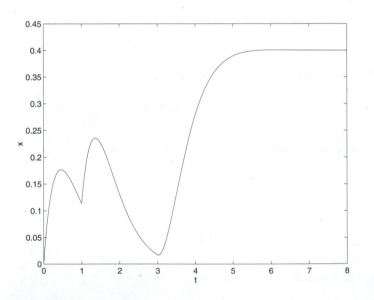

Figure 16.4
The solution of the initial value problem (16.4.4)

Note that the solution to (16.4.4) tends asymptotically to $x = 0.4$ for large t. This behavior can be explained as follows: For large t, equation (16.4.4) becomes

$$\ddot{x} + 4\dot{x} + 5x = 2.$$

The only equilibrium (or time independent solution) to this equation is $x(t) = 0.4$, and each solution of (16.4.4) tends to that equilibrium for large t.

HAND EXERCISES

1. Consider the function

$$z(t) = H_1(t)(1 - \cos(2(t-1))) = \begin{cases} 0 & t \le 1 \\ 1 - \cos(2(t-1)) & t > 1. \end{cases}$$

Show that $z(t)$ is continuous at $t = 1$ by verifying that

$$\lim_{t \to 1^+} z(t) = \lim_{t \to 1^-} z(t).$$

Show that $z(t)$ is differentiable and that $\dot{z}(1) = 0$ by verifying that

$$\lim_{h \to 0^+} \frac{z(1+h) - z(1)}{h} = 0 = \lim_{h \to 0^-} \frac{z(1+h) - z(1)}{h}.$$

2. Reconfirm that the solution of the initial value problem

$$\ddot{x} + 4x = H_1(t); \quad x(0) = 1, \quad \dot{x}(0) = 0,$$

is given by

$$x(t) = \frac{1}{4} H_1(t)(1 - \cos(2(t-1))) + \cos(2t).$$

Use the methods of Chapter 15 and proceed as follows:
   (a) Find the solution $x_1(t)$ of the initial value problem

$$\ddot{x} + 4x = 0; \quad x(0) = 1, \quad \dot{x}(0) = 0,$$

on the time interval $t \in [0, 1]$.
   (b) Find the general solution $x_2(t)$ of the second-order differential equation $\ddot{x} + 4x = 1$.
   (c) Adjust the parameters in $x_2(t)$ such that the function

$$x(t) = \begin{cases} x_1(t) & t \in [0, 1] \\ x_2(t) & t > 1 \end{cases}$$

is differentiable at $t = 1$.

COMPUTER EXERCISES

3. Use the MATLAB command ode45 to compute the solution of the grand finale initial value problem (16.4.4). More precisely, write (16.4.4) as a nonautonomous first-order system of ODEs by setting $y = \dot{x}$ and obtaining

$$\begin{aligned} \dot{x} &= y \\ \dot{y} &= -5x - 4y + \delta_1(t) + 2H_3(t). \end{aligned} \qquad (16.4.5)^*$$

There is a numerical difficulty when attempting to compute delta functions. A standard way around this difficulty is to approximate the Dirac delta function $\delta_1(t)$ using (16.2.4) with $h = 0.01$. We use the MATLAB m-file f16_4_5.m for the numerical computations:

```
function y = f16_4_5(t,x)
    h = 0.01;
y(1) = x(2);
y(2) = -4*x(2)-5*x(1);
if (abs(t-(1-h/2)) <= h/2)
        y(2) = y(2) + 1/h;
end
if (t >= 3)
        y(2) = y(2) + 2;
end
y = [y(1) y(2)]';
```

Using ode45, compute the solution on the time interval [0, 8] using the command

```
[t,x] = ode45('f16_4_5',[0 8],[0,1]);
```

Plot the first component of x by typing plot(t,x(:,1)), and observe the difference between the numerically computed solution and the exact solution, whose graph is plotted in Figure 16.4.

Next, set the relative error tolerance 1e−8 using the command

```
options = odeset('RelTol',1e\(-\)8);
```

and compute the solution using the command

```
[t,x] = ode45('f16_4_5',[0 8],[0,1],options);
```

Compare the results of this numerical integration with the theoretically exact result in Figure 16.4. Try to explain why the results differ for different tolerances.

16.5 RLC CIRCUITS

RLC circuits provide an excellent example of a physical system that is well modeled by a second-order linear differential equation that is periodically forced by a discontinuous function. Consider the electrical circuit shown in Figure 16.5. The circuit consists of a resistor with resistance R, a coil with inductance L, a capacitor with capacitance C, and a voltage source producing a time-dependent voltage $V(t)$.

We describe the behavior of the circuit by the voltage drop $x(t)$ at the capacitor. Kirchhoff's laws for electrical circuits show that $x(t)$ satisfies the second-order differential equation

$$\frac{1}{CL} V(t) = \frac{d^2x}{dt^2}(t) + \frac{R}{L}\frac{dx}{dt}(t) + \frac{1}{CL} x(t), \qquad (16.5.1)$$

as we now explain. *Kirchhoff's Voltage Law* states that at each instant of time the voltage $V(t)$ produced at the source is equal to the sum of the voltage drops at the three

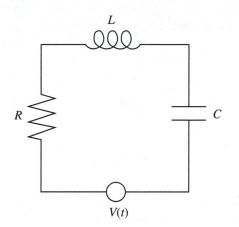

Figure 16.5
An RLC circuit with resistance R, inductance L,
capacitance C, and a voltage source $V(t)$

elements of the circuit. So, if we denote the voltage drops at the coil by V_{coil} and at the resistor by $V_{resistor}$, then we have

$$V(t) = V_{coil}(t) + V_{resistor}(t) + x(t), \tag{16.5.2}$$

recalling that $x(t)$ is the voltage drop at the capacitor.

Let $I(t)$ be the current through the system. Then:

(a) The voltage drop in a capacitor is proportional to the charge difference $Q_{capacitor}$ between the two plates:

$$x(t) = \frac{1}{C} Q_{capacitor},$$

where C is the *capacitance* measured in farads. The charge difference itself is related to the current by

$$I = \frac{dQ_{capacitor}}{dt} = C\frac{dx}{dt}.$$

(b) The voltage drop at a resistor is proportional to the current:

$$V_{resistor} = RI = RC\frac{dx}{dt},$$

where R is the *resistance* measured in ohms.

(c) The voltage drop in a coil is given by *Faraday's Law*; the drop is proportional to the rate of change of the current:

$$V_{coil} = L\frac{dI}{dt} = LC\frac{d^2x}{dt^2},$$

where L is the inductance of the coil measured in henrys.

Combining (a, b, c) with (16.5.2), we obtain the differential equation

$$V(t) = LC\ddot{x} + RC\dot{x} + x.$$

After dividing by LC, we obtain the second-order equation (16.5.1), and on setting

$$a = \frac{R}{L}, \quad b = \frac{1}{CL}, \quad \text{and} \quad g(t) = \frac{1}{CL} V(t),$$

we have an initial value problem in the form (16.1.6).

A typical input at the voltage source (associated with alternating current) is a $2T$-periodic square wave with amplitude $A > 0$ defined by periodicity and

$$V(t) = \begin{cases} A & 0 < t \leq T \\ -A & T < t \leq 2T. \end{cases}$$

See Figure 16.6. We will see how to use Laplace transforms to solve second-order equations with a discontinuous forcing of this type. Indeed, it is the possibility of using Laplace transforms to solve linear equations with piecewise smooth forcing terms that is the main strength of Laplace transforms.

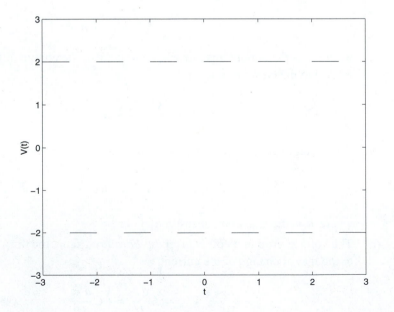

Figure 16.6
A 1-periodic square wave $V(t)$ of amplitude $A = 2$

Piecewise Smooth Periodic Forcing

We now show how Laplace transforms can be used to solve second-order equations with periodic discontinuous forcing. As an example we consider the equation (16.4.1), where the external force $g(t)$ is given by

$$g(t) = \begin{cases} 1 & \text{for } t \in [2k, 2k+1] \\ 0 & \text{for } t \in [2k+1, 2k+2] \end{cases} \qquad (k = 0, 1, 2, \ldots). \qquad \textbf{(16.5.3)}$$

We begin by computing the Laplace transform of $g(t)$. Using the definition, we obtain

$$\mathcal{L}[g] = \int_0^\infty e^{-st} g(t)\, dt$$

$$= \sum_{k=0}^\infty \int_k^{k+1} e^{-st} g(t)\, dt$$

$$= \sum_{k=0}^\infty \int_{2k}^{2k+1} e^{-st}\, dt,$$

since $g(t)$ vanishes for $t \in [2k+1, 2k+2]$. Performing the integrations leads to

$$\mathcal{L}[g] = \frac{1}{s} \sum_{k=0}^\infty \left(e^{-2sk} - e^{-s(2k+1)} \right)$$

$$= \frac{1}{s} \sum_{k=0}^\infty (1 - e^{-s}) e^{-2sk}$$

$$= \frac{1}{s}(1 - e^{-s}) \frac{1}{1 - e^{-2s}}$$

$$= \frac{1}{s(1 + e^{-s})}.$$

Hence an application of the Laplace transform to the differential equation (16.4.1) gives

$$s^2 \mathcal{L}[x] - s + 4\mathcal{L}[x] = \mathcal{L}[g] = \frac{1}{s(1 + e^{-s})},$$

and solving this for $\mathcal{L}[x]$ leads to

$$\mathcal{L}[x] = \frac{1}{s(s^2 + 4)(1 + e^{-s})} + \frac{s}{s^2 + 4}.$$

It remains to compute the inverse Laplace transform of the first term. For this we observe that for real numbers x, with $|x| < 1$, the following identity holds:

$$\frac{1}{1 + x} = 1 - x + x^2 - x^3 + x^4 - \cdots = \sum_{k=0}^\infty (-1)^k x^k.$$

Hence, for positive s,

$$\frac{1}{1+e^{-s}} = \sum_{k=0}^{\infty} (-1)^k e^{-sk},$$

and it follows that

$$\mathcal{L}[x] = \frac{1}{s(s^2+4)} \sum_{k=0}^{\infty} (-1)^k e^{-sk} + \frac{s}{s^2+4}$$

$$= \frac{1}{4}\left(\frac{1}{s} - \frac{s}{s^2+4}\right) \sum_{k=0}^{\infty} (-1)^k e^{-sk} + \frac{s}{s^2+4}$$

$$= \frac{1}{4}\left(\frac{1}{s} + \frac{3s}{s^2+4} + \left(\frac{1}{s} - \frac{s}{s^2+4}\right) \sum_{k=1}^{\infty} (-1)^k e^{-sk}\right)$$

$$= \frac{1}{4}\left(\frac{1}{s} + \frac{3s}{s^2+4} + \sum_{k=1}^{\infty} (-1)^k \frac{e^{-sk}}{s} - \sum_{k=1}^{\infty} (-1)^k \frac{e^{-sk}s}{s^2+4}\right).$$

Now we have derived an expression for $\mathcal{L}[x]$ in which we can compute the inverse Laplace transform for each of the individual summands. Using Table 1 and Proposition 16.2.5, we obtain

$$x(t) = \frac{1}{4}\left(1 + 3\cos(2t) + \sum_{k=1}^{\infty} (-1)^k H_k(t) - \sum_{k=1}^{\infty} (-1)^k H_k(t)\cos(2(t-k))\right)$$

$$= \frac{1}{4}\left(1 + 3\cos(2t) + \sum_{k=1}^{\infty} (-1)^k H_k(t)(1 - \cos(2(t-k)))\right)$$

$$= \frac{1}{4}\left(1 + 3\cos(2t) + 2\sum_{k=1}^{\infty} (-1)^k H_k(t)\sin^2(t-k)\right).$$

In Figure 16.7 we show the solution for $t \in [0, 20]$. Observe that this solution is certainly not periodic in time. Indeed, since the internal frequency of equation (16.4.1) is $1/\pi$ and the frequency of the external periodic forcing is $1/2$, we expect to obtain a quasiperiodic motion.

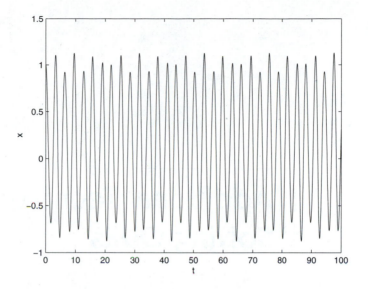

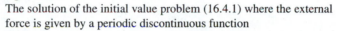

Figure 16.7
The solution of the initial value problem (16.4.1) where the external
force is given by a periodic discontinuous function

Additional Techniques for Solving ODEs

In Chapters 14, 15, and 16 we developed analytic methods for finding solutions to *linear differential equations* with constant coefficients and with forcing. In this chapter we show how to find closed form solutions to certain types of nonlinear equations as well as to certain linear equations with time-dependent coefficients.

Most nonlinear differential equations cannot be solved analytically, and for these equations one has to rely either on advanced theory or on numerical methods. There are, however, specific types of ordinary differential equations whose solution by hand is possible. In this chapter we describe several of these types. In Sections 17.1 and 17.2 we solve forced nonconstant coefficient linear differential equations and systems using *variation of parameters*. In Section 17.4 we show how to solve higher order equations by reduction to first-order systems and by a variant of variation of parameters called *reduction of order*. Sometimes it is possible to transform a nonlinear equation into an equation that can be treated by one of the methods mentioned above. This is accomplished by substituting another function for the function $x(t)$ in a suitable way. *Simplifications by substitution* are discussed in Section 17.5.

The chapter ends with a discussion of two types of differential equations whose solutions lie on level curves of a real-valued function: nonautonomous *exact differential equations* and autonomous *Hamiltonian systems*. Exact equations are treated in Section 17.6. Hamiltonian systems, which arise naturally in mechanical systems, are treated in Section 17.7.

17.1   NONCONSTANT COEFFICIENT LINEAR EQUATIONS

The simplest nonconstant coefficient homogeneous linear differential equation is

$$\frac{dx}{dt} = a(t)x. \qquad \textbf{(17.1.1)}$$

This equation does not have constant coefficients, since the coefficient a depends on t. The equation is linear because linear combinations of solutions are solutions. Note that (17.1.1) is of the form (4.4.10) and can be solved by separation of variables.

If $x(t_0) = 0$, then $x(t) = 0$ is the solution to the initial value problem (17.1.1). To solve the initial value problem $x(t_0) = x_0$ for (17.1.1) when $x_0 \neq 0$, we apply the technique of separation of variables to this equation, which yields

$$\ln |x| = H(t) + C,$$

where $H(t)$ is the definite integral

$$H(t) = \int_{t_0}^{t} a(\tau)\, d\tau. \qquad \textbf{(17.1.2)}$$

Exponentiation implies that

$$x(t) = K e^{H(t)}$$

for an appropriate scalar K. Indeed, if $x(t_0) = x_0$, then

$$x_0 = x(t_0) = K e^{H(t_0)} = K e^0 = K.$$

We have shown that the function

$$x(t) = x_0 e^{H(t)} \qquad \textbf{(17.1.3)}$$

is the unique solution to the initial value problem (17.1.1), where $x(t_0) = x_0$.

An Example of a Nonconstant Coefficient Equation

We illustrate (17.1.3) by an example. Solve the initial value problem

$$\frac{dx}{dt} = -\frac{x}{t}$$
$$x(2) = 5.$$

Since $a(t) = -\frac{1}{t}$, we can use (17.1.2) to compute

$$H(t) = -\int_{2}^{t} \frac{1}{\tau}\, d\tau = -\ln t + \ln 2 = \ln\left(\frac{2}{t}\right).$$

Then the solution (17.1.3) is given by

$$x(t) = 5e^{\ln(2/t)} = \frac{10}{t}.$$

The Inhomogeneous Equation

Next we consider forced linear differential equations of the type

$$\frac{dx}{dt} = a(t)x + g(t), \tag{17.1.4}$$

where a and g are continuous functions of t. The differential equation (17.1.4) is homogeneous when $g = 0$ and inhomogeneous otherwise. As is always the case, we find the general solution to the inhomogeneous equation by adding the general solution to the homogeneous equation (that we can find by separation of variables) to a particular solution to the inhomogeneous equation.

Two Examples of Solutions of Inhomogeneous Equations

(a) Find all solutions of the differential equation

$$\frac{dx}{dt} = 3t^2x + 3t^2. \tag{17.1.5}$$

By inspection, the constant function $x_p(t) = -1$ is a particular solution of (17.1.5). Since all solutions of the homogeneous equation are of the form Ke^{t^3} for some real constant K, it follows that the general solution of the inhomogeneous equation is

$$x(t) = -1 + Ke^{t^3}.$$

(b) Find all solutions of the differential equation

$$\frac{dx}{dt} = \left(\frac{5}{t} - t\right)x + t^6. \tag{17.1.6}$$

A calculation shows that $x_p(t) = t^5$ is a particular solution of (17.1.6). Since all solutions to the homogeneous equation are of the form $Kt^5e^{-t^2/2}$ for some real constant K, it follows that the general solution of the inhomogeneous equation is

$$x(t) = t^5 + Kt^5e^{-t^2/2}.$$

Inhomogeneous Equations and Variation of Parameters

In examples (17.1.5) and (17.1.6) we guessed or were given a particular solution to the inhomogeneous equation. Now we discuss a method for finding a particular solution to the inhomogeneous equation (17.1.4) when $g(t)$ is a nonzero function. This technique for finding a particular solution is called *variation of parameters*.

We already know by separation of variables that every solution of the *homogeneous* equation ($g = 0$) has the form $Ke^{H(t)}$, where $dH/dt = a$. The idea behind variation of parameters is to allow K to depend on t. More precisely, assume that a solution $x(t)$ has the form

$$x(t) = c(t)e^{H(t)},$$

where c is a differentiable function of t. Substituting $x(t)$ in (17.1.4), using the product rule for differentiation, and suppressing the explicit dependence of functions on t lead to the identity

$$\frac{dc}{dt} e^H + ce^H a = ace^H + g,$$

which simplifies to

$$\frac{dc}{dt} = ge^{-H}.$$

This last equation may be solved for $c(t)$ by integration, obtaining

$$c(t) = \int g(\tau)e^{-H(\tau)} \, d\tau. \qquad (17.1.7)$$

Uniqueness of solutions to initial value problems implies the main result of this section.

Theorem 17.1.1 (Variation of parameters) *The unique solution to the initial value problem*

$$\frac{dx}{dt} = a(t)x + g(t)$$
$$x(t_0) = x_0$$

is

$$x(t) = c(t)e^{H(t)},$$

where

$$H(t) = \int_{t_0}^{t} a(\tau) \, d\tau \quad and \quad c(t) = \int_{t_0}^{t} g(\tau)e^{-H(\tau)} \, d\tau + x_0.$$

Two Examples of Variation of Parameters

(a) Solve the initial value problem

$$\frac{dx}{dt} = -\frac{x}{t} + \frac{2}{t^4}$$

$$x(1) = 0.$$

(17.1.8)

Compute

$$H(t) = -\int_1^t \frac{1}{\tau} \, d\tau = -\ln t + \ln 1 = -\ln t.$$

Hence

$$c(t) = \int_1^t \frac{2}{\tau^4} e^{\ln \tau} \, d\tau + 0 = \int_1^t \frac{2}{\tau^3} \, d\tau = -\frac{1}{t^2} + 1,$$

which implies that

$$x(t) = \left(1 - \frac{1}{t^2}\right)\frac{1}{t}$$

(17.1.9)

is the solution. For comparison we show the graph of (17.1.9) in Figure 17.1(left) and the result of a numerical computation of the initial value problem using dfield5 in Figure 17.1 (right).

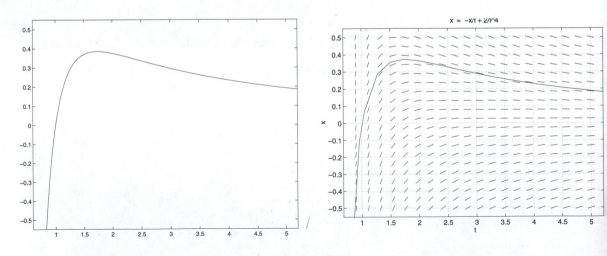

Figure 17.1
Left: Graph of solution (17.1.9) to equation (17.1.8). *Right*: The time series for solution to (17.1.8) with initial condition $x(1) = 0$ using dfield5.

(b) Solve the initial value problem

$$\frac{dx}{dt} = 4x + \cos t + e^{2t}$$
$$x(0) = 1.$$

Here $H(t) = 4t$ and

$$c(t) = \int_0^t g(\tau)e^{-4\tau} \, d\tau + 1$$

$$= \int_0^t \left(\cos \tau + e^{2\tau}\right) e^{-4\tau} \, d\tau + 1$$

$$= \frac{1}{17}\left(e^{-4t}(\sin t - 4\cos t) + 4\right) + \frac{1}{2}\left(1 - e^{-2t}\right) + 1.$$

Thus the solution is

$$x(t) = c(t)e^{4t} = \frac{1}{17}(\sin t - 4\cos t + 4e^{4t}) - \frac{1}{2}e^{2t} + \frac{3}{2}e^{4t}.$$

HAND EXERCISES

In Exercises 1–6, decide whether or not the given differential equation is linear. If it is linear, specify whether the equation is homogeneous or inhomogeneous.

1. $\dfrac{dx}{dt} = 0$

2. $\dfrac{dx}{dt} = x^2 \cos t$

3. $\dfrac{dx}{dt} = x + t^2$

4. $\dfrac{dx}{dt} = \dfrac{t}{x} - \dfrac{1}{t}$

5. $(t^2 + 1)\dfrac{dx}{dt} = x + \sin t$

6. $x\dfrac{dx}{dt} = \cos t$

In Exercises 7–10, solve each initial value problem by variation of parameters.

7. $\dfrac{dx}{dt} = t^2 x + t^2;$ $x(1) = 1$

8. $\dfrac{dx}{dt} = x + 2t;$ $x(0) = -1$

9. $\dfrac{dx}{dt} = \dfrac{t}{t^2 + 1}x + \sin(t)\sqrt{t^2 + 1};$ $x(0) = 2$

10. $\dfrac{dx}{dt} = 2x + \dfrac{1}{t}e^{2t};$ $x(1) = 4$

COMPUTER EXERCISES

In Exercises 11–13, use dfield5 to compute solutions to the given linear differential equations. What is the asymptotic behavior of the solutions as t tends to infinity? Use variation of parameters to explain the behavior.

11. $\dfrac{dx}{dt} = -2x + t$

12. $\dfrac{dx}{dt} = -2x + t^2$

13. $\dfrac{dx}{dt} = -2x + \sin t$

17.2 VARIATION OF PARAMETERS FOR SYSTEMS

In this section we consider solving inhomogeneous systems of linear differential equations of the form

$$\frac{dX}{dt} = A(t)X + G(t), \tag{17.2.1}$$

where $A(t) = (a_{ij}(t))$ is an $n \times n$ matrix and $G(t) = (g_1(t), \ldots, g_n(t))^t$ is a vector of continuous functions. The system is *homogeneous* when $G(t) = 0$.

We divide this discussion into two parts: finding the general solution of the homogeneous equation and finding a particular solution to the inhomogeneous equation using variation of parameters.

Homogeneous Nonconstant Coefficient Systems

From the theory of constant coefficient systems of linear differential equations, we know abstractly how to write a basis of solutions to the homogeneous system $\dot{X} = AX$, where A is an $n \times n$ constant matrix. That basis is

$$X_j(t) = e^{tA} v_j,$$

where $\{v_1, \ldots, v_n\}$ is a basis of $\mathbf{R}^n$. For the homogeneous system of n nonconstant coefficient linear differential equations

$$\dot{X} = A(t)X \tag{17.2.2}$$

a similar statement is true. It should be noted, however, that only in special cases can (17.2.2) be solved in closed form.

Proposition 17.2.1 *Let $v_1, \ldots, v_n$ be a basis for $\mathbf{R}^n$. Then there exist solutions*

$$X_1(t), \ldots, X_n(t)$$

of (17.2.2) satisfying $X_j(0) = v_j$ and these solutions form a basis of solutions to (17.2.2).

Proof: The theory of differential equations implies that there exists a unique solution to (17.2.2) satisfying the initial condition $X(0) = X_0$ for any $X_0 \in \mathbf{R}^n$. Therefore there exist solutions $X_j(t)$ such that $X_1(0) = v_j$. We claim that these solutions form a basis for the vector space of all solutions to (17.2.2).

Let $X(t)$ be any solution to (17.2.2) with initial condition $X(0) = X_0$. Since $\{X_1(0), \ldots, X_n(0)\}$ is a basis for $\mathbf{R}^n$, we can find scalars α_j such that

$$X_0 = \alpha_1 X_1(0) + \cdots + \alpha_n X_n(0).$$

Linearity of (17.2.2) implies that

$$\alpha_1 X_1(t) + \cdots + \alpha_n X_n(t)$$

is a solution. Therefore uniqueness of solutions to (17.2.2) implies that

$$X(t) = \alpha_1 X_1(t) + \cdots + \alpha_n X_n(t)$$

for all time t, which proves the proposition.   ◆

Existence of solutions to the initial value problem $X(t_0) = X_0$ also guarantees that the vectors $X_1(t_0), \ldots, X_n(t_0)$ must be linearly independent. Indeed, since
$X_1(t), \ldots, X_n(t)$ is a basis of solutions to (17.2.2), we must be able to write

$$X_0 = \alpha_1 X_1(t_0) + \cdots + \alpha_n X_n(t_0)$$

for scalars $\alpha_1, \ldots, \alpha_n$ in order for a solution to this initial value problem to exist. That
is, the vectors $X_1(t_0), \ldots, X_n(t_0)$ must span $\mathbf{R}^n$. Hence they are linearly independent.
We have proved the following:

Lemma 17.2.2   *Let $X_1, \ldots, X_n$ be a linearly independent set of solutions to (17.2.2).
Then, for every t, the matrix*

$$Y(t) = \big(X_1(t)| \cdots |X_n(t)\big) \tag{17.2.3}$$

is invertible.

Proof: Proposition 17.2.1 implies that the columns of $Y(t)$ are linearly independent.
Hence $Y(t)$ is invertible.   ◆

Rather than relying on the general existence and uniqueness theory for systems of
differential equations, there is a computationally direct way of proving Lemma 17.2.2.
It is possible to show that the determinant of the matrix $Y(t)$ in (17.2.3) is nonzero and
hence $Y(t)$ is invertible. This approach is carried out in Section 17.3.

The Theory of Variation of Parameters

Let $X_1(t), \ldots, X_n(t)$ be a basis of solutions to the homogeneous system $\dot{X} = A(t)X$.
In the method of variation of parameters we look for solutions to the inhomogeneous
system (17.2.1) of the form

$$X(t) = c_1(t)X_1(t) + \cdots + c_n(t)X_n(t),$$

where $c_1(t), \ldots, c_n(t)$ are differentiable functions. In order to find a method for determining the functions c_j, we assume that X is a solution to (17.2.1).

Use the product rule to compute

$$\frac{dX}{dt} = \sum_{j=1}^{n} \left(c_j \frac{dX_j}{dt} + \frac{dc_j}{dt} X_j \right)$$

$$= \sum_{j=1}^{n} \left(c_j A X_j + \frac{dc_j}{dt} X_j \right)$$

$$= AX + \sum_{j=1}^{n} \frac{dc_j}{dt} X_j.$$

It follows that X is a solution of (17.2.1) if and only if

$$G(t) = \frac{dc_1}{dt}(t) X_1(t) + \cdots + \frac{dc_n}{dt}(t) X_n(t). \qquad (17.2.4)$$

So variation of parameters works if we can find functions c_j that satisfy (17.2.4).

We claim that it is always possible to find functions $d_j(t)$ so that

$$G(t) = d_1(t) X_1(t) + \cdots + d_n(t) X_n(t). \qquad (17.2.5)$$

To verify (17.2.5), note that (17.2.5) can be rewritten in matrix form as

$$Y(t) D(t) = G(t),$$

where $Y(t)$ is defined in (17.2.3) and

$$D(t) = (d_1(t), \ldots, d_n(t))^t.$$

Using the invertibility of $Y(t)$, as proved in Lemma 17.2.2, we find that

$$D(t) = Y(t)^{-1} G(t).$$

It now follows that (17.2.4) is satisfied by integrating the differential equations $\dot{c}_j = d_j$, where the functions $d_j(t)$ are defined in (17.2.5), to find the functions c_j.

We summarize this discussion as follows:

Theorem 17.2.3 (Variation of parameters: systems)   *Consider*

$$\frac{dX}{dt} = A(t)X + G(t) \qquad (17.2.6)$$
$$X(t_0) = X_0.$$

(a) *Let $X_1(t), \ldots, X_n(t)$ be a basis of solutions to the homogeneous differential equation $\dot{X} = A(t)X$.*

(b) *Let $Y(t)$ be defined as in (17.2.3) and define $D(t) = (d_1(t), \ldots, d_n(t))^t$ by*

$$D(t) = Y(t)^{-1} G(t). \qquad (17.2.7)$$

(c)   *Let the functions* $c_1(t), \ldots, c_n(t)$ *satisfy the initial value problem*

$$\frac{dc_j}{dt} = d_j \quad and \quad c_1(t_0)X_1(t_0) + \cdots + c_n(t_0)X_n(t_0) = X_0. \qquad \textbf{(17.2.8)}$$

Then

$$X(t) = c_1(t)X_1(t) + \cdots + c_n(t)X_n(t)$$

is the unique solution of (17.2.6).

Note that when $n \geq 3$, the hand computation of $Y(t)^{-1}$ can be painful.

Examples with Constant Coefficient Matrix A When $n = 2$

We consider two examples of variation of parameters applied to forced systems of differential equations of the form

$$\dot{X} = AX + G(t).$$

An Example with Two Real Simple Eigenvalues

Consider the initial value problem (17.2.6), where

$$A = \begin{pmatrix} -1 & -8 \\ -16 & 7 \end{pmatrix}, \quad G(t) = \begin{pmatrix} 1-t \\ -2-t \end{pmatrix}, \quad X_0 = \begin{pmatrix} -1 \\ -1 \end{pmatrix}. \qquad \textbf{(17.2.9)}$$

Variation of parameters proceeds in four steps:

Step 1: Solutions to the homogeneous equation.   The eigenvalues and eigenvectors of A can be found either by using MATLAB or by hand. The eigenvalues are $\lambda_1 = -9$ and $\lambda_2 = 15$, and the associated eigenvectors are $v_1 = (1, 1)^t$ and $v_2 = (1, -2)^t$. Therefore solutions of the homogeneous equation are

$$X_1(t) = e^{-9t} \begin{pmatrix} 1 \\ 1 \end{pmatrix} \quad and \quad X_2(t) = e^{15t} \begin{pmatrix} 1 \\ -2 \end{pmatrix}.$$

Step 2: Computation of the vector D(t).   We compute the inverse

$$Y(t)^{-1} = \begin{pmatrix} e^{-9t} & e^{15t} \\ e^{-9t} & -2e^{15t} \end{pmatrix}^{-1} = -\frac{1}{3e^{6t}} \begin{pmatrix} -2e^{15t} & -e^{15t} \\ -e^{-9t} & e^{-9t} \end{pmatrix} = \frac{1}{3} \begin{pmatrix} 2e^{9t} & e^{9t} \\ e^{-15t} & -e^{-15t} \end{pmatrix}.$$

The vector $D(t)$ is

$$D(t) = Y(t)^{-1}G(t) = \frac{1}{3} \begin{pmatrix} 2e^{9t} & e^{9t} \\ e^{-15t} & -e^{-15t} \end{pmatrix} \begin{pmatrix} 1-t \\ -2-t \end{pmatrix} = \begin{pmatrix} -te^{9t} \\ e^{-15t} \end{pmatrix}.$$

Step 3: Computation of the functions $c_j(t)$.   By assumption,

$$X_0 = c_1(0)X_1(0) + c_2(0)X_2(0).$$

Therefore

$$\begin{pmatrix} -1 \\ -1 \end{pmatrix} = c_1(0)\begin{pmatrix} 1 \\ 1 \end{pmatrix} + c_2(0)\begin{pmatrix} 1 \\ -2 \end{pmatrix},$$

from which it follows that $c_1(0) = -1$ and $c_2(0) = 0$. Since

$$\dot c_1 = d_1 = -te^{9t} \quad \text{and} \quad \dot c_2 = d_2 = e^{-15t},$$

it follows that

$$c_1(t) = \int_0^t (-\tau)e^{9\tau}\, d\tau - 1 = -\frac{1}{9}\left(t - \frac{1}{9}\right)e^{9t} - \frac{82}{81}$$

$$c_2(t) = \int_0^t e^{-15t}\, d\tau = \frac{1}{15}\left(1 - e^{-15t}\right).$$

Step 4: Write out the solution.   The solution $X(t)$ of the initial value problem (17.2.9) is now given by

$$X(t) = c_1(t)e^{-9t}\begin{pmatrix} 1 \\ 1 \end{pmatrix} + c_2(t)e^{15t}\begin{pmatrix} 1 \\ -2 \end{pmatrix}$$

$$= \begin{pmatrix} -\dfrac{1}{9}\left(t - \dfrac{1}{9}\right) - \dfrac{82}{81}e^{-9t} + \dfrac{1}{15}\left(e^{15t} - 1\right) \\[2mm] -\dfrac{1}{9}\left(t - \dfrac{1}{9}\right) - \dfrac{82}{81}e^{-9t} - \dfrac{2}{15}\left(e^{15t} - 1\right) \end{pmatrix}.$$

An Example with a Nontrivial Jordan Block

Solve the system

$$\dot X = AX + (t, 1)^t \tag{17.2.10}$$
$$X(0) = (1, -1)^t,$$

where

$$A = \begin{pmatrix} 2 & 1 \\ 0 & 2 \end{pmatrix}.$$

Step 1:   A basis for solutions to the homogeneous equation $\dot X = AX$ is

$$X_1(t) = e^{2t}\begin{pmatrix} 1 \\ 0 \end{pmatrix} \quad \text{and} \quad X_2(t) = e^{2t}\begin{pmatrix} t \\ 1 \end{pmatrix}.$$

Step 2:   We compute

$$D(t) = e^{-2t} \begin{pmatrix} 1 & t \\ 0 & 1 \end{pmatrix}^{-1} G(t) = e^{-2t} \begin{pmatrix} 1 & -t \\ 0 & 1 \end{pmatrix} \begin{pmatrix} t \\ 1 \end{pmatrix} = e^{-2t} \begin{pmatrix} 0 \\ 1 \end{pmatrix}.$$

Step 3:   Note that

$$X(0) = \begin{pmatrix} 1 \\ -1 \end{pmatrix} = c_1(0) \begin{pmatrix} 1 \\ 0 \end{pmatrix} + c_2(0) \begin{pmatrix} 0 \\ 1 \end{pmatrix}.$$

So $c_1(0) = 1$ and $c_2(0) = -1$. Now solve the differential equations

$$\dot{c}_1 = d_1(t) = 0$$
$$\dot{c}_2 = d_2(t) = e^{-2t},$$

obtaining

$$c_1(t) = 1$$
$$c_2(t) = -\frac{1}{2}(e^{-2t} + 1).$$

Step 4:   It follows from Theorem 17.2.3 that the solution to (17.2.10) is

$$X(t) = c_1(t)X_1(t) + c_2(t)X_2(t) = -\frac{1}{2} \begin{pmatrix} t(1 + e^{2t}) - 2e^{2t} \\ e^{2t} + 1 \end{pmatrix}.$$

HAND EXERCISES

In Exercises 1 and 2, solve the given initial value problems by variation of parameters.

1. $\dfrac{dX}{dt} = \dfrac{1}{2} \begin{pmatrix} 1 & 1 \\ 1 & 1 \end{pmatrix} X + \begin{pmatrix} e^t \\ 0 \end{pmatrix};$ $X(0) = \begin{pmatrix} 1 \\ -1 \end{pmatrix}$

2. $\dfrac{dX}{dt} = \begin{pmatrix} 1 & 3 \\ 3 & 1 \end{pmatrix} X + \begin{pmatrix} 2t \\ t \end{pmatrix};$ $X(0) = \begin{pmatrix} 0 \\ 1 \end{pmatrix}$

3. Find the general solution to the system of differential equations

$$\frac{dX}{dt} = \begin{pmatrix} -\dfrac{1}{t} & \dfrac{1}{t^2} \\ 1 & 0 \end{pmatrix} X + \begin{pmatrix} 1 \\ 3t \end{pmatrix}.$$

Hint   Verify that

$$X_1(t) = \begin{pmatrix} 1 \\ t \end{pmatrix} \quad \text{and} \quad X_2(t) = \begin{pmatrix} \dfrac{1}{t^2} \\ -\dfrac{1}{t} \end{pmatrix}$$

are solutions to the homogeneous equation and then use variation of parameters.

4. Find the general solution to the system of differential equations

$$\frac{dX}{dt} = \begin{pmatrix} 0 & 1 \\ 3 & 1 \\ t^2 & t \end{pmatrix} X + \begin{pmatrix} t \\ 2t \end{pmatrix}.$$

Hint Verify that

$$X_1(t) = \begin{pmatrix} t^3 \\ 3t^2 \end{pmatrix} \quad \text{and} \quad X_2(t) = \begin{pmatrix} \dfrac{1}{t} \\ -\dfrac{1}{t^2} \end{pmatrix}$$

are solutions to the homogeneous equation and then use variation of parameters.

17.3 THE WRONSKIAN

We return to the homogeneous system of linear differential equations

$$\frac{dX}{dt} = A(t)X, \tag{17.3.1}$$

where $A(t) = (a_{ij}(t))$ is an $n \times n$ matrix. Let $X_1, \ldots, X_n$ be a linearly independent set of solutions to (17.3.1) and let

$$Y(t) = \left(X_1(t) | \cdots | X_n(t) \right). \tag{17.3.2}$$

Using existence and uniqueness of solutions to the initial problem for (17.3.1), we showed in Lemma 17.2.2 that $Y(t)$ is an invertible matrix for every time t. In this section we prove this same result by explicitly showing that the determinant of $Y(t)$ is always nonzero.

Define the *Wronskian* to be

$$W(t) = \det Y(t).$$

Theorem 17.3.1   *Let $X_1, \ldots, X_n$ be solutions to (17.2.2) such that the vectors $X_j(0)$ form a basis of $\mathbf{R}^n$. Let $Y(t)$ be the matrix defined in (17.2.3). Then*

$$W(t) = e^{\int_0^t tr(A(\tau)) \, d\tau} W(0). \tag{17.3.3}$$

It follows directly from (17.3.3) that the determinant of $Y(t)$ is nonzero when the determinant of $Y(0)$ is nonzero. But $\det Y(0) \neq 0$ since the vectors $X_j(0)$ form a basis of $\mathbf{R}^n$.

We prove Theorem 17.3.1 in two important special cases: linear constant coefficient systems and linear nonconstant 2×2 systems. The proof for constant coefficient systems is based on Jordan normal forms, while the proof for 2×2 systems is based on solving a separable differential equation for the Wronskian itself. It is this latter proof that generalizes to a proof of the theorem.

Wronskians for Constant Coefficient Systems

We begin by interpreting the Wronskian directly in terms of the constant coefficient matrix A. Note that $X_j(t) = e^{tA}e_j$ is just the jth column of the matrix e^{tA}. It follows that

$$W(t) = \det\ e^{tA}.$$

Lemma 17.3.2   *Let A be an n × n matrix. Then*

$$\det\ e^A = e^{tr(A)}.$$

Proof: This result is proved using Jordan normal forms. To see why normal form theory is relevant, suppose that A and B are similar matrices. Then e^A and e^B are similar matrices, and $\text{tr}(A) = \text{tr}(B)$ and $\det\ e^A = \det\ e^B$. So if we can show that the lemma is valid for matrices in Jordan normal form, then the lemma is valid for all matrices.

Suppose that the matrix J is a $k \times k$ Jordan block matrix associated with the eigenvalue λ. Then J is upper triangular and the diagonal entries of J all equal λ. It follows that $\text{tr}(J) = k\lambda$. It also follows that e^J is an upper triangular matrix whose diagonal entries all equal e^λ. Hence

$$\det\ e^J = \left(e^\lambda\right)^k = e^{k\lambda} = e^{\text{tr}(J)}.$$

So the lemma is valid for Jordan block matrices.

Next suppose that A is block diagonal; that is

$$A = \begin{pmatrix} B & 0 \\ 0 & C \end{pmatrix}.$$

We claim that if the lemma is valid for matrices B and C, then it is valid for the matrix A. To see this observe that

$$\text{tr}(A) = \text{tr}(B) + \text{tr}(C)$$

and that

$$e^A = \begin{pmatrix} e^B & 0 \\ 0 & e^C \end{pmatrix}.$$

Hence

$$\det\ e^A = \det\ e^B \det\ e^C = e^{\text{tr}(B)}e^{\text{tr}(C)},$$

by assumption. It follows that

$$\det e^A = e^{\text{tr}(B)+\text{tr}(C)} = e^{\text{tr}(A)},$$

as desired. By induction, the lemma is valid for Jordan normal form matrices and hence for all matrices. ◆

Wronskians for Planar Systems

In the time-dependent case we verify Theorem 17.3.1 for only 2×2 systems, as this substantially simplifies the discussion. Let

$$A(t) = \begin{pmatrix} a_{11}(t) & a_{12}(t) \\ a_{21}(t) & a_{22}(t) \end{pmatrix}$$

and let

$$X_1(t) = \begin{pmatrix} x_1(t) \\ y_1(t) \end{pmatrix} \quad \text{and} \quad X_2(t) = \begin{pmatrix} x_2(t) \\ y_2(t) \end{pmatrix}$$

be solutions of (17.2.2). It follows that

$$\begin{aligned}
\dot{x}_1 &= a_{11}x_1 + a_{12}y_1 \\
\dot{y}_1 &= a_{21}x_1 + a_{22}y_1 \\
\dot{x}_2 &= a_{11}x_2 + a_{12}y_2 \\
\dot{y}_2 &= a_{21}x_2 + a_{22}y_2.
\end{aligned} \tag{17.3.4}$$

In this notation

$$Y(t) = \begin{pmatrix} x_1(t) & x_2(t) \\ y_1(t) & y_2(t) \end{pmatrix}$$

and

$$W(t) = x_1(t)y_2(t) - x_2(t)y_1(t).$$

We claim that

$$\dot{W} = \text{tr}(A)W.$$

If so, we can use separation of variables to solve this differential equation, obtaining

$$\ln|W| = \int \text{tr}(A(t))\, dt$$

from which the proof of Theorem 17.3.1 follows.

Proof: Use the product rule to compute

$$\dot{W} = \dot{x}_1 y_2 + x_1 \dot{y}_2 - \dot{x}_2 y_1 - x_2 \dot{y}_1.$$

Now substitute (17.3.4) to see that

$$\dot{W} = (a_{11} x_1 + a_{12} y_1) y_2 + x_1 (a_{21} x_2 + a_{22} y_2) - (a_{11} x_2 + a_{12} y_2) y_1 - x_2 (a_{21} x_1 + a_{22} y_1),$$

from which it follows that

$$\dot{W} = (a_{11} + a_{22}) W = \operatorname{tr}(A) W,$$

as claimed. ◆

HAND EXERCISES

In Exercises 1–4, compute the Wronskian $W(t)$ for each matrix.

1. $Y(t) = \begin{pmatrix} t & 1 \\ -1 & t \end{pmatrix}$

2. $Y(t) = \begin{pmatrix} \cos t & -\sin t \\ \sin t & \cos t \end{pmatrix}$

3. $Y(t) = \begin{pmatrix} 0 & t & e^t \\ 1 & t & te^t \\ t & t & t^2 e^t \end{pmatrix}$

4. $Y(t) = \begin{pmatrix} 1 & 0 & 1 \\ 1 + \sin t & 1 & e^t \\ -4 + \cos t & t & te^t \end{pmatrix}$

5. Use (17.3.3) to verify that the Wronskian for solutions to the linear constant coefficient system of ODE $\dot{X} = AX$ is

$$W(t) = e^{\operatorname{tr}(A)t} W(0). \tag{17.3.5}$$

In Exercises 6–8, verify (17.3.5) for the given matrix A.

Hint First you need to find linearly independent solutions to the system $\dot{X} = AX$.

6. $A = \begin{pmatrix} 0 & -1 \\ 1 & 0 \end{pmatrix}$

7. $A = \begin{pmatrix} 2 & 3 \\ 3 & 2 \end{pmatrix}$

8. $A = \begin{pmatrix} 2 & -2 \\ 4 & -3 \end{pmatrix}$

17.4 HIGHER ORDER EQUATIONS

Using the same trick as in the case of equations with constant coefficients, we can rewrite a higher order linear equation with nonconstant coefficients as a first-order system of linear equations. This way we can also use the method of variation of parameters for the solution of higher order equations. In fact, consider a general linear differential equation of order n:

$$\frac{d^n x}{dt^n} + a_{n-1}(t) \frac{d^{n-1} x}{dt^{n-1}} + \cdots + a_1(t) \frac{dx}{dt} + a_0(t) x = g(t). \tag{17.4.1}$$

As in Section 15.2 we define the functions $x_1(t), \ldots, x_n(t)$ by

$$x_1(t) = x(t), \quad x_2(t) = \frac{dx}{dt}(t), \quad \ldots, \quad x_n(t) = \frac{d^{n-1}x}{dt^{n-1}}(t)$$

and see that (17.4.1) is equivalent to

$$\frac{dx_j}{dt} = x_{j+1} \quad (j = 1, \ldots, n-1)$$

$$\frac{dx_n}{dt} = g(t) - a_0(t)x_1 - \cdots - a_{n-1}(t)x_n.$$

Introducing the vectors

$$X(t) = (x_1(t), \ldots, x_n(t))^t \quad \text{and} \quad G(t) = (0, \ldots, 0, g(t))^t,$$

we conclude that (17.4.1) is equivalent to the first-order system

$$\frac{dX}{dt} = A(t)X + G(t),$$

where $A(t)$ is the matrix

$$A(t) = \begin{pmatrix} 0 & 1 & 0 & \cdots & 0 \\ 0 & 0 & 1 & \cdots & 0 \\ \vdots & \vdots & \vdots & & \vdots \\ 0 & 0 & 0 & \cdots & 1 \\ -a_0(t) & -a_1(t) & -a_2(t) & \cdots & -a_{n-1}(t) \end{pmatrix}.$$

In particular, $x_1(t)$ is a solution of (17.4.1) if $X(t) = (x_1(t), \ldots, x_n(t))^t$ is a solution of the system $\dot{X} = A(t)X + G(t)$, which can in principle be solved by variation of parameters.

An Example of Second Order

As an example we find a solution to the initial value problem

$$\ddot{x} - \frac{2}{t}\dot{x} + \frac{2}{t^2}x = t \tag{17.4.2}$$

$$x(1) = 0, \quad \dot{x}(1) = 1.$$

This second-order differential equation is equivalent to the system

$$\dot{X} = A(t)X + G(t)$$

$$X(1) = (0, 1)^t, \tag{17.4.3}$$

where

$$A(t) = \begin{pmatrix} 0 & 1 \\ -\dfrac{2}{t^2} & \dfrac{2}{t} \end{pmatrix} \quad \text{and} \quad G(t) = \begin{pmatrix} 0 \\ t \end{pmatrix}.$$

Step 1:   A basis of solutions to the homogeneous equation $\dot{X} = A(t)X$ is

$$X_1(t) = \begin{pmatrix} t \\ 1 \end{pmatrix} \quad \text{and} \quad X_2(t) = \begin{pmatrix} t^2 \\ 2t \end{pmatrix}.$$

Step 2:   The inverse of $Y(t)$ is given by

$$Y(t)^{-1} = \frac{1}{t^2} \begin{pmatrix} 2t & -t^2 \\ -1 & t \end{pmatrix}.$$

Hence we obtain

$$D(t) = Y(t)^{-1}G(t) = \frac{1}{t^2} \begin{pmatrix} 2t & -t^2 \\ -1 & t \end{pmatrix} \begin{pmatrix} 0 \\ t \end{pmatrix} = \begin{pmatrix} -t \\ 1 \end{pmatrix}.$$

Step 3:   Note that

$$X(1) = \begin{pmatrix} 0 \\ 1 \end{pmatrix} = c_1(1) \begin{pmatrix} 1 \\ 1 \end{pmatrix} + c_2(1) \begin{pmatrix} 1 \\ 2 \end{pmatrix}.$$

So $c_1(1) = -1$ and $c_2(1) = 1$. Now solve the differential equations

$$\dot{c}_1 = d_1(t) = -t$$
$$\dot{c}_2 = d_2(t) = 1$$

with these initial conditions, obtaining

$$c_1(t) = -\frac{1}{2}(t^2 + 1)$$
$$c_2(t) = t.$$

Step 4:   Theorem 17.2.3 implies that the solution to (17.4.3) is

$$X(t) = c_1(t)X_1(t) + c_2(t)X_2(t) = \begin{pmatrix} \dfrac{t}{2}(t^2 - 1) \\ \dfrac{1}{2}(3t^2 - 1) \end{pmatrix},$$

and hence the solution of (17.4.2) is given by

$$x(t) = \frac{t}{2}(t^2 - 1).$$

An Example of an Electrical Circuit

It is rare, however, that variation of parameters can actually be used to find a closed form solution to an inhomogeneous higher order linear differential equation with variable coefficients. In such instances it is best to resort to numerical integration of the first-order system obtained from the higher order equation.

We consider here the temporal behavior of an RLC circuit introduced in Section 16.5. This circuit is described by the differential equation

$$\ddot{x} + \frac{R}{L}\dot{x} + \frac{1}{CL}x = \frac{1}{CL}V(t),$$

where $x(t)$ is the voltage drop at the capacitor, R is the resistance, L is the inductance, C is the capacitance, and a voltage source is producing a time-dependent voltage $V(t)$, see (16.5.1). Now we assume that the circuit additionally contains a microphone that has a time-dependent resistance $R_{\mathrm{mic}}(t)$ of the form

$$R_{\mathrm{mic}}(t) = R_0 + R_1 \cos(\omega t).$$

This corresponds to the fact that a periodic signal of frequency $2\pi/\omega$ is entering the microphone. Then the differential equation for the electrical circuit becomes

$$\ddot{x} + \frac{R + R_{\mathrm{mic}}(t)}{L}\dot{x} + \frac{1}{CL}x = \frac{1}{CL}V(t). \tag{17.4.4}$$

For simplicity we set

$$R = 0 \quad \text{and} \quad C = L = R_0 = R_1 = \omega = 1$$

and assume that the voltage source is producing the constant voltage

$$V(t) = 1.$$

Thus we obtain the following linear differential equation of second order:

$$\ddot{x} + (1 + \cos t)\dot{x} + x = 1,$$

which is equivalent to the system $\dot{X} = A(t)X + G(t)$, where

$$A(t) = \begin{pmatrix} 0 & 1 \\ -1 & -(1 + \cos t) \end{pmatrix} \quad \text{and} \quad G(t) = \begin{pmatrix} 0 \\ 1 \end{pmatrix}. \tag{17.4.5}*$$

It is not obvious how to find the general solution of this equation by hand. In fact, it is not even clear how to construct solutions of the homogeneous equation. Hence we have to rely on numerical methods and compute solutions using ode45 in MATLAB. The right-hand side in (17.4.5) is stored in the function m-file f17_4_5.m. For completeness,

```
function f = f17_4_5(t,x)
A = [0  1; -1  -(1+cos(t))];
f = A*x + [0; 1];
```

We compute the solution of (17.4.5) with initial condition $X(0) = (1, 1)'$ for $t \in [0, 30]$ by typing

```
[t,x] = ode45('f17_4_5',[0 30],[1,1]');
```

The result of this computation is shown in Figure 17.2. It can be observed that $x_1(t)$ tends to 1 and $x_2(t)$ tends to 0 for increasing time t. In other words, the solution approaches the time-independent solution $X(t) = (1, 0)'$ of (17.4.5).

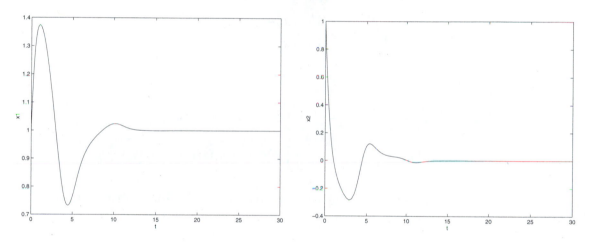

Figure 17.2
The two components of the solution of (17.4.5) for $t \in [0, 30]$ with initial condition $X(0) = (1, 1)'$

Reduction of Order

In order to find solutions to a higher order linear differential equation, we have rewritten the higher order equation as a first-order system of linear equations, and then we applied variation of parameters to that system. There is another method known as *reduction of order* that uses the same idea for the construction principle of the solution—but this idea is applied directly to the higher order equation. We remark that *reduction of order* can sometimes work even when a method like undetermined coefficients fails, since its applicability does not depend on the specific type of the inhomogeneity. However, as in the case of variation of parameters, the practical use of this technique is very limited. The reason is that the integrations that are involved in the computations can rarely be performed by hand.

We illustrate this method by the second-order linear differential equation

$$\ddot{x} + a_1(t)\dot{x} + a_0(t)x = g(t). \tag{17.4.6}$$

Roughly speaking, reduction of order is a way to reduce the second-order inhomogeneous equation to an inhomogeneous first-order equation.

Suppose that $x_h(t)$ is a solution of the homogeneous second-order equation $\ddot{x} + a_1(t)\dot{x} + a_0(t)x = 0$. In analogy to the method of variation of parameters, we try to find a solution $x(t)$ of the inhomogeneous equation (17.4.6) of the form

$$x(t) = c(t)x_h(t),$$

where $c(t)$ is a smooth function. To determine $c(t)$ we substitute $x(t)$ into the inhomogeneous equation (17.4.6). Before proceeding, we compute the derivatives of $x(t)$:

$$\dot{x} = \dot{c}x_h + c\dot{x}_h$$
$$\ddot{x} = \ddot{c}x_h + 2\dot{c}\dot{x}_h + c\ddot{x}_h.$$

Next we substitute $\dot{x}$ and $\ddot{x}$ into (17.4.6) and find

$$\ddot{c}x_h + 2\dot{c}\dot{x}_h + c\ddot{x}_h + a_1(t)\left(\dot{c}x_h + c\dot{x}_h\right) + a_0(t)cx_h =$$

$$x_h\left(\ddot{c} + \dot{c}\left(2\frac{\dot{x}_h}{x_h} + a_1(t)\right)\right) + c(\ddot{x}_h + a_1(t)\dot{x}_h + a_0(t)x_h) = g(t).$$

Since x_h is a solution of the homogeneous equation we have that $\ddot{x}_h + a_1(t)\dot{x}_h + a_0(t)x_h = 0$. After division by x_h, the function $c(t)$ satisfies

$$\ddot{c} + \dot{c}\left(2\frac{\dot{x}_h}{x_h} + a_1(t)\right) = \frac{g(t)}{x_h}.$$

Introducing $y(t) = \dot{c}(t)$, we arrive at the differential equation

$$\frac{dy}{dt} = -y\left(2\frac{\dot{x}_h}{x_h} + a_1(t)\right) + \frac{g(t)}{x_h},$$

which is linear and of first order. If we can find a solution $y(t)$ of this equation, then we can compute $c(t)$ by integration and we have constructed a particular solution $x(t) = c(t)x_h(t)$ of the inhomogeneous equation (17.4.6). Thus:

Theorem 17.4.1 (Reduction of order) *Consider the inhomogeneous linear ODE of second order (17.4.6).*

(a) Let $x_h(t)$ be a nonzero solution of the homogeneous equation

$$\ddot{x} + a_1(t)\dot{x} + a_0(t)x = 0.$$

(b)   Let $c(t)$ be a function such that its derivative $\dot{c}(t)$ is a solution of the linear differential equation

$$\frac{dy}{dt} = -\left(\frac{2\dot{x}_h(t) + a_1(t)x_h(t)}{x_h(t)}\right)y + \frac{g(t)}{x_h(t)}. \qquad \textbf{(17.4.7)}$$

Then $x_p(t) = c(t)x_h(t)$ is a particular solution of the second-order inhomogeneous equation (17.4.6).

Observe that (17.4.7) is again a scalar first-order differential equation that can in principle be solved by variation of parameters (see Theorem 17.1.1). Indeed, to see this one has to substitute

$$-\left(\frac{2\dot{x}_h(t) + a_1(t)x_h(t)}{x_h(t)}\right) \quad \text{and} \quad \frac{g(t)}{x_h(t)}$$

for

$$a(t) \quad \text{and} \quad g(t)$$

in (17.1.4).

A Specific Case: Constant Coefficients and Real Eigenvalues

As a special case of Theorem 17.4.1, we suppose that the coefficients a_0 and a_1 do not depend on t and that λ is a real root of the characteristic polynomial of the homogeneous equation $\ddot{x} + a_1\dot{x} + a_0 x = 0$. Then we can choose $x_h(t) = e^{\lambda t}$, and (17.4.7) becomes

$$\frac{dy}{dt} = -(2\lambda + a_1)y + e^{-\lambda t}g(t). \qquad \textbf{(17.4.8)}$$

If $\dot{c}(t)$ is a solution to (17.4.8), then $x(t) = c(t)e^{\lambda t}$ is a particular solution of the second-order inhomogeneous equation (17.4.6).

As an example, apply reduction of order to find a solution to the inhomogeneous ODE

$$\ddot{x} - 2\dot{x} + x = \frac{e^t}{t^3}. \qquad \textbf{(17.4.9)}$$

Observe that this equation cannot be solved by the method of undetermined coefficients since the right-hand side in (17.4.9) is not a solution of a homogeneous linear differential equation with constant coefficients.

The characteristic polynomial of (17.4.9) has the double eigenvalue $\lambda = 1$. Since $a_1 = -2$ and $g(t) = e^t/t^3$, (17.4.8) takes the form

$$\frac{dy}{dt} = -(2\lambda - 2)y + e^{-\lambda t}\left(\frac{e^t}{t^3}\right) = \frac{1}{t^3}.$$

Hence

$$y(t) = -\frac{1}{2t^2},$$

and integrating $y(t)$ leads to

$$c(t) = \int y(t)\, dt = \frac{1}{2t}.$$

Now Theorem 17.4.1 guarantees that

$$x(t) = c(t)\, e^t = \frac{e^t}{2t}$$

is a solution to (17.4.9). The other solutions to the inhomogeneous equation are found by adding the general solution to the homogeneous equation.

HAND EXERCISES

In Exercises 1 and 2, transform the given second-order ODE into a first-order system and then solve the initial value problem by variation of parameters.

1. $\ddot{x} - \dfrac{6}{t^2} x = 14 t^3; \quad x(-1) = -1, \quad \dot{x}(-1) = 5$

Hint The linearly independent solutions of the homogeneous equation are

$$x_h^1(t) = t^3 \quad \text{and} \quad x_h^2(t) = \frac{1}{t^2}.$$

2. $\ddot{x} + \dfrac{2}{t} \dot{x} - \dfrac{2}{t^2} x = 4; \quad x(1) = 1, \quad \dot{x}(1) = 2$

Hint The linearly independent solutions of the homogeneous equation are

$$x_h^1(t) = t \quad \text{and} \quad x_h^2(t) = \frac{1}{t^2}.$$

3. Use reduction of order to find a solution to the equation

$$\ddot{x} + 3\dot{x} + 2x = t. \qquad\qquad (17.4.10)$$

Compare the level of effort with the solution obtained using the method of undetermined coefficients.

4. Consider the second-order differential equation

$$\frac{d^2 x}{dt^2} + p(t)\frac{dx}{dt} + q(t)x = g(t). \qquad\qquad (17.4.11)$$

Suppose that

$$x_1(t) = 1, \quad x_2(t) = 1 + t, \quad \text{and} \quad x_3(t) = 1 + t + t^2$$

are all solutions to (17.4.11). Then determine $p(t)$, $q(t)$, and $g(t)$.

5. Set $R = 0$ and $C = L = 1$ in the electrical circuit equation (17.4.4). Show that this equation may be rewritten as the first-order system

$$\begin{aligned} \dot{x} &= y \\ \dot{y} &= -x - R_{\text{mic}}(t)y + V(t). \end{aligned} \tag{17.4.12}$$

COMPUTER EXERCISES

In Exercises 6–9, use MATLAB to find solutions for the electrical circuit equation (17.4.4). In each exercise, set $R = 0$ and $C = L = 1$ in addition to the specified information, and use the first-order system (17.4.12).

6. Parameters for the circuit: $R_{\text{mic}}(t) = 1 + \cos(5t)$ and $V(t) = 1$;
initial conditions and time interval: $x(0) = 1$, $\dot{x}(0) = 0.9$, and $t \in [0, 20]$

7. Parameters for the circuit: $R_{\text{mic}}(t) = 1 + \sin t$ and $V(t) = \sin(2t)$;
initial conditions and time interval: $x(0) = 2$, $\dot{x}(0) = 1$, and $t \in [0, 40]$

8. Parameters for the circuit: $R_{\text{mic}}(t) = 1 + \cos t$ and $V(t) = 0.4 + e^{-t}\sin(3t)$;
initial conditions and time interval: $x(0) = 1$, $\dot{x}(0) = -1.5$, and $t \in [0, 30]$

9. Parameters for the circuit: $R_{\text{mic}}(t) = 0.02 + \sin t$ and $V(t) = 1$;
initial conditions and time interval: $x(0) = 1.2$, $\dot{x}(0) = 1.1$, and $t \in [0, 60]$

17.5 SIMPLIFICATION BY SUBSTITUTION

So far in this chapter—as well as in Chapter 15—we have seen how to find solutions of ordinary differential equations that have special forms. (For instance, separation of variables can be applied to solve equations of the form $dx/dt = g(x)h(t)$.) However, "most" differential equations are not of the form needed to apply one of these techniques. But sometimes it is possible to transform the equation into such a form. This is accomplished by substituting a new function for $x(t)$, so that the differential equation for this new function has a simpler form. We illustrate this procedure in two cases.

Homogeneous Coefficients

Differential equations of the form

$$\frac{dx}{dt} = F\left(\frac{x}{t}\right), \tag{17.5.1}$$

where $F : \mathbf{R} \to \mathbf{R}$ is continuous, are said to have *homogeneous coefficients*. In general, none of the techniques described so far can be applied to this equation.

Since the function $x(t)/t$ appears as the argument of F in the equation, it seems plausible to write down (17.5.1) in terms of this function. Having this in mind, define

$$v(t) = \frac{x(t)}{t}.$$

Then $x(t) = tv(t)$ and (17.5.1) becomes

$$v(t) + t\frac{dv}{dt}(t) = F(v(t)).$$

When $t \neq 0$, this equation is equivalent to

$$\frac{dv}{dt} = \frac{F(v) - v}{t}. \tag{17.5.2}$$

We have arrived at an equation to which we can apply separation of variables. Once a solution $v(t)$ of (17.5.2) is found, we obtain a solution of (17.5.1) by setting

$$x(t) = tv(t).$$

If an initial condition $x(t_0) = x_0$ is specified in (17.5.1), then we have to solve (17.5.2) with the transformed initial condition $v(t_0) = x_0/t_0$.

An Example

Consider the initial value problem

$$\frac{dx}{dt} = \frac{2t^2 + x^2}{tx} \tag{17.5.3}$$
$$x(2) = 6.$$

Since the right-hand side of this equation can be written as

$$F\left(\frac{x}{t}\right) = 2\frac{t}{x} + \frac{x}{t},$$

we see that (17.5.3) has homogeneous coefficients. We set $v(t) = x(t)/t$. By (17.5.2) we have to solve the initial value problem

$$\frac{dv}{dt} = \frac{F(v) - v}{t} = \frac{2\frac{1}{v} + v - v}{t} = \frac{2}{tv}$$
$$v(2) = 6/2 = 3.$$

Separation of variables shows that $v(t)$ has to satisfy

$$\frac{v^2}{2} = \ln(t) - \ln(2) + \frac{9}{2}$$

(see Section 4.4). Therefore

$$v(t) = \sqrt{9 + 2\ln\frac{t}{2}}.$$

Finally, we obtain the solution

$$x(t) = tv(t) = t\sqrt{9 + 2\ln\frac{t}{2}}.$$

Bernoulli's Equation

Let $r, s : \mathbf{R} \to \mathbf{R}$ be continuous functions, and let $p \in \mathbf{R}$ be a real number. Then an equation of the form

$$\frac{dx}{dt} = r(t)x + s(t)x^p \tag{17.5.4}$$

is called a *Bernoulli equation*. For $p = 0$ or $p = 1$ the equation is linear and we can, in principle, find all the solutions by variation of parameters (see Chapter 15). Hence we assume that

$$p \neq 0, 1.$$

The idea is to substitute $x(t)$ in such a way that also for these values of p (17.5.4) becomes linear in the new function $v(t)$. We try the guess

$$v(t) = x(t)^{1/\alpha},$$

where we choose the real constant $\alpha \neq 0$ later to simplify the transformed equation. Using the chain rule on $x(t) = v(t)^\alpha$, we compute

$$\frac{dx}{dt} = \alpha v^{\alpha-1}\frac{dv}{dt}.$$

Substitution into (17.5.4) yields

$$\alpha v^{\alpha-1}\frac{dv}{dt} = r(t)v^\alpha + s(t)v^{p\alpha}.$$

Thus

$$\frac{dv}{dt} = \frac{1}{\alpha}r(t)v + \frac{1}{\alpha}s(t)v^{p\alpha-\alpha+1}.$$

To simplify this equation, choose the constant α so that $p\alpha - \alpha + 1 = 0$; that is, set

$$\alpha = \frac{1}{1-p}.$$

Then we arrive at the *linear* equation

$$\frac{dv}{dt} = (1-p)r(t)v + (1-p)s(t), \tag{17.5.5}$$

which can, in principle, be solved by variation of parameters. If this can be done, then we obtain a solution $x(t)$ of (17.5.4) by setting

$$x(t) = (v(t))^{\frac{1}{1-p}}.$$

If an initial condition $x(t_0) = x_0$ is specified in (17.5.4), then we have to solve (17.5.5) with the transformed initial condition $v(t_0) = x_0^{1/\alpha} = x_0^{1-p}$.

An Example

Consider the initial value problem

$$\frac{dx}{dt} = 3t^2 x + 3t^2 x^{2/3}$$
$$x(0) = 27.$$

The differential equation is a Bernoulli equation with

$$r(t) = 3t^2, \quad s(t) = 3t^2, \quad \text{and} \quad p = \frac{2}{3}.$$

Hence we have to find a solution of the linear initial value problem (see (17.5.5))

$$\frac{dv}{dt} = t^2 v + t^2$$
$$v(0) = 27^{1/3} = 3.$$

The solution to this differential equation can be obtained by variation of parameters and is

$$v(t) = -1 + 4e^{t^3/3}.$$

Therefore a solution of the Bernoulli equation is given by

$$x(t) = v(t)^{\frac{1}{1-p}} = \left(-1 + 4e^{t^3/3}\right)^3.$$

HAND EXERCISES

In Exercises 1–5, decide whether the given differential equation has homogeneous coefficients or is a Bernoulli equation.

1. $\dfrac{dx}{dt} = \dfrac{x^2}{t^3}$

2. $\dfrac{dx}{dt} = t$

3. $\dfrac{dx}{dt} = \dfrac{\cos t}{x^4}\sqrt{t + x^2}$

4. $\dfrac{dx}{dt} = \dfrac{t^2}{\sqrt{x}} + x \sin t$

5. $\dfrac{dx}{dt} = \dfrac{x}{t} + \dfrac{\sqrt{t}}{\sqrt{x}}$

In Exercises 6–9, solve each initial value problem by an appropriate solution technique.

6. $\dfrac{dx}{dt} = \sec\left(\dfrac{x}{t}\right) + \dfrac{x}{t}$, where $x(1) = \pi$

7. $\dfrac{dx}{dt} = x + x^2$, where $x(2) = 1$

8. $\dfrac{dx}{dt} = -\dfrac{x}{t} - t^3 x^3$, where $x(1) = 1$

9. $\dfrac{dx}{dt} = \dfrac{x(x + t)}{t^2}$, where $x(1) = 1$

10. We set $v(t) = x(t)^{1-p}$ to transform the Bernoulli equation (17.5.4) into the equation

$$\frac{dv}{dt} = (1 - p)r(t)v + (1 - p)s(t).$$

Now set $w(t) = v(\beta t)$ and show that β can be chosen so that $w(t)$ is a solution to the linear differential equation

$$\frac{dw}{dt} = r\left(\frac{t}{1 - p}\right)w + s\left(\frac{t}{1 - p}\right).$$

17.6   EXACT DIFFERENTIAL EQUATIONS

We illustrate the idea behind exact equations with the following example:

$$\frac{dx}{dt} = \frac{x - 1}{2x - t}. \qquad (17.6.1)$$

Since the right-hand side of (17.6.1) is neither of the form $g(x)h(t)$ nor of the form $F(x/t)$, it cannot be solved either by separation of variables or by substitution. However, it is possible to determine all solutions of (17.6.1), as we now explain.

We may rewrite (17.6.1) as

$$(2x - t)\frac{dx}{dt} = (x - 1). \qquad (17.6.2)$$

Suppose that $x(t)$ is a solution to (17.6.1) and let $y(t) = x(t)^2 - x(t)t + t$. Then differentiation shows that

$$\dot{y} = 2x\dot{x} - \dot{x}t - x + 1 = (2x - t)\dot{x} - (x - 1) = 0,$$

with the last equality coming from (17.6.2). Thus, if $x(t)$ is a solution, then the function $y(t) = x(t)^2 - x(t)t + t$ is constant. It follows that for each solution $x(t)$ there is a real constant c such that

$$x(t)^2 - x(t)t + t = c. \qquad (17.6.3)$$

Since (17.6.3) is quadratic in x, the quadratic formula determines two solutions for each value of c—namely,

$$x_\pm(t) = \frac{t \pm \sqrt{t^2 - 4t + 4c}}{2}. \qquad (17.6.4)$$

When $c = 1$, the radical in (17.6.4) is a perfect square and the two solutions to (17.6.1) are:

$$x_+(t) = t - 1 \quad \text{and} \quad x_-(t) = 1,$$

as can readily be checked.

Solutions Lie on Level Contours

To derive a geometric interpretation for the identity (17.6.3), define the function $F : \mathbf{R}^2 \to \mathbf{R}$ by

$$F(t, x) = x^2 - xt + t.$$

A *level set* or *contour* of F is defined to be the set of points in the (t, x)-plane for which

$$F(t, x) = c$$

for some real constant c. Thus (17.6.3) can now be restated as follows: Solutions $x(t)$ of the differential equation (17.6.1) lie on the level sets of F, since $F(t, x(t)) = c$.

Drawing Contours Using MATLAB

We use the MATLAB command contour to illustrate how this information helps us to visualize all the solutions. Type

```
[t,x] = meshgrid(-1.5:0.1:1.5,-1.5:0.1:1.5);
F = x.^2 - x.*t + t;
```

After these commands the data for the surface defined by the function F in the square $[-1.5, 1.5] \times [-1.5, 1.5]$ are stored in the MATLAB variable F. The command contour(F) allows us to display the level sets of F. To have the correct scales on the axes, type

```
contour(t,x,F)
```

Now the contour lines—the level sets corresponding to different levels c—are displayed. Moreover, we want to know to which levels the curves in that picture belong. Suppose that we are interested in the level sets corresponding to

$$c \in \{-2, -1, 0, 1, 2, 3, 4, 5\}.$$

Then we obtain this information by typing

```
cs = contour(t,x,F,[-2,-1,0,1,2,3,4,5]);
clabel(cs)
xlabel('t')
ylabel('x')
```

The command `clabel` allows us to label the different contour lines by their actual level. The desired information is displayed in Figure 17.3.

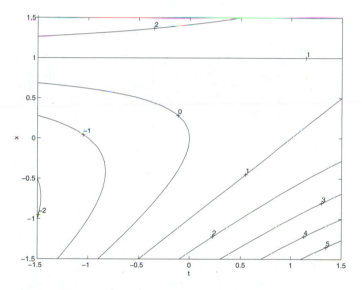

Figure 17.3
Contour lines of $F(t, x) = x^2 - xt + t$ for
$(t, x) \in [-1.5, 1.5] \times [-1.5, 1.5]$

Note that the straight line solutions $x(t) = 1$ and $x(t) = t - 1$ correspond to the level set for $c = 1$, as expected. We can also see from Figure 17.3 that the contour lines for the levels $c = 0, -1, -2$ have a turning point; that is, they cannot be parameterized entirely by t.

Setting $c = 0$ in (17.6.4), we find the expressions

$$x_+(t) = \frac{1}{2} \left(t + \sqrt{t(t-4)} \right) \quad \text{and} \quad x_-(t) = \frac{1}{2} \left(t - \sqrt{t(t-4)} \right).$$

These functions are not real-valued for $t \in (0, 4)$. In particular, they cannot correspond to solutions of (17.6.1) for these values of t. This provides an explanation for the existence of the turning point at $(t, x) = (0, 0)$ and for why the solutions $x_+(t)$ and $x_-(t)$ "collide" at this point.

Finally, we use dfield5 to confirm our theoretical discussion. We set the minimum and maximum values of t and x in The display window as in Figure 17.3 and start the computation in

$$(t_0, x_0) = (-0.5, x_-(-0.5)) = (-0.5, -1)$$

using the Keyboard input. The result—which is in good agreement with the contour plot—is shown in Figure 17.4. Observe that dfield5 will not stop the computation of the forward orbit on its own, and one has to use the Stop button to stop the numerical solution. The reason is that the program encounters numerical difficulties approaching the turning point $(0, 0)$, since the right-hand side of the differential equation is not defined at $(0, 0)$.

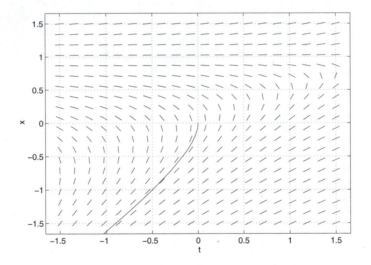

Figure 17.4
The solution of (17.6.1) for the initial value $(t_0, x_0) = (-0.5, -1)$
computed by dfield5

The General Definition of Exact Differential Equations

After this motivating example, we consider the general situation. Suppose that we have an ordinary differential equation of the form

$$\frac{dx}{dt} = \frac{G(t, x)}{H(t, x)}, \tag{17.6.5}$$

where G and H are real-valued continuous functions. In (17.6.1), $G(t, x) = x - 1$ and $H(t, x) = 2x - t$. The differential equation (17.6.5) is *exact* if there is a function $F(t, x)$ such that

$$\frac{\partial F}{\partial t} = -G \quad \text{and} \quad \frac{\partial F}{\partial x} = H. \tag{17.6.6}$$

Indeed, the differential equation (17.6.1) is exact—just set $F(t, x) = x^2 - xt + t$.

Supposing that the differential equation (17.6.5) is exact, proceed as in the example by rewriting (17.6.5) as

$$H(t, x)\frac{dx}{dt} - G(t, x) = 0. \tag{17.6.7}$$

Let $x(t)$ be a solution to (17.6.1) and use the chain rule, (17.6.6), and (17.6.7) to obtain

$$\frac{d}{dt}F(t, x(t)) = \frac{\partial F}{\partial x}\frac{dx}{dt} + \frac{\partial F}{\partial t} = H\frac{dx}{dt} - G = 0.$$

Hence solutions of (17.6.5) must lie on a level set of F defined by $F(t, x) = c$. We have proved:

Theorem 17.6.1   *Assume that the differential equation (17.6.5) is exact. Then, for any solution $x(t)$ of (17.6.5) with $x(t_0) = x_0$, there is a constant c such that for all t close to t_0, $F(t, x(t)) = c$.*

On the Existence of *F*

In order to solve exact differential equations, we must answer two questions:

- When does a function F exist that satisfies the conditions in (17.6.6)?
- If the function F exists, then how can we compute it?

The following proposition, based on the equality of mixed partial derivatives, gives a necessary condition for having a positive answer to the first question.

Proposition 17.6.2   *Let $G, H : \mathbf{R}^2 \to \mathbf{R}$ be differentiable functions such that $\partial G/\partial x$ and $\partial H/\partial t$ are continuous. Suppose that the corresponding differential equation (17.6.5) is exact. Then*

$$-\frac{\partial G}{\partial x}(t, x) = \frac{\partial H}{\partial t}(t, x). \tag{17.6.8}$$

Proof: Using (17.6.6), we compute

$$\frac{\partial^2 F}{\partial x \partial t} = -\frac{\partial G}{\partial x}(t, x) \quad \text{and} \quad \frac{\partial^2 F}{\partial t \partial x} = \frac{\partial H}{\partial t}(t, x).$$

Since $\partial G/\partial x$ and $\partial H/\partial t$ are continuous, equality of mixed partial derivatives holds, and

$$-\frac{\partial G}{\partial x} = \frac{\partial^2 F}{\partial x \partial t} = \frac{\partial^2 F}{\partial t \partial x} = \frac{\partial H}{\partial t},$$

as desired.   ◆

Remark 17.6.3   *It can be shown that criterion (17.6.8) is also sufficient for the exactness of the corresponding differential equation, if this criterion is satisfied in a region in the (t, x)-plane having no holes. The proof of this result can be found in any text on multidimensional calculus.*

Two Examples

We use two examples to illustrate the test for exactness of the underlying differential equation given by (17.6.8).

(**a**)   Consider (17.6.1); that is, suppose $G(t, x) = x - 1$ and $H(t, x) = 2x - t$. Then

$$-\frac{\partial G}{\partial x} = -1 = \frac{\partial H}{\partial t},$$

and according to Remark 17.6.3, a function F with the desired properties may be found.

(**b**)   Suppose that $G(t, x) = 1$ and $H(t, x) = t$. Then

$$-\frac{\partial G}{\partial x}(t, x) = 0 \neq 1 = \frac{\partial H}{\partial t}(t, x),$$

and by Proposition 17.6.2 a function F satisfying (17.6.6) cannot exist.

On the Computation of F

Next, consider the second question: How can F be computed when it exists? By (17.6.6) we have

$$\frac{\partial F}{\partial t}(t, x) = -G(t, x).$$

Therefore there is a differentiable function $g : \mathbf{R} \to \mathbf{R}$ such that

$$F(t, x) = -\Gamma(t, x) + g(x), \qquad \text{(17.6.9)}$$

where

$$\Gamma(t, x) = \int G(\tau, x)\, d\tau$$

is an indefinite integral of G with respect to the variable t. Condition (17.6.6) also implies that

$$H(t, x) = \frac{\partial F}{\partial x}(t, x) = -\frac{\partial}{\partial x}\Gamma(t, x) + g'(x);$$

that is

$$g'(x) = \frac{\partial}{\partial x}\Gamma(t, x) + H(t, x). \tag{17.6.10}$$

Equations (17.6.10) and (17.6.9) can now be used to compute the function F as long as the corresponding integrations can be performed.

Equivalently, we can first integrate the condition $F_x = H$ to obtain

$$F(t, x) = \Omega(t, x) + h(t), \tag{17.6.11}$$

where $\Omega(t, x) = \int H(t, x)dx$ is an indefinite integral of H with respect to x. Then (17.6.6) leads to

$$h'(t) = -\frac{\partial}{\partial t}\Omega(t, x) - G(t, x).$$

In practice, we prefer one way to the other depending on which integrations are easier to perform.

Two Examples

(a) Reconsider (17.6.1); that is, suppose $G(t, x) = x - 1$ and $H(t, x) = 2x - t$. We can choose

$$\Gamma(t, x) = \int G(\tau, y)\, d\tau = tx - t,$$

and (17.6.10) becomes

$$g'(x) = t + H(t, x) = 2x.$$

We can then choose $g(x) = x^2$ and, by (17.6.9), obtain

$$F(t, x) = -\Gamma(t, x) + g(x) = -tx + t + x^2.$$

(b) Next, consider the differential equation

$$\frac{dx}{dt} = \frac{2t(1 - x^2) - x^6}{x(2t^2 + 3x + 6tx^4)}. \tag{17.6.12}$$

In our notation

$$G(t, x) = 2t(1 - x^2) - x^6 \quad \text{and} \quad H(t, x) = x(2t^2 + 3x + 6tx^4).$$

Therefore

$$-\frac{\partial G}{\partial x} = 4tx + 6x^5 = \frac{\partial H}{\partial t}.$$

Hence the equation is exact (see Proposition 17.6.2 and Remark 17.6.3).

It is easier to compute $\Gamma(t, x) = \int G(\tau, x)\, d\tau$ than $\Omega(t, x) = \int H(s, x)dx$. We choose

$$\Gamma(t, x) = t^2(1 - x^2) - tx^6,$$

and (17.6.10) becomes

$$g'(x) = -(2t^2x + 6tx^5) + x(2t^2 + 3x + 6tx^4) = 3x^2.$$

Now we use (17.6.9) to obtain

$$F(t, x) = t^2(x^2 - 1) + tx^6 + x^3.$$

Solutions of the differential equation (17.6.12) lie on level sets of F. In particular, the solution $x(t)$ determined by the initial condition $x(2) = 1$ satisfies

$$F(t, x) = t^2(x^2 - 1) + tx^6 + x^3 = 3,$$

since $F(2, 1) = 3$. To see the qualitative behavior of the solutions, we use the following sequence of MATLAB commands:

```
[t,x] = meshgrid(1:0.02:3,0.5:0.01:1.5);
F   = t.^2.*(x.^2-1) + t.*x.^6 + x.^3;
cs = contour(t,x,F,[0,3,6,9,12]);
clabel(cs)
hold on
plot(2,1,'o')
xlabel('t')
ylabel('x')
```

The result is shown in Figure 17.5.

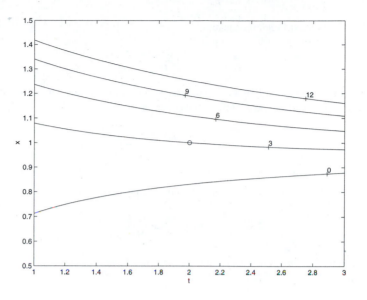

Figure 17.5
Solutions of the differential equation (17.6.12) corresponding to the
levels $c = 0, 3, 6, 9, 12$ in the rectangle $[1, 3] \times [0.5, 1.5]$

HAND EXERCISES

In Exercises 1 and 2, determine whether or not the given differential equation is exact.

1. $\dfrac{dx}{dt} = \dfrac{\cos x}{t \sin x}$

2. $\dfrac{dx}{dt} = \dfrac{\cos t}{x \sin t}$

3. Verify that the differential equation

$$\frac{dx}{dt} = \frac{2t - x}{t + 1}$$

is exact and find a solution with initial value $x(0) = 3$.

4. Verify that the differential equation

$$\frac{dx}{dt} = \frac{x}{x - t}$$

is exact and find a solution with initial value $x(2) = 3$. Compare your answer with Exercise 19 in
Section 4.2.

5. Show that the differential equation

$$\frac{dx}{dt} = -\frac{1 + 3tx}{t^2} \tag{17.6.13}$$

is not exact. Now multiply both the numerator and denominator of the right-side of (17.6.13) by t
and show that this "new" differential equation

$$\frac{dx}{dt} = -\frac{t + 3t^2 x}{t^3}$$

is exact. Now find all solutions to (17.6.13).

6. Show that the differential equation

$$\frac{dx}{dt} = -\frac{3x}{2t} \tag{17.6.14}$$

is not exact. Now multiply both the numerator and denominator of the right-side of (17.6.14) by $t^2 x$ and show that this "new" differential equation

$$\frac{dx}{dt} = -\frac{3t^2 x^2}{2t^3 x}$$

is exact. Now find all solutions to (17.6.14).

7. Let $G, H : \mathbf{R}^2 \to \mathbf{R}$ be continuous functions and suppose that the differential equation

$$\frac{dx}{dt} = \frac{G(t, x)}{H(t, x)} \tag{17.6.15}$$

is **not** exact. An *integrating factor* for (17.6.15) is a function $\rho : \mathbf{R}^2 \to \mathbf{R}$ such that the differential equation

$$\frac{dx}{dt} = \frac{\rho(t, x)G(t, x)}{\rho(t, x)H(t, x)}$$

is exact. Suppose that there exists an integrating factor $\rho(t)$ for (17.6.15) that is a function of t alone. Show that ρ satisfies

$$\frac{d\rho}{dt} = -\frac{\rho}{H}\left(\frac{\partial G}{\partial x} + \frac{\partial H}{\partial t}\right).$$

8. Use the result of Exercise 7 to find an integrating factor for the differential equation

$$\frac{dx}{dt} = -\frac{8x + 5t}{4t},$$

and then determine all solutions of this differential equation.

COMPUTER EXERCISES

9. Consider the differential equation

$$\frac{dx}{dt} = -\frac{2tx + \cos x}{t(t - \sin x)}.$$

Show that this differential equation is exact. Then use the command `contour` in MATLAB to find solutions. For the display, use the rectangle $[-2, 2] \times [-2, 2]$ in the (t, x)-plane and show contour lines for the different levels $c = -4, -3, -2, -1, 0, 1, 2, 3, 4$. Mark the solution that satisfies the initial condition $x(1) = 0$ by a circle.

Hint Use the same procedure that was used to create Figure 17.5.

17.7 HAMILTONIAN SYSTEMS

In the previous sections we have considered solution techniques that can be applied to *scalar* differential equations. In this section we present *planar* ODEs of a specific structure for which—similar to the case of exact differential equations, see Section 17.6—solutions can be identified as level curves of a certain function.

An autonomous planar system of differential equations

$$\dot{x} = f(x, y)$$
$$\dot{y} = g(x, y)$$

(17.7.1)

is *Hamiltonian* if there exists a real-valued function $H(x, y)$ such that

$$f(x, y) = \frac{\partial H}{\partial y}(x, y) \quad \text{and} \quad g(x, y) = -\frac{\partial H}{\partial x}(x, y).$$

The function H is called the *Hamiltonian* of the system (17.7.1). By the following result, Hamiltonian systems can be solved in a way analogous to exact systems.

Theorem 17.7.1 *Every solution trajectory of the Hamiltonian system (17.7.1) lies on a level curve of the associated Hamiltonian H.*

Proof: Let $(x(t), y(t))$ be a solution trajectory of (17.7.1). We need to verify that $H(x(t), y(t))$ is constant in t. This is most easily accomplished by showing that

$$\frac{d}{dt} H(x(t), y(t)) = 0.$$

(17.7.2)

Evaluate the left-hand side of (17.7.2) by using the chain rule:

$$\frac{d}{dt} H(x(t), y(t)) = \frac{\partial H}{\partial x} \frac{dx}{dt} + \frac{\partial H}{\partial y} \frac{dy}{dt}$$
$$= -g \frac{dx}{dt} + f \frac{dy}{dt},$$

where the last equality follows from the definition of the Hamiltonian. Using the fact that $(x(t), y(t))$ is a solution of (17.7.1) leads to

$$\frac{d}{dt} H(x(t), y(t)) = -gf + fg = 0,$$

as desired. ◆

Potential Systems

The system of differential equations

$$\dot{x} = y$$
$$\dot{y} = -V'(x) \tag{17.7.3}$$

is a Hamiltonian system for any *potential* function $V(x)$, where the Hamiltonian is

$$H(x, y) = \frac{1}{2} y^2 + V(x).$$

To verify this point, just check that

$$-\frac{\partial H}{\partial x} = -V'(x) \quad \text{and} \quad \frac{\partial H}{\partial y} = y.$$

In particular, the linear center is a Hamiltonian system; just take $V(x) = \frac{1}{2}x^2$ so that

$$H(x, y) = \frac{1}{2}(x^2 + y^2).$$

Note that the level curves of this Hamiltonian are just circles centered at the origin. This observation provides a second verification that trajectories of this center lie on circles.

A more interesting example occurs if we take the potential function equal to

$$V(x) = \frac{1}{4}x^4 - \frac{1}{2}x^2.$$

The corresponding Hamiltonian system is

$$\dot{x} = y$$
$$\dot{y} = x - x^3 \tag{17.7.4*}$$

since $V'(x) = x^3 - x$. It is easy to verify that (17.7.4) has three equilibria at $(1, 0)$, $(0, 0)$, and $(-1, 0)$.

The Jacobian of (17.7.3) is

$$J = \begin{pmatrix} 0 & 1 \\ -V''(x) & 0 \end{pmatrix}.$$

Thus equilibria of (17.7.3) are saddles if $V'' < 0$ or centers if $V'' > 0$. Indeed, this dichotomy is generally valid for Hamiltonian systems, and Hamiltonian systems are *not* usually Morse-Smale. For instance, in (17.7.4)

$$V''(x) = 3x^2 - 1$$

and the origin is a saddle while the other two equilibria are centers.

For planar Hamiltonian systems, it is always the case (this is a theorem similar to the Hopf bifurcation theorem) that centers are surrounded by a continuous family of periodic trajectories. Using `pplane5`, we can compute the phase portrait for this potential system and verify the existence of continuous families of periodic trajectories. The result is presented in Figure 17.6. Note the numerical evidence for the existence of two homoclinic trajectories.

We remark that the system (17.7.3) is just the one obtained from the second-order equation

$$\ddot{x} + V(x) = 0 \qquad\qquad \textbf{(17.7.5)}$$

by the trick of setting $y = \dot{x}$ to obtain a planar system.

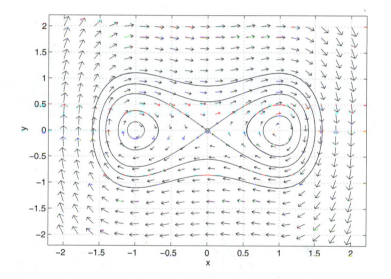

Figure 17.6
Phase portrait of (17.7.4)

Newton's Second Law and Potential Systems

There are several related interpretations of the potential V in (17.7.3) all based on Newton's second law of motion. In particular, we can rewrite (17.7.5) as

$$\ddot{x} = -V(x)$$

and interpret $-V$ as the force acting on a particle of unit mass. In this interpretation it is assumed that the force just depends on the value of x.

We consider three examples:

(a)   Suppose that x is the position of a unit mass on a vertical line moving under the influence of gravity. Then $V(x) = g$, where g is the gravitational constant.

(b)   Suppose that x is the distance of a particle of unit mass from the sun. The gravitational attraction of the sun is given by the inverse square law. If we assume that the sun is at the origin, then $V(x) = 1/x^2$.

(c)   Consider a pendulum with a unit mass attached to its end. Assume that the pendulum itself is idealized to have unit length and zero mass. The motion of an ideal pendulum is driven only by gravity. To derive the force acting on the pendulum, let x be the angle that the pendulum makes with the vertical, as in Figure 17.7. In that figure, the angle $z = (\pi/2) - x$. Then the force on the pendulum is $g \cos z = g \sin x$, and the pendulum equation is

$$\dot{x} = y$$
$$\dot{y} = -g \sin x. \qquad\qquad \textbf{(17.7.6)}^*$$

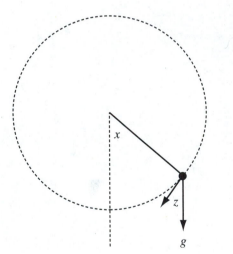

Figure 17.7
Pendulum geometry

The phase portrait for the pendulum equations is shown in Figure 17.8.

The Two-Body Problem

As a final example, consider the motion of two bodies with masses m_1 and m_2 under the influence of an attractive inverse square law of force. Both bodies move in three-dimensional space and therefore their motion would in principle be described by a system of three second-order ODEs. However, by considering the motion inside an appropriate moving frame, we can neglect both the translational motion in space and

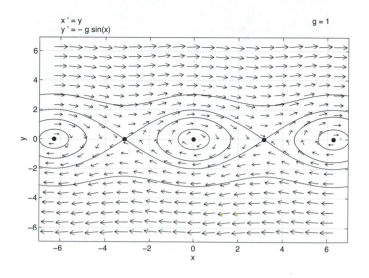

Figure 17.8
Phase portrait of the pendulum equation (17.7.6)

the rotation of the two bodies around their common center of mass. If this is done, then the relative motion of the two bodies is described by the single second-order equation

$$m\ddot{x} = -\frac{1}{x^2} + \frac{\ell^2}{mx^3}. \tag{17.7.7}$$

Here $x(t)$ is the distance between the two bodies at time t, ℓ is the (constant) *angular momentum* related to their rotational motion, and $m = m_1 m_2/(m_1 + m_2)$ is the *reduced mass*. We can rewrite (17.7.7) in the form $\ddot{x} + V(x) = 0$, where

$$V(x) = \frac{1}{mx^2} - \frac{\ell^2}{m^2 x^3}.$$

For two unit masses $m_1 = m_2 = 1$ (that is, $m = \frac{1}{2}$) with angular momentum $\ell = 1$, we can rewrite the second-order equation as the first-order system

$$\begin{aligned} \dot{x} &= y \\ \dot{y} &= -\frac{2}{x^2} + \frac{4}{x^3}. \end{aligned} \tag{17.7.8}*$$

The phase portrait of (17.7.8) is shown in Figure 17.9. In physical space, the steady-state solution $(x, y) = (2, 0)$ corresponds to a circular motion of the two bodies around their common center of mass and the periodic solutions correspond to motions on ellipsoids.

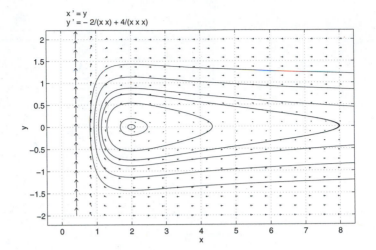

Figure 17.9
Phase portrait of the equation (17.7.8)

HAND EXERCISES

In Exercises 1–4, write the system of differential equations corresponding to each of the given Hamiltonians.

1. $H(x, y) = x + 3y + 2$

2. $H(x, y) = y^3$

3. $H(x, y) = 1 + \sin x \cos y$

4. $H(x, y) = xy^2 - x + x \cos y$

In Exercises 5–8, decide whether or not the given planar systems of differential equations are Hamiltonian.

5. $\begin{aligned} \dot{x} &= 0 \\ \dot{y} &= -1 \end{aligned}$

6. $\begin{aligned} \dot{x} &= x \\ \dot{y} &= y \end{aligned}$

7. $\begin{aligned} \dot{x} &= x - y^2 \\ \dot{y} &= -y - x^2 \end{aligned}$

8. $\begin{aligned} \dot{x} &= x \cos y \\ \dot{y} &= -\sin y \end{aligned}$

9. Let J be the Jacobian matrix of an equilibrium in a planar Hamiltonian system. Show that $\mathrm{tr}(J) = 0$. If $\det(J) \neq 0$, what are the possible types for this equilibrium?

10. Determine the equilibria in the pendulum equations (17.7.6). Refer to the phase portrait of the ideal pendulum in Figure 17.8 and describe in words the motion of the pendulum corresponding to a periodic solution surrounding one of the centers. Describe the motion of the pendulum corresponding to the initial condition $(x(0), y(0)) = (0, 2)$.

11. Consider the Hamiltonian $H(x, y) = x^3 + x^2 - y^2$. Show that the associated Hamiltonian system has a homoclinic trajectory.
Hint   Consider the level curve $H(x, y) = 0$.

12. Consider the Hamiltonian $H(x, y) = -\frac{1}{4}x^4 + \frac{1}{2}x^2 + y^2$. Show that the associated Hamiltonian system has saddle points at $(\pm 1, 0)$ and heteroclinic trajectories connecting these saddle points.
Hint   Consider the level curve $H(x, y) = \frac{1}{4}$.

CHAPTER

18

Numerical Solutions of ODEs

In general it is difficult, if not impossible, to find solutions to differential equations analytically. When this happens we rely on the numerical approximation of solutions, and in this chapter we discuss how numerical solutions to initial value problems of the form

$$\frac{dx}{dt}(t) = f(t, x(t))$$

$$x(t_0) = x_0$$

are found. This discussion provides insight into the numerical techniques used in the MATLAB programs dfield5, pline, and pplane5.

The numerical schemes that we consider approximate solutions up to a specified accuracy, and in Section 18.1 we describe the basic ideas underlying the construction of such schemes. In this section we see that in general there is an error between the numerical approximation and the analytic solution of the underlying initial value problem. One of the main tasks in the numerical analysis of ODEs is to derive bounds on this error, and in Section 18.2 we derive these bounds for the simplest numerical scheme, *Euler's method*. Finally, in Section 18.3 we generalize this treatment and indicate how to derive error bounds for arbitrary numerical schemes. In particular it turns out that in general the *fourth-order Runge-Kutta method* leads to much more accurate numerical approximations than does Euler's method.

18.1   A DESCRIPTION OF NUMERICAL METHODS

By definition, derivatives are limits of Newtonian quotients; approximating this limit is one of the basic ideas in the construction of numerical methods. More precisely, let x be the solution of the initial value problem $\dot{x} = f(t, x)$, $x(t_0) = x_0$. The derivative of x at time t is the limit

$$\frac{dx}{dt}(t) = \lim_{h \to 0} \frac{x(t + h) - x(t)}{h}.$$

Hence we expect

$$x(t + h) \approx x(t) + h\frac{dx}{dt}(t) = x(t) + hf(t, x(t)) \qquad \textbf{(18.1.1)}$$

to be a good approximation of x at time $t + h$ for small h. Indeed, by Taylor's formula,

$$
\begin{aligned}
x(t + h) &= x(t) + h\frac{dx}{dt}(t) + \frac{h^2}{2}\frac{d^2x}{dt^2}(t + \theta h) \\
&= x(t) + hf(t, x(t)) + \frac{h^2}{2}\frac{d^2x}{dt^2}(t + \theta h)
\end{aligned}
\qquad \textbf{(18.1.2)}
$$

with an appropriate $0 < \theta < 1$. Thus the error made in the approximation in (18.1.1) is the size h^2 as long as the second derivative of $x(t)$ is bounded. Consequently, if we know the value of the solution x at time t, then we can use the right-hand side f in the differential equation to compute an approximation of the solution x at time $t + h$.

Euler's Method

Since $f(t, x(t))$ is the derivative of x at time t, there is a simple geometric interpretation of (18.1.1) (see Figure 18.1). Recall that the tangent line to the graph of the function $x(t)$ at the point $(t, x(t))$ is the line going through the point $(t, x(t))$ whose slope is $\dot{x}(t) = f(t, x(t))$. The approximation of x at time $t + h$ is just the one given by the value of that tangent line at $t + h$. The numerical method that is based on (18.1.1) is called *Euler's method*. It seems evident that smaller values of h should lead to a more accurate approximation of the solution x on the interval $[t, t + h]$. On the other hand, note that simultaneously the length of the interval $[t, t + h]$ is shrinking and that more approximations are needed to approximate a solution on a fixed time interval.

Concretely, Euler's method produces a sequence of approximations to a solution of an initial value problem. To understand this point, begin the numerical integration at the exact initial value (t_0, x_0). Choose a *step size* $h > 0$. Use (18.1.1) to obtain after one *integration step* the approximation $x(t_1) \approx x_1$, where

$$t_1 = t_0 + h \quad \text{and} \quad x_1 = x_0 + hf(t_0, x_0).$$

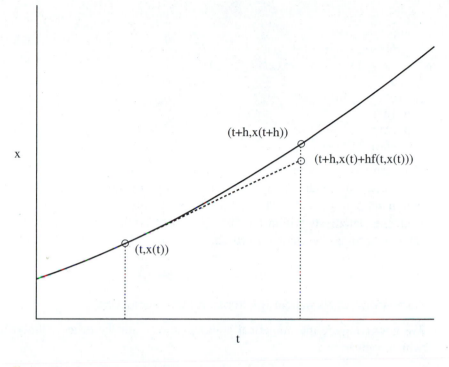

Figure 18.1
Illustration of one step in Euler's method for $h = 0.2$

Continuing this process, construct a sequence

$$t_{k+1} = t_k + h \quad \text{and} \quad x_{k+1} = x_k + hf(t_k, x_k) \qquad \textbf{(18.1.3)}$$

for $k = 0, 1, \ldots, K - 1$, where K is the total number of steps that are performed in the numerical approximation.

An Example of Euler's Method

We illustrate how Euler's method works for the example

$$\frac{dx}{dt} = x + t$$

$$x(1) = 2 \qquad \textbf{(18.1.4)}$$

by using MATLAB to compute an approximation to $x(3)$. Suppose that we set the step size to $h = 0.2$. Then the number of steps needed to reach $t = 3$ is $K = 10$. The code for computing the sequence in (18.1.3) is:

```
h    = 0.2;
t(1) = 1;
x(1) = 2;
```

```
K    = 10;
for k = 1:K
    t(k+1) = t(k)+h;
    x(k+1) = x(k)+h*(x(k)+t(k));
end
plot(t,x,'o')
hold on
plot(t,x,'- -')
xlabel('(a)')
```

The result is shown in Figure 18.2(a). Note that the 'o' in the MATLAB command plot(t,x,'o') puts o's at the 11 numerically computed points, and the '--' in the command plot(t,x,'--') interpolates dashed lines between successive o's. Moreover, the statements within the for loop—that is, the lines from for k = 1:K to end—reproduce the iteration procedure given in (18.1.3). (In MATLAB the index of a vector is not allowed to be 0. Therefore, the for loop is programmed to run from $k = 1, \ldots, K$ rather than from $k = 0, \ldots, K - 1$.)

Comparison of Numerical Solutions with Exact Solutions

We now compare the numerical approximation with the exact solution of (18.1.4), which is given by

$$x(t) = 4e^{t-1} - t - 1.$$

This comparison is made in Figure 18.2(b). The second curve in the figure is obtained using the additional commands

```
s = 1 : 0.01 : 3;
y = 4*exp(s-1)-s-1;
plot(s,y)
xlabel('(b)')
```

Note that in that figure the scale on the vertical axis is different from the one in Figure 18.2(a).

In this example we see that after just ten steps, the numerical solution of the initial value problem by Euler's method has led to a significant error; see Figure 18.2(b). There are two ways to proceed. Either we can use a smaller step size in Euler's method in an attempt to improve accuracy, or we can develop numerical methods that give more precise results for the same step size. For an illustration of the first approach, we show in Figure 18.3 an approximation with Euler's method using a step size $h = 0.05$. It can be seen that the error is now indeed much smaller but that also quite a few iterations of Euler's method are needed to produce this result. Therefore the latter approach has generally proved preferable, and one idea for improving accuracy is to replace $f(t_k, x_k)$ in the right-hand side of (18.1.3) by another expression in ways that we now explain.

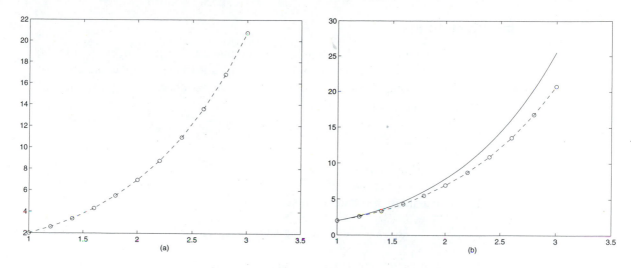

Figure 18.2

Approximation of the solution of (18.1.4) by Euler's method. (a) Circles mark the computed points; dashed lines are linear interpolations between the o's. (b) A comparison is made to the exact solution pictured as a solid line.

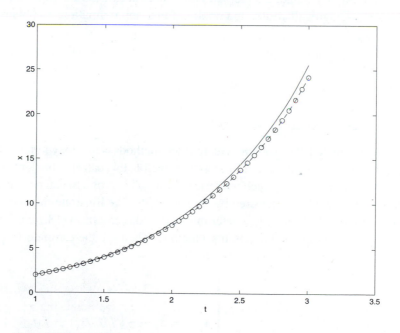

Figure 18.3

Approximation of the solution of (18.1.4) by Euler's method with step size $h = 0.05$. (Marks and lines as in Figure 18.2.)

Implicit Methods

In Euler's method, the approximation x_{k+1} is computed using the tangent line to the graph of the solution at the point (t_k, x_k). Another idea is to approximate the graph of $x(t)$ by the line passing through the point (t_k, x_k) whose slope is $f(t_{k+1}, x_{k+1})$. This idea leads to the numerical approximation

$$t_{k+1} = t_k + h \quad \text{and} \quad x_{k+1} = x_k + hf(t_{k+1}, x_{k+1}),$$

which is known as the *implicit Euler method*. One difficulty with this method is that we do not yet know what the value of x_{k+1} is. However, it is *implicitly* defined in the sense that we can think of the iteration step

$$x_{k+1} = x_k + hf(t_{k+1}, x_{k+1}) \tag{18.1.5}$$

as an equation in the unknown x_{k+1} and use this nonlinear equation to solve for x_{k+1}. We remark that there are ODEs (e.g., so-called *stiff* equations) for which implicit schemes typically produce more reliable results. On the other hand, implicit methods require considerable additional numerical effort at each time step in order to solve the equations for x_{k+1}.

An implicit numerical method that is more accurate than implicit Euler is obtained by just averaging the Euler and implicit Euler approximations to obtain

$$t_{k+1} = t_k + h \quad \text{and} \quad x_{k+1} = x_k + \frac{h}{2}\Big(f(t_k, x_k) + f(t_{k+1}, x_{k+1})\Big). \tag{18.1.6}$$

This method is called the *implicit trapezoidal rule*.

The Modified Euler Method

As we discussed, the problem with implicit methods is that they require the solution of a nonlinear equation at each time step. In principle this can also be done numerically, but such solutions require much numerical effort. This problem can be overcome partly by using a clever idea discovered by Runge (1895). We illustrate this idea on the implicit trapezoidal rule. Rather than determining x_{k+1} directly from (18.1.6), first estimate x_{k+1} by y_{k+1} using Euler's (tangent line) method. Then use the estimate y_{k+1} in the implicit trapezoidal rule. In formulas we obtain

$$t_{k+1} = t_k + h \quad \text{and} \quad \begin{cases} y_{k+1} = x_k + hf(t_k, x_k) \\[2mm] x_{k+1} = x_k + \dfrac{h}{2}\Big(f(t_k, x_k) + f(t_{k+1}, y_{k+1})\Big) \end{cases}$$

or, in one line,

$$t_{k+1} = t_k + h \quad \text{and} \quad x_{k+1} = x_k + \frac{h}{2}\Big(f(t_k, x_k) + f(t_{k+1}, x_k + hf(t_k, x_k))\Big). \tag{18.1.7}$$

The resulting numerical method is called the *modified Euler method.*

Let us use MATLAB to solve the initial value problem (18.1.4) by the modified Euler method. Using the same data as in the previous example, we type

```
h    = 0.2;
t(1) = 1;
x(1) = 2;
K    = 10;
for k = 1:K
    t(k+1) = t(k)+h;
    y(k)   = x(k)+h*(x(k)+t(k));
    x(k+1) = x(k)+(h/2)*(x(k)+t(k)+y(k)+t(k+1));
end
plot(t,x,'o')
hold on
plot(t,x,'- -')
s = 1 : 0.01 : 3;
y = 4*exp(s-1)-s-1;
plot(s,y)
xlabel('(a)')
```

to obtain the illustration in Figure 18.4(a). We see that the approximation can hardly be distinguished from the exact solution. Even if we double the step size to $h = 0.4$ and reduce the number of steps to $K = 5$, the approximation is still acceptable. See Figure 18.4(b).

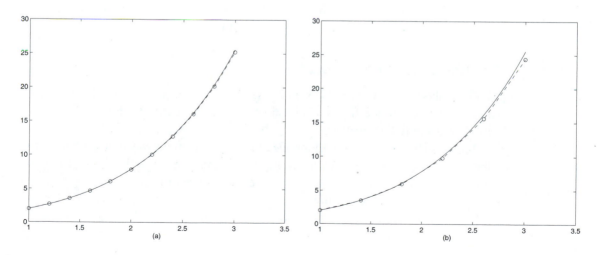

Figure 18.4

Approximations of the solution of (18.1.4) by the modified Euler method with step sizes: (a) $h = 0.2$; (b) $h = 0.4$. The exact solution is shown with a solid line.

Fourth-Order Runge-Kutta Method

The idea in the construction of the modified Euler method is to obtain a better approximation of the solution by using more information on the underlying differential equation. This is realized in the scheme by taking the arithmetic average of two evaluations of the right-hand side f at different points rather than at just one point as in the standard Euler methods.

In the frequently used *fourth-order Runge-Kutta method*, four different evaluations of f are taken into account in the computation of the next iterate. The method is given as follows: Set

$$f_1 = f(t_k, x_k)$$
$$f_2 = f\left(t_k + \frac{h}{2}, x_k + \frac{h}{2} f_1\right)$$
$$f_3 = f\left(t_k + \frac{h}{2}, x_k + \frac{h}{2} f_2\right)$$
$$f_4 = f(t_k + h, x_k + h f_3),$$

and define the new approximation by

$$x_{k+1} = x_k + \frac{h}{6}(f_1 + 2f_2 + 2f_3 + f_4).$$

By this method we can approximate solutions of differential equations in a very precise way. We illustrate this by an application to the initial value problem (18.1.4). In Figure 18.5 we show the approximation obtained with the step size $h = 0.6$ for $K = 20$ steps. The result is particularly convincing; observe that in contrast to Figures 18.2–18.4 the solution is approximated on a significantly larger interval.

Systems of Differential Equations

All the methods we have introduced in this section can also be used to find numerically solutions of systems of differential equations. We illustrate this fact for systems of two equations using Euler's method.

In the scalar case the underlying idea for Euler's method was to approximate the solution of the differential equation over a short interval by a line that is tangent to the solution curve (see Figure 18.1). We now use the same approach by applying this idea to each component of the solution.

Suppose that $(x(t), y(t))$ is a solution of the initial value problem

$$\frac{dx}{dt} = f(t, x, y)$$

$$\frac{dy}{dt} = g(t, x, y),$$

$$(18.1.8)$$

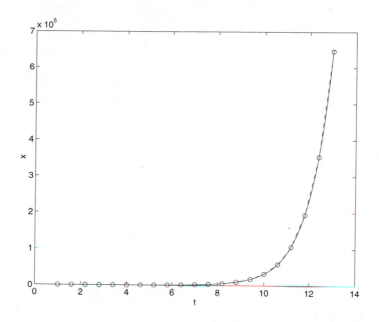

Figure 18.5
Approximation of the solution of (18.1.4) by the fourth-order
Runge-Kutta method. The exact solution is superimposed (solid line).

where $(x(t_0), y(t_0)) = (x_0, y_0)$. As in (18.1.1) we obtain for the components $x(t)$ and $y(t)$ by using the differential equation

$$x(t + h) \approx x(t) + h\frac{dx}{dt}(t) = x(t) + hf(t, x(t), y(t))$$

$$y(t + h) \approx y(t) + h\frac{dy}{dt}(t) = y(t) + hg(t, x(t), y(t)).$$

Again $h > 0$ is called the step size. We can now proceed as in the case of a scalar differential equation and obtain in analogy to (18.1.3),

$$t_{k+1} = t_k + h \quad \text{and} \quad \begin{cases} x_{k+1} = x_k + hf(t_k, x_k, y_k) \\ y_{k+1} = y_k + hg(t_k, x_k, y_k) \end{cases} \quad (k = 0, 1, \dots, K - 1).$$

In vector notation, Euler's method applied to the initial value problem (18.1.8) can be written as

$$\begin{pmatrix} x_{k+1} \\ y_{k+1} \end{pmatrix} = \begin{pmatrix} x_k \\ y_k \end{pmatrix} + h\begin{pmatrix} f(t_k, x_k, y_k) \\ g(t_k, x_k, y_k) \end{pmatrix}$$
$$t_{k+1} = t_k + h \tag{18.1.9}$$

for $k = 0, 1, \dots, K$.

Similarly, the implicit Euler method takes the form

$$\begin{pmatrix} x_{k+1} \\ y_{k+1} \end{pmatrix} = \begin{pmatrix} x_k \\ y_k \end{pmatrix} + h \begin{pmatrix} f(t_{k+1}, x_{k+1}, y_{k+1}) \\ g(t_{k+1}, x_{k+1}, y_{k+1}) \end{pmatrix}$$
$$t_{k+1} = t_k + h.$$

Let us use MATLAB to compute an approximation of the initial value problem

$$\frac{dx}{dt} = y - 3t$$

$$\frac{dy}{dt} = y + x^2,$$

(18.1.10)

where $(x(1), y(1)) = (0, 2)$. We specify the step size $h = 0.05$ and find an approximation on the interval $[1, 3]$ by typing

```
h     = 0.05;
t(1) = 1;
x(1) = 0;
y(1) = 2;
K     = 40;
for k = 1:K
   t(k+1) = t(k)+h;
   x(k+1) = x(k)+h*(y(k)-3*t(k));
   y(k+1) = y(k)+h*(y(k)+x(k)^2);
end
plot(x,y,'o')
hold on
plot(x,y,'- -')
xlabel('x')
ylabel('y')
```

The result is shown in Figure 18.6. Observe that the solution $(x(t), y(t))$ is graphed in the xy-plane. In particular, the variable t does not appear in the figure, and this is the reason the single steps in the approximation do not seem to be equally spaced. However, in Figure 18.7 we show graphs of the approximations of the time series x vs. t and y vs. t. Here it can be seen that the step size is indeed constant.

There are several ways to improve the accuracy of numerical schemes approximating solutions of initial value problems. One very successful method is to adapt the step size h in each step of the numerical approximation. We illustrate this strategy in Appendix 18.4.

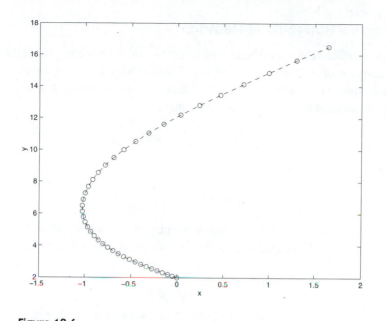

Figure 18.6
Approximation of the solution of (18.1.10) by Euler's method with step size $h = 0.05$. The single points of the approximation are marked by circles.

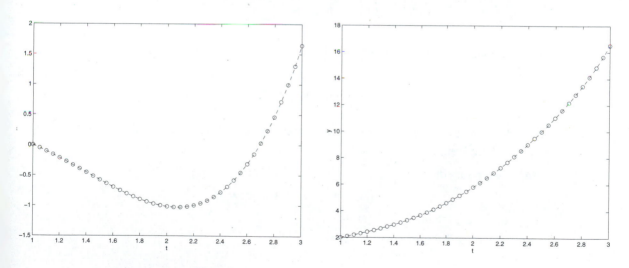

Figure 18.7
Approximation of the x and y components of the solution of (18.1.10) by Euler's method with step size $h = 0.05$

General Runge-Kutta Methods

The modified Euler method as well as the fourth-order Runge-Kutta method are specific examples of a general class of numerical schemes for the solution of initial value problems. These schemes are called *Runge-Kutta methods*. A general explicit Runge-Kutta method is fully described by numbers

$$b_1, \ldots, b_s \quad \text{and} \quad c_1, \ldots, c_s$$

and

$$a_{pq}, \quad p = 2, \ldots, s, \quad q = 1, \ldots, p - 1.$$

With these numbers the kth step in the numerical solution of an initial value problem using a Runge-Kutta method can be described as follows: Given (t_k, x_k), first evaluate f for s times:

$$f_1 = f\left(t_k + c_1 h, x_k\right)$$
$$f_2 = f\left(t_k + c_2 h, x_k + h a_{21} f_1\right)$$
$$f_3 = f\left(t_k + c_3 h, x_k + h(a_{31} f_1 + a_{32} f_2)\right)$$
$$\vdots$$
$$f_s = f\left(t_k + c_s h, x_k + h \sum_{j=1}^{s-1} a_{ij} f_j\right).$$

Second, obtain the next approximation by a weighted average of $f_1, \ldots, f_s$:

$$t_{k+1} = t_k + h \quad \text{and} \quad x_{k+1} = x_k + h \sum_{i=1}^{s} b_i f_i.$$

We now show that—except for the implicit schemes—all the methods that we have discussed are Runge-Kutta methods.

(a) *Euler's method:* Set

$$s = 1, \quad b_1 = 1, \quad c_1 = 0,$$

and there are no a_{pq}s.

(b) *The modified Euler method:* Set

$$s = 2, \quad b_1 = b_2 = \frac{1}{2}, \quad c_1 = 0, \quad c_2 = 1, \quad a_{21} = 1.$$

Indeed, from these data,

$$f_1 = f(t_k, x_k)$$
$$f_2 = f(t_{k+1}, x_k + h f_1) = f(t_{k+1}, x_k + h f(t_k, x_k))$$

and

$$x_{k+1} = x_k + h \left(\frac{1}{2} f_1 + \frac{1}{2} f_2 \right)$$

$$= x_k + \frac{h}{2} \Big(f(t_k, x_k) + f(t_{k+1}, x_k + hf(t_k, x_k)) \Big).$$

Hence we have obtained (18.1.7).

(c) *Fourth-order Runge-Kutta method:* Set

$$s = 4, \quad b_1 = b_4 = \frac{1}{6}, \quad b_2 = b_3 = \frac{2}{6}, \quad c_1 = 0, \quad c_2 = c_3 = \frac{1}{2}, \quad c_4 = 1,$$

and

$$a_{21} = \frac{1}{2}, \quad a_{31} = 0, \quad a_{32} = \frac{1}{2}, \quad a_{41} = 0, \quad a_{42} = 0, \quad a_{43} = 1.$$

HAND EXERCISES

1. Consider the initial value problem

$$\frac{dx}{dt}(t) = -2$$
$$x(t_0) = 0.$$

Show that in this case an application of Euler's method leads to exact results for all step sizes h; that is, $x_k = x(t_k)$ for all k.

2. Write down the implicit trapezoidal rule for the initial value problem (18.1.8).

In Exercises 3–5, decide whether or not the following numerical schemes are Runge-Kutta methods.

3. $x_{k+1} = x_k + hf(t_k + \frac{h}{2}, x_k + \frac{h}{2} f(t_k, x_k))$

4. $x_{k+1} = x_k + hf(t_k, x_k + t_k)$

5. $x_{k+1} = x_k + \frac{h}{4}(f(t_k, x_k) + 3f(t_k + \frac{2h}{3}, x_k + \frac{2h}{3} f(t_k, x_k)))$

6. Consider the initial value problem

$$\frac{dx}{dt}(t) = 1$$
$$x(t_0) = 0.$$

(a) Suppose that a Runge-Kutta method leads to exact results for all step sizes h; that is, $x_k = x(t_k)$. What does this imply for the numbers $b_1, \ldots, b_s$?

(b) Verify that the relation on the numbers $b_1, \ldots, b_s$ derived in (a) is satisfied for Euler's method, for the modified Euler method, and for the fourth-order Runge-Kutta method.

7. Use formula (18.1.2) to show that the first step in the numerical approximations shown in Figures 18.2(b) and 18.3 have to lie below the graph of the exact solution $x(t)$.

COMPUTER EXERCISES

8. Use MATLAB to reproduce the result displayed in Figure 18.3.

In Exercises 9–11, solve the given initial value problem by the explicit and the modified Euler method. Approximate the solution on the given interval and use the prescribed step size h.

9. $\dot{x} = 1$, $x(2) = 3$, step size $h = 0.1$, interval $[2, 4]$

10. $\dot{x} = x^2$, $x(1) = 0.5$, step size $h = 0.2$, interval $[1, 3]$

11. $\dot{x} = tx$, $x(1) = -1$, step size $h = 0.05$, interval $[1, 3]$

12. Use MATLAB to solve the initial value problem

$$\frac{dx}{dt} = x, \quad x(0) = 1$$

(a) by the implicit Euler method, and (b) by the implicit trapezoidal rule on the interval $[0, 2]$. In each case use the step size $h = 0.1$.

Hint Before working with MATLAB, use this specific differential equation to solve explicitly for x_{k+1} in (18.1.5) and (18.1.6).

13. Program the fourth-order Runge-Kutta method in MATLAB to verify the numerical computation in Figure 18.5. Compare the results to those obtained by ode45.

14. Use MATLAB to solve the initial value problem

$$\frac{dx}{dt} = -y$$
$$\frac{dy}{dt} = x,$$

where $(x(1), y(1)) = (1, 0)$. Find an approximation on the interval $[0, 12]$ with a step size $h = 0.1$ using (a) Euler's method, and (b) the implicit Euler method. Write down the explicit solution of the initial value problem and compare this with the numerical results.

18.2 ERROR BOUNDS FOR EULER'S METHOD

The numerical results of the preceding section indicate that the fourth-order Runge-Kutta method leads to more reliable results than Euler's method in the sense that the solution of the initial value problem (18.1.4) is much better approximated (see Figures 18.2 and 18.5). The purpose of the following sections is to derive error bounds for some numerical methods. These error bounds allow us to compare the accuracy of different methods when solving initial value problems.

To motivate the general treatment, let us explicitly compute the error of a specific numerical method. We apply the "simplest" method, Euler's method, to the "simplest" initial value problem that is not solved exactly by Euler's method:

$$\frac{dx}{dt} = x \tag{18.2.1}$$
$$x(0) = 1.$$

More precisely, we approximate the solution $x(t) = e^t$ on the interval $[0, T]$ with step size $h = T/K$, so that the numerical approximation consists of $K + 1$ points. With $t_0 = 0$ and $x_0 = 1$, Euler's method (18.1.3) takes the form

$$t_{k+1} = t_k + h$$
$$x_{k+1} = x_k + hx_k = (1 + h)x_k,$$

where $k = 0, 1, \ldots, K - 1$.

There are two essentially different types of error that are both relevant: the *local* and the *global discretization error*. Roughly speaking, the local discretization error is the error that is made by one single step in the numerical integration, whereas the global error is the error that is made on the whole time interval in the course of the approximation.

Local Error for Euler's Method

First we discuss the local error for Euler's method. We assume that the numerical solution is exact up to step k; that is, in our case we start in $x(t_k) = e^{t_k}$. Then the local discretization error $\delta(k + 1)$ is given by the error made in the following step:

$$\delta(k + 1) = x(t_{k+1}) - (x(t_k) + hx(t_k)) = e^{t_{k+1}} - (1 + h)e^{t_k}. \tag{18.2.2}$$

For instance, since $t_0 = 0$ and $t_1 = h$,

$$\delta(1) = x(t_1) - (e^{t_0} + he^{t_0}) = e^h - (1 + h).$$

In general $t_k = kh$, and we obtain from (18.2.2)

$$\delta(k + 1) = e^{t_{k+1}} - (1 + h)e^{t_k} = e^{(k+1)h} - (1 + h)e^{kh} = e^{kh}(e^h - (1 + h)).$$

The exponential function has the expansion

$$e^h = 1 + h + \frac{h^2}{2} + \frac{h^3}{6} + \cdots \quad \text{for all } h \geq 0, \tag{18.2.3}$$

and therefore it follows that

$$e^h - (1 + h) = \frac{h^2}{2} + \frac{h^3}{6} + \cdots = \frac{h^2}{2}\left[1 + \frac{h}{3} + \frac{h^2}{12} + \cdots\right] \leq \frac{h^2}{2}e^h.$$

Since $(k + 1)h = (k + 1)/K \leq T$ for $k = 0, 1, \ldots, K - 1$, we finally have the desired bound

$$\delta(k + 1) = e^{kh}(e^h - (1 + h)) \leq e^{(k+1)h}\left(\frac{h^2}{2}\right) \leq e^T \frac{h^2}{2} \quad \text{for all } h \geq 0. \tag{18.2.4}$$

Observe that we have obtained a bound for the local discretization error that depends just on the step size h and not on the actual step k.

Global Error for Euler's Method

We now consider the global discretization error after k steps. It is defined by

$$\epsilon(k) = x(t_k) - x_k, \quad k = 0, 1, \ldots, K.$$

The basic trick in the computation of a bound for $|\epsilon(k)|$ is to derive an equation for the evolution of this error while k is varied. We do this as follows: By (18.2.2) we have for $k = 1, 2, \ldots, K$,

$$x(t_k) = (1 + h)x(t_{k-1}) + \delta(k),$$

and, on the other hand, Euler's method applied to (18.2.1) is given by

$$x_k = (1 + h)x_{k-1}.$$

Subtracting these two equations from each other, we obtain

$$x(t_k) - x_k = (1 + h)(x(t_{k-1}) - x_{k-1}) + \delta(k)$$

and therefore with the bound (18.2.4) for the local error

$$\begin{aligned}
|\epsilon(k)| = |x(t_k) - x_k| &= |(1 + h)(x(t_{k-1}) - x_{k-1}) + \delta(k)| \\
&\leq (1 + h)|\epsilon(k - 1)| + \delta_h
\end{aligned}$$

with

$$\delta_h = e^T \frac{h^2}{2}. \tag{18.2.5}$$

We have accomplished our first goal: The global discretization error at step k is expressed in terms of the global discretization error at step $k - 1$ in combination with a bound for the local discretization error. This allows us to apply this formula repeatedly until k is 0:

$$\begin{aligned}
|\epsilon(k)| &\leq (1 + h)|\epsilon(k - 1)| + \delta_h \\
&\leq (1 + h)[(1 + h)|\epsilon(k - 2)| + \delta_h] + \delta_h \\
&= (1 + h)^2|\epsilon(k - 2)| + ((1 + h) + 1)\delta_h \\
&\leq (1 + h)^2[(1 + h)|\epsilon(k - 3)| + \delta_h] + ((1 + h) + 1)\delta_h \\
&= (1 + h)^3|\epsilon(k - 3)| + ((1 + h)^2 + (1 + h) + 1)\delta_h \\
&\vdots \\
&\leq (1 + h)^k|\epsilon(0)| + ((1 + h)^{k-1} + \cdots + (1 + h) + 1)\delta_h.
\end{aligned} \tag{18.2.6}$$

But $\epsilon(0) = x(t_0) - x_0 = 0$, and using the formula

$$\alpha^{k-1} + \alpha^{k-2} + \cdots + \alpha + 1 = \frac{\alpha^k - 1}{\alpha - 1}, \quad \alpha \neq 1,$$

with $\alpha = 1 + h$, we obtain

$$|\epsilon(k)| \leq \frac{(1+h)^k - 1}{h} \delta_h.$$

By (18.2.3) $1 + h \leq e^h$ and we finally obtain the desired bound on the global discretization error for Euler's method applied to (18.2.1):

$$
\begin{aligned}
|\epsilon(k)| &\leq \frac{1}{h}(e^{kh} - 1)\delta_h = \frac{1}{h}(e^{kh} - 1)e^T \frac{h^2}{2} \\
&= e^T(e^{kh} - 1)\frac{h}{2} \leq e^T(e^T - 1)\frac{h}{2}.
\end{aligned}
\tag{18.2.7}
$$

We summarize our computations in the following proposition:

Proposition 18.2.1   *Let x_k for $0 \leq k \leq K$ be the numbers generated by Euler's method applied to the initial value problem*

$$\frac{dx}{dt} = x, \quad x(0) = 1,$$

on the interval $[0, T]$ with step size $h = T/K$ and such that $x_0 = x(0) = 1$. Then:

 (a)   *The local discretization error $\delta(k)$ satisfies*

$$\delta(k) \leq e^T \frac{h^2}{2}.$$

 (b)   *The global discretization error $\epsilon(k)$ satisfies*

$$|\epsilon(k)| \leq e^T(e^{kh} - 1)\frac{h}{2} \leq e^T(e^T - 1)\frac{h}{2}.$$

 (c)   *Euler's method converges to the solution $x(t) = e^t$ of the initial value problem on $[0, T]$ if the step size h tends to 0.*

Proof: The statements (a) and (b) are precisely (18.2.4) and (18.2.7). Moreover, (c) follows from (18.2.7) since the right-hand side in

$$|x(t_k) - x_k| \leq e^T(e^T - 1)\frac{h}{2}$$

becomes arbitrarily small for $h \to 0$ and uniformly in k.   ◆

Verification of Error Analysis Using MATLAB

Let us verify the estimate in (18.2.7) using MATLAB. For a numerical approximation of the solution of the initial value problem (18.2.1) with step size $h = 0.1$ on the interval $[0, 1]$, we type

```
h       = 0.1;
t(1)    = 0;
x(1)    = 1;
err(1)  = 0;
est(1)  = 0;
K       = 1/h;
for k = 1:K
    t(k+1)  = t(k)+h;
    x(k+1)  = (1+h)*x(k);
   err(k+1) = exp(t(k+1))-x(k+1);
   est(k+1) = exp(1)*(exp(k*h)-1)*h/2;
end
plot(t,err,'+')
hold on
plot(t,est,'x')
xlabel('(a)')
```

The result is shown in Figure 18.8(a). It can be seen that, indeed, we have obtained an upper bound for the global discretization error in (18.2.7). In the proof of part (c) of Proposition 18.2.1 we used the fact that this upper bound tends to 0 with decreasing step size h. This fact is illustrated in Figure 18.8(b), where we have changed the step size h to 0.02.

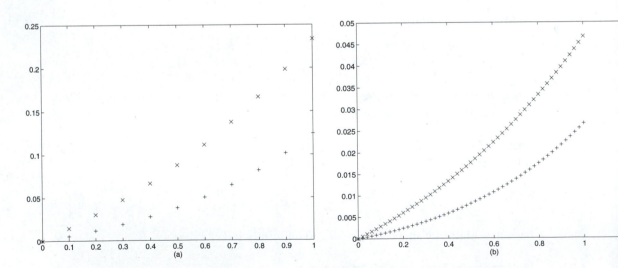

Figure 18.8
Comparison of the global discretization error (marked by $+$) with its estimate given in (18.2.7) (marked by $\times$) for the step sizes (a) $h = 0.1$ and (b) $h = 0.02$

Explicit Computation of Error Bounds

Another important consequence of Proposition 18.2.1 is that it allows us to compute the solution of the initial value problem (18.2.1) on a given interval $[0, T]$ up to a prescribed accuracy. Indeed, we just have to use the estimate (18.2.7) on the global discretization error.

For an illustration of this fact, suppose that we want to approximate a solution of (18.2.1) on the interval $t \in [0, 2]$ where the maximal global discretization error is less than 0.01. It follows from (18.2.7) that this is certainly the case if the step size h is chosen such that

$$e^2(e^2 - 1)\frac{h}{2} = 0.01$$

or, equivalently,

$$h = \frac{0.02}{e^2(e^2 - 1)} \approx 0.000424.$$

Indeed, if this is the case, then we find with (18.2.7)

$$|\epsilon(k)| \le e^2(e^{kh} - 1)\frac{h}{2} \le e^2(e^2 - 1)\frac{h}{2} = 0.01.$$

The result of the corresponding MATLAB computation of the global discretization error is shown in Figure 18.9.

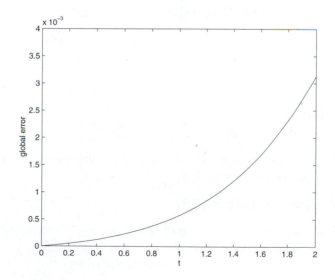

Figure 18.9
The global discretization error for the solution of (18.2.1) on the interval $[0, 2]$ with step size $h = 0.000424$

HAND EXERCISES

In Exercises 1–3, compute bounds on the local and global errors for Euler's method applied to the given initial value problem. Perform the same steps as in the treatment of the initial value problem (18.2.1) in the text.

1. $\frac{dx}{dt} = 3x$; $x(0) = 1$ 2. $\frac{dx}{dt} = x$; $x(0) = 2$ 3. $\frac{dx}{dt} = 3x$; $x(0) = 2$

4. Apply the implicit Euler method to the initial value problem (18.2.1). Then determine a formula for the local discretization error that is analogous to (18.2.2).

Hint   Before proceeding as for Euler's method, solve for x_{k+1} in (18.1.5) in this specific case.

COMPUTER EXERCISES

In Exercises 5 and 6, determine a step size h such that the global discretization error for the solution of (18.2.1) on the given interval using Euler's method is less than the given absolute tolerance. Verify your results by a computation of the global discretization error using MATLAB.

5. Interval [0, 4] and absolute tolerance 2

6. Interval [0, 5] and absolute tolerance 10

7. Suppose that you want to solve (18.2.1) by Euler's method with a fixed step size $h = 0.005$ on the interval [0, T]. How large can you choose T so that the global discretization error does not exceed 0.05? Confirm your result using MATLAB.

18.3   LOCAL AND GLOBAL ERROR BOUNDS

We now want to analyze the error of a numerical method applied to an initial value problem of the form

$$\frac{dx}{dt} = f(t, x)$$

$$x(t_0) = x_0.$$

Before introducing the general concept, we recall the crucial points in the error analysis that we performed in Section 18.2.

Error Analysis from Section 18.2

We introduced two different types of error: the *local discretization error* $\delta(k)$ and the *global discretization error* $\epsilon(k)$.

For Euler's method, the local error is bounded by

$$\delta(k) \leq \delta_h \leq C_E h^2,$$

where $C_E > 0$ is a constant depending on the length of the interval [0, T] on which the solution is approximated; see (18.2.4). (For example, $C_E = e^T / 2$.) In particular, on the interval $[0, T]$, the local error goes to 0 at least as fast as a quadratic function in the step size h.

For Euler's method the global error is bounded by

$$|\epsilon(k)| \leq Dh,$$

with a constant $D > 0$ that again depends on just T; see (18.2.7). Hence the global error tends to 0 at least as fast as a linear function in the step size h. Roughly speaking, one power is lost going from the local to the global error, and this power is lost while going through the estimate in (18.2.6). The same phenomenon occurs in the general error analysis.

A General Form for a Numerical Method

Let $x(t)$ be the solution of the initial value problem. From now on we represent an explicit numerical method by a function Φ as follows: For $k = 0, 1, \ldots, K - 1$ we have $t_k = t_0 + kh$, $x_0 = x(t_0)$, and

$$x_{k+1} = x_k + h\Phi(t_k, x_k, h).$$

In Euler's method (18.1.3),

$$\Phi(t_k, x_k, h) = f(t_k, x_k).$$

In particular, in this case, Φ does not depend on the step size h. In the modified Euler method (18.1.7),

$$\Phi(t_k, x_k, h) = \frac{1}{2}\Big(f(t_k, x_k) + f(t_k + h, x_k + hf(t_k, x_k))\Big).$$

A General Form for Errors

Next we introduce the two different types of error for a numerical method given by Φ.

Definition 18.3.1
  (a)   The local discretization error $\delta(k + 1)$ *is defined as*

$$\delta(k + 1) = x(t_{k+1}) - (x(t_k) + h\Phi(t_k, x(t_k), h)),$$

  where $k = 0, 1, \ldots, K - 1$.
  (b)   The global discretization error $\epsilon(k)$ *is given by*

$$\epsilon(k) = x(t_k) - x_k,$$

  where $k = 0, 1, \ldots, K$.

A Theorem on Global Discretization Errors

The purpose is to find a bound on the global error by the same technique as in Section 18.2. For this we need two assumptions:

(i) In the error estimates for Euler's method it was very convenient to use a bound for the local error that is independent of the actual step k. Hence we assume that there is a constant $\delta_h > 0$ such that

$$|\delta(k+1)| \leq \delta_h \quad \text{for all } k = 0, 1, \ldots, K - 1. \qquad \textbf{(18.3.1)}$$

In concrete examples this fact can be guaranteed by the differentiability of the right-hand side f in the initial value problem.

(ii) We also need an additional assumption on the function Φ: We assume that there is a constant $L > 0$ such that for all t, h, y, z,

$$|\Phi(t, y, h) - \Phi(t, z, h)| \leq L|y - z|. \qquad \textbf{(18.3.2)}$$

For Euler's method $\Phi(t, x, h) = f(t, x)$, and therefore (18.3.2) is satisfied if f is (globally) Lipschitz continuous in x.

We are now in the position to derive a bound for the global discretization error proceeding completely analogous to Section 18.2. Observe that by Definition 18.3.1(a),

$$x(t_k) = x(t_{k-1}) + h\Phi(t_{k-1}, x(t_{k-1}), h) + \delta(k).$$

On the other hand, the numerical method can be written as

$$x_k = x_{k-1} + h\Phi(t_{k-1}, x_{k-1}, h).$$

Subtracting these two equations from each other and using (18.3.2) and (18.3.1), we obtain

$$\begin{aligned}
|\epsilon(k)| &= |x(t_k) - x_k| \\
&= |x(t_{k-1}) - x_{k-1} + h[\Phi(t_{k-1}, x(t_{k-1}), h) - \Phi(t_{k-1}, x_{k-1}, h)] + \delta(k)| \\
&\leq |x(t_{k-1}) - x_{k-1}| + hL|x(t_{k-1}) - x_{k-1}| + |\delta(k)| \\
&\leq (1 + hL)|x(t_{k-1}) - x_{k-1}| + \delta_h \\
&= (1 + hL)|\epsilon(k-1)| + \delta_h.
\end{aligned}$$

This inequality is of the same type as the inequality in the first line in (18.2.6)—one just has to replace the first h by hL. Hence we can repeat the estimate in (18.2.6) and obtain

$$|\epsilon(k)| \leq \frac{(1 + hL)^k - 1}{hL}\delta_h.$$

Since $1 + hL \leq e^{hL}$, we have proved the following result:

Theorem 18.3.2   *Suppose that (18.3.1) and (18.3.2) hold. Then the global discretization error of the numerical method given by the function Φ satisfies*

$$|\epsilon(k)| \leq \left(\frac{e^{khL} - 1}{L}\right) \frac{\delta_h}{h}. \tag{18.3.3}$$

An Example Using Euler's Method

The estimate (18.3.3) indicates that the constant L plays an important role in the size of the global discretization error. Indeed, let us illustrate this fact by applying Euler's method to the initial value problem

$$\frac{dx}{dt} = \lambda x \tag{18.3.4}$$
$$x(0) = 1$$

on the interval $[0, T]$. In this case $L = \lambda$, since for Euler's method

$$|\Phi(t, y, h) - \Phi(t, z, h)| = |f(t, y) - f(t, z)| = |\lambda y - \lambda z| = \lambda |y - z|.$$

Moreover, since we know the exact solution, we may compute δ_h for this case as follows: We have

$$\delta(k + 1) = e^{\lambda t_{k+1}} - (e^{\lambda t_k} + \lambda h e^{\lambda t_k}) = e^{\lambda k h}(e^{\lambda h} - (1 + \lambda h)),$$

and since $e^{\lambda h} - (1 + \lambda h) \leq e^{\lambda h}((\lambda h)^2/2)$, we obtain (see also (18.2.4))

$$\delta(k + 1) \leq e^{\lambda k h} e^{\lambda h} \frac{(\lambda h)^2}{2} \leq e^{\lambda T} \frac{(\lambda h)^2}{2}.$$

Therefore we can choose

$$\delta_h = \lambda^2 e^{\lambda T} \frac{h^2}{2}. \tag{18.3.5}$$

In fact, observe that for $\lambda = 1$ we recover (18.2.5).

Proceeding as in Section 18.2, we set $\lambda = 2$ and compute the global discretization error and its bound in (18.3.3) by MATLAB on the interval $[0, 1]$ for the step size $h = 0.1$. Concretely we type

```
h      = 0.1;
L      = 2;
t(1)   = 0;
x(1)   = 1;
err(1) = 0;
est(1) = 0;
K      = 1/h;
```

```
for k = 1:K
    t(k+1) = t(k)+h;
    x(k+1) = (1+L*h)*x(k);
  err(k+1) = exp(L*t(k+1))-x(k+1);
  est(k+1) = exp(L)*(exp(L*k*h)-1)*L*h/2;
end
plot(t,err,'+')
hold on
plot(t,est,'x')
```

The result is shown in Figure 18.10. A comparison with Figure 18.8 shows that the error is indeed significantly bigger, reflecting the fact that here $L = 2$ whereas in the example in Section 18.2 this constant was 1.

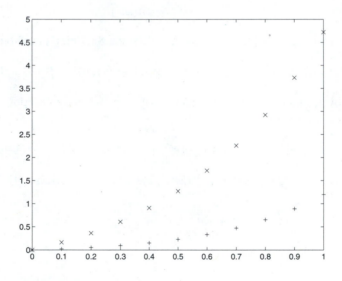

Figure 18.10
The global discretization error (marked by +) and its bound given in (18.3.3) (marked by ×) for the step size $h = 0.1$

A Bound on Local Discretization Errors

It remains to demonstrate how to determine a bound δ_h on the local discretization error $\delta(k)$ for a given numerical method applied to the initial value problem

$$\frac{dx}{dt} = f(t, x)$$

$$x(t_0) = x_0$$

on the interval $[t_0, t_e]$. Again we consider Euler's method. The crucial tool for the computation of the local error is a Taylor expansion of the solution combined with the fact that the derivatives of $x(t)$ can be written in terms of the function f by the differential equation. Using this idea, we expand $x(t_{k+1}) = x(t_k + h)$ up to second order by

$$x(t_{k+1}) = x(t_k) + h \frac{dx}{dt}(t_k) + \frac{h^2}{2} \frac{d^2 x}{dt^2}(t_k + \theta h)$$

$$= x(t_k) + h f(t_k, x(t_k)) + \frac{h^2}{2} \frac{d^2 x}{dt^2}(t_k + \theta h)$$

with an appropriate $0 < \theta < 1$. Substitution into the expression for the local error leads to

$$\delta(k+1) = x(t_{k+1}) - (x(t_k) + h f(t_k, x(t_k)))$$

$$= \frac{h^2}{2} \frac{d^2 x}{dt^2}(t_k + \theta h).$$

Hence, choosing a constant C_E by

$$C_E = \frac{1}{2} \max \left\{ \left| \frac{d^2 x}{dt^2}(s) \right|, \quad t_0 \leq s \leq t_e \right\},$$

we find that $\delta(k+1) \leq \delta_h$ for all k if we set

$$\delta_h = C_E h^2.$$

Since the solution $x(t)$ of the initial value problem is in general not known, it is more appropriate to find a bound for the constant C_E directly from the right-hand side f in the differential equation. We now outline a procedure for how this can be accomplished.

We have $(dx/dt)(t) = f(t, x(t))$ and it follows that

$$\frac{d^2 x}{dt^2}(t) = \frac{\partial f}{\partial t}(t, x(t)) + \frac{\partial f}{\partial x}(t, x(t)) \frac{dx}{dt}(t)$$

$$= \frac{\partial f}{\partial t}(t, x(t)) + \frac{\partial f}{\partial x}(t, x(t)) f(t, x(t)). \tag{18.3.6}$$

Hence if a solution has to be computed for $t_0 \leq t \leq t_e$, and the values of the solution certainly are in the range $x_\ell \leq x \leq x_u$, then

$$C_E \leq \frac{1}{2} \max \left\{ \left| \frac{\partial f}{\partial t}(s, y) + \frac{\partial f}{\partial x}(s, y) f(s, y) \right|, \quad t_0 \leq s \leq t_e, \quad x_\ell \leq y \leq x_u \right\}. \tag{18.3.7}$$

As an example of (18.3.7) we approximate the solution of the initial value problem

$$\frac{dx}{dt} = 5x$$

$$x(0) = 1$$

on the interval $[0, T]$. Here $f(t, x) = 5x$ and therefore

$$\frac{\partial f}{\partial t}(t, x) = 0 \quad \text{and} \quad \frac{\partial f}{\partial x}(t, x) = 5.$$

From the ODE we see that the solution is monotonically increasing, and thus we find that

$$C_E \leq \frac{1}{2} \max \{|0 + 5 \cdot 5y|, \quad y = x(T)\} = \frac{25}{2} x(T).$$

Therefore we can choose

$$\delta_h = \frac{25}{2} \bar{x} h^2$$

for any $\bar{x}$ such that $\bar{x} \geq x(T)$. In particular, we have confirmed the result we have previously obtained; see (18.3.5).

The computation of the local discretization error using Taylor expansions is quite tedious. Therefore we just remark here that for the modified Euler method, the local error can be bounded by a function of the form

$$\delta_h = C_M h^3,$$

and for the fourth-order Runge-Kutta method, the local discretization error is bounded by

$$\delta_h = C_R h^5.$$

Here C_M and C_R are positive constants.

A Bound on Global Discretization Errors

Once the local discretization error is bounded, we can use Theorem 18.3.2 to obtain a bound on global discretization error. In particular, we can prove:

Proposition 18.3.3   *Suppose that (18.3.2) holds. Then there are constants C_E, C_M, and C_R such that the global discretization error of:*

  (a)   Euler's method is bounded by

$$|\epsilon(k)| \leq \frac{e^{khL} - 1}{L} C_E h;$$

(b)   *the modified Euler method is bounded by*

$$|\epsilon(k)| \le \frac{e^{khL} - 1}{L} C_M h^2;$$

(c)   *the fourth-order Runge-Kutta method is bounded by*

$$|\epsilon(k)| \le \frac{e^{khL} - 1}{L} C_R h^4.$$

In view of this proposition it is clear that the fourth-order Runge-Kutta method is much better than Euler's method or the modified Euler method: The reason is that for small step sizes h, the value of h^4 is much smaller than h^2 or even h itself. Hence the global discretization error of the fourth-order Runge-Kutta method is, for reasonably small step sizes, much smaller than for the other two methods. In addition, it is also evident how the *fourth-order* Runge-Kutta method got its name.

Specifying the Tolerance

In principle, we can use Proposition 18.3.3 to find a step size for which the global discretization error is not bigger than a specified tolerance. To illustrate this point, consider the initial value problem (see also (18.1.4))

$$\frac{dx}{dt} = x + t$$

$$x(1) = 2.$$

The aim is to find a step size h such that Euler's method is approximating the solution on the interval $[1, 3]$ up to a global discretization error smaller than 0.1. By Proposition 18.3.3 this can be guaranteed if h is chosen so that

$$\frac{e^{khL} - 1}{L} C_E h < 0.1.$$

For this example $L = 1$, since

$$|y + t - (z + t)| = |y - z| = 1 \cdot |y - z|.$$

We now find a bound for the constant C_E. We compute

$$\frac{\partial f}{\partial t}(s, y) = \frac{\partial f}{\partial x}(s, y) = 1.$$

We see from the ODE that the solution is monotonically increasing. Suppose that we know that for $t \in [1, 3]$, the solution $x(t)$ is bounded by 40 from above. Using (18.3.7), we obtain an estimate for C_E by

$$C_E \leq \frac{1}{2} \max \{|1 + 1 \cdot f(s, y)|, \quad 1 \leq s \leq 3, \quad 2 \leq y \leq 40\} = \frac{44}{2} = 22.$$

Since $e^{kh} \leq e^{t_e - t_0} = e^2$, we can choose an h such that

$$(e^2 - 1)22h < 0.1 \quad \Longleftrightarrow \quad h < \frac{1}{220(e^2 - 1)} \approx 0.00071.$$

Indeed, computing a solution with MATLAB by

```
h      = 0.0007;
t(1)   = 1;
x(1)   = 2;
K      = round(2/h);
for k = 1:K
    t(k+1) = t(k)+h;
    x(k+1) = x(k)+h*(x(k)+t(k));
end
plot(t,x)
xlabel('t')
ylabel('x')
```

we obtain the result in Figure 18.11. (The MATLAB command round rounds a number toward the nearest integer.) The outcome cannot be distinguished from the exact solution shown in Figure 18.2(b).

HAND EXERCISES

1. Determine the function $\Phi = \Phi(t_k, x_k, h)$ which belongs to the fourth-order Runge-Kutta method.

In Exercises 2 and 3, find an estimate for the constant C_E of the form $C_E \leq Kx(t_0)$, where $x(t)$ is the solution of the specified initial value problem on $[0, T]$ and $t_0 \in [0, T]$.

2. $\dfrac{dx}{dt} = 3x$, where $x(0) = 2$

3. $\dfrac{dx}{dt} = -x$, where $x(0) = 1$

COMPUTER EXERCISES

4. Set $\lambda = 0.2$ and the step size $h = 0.1$. Use MATLAB to compute the global discretization error and its bound in (18.3.3) on the interval $[0, 1]$. Compare the result with the ones obtained in Figures 18.8 and 18.10.

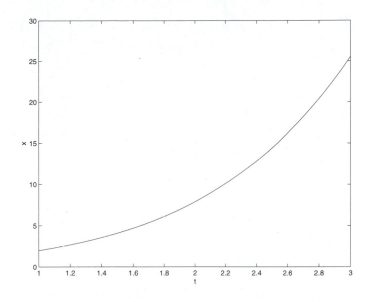

Figure 18.11
The numerical solution obtained by Euler's method for the step size
$h = 0.0007$

18.4   APPENDIX: VARIABLE STEP METHODS

The idea underlying variable step methods is to perform *step length controls* rather than using steps of fixed length. The strategy is to vary the step size in each step of the integration so that a fixed error bound is always maintained.

More concretely, suppose that we can estimate in step k the error that is made in the numerical solution. Then we can use this estimate to find an appropriate step size h_{k+1} for the next step. This principle has the advantage that large step sizes are used when the error bound is small and small step sizes are used when the error bound is large. The MATLAB command ode45 is based on explicit Runge-Kutta methods, and it uses step length control to solve initial value problems numerically.

Rather than explaining step length control mechanisms in detail, we give an example illustrating such a mechanism used in ode45. Consider the initial value problem

$$\frac{dx}{dt}(t) = -e^{10t}x^2$$
$$x(0) = 3. \tag{18.4.1}*$$

Store the right-hand side in the m-file f18_4_1.m,

```
function f = f18_4_1(t,x)
f = -exp(10*t)*x*x;
```

and compute an approximation of the solution on the interval $t \in [0, 1.2]$:

```
[t,x] = ode45('f18_4_1',[0 1.2],3);
plot(t,x)
xlabel('t')
ylabel('x')
```

The result is shown in Figure 18.12(a). In particular, note that the solution is almost linear for $t \in [0, 0.2]$ and indistinguishable from 0 for $t > 0.8$. Since the behavior of linear functions is easy to predict, we expect that the step length control mechanism in ode45 will lead to larger step sizes in those intervals. To look at the step lengths, we have to compute the differences of the consecutive entries in t. In MATLAB this can be done with the command diff. Typing

```
dt  = diff(t);
tdt = t(2:length(t));
plot(tdt,dt,'o')
xlabel('t')
ylabel('step lengths')
```

we obtain the result shown in Figure 18.12(b). At the beginning of the numerical solution—within the region where the solution shows almost linear behavior—the step sizes are quite large. Then the step sizes decrease until the almost constant part is detected by the step length control mechanism and the step sizes increase again. At the end the step size becomes smaller, since the numerical approximation ends at $t_e = 1.2$ and the step sizes have to be adjusted to this value. Altogether ode45 qualitatively behaves as expected.

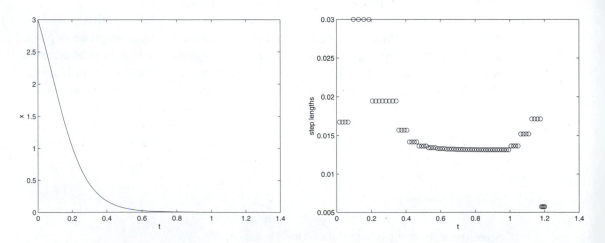

Figure 18.12
(a) Numerical solution of the initial value problem (18.4.1) using ode45; (b) the different step sizes used by ode45 in the numerical approximation.

MATLAB Commands

† indicates an laode toolbox command not found in MATLAB

CHAPTER 1: PRELIMINARIES

Editing and Number Commands

quit   Ends MATLAB session

;   (a) At end of line the semicolon suppresses echo printing

    (b) When an array is entered the semicolon indicates a new row

↑   Displays previous MATLAB command

[]   Indicates the beginning and the end of a vector or a matrix

x=y   Assigns x the value of y

x(j)   Recalls jth entry of vector x

A(i,j)   Recalls the entry in the ith row, jth column of matrix A

A(i,:)   Recalls ith row of matrix A

A(:,j)   Recalls jth column of matrix A

Vector Commands

norm(x)   The norm or length of a vector x

dot(x,y)   Computes the dot product of vectors x and y

†addvec(x,y)   Graphics display of vector addition in the plane

†addvec3(x,y)   Graphics display of vector addition in three dimensions

Matrix Commands

A'	(Conjugate) transpose of matrix
zeros(m,n)	Creates an $m \times n$ matrix all of whose entries equal 0
zeros(n)	Creates an $n \times n$ matrix all of whose entries equal 0
diag(x)	Creates an $n \times n$ diagonal matrix whose diagonal entries are the components of the vector $x \in \mathbf{R}^n$
eye(n)	Creates an $n \times n$ identity matrix

Special Numbers in MATLAB

pi	The number $\pi = 3.1415\ldots$
acos(a)	The inverse cosine of the number a

CHAPTER 2: SOLVING LINEAR EQUATIONS

Editing and Number Commands

format	Changes the numbers display format to standard five-digit format
format long	Changes display format to 15 digits
format rational	Changes display format to rational numbers
format short e	Changes display to five digit floating point numbers

Vector Commands

x.*y	Componentwise multiplication of the vectors x and y
x./y	Componentwise division of the vectors x and y
x.^y	Componentwise exponentiation of the vectors x and y

Matrix Commands

A([i j],:) = A([j i],:)	Swaps ith and jth rows of matrix A
A\b	Solves the system of linear equations associated with the augmented matrix $(A\|b)$
x = linspace(xmin,xmax,N)	Generates a vector x whose entries are N equally spaced points from xmin to xmax
x = xmin:xstep:xmax	Generates a vector whose entries are equally spaced points from xmin to xmax with stepsize xstep
[x,y] = meshgrid(XMIN:XSTEP:XMAX,YMIN:YSTEP:YMAX);	Generates two vectors x and y; the entries of x are values from XMIN to XMAX in steps of XSTEP, similarly for y.
rand(m,n)	Generates an $m \times n$ matrix whose entries are randomly and uniformly chosen from the interval $[0, 1]$
rref(A)	Returns the reduced row echelon form of the $m \times n$ matrix A
rank(A)	Returns the rank of the $m \times n$ matrix A

Graphics Commands

`plot(x,y)`	Plots a curve by connecting the points $(x(i), y(i))$ in sequence
`xlabel('labelx')`	Prints `labelx` along the x-axis
`ylabel('labely')`	Prints `labely` along the y-axis
`surf(x,y,z)`	Plots a three-dimensional graph of $z(j)$ as a function of $x(j)$ and $y(j)$
`hold on`	Instructs MATLAB to *add* new graphics to the previous figure
`hold off`	Instructs MATLAB to *clear* figure when new graphics are generated
`grid`	Toggles grid lines on a figure
`axis('equal')`	Forces MATLAB to use equal x and y dimensions
`view([a b c])`	Sets viewpoint from which an observer sees the current three-dimensional plot
`zoom`	Zooms in and out on two-dimensional plot; on each mouse click, axes change by a factor of 2

Special Numbers and Functions in MATLAB

`exp(x)`	The number e^x, where $e = \exp(1) = 2.7182\ldots$
`sqrt(x)`	The number $\sqrt{x}$
`i`	The number $\sqrt{-1}$

CHAPTER 3: MATRICES AND LINEARITY

Matrix Commands

`A*x`	Performs the matrix vector product of the matrix A with the vector x
`A*B`	Performs the matrix product of the matrices A and B
`size(A)`	Determines the numbers of rows and columns of a matrix A
`inv(A)`	Computes the inverse of a matrix A

Program for Matrix Mappings

†map	Allows the graphic exploration of planar matrix mappings

CHAPTER 4: SOLVING ORDINARY DIFFERENTIAL EQUATIONS

Special Functions in MATLAB

`sin(x)`	The number $\sin(x)$
`cos(x)`	The number $\cos(x)$

Matrix Commands

`eig(A)`	Computes the eigenvalues of the matrix A
`null(A)`	Computes the solutions to the homogeneous equation $Ax = 0$

Programs for the Solution of ODEs

†dfield5	Displays graphs of solutions to differential equations
†pline	Dynamic illustration of phase line plots for single autonomous differential equations
†pplane5	Displays phase space plots for systems of autonomous differential equations

CHAPTER 6: CLOSED FORM SOLUTIONS FOR PLANAR ODEs

Matrix Commands

expm(A)	Computes the matrix exponential of the matrix A

Functions in MATLAB

prod(1:n)	Computes the product of the integers $1, \ldots, n$

CHAPTER 8: DETERMINANTS AND EIGENVALUES

Matrix Commands

det(A)	Computes the determinant of the matrix A
poly(A)	Returns the characteristic polynomial of the matrix A
sum(v)	Computes the sum of the components of the vector v
trace(A)	Computes the trace of the matrix A

CHAPTER 9: LINEAR MAPS AND CHANGES OF COORDINATES

Vector Commands

†bcoord	Geometric illustration of planar coordinates by vector addition
†ccoord	Geometric illustration of coordinates relative to two bases

CHAPTER 10: ORTHOGONALITY

Matrix Commands

orth(A)	Computes an orthonormal basis for the column space of the matrix A
[Q,R] = qr(A,0)	Computes the QR decomposition of the matrix A

Graphics Commands

axis([xmin,xmax,ymin,ymax])	In a two-dimensional plot forces MATLAB to use the intervals [xmin,xmax] and [ymin,ymax] on the x- and y-axes, respectively
plot(x,y,'o')	Same as plot but now the points $(x(i), y(i))$ are marked by circles and no longer connected in sequence

CHAPTER 11: AUTONOMOUS PLANAR NONLINEAR SYSTEMS

Matrix Commands

`[V,D] = eig(A)`   Computes eigenvectors and eigenvalues of the matrix A

CHAPTER 13: MATRIX NORMAL FORMS

Vector Commands

`real(v)`   Returns the vector of the real parts of the components of the vector v

`imag(v)`   Returns the vector of the imaginary parts of the components of the vector v

CHAPTER 14: HIGHER DIMENSIONAL SYSTEMS

Commands for the Solution of Initial Value Problems

`[t,x]=ode45('fun',[t0 te],x0)`

Computes the solution to differential equation with right-hand side `fun` on interval `[t0 te]` with the initial condition `x0` at time `t0`

`odeset`   Displays a list of options that can be used in `ode45`

`lorenz`   Displays a dynamic simulation of a solution to the Lorenz equations

Graphics Commands

`subplot(m,n,p)`   Activates the pth subfigure in a matrix of $m \times n$ subfigures

`plot3(x,y,z)`   Plots curve in three-dimensional space connecting the points $(x(i), y(i), z(i))$ in sequence

`zlabel('labelz')`   Prints `labelz` along the z-axis

`clf`   Clears the previous graphics

Special Functions in MATLAB

`abs(v)`   Computes the absolute value of the components of the vector `v` and returns the answer in a vector of the same length

CHAPTER 15: LINEAR DIFFERENTIAL EQUATIONS

Commands for Polynomials

`roots(a)`   Computes the roots of the polynomial with coefficients specified in the vector `a`

CHAPTER 16: LAPLACE TRANSFORMS

Commands for Polynomials

residue(p,q)   Determines partial fraction expansion of p/q, where p and q are polynomials

CHAPTER 17: ADDITIONAL TECHNIQUES FOR SOLVING ODEs

Graphics Commands

contour(F)   Plots contour lines of the function F

contour(x,y,F)   Plots contour lines of the function F where the axis scales are given by x and y

clabel(c)   Labels contour lines obtained by contour by their actual levels

CHAPTER 18: NUMERICAL SOLUTIONS OF ODES

Graphics Commands

plot(x,y,'--')   Plots a curve by connecting the points $(x(i), y(i))$ in sequence with a dashed line

plot(x,y,'+')   Plots a curve by connecting the points $(x(i), y(i))$ in sequence and marks each point with a $+$

plot(x,y,'x')   Plots a curve by connecting the points $(x(i), y(i))$ in sequence and marks each point with an x

MATLAB Function

round(x)   Rounds the number x toward the nearest integer

Vector Commands

diff(v)   Computes the differences of consecutive entries in the vector v

length(v)   Returns the number of components of the vector v

Programming Commands

for k = 1:K   The MATLAB commands between for k = 1:K and end are done K times, where k varies from $1, 2, \ldots, K$.

  MATLAB commands

end

Answers to Selected Odd-Numbered Problems

CHAPTER 1: PRELIMINARIES

Section 1.1: Vectors and Matrices

1. $(3, 2, 2)$

3. $(1, -1, 9)$

5. $(5, 0)$

7. not possible

9. not possible

11. $\begin{pmatrix} 4 & -4 \\ -11 & 11 \end{pmatrix}$

Section 1.2: MATLAB

1. (a) 11;   (b) $\begin{pmatrix} 15 \\ 3 \\ 24 \end{pmatrix}$;   (c) $(-6, -2, -6, -4)$;   (d) $\begin{pmatrix} 15 \\ 0 \\ 18 \end{pmatrix}$

3. $(23.1640, -3.5620, -12.8215)$

5. $\begin{pmatrix} -14.0300 & -5.8470 & 7.0600 \\ -9.7600 & 11.0570 & -9.6600 \end{pmatrix}$

Section 1.3: Special Kinds of Matrices

1. symmetric

3. symmetric

5. symmetric

7. strictly upper triangular

9. not upper triangular

11. three

13. mn

15. $n(n + 1)/2$

17. true

19. false

21. $A^t = (3)$

Section 1.4: The Geometry of Vector Operations

1. $\|x\| = 3$

3. $\|x\| = \sqrt{3}$

5. perpendicular

7. not perpendicular

9. $a = 10/3$

11. $x \cdot y = 4$; $\cos\theta = 2/\sqrt{5}$

13. $x \cdot y = 13$; $\cos\theta = 13/6\sqrt{5}$

15. $x \cdot y = 31$; $\cos\theta = 31/\sqrt{1410} \approx 0.8256$

19. $(0.1244, 0.8397, -0.4167, 0.3253)$.

21. $15.5570°$

23. $124.7286°$

25. $\sqrt{147} \approx 12.1244$

CHAPTER 2: SOLVING LINEAR EQUATIONS

Section 2.1: Systems of Linear Equations and Matrices

1. $(x, y) = (2, 4)$

3. $(x, y) = (-4, 1)$

5. (a) has an infinite number of solutions;   (b) has no solutions.

7. (a) $p(x) = -x^2 + 5x + 1$

11.
```
   ans =
    -12.0495
     -0.8889
      7.8384
```

Section 2.2: The Geometry of Low-Dimensional Solutions

1. $2x + 3y + z = -5$

3. $z = x$

5. (a) $u = (2, 2, 1)$;   (b) $v = (1, 1, 2)$;   (c) $\cos\theta = 2/\sqrt{6}$; $\theta = 35.2644°$

7. $(x, y) \approx (2.15, -1.54)$

9. $(x, y, z) = (1, 3, -1)$

11. The function has three relative maxima.

Section 2.3: Gaussian Elimination

1. not in reduced echelon form

3. not in reduced echelon form

5. The 1st, 3rd, and 5th columns contain pivots. The system is inconsistent; no solutions.

9. inconsistent; no solutions

11. (a) Infinitely many solutions; (b) one variable can be assigned arbitrary values.

13. unique solution

15. linear

17. not linear

19. not linear

21. consistent

23. The row echelon form is:

```
A =
      0   1.0000   2.0000   1.0000  14.0000  21.0000        0  -1.0000
      0        0        0   1.0000   3.0000   5.0000        0   9.0000
      0        0        0        0   1.0000  -0.5000        0  -4.7143
      0        0        0        0        0   1.0000        0   0.3457
      0        0        0        0        0        0        0   1.0000
      0        0        0        0        0        0        0        0
```

Section 2.4: Reduction to Echelon Form

1. The reduced echelon form of the matrix A is $\begin{pmatrix} 1 & 2 & 0 & 4 \\ 0 & 0 & 1 & 2 \\ 0 & 0 & 0 & 0 \end{pmatrix}$; $\text{rank}(A) = 2$.

3. four

5. consistent; three parameters

7. inconsistent

9. 1

11. 2

Section 2.5: Linear Equations with Special Coefficients

1. $\begin{pmatrix} x_1 \\ x_2 \end{pmatrix} = \begin{pmatrix} \frac{3}{2} - \frac{1}{2}i \\ -\frac{1}{2} - \frac{1}{2}i \end{pmatrix}$

3. $\begin{pmatrix} x_1 \\ x_2 \end{pmatrix} = \begin{pmatrix} \frac{1}{2} \\ -\frac{1}{2} \end{pmatrix}$

5.   A\b=
        0.3006+ 0.2462i
       −0.6116+ 0.0751i

CHAPTER 3: MATRICES AND LINEARITY

Section 3.1: Matrix Multiplication of Vectors

1. $\begin{pmatrix} 4 \\ -11 \end{pmatrix}$

3. $\begin{pmatrix} 6 \\ -10 \end{pmatrix}$

5. (13)

9. $\begin{pmatrix} 2 & 3 & -2 \\ 6 & 0 & -5 \end{pmatrix} \begin{pmatrix} x_1 \\ x_2 \\ x_3 \end{pmatrix} = \begin{pmatrix} 4 \\ 1 \end{pmatrix}$

11. $A = \begin{pmatrix} 3 & 1 \\ -5 & 4 \end{pmatrix}$

13. No upper triangular matrix satisfies (3.1.6), but any symmetric matrix of the form $\begin{pmatrix} 1 & 2 \\ 2 & a_{22} \end{pmatrix}$ satisfies (3.1.6).

15.   b =
        103.5000
        175.8000
       −296.9000
       −450.1000
        197.4000
        656.6000
        412.4000

17.   A\b=
       −2.3828
       −1.0682
        0.1794

Section 3.2: Matrix Mappings

1. $(x_1, 0)^t$

3. $(x_1, 3x_1)^t$

5. $R_{(-45°)} = \begin{pmatrix} \frac{1}{\sqrt{2}} & \frac{1}{\sqrt{2}} \\ -\frac{1}{\sqrt{2}} & \frac{1}{\sqrt{2}} \end{pmatrix}$

7. $\begin{pmatrix} 1 & 0 \\ 0 & -1 \end{pmatrix}$

9. $\begin{pmatrix} 0 & 1 \\ 1 & 0 \end{pmatrix}$

11. $R_{90°} = \begin{pmatrix} 0 & -1 \\ 1 & 0 \end{pmatrix}$

15. A maps $x = (1, 1)^t$ to twice its length and $x = s(0, 1)^t$ to half its length.

17. C maps $x = (1, 0)$ to twice its length and $x = (1, 2)$ to $-\frac{1}{2}$ times its length.

19. B rotates the plane by $\theta \approx 3.0585$ counterclockwise and dilatates it by a factor of $c = \sqrt{5.8} \approx 2.4083$.

21. A rotates the plane $30°$ clockwise.

23. C reflects the plane across the line $y = x$.

25. E maps (x, y) to a point on the line $y = x$; that point is $(\frac{x+y}{2}, \frac{x+y}{2})$.

Section 3.3: Linearity

1. (a) $(-5, 11)$;   (b) $(6, 8, -16)$;   (c) $(21, 7, -10, -2)$

3. $\alpha = -\frac{7}{5}; \beta = -\frac{3}{5}$

5. $\alpha = \frac{5}{13}\gamma + \frac{7}{13}$ and $\beta = -\frac{14}{13}\gamma - \frac{4}{13}$

7. not linear

9. not linear

11. $A = \begin{pmatrix} 0 & -1 & -1 \\ 1 & 0 & 2 \\ 1 & -2 & 0 \end{pmatrix}$

13. $A = \begin{pmatrix} 0 & 1 & 0 \\ 0 & 0 & 1 \\ 1 & 0 & 0 \end{pmatrix}$

17. The mapping rotates a 2-vector $90°$ clockwise and then it halves its length.

Section 3.4: The Principle of Superposition

1. (a) $v_1 = \begin{pmatrix} -1 \\ 1 \\ 0 \end{pmatrix}$; $v_2 = \begin{pmatrix} -1 \\ 0 \\ 1 \end{pmatrix}$;   (b) $w_1 = \begin{pmatrix} 1 \\ 0 \\ -1 \end{pmatrix}$; $w_2 = \begin{pmatrix} 0 \\ 1 \\ -1 \end{pmatrix}$

3. (a) $s\begin{pmatrix} -11 \\ 7 \\ 1 \end{pmatrix}$;   (b) $\begin{pmatrix} 1 \\ 1 \\ 1 \end{pmatrix}$;   (c) $\begin{pmatrix} 1 \\ 1 \\ 1 \end{pmatrix} + s\begin{pmatrix} -11 \\ 7 \\ 1 \end{pmatrix}$

Section 3.5: Composition and Multiplication of Matrices

1. $AB = \begin{pmatrix} -2 & 0 \\ 7 & -1 \end{pmatrix}$; $BA = \begin{pmatrix} -2 & 0 \\ 5 & -1 \end{pmatrix}$

3. Neither AB nor BA is defined.

5. $\begin{pmatrix} -11 & 8 \\ -3 & 2 \end{pmatrix}$

7. $\begin{pmatrix} -4 & 13 & 3 \\ -12 & 11 & -11 \\ 3 & -1 & 4 \end{pmatrix}$

9. $B = \begin{pmatrix} b_{11} & 0 \\ 0 & b_{22} \end{pmatrix}$

11. $\begin{pmatrix} 10 & -5 & 15 \\ -5 & 6 & -4 \\ 15 & -4 & 26 \end{pmatrix}$

13. Neither AB nor BA is defined.

Section 3.6: Properties of Matrix Multiplication

3. $B = \begin{pmatrix} 1 & 1 & \frac{1}{2} \\ 0 & 1 & 1 \\ 0 & 0 & 1 \end{pmatrix}$; $C = \begin{pmatrix} 1 & t & \frac{t^2}{2} \\ 0 & 1 & t \\ 0 & 0 & 1 \end{pmatrix}$

Section 3.7: Solving Linear Systems and Inverses

3. $a \neq 0$ and $b \neq 0$

5. $A^{-1} = \dfrac{1}{10} \begin{pmatrix} -8 & 32 & -9 \\ 2 & 2 & 1 \\ 2 & -8 & 1 \end{pmatrix}$

9. A is invertible for any a, b, and c, and $A^{-1} = \begin{pmatrix} 1 & -a & -b+ac \\ 0 & 1 & -c \\ 0 & 0 & 1 \end{pmatrix}$.

11. Type N = [B eye(4)] in MATLAB and then row reduce N to obtain

```
ans =
 1.0000        0         0         0   -1.5714   -0.4286         0    1.4286
      0   1.0000         0         0    0.7429    0.0571    0.2000   -0.4571
      0        0    1.0000         0   -0.9143    0.3143   -0.4000    0.4857
      0        0         0    1.0000   -0.6000   -0.2000   -0.2000    0.6000
```

Section 3.8: Determinants of 2 × 2 Matrices

1. $\begin{pmatrix} 2 & -1 \\ -3 & 2 \end{pmatrix}$

9. $y = \frac{29}{11}$

11. A is invertible; $\det(A) = 4$.

13. A is not invertible; $\det(A) = 0$

CHAPTER 4: SOLVING ORDINARY DIFFERENTIAL EQUATIONS

Section 4.1: A Single Differential Equation

1. $x(t) = \dfrac{5 - \cos(2t)}{2}$

3. $x(t) = 2 - \dfrac{1}{t}$

5. $x_1(t)$ is a solution; $x_2(t)$ is not a solution.

7. $x_1(t)$ is not a solution; $x_2(t)$ is a solution.

9. $t_1 = -\frac{1}{3}\ln(0.5)$

11. $P_{\text{instant}}(1.5) = \$11{,}190.72$; $P_{\text{monthly}}(1.5) = \$11{,}186.81$

13. If the population doubles every 50 years, then $r = \frac{1}{50}(\ln 2)$. If the population doubles every 25 years, then $r = \frac{1}{25}(\ln 2)$.

15. 11:23 am

17. The graph is shown below.   **19.** The graph is shown below.

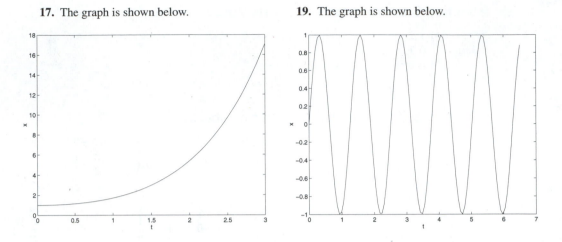

21. After two years, Statewide returns more interest than Intrastate. After one year, Intrastate returns more.

Section 4.2: Graphing Solutions to Differential Equations

1. increasing   **3.** increasing   **9.** nonautonomous   **11.** autonomous

17. The minimum value is ≈ -1.39.

Section 4.3: Phase Space Pictures and Equilibria

1. $0.055 \leq x(0) \leq 0.109$ **3.** $\lambda > 0$

7. $x = 4$ and $x = -2$ are unstable equilibria, and $x = 0$ is stable.

9. $x = -3 + 2\sqrt{2}$ is an unstable equilibrium, and $x = -3 - 2\sqrt{2}$ is stable.

13. autonomous   **15.** nonautonomous

Section 4.4: Separation of Variables

1. Separation of variables can be used; set $g(x) = x^{1/2}$ and $h(t) = \cos t$.

3. Separation of variables cannot be used.

5. $x(t) = -e^{-3t}$ **7.** $x(t) = \frac{1}{9}(2t^{3/2} + 1)^2$

9. $x(t) = Ke^{5t} - \frac{2}{5}$ for some $K \geq 0$; $x(t) = x_0 = -\frac{2}{5}$ is a constant solution.

Section 4.5: Uncoupled Linear Systems of Two Equations

1. $(0, 0)$ and $(1, 1)$ **3.** The origin is a saddle.

5. The origin is not a saddle, a source, or a sink.

9. Trajectories approach the origin tangent to the y-axis if $A < D < 0$.

11. Identical initial conditions have identical trajectories for the two systems.

Section 4.6: Coupled Linear Systems

1. (a) Trajectories converge to the origin when $D < 0$; (b) trajectories move away from the origin when $D > 0$; (c) trajectories form circles around the origin when $D = 0$.

5. Both function pairs are solutions.

7. $(e^t \sin t, e^t \cos t)$ is a solution, whereas $(3e^t, -2e^t)$ is not.

Section 4.7: The Initial Value Problem and Eigenvectors

1. $\dfrac{dX}{dt} = \begin{pmatrix} 4 & 5 \\ 2 & -3 \end{pmatrix} X$, where $X(t) = (x_1(t), x_2(t))^t$.

5. $(1, 1)^t$ is an eigenvector of C with eigenvalue $a + b$; $(1, -1)^t$ is an eigenvector with eigenvalue $a - b$.

9. $(-1, 2)^t$ is an eigenvector of B with eigenvalue 6, and $(-2, 1)^t$ is an eigenvector with eigenvalue 3.

11. (a) $\alpha = 1$; $\beta = 1$

Section 4.8: Eigenvalues of 2 × 2 Matrices

1. $\lambda = 5$ or $\lambda = -1$

3. The determinant is 23, the trace is 7, and the characteristic polynomial is $\lambda^2 - 7\lambda + 23$.

5. The determinant is 0, the trace is 9, and the characteristic polynomial is $\lambda^2 - 9\lambda$.

7. The eigenvalues are $\frac{-3+\sqrt{17}}{2}$ and $\frac{-3-\sqrt{17}}{2}$.

13. The eigenvectors are $(2, 1)^t$ with eigenvalue 4, and $(1, -1)^t$ with eigenvalue -2.

15. $\lambda \approx 2.5291 \pm 2.1111i$

Section 4.9: Initial Value Problems Revisited

1. $X(t) = 4e^{2t} \begin{pmatrix} 1 \\ 1 \end{pmatrix} - 3e^t \begin{pmatrix} 1 \\ 0 \end{pmatrix}$

3. $X(t) = \frac{8}{3}e^t \begin{pmatrix} 1 \\ 2 \end{pmatrix} - \frac{5}{3}e^{-2t} \begin{pmatrix} 2 \\ 1 \end{pmatrix}$

5. $X(t) = \frac{1}{5}e^{-t} \begin{pmatrix} 1 \\ 2 \end{pmatrix} + \frac{2}{5}e^{4t} \begin{pmatrix} 2 \\ -1 \end{pmatrix}$

7. $X(t) \approx 1.0860e^{-1.8035t} \begin{pmatrix} -0.9005 \\ 0.4348 \end{pmatrix} - 2.4527e^{4.4835t} \begin{pmatrix} -0.8880 \\ -0.4598 \end{pmatrix}$

9. $X(0.5) = (1.621, 0.291)^t$; the two methods agree to three decimal places.

Section 4.10: Markov Chains

3. The matrix Q is not a Markov matrix, since there is no positive integer k for which the 2nd row, 1st column of $Q^k > 0$.

5. $\begin{pmatrix} \frac{1}{3} & 0 & \frac{1}{3} & \frac{1}{3} & 0 \\ \frac{1}{3} & \frac{1}{3} & \frac{1}{3} & 0 & 0 \\ 0 & 0 & \frac{1}{2} & \frac{1}{2} & 0 \\ 0 & 0 & 0 & \frac{1}{2} & \frac{1}{2} \\ \frac{1}{4} & \frac{1}{4} & 0 & \frac{1}{4} & \frac{1}{4} \end{pmatrix}$

7. (a) $\text{PDOG} = \begin{pmatrix} 0 & \frac{1}{3} & \frac{1}{3} & \frac{1}{3} \\ 1 & 0 & 0 & 0 \\ 0 & 0 & 0 & 1 \\ 0 & \frac{1}{2} & \frac{1}{2} & 0 \end{pmatrix}$;   (b) $\frac{7}{36}$;

(c) 0;   (d) After a large number of steps, there will be approximately 23 dogs in each of rooms 1, 2, and 3, and 31 dogs in room 4.

9. (a) 11.62%;   (b) 14.07%;   (c) $(P^t)^4 \begin{pmatrix} 20 \\ 20 \\ 20 \\ 20 \\ 20 \end{pmatrix} = \begin{pmatrix} 14.1466 \\ 22.1620 \\ 21.2874 \\ 30.0624 \\ 12.3417 \end{pmatrix}$;   (d) An eigenvector of P^t

with eigenvalue 1 is $V = \begin{pmatrix} 0.1408 \\ 0.2195 \\ 0.2154 \\ 0.3032 \\ 0.1211 \end{pmatrix}$.

11. 1.53%

CHAPTER 5: VECTOR SPACES

Section 5.1: Vector Spaces and Subspaces

3. V_1 and V_3 are identical.   **5.** not a subspace   **7.** subspace

9. subspace   **11.** not a subspace   **13.** subspace

15. not a subspace

17. subspace when $c = 0$; not a subspace when $c \neq 0$

Section 5.2: Construction of Subspaces

1. span$\{(1, 0, -4)^t, (0, 1, 2)^t\}$ **3.** span$\{(1, 0, -1)^t, (0, 1, -1)^t\}$

5. span$\{(-2, 1, 0, 0, 0)^t, (-1, 0, -4, 1, 0)^t\}$ **7.** span$\{(-2, -1, 1)^t\}$

9. $\begin{pmatrix} 1 & 1 & 0 & 0 \\ 0 & 0 & 1 & 1 \end{pmatrix}$ **11.** $(2, 20, 0)^t = -4w_1 + 6w_2$

13. $t^4 \notin W$

15. $y(t) = 0.5t^2 \in W$, but $\{y(t), x_2(t)\}$ does not span W.

Section 5.3: Spanning Sets and MATLAB

1. span$\{(0.3225, 0.8931, -0.0992, 0.2977)^t, (0, -0.1961, 0.5883, 0.7845)^t\}$

3. $\text{span}\{(-0.8452, -0.1690, 0.5071)^t\}$ **5.** $\text{span}\{(-1, -3, 1, 0)^t, (\frac{3}{4}, 2, 0, 1)^t\}$

7. $v_2 \notin W$

Section 5.4: Linear Dependence and Linear Independence

3. linearly dependent   **9.** linearly dependent

Section 5.5: Dimension and Bases

3. $\{(1, 1, 1, 0), (-2, -2, 0, 1)\}$ is a basis; the dimension is two.

5. $\dim \mathcal{P}_2 = 3$; $\dim \mathcal{P}_n = n + 1$

Section 5.6: The Proof of the Main Theorem

1. a plane with $N = n_3(-\frac{3}{2}, 1, 1)$ **3.** a plane with $N = n_3(0, 0, 1)$

5. (a) 5;   (c) $5 - r$;   (d) $5 - r$ **9.** (a) $\lambda \neq 2$; (b) $\lambda = 2$

11. $\{(1, 0, 0, -\frac{1}{2}, \frac{3}{2}), (0, 1, 0, \frac{1}{2}, -\frac{1}{2}), (0, 0, 1, \frac{1}{2}, \frac{3}{2})\}$

CHAPTER 6: CLOSED FORM SOLUTIONS FOR PLANAR ODEs

Section 6.1: The Initial Value Problem

1. The associated eigenvalues are 2 and -1.   **3.** $2e^{2t}\begin{pmatrix} 2 \\ -3 \end{pmatrix} + e^{-t}\begin{pmatrix} -7 \\ 11 \end{pmatrix}$

5. The associated eigenvectors are $(1, 1)^t$ and $(1, -1)^t$.

7. $4\begin{pmatrix} 1 \\ 1 \end{pmatrix} - 2e^{2t}\begin{pmatrix} 1 \\ -1 \end{pmatrix}$ **11.** $(-\sin t, \cos t)$ **13.** $e^{-2t}(1, 0)$

15. The associated eigenvalues are $-1, -2,$ and -3.

17. $-2e^{-2t}\begin{pmatrix} 2 \\ -1 \\ 2 \end{pmatrix} + e^{-3t}\begin{pmatrix} 6 \\ -3 \\ 7 \end{pmatrix}$.

19. (b) The eigenvalues are 2 and -1;   (c) $2e^{2t}\begin{pmatrix} 1 \\ -1 \end{pmatrix} + e^{-t}\begin{pmatrix} 2 \\ 1 \end{pmatrix}$

Section 6.2: Closed Form Solutions by the Direct Method

3. $3\cos^2 \theta \sin \theta - \sin^3 \theta$

5. $X(t) = \alpha_1 \begin{pmatrix} 5e^{2t}\cos(3t) \\ e^{2t}(2\cos(3t) + \sin(3t)) \end{pmatrix} + \alpha_2 \begin{pmatrix} 5e^{2t}\sin(3t) \\ e^{2t}(2\sin(3t) - \cos(3t)) \end{pmatrix}$

7. $X(t) = \alpha e^{-2t}\begin{pmatrix} 2 \\ 1 \end{pmatrix} + \beta e^{-2t}\begin{pmatrix} 2t + 1 \\ t + 1 \end{pmatrix}$

Section 6.3: Solutions Using Matrix Exponentials

1. $m = 7$

3. One example is $C_1 = \begin{pmatrix} 1 & -2 \\ 3 & 1 \end{pmatrix}$ and $C_2 = \begin{pmatrix} -2 & 3 \\ -1 & -2 \end{pmatrix}$.

5. $e^{tC} = \begin{pmatrix} 1 & t & t^2/2 \\ 0 & 1 & t \\ 0 & 0 & 1 \end{pmatrix}$

Section 6.4: Linear Normal Form Planar Systems

1. $\begin{pmatrix} x(t) \\ y(t) \end{pmatrix} = \begin{pmatrix} e^{2t}(\cos(3t) - 2\sin(3t)) \\ -e^{2t}(\sin(3t) + 2\cos(2t)) \end{pmatrix}$

3. $e^{tC} = e^{2t}(I_n + tA + \frac{t^2}{2}A^2)$

Section 6.5: Similar Matrices

3. not similar

7. $e^A = e^2 \begin{pmatrix} 2 & -1 \\ 1 & 0 \end{pmatrix}$

Section 6.6: Formulas for Matrix Exponentials

1. $X(t) = -3e^t \begin{pmatrix} 1 \\ 1 \end{pmatrix} + e^{2t} \begin{pmatrix} 1 \\ 2 \end{pmatrix}$

3. $X(t) = e^t \begin{pmatrix} 2\sin t + \cos t \\ -3\sin t + \cos t \end{pmatrix}$

Section 6.7: Second-Order Equations

1. $t \approx 1.63$ seconds

3. $x(t) = \alpha e^t + \beta e^{-3t}$

5. $x(t) = e^{-t}(\alpha \cos t + \beta \sin t)$

9. false

CHAPTER 7: QUALITATIVE THEORY OF PLANAR ODES

Section 7.1: Sinks, Saddles, and Sources

1. not asymptotically stable

3. not asymptotically stable

5. saddle

7. source

9. source

11. saddle

13. sink

15. (a) $P \approx \begin{pmatrix} 7.1063 & 0 \\ 0 & 0.7106 \end{pmatrix}$; P stretches the x-coordinate of a vector and shrinks the y-coordinate. (b) The solutions of $\dot{X} = CX$ are obtained from the solutions of $\dot{X} = BX$ by stretching the x-coordinate by a factor of 7.1063 and the y-coordinate by a factor of 0.7106.

Section 7.2: Phase Portraits of Sinks

1. saddle

3. saddle

5. A possible matrix is $\begin{pmatrix} -0.5 & -1 \\ 1 & -0.5 \end{pmatrix}$.

7. A possible matrix is $\begin{pmatrix} -1 & 0 \\ 1 & -2 \end{pmatrix}$.

9. (a) no;   (b) none

Section 7.3: Phase Portraits of Nonhyperbolic Systems

1. One such matrix is $\begin{pmatrix} 0 & -1 \\ 1 & 0 \end{pmatrix}$.

5. asymptotically stable for A and unstable for B, C, and D

7. $\det(A) > 0$, $\det(D) > 0$, $\det(B) < 0$, $\det(C) = 0$

9. hyperbolic; spiral source

11. hyperbolic; improper nodal source

13. not hyperbolic; center

15. hyperbolic; saddle

CHAPTER 8: DETERMINANTS AND EIGENVALUES

Section 8.1: Determinants

1. -28

3. 14

7. -4

9. (a) 1 and $\frac{-1}{3}$;   (b) yes

13. $B_{11} = \begin{pmatrix} 7 & -2 & 10 \\ 0 & 0 & -1 \\ 4 & 2 & -10 \end{pmatrix}$; $B_{23} = \begin{pmatrix} 0 & 2 & 5 \\ 0 & 0 & -1 \\ 3 & 4 & -10 \end{pmatrix}$; $B_{43} = \begin{pmatrix} 0 & 2 & 5 \\ -1 & 7 & 10 \\ 0 & 0 & -1 \end{pmatrix}$

Section 8.2: Eigenvalues

1. $p_A(\lambda) = -\lambda^3 + 2\lambda^2 + \lambda - 2$; the eigenvalues are 1, -1, and 2.

3. span $\{(-1, 1, 0)^t, (1, 0, 1)^t\}$

5. (a) The eigenvalues are 3 and -2 with corresponding eigenvectors $(1, -1)^t$ and $(1, -2)^t$; (c) $(2x_1 + x_2, -x_1 - x_2)$.

9. false

11. (a) The eigenvalues are -0.5861 ± 20.2517, -12.9416, -9.1033, and 5.2171. The trace is -18. The characteristic polynomial is $\lambda^5 + 18\lambda^4 + 433\lambda^3 + 6{,}296\lambda^2 + 429\lambda - 252{,}292$.

13. B is the zero matrix.

CHAPTER 9: LINEAR MAPS AND CHANGES OF COORDINATES

Section 9.1: Linear Mappings and Bases

1. $A = \begin{pmatrix} -7 & -11 & 3 \\ -4 & -7 & 2 \end{pmatrix}$

Section 9.2: Row Rank Equals Column Rank

1. The possible choices are $\alpha_3(-1, -1, 1)$ and $\beta_3(-\frac{7}{5}, -\frac{9}{5}, 1)$.

3. (a) $\{(1, 0, 1, 0), (0, 1, -1, 0), (0, 0, 0, 1)\}$ is a basis for the row space of A; the row rank of A is 3.   (b) The column rank of A is 3; $\{(1, 0, 0), (0, 1, 0), (0, 0, 1)\}$ is a basis for the column space of A.   (c) $\{(-1, 1, 1, 0)\}$ is a basis for the null space; the nullity of A is 1. (d) The null space is trivial and the nullity of A^t is 0.

Section 9.3: Vectors and Matrices in Coordinates

1. $[v]_W = (7, 4)$ **5.** $[v]_W = (-2, 2, -1)$

7. $[L]_W$ is diagonal in the basis $W = \left\{ \begin{pmatrix} 1 \\ 2 \end{pmatrix}, \begin{pmatrix} 2 \\ 3 \end{pmatrix} \right\}$.

Section 9.4: Matrices of Linear Maps on a Vector Space

1. $C_{WZ} = \begin{pmatrix} 2 & 3 \\ -1 & -2 \end{pmatrix}$

5. (a) A fixes w_1, moves w_2 to w_3, and moves w_3 to $-w_2$; (b) $[L_A]_W = \begin{pmatrix} 1 & 0 & 0 \\ 0 & 0 & -1 \\ 0 & 1 & 0 \end{pmatrix}$; (c) $[L_A]_W$

fixes e_1, moves e_2 to e_3, and moves e_3 to $-e_2$.

CHAPTER 10: ORTHOGONALITY

Section 10.1: Orthonormal Bases

1. $\frac{1}{\sqrt{3}}(1, 1, -1)$ and $\frac{1}{\sqrt{2}}(0, 1, 1)$

Section 10.2: Least Squares Approximations

1. $\frac{1}{5}(3, 4)$ and $\frac{1}{5}(-4, 3)$

Section 10.3: Least Squares Fitting of Data

1. (a) $m \approx 0.4084$ and $b \approx 0.9603$, where m and b are in billions.   (b) In 1910 $P \approx 1369$ million people.   (c) The prediction for 2000 is likely to be low.

3. Let R be the number of days in the year with precipitation and let s be the percentage of sunny hours to daylight hours. Then the best linear estimate of the relationship between the two is $R \approx 199.2 - 156.6s$.

Section 10.4: Symmetric Matrices

3. The eigenvectors are $(1, 1)$ and $(1, -1)$; the eigenvalues are 4 and -2.

5. The eigenvectors are $(1, 1)$ and $(1, -1)$; the eigenvalues of C are -2 and 2.01.

Section 10.5: Orthogonal Matrices and *QR* Decompositions

1. not orthogonal

3. orthogonal

5. not orthogonal

7. $H = \begin{pmatrix} 0 & -1 \\ -1 & 0 \end{pmatrix}$

9. $H = \dfrac{1}{27} \begin{pmatrix} 25 & 2 & 10 \\ 2 & 25 & -10 \\ 10 & -10 & -50 \end{pmatrix}$

11. $\begin{pmatrix} -\frac{3}{5} & \frac{4}{5} \\ \frac{4}{5} & \frac{3}{5} \end{pmatrix}$

13. The orthonormal basis generated by the command [Q R] = qr(A,0) is:

```
v1 =          v2 =
  -0.7071        0.7071
   0.7071        0.7071
```

15. The orthonormal basis generated by the command [Q R] = qr(A,0) is:

```
v1 =          v2 =          v3 =
  -0.2673        0.0514       -0.9623
   0.5345       -0.8230       -0.1925
  -0.8018       -0.5658        0.1925
```

17.
```
  H1 =                                          H2 =
   0.6245  -0.7220  -0.2744   0.1155     0.2807   0.6679  -0.3083  -0.6165
  -0.7220  -0.3885  -0.5276   0.2222     0.6679   0.3798   0.2862   0.5725
  -0.2744  -0.5276   0.7995   0.0844    -0.3083   0.2862   0.8679  -0.2642
   0.1155   0.2222   0.0844   0.9645    -0.6165   0.5725  -0.2642   0.4716
   H =
  -0.2935   0.1305  -0.6678  -0.6714
  -0.4365  -0.6536  -0.4053   0.4669
  -0.7279  -0.1065   0.6051  -0.3043
  -0.4398   0.7378  -0.1536   0.4885
```

CHAPTER 11: AUTONOMOUS PLANAR NONLINEAR SYSTEMS

Section 11.1: Introduction

1. There are three equilibria in the interval $[-3, 1]$; $x \approx -2.910$ and $x \approx 0.925$ are asymptotically stable, and $x \approx -0.408$ is unstable.

3. A valid initial condition is $X(0) = (1, 4)$.   **5.** A valid initial condition is $X(0) = (1, 2)$.

Section 11.2: Equilibria and Linearization

1. (b) $\lim\limits_{t \to \infty} x(t) = -\infty$

3. (b) If there exists a positive equilibrium p_e, then $\lim\limits_{p \to \infty} p(t) = p_e$. If no positive equilibrium exists, then $\lim\limits_{p \to \infty} p(t) = \infty$.

5. spiral source

7. not hyperbolic

9. (a) $(1, 0)$ and $(0, 1)$;   (b) $(1, 0)$ is a saddle; $(0, 1)$ is a spiral sink.

11. (a) $(0, 0)$, $(\sqrt{2.2}, 0)$, and $(-\sqrt{2.2}, 0)$; (c) $(\sqrt{2.2}, 0)$ is not hyperbolic when $a = \frac{-2.1}{\sqrt{2.2}}$; $(-\sqrt{2.2}, 0)$ is not hyperbolic when $a = \frac{2.1}{\sqrt{2.2}}$; (d) $(\sqrt{2.2}, 0)$ is asymptotically stable when $a < -\frac{2.1}{\sqrt{2.2}}$, and $(-\sqrt{2.2}, 0)$ is asymptotically stable when $a > \frac{2.1}{\sqrt{2.2}}$. There is no value of a for which both equilibria are stable. $(0, 0)$ is not stable.

13. (a) The Jacobian matrix at the origin is $\begin{pmatrix} 0 & 1 \\ -1 & 0 \end{pmatrix}$; its eigenvalues are $\pm i$; and the origin is a center.

15. (b) The eigenvectors are $v_1 = (0.4723, 0.8814)$ and $v_2 = (0.9911, 0.1328)$. The trajectories approach the origin in the direction of v_2.

Section 11.3: Periodic Solutions

1. two limit cycles: $r = 1$ and $r = \sqrt{3}$ **3.** two limit cycles: $r = \sqrt{2}$ and $r = \sqrt{3}$

7. nine

Section 11.4: Stylized Phase Portraits

3. The equilibria and their types are:

```
(1.4650,  1.3590)       Spiral source
(-1.3651, -1.4635)      Spiral sink
(1.4650, -1.3590)       Saddle point
(-1.3651,  1.4635)      Saddle point
```

There is a limit cycle surrounding a spiral source at $(1.4650, 1.3590)$ and a limit cycle surrounding a spiral sink at $(-1.3651, -1.4635)$.

5. (a) five; (b) two

CHAPTER 12: BIFURCATION THEORY

Section 12.1: Two-Species Population Models

1. Predators eat prey faster in (12.1.8) than in (12.1.9).

3. For $\mu = 6$, there is a center at $e_4 = (1, 8)$.

5. Two equilibria: $(0, 0)$ and $(-\mu_2/\sigma_2, -\mu_1/\sigma_1)$; the second equilibrium is in the first quadrant when $\mu_1 > 0, \mu_2 < 0, \sigma_1 < 0, \sigma_2 > 0$.

Section 12.2: Examples of Bifurcations

3. The maximum number of equilibria is four, which occurs when $8 < \rho < 13$.

5. $(2, 1)$

9. (b) $J_{(x,y)} = \begin{pmatrix} 0 & 1 \\ 2x + y & 1 + x \end{pmatrix}$. At the equilibria $(\pm\sqrt{\rho}, 0)$,

$$J_{(\sqrt{\rho}, 0)} = \begin{pmatrix} 0 & 1 \\ 2\sqrt{\rho} & 1 + \sqrt{\rho} \end{pmatrix} \quad \text{and} \quad J_{(-\sqrt{\rho}, 0)} = \begin{pmatrix} 0 & 1 \\ -2\sqrt{\rho} & 1 - \sqrt{\rho} \end{pmatrix}.$$

$(\sqrt{\rho}, 0)$ is a saddle. $(-\sqrt{\rho}, 0)$ is a sink when $\rho > 1$ and is a spiral when $1 - 10\sqrt{\rho} + \rho < 0$—that is, when $1 < \rho < 97$.

11. homoclinic bifurcation   **13.** $\rho > 0$

Section 12.4: The Remaining Global Bifurcations

1. A saddle-node bifurcation of periodic solutions occurs at $\rho \approx 0.995$. There are two limit cycles when $\rho = 0.7$ and no periodic solutions when $\rho = 1.1$.

5. The system appears to have an infinite number of limit cycles as the size of the computation area increases.

7. a saddle near $(0.4, -1.2)$ and a spiral source near $(2.0, 1.9)$

9. Hopf bifurcation; the spiral source becomes a spiral sink and the periodic solution disappears.

11. There are saddle points at $(-\rho \pm \sqrt{\rho^2 + 4}, 0)$ and a spiral sink at the origin.

13. heteroclinic bifurcation at $\rho \approx 0.141$

Section 12.5: Saddle-Node Bifurcations Revisited

3. two equilibria near $x = 3$ when $\rho > 1$

Section 12.6: Hopf Bifurcations Revisited

1. Hopf bifurcation occurs at the origin when $\rho = 0$; the eigenvalue crossing condition is satisfied.

CHAPTER 13: MATRIX NORMAL FORMS

Section 13.1: Real Diagonalizable Matrices

1. (a) The eigenvalues are 3 and -3 with eigenvectors $v_1 = (1, 1)^t$ and $v_2 = (1, -1)^t$; (b) $S = (v_1|v_2)$.

3. The eigenvalues are $\lambda_1 = 1$ and $\lambda_2 = -1$. The eigenvector associated with λ_1 is $v_1 = (1, 1, 1)^t$. There are two eigenvectors associated with λ_2: $v_2 = (1, 0, 0)^t$ and $v_3 = (0, 1, 2)^t$. $S = (v_1|v_2|v_3)$.

11. The eigenvalues of C are

```
ans =
   -4.0000
  -12.0000
   -8.0000
  -16.0000
```

and

```
S =
    0.5314   -0.5547    0.0000    0.4082
   -0.4871    0.5547   -0.4082   -0.8165
    0.6199   -0.5547    0.8165    0.4082
   -0.3100    0.2774   -0.4082    0.0000
```

13. not real diagonalizable

Section 13.2: Simple Complex Eigenvalues

1. $T = \begin{pmatrix} 1-i & 1+i \\ 2i & -2i \end{pmatrix}$; $S = \begin{pmatrix} 1 & -1 \\ 0 & 2 \end{pmatrix}$

3. A rotates plane $45°$ counterclockwise and expands by $\sqrt{2}$.

5. The matrices are:

T =

0.9690	0.0197 + 0.3253i	0.0197 − 0.3253i
0.1840	0.0506 − 0.5592i	0.0506 + 0.5592i
0.1647	−0.4757 − 0.5935i	−0.4757 + 0.5935i

S =

0.9690	0.0197	0.3253
0.1840	0.0506	−0.5592
0.1647	−0.4757	−0.5935

7. The matrices are:

T =

Columns 1 through 4

−0.1933−0.2068i	−0.1933+0.2068i	−0.6791+0.5708i	−0.6791−0.5708i
−0.0362+0.4192i	−0.0362−0.4192i	0.2735−0.3037i	0.2735+0.3037i
0.4084+0.1620i	0.4084−0.1620i	0.0881+0.0243i	0.0881−0.0243i
−0.0000−0.0000i	−0.0000+0.0000i	−0.0000+0.0000i	−0.0000−0.0000i
−0.1933−0.2068i	−0.1933+0.2068i	−0.1321−0.0365i	−0.1321+0.0365i
0.2657−0.6317i	0.2657+0.6317i	0.1321+0.0365i	0.1321−0.0365i

Columns 5 through 6

0.4205−0.1238i	0.4205+0.1238i
0.0855+0.2601i	0.0855−0.2601i
−0.1639−0.1479i	−0.1639+0.1479i
−0.5203+0.1710i	−0.5203−0.1710i
0.4205−0.1238i	0.4205+0.1238i
−0.4205+0.1238i	−0.4205−0.1238i

S =

−0.1933	−0.2068	−0.6791	0.5708	0.4205	−0.1238
−0.0362	0.4192	0.2735	−0.3037	0.0855	0.2601
0.4084	0.1620	0.0881	0.0243	−0.1639	−0.1479
−0.0000	−0.0000	−0.0000	0.0000	−0.5203	0.1710
−0.1933	−0.2068	−0.1321	−0.0365	0.4205	−0.1238
0.2657	−0.6317	0.1321	0.0365	−0.4205	0.1238

Section 13.3: Multiplicity and Generalized Eigenvectors

1. The eigenvalues of matrix A are:

Eigenvalue	Algebraic Multiplicity	Geometric Multiplicity
2	1	1
3	2	1
4	1	1

3. The eigenvalues of matrix C are:

Eigenvalue	Algebraic Multiplicity	Geometric Multiplicity
-1	3	2
1	1	1

5. $v_1 = (-1, 1)$ and $v_2 = (0, 1)$

7. $v_1 = (9, 1, -1)$, $v_2 = (-2, 0, 1)$, and $v_3 = (9, 1, -2)$

9. The eigenvalue 2 has algebraic multiplicity 4 and geometric multiplicity 1.

Section 13.4: The Jordan Normal Form Theorem

1. Two such matrices are $\begin{pmatrix} 2 & 1 & 0 & 0 \\ 0 & 2 & 0 & 0 \\ 0 & 0 & 2 & 1 \\ 0 & 0 & 0 & 2 \end{pmatrix}$ and $\begin{pmatrix} 2 & 1 & 0 & 0 \\ 0 & 2 & 1 & 0 \\ 0 & 0 & 2 & 0 \\ 0 & 0 & 0 & 2 \end{pmatrix}$.

3. There are six different Jordan normal form matrices.

5. $\begin{pmatrix} \dfrac{3+\sqrt{17}}{2} & 0 \\ 0 & \dfrac{3-\sqrt{17}}{2} \end{pmatrix}$

7. $\begin{pmatrix} -1 & 0 & 0 \\ 0 & 2 & 1 \\ 0 & 0 & 2 \end{pmatrix}$

9. $\begin{pmatrix} 1 & 1 & 0 \\ 0 & 1 & 1 \\ 0 & 0 & 1 \end{pmatrix}$

11. $\begin{pmatrix} e^t & 0 & 0 \\ 0 & e^{-t} & te^{-t} \\ 0 & 0 & e^{-t} \end{pmatrix}$

15. (a) $\begin{pmatrix} 3 & 0 & 0 & 0 \\ 0 & 1 & 0 & 0 \\ 0 & 0 & 0 & 0 \\ 0 & 0 & 0 & -1 \end{pmatrix}$;

(b)

$S =$

```
-0.1387   -0.1543   -0.0000   -0.5774
 0.1387   -0.3086   -0.4082    0.0000
 0.1387    0.9258    0.8165    0.5774
-0.9707   -0.1543    0.4082   -0.5774
```

17. (a) $J = \begin{pmatrix} 0 & 2 & 0 & 0 \\ -2 & 0 & 0 & 0 \\ 0 & 0 & -2 & 1 \\ 0 & 0 & -1 & -2 \end{pmatrix}$;

(b)

$T =$

```
-0.2118   -0.0456    0.2211    0.0060
-0.8548   -0.2507    0.8762    0.1803
-0.3555   -0.0988    0.3529    0.0669
-0.1437   -0.0531    0.1440    0.0344
```

19. (a)
$$\begin{pmatrix} -1 & 1 & 0 & 0 \\ 0 & -1 & 1 & 0 \\ 0 & 0 & -1 & 0 \\ 0 & 0 & 0 & 1 \end{pmatrix};$$

(b)

$$S = \begin{array}{cccc} -1 & 1 & 0 & 0 \\ -1 & 1 & 0 & 1 \\ -1 & 0 & 1 & 1 \\ -1 & 0 & 0 & 1 \end{array}$$

CHAPTER 14: HIGHER DIMENSIONAL SYSTEMS

Section 14.1: Linear Systems in Jordan Normal Form

1.
$$\begin{pmatrix} e^{2t}(t-1) \\ e^{2t} \\ e^{t}(2\cos t + 3\sin t) \\ e^{t}(2\sin t - 3\cos t) \end{pmatrix}$$

3. $X(t) = \begin{pmatrix} t + \frac{t^2}{2} \\ 1 + t \\ 1 \end{pmatrix}$; $\|X(0)\| = \sqrt{2} \approx 1.4142$; $\|X(2)\| = e^{-1}\sqrt{26} \approx 1.8758$

7.
$$\left(e^{-t}\begin{pmatrix} 1 & t & \frac{1}{2}t^2 \\ 0 & 1 & t \\ 0 & 0 & 1 \end{pmatrix} \quad 0 \\ \quad 0 \quad e^{2t}\begin{pmatrix} 1 & t \\ 0 & 1 \end{pmatrix} \right)$$

9. $e^{t}\begin{pmatrix} 3 & -4 & 1 \\ -6 & 11 & 0 \\ 15 & -32 & 0 \end{pmatrix}\begin{pmatrix} 1 & t & \frac{1}{2}t^2 \\ 0 & 1 & t \\ 0 & 0 & 1 \end{pmatrix}\begin{pmatrix} 0.4074 \\ 0.2222 \\ 0.6667 \end{pmatrix}$

11. Solutions are linear combinations of the functions e^{5t} and $t^j e^t$ for $j = 0, 1, 2$.

Section 14.2: Qualitative Theory near Equilibria

1. $\begin{pmatrix} -10 & -2 & -1 \\ 9 & 1 & 4 \\ -14 & 20 & 1 \end{pmatrix}$ **3.** not stable

5. $(1, 1, 1)$ and $(3, -3, 9)$; the equilibrium $(1, 1, 1)$ has three unstable directions, whereas the equilibrium $(3, -3, 9)$ has two unstable directions and one stable direction.

7. asymptotically stable   **9.** asymptotically stable

11. four unstable and one stable direction

Section 14.3: MATLAB ode45 in One Dimension

1. Create the file exode3_1.m containing the text

```
function f1 = exode3_1(t)
f1 = sin(t) - t^3;
```

3. Create the file exode3_3.m containing the text

```
function f3 = exode3_3(x,y)
f3 = [x*y - 1, x + 2*y];
```

5. Create the file exode3_5.m containing the text

```
function f5 = exode3_5(u)
f5 = [u(1) - 2*u(3); u(1)*u(2)*u(3)];
```

Section 14.4: Higher Dimensional Systems Using ode45

1. stable   **3.** asymptotically stable   **5.** not asymptotically stable

Section 14.5: Quasiperiodic Motions and Tori

3. no   **5.** no

Section 14.6: Chaos and the Lorenz Equation

1. Unstable equilibria at $(0, 0, 0)$, $(6\sqrt{2}, 6\sqrt{2}, 27)$, and $(-6\sqrt{2}, -6\sqrt{2}, 27)$.

3. Unstable equilibria at $(0, 0, 0)$, $(\sqrt{60}, 0, -\sqrt{60})$, and $(-\sqrt{60}, 0, \sqrt{60})$. In forward time, trajectories oscillate between circuits around the two nonzero equilibria.

5. asymptotic to an equilibrium   **7.** new form of asymptotic behavior

9. asymptotic to chaotic   **11.** asymptotic to periodic

CHAPTER 15: LINEAR DIFFERENTIAL EQUATIONS

Section 15.1: Solving Systems in Original Coordinates

1. $X_1(t) = e^{2t} \begin{pmatrix} 1 \\ 0 \end{pmatrix}$; $X_2(t) = e^{2t} \begin{pmatrix} t \\ 1 \end{pmatrix}$

3. $X_1(t) = e^{2t} \begin{pmatrix} 1 \\ -1 \end{pmatrix}$; $X_2(t) = e^{2t} \begin{pmatrix} t - 1 \\ -t \end{pmatrix}$

5. $-\dfrac{1}{2} \begin{pmatrix} e^t + e^{-t} & e^t - e^{-t} \\ e^t - e^{-t} & e^t + e^{-t} \end{pmatrix}$

7. $-\begin{pmatrix} e^{3t} & e^{3t} - e^t \\ 0 & e^t \end{pmatrix}$

9. $\begin{pmatrix} -4e^{-t} + 3 + 2e^t \\ e^{-t} + e^t \\ -6e^{-t} + 3 + 2e^t \end{pmatrix}$

11. $\begin{pmatrix} 1 & t \\ 0 & 1 \end{pmatrix}$

13. $\dfrac{1}{4} \begin{pmatrix} 1 & -1 & -2 + 2t \\ -1 & 1 & -2t \\ 0 & 0 & 2 \end{pmatrix} + e^{2t} \begin{pmatrix} 2 & 2 & 2 \\ 2 & 2 & 2 \\ 0 & 0 & 0 \end{pmatrix}$

Section 15.2: Higher Order Equations

1. $\begin{cases} \dot{x} = y \\ \dot{y} = -x - 5y \end{cases}$

3. $\begin{cases} \dot{x} = y \\ \dot{y} = z \\ \dot{z} = 10x + 4z \end{cases}$

5. $\alpha_1 e^{4t} + \alpha_2 e^{-t}$

7. $\alpha_1 e^{3t} + \alpha_2 t e^{3t}$

9. $\alpha_1 + \alpha_2 \cos(2t) + \alpha_3 \sin(2t)$

11. $\alpha_1 e^{3t} + \alpha_2 e^{2t} + \alpha_3 e^t$

13. $\alpha_1 e^{2t} + \alpha_2 e^{-t} + \alpha_3 e^{-2t}$

15. $\alpha_1 e^{2t} + \alpha_2 e^{-2t} + \alpha_3 \cos t + \alpha_4 \sin t$

Section 15.3: Linear Differential Operators

1. (a) $\sin(3t) + 3\cos(3t)$; (b) $2te^{-t}$

3. (a) $104 \sin(3t)$; (b) $(4t^2 + 8)e^{-t}$

5. $\frac{dx}{dt} + x = 0; \lambda + 1; \alpha_1 e^{-t}$

7. $-4\frac{d^5 x}{dt^5} + 4\frac{d^4 x}{dt^4} - 8\frac{d^2 x}{dt^2} = 0; \lambda^2(\lambda + 1)(\lambda^2 - 2\lambda + 2);$
$\alpha_1 + \alpha_2 t + \alpha_3 e^{-t} + \alpha_4 e^t \cos t + \alpha_5 e^t \sin t$

11. $ae^t \cos t \begin{pmatrix} 1 \\ 1 \end{pmatrix} + be^t \sin t \begin{pmatrix} 1 \\ -1 \end{pmatrix}$

13. $ae^{2t} \begin{pmatrix} 1 \\ 0 \end{pmatrix} + be^t \begin{pmatrix} 1 \\ 1 \end{pmatrix}$

15. $\frac{d^4 x}{dt^4} + 4\frac{d^2 x}{dt^2} - 6\frac{dx}{dt} = 0$; roots 0, 1.1347, and $-0.5674 \pm 2.2284i$;
$\alpha_1 + \alpha_2 e^{1.1347t} + \alpha_3 e^{-0.5674t} \cos(2.2284t) + \alpha_4 e^{-0.5674t} \sin(2.2284t)$

Section 15.4: Undetermined Coefficients

1. $\frac{d^2 x}{dt^2} - 8\frac{dx}{dt} + 41x$

3. $\frac{d^3 x}{dt^3} - 6\frac{d^2 x}{dt^2} + 12\frac{dx}{dt} - 8x$

5. $\frac{1}{3}\cos t + \frac{1}{6}\sin t$

7. $\frac{1}{6}t^2 e^{-t}$

9. $\frac{1}{4}(t\sin t - t^2 \cos t)$

11. $-4te^{-t}\cos t$

13. $\frac{1}{72}(-3e^{2t} + 6te^{2t} + 3e^{-t}\cos(\sqrt{3}t) + \sqrt{3}e^{-t}\sin(\sqrt{3}t))$

15. $\frac{1}{2}e^{t-1} + \frac{1}{2}e^{-t+1} - 1$

Section 15.5: Periodic Forcing and Resonance

1. $\alpha_1 \cos(\sqrt{\kappa}t) + \alpha_2 \sin(\sqrt{\kappa}t) + \begin{cases} \frac{A}{1-\omega^2}\cos(\omega t) & \omega \neq \sqrt{\kappa} \\ \frac{A}{2\sqrt{\kappa}}t\cos(\sqrt{\kappa}t) & \omega = \sqrt{\kappa} \end{cases}$

5. beats

7. resonance

CHAPTER 16: LAPLACE TRANSFORMS

Section 16.1: The Method of Laplace Transforms

1. $\frac{1}{s-3} - \frac{2}{s+4}$

5. $4e^{-3t} - 3e^{-2t}$

7. $-\frac{1}{5}(e^{4t} + 9e^{-t})$

Section 16.2: Laplace Transforms and Their Computation

1. $4\left(\dfrac{s\cos(1)+\sin(1)}{s^2+1}\right)$

3. $\dfrac{9s^2-6s+2}{s^3}$

5. $2e^{-t}\cos(2t)-2e^{-t}\sin(2t)$

7. $\frac{1}{13}\left(1+e^{3t}\cos(2t)-8e^{3t}\sin(2t)\right)$

9. $\frac{1}{30}(35e^{2t}-6e^t+e^{-4t})$

Section 16.3: Partial Fractions

1. $e^{2t}+\frac{2}{3}e^{-3t}-\frac{5}{3}$

3. $\frac{2}{s-2}$

5. $\dfrac{-\frac{3}{4}i}{s-2i}+\dfrac{\frac{3}{4}i}{s+2i}$

7. $-\dfrac{4}{s-1}+s-1$

CHAPTER 17: ADDITIONAL TECHNIQUES FOR SOLVING ODES

Section 17.1: Nonconstant Coefficient Linear Equations

1. linear and homogeneous

3. linear and inhomogeneous

5. linear and inhomogeneous

7. $2e^{\frac{1}{3}(t^3-1)}-1$

9. $(3-\cos t)\sqrt{t^2+1}$

Section 17.2: Variation of Parameters for Systems

1. $\frac{1}{2}\begin{pmatrix} te^t+e^t+1 \\ te^t-e^t-1 \end{pmatrix}$

3. $X(t)=(\alpha_1+2t)X_1(t)+\left(\alpha_2-\frac{1}{3}t^3\right)X_2(t)$

Section 17.3: The Wronskian

1. t^2+1

3. $te^t(1-t)$

7. $-2e^{4t}$

Section 17.4: Higher Order Equations

1. t^5

3. $x_p(t)=\frac{1}{2}(t-\frac{3}{2})$

Section 17.5: Simplification by Substitution

1. Bernoulli

3. Bernoulli

5. homogeneous coefficients and Bernoulli

7. $\dfrac{t}{4-t}$

9. $\dfrac{t}{1-\ln|t|}$

Section 17.6: Exact Differential Equations

1. exact

3. $\dfrac{3+t^2}{1+t}$

5. $\dfrac{C}{t^3}-\dfrac{1}{2t}$

Section 17.7: Hamiltonian Systems

1. $\begin{cases} \dot{x} = 3 \\ \dot{y} = -1 \end{cases}$

3. $\begin{cases} \dot{x} = -\sin x \sin y \\ \dot{y} = -\cos x \cos y \end{cases}$

5. yes

7. yes

9. The equilibria are saddles or centers

CHAPTER 18: NUMERICAL SOLUTIONS OF ODES

Section 18.1: A Description of Numerical Methods

3. yes

5. yes

Section 18.2: Error Bounds for Euler's Method

1. local discretization error: $\delta(k+1) \leq e^{3T} \frac{(3h)^2}{2}$ for all $h \geq 0$; global discretization error: $|\epsilon(k)| \leq e^{3T}(e^{3T}-1)\frac{3h}{2}$

3. local discretization error: $\delta(k+1) \leq 2e^{3T} \frac{(3h)^2}{2}$ for all $h \geq 0$; global discretization error: $|\epsilon(k)| \leq 2e^{3T}(e^{3T}-1)\frac{3h}{2}$

5. $h = 0.0014$

7. $T = 1.609$

Section 18.3: Local and Global Error Bounds

1.

$$
\begin{aligned}
\Phi(t_k, x_k, h) = \frac{1}{6}\Big[& f(t_k, x_k) \\
& +2f\left(t_k + \frac{h}{2}, x_k + \frac{h}{2}f(t_k, x_k)\right) \\
& +2f\left(t_k + \frac{h}{2}, x_k + \frac{h}{2}f\left(t_k + \frac{h}{2}, x_k + \frac{h}{2}f(t_k, x_k)\right)\right) \\
& + f\left(t_k + h, x_k + hf\left(t_k + \frac{h}{2}, x_k + \frac{h}{2}f\left(t_k + \frac{h}{2}, x_k + \frac{h}{2}f(t_k, x_k)\right)\right)\right)\Big]
\end{aligned}
$$

3. $C_E \leq \frac{1}{2}x(0) = \frac{1}{2}$

Index

INSTRUCTIONS FOR DOWNLOADING FROM FTP OR HTTP SITES

You may access the `laode` toolbox files by using ftp on a web browser or by using anonymous ftp protocols (fetch on MACs). Point your web browser to: *ftp://ftp.mathworks.com/pub/books/golubitsky*

Anonymous ftp:	*ftp://ftp.mathworks.com*
Name:	anonymous
Password:	"e-mail address"
	cd pub/books/golubitsky

Or, access the files at the University of Houston by pointing your web browser to: *ftp://ftp.math.uh.edu/pub/laode*

Anonymous ftp:	*ftp://ftp.math.uh.edu*
Name:	anonymous
Password:	"e-mail address"
	cd pub/laode

You may want to browse the Brooks/Cole Math Resources Center at *http://www.brookscole.com/math* where you will also find these files.

We are maintaining these sites for three purposes:

1. To allow access to the laode files for those who cannot use the included CD-ROM
2. To provide up-dated files that conform with changes to future versions of MATLAB
3. To provide corrections and a list of known bugs

Each site is structured as follows. There are three README files (README_UNIX, README_PC, README_MAC) and three archive files (LAODE_UNIX.tar.gz, LAODEPC.exe, LAODEMAC.sea). Begin by reading the README_ file for your platform. The README_ files contain instructions on how to download the `laode` toolbox files on each of the three major platforms and how to extract the archived files.

After downloading and extracting the files that are appropriate for your computer platform, please read the downloaded README file (README.unix, README.pc, README.mac) and follow the instructions for installing the `laode` toolbox so that the needed files will be available on your computer from inside of MATLAB.

For technical support call: 1-800-423-0563 or e-mail: support@kdc.com